PRACTICAL ENHANCED RESERVOIR ENGINEERING

PRACTICAL ENHANCED RESERVOIR ENGINEERING

Assisted with Simulation Software

Abdus Satter, Ph.D.
Ghulam M. Iqbal, Ph.D., P.E.
James L. Buchwalter, Ph.D., P.E.

Copyright © 2008 by
PennWell Corporation
1421 South Sheridan Road
Tulsa, Oklahoma 74112-6600 USA

800.752.9764
+1.918.831.9421
sales@pennwell.com
www.pennwellbooks.com
www.pennwell.com

Marketing Manager: Julie Simmons
National Account Executive: Francie Halcomb
Director: Mary McGee
Managing Editor: Marla Patterson
Production/Operations Manager: Traci Huntsman
Production Editor: Tony Quinn
Book Designer: Susan E. Ormston Thompson
Cover Designer: Charles Thomas

Library of Congress Cataloging-in-Publication Data

Satter, Abdus.

 Practical enhanced reservoir engineering : assisted with simulation software / Abdus Satter, Ghulam M. Iqbal, and James L. Buchwalter.

     p. cm.

 ISBN-13: 978-1-59370-056-0 (hardcover)

 ISBN-10: 1-59370-056-3 (hardcover)

 1. Petroleum--Geology--Data processing. 2. Petroleum engineering--Data processing. 3. Oil reservoir engineering--Data processing. I. Iqbal, Ghulam M. II. Buchwalter, James L. III. Title.

 TN870.5.S28 2007

 622'.3382--dc22

                                                                                           2007011741

Printed in the United States of America

1 2 3 4 5 12 11 10 09 08

Contents

PREFACE

Practical Enhanced Reservoir Engineering—Assisted with Simulation Software is written to modernize and bring up-to-date petroleum reservoir engineering for college students. It is designed to prepare graduates to play an active and important role throughout the reservoir life cycle in the various phases of the reservoir management process with fellow geoscientists as members of the asset team. Teamwork is more important than ever, as we need to manage our reservoirs in a way that will make our projects profitable and our companies successful.

This book is not just a usual college textbook, but a modern and very practical guide with reservoir engineering fundamentals, advanced reservoir-related topics, and reservoir simulation fundamentals, problems, and case studies from around the world. The graduates can use these in their profession on a daily basis.

Reservoir engineering is the heart of petroleum engineering. In essence, reservoir engineering deals with the flow of oil, gas, and water through porous media, and the associated recovery efficiencies. Along with basic reservoir engineering, students will gain additional and advanced knowledge to play an active and important role throughout the reservoir life cycle, including discovery, delineation, development, production, and abandonment. Students will also be equipped to understand the various phases of the reservoir management process, including setting a strategy, developing a plan, and implementing, monitoring, evaluating, and completing it. In the digital age, petroleum fields are increasingly viewed as "digital fields," "smart fields," or "e-fields." This has occurred as wells have been transformed into next-generation "smart wells," coupled with robust information management systems. The vision is to attain real-time or near–real-time control on the assets, including continuous optimization of oil and gas production and maximization of recovery. Reservoir engineers are involved now not only in deterministic but also probabilistic methods, economics, recovery processes, and reserves estimation. Stand-alone studies in reservoir engineering and management are transitioned into integrated modeling.

In writing this book, the authors bring their lifelong experience and expertise in reservoir engineering and simulation techniques and practice. Reservoir simulation techniques play a very important role in enhancing basic reservoir engineering concepts and practice. Thus, applications of reservoir simulation methods are included throughout the various chapters of this book.

This practical book will consider the functions of reservoir engineers and how they analyze, think, and work in real-life situations. It presents the following:

- Rock and fluid properties, fluid flow principles, and reservoir performance analysis techniques; also, classical analyses in reservoir engineering, including volumetric, decline curve, and material balance studies. Most techniques are illustrated with the aid of software tools available in the industry.

- New topics such as well test analysis, reserves, reservoir economics, risk and uncertainties, probabilistic methods, and recovery processes, including waterflood and enhanced recovery processes such as thermal, chemical, and miscible floods. Recovery techniques from unconventional resources, such as oil sands, are also discussed.

- Fundamentals, applications, and value of reservoir simulation models.

- Probability analyses of hydrocarbons in place, well production, and petroleum reserves.

- Operational problems encountered by reservoir engineers and specific solution strategies to augment the recovery of oil and gas in a variety of circumstances, including marginal, matured, low permeability, stratified, and fractured reservoirs. Applications of smart well technology are also presented.

- Assignment of class projects in which the students have the opportunity to apply what they have learned to treat their problems.

Throughout the book, class exercises are designed to encourage the students to review published literature in order to learn about how real-life reservoirs are managed effectively. Furthermore, students are required to formulate strategies based on valid assumptions in the absence of necessary data, as frequently is the case in the reservoir engineering profession.

The book is designed to aid students and professionals alike in playing an active and important role throughout the reservoir life cycle in the various phases of the reservoir management process with fellow geoscientists and others as asset team members in the project.

We are confident this book will serve students and the industry well.

Abdus Satter, Ghulam Iqbal and Jim Buchwalter

ACKNOWLEDGMENTS

The authors would like to thank our many industry colleagues, coworkers, and students, from whom we learned the various aspects of reservoir engineering as practiced worldwide and the need for computer-assisted reservoir simulation software. In particular, we acknowledge the assistance of the following companies for allowing us to use their software: Fekete Associates for RTA, Gemini Solutions, Inc. for Merlin, IHS Inc. for OilWat/GasWat, Palisade Corp. for @Risk, Weatherford Company for PanSystem, and, in addition, Core Laboratories, Inc., through John Dacy for laboratory fluid properties data. In particular, we acknowledge the contributions of all the persons in the preparation of the manuscript, including the graphics designers Elizabeth Marroquin, Shahriar Arif, and Mahmudul Islam. Sara and Raya Iqbal aided in preparing certain figures and tables.

Special thanks and appreciations go to our wives: Yolanda Satter, Jesmin Iqbal, Betty Buchwalter, and our families for their patience, understanding, and encouragement during the long period of planning and writing this book.

1 · Introduction

Reservoir engineering is the heart of petroleum engineering. In the 1930s and 1940s, reservoir engineering evolved as a separate and important discipline of petroleum engineering. In essence, reservoir engineering deals with the flow of oil, gas, and water through porous media in rocks and also with the associated recovery efficiencies. Reservoir engineers play an active and important role throughout the reservoir life cycle and in the various phases of the reservoir management process.

Before 1970, reservoir engineering was considered to be the most important technical function in reservoir management. Since then, the value of synergism between engineering and the geosciences—geology, geophysics, petrophysics, and geostatistics—has been recognized. Furthermore, in recent years, integration and teamwork involving multidisciplinary professionals, tools, technologies, and data are considered essential for successful reservoir management.[1,2]

Many reservoir engineering books have been published in the past 50 years.[3–18] In addition, several reservoir simulation books have been published since the 1970s with the advent of digital computing.[19–26]

In writing this book, the authors intend to share their lifelong experience and expertise in reservoir engineering, including reservoir simulation techniques and reservoir management practices. The goal is to present a comprehensive book, starting from basic principles and leading to real-life reservoir management aided by simulation and other software tools. This practical book will consider the functions of reservoir engineers and how they analyze, think, plan, and work in real-life situations. It will present the following:

- Rock and fluid properties, fluid flow principles, well test analysis, and reservoir performance analysis techniques
- New topics such as reserves, reservoir economics, risk and uncertainties, probabilistic methods, and recovery processes
- The role of reservoir simulation models in enhancing basic reservoir engineering concepts and practice

Computer-based tools, including reservoir simulators and related software tools, are used extensively in this book to illustrate various concepts in reservoir engineering.

The learning objectives of this chapter are:
- Elements of petroleum reservoirs
- Composition of petroleum
- Origin, accumulation, and migration of petroleum
- History of reservoir engineering
- Reservoir life cycle
- Reservoir management goal
- Reservoir management process
- Reservoir engineers' functions

The scope, objectives, and organization of this book will be presented in the following sections.

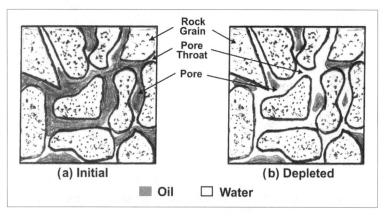

Fig. 1–1. Microscopic view of typical reservoir rock showing grains, pores, and fluid saturation distribution (a) at discovery of the reservoir and (b) following oil production by water encroachment (or injection) into the reservoir. Note the replacement of oil by water in the microscopic pore spaces during the reservoir life cycle.

Petroleum Reservoirs

Petroleum reservoirs consist of underground rocks with porosity, permeability, and trap. Porosity and permeability influence a reservoir's ability to accumulate and produce hydrocarbons. A petroleum trap is any barrier to the upward or lateral movement of oil and gas, allowing them to accumulate. The reservoir rocks contain oil or gas, or both, along with in-situ waters.

Rocks are composed of grains, pores, and cementing materials. Porosity (percentage of bulk volume) or the pore space provides the storage capacity of the rock. Permeability is a measure of the rock's ability to transmit fluid through pores.

Figure 1–1 presents typical views of petroleum-bearing porous rock in a reservoir. Microscopic pores are interdispersed in grains of the rock. Pore spaces are occupied by fluids, oil, and formation water under elevated pressure and temperature. Continuous channels exist in the porous network through which oil can flow towards the wells under the pressure gradient. A time-lapse study of oil production at a microscopic level would reveal the following:

a. At the discovery of an oil reservoir, large bodies of petroleum fluids (oil and gas) are found in the microscopic pore network. Oil migrated from a source rock, displaced water from the pores in the reservoir rock, and became trapped under favorable geologic conditions millions of years ago.

b. One of the major oil production mechanisms is based on water encroachment from the adjacent aquifer. Invading water from the aquifer drives a portion of the oil towards the wells. Consequently, pores in the rock become depleted in oil, accompanied by an increase in water saturation. Other mechanisms of production, including various improved and enhanced recovery processes, will be discussed in chapters 8, 16, and 17.

The amount of oil and gas in a reservoir, visualization of reservoir dynamics, design of recovery processes and quantities of commercially producible petroleum, referred to as reserves, are of prime importance to reservoir engineers. The topic of petroleum reserves will be treated in chapter 15.

Reservoir Geology

Reservoir rocks, mostly sedimentary in origin, are classified as the following:

• **Clastic rocks.** These reservoir rocks are formed from preexisting rocks by erosion, transformation, and deposition. These include sands, sandstones, and conglomerates, and less importantly, siltstones and shales.

• **Carbonate rocks.** These rocks are formed from organic constituents and chemical precipitates and include dolomites, reef rocks, limestones, and chalks.

Movement of fluids through the network of microscopic pores is controlled by rock and fluid properties, as described in chapters 2 and 3. Water injection into petroleum reservoirs in order to augment recovery is discussed in chapter 16.

Hydrocarbons in a reservoir may accumulate and be "trapped," or prevented from escaping, by various mechanisms, as illustrated in Figure 1–2 and 1–3. Hydrocarbon accumulations in a reservoir may be due to structural or stratigraphic traps, or a combination of both. Structural traps of hydrocarbons occur after deposition of the rock due to tectonic activity, such as faulting (fig. 1–2) or folding of the rock units. Often multiple tectonic events have occurred related to the hydrocarbon

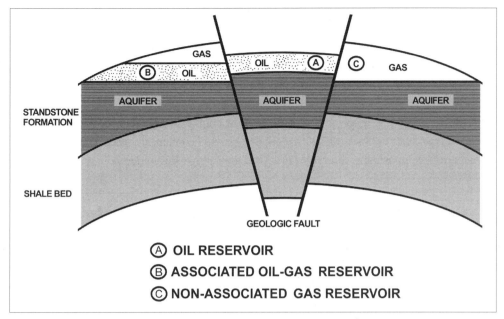

Fig. 1–2. Occurrence of oil and gas reservoirs in anticlinal traps that are faulted

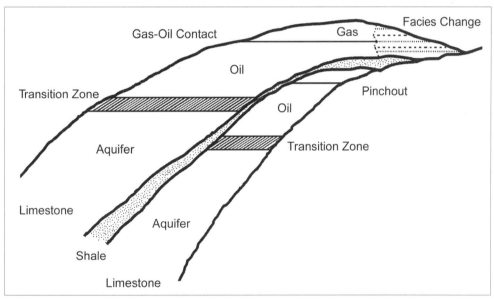

Fig. 1–3. Formation of geologic traps due to facies changes and unconformities. Included in the illustration are gas/oil contacts and transition zones between oil and water found in reservoirs where oil is overlain by free gas.

accumulation. Anticlinal traps, caused by an upward folding of the rock, are a type of structural trap responsible for most hydrocarbon accumulations worldwide. Hydrocarbon accumulations may also be found in stratigraphic traps, which result from the variations in lithology or stratigraphy or both. Stratigraphic traps may be due to depositional occurrences, such as a reef formation, alluvial deposits, or submarine turbidite deposits, among others. They may also be caused by diagenesis, facies change, and unconformities. Figure 1–3 shows an example of traps formed due to lateral changes in rock properties and unconformities. Combinations of structural and stratigraphic traps are possible. Finally, hydrodynamic traps also may allow accumulations of hydrocarbons. Hydrodynamic traps are associated with other types of traps, and they occur due to the movement of water in the formation in a manner that prevents escape of the associated hydrocarbons.

Petroleum Composition

Petroleum is a naturally occurring hydrocarbon mixture that includes crude oil, condensate, dry gas, tar, and bitumen. Chemically, it is composed largely of carbon and hydrogen, with impurities such as helium, nitrogen, oxygen, carbon dioxide, and sulfide dioxide, etc. Whether it occurs as liquid or gas depends upon its composition and the reservoir pressure and temperature.

The original pressure and temperature of a reservoir and the composition of its hydrocarbon system determine the reservoir or petroleum type, such as black oil, volatile oil, dry gas, and condensate reservoirs.

Chapter 3 reviews fluid compositions, reservoir types, and their properties in detail.

Origin, Accumulation, and Migration of Petroleum

It is important to have some knowledge of the origin, migration, and accumulation of petroleum, thus a short discussion follows.

Geologists generally believe that petroleum was originated in organic-rich, fine-grained sediments, commonly known as source beds (2%–10% organic carbon by weight). Plants convert carbon dioxide and water into oxygen and carbohydrates by photosynthesis. In ancient times (tens or hundreds of millions of years ago), the sediments along with the remnants of plants were buried and subjected to high pressure and temperature in the subsurface environment. However, an alternate theory of the origin of petroleum, known as abiogenic theory, postulates that hydrocarbons were trapped beneath as the earth was formed.

Rock containing kerogen, an organic compound based on the remnants of plants, is referred to as source rock. Kerogen occurs as humic and sapropelic types. Fluvial sediments tend to be enriched in humic kerogen, whereas deep marine sediments and some lake deposits tend to be enriched in sapropelic-type kerogen. Many shallow marine source beds contained significant amounts of both types of kerogen.

Kerogen is transformed into petroleum by deep burial and subsequent subjection to increasing pressure and temperature through millions of years. Humic kerogen yields mainly natural gas under thermal maturation, whereas sapropelic-rich sediments yield primarily oil.

Petroleum fluids formed from kerogen are eventually expelled from the pores of source material due to intense pressure. Subsequent migration of fluids occurs vertically thousands of feet or laterally tens of miles from the source beds to the overlying porous and permeable beds. This migration of fluids involves a complex interplay of buoyancy, capillary forces, and hydrodynamics.

The migrating hydrocarbon fluids, displacing in-situ water, are eventually trapped and accumulated under structural or stratigraphic traps as discussed above.

Deposition of sediments occurs under marine and nonmarine (barrier island, river channel, delta, and desert) environments. The depositional environment influences the occurrence of hydrocarbons and the size, shape, and properties of the reservoir rock.

History of Reservoir Engineering

The Drake Well, bearing the name of its supervisor, Col. Edwin Drake, was the very first well drilled purposely for oil in the United States. It was drilled in 1859 to a depth of 69 ft near Titusville, Pennsylvania, and produced about 25 barrels (bbl) per day, which sold for $20/bbl. Following the success of the Drake Well, petroleum production and processing rapidly grew into a major industry in the United States.

The birth of petroleum technology began in a real sense in 1914 when petroleum geologists started "well-sitting," as it is known today. By the mid-1930s, a new science in petroleum production engineering began to evolve. In the late 1940s, reservoir engineering technology became available on a practical and widespread basis.

In the 1850s, even before the drilling of the Drake Well, Henry Darcy, a civil engineer in Dijon, France, was studying methods of purifying drinking water. In the process, he experimentally found the law of fluid flow through porous sands. Since then, Darcy's law for single-phase linear flow has been extended to account for multiphase, multidimensional flow, providing the foundation of reservoir engineering.

Reservoir engineering became a powerful and well-defined branch of petroleum engineering due in part to the development of material balance principles in the 1930s. Other subsequent developments were the frontal advance theory, water influx effects and functions in the 1940s, and reservoir simulation theory and applications in the late 1940s and early 1950s. Further significant progress resulted from various well testing techniques and analyses in the 1950s and beyond.

Significant advancements have been made to characterize petroleum reservoirs more accurately. Various techniques have been applied to augment reservoir performance under different operating plans.

The ongoing major challenge in reservoir engineering is to maximize economic recovery of oil and gas. This has led to waterflooding and thermal and nonthermal recovery processes for oil, and gas cycling for condensate reservoirs.

Reservoir Life Cycle

In modern times, a reservoir's life cycle (fig. 1–4) from cradle to grave consists of many phases, which include exploration, discovery, delineation, development, production, and abandonment.

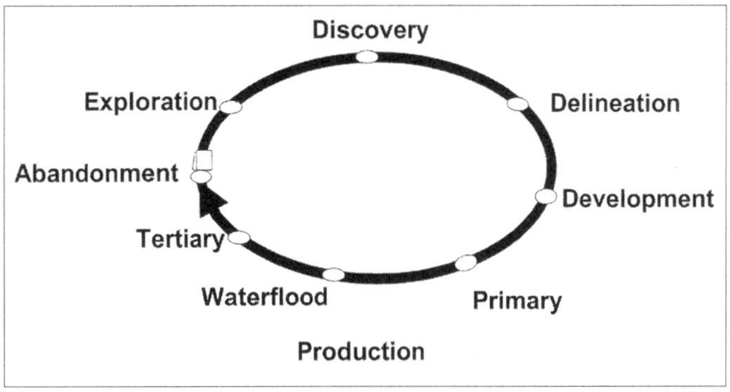

Exploration. Geologists and geophysicists are involved in exploration and contribute to reservoir definition. This includes depth, structure, stratigraphy, fractures, faults, size, aquifer system, and the location of the prospect reservoir.

Discovery. Hopefully, drilling into a prospect location will yield a discovery. Drilling engineers, petrophysicists, and reservoir engineers contribute to locating producible formations with pay thickness, porosity, oil saturation, reservoir pressure, and probable producing rates.

Fig. 1–4. Reservoir life cycle. *Source: A. Satter, J. E. Varnon, and M. T. Hoang. 1992. Reservoir management: technical perspective. SPE Paper #22350. SPE International Meeting on Petroleum Engineering, Beijing, China, March 24-27. © Society of Petroleum Engineers. Reprinted with permission.*

Delineation. Drilling additional wells delineates reservoir size and extent. Drilling engineers, petrophysicists, and reservoir engineers are again involved. Additional data on reservoir continuity and variations in pay thickness, porosity, oil saturation, and reservoir pressure is collected. Normally, one of the wells is cored, and the cores are analyzed in the laboratory for porosity, absolute permeability, relative permeability, and spectrographic characteristics. Oil, gas, and water properties, such as gas solubility, formation volume factor, compressibility, and viscosity, are determined by analyzing the reservoir fluid samples.

Development. Reservoir, drilling, operation, and facilities engineers are mainly involved in developing the field using an economically viable number of wells and spacings.

Production. This includes primary, secondary, and even enhanced oil recovery (EOR) processes. Primary production from oil or gas reservoirs is obtained at the expense of the natural reservoir energy. Secondary recovery from oil reservoirs is made by injecting fluids to augment natural energy. This is attained by gas injection, waterflooding, and gas-water combination floods. Enhanced oil recovery processes include thermal, chemical, and miscible floods. These are employed by using an external source of energy to recover oil that cannot be produced economically by conventional primary and secondary means.

Abandonment. Oil or gas fields are abandoned when no more recovery can be obtained economically.

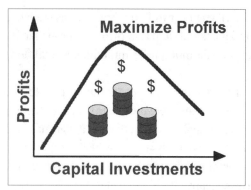

Fig. 1-5. Reservoir management goal. *Source: A. Satter, J. E. Varnon, and M. T. Hoang. 1992. Reservoir management: technical perspective. SPE Paper #22350. SPE International Meeting on Petroleum Engineering, Beijing, China, March 24-27. © Society of Petroleum Engineers. Reprinted with permission.*

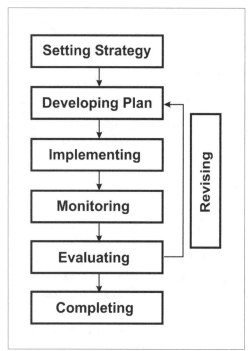

Fig. 1-6. Reservoir management process. *Source: A. Satter, J. E. Varnon, and M. T. Hoang. 1992. Reservoir management: technical perspective. SPE Paper #22350. SPE International Meeting on Petroleum Engineering, Beijing, China, March 24-27. © Society of Petroleum Engineers. Reprinted with permission.*

Geologists, geophysicists, petrophysicists, and engineers, with their tools, technologies, and available information about the reservoir, work together throughout the reservoir life cycle. Multidisciplinary professionals working as an integrated team form the key to the successful operation of the reservoir.

Reservoir Management Goal

The goal of reservoir management is to maximize profitability or net present value of the asset (petroleum reserves) while minimizing capital investments and operating costs (fig. 1–5). Utilizing the proven reservoir management process, and maximizing the use of the company's resources, such as professionals, technologies, tools, and data, can achieve success. Successful reservoir management requires synergy and well-coordinated team efforts.

Reservoir Management Process

The modern reservoir management process involves setting a strategy, developing a plan, and implementing, monitoring, evaluating, and completing it (fig. 1–6). None of the components of the process is independent of the others. Integration of all these is essential for successful reservoir operation and management. It is a dynamic and ongoing process. It must be kept in mind that every field is unique due to the nature of petroleum fluids, rock characteristics, geological setting, and reservoir size and prevailing pressure, among other aspects. These usually require state-of-the art technology and unique management strategies to succeed.

Sound reservoir management requires constant monitoring and surveillance of the reservoir performance as a whole. This must be done in order to determine if the reservoir performance is conforming to the management plan. The major areas of monitoring and surveillance involving data acquisition and management include the following:

- Oil, water, and gas production
- Gas and water injection
- Static and flowing bottomhole pressures
- Production and injection well tests
- Well injection and production profiles
- Analysis of produced fluids
- Any others aiding surveillance of the reservoir, including seismic surveys

The coordinated efforts of the various functional groups working on the project are needed to successfully carry out the monitoring and surveillance program. Monitoring and surveillance information is integrated with the reservoir simulation model to predict the future performance of the reservoir. As additional data becomes available, the reservoir management plan is refined and implemented with appropriate changes. These could include well recompletion, infill drilling, and adjustment of well injection and production rates, to name a few. Revision of plans and strategies is needed when the reservoir performance does not conform to the operating plan or when conditions change.

In the traditional approach, data analysis by geophysicists, geologists, and engineers was sequential rather than integrated. Each group of professionals worked in isolation to analyze relevant data and conceptualize the reservoir based on their expertise in one single discipline. As the reservoir management process evolved, the value of asset teams became apparent. Multidisciplinary professionals working with their technologies, tools, and data communicate directly with each other before conclusions are drawn and reservoir management decisions are made. The success of the operation can be ensured when the professionals work as an integrated team rather than as a relay team.

Reservoir Engineers' Functions

The first priority in reservoir engineering is to estimate the original hydrocarbon in place (OHCIP), followed by determination of reserves, field development strategy, production rates, reservoir monitoring plan, and economic life. Ideally, the reservoir engineers are involved in working with an integrated team of geologists, geophysicists, petrophysicists, and engineers from other disciplines. The reservoir team is responsible for achieving goals set out by the management that may include field development and production enhancement. It could also include providing solutions for reservoir and individual well issues, in addition to cost management. State-of-the-art techniques, based on global know-how, are employed to help attain these goals.

The functions of reservoir engineers involve the following:

1. Working with an integrated team
2. Setting project objective(s)
3. Collecting, analyzing, validating, and managing data related to the project
4. Developing plans and project performance
5. Evaluating project economics
6. Obtaining management approval for the project
7. Implementing, monitoring, and evaluating the project performance

Further discussions are presented in Figure 1–7.

Scope and Objectives

The goal has been to present a comprehensive book, starting from the basic principles and leading to successful management throughout the reservoir life cycle, aided by simulation.

The book is designed for graduate and undergraduate students, as well as for practicing engineers. Geoscientists can also benefit from this book, especially in the areas of data integration, reservoir model development, and application of innovative techniques in order to efficiently manage oil and gas fields. The reservoir engineering concepts, methodology, and applications are treated in such a logical sequence that academia and industry professionals alike can benefit from it. The learning objectives are cited at the introduction of various topics in the book, while key points focus on core concepts in a clear and concise manner. Most chapters are concluded with a section summarizing the topics discussed. The last chapter presents several real-life problems that reservoir engineers typically face in their profession.

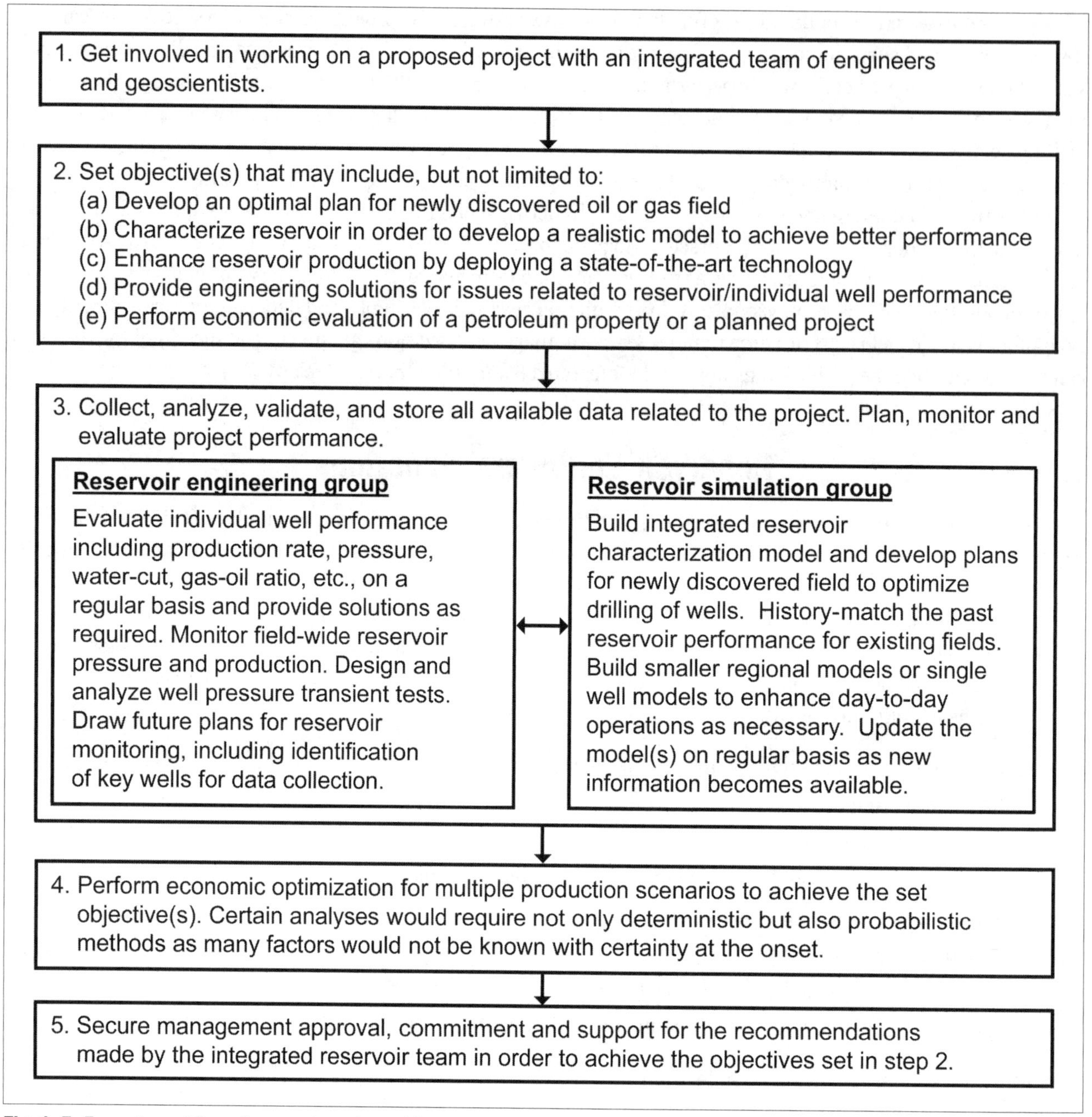

1. Get involved in working on a proposed project with an integrated team of engineers and geoscientists.

2. Set objective(s) that may include, but not limited to:
   (a) Develop an optimal plan for newly discovered oil or gas field
   (b) Characterize reservoir in order to develop a realistic model to achieve better performance
   (c) Enhance reservoir production by deploying a state-of-the-art technology
   (d) Provide engineering solutions for issues related to reservoir/individual well performance
   (e) Perform economic evaluation of a petroleum property or a planned project

3. Collect, analyze, validate, and store all available data related to the project. Plan, monitor and evaluate project performance.

   Reservoir engineering group

   Evaluate individual well performance including production rate, pressure, water-cut, gas-oil ratio, etc., on a regular basis and provide solutions as required. Monitor field-wide reservoir pressure and production. Design and analyze well pressure transient tests. Draw future plans for reservoir monitoring, including identification of key wells for data collection.

   Reservoir simulation group

   Build integrated reservoir characterization model and develop plans for newly discovered field to optimize drilling of wells. History-match the past reservoir performance for existing fields. Build smaller regional models or single well models to enhance day-to-day operations as necessary. Update the model(s) on regular basis as new information becomes available.

4. Perform economic optimization for multiple production scenarios to achieve the set objective(s). Certain analyses would require not only deterministic but also probabilistic methods as many factors would not be known with certainty at the onset.

5. Secure management approval, commitment and support for the recommendations made by the integrated reservoir team in order to achieve the objectives set in step 2.

Fig. 1–7. Example workflow of reservoir engineers' functions

Organization of the Book

Reservoir engineering is the heart of petroleum engineering. It plays a very important role in reservoir management. This book presents the following:

- Chapter 1: Introduction. Introduces petroleum reservoirs, petroleum composition, origin, accumulation, and migration, the history of reservoir engineering, reservoir life cycle and management, and functions of reservoir engineers.

- Chapters 2 and 3: Rock Characteristics, Significance in Petroleum Reservoirs, and Applications; Fundamentals of Reservoir Fluid Properties, Phase Behavior, and Applications. Reservoir rock and fluid properties: definitions and significance in reservoir performance.

- Chapter 4: Fundamentals of Fluid Flow in Petroleum Reservoirs and Applications. Steady and unsteady state.

- Chapter 5: Transient Well Pressure Analysis and Applications.
- Chapters 6 and 7: Fundamentals of Data Acquisition, Analysis, Management, and Applications; Integration of Geosciences and Engineering Models.
- Chapters 8–14: Evaluation of Primary Reservoir Performance; Empirical Methods for Reservoir Performance Analysis and Applications; Volumetric Methods for Performance Analysis and Applications; Decline Curve Analysis and Applications; Material Balance Methods and Applications; Reservoir Simulation Fundamentals; and Reservoir Simulation Model Applications. Includes primary reservoir performance analysis and techniques, and forecasts: volumetric, empirical, decline curve, material balance, and reservoir simulation methods.
- Chapter 15: Fundamentals of Oil and Gas Reserves and Applications. Proved, probable, and possible reserves.
- Chapter 16: Improved Recovery Processes: Fundamentals of Waterflooding and Applications.
- Chapter 17: Improved Recovery Processes: Enhanced Oil Recovery and Applications. Thermal, miscible, and chemical floods.
- Chapter 18: Fundamentals of Petroleum Economics, Integrated Modeling, and Risk and Uncertainty Analysis.
- Chapter 19: Operational Issues in Reservoir Development and Management.
- Chapter 20: Class Projects, each requiring one or more weeks of team effort.
- Glossary of Selected Key Terms. A selected list of reservoir-related key words, with brief explanations. When a key word related to reservoir engineering appears to be unfamiliar in the course of study, students are encouraged to consult the glossary.

Tables of lists of symbols, unit conversions, and selected acronyms are included at the end of the chapter.

Summing Up

The goal of this book is to make a comprehensive presentation, starting from the basic principles and leading to real-life reservoir management aided by simulation and other software tools.

Petroleum reservoirs. Petroleum reservoirs consist of underground rocks with porosity, permeability, and trap. Porosity provides storage capacity to hold petroleum. The rock's ability to conduct fluids depends on its permeability, while traps, such as structural or stratigraphic, prevent the movements of the fluids.

Composition of petroleum. Petroleum is a naturally occurring hydrocarbon mixture with impurities. Whether it occurs as crude oil, condensate, dry gas, tar, or bitumen depends upon its composition and the reservoir pressure and temperature.

Origin, accumulation, and migration of petroleum. Geologists generally believe that petroleum originated in organic-rich, fine-grained sediments. Hydrocarbon fluids migrating vertically or laterally are eventually trapped and accumulated under structural or stratigraphic barriers.

Depositions of sediments under marine and nonmarine environments. The depositional environment influences the occurrence of hydrocarbons and size, shape, and properties of the reservoir rock.

History of reservoir engineering. Even before the drilling of the Drake Well, Henry Darcy, a civil engineer in Dijon, France, proposed the law of fluid flow through porous sands. Since then, Darcy's law for single-phase linear flow has been extended to account for multiphase, multidimensional flow, providing the foundation of reservoir engineering. Various advancements have been made to characterize reservoirs more accurately and to apply various techniques to augment reservoir performance under different operating plans. The ongoing major challenge in reservoir engineering is to maximize economic recovery of oil and gas. This has led to waterflooding and thermal and nonthermal recovery processes for oil, and gas cycling for condensate reservoirs.

Reservoir life cycle. In modern days, the reservoir life cycle consists of exploration, discovery, delineation, development, production, and eventually abandonment.

Reservoir management goal. The goal of reservoir management is to maximize profitability while minimizing capital investments and operating costs. Utilizing the proven reservoir management process and maximizing use of the company's resources can lead to success. Multidisciplinary professionals working as an integrated team are the key to successful operation of reservoirs.

Reservoir management process. The modern reservoir management process involves setting a strategy, developing a plan, and implementing, monitoring, evaluating, and completing it. None of the components of the process is independent of the others. Integration of all these is essential for successful reservoir operation and management. It is a dynamic and ongoing process.

Reservoir engineers' functions. The first priority in reservoir engineering is to estimate the original hydrocarbon in place, followed by determination of reserves, field development strategy, production rates, reservoir monitoring plan, and economic life. Reservoir engineers are involved throughout the reservoir life cycle in data acquisition, analysis, and management. They are also involved with the integrated reservoir model, production and reserves forecasts, evaluation of uncertainties, implementation of new technologies, economic optimization, and management approval. Reservoir engineers need to work with multidisciplinary professionals to achieve the company's reservoir management goals.

Class Assignments

Questions

1. Name the essential elements of petroleum reservoirs.

2. What is the composition of petroleum found in subsurface reservoirs?

3. What is the origin of petroleum, and how is it formed?

4. Do the origin, migration, and accumulation of petroleum occur at the same time and in the same location? Briefly describe the geologic processes responsible for the above.

5. What is the goal of reservoir management, and how can it be achieved?

6. Define the reservoir management process. Is it static or dynamic?

7. Is the reservoir management process applicable to health management, financial management, or any other kind of management process?

8. What is an integrated reservoir model? Why it is important?

9. Describe briefly the functions of reservoir engineers. Why is teamwork crucial in reservoir studies?

10. What is reservoir simulation? How does it aid in managing a reservoir?

Exercises

1.1. Arrange the reservoir life cycle in correct sequence: delineation, production, discovery, development, exploration, and abandonment.

1.2. Based on a literature survey, describe an interesting reservoir engineering project. Include the following in brief:
   - Name, location, and size of the field
   - Reservoir characteristics and properties of oil
   - Objective of the reservoir engineering project
   - Planning and implementation
   - Measure of success
   - Lessons learned

1.3. Describe a petroleum basin chosen from literature. Include in the basin description the following:
- Geologic age of the basin
- Year discovered
- Characteristics of the rock
- Trapping mechanism
- Characteristics of the crude oil
- Estimated number of fields
- Development history and production statistics

References

Reservoir engineering

1. Satter, A., and G. C. Thakur. 1994. *Integrated Petroleum Reservoir Management—A Team Approach*. Tulsa: PennWell Books.
2. Satter, A., J. E. Varnon, and M. T. Hoang. 1992. Reservoir management: technical perspective. SPE Paper #22350. SPE International Meeting on Petroleum Engineering, Beijing, China, March 24–27.
3. Calhoun, J. C. 1953. *Fundamentals of Reservoir Engineering*. Norman: University of Oklahoma Press.
4. Pirson, S. J. 1958. *Oil Reservoir Engineering*. New York: McGraw-Hill Book Company.
5. Craft, B. C., and M. F. Hawkins. 1959. *Applied Petroleum Reservoir Engineering*. Englewood Cliffs, NJ: Prentice-Hall.
6. Amyx, J. A., D. M. Bass, and R. L. Whiting. 1960. *Petroleum Reservoir Engineering—Physical Properties*. New York: McGraw-Hill Book Company, Inc.
7. Cole, F. W. 1969. *Reservoir Engineering Manual*. Houston: Gulf Publishing Co.
8. Clark, N. J. 1969. *Elements of Petroleum Reservoirs*. Dallas: Society of Petroleum Engineers of AIME.
9. Slider, H. C. 1976. *Practical Petroleum Engineering Methods*. Tulsa: Petroleum Publishing Co.
10. Dake, L. P. 1978. *Fundamentals of Reservoir Engineering*. Amsterdam: Elsevier Scientific Publishing Co.
11. Slider, H. C. 1983. *Worldwide Practical Petroleum Reservoir Engineering Methods*. 2nd ed. Tulsa: PennWell Books.
12. Dake, L. R. 1978. *The Practice of Reservoir Engineering*. Amsterdam: Elsevier Scientific Publishing Co.
13. Mian, M. A. 1992. *Petroleum Engineering Handbook for the Practicing Engineer*. Tulsa: PennWell Books.
14. Dake, L. R. 2001. *The Practice of Reservoir Engineering*. Rev. ed. Amsterdam: Elsevier Scientific Publishing Co.
15. Lee, W. J., and R. A. Wattenbarger. 2002. *Gas Reservoir Engineering*. Richardson, TX: Society of Petroleum Engineers.
16. Towler, B. F. 2002. *Fundamental Principles of Reservoir Engineering*. Richardson, TX: Society of Petroleum Engineers.
17. Ahmed, T. 2001. *Reservoir Engineering Handbook*. 2nd ed. Houston: Gulf Publishing Co.
18. Ahmed, T., and P. McKinney. 2004. *Advanced Reservoir Engineering*. Houston: Gulf Publishing Co.

Reservoir simulation

19. Crichlow, H. B. 1977. *Modern Reservoir Engineering: A Simulation Approach*. Englewood Cliffs, NJ: Prentice-Hall.
20. Aziz, K., and A. Settari. 1979. *Petroleum Reservoir Simulation*. New York: Applied Science Publishers, Ltd. Reprinted in 1990 and available through the Society of Petroleum Engineers.
21. Mattax, C. C., and R. L. Dalton. 1990. *Reservoir Simulation*. Richardson, TX: Society of Petroleum Engineers.
22. Turgay, E., J. H. Abou-Kassem, and G. R. King. 2001. *Basic Applied Reservoir Simulation*. Richardson, TX: Society of Petroleum Engineers.
23. Fanchi, J. R. 2006. *Principles of Applied Reservoir Simulation*. 3rd ed. Houston: Gulf Publishing Co.
24. Carlson, M. 2003. *Practical Reservoir Simulation*. Richardson, TX: Society of Petroleum Engineers.
25. Abou-Kassem, J. H., S. M. Farouq Ali, and M. R. Islam. 2006. *Petroleum Reservoir Simulations*. Houston: Gulf Publishing Co.
26. Thomas, G. W. 1982. *Principles of Reservoir Hydrocarbon Simulation*. Boston: International Human Resources Development Corp.

Tables—SPE Symbols, Unit Conversions, and Acronyms

Table 1–1. List of selected SPE symbols

Letter Symbol	Reserve SPE Letter Symbol	Quantity/Description	Dimensions
a		Decline factor, nominal	
b	Y	Intercept	various
B	F	Formation volume factor	
c	k, κ	Compressibility	Lt^2/m
C	n_C	Components, number of	
C		Water drive constant	L^4t^2/m
C_{fD}		Fracture conductivity, dimensionless	
C_L	C_L, n_L	Condensate or natural gas liquids content	various
d		Decline factor, effective	
d	L_d, L_2	Distance between adjacent rows of injection and production wells	L
D		Deliverability (gas well)	L^3/t
D	y, H	Depth	L
D	μ, δ	Diffusion coefficient	L^2/t
D		Discount factor, general	
e	i	Influx (encroachment) rate	L^3/t
e^z	exp z	Exponential function	
E	η, e	Efficiency	
E	U	Energy	mL^2/t^2
E_A	η_A, e_A	Areal sweep efficiency	
E_D	η_D, e_D	Displacement efficiency	
$-Ei(-x)$		Exponential integral	
$Ei(x)$		Exponential integral, modified	
E_I	η_I, e_I	Invasion or vertical sweep efficiency	
E_R	η_R, e_R	Efficiency, overall reservoir recovery	
E_V	η_V, e_V	Volumetric sweep efficiency	
f	F	Fraction	
g	γ	Gradient	various
G	g	Gas in place in reservoir, total initial	L^3
G_L	g_L	Condensate liquids in place in reservoir, initial	L^3
h	d, e	Height (other than elevation)	L
h	d, e	Thickness	L
i		Injection rate	L^3/t
i		Interest rate	$1/t$
i_R		Rate of return (earning power)	
i_w		water injection rate	L^3/t
I	i	Injectivity index	L^4t/m
j	ω	Reciprocal permeability	$1/L^2$
J	j	Productivity index	L^4t/m
k	K	Permeability, absolute (fluid flow)	L^2
K	k, F_{eq}	Equilibrium ratio (y/x)	
K_{ani}	M_{ani}	Anisotropy coefficient	
ln		Natural logarithm, base e	
log		Common logarithm, base 10	
log_a		Logarithm, base a	
L	n_L	Moles of liquid phase	
L_f	x_f	Fracture half-length	L
m		Mass	m
m		Cementation (porosity) exponent	
m		Ratio of initial reservoir free-gas volume to initial reservoir oil volume	
m	A	Slope	various
m		Initial gascap volume as a fraction of oil volume	
M		Mobility ratio, general	
M		Molecular weight	m
M	$m_{\theta D}$	Slope, interval transit time vs. density	tL^2/m
n	N	Density (number/unit volume)	$1/L^3$
n		Exponent of backpressure curve, gas well	
n		Saturation exponent	
n_t	N_t	Moles, number of, total	
N	n	Oil in place, initial	L^3
N_p		Cumulative oil production	L^3
O		Operating expense	various
p	P	Pressure	m/Lt^2
P		Phases, number of	
p_b	p_s, P_b, P_s	Bubblepoint pressure (saturation)	mL/t^2
p_c	P_C, p_C	Capillary pressure	mL/t^2
P_{PV}		Discounted cash flow	
q	Q	Production rate or flow rate	L^3/t
Q_i	q_i	Pore volumes of injected fluid, cumulative, dimensionless	
r	R	Radius	L
R	F_g, F_{go}	Gas/oil ratio, producing	
R_s	F_{gs}, F_{gos}	Gas solubility (solution GOR)	
s	L	Displacement	L
s	S, σ	Skin effect	
s^2		Variance of a random variable, estimated	
S	ρ, s	Saturation	
S	s, σ	Storage, storage capacity	various
t_{ma}	Δt_{ma}	Matrix interval transit time	t/L
t_p	τ_p	Time well on production prior to shut-in	t

Table 1–1. cont.

Letter Symbol	Reserve SPE Letter Symbol	Quantity/ Description	Dimensions
T	θ	Temperature	T
T	T	Transmissivity, transmissibility	various
v	V, u	Velocity	L/t
V	R	Gross revenue ("value"), total	M
V	n_v	Moles of vapor phase	
w	m	Mass flow rate	m/t
W	w	Initial water in place	L^3
W	w, G	Weight (gravitational)	mL/t^2
W_e		Cumulative natural water influx	
W_i		Cumulative water injection	L^3
x		Mole fraction of a component in liquid phase	
y	f	Holdup (fraction of pipe volume filled by oil or water)	
y		Mole fraction of a component in vapor phase	
z	Z	Gas compressibility factor (deviation factor)	
Z	D, h	Elevation (referred to datum)	L
α	β, γ	Angle	1/L
α	α, η_h	Heat or thermal diffusivity	L^2/t
γ	s, F_s	Specific gravity	
γ	k	Specific heat ratio	
δ	Δ	Decrement	various
δ		Deviation, hole	
Δ		Difference or difference operator	
η		Hydraulic diffusivity	L^2/t
θ	β, γ	Angle	
θ	α_d	Angle of dip	
λ		Mobility	L^3t/m
μ		Viscosity	m/Lt
ρ	D	Density	m/L^3
σ		Standard deviation of a random variable	
σ	y, γ	Surface tension, interfacial	m/t^2
τ		Tortuosity	
ϕ	f, ε	Porosity	
Φ	β_d	Dip, azimuth of	
Φ	f	Fluid potential or potential function	various
Ψ		Dispersion factor	
Ψ		Stream function	various
Ψ		Angular frequency	1/t

Table 1–2. Additional symbols used

Symbol	Definition	Unit
a	Parameter for mole attraction	
b	Parameter for mole repulsion	
C	Coefficient of wellbore storage	$L^3/(mL/t^2)$
C_A	Shape factor	
D	Turbulent flow factor	
E	Gas expansion factor	L^3/L^3
E_g	Expansion of gascap gas	L^3/L^3
E_o	Expansion of oil and original gas in solution	L^3/L^3
f	Normal probability distribution of x	
F	Non-Darcy flow coefficient	
F	Underground withdrawal volume	L^3
f_w	Fractional flow, water displacing oil	
i	Internal rate of return	
M_c	Molecular weight of condensate	
n	Number of trials	
N_{ca}	Capillary number	
p	Probability of success	
R_n	Random number between 0 and 1	
s	Distance	L
S	Aquifer function	
U	Aquifer constant	
V	Dykstra-Parsons permeability variation factor	
W	Fracture width	L
x	Number of successful outcomes	
y	Reduced density factor	
∂	Differential operator	
ε	Correction factor for pseudocritical properties	

Table 1–3. Unit conversion: SI metric to oilfield units

Quantity	Conversion[a]		Unit
Pressure (reservoir, fluid, wellbore, capillary, standard, etc.)	$kPa \times 0.14504$	=	psia
	$MPa \times 145.04$	=	psia
	$kPa \times 4.01865$	=	in. H_2O (60°F)
Pressure gradient	$Pa/m \times 0.04421$	=	psi/ft
Length	$km \times 0.62137$	=	mile
Liquid head	$mm \times 0.03937$	=	in.
Area	$m^2 \times 2.47105 \times 10^{-4}$	=	acre
	$km^2 \times 0.3861$	=	sq. mile
Formation thickness	$m \times 3.2808$	=	ft
Reservoir volume	$m^3 \times 8.10708 \times 10^{-4}$	=	acre-ft
Fluid volume	$m^3 \times 6.28981$	=	bbl
	$m^3 \times 35.31466$	=	cft
	$m^3 \times 264.172$	=	U.S. gal
Fluid viscosity	$Pa.s \times 1000$	=	cp
Oil gravity	$\frac{141.5}{sp.\ gr.} - 131.5$	=	°API
Fluid density	$kg/m^3 \times 0.062428$	=	lb_m/cft
	$kg/m^3 \times 0.008345$	=	$lb_m/US\ gal\ (ppg)$
Specific volume	$m^3/kmol \times 16.01846$	=	ft³/lb-mol
Fluid or rock compressibility	$Pa^{-1} \times 6894.6497$	=	psi^{-1}
Fluid velocity	$m/s \times 3.2808$	=	ft/s
Liquid flow rate	$m^3/d \times 6.28981$	=	bbl/d
Gas flow rate	$m^3/d \times 35.31467$	=	cft/d
Gas/oil ratio	m^3/m^3 (std.) $\times 5.5519$	=	scf/stb
Productivity index (PI)	$m^3/kPa.d \times 43.367$	=	bbl/d-psi
Specific productivity index	$m^3/kPa.d.m \times 13.218$	=	bbl/d-psi-ft
Permeability	$\mu m^2 \times 1.01325$	=	darcy
Permeability-thickness	$mD\text{-}m \times 3.2808$	=	mD-ft
Fluid mobility	$\mu m^2/Pa.s \times 1.01325 \times 10^{-3}$	=	mD/cp
Oil recovery per unit volume	$m^3/m^3 \times 0.77583$	=	bbl/ac-ft
Concentration	$mg/kg \times 1.0$	=	wt ppm
Particle size	$\mu m \times 1.0$	=	micron
Surface or interfacial tension	$mN/m \times 1.0$	=	dyne/cm
Temperature	°C $\times 1.8 + 32$	=	°F
Temperature gradient	$mK/m \times 0.054864$	=	°F/100 ft
Heat flux	$kW/m^2 \times 316.998$	=	Btu/(hr-ft²)
Heat transfer coefficient	$kW/(m^2.k) \times 176.11$	=	Btu/(hr-ft³-°F)
Thermal conductivity	$W/(m.k) \times 0.57779$	=	Btu/(hr-ft²-°F/ft)

[a] Conversions are approximate in most cases.

Note: To convert a reservoir pressure of 20,000 kPa or 20 MPa to oilfield units, multiply by a factor of 0.14504 as follows: (20,000 kPa) $\times$ (0.14504 psia/kPa) = 2,900.8 psia; Similarly, (20MPa) $\times$ (145.04 psia/MPa) = 2,900.8 psia.

Table 1–4. List of selected acronyms

Acronym	Description	Acronym	Description
AAPG	American Association of Petroleum Geologists	ER, E_R	Recovery efficiency
acre-ft, ac-ft	Acre-feet	ESP	Electric submersible pump
AGA	American Gas Association	ETR	Early time region
AIME	American Institute of Mining, Metallurgical, and Petroleum Engineers	EV	Expected value
		FAL	Formation analysis log
AOFP	Absolute open flow potential	FAWAG	Foam-assisted water alternating gas (injection)
API	American Petroleum Institute	FFM	Full-field (simulation) model
ASTM	American Society for Testing and Materials	FVF	Formation volume factor
bbl	Barrel	FWL	Free water level
bcf	Billion cubic feet	GIS	Geographic information system
BFIT	Before federal income taxes	GLR	Gas/liquid ratio
BHA	Bottomhole assembly	GOM	Gulf of Mexico
BHCT	Bottomhole circulating temperature	GOR	Gas/oil ratio
BHFP	Bottomhole flowing pressure	GR	Gamma ray
BHP	Bottomhole pressure	GTL	Gas-to-liquids (technology)
BHST	Bottomhole static temperature	GPSA	Gas Producers Suppliers Association
BOE	Barrels of oil equivalent	GWC	Gas/water contact
BOP	Blowout preventer	HC	Hydrocarbon
bopd	Barrels of oil per day	HCIP	Hydrocarbon in place
BS&W	Basic sediment & water (analysis)	HCPV	Hydrocarbon pore volume
BSCWEPD	Barrels of steam (as cold water equivalent) per day	IGIP	Initial gas in place
Btu	British thermal unit	IMPES	Implicit pressure-explicit saturation (method)
BWPD	Barrels of water per day	IOR	Improved oil recovery
CAPEX	Capital expenditure	IPR	Inflow performance relationship
CBM	Coalbed methane	IRR	Internal rate of return
CCE	Constant composition expansion (test)	ISO	International Standards Organization
CDF	Cumulative (probability) distribution function	IWS	Intelligent well system
CGR	Condensate/gas ratio	kPa	Kilo Pascal
CHOPS	Cold heavy oil production with sand	LNG	Liquefied natural gas
CNG	Compressed natural gas	LPG	Liquefied petroleum gas
C/O	Carbon-oxygen log	LTR	Late time region
COFCAW	Combination of forward combustion and waterflooding	LWD	Logging while drilling
		M	10^3 (Oil and gas industry convention)
CVD	Constant volume depletion (test)	MBH	Matthews-Brons-Hazebroek
DCF	Discounted cash flow	Mbo	Thousand barrels of oil
DCFROI	Discounted cash flow return on investment	Mcf	Thousand cubic feet
DF	Discount factor	MD	Measured depth
DHPV	Displaceable hydrocarbon pore volume	md, mD	Millidarcy (unit of rock permeability)
DST	Drillstem test	MDH	Miller-Dyes-Hutchinson
DTS	Distributed temperature sensing	MDT	Modular Dynamics Tool (trademark of Schlumberger)
EF	Escalation factor		
EIA	Energy Information Administration	MEOR	Microbial enhanced oil recovery
EOR	Enhanced oil recovery	MMMbo	Billion barrels of oil
EOS	Equation of state	MMP	Minimum miscible pressure

Table 1–4. cont.

Acronym	Description	Acronym	Description
MRC	Maximum reservoir contact (well)	rb	Reservoir barrels
MMscfd	Million standard cubic feet per day	RF	Recovery factor
MPa	Mega pascal (10^6 pascal)	RFT	Repeat Formation Tester (trademark of Schlumberger)
MPY	Mils per year (thousandths of inch per year—measure of corrosion)	ROP	Rate of penetration
Mscfd	Thousand standard cubic feet per day	ROS	Remaining oil saturation
MSDS	Material safety data sheet	RQI	Reservoir quality index
MTR	Middle time region	SAGD	Steam-assisted gravity drive
MWD	Measurements while drilling	SBHP	Static bottomhole pressure
NAF	Acre-feet of net pay (reservoir volume)	SCAL	Special core analysis
NGL	Natural gas liquids	SCF	Standard cubic feet
NMR	Nuclear magnetic resonance	SEC	U.S. Securities and Exchange Commission
NPV	Net present value	SI	International System of Units
NTG	Net to gross (thickness ratio)	SP	Spontaneous potential
OECD	Organization for Economic Cooperation and Development	SPE	Society of Petroleum Engineers
		SPEE	Society of Petroleum Evaluation Engineers
OGIP	Original gas in place	SPWLA	Society of Petrophysicists and Well Log Analysts
OGJ	*Oil & Gas Journal*	SS	Subsea
OGR	Oil/gas ratio	SSP	Static spontaneous potential
OHCIP	Original hydrocarbon in place	SSSV	Subsurface safety valve
OOIP	Original oil in place	stb	Stock-tank barrels
OPEC	Organization of the Petroleum Exporting Countries	stbd	Stock-tank barrels per day
		STOIIP	Stock-tank oil initially in place
OWC	Oil/water contact	SWAG	Simultaneous water and gas (injection)
P&A	Plugged and abandoned (well)	t	Tonne (metric ton)
PBU	Pressure buildup (test)	tcf	Trillion cubic feet
PDF	Probability distribution function	TDS	Total dissolved solids
PDG	Permanent downhole gauge	TDT	Thermal (neutron) decay time log
PFO	Pressure falloff (test)	THAI	Toe-to-heel air injection
PI	Productivity index	TOC	Total organic carbon
PLT	Production logging tool	TVD	True vertical depth
PRMS	Petroleum Resources Management System	V	Permeability variation factor
psi	Pounds per square inch	VAPEX	Vapor extraction (recovery process)
psia	Pounds per square inch absolute	VSP	Vertical seismic profile
psig	Pounds per square inch gauge	WAF	Well allocation factor
PV	Pore volume	WAG	Water alternating gas (injection)
PVFI	Pore volume of fluid injected	WC	Water cut
PVT	Pressure-volume-temperature (data, correlation, analysis, or cell)	WHP	Wellhead pressure
		WHT	Wellhead temperature
PVWI	Pore volume of water injected	WOR	Water/oil ratio
PW	Present worth	WPC	World Petroleum Council
PWNP	Present worth net profit		

2 · Rock Characteristics, Significance in Petroleum Reservoirs and Applications

Introduction

Rock and fluid properties are the building blocks in any reservoir engineering study that lead to the formulation of a successful reservoir management strategy. Sometimes the study involves the estimation of oil and gas reserves based on a simple analytical approach, as demonstrated in this chapter. In other instances, reservoir performance prediction is accomplished by robust multiphase, multidimensional simulation models, as discussed in chapter 13. Regardless of the study and related complexity, the reservoir engineer must have a sound understanding of the rock properties involved. What is more important is the knowledge of the variability of rock properties throughout the reservoir and how heterogeneous reservoirs perform in the real world.

It is a common observation that rock properties vary from one location to another in the reservoir, often impacting reservoir performance. Some reservoir analyses are based on the assumption that a reservoir is homogeneous and isotropic, implying that the rock properties are nonvariant and uniform in all directions. However, such idealized conditions are seldom encountered in the field. Various geologic and geochemical processes leave imprints on a reservoir over millions of years, leading to the occurrence of reservoir heterogeneities that are largely unknown prior to oil and gas production. For example, the occurrence of a few fractures, an unknown geological barrier, or multiple zones of rocks having dissimilar properties can alter the reservoir performance markedly. The reservoir engineer's perception about how to develop, produce, and manage the field throughout its active life could be changed accordingly. Rock properties also may be altered as the reservoir is produced. Reservoir fluid properties are described in chapter 3.

Worldwide studies of rock properties indicate that all reservoirs are basically heterogeneous as well as unique in character. In a heterogeneous geologic formation, rock properties vary from one location to another, sometimes drastically, within a short vertical or horizontal interval. Petroleum reservoirs are unique, and patterns and trends related to the rock characteristics observed in one reservoir cannot be readily assumed to be the same as another. This is true even when the two reservoirs are situated in the same geographical region and geologic setting. In most cases, a large number of wells must be drilled and produced in order to gain adequate knowledge of rock properties and their influence on overall reservoir behavior. In the early stages of field development, very few wells have been drilled, and information regarding vital rock characteristics is very limited. It is interesting to note that critical rock heterogeneities may become apparent only at the time water or gas injection is initiated in the reservoir to augment oil recovery.

Reservoir engineers, often working as a team with geosciences professionals, face a monumental challenge in successfully developing the reservoir based on their experience, ingenuity, and a bit of luck. The effect of rock properties in influencing oil and gas recovery from a specific reservoir is a continuous learning process. A myriad of data obtained throughout the reservoir life cycle needs to be analyzed and integrated. This must be done to achieve the common goal of adding value to proven reserves.

The basic properties of rocks can be classified as the following:
1. **Skeletal.** The "skeleton" of the rocks is influenced by the depositional environment and various earth processes following deposition.
2. **Dynamic.** This relates to the interaction of the rocks and fluids in the reservoir.

Skeletal properties of interest to reservoir engineers include porosity, pore size distribution, compressibility, and absolute permeability of the rock. Dynamic or interaction properties of rock are influenced by the nature and interaction between fluids, as well as between the fluid and the rock surface. These include wettability, capillary pressure, saturation, and relative permeability.

This chapter is devoted to learning about the following:

- Reservoir rock types
- Skeletal rock properties
  - Porosity
  - Permeability
  - Formation compressibility
- Dynamic rock properties
  - Reservoir fluid saturation
  - Interfacial tension
  - Wettability
  - Capillary pressure
  - Leverett J function
  - Capillary number
  - Relative permeability
  - Formation transmissibility and storavity
- Measures of rock heterogeneity
- Reservoir characterization
- Oil and gas in place estimation
- Petroleum reserves
- Sources of rock properties data
  - Core analysis
  - Log analysis
  - Well test analysis
  - Geosciences data
  - Emerging technology
- Examples
- Applications
- Class problems

Reservoir Rock Types

As mentioned in chapter 1, reservoir rocks are classified as the following:

- **Clastic rocks.** Formed from preexisting rocks by erosion, transformation, and deposition. These include sands, sandstones, and conglomerates, and, less importantly, siltstones and shales.
- **Carbonate rocks.** Formed from organic constituents and chemical precipitates. These include dolomites, reef rocks, limestones, and chalks.

Sandstone rocks are aggregates of particles or fragments of minerals or older rocks. They come in gradation ranging from clean to dirty, sucrosic or coarse to very fine grained, white to black, unconsolidated to very consolidated or cemented, shale-free to very shaly, and no lime to limey. Most hydrocarbon-bearing geologic formations are sedimentary in origin.

Some reservoir rocks are composed chiefly of chemical or biochemical precipitates. They are carbonate sediments, mostly limestones and dolomites. Their physical characteristics range from spongy to chalky, cavernous, or vuggy. They may be classified as oolitic, oolicastic, fractured, crystalline, or reef minerals. The main biochemical agents in forming limestones are algae, bacteria, foraminifera, corals, bryozoa, brachiopods, and molluska.

In essence, petroleum reservoirs are broadly classified as either sandstone or carbonate. Again, most carbonate reservoirs consist of limestone, dolomite, or chalk formations. It is noteworthy that the majority of giant petroleum reservoirs in the world, many of which are concentrated in the Middle East, are composed of carbonate rocks. Multiple stratigraphic sequences bearing petroleum fluid in the form of oil or gas, or both, separated by relatively thin shale beds, are commonplace in petroleum reservoirs. Some shales and granite have been found to contain commercial quantities of petroleum. However, these occurrences are rare.

Skeletal Properties of Reservoir Rocks

Petroleum reservoirs have the ability to contain hydrocarbon fluids (oil and gas) in the microscopic pores of geologic formations and to transmit the fluids under certain driving forces. This ability is related to the skeletal properties of the geologic formation, including bulk porosity and absolute permeability. Skeletal rock properties are largely shaped during the depositional period or thereafter, over spans of millions of years.

Porosity

Geologic formations containing petroleum are characterized by the presence of a microscopic pore network in which oil, gas, and water may coexist (fig. 2–1). Pores are microscopic void spaces in between minute grains or particles that constitute the rock. The grains are of various shapes, sizes, and patterns that dictate the nature of the pore channels through which subsurface fluids are able to move. In certain other instances, oil and gas are found in a network of fractures, fissures, or cavities of formations. Rock porosity is a measure of the pore volume of the rock over its bulk volume.

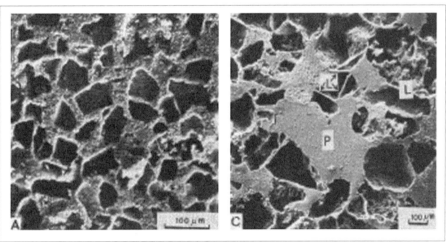

Fig. 2–1. Microscopic photos of core specimens. Dark-colored grains of porous rock amid a light-colored porous network are observed. The photo on the left exhibits well-sorted grains of sandstone and primary porosity. Pore throat diameter is relatively large compared to pore volume, leading to high potential recovery. In the next photo, certain pores are found to be larger than grains. Secondary porosity appears to have developed by dissolution of carbonate and feldspar. However, pore throat diameter is relatively small compared to pore volume. Laboratory studies indicated that oil recovery was lower from the core having secondary porosity. *Source: N. C. Wardlaw and J. P. Cassan. 1989. Oil recovery efficiency and the rock-pore properties of some sandstone reservoirs. In Reservoir Characterization. Vol. I. SPE Reprint Series no. 27. p. 214. © Society of Petroleum Engineers. Reprinted with permission.*

Porosity of a rock can be expressed as follows:

$$\emptyset = \frac{\text{volume of pore spaces in rock}}{\text{bulk volume of rock}}$$

(2.1)

The unit of porosity is dimensionless. In oilfield literature, porosity is also reported in percent of the bulk volume of the rock. If a core sample has a volume of 10 cubic inches (in.^{3}), and the total void space in the porous network of the core is measured as 2 in.^{3}, the core is said to have a porosity of 0.20 or 20%. The grain volume of the sample, defined as the volume occupied by the grains of rock, would be the difference, namely, 8 in.^{3} in this case. In other words, the bulk volume of the rock is made up of two components: (a) volume occupied by the grains, and (b) pore spaces between the grains. Oil reservoirs that are capable of commercial production usually exhibit porosity values ranging from 5% to 35% or higher. Gas reservoirs are found to produce from formations having even lower porosity. Certain naturally fractured oil- and gas-bearing formations may have significantly low or negligible porosity.

Absolute porosity and effective porosity

Due to various geologic processes that characterize a geologic formation on a microscopic scale, including deposition of grains and cementation between them, not all the pore spaces are interconnected throughout the reservoir. It is quite likely that the total pore volume is made up of two classes of pores. Certain pores are interconnected, forming continuous channels for fluid flow, while others are isolated due to excessive cementation or bonding between surrounding grains. The pores that are not part of a continuous channel network do not contribute to oil and gas production. These are not considered in the estimation of petroleum reserves. Consequently, the measure of interconnected pore volume as opposed to total pore volume is of more interest to reservoir engineers in estimating producible fluid volume and analyzing reservoir performance. This leads to the differentiation between absolute porosity and effective porosity as follows:

$$\phi_{abs} = \frac{\text{volume of connected and non-connected pores}}{\text{bulk volume of porous rock}} \qquad (2.2)$$

$$\phi_{eff} = \frac{\text{volume of interconnected pores}}{\text{bulk volume of porous rock}} \qquad (2.3)$$

where

ϕ_{abs} = absolute porosity, dimensionless, and

ϕ_{eff} = effective porosity, dimensionless.

The effective porosity of a rock is equal to the absolute porosity in reservoirs where all of the pores are interconnected to form fluid flow channels that are essentially continuous. However, in geologic formations where certain pore spaces are not interconnected, and microscopic flow channels are discontinuous, the effective porosity would be less than the absolute porosity. Absolute and effective porosity values of reservoir rock samples are routinely determined in the laboratory.

Primary porosity and secondary porosity

The porosity that initially develops in a reservoir rock during its deposition in prehistoric times is known as primary porosity. Void spaces that exist from the time of deposition between grains and crystals of the rock are common examples of primary porosity. However, secondary porosity may develop following original deposition due to various geological and geochemical processes, leading to significant alteration in rock characteristics. Examples of secondary porosity are vugs, or the cavities that are typically observed in limestone formations. Circulation of certain solutions, dolomitization of carbonate rocks, and development of fractures in the rock matrix may lead to secondary or induced porosity of the rock. The presence of secondary porosity adds to the heterogeneity of the reservoir rock and may influence the flow of fluids in the reservoir. Consequently, the actual reservoir performance may depart significantly from the case where the formation is assumed to have primary porosity only. It is further noted that secondary porosity may or may not be detected in the case of limited information obtained from the reservoir.

Factors affecting porosity

The porosity of a rock can be affected by a host of factors during deposition, as well as in the long periods following deposition. Shape, angularity, and packing and sorting of grains in a rock would dictate the volume of void space in the rock during deposition. As mentioned in the preceding section, the extent of certain processes after deposition, such as leaching, dolomitization, and fracture inducement, may lead to the development of secondary porosity. Limestones, dolomites, and shales are more frequently subjected to secondary or induced porosity.

Pore geometry, and its random or repeatable pattern throughout the formation, will influence virtually all other rock properties directly or indirectly. Of particular interest are pore size distribution, pore diameter, pore throat size, and the extent of interconnected pores in the flow network. In fact, flow through pore throats is identified as a key area for future studies that may lead to enhancement of petroleum fluid transport through the constricted pathways.

Literature review indicates that several mathematical models, ranging from simple to highly complex, have been developed that can relate various rock properties to the factors mentioned above. Certain carbonate reservoirs appear to be more difficult to model due to their highly heterogeneous nature.

Sources of porosity data

Porosity values in a reservoir can be obtained by laboratory studies of core samples, wireline porosity logs (acoustic and neutron-porosity), logging-while-drilling (LWD) tools, and seismic studies. Laboratory determination of porosity is described in the following. Wireline logs and seismic surveys are described briefly in later sections.

A logging-while-drilling tool consists of arrays of sensors to record various rock properties down the hole, such as resistivity, porosity, density, and gamma ray emission. The tool is placed above the drill collar during drilling. Collected data is stored in the tool and then transmitted to the surface through pressure pulses in the mud system. The tool is part of a measurements-while-drilling (MWD) system designed to monitor pressure, temperature, and wellbore trajectory.

Geostatistical methods are utilized to build a porosity model of a reservoir, which attempts to predict porosity values away from wells based on well data, rock type, and degree of uncertainty.

Porosity measurement of core samples

The absolute or total porosity of a core can be determined by comparing its volume before and after crushing. All pore spaces that exist in the core, regardless of whether they are interconnected or isolated, reduce to zero in the process. Equation 2.2 is then used to calculate the total porosity of the rock sample.

In contrast, the effective porosity can be determined by allowing a fluid of known density to enter the empty pores of a dry core. The volume of fluid that enters the core is readily known from the increase in weight of the saturated core and the density of the fluid. Needless to say, the fluid can enter only the interconnected pores of the rock, and Equation 2.3 can be used to calculate effective porosity.

Example 2.1. Calculation of the grain volume of a sand pack. Calculate the grain volume and porosity of a sand pack. Available data is given as follows: length = 20 in.; diameter = 2.4 in.; weight of dry sand = 6.1 lbs; and the specific gravity of the dry sand = 2.6 (water = 1.0).

Solution:

$$\text{Bulk volume of sand pack } (V_b) = \left[\frac{\pi}{4} (2.4)^2 \text{ in}^2 \, (20) \text{ in} \right] \left[\frac{1}{12^3} \frac{\text{ft}^3}{\text{in}^3} \right]$$
$$= 0.05236 \text{ ft}^3$$

Since the density of the sand pack is determined by its mass over its volume, the grain volume of the sand pack can be calculated as given in the following:

$$\text{Grain volume of the sand pack } (V_g) = (6.1 \text{ lb}) \frac{1}{62.4 \times 2.6} \frac{\text{ft}^3}{\text{lb}}$$
$$= 0.0376 \text{ ft}^3$$

Finally, the porosity of the sand pack is calculated based on bulk volume and grain volume:

$$\text{Porosity of core } (\emptyset) = \frac{\text{Bulk volume} - \text{grain volume}}{\text{Bulk volume}}$$
$$= \frac{0.05236 - 0.0376}{0.05236}$$
$$= 0.2819 \text{ or } 28.19\%$$

Example 2.2. Computation of grain density and porosity of a core sample. Calculate the porosity and grain density of a core. Can the information obtained from a core analysis be used in estimating oil and gas in place? Available core and fluid data include the following: diameter of core = 3.8 cm; length of core = 10.0 cm; dry weight of core = 275 g; weight of 100% brine-saturated core = 295 g; and brine density = 1.05 g/cm^3.

Solution: Since the core is 100% saturated with brine, the pore volume of the core can be calculated based on the volume of brine:

$$\text{Pore volume of core } (V_p) = (295 - 275 \text{ gm}) \left(\frac{1}{1.05} \frac{cm^3}{g} \right)$$

$$= 19.0476 \text{ cm}^3$$

$$\text{Bulk volume of core } (V_b) = \frac{\pi}{4} (3.8 \text{ cm})^2 (10 \text{ cm})$$

$$= 113.411 \text{ cm}^3$$

$$\text{Porosity of core } (\phi) \quad = \frac{19.0476 \text{ cm}^3}{113.411 \text{ cm}^3}$$

$$= 0.168 \text{ or } 16.8\%$$

$$\text{Grain density } (\rho_g) \quad = \frac{275 \text{ gm}}{(113.411 - 19.0476) \text{ cc}}$$

$$= 2.914 \text{ g/cm}^3$$

One of the principal applications of porosity determination involves the estimation of original hydrocarbon in place (OHCIP) as illustrated later. Log, core, and seismic studies based on porosity values are integrated to estimate oil and gas in place. Geostatistical methods are often utilized to predict porosity and other rock properties away from the wells where direct measurement is not possible. As more information on porosity and related reservoir characteristics is obtained by further developing the field, the accuracy of the estimation improves.

Cutoff porosity and net thickness of a reservoir

Many reservoirs are encountered where porosity is rather low in certain vertical sequences of the geologic formation. Certain vertical sections of formation exhibit shaliness which is nonproductive. Variations in porosity and other rock properties do occur in a reservoir due to changes in the depositional environment millions of years ago. Changes in the rock fabric after deposition are also observed. Petroleum fluids occupying the smaller pores do not contribute to production in any significant volume. It is a common practice in the industry to use a cutoff value for porosity in reservoir studies. Depending on reservoir characteristics, typical porosity cutoff points around 5% are used in oil reservoirs. Cutoff porosity is usually determined by considering only the portion of the formation having relatively high porosity that would facilitate oil and gas production in commercial quantities. Log studies conducted in various wells lead to the determination of the cutoff porosity value in a reservoir.

The concept of cutoff porosity leads to the introduction of net thickness as opposed to gross thickness of a reservoir in estimating oil and gas reserves. Net thickness represents the portion of the hydrocarbon-bearing formation that can be produced by conventional means where porosity is relatively high. Typical values for the net to gross thickness (NTG) ratio may be about 0.95 or less in petroleum reservoirs. In addition to low porosity, the net pay thickness of a reservoir is influenced by the possible existence of poor permeability and relatively high water saturation in the pores. Changes in lithology due to shaliness are mentioned earlier. Cutoff values assigned to the above properties are also commonplace. Rock permeability and water saturation are discussed later in the chapter.

High cutoff values of porosity, permeability, and water saturation, along with low values of net to gross thickness in a reservoir, lead to a decrease in petroleum reserves.

Key points—porosity

The important points to keep in mind about porosity include the following:

1. The reservoir rock must have a network of interconnected pores, or a finite effective porosity, in order to hold the petroleum fluid that is eventually produced. However, there are some exceptions, such as when the fluid is contained in a network of fractures or other contraptions.

2. The pore geometry and its random or repeatable pattern throughout the rock, as dictated by the size, shape, and sorting of the rock particles or grains, influence virtually all other rock properties.

3. The effective porosity of the rock could be lower than the absolute porosity, as not all the pores form continuous channels to transmit petroleum fluids towards the wellbore.

4. In carbonate formations, secondary porosity that may develop after deposition adds to reservoir heterogeneity and complexity.

5. The net thickness of a reservoir, as opposed to its gross thickness, is used in estimating the petroleum reserve. Net thickness depends on the cutoff value of porosity. Cutoff porosity represents a threshold value below which the formation does not contribute to production. Additionally, poor permeability and high water saturation also influence net pay, and cutoff values are assigned for these properties.

6. Porosity data is primarily obtained from log and core studies. It is also obtained from logging while drilling tools. Adequate knowledge of porosity over the entire reservoir, along with knowledge of fluid saturation, are essential in estimating initial oil and gas in place. Geostatistical methods are usually employed to model variations in porosity throughout the reservoir.

Areal, vertical, and volumetric averages of rock properties

Knowledge of basic rock properties can lead to the estimation of the total volume of oil or gas in a reservoir. This will be illustrated in Examples 2.9 and 2.10. It must be emphasized that the porosity and other rock properties used in estimating subsurface petroleum volumes must represent the reservoir properties as accurately as possible. Besides calculating the arithmetic mean of several values of porosity in a straightforward manner, more accurate approaches based on reservoir area, thickness, or volume may be employed to estimate average porosity. Vertical and areal variations of porosity and fluid saturation, as obtained from several wells, need to be known. Higher accuracy in computation is usually achieved when many wells are drilled, and a large amount of data becomes available.

A classical equation to calculate the average value of rock porosity is given in the following:

$$\emptyset_{avg} = \frac{\sum \emptyset_k x_k}{\sum x_k}$$ (2.4)

where

$\emptyset_{avg}$ = estimated average porosity of the reservoir,

x_k = reservoir area, thickness, or volume assigned to $\emptyset_k$ to obtain areal, vertical, or volumetric averages of rock porosity, respectively, and

$k = 1$ to n, n being the total number of data points.

Fluid saturation is a dynamic property of rock discussed later in the chapter. However, it is not out of place to mention that certain other rock properties such as fluid saturation are averaged in a similar manner. The average value of initial fluid saturation in the reservoir can be estimated in a manner similar to Equation 2.4. For example, the average connate water saturation is found as in the following:

$$S_{wc, avg} = \frac{\sum S_{wc} x_k}{\sum x_k}$$ (2.5)

Connate water saturation indicates the minimum value of saturation of the formation water that could not be expelled from the pores during oil or gas migration.

Example 2.3. Thickness-weighted average porosity of a formation. Compute the average porosity of a geologic formation having a total thickness of 9 ft. The necessary data is obtained from electric logs and is tabulated in the following:

Thickness, ft	Porosity, fraction
0.5	0.10
0.5	0.12
1.0	0.14
2.0	0.16
3.5	0.18
1.5	0.20

Compare the result with the arithmetic average value.

Solution: The thickness-weighted average porosity can be estimated based on the general expression as in Equation 2.4. In this case, x_k is the portion of reservoir thickness associated with each porosity value. Hence, Equation 2.4 becomes the following:

$$\emptyset_{avg(h)} = \frac{\Sigma \, \emptyset_k \, h_k}{\Sigma \, h_k} \tag{2.6}$$

where

h_k = reservoir thickness associated with k^{th} value of porosity.

Substituting the values gives the following:

$$\emptyset_{avg(h)} = \frac{[0.10][0.5]+[0.12][0.5]+[0.14][1.0]+[0.16][2.0]+[0.18][3.5]+[0.20][1.5]}{0.5+0.5+1.0+2.0+3.5+1.5}$$

$$= 0.1667 \text{ or } 16.67\%$$

The arithmetic average or mean value of porosity is found by the following:

$$\emptyset_m = \frac{\Sigma \, \emptyset_k}{n} \tag{2.7}$$

$$= \frac{0.10+0.12+0.14+0.16+0.18+0.20}{6}$$

$$= 0.15 \text{ or } 15\%$$

Equation 2.7 assumes that the values of porosity used in this example are uniformly distributed in the vertical sequence. In the case of a large reservoir, original hydrocarbon in place would be underestimated substantially with an incorrectly averaged porosity, although the thickness-weighted value appears to be only slightly higher. Further accuracy could be attained by calculating the volumetric average of porosity if a large amount of data is available. In this case, Equation 2.4 takes the following form:

$$\emptyset_{avg(Ah)} = \frac{\Sigma \, \emptyset_k \, A_k \, h_k}{\Sigma A_k \, h_k} \tag{2.8}$$

where

A_k = area of reservoir associated with k^{th} value of porosity.

Estimation of pore and hydrocarbon volumes

Reservoir engineers are interested in knowledge of porosity in determining the following, among others:

- Pore volume of rock
- Hydrocarbon volume, initial oil or gas in place
- Movable hydrocarbon volume, recovery

Pore volumes are obtained by multiplying porosity with the bulk volume of the reservoir rock. However, both need to be known with reasonable accuracy based on geophysical, geologic, petrophysical, and well tests, and on other studies.

Hydrocarbon pore volume can be obtained based on pore volume and the amount of petroleum fluid in the pores. An estimate of original oil or gas in place (the volume of hydrocarbon fluids in reservoir) is critically dependent on the porosity distribution in the reservoir. During the appraisal phase, reservoir data is very limited. Only approximate estimates of hydrocarbon volume can be obtained based on geophysical data and information obtained from exploratory wells.

Estimation of hydrocarbon volume that can be moved or produced is accomplished by laboratory analysis, available correlations, and reservoir model simulation. Estimations can also rely on prior experience and a review of worldwide trends in the recovery of petroleum. These are discussed in later chapters of the book. However, movable hydrocarbon volume is primarily dependent on permeability and relative permeability of the rock, among other factors described later in the chapter.

Uncertainties in porosity and other rock property data

Rock properties are seldom known accurately in all locations of the reservoir. Values of porosity and other properties may be estimated between wells by geostatistical modeling. Sometimes referred to as stochastic modeling, this method involves varying the properties within certain bounds dictated by well data and rock facies. The geostatistical method is discussed briefly later in this chapter and also in chapter 7. The net result of geostatistical modeling is the generation of multiple realizations of the reservoir description in the face of uncertainty.

The probability distribution of original hydrocarbon in place based on a range of values is generated using a Monte Carlo simulation rather than attempting to calculate a single number. (An example of this is shown in chapter 9.) When the field is developed, a large volume of data related to core, log, production, and well testing, and other factors, becomes available. A number of analytic methods can be employed to verify the accuracy of the original oil or gas in place estimation. These methods, including decline curve analysis and material balance, are described in later chapters with the aid of software applications.

In the preceding sections, porosity, the property of rock that is essential in providing storage space for petroleum fluids in a reservoir, has been discussed. Now attention can be given to the rock property that is instrumental in producing oil and gas from the reservoir.

Permeability of reservoir rock

Permeability is a measure of the capability of a porous medium to transmit fluid through a network of microscopic channels under a certain driving force. In subsurface porous media, the driving force originates from the pressure differential that exists between two points in the flow path of the subsurface fluid. Reservoir pressure at the location of a producing well is significantly lower than at other areas where no well is drilled. The driving force could be either natural or created by engineering design involving fluid injection through certain wells placed optimally to achieve the best results. Reservoir rocks are found to be permeable in both horizontal and vertical directions. Among a myriad of factors related to geological and geochemical processes, the permeability of a rock is influenced by the size, shape, configuration, and connectivity of a porous network.

When a new reservoir is discovered, rock permeability is one of the most valuable characteristics the reservoir engineer seeks to determine. General visualization of the future performance of a new reservoir becomes apparent when information on the range and trend in rock permeability is available. The following cases illustrate the point:

- Large reservoirs with "good" permeability, operating under favorable conditions, usually lead to a high rate of oil and gas production for a long period if the geologic formation is not highly heterogeneous. Not unexpectedly, the ultimate oil recovery from these reservoirs is relatively high.

- In contrast, a reservoir having low to very low permeability may not produce commercially for a long period. It may warrant drilling of closely spaced producers and injectors to operate, among other management strategies, such as hydraulically fracturing the formation and creating conductive pathways.

- Rocks exhibiting significant variations in permeability in various subzones within the same formation may lead to poor oil recovery during water or gas injection. The injected fluid tends to bypass oil in low permeability zones.

Spatial trends in rock permeability

Many reservoirs exhibit a trend in which good permeability is found in the central location or at the crest of the reservoir. However, gradual degradation of rock permeability is encountered as wells are drilled closer to the reservoir boundaries. Sealing faults, facies changes, or other barriers run through some reservoirs, causing rock permeability (and reservoir connectivity) to alter in an abrupt manner. Certain reservoirs, specifically carbonate reservoirs, may exhibit wide variations in permeability at various depths. Such variations can include the presence of high permeability streaks, as evident from cores obtained from several wells.

Absolute permeability

In this section, the absolute permeability of a rock unit, a skeletal property, is discussed. The effective permeability, however, is a dynamic property of rock indicative of the ability of a specific fluid to flow when multiple fluid phases are present in the porous network as their saturations change during production. Relative permeability to a fluid is a ratio of effective permeability to absolute permeability. Effective permeability and relative permeability are discussed later in this chapter. Knowledge of absolute permeability, among other properties, is traditionally obtained from laboratory studies on core samples collected in newly drilled wells. Various sources of permeability data utilized in reservoir studies are listed later in the chapter.

Measure of rock permeability—Darcy's law

In 1856, French engineer Henry Darcy conducted an experiment in order to study fluid flow behavior through a bed of packed sand particles, emulating a subsurface aquifer

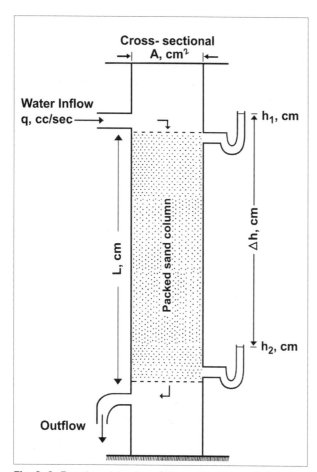

Fig. 2–2. Experimental setup of Henry Darcy to study fluid flow characteristics through a packed bed

or petroleum reservoir (fig. 2–2). He observed that the volumetric flow rate of water through the packed bed is a function of (a) the dimension of the porous medium, and (b) the difference in hydraulic head, as given in the following:[1,2]

$$q = K A \frac{h_1 - h_2}{L}$$

(2.9)

where

q = volumetric flow rate, cm^3/sec,
K = constant of proportionality for the medium; hydraulic conductivity,
A = cross-sectional area of flow, cm^2,
h_1, h_2 = hydraulic head at points 1 and 2, respectively, and
L = length of the porous medium, cm.

Darcy's law is found to be valid for other fluids, such as reservoir oil and gas, when the above equation is modified to include (i) the viscosity of the fluid and (ii) inclined flow in a dipping reservoir, as follows:[3]

$$v = \frac{q}{A} = -\frac{k}{\mu} \left[\frac{\partial P}{\partial L} - 0.433 \, \gamma \, \text{Cos} \, \alpha \right]$$

(2.10)

where

v = apparent fluid velocity, cm/sec,
A = cross-sectional area of flow, cm^2,
k = permeability of a porous medium, darcies,
μ = fluid viscosity, centipoise (cp),
$\frac{\partial P}{\partial L}$ = pressure gradient over the length of the flow path, atm/cm,
γ = fluid specific gravity (water = 1), and
α = angle of dip measured counterclockwise between the vertical direction downward and the inclined plane of the fluid flow.

In Equation 2.10, a negative sign appears on the right side as the fluid flows counter to the higher pressure. A porous medium would have a permeability of 1 darcy when a fluid having viscosity of 1 cp flows at a rate of 1 cm^3/sec under a pressure gradient of 1 atm/cm. The following expressions of darcy in familiar units are worth noting:

$$\text{darcy (D)} = \frac{(cc/sec)(cp)}{(sq \, cm)(atm)/cm}$$

$$= 1.127 \frac{(bbl/d)(cp)}{(sq \, ft)(psi)/ft}$$

In oil and gas reservoirs, the value of permeability is usually less than 1 darcy. A more convenient unit of rock permeability is a millidarcy, abbreviated in the literature as md or mD. Note that 1 darcy is equal to 1,000 millidarcies following the conventions of the metric system. Although the reservoir permeability is found to vary from less than 1 millidarcy to several darcies, producing oil reservoirs operate between a few to a few hundred millidarcies in many instances. Since gas requires much less driving force to move in the porous media, certain gas reservoirs may produce economically when the rock permeability is much less. In general, an oil reservoir having very low permeability (in the low single digits of millidarcies) may not be viewed as a good candidate for substantial production over a long period. One notable exception occurs with a fractured reservoir having low matrix permeability. In this case, fluid flow may occur predominantly through a network of highly conductive microchannels.

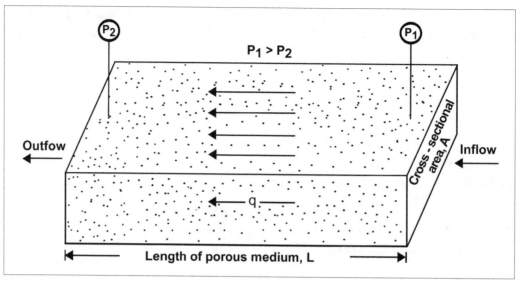

Fig. 2–3. Linear flow of fluid through porous media

Linear fluid flow through porous media

This review of Darcy's equation begins by considering the simplest case first, i.e., 1-D linear horizontal flow in homogeneous porous media. Rearranging Equation 2.10, and integrating between the limits of fluid pressure (p_2, p_1) over the length of flow path (L, 0) and noting that $\alpha=0$, a relationship between the observed pressure drop and resulting flowrate for a linear flow system (fig. 2–3) can be obtained as follows:

$$q \int_0^L \partial L = -\frac{kA}{\mu} \int_{p_1}^{p_2} \partial p$$

$$qL = \frac{kA}{\mu}(p_2 - p_1) \tag{2.11}$$

$$q = \frac{kA}{\mu L}\Delta p \tag{2.12}$$

where

L = length of linear flow path, cm, and

Δp = pressure drop ($p_1 - p_2$) of flowing fluid over length L, atm.

Note that fluid pressure decreases along the flow path, and the negative sign in Equation 2.11 is eliminated. Finally, an expression for permeability can be obtained by rearranging the above equation:

$$k = \frac{q\mu L}{A\Delta p} \tag{2.13}$$

The above equation is frequently used in the laboratory to estimate the permeability of a core sample. The important assumptions are that the fluid is incompressible and flow is steady state. In addition, the following conditions must be met:

1. Fluid flow occurs in a horizontal direction.
2. Flow occurs in a laminar regime without any turbulence effects.
3. Only one fluid is present in the system occupying the entire pore space.
4. There is no chemical reaction between the rock and the fluid.

In oilfield units, Equation 2.13 takes the following form:

$$q = 1.127 \times 10^{-3}\frac{kA}{\mu L}\Delta p \tag{2.14}$$

where
  q is given in barrels per day (bbl/d),
  k is given in millidarcies (mD),
  A is given in square feet (ft^2),
  μ is given in centipoise (cp),
  L is given in feet (ft), and
  Δp is given in pounds per square inch (psi).

Permeability derived by the aforementioned equation is referred to as the absolute permeability of the rock, as only one fluid phase completely saturates the porous medium. However, sometimes more than one fluid is present in the rock pores, such as oil and water. Then effective and relative permeability values for the individual fluid phases need to be known to calculate the flow of individual fluid phase. This topic is discussed with the dynamic properties of rock, the properties being dependent on the fluid saturation of the rock's pores.

A common method for laboratory measurement of permeability involves utilization of air as a flowing fluid through the core. Permeability measured in this manner is referred to as permeability to air or simply air permeability of the core sample. Measurement of air permeability is relatively less time-consuming and is more convenient. The equation to calculate air permeability in a linear plug of a porous medium typically simulated in a laboratory is given as follows:[4,5]

$$k_{air} = \frac{q_a \, p_a \, \mu \, L}{p_m \, (p_1 - p_2) A} \tag{2.15}$$

where
  q_a = flow rate of air based on atmospheric pressure, cm^3/sec,
  p_a = atmospheric pressure, atm, and
  p_m = mean pressure between two ends of a linear plug, atm.

The mean pressure can be calculated as $\frac{p_1 + p_2}{2}$. The derivation of Equation 2.15 is left as an exercise at the end of the chapter.

Example 2.4. Computation of core permeability based on flow of air. Calculate the permeability of the core in Example 2.1. The available data is given as follows: air flow rate = 30 cm^3/sec; inlet pressure = 10 psig; outlet pressure = 1 atm (14.7 psia); and air viscosity at elevated temperature = 0.0198 cp.

Solution:

Inlet pressure $(p_1) = \dfrac{10 + 14.7}{14.7}$

$\qquad\qquad = 1.68$ atm

Mean pressure $(p_m) = \dfrac{1.68 + 1.0}{2}$

$\qquad\qquad = 1.34$ atm

Length of core (L) = (20 in) $\left(2.54 \dfrac{cm}{in}\right)$

$\qquad\qquad = 50.8$ cm

Cross-sectional flow area (A) = $\dfrac{\pi}{4}$ (2.4 in)^2 $\left(2.54 \dfrac{cm}{in}\right)^2$

$\qquad\qquad = 29.186$ cm^2

Permeability to air in core $(k_{air}) = \dfrac{[30][1.0][0.0198][50.8]}{[1.34][0.68][29.186]}$

$\qquad\qquad = 1.135$ darcies or 1,135 mD

Linear flow in a stratified system

Due to changes in the depositional environment over the geologic time scale, many reservoirs are composed of multiple layers having distinct characteristics, including rock permeability. The average permeability of a layered system can be estimated by noting the following:

1. The total flow rate through the layered system is the sum of the fluid flow rate in each layer.

2. The pressure drop experienced in each layer is the same under steady-state conditions.

Fluid flow also can be considered through a number (n) of distinct layers having the same width but dissimilar thicknesses in a porous medium. (Note: all the assumptions inherent in Darcy's law are considered valid.) Furthermore, no crossflow due to vertical permeability is assumed to take place between adjacent layers. Equation 2.13 can then be applied to compute the volumetric flow rate of individual layers as well as of the total system, as in the following:

$$q_t = \Sigma q_i \tag{2.16}$$

$$q_i = \frac{k_i A_i}{\mu L} \Delta p \tag{2.17}$$

$$q_t = \frac{k_{system} A_t}{\mu L} \Delta p = \Sigma \frac{k_i A_i}{\mu L} \Delta p \tag{2.18}$$

where

q_t = total flow rate through the layered system,
q_i = flow rate through i^{th} layer,
k_{system} = average permeability of the layered system,
k_i = permeability of ith layer,
A_t = total cross-sectional flow area of all the layers, and
A_i = cross-sectional flow area of the individual layers.

Since all the layers have the same width (w) but different thicknesses (h_i), it can be noted:

$$A_t = h_t w \tag{2.19}$$

$$A_i = h_i w \tag{2.20}$$

Combining Equations 2.17 through 2.20, the following can be obtained:

$$\frac{q_i}{q_t} = \frac{k_i h_i}{k_{system} h_t} \tag{2.21}$$

Equation 2.21 represents an important concept for a stratified flow system. The contribution of fluid flow of an individual layer is proportional to its permeability-thickness product. The above is based on the assumption of 1-D horizontal flow of a fluid. However, in a large number of stratified reservoirs, interlayer communication of fluids (vertical crossflow) occurs in varying degrees. This must be considered in order to predict the overall fluid flow behavior.

Thickness-weighted average of permeability

The thickness-weighted average permeability of the entire system can be obtained by simplifying Equation 2.18, as given in the following:

$$k_{system} h_t = \Sigma k_i h_i \tag{2.22}$$

$$k_{system} = \frac{\Sigma k_i h_i}{h_t} \tag{2.23}$$

Linear flow in a composite system

In addition to multiple layering in a vertical direction, reservoir heterogeneity can be observed in which the fluid flowing horizontally toward the producer encounters zones of differing rock permeabilities. The reservoir can be visualized to have a series of zones in a lateral direction. The average permeability, referred to as the harmonic average, can be obtained by noting that the pressure drop experienced by the flowing fluid across the entire system is the sum of the individual pressure drops in each zone having dissimilar permeability. As can be concluded from Darcy's law, the pressure drop will be less in a zone of relatively high permeability, and vice versa.

$$\Delta p_{system} = \Sigma \, \Delta p_i \tag{2.24}$$

Harmonic average of reservoir permeability

Based on Equation 2.13 and 2.24, the pressure drops can be expressed in terms of rock permeabilities, as in the following:

$$\frac{q\mu L_{system}}{k_{system} A} = \Sigma \, \frac{q\mu L_i}{k_i A_i} \tag{2.25}$$

$$k_{system} = \frac{L_{system}}{\Sigma \, (L_i/k_i)} \tag{2.26}$$

where
   L_i = length of ith zone in the series, and
   k_i = permeability of ith zone in the series.

Geometric average of reservoir permeability

Warren and Price proposed that the geometric average of the permeabilities obtained from a number of core samples could be utilized to represent the reservoir in certain heterogeneous cases.[6] The geometric average of permeability based on core analysis is given in the following:

$$k_{system} = \exp \left[\frac{\Sigma \, h_i \ln \, (k_i)}{\Sigma \, h_i} \right] \tag{2.27}$$

Example 2.5. Individual and average layer permeabilities of a layered system. Consider the linear flow of oil through a porous medium comprised of two layers. Calculate the following:

(a) The average permeability of the entire system

(b) The individual layer contribution to volumetric flow.

Assume that Darcy's law is valid under the prevailing conditions and there is no crossflow between the layers. The following data is given: $k_1 = 100$ mD; $k_2 = 50$ mD; $h_1 = 5$ ft; $h_2 = 10$ ft; L = 200 ft; w = 150 ft; $\Delta p = 125$ psi; and $\mu = 2.1$ cp.

Solution:

(a) The average permeability of the two-layer system is calculated from Equation 2.23, as permeability and thickness of the layers are known:

$$k_{system} = \frac{[100][5] + [50][10]}{5 + 10}$$

$$= 66.67 \text{ mD}$$

(b) The flow rate through layers 1 and 2 can now be calculated based on Equation 2.14:

$$q_1 = 1.127 \times 10^{-3} \frac{[100][150][5]}{[2.1][200]} [125] = 25.16 \text{ bbl/day}$$

$$q_2 = 1.127 \times 10^{-3} \frac{[50][150][10]}{[2.1][200]} [125] = 25.16 \text{ bbl/day}$$

Although the permeability values of the two layers are very different, each layer's contribution to the total flow through the system is the same. This is due to the fact that the permeability-thickness product (kh) of the two layers is the same.

As a final check, the volumetric flow rate is computed through the entire system based on the average value of permeability:

$$q_t = 1.127 \times 10^{-3} \frac{[66.67][150][5+10]}{[2.1][200]} [125] = 50.32 \text{ bbl/day}$$

It is worth noting that due to the uncertainties inherent in reservoir description, rock permeability (millidarcies, mD), and calculated fluid flow rate (stock-tank barrels per day, stb/d) are not usually reported up to two decimal places. A notable exception occurs with tight gas reservoirs, where rock permeability in microdarcies (μD) may be encountered.

Radial flow equation predicting well rate

Fluid flow in the immediate vicinity of a vertical oil or gas well is predominantly radial, as observed in Figure 2–4. Darcy's equation can be recast in radial form in order to estimate fluid flow rate through a well when related information is known. This input includes the permeability of the rock, reservoir fluid properties and pressure profile, and formation thickness. When the fluid flow geometry is radial, fluid pressure decreases as the radial distance r deceases from reservoir boundary toward the wellbore. Hence, the need for a negative sign is eliminated and Equation 2.13 takes the following form:

$$q = \frac{kA}{\mu} \frac{\partial p}{\partial r} \qquad (2.28)$$

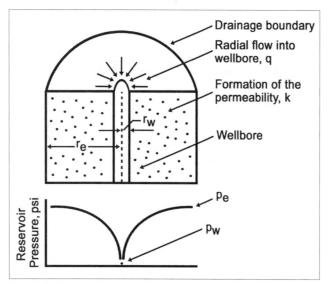

Fig. 2–4. Schematic of radial flow pattern at a producing well

The next step is to integrate between the limits of fluid pressure (p_e, p_w) at the outer radius of the drainage area and the radius of wellbore (r_e, r_w), respectively. Noting that cross-sectional flow area is $2\pi rh$ for radial flow, the following can be derived:[7,8]

$$q \int_{r_w}^{r_e} \frac{\partial r}{r} = \frac{2\pi kh}{\mu} \int_{p_w}^{p_e} \partial p$$

$$q = \frac{2\pi kh (p_e - p_w)}{\mu \ln (r_e/r_w)} \qquad (2.29)$$

In oilfield units, the following can be shown:[9]

$$q = \frac{7.08 \times 10^{-3} kh (p_e - p_w)}{\mu B_o \ln (r_e/r_w)} \qquad (2.30)$$

where

q is given in stb/d, and

B_o is given in rb/stb.

The equation above is based on the following assumptions:

1. Fluid flow is steady state and does not change with time.
2. Only one fluid occupies the entire pore space.
3. The reservoir is homogeneous.
4. The drainage radius is known accurately.

Although such ideal conditions do not exist in reality, Equation 2.30 nevertheless aids in visualizing the physics of fluid flow surrounding the wells. B_o is the oil formation volume factor defined in chapter 3.

Similarly, in the case of a gas reservoir, the following equation for flow can be derived:

$$q_{sc} = \frac{19.88 \times 10^{-3} \, T_{sc} \, k \, h \, (p_e^2 - p_w^2)}{p_{sc} \, T \, (z \, \mu) \, \ln(r_e/r_w)} \qquad (2.31)$$

The above equation is valid for relatively low reservoir pressure (< 2,000 psia), as the product $z\mu$ is assumed to be constant. A pseudopressure function, described in chapters 4 and 5, is used to treat the flow of gas in wide ranges of reservoir pressure. In Equation 2.31, T is the reservoir temperature and z is the gas deviation factor described in chapter 3. Furthermore, p_{sc} and T_{sc} are pressure and temperature at standard conditions, respectively.

Radial flow in a composite system

The harmonic average of permeability can be estimated for porous media having zones of dissimilar permeabilities in the radial direction. In the immediate area surrounding a producer or injector, a zone of altered permeability may exist. This could be due to the occurrence of various near-wellbore phenomena, such as fines migration, clay swelling, drilling fluid invasion, or an acidizing operation. The zone is visualized to be roughly circular in most cases, extending a few inches to several feet into the reservoir. Following the procedure to calculate the harmonic average in a lateral direction shown earlier, the following can be obtained in the case of radial flow that encounters concentric zones of varying permeability:

$$k_{system} = \frac{\ln(r_e/r_w)}{\sum [\ln(r_i/r_{i-1})/k_i]} \qquad (2.32)$$

where
  r_i is the radius of ith layer, and
  k_i is the permeability of ith layer.

Significance of Darcy's law

Darcy's law provides a simple, yet powerful, tool to reservoir engineers in visualizing and evaluating the factors that affect fluid flow in porous media, allowing them to draw useful inferences. Inspection of Equations 2.10 and 2.13 suggests the following:

1. Fluid flow in a porous medium takes place in a direction opposite to increasing pressure. In other words, fluid is driven toward the producing wells due to diminished reservoir pressure near the wellbore.
2. Reservoirs having high permeability would produce with relative ease, and better recovery is usually expected. If all other factors are the same, a relatively high pressure gradient (more energy) would be required to attain a similar rate of production in a low permeability formation.
3. The fluid flow rate is inversely proportional to its viscosity. This indicates that the gas phase, having much less viscosity, will flow at a faster rate than the oil phase in the same formation. Similarly, water is expected to be more mobile in a porous medium, as it is less viscous than oil. Many of the challenges facing the reservoir engineer in managing a reservoir are related to these phenomena.
4. Industry-wide reservoir strategies that aim for better oil recovery are readily understood by considering Darcy's law. As oil and gas are produced, the pressure gradient in the reservoir decreases in order to attain equilibrium.

This is expected when pressure support from the surrounding aquifer is either nonexistent or is limited. At the end of primary production, an improved petroleum recovery program can be initiated by adding additional driving energy to the remaining fluid. This is achieved by injecting a fluid in the reservoir and recreating a high pressure gradient. In thermal recovery processes, the viscosity of heavy crude is lowered by adding thermal energy to augment recovery.

It is again emphasized that conceptual application of Darcy's law greatly aids in visualizing reservoir performance in qualitative sense. This is true even before the reservoir engineering team embarks on detailed studies.

Klinkenberg effect

When gas is used to determine permeability in a porous medium, it is observed that the measured apparent permeability varies with different gases due to a "slippage effect." This is called the Klinkenberg effect.[10] The observed permeability values are larger than those obtained with a liquid and are found to be a function of the reciprocal of mean pressure. The relationship between liquid permeability and gas permeability can be expressed as the following:

$$k_L = \frac{k_g}{1 + b/P_m} \tag{2.33}$$

where

k_L = permeability to liquid when the core is 100% saturated with the liquid, millidarcies,

k_g = permeability to gas when the core is 100% saturated with the gas, millidarcies,

b = Klinkenberg factor, and

p_m = mean flowing pressure at which the gas permeability is measured, atm.

In the laboratory, the liquid permeability of a core can be obtained by determining air permeability at a series of points by varying the mean flowing pressure as shown in Figure 2–5.

Factors and processes affecting rock permeability

The permeability of a rock unit may depend on a variety of factors, including:

1. Size, shape, composition, and orientation of grains that influence pore geometry

2. Degree of cementation and shaliness

3. Presence of fractures and fissures, among other factors[11]

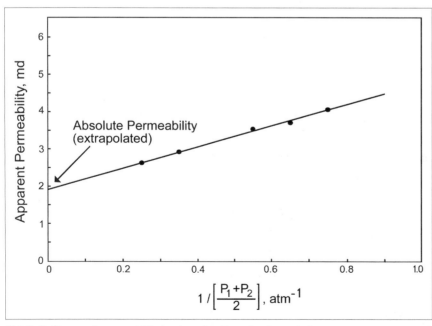

Fig. 2–5. Observed permeability to air as function of reciprocal of mean pressure

Two rock samples having similar porosities can have permeabilities that differ by one order of magnitude or even more, due to the degree of cementation between grains, among other factors. Important controlling factors in determining rock permeability include critical pore diameter, pore throat size, and tortuosity of the porous media. These allow fluid flow with varying degrees of ease or difficulty.

The depositional environment varies notably in space and time, which leads to significant variations in rock permeability, both laterally and vertically. Microscopic core studies reveal that both horizontal and vertical permeability would be high in reservoirs with large and well-rounded grains. In contrast, core samples made up of smaller and irregular-shaped grains would lead to poor permeability, and the effect could be more pronounced in the vertical direction. Moreover, the presence of clay or shale in the rock matrix is detrimental to permeability. It should be noted that shale is impermeable for all practical purposes.

In the period after deposition, various geological and geochemical processes can influence rock permeability. Any fractures or fissures that may develop in a geologic formation due to certain stresses could significantly enhance its permeability. In carbonate rocks, the process of leaching by mineral solutions flowing through the porous network may lead to permeability enhancement. Compaction of pores as reservoir pressure declines may lead to permeability damage.

Rock permeability correlations

Porosity values are obtained in various locations of a reservoir from wireline logs with relative ease. The same cannot be said about obtaining and validating corresponding permeability values. Determining permeability usually requires extensive petrophysical studies, well tests, and production history matching based on computer models. Moreover, variation of rock permeability in a geologic formation is usually more pronounced than that of porosity. Reservoir engineers need to get as much information as possible about rock permeability throughout the reservoir, as it primarily controls well flow rates. Reservoir simulation models require permeability values to be defined with good accuracy in order to predict reservoir performance. Consequently, several approaches have been developed in the industry in an attempt to correlate permeability with porosity and other rock characteristics. These models or correlations are often reservoir- or field-specific, and are dependent on lithofacies and rock groups.

Basic models involve plotting of permeability versus porosity values as obtained from petrophysical studies on a semilog scale, followed by identification of a detectable trend. The variation in permeability is likely to be more significant than the variation in porosity from one point to another in a geologic formation. As a result, a logarithmic scale is commonly used for permeability values against porosity. Less heterogeneous or near-homogeneous lithologic units in a sandstone reservoir may exhibit a definitive trend between porosity and permeability.

A general relationship between porosity and permeability can be derived as follows:

$$\log_{10} k = m\emptyset + c \tag{2.34}$$

where
m = gradient, and
c = a constant.

If two sets of permeability and porosity data in a reservoir are known, the values of gradient (m) and a constant (c) can be determined. For example, one could consider the following data obtained from cores in a formation:

Porosity, fraction	Permeability, mD
0.24	20
0.25	30
0.28	97
0.29	144

Based on these values, the gradient and intercept can be determined in Equation 2.34, as given in the following:

$$\log k = 17.1\phi - 2.8 \tag{2.35}$$

In a reservoir where extensive porosity and permeability data is available, porosity is plotted against the log of permeability in order to obtain a least squares fit. Large deviations in values of permeability from the plotted line readily point to significant heterogeneities that may exist in the reservoir.

The reservoir engineer must recall the uniqueness of the reservoir rock's properties, including porosity-permeability relationships, when relying on available correlations in the absence of extensive field data. For example, a carbonate reservoir having secondary porosity, stratification, and high permeability streaks could be difficult to correlate or model in terms of its porosity-permeability relationship.

Figure 2–6 shows the results of a study based on a large number of core samples obtained from carbonate reservoirs where the grain size of the porous rock is correlated to its permeability. The rock samples exhibited uniform cementation

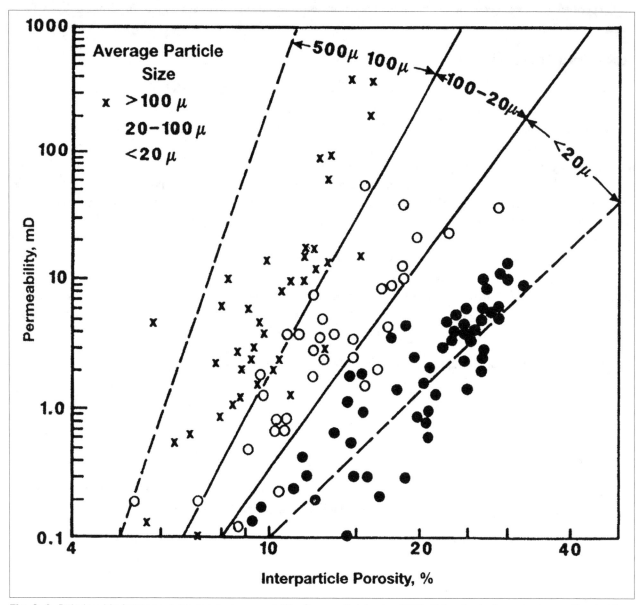

Fig. 2–6. Relationship between grain size and permeability. *Source: F. J. Lucia. 1989. Petrophysical parameters estimated from descriptions of carbonate rocks: a field classification of carbonate pore space. Reservoir Characterization. Vol. I. SPE Reprint Series no. 27. p. 89. © Society of Petroleum Engineers. Reprinted with permission.*

and had no vugs. For the same porosity, larger particle size may lead to greater permeability. Generally speaking, rock permeability increases with increasing porosity, as depicted in Figure 2–6. Higher porosity values (15% or greater) are associated with relatively high permeability when most of the porous network is interconnected and contributes to flow. Formations having low porosity are not expected to be good conductors of fluids unless high permeability streaks or fractures are present. High permeability streaks are described later.

Certain porosity-permeability correlations incorporate factors like grain size, shape, and sorting, degree of cementation, pore throat diameter, formation factor, and connectivity among pores. These factors usually influence the permeability of the rock in a complex manner. For similar rock porosities, permeability can vary significantly from one geologic formation to another due to cementation between the grains and the tortuosity of the pores, as noted earlier. Rigorous mathematical models that attempt to predict permeability values from porosity information may involve multidisciplinary studies based on geophysical information and geostatistical studies, among others. It must be emphasized again that the porosity-permeability correlations are essentially reservoir specific. They cannot be applied from one petroleum basin to another or from one reservoir to another in the same basin. They cannot even be applied from one lithologic unit to another in the same reservoir when significant heterogeneities are present (fig. 2–7).

Besides those already presented, correlations are available to estimate rock permeability as a function of both porosity and connate or formation water saturation. These correlations indicate that rock permeability varies inversely with

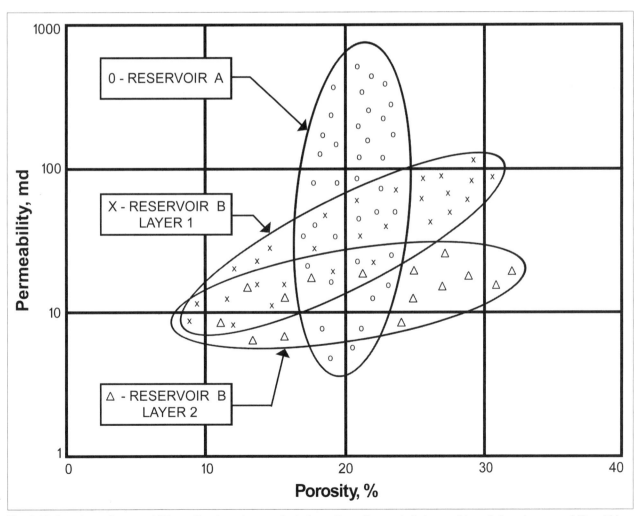

Fig. 2–7. Porosity and permeability profiles in various geologic formations. Reservoir A shows wide variations in permeability within a narrow range of porosity, indicating severe heterogeneity. Zones in the same reservoir may indicate distinctly different trends in permeability, depending on variations in depositional environment. Layers 1 and 2 in Reservoir B illustrate this phenomenon.

connate water saturation, i.e., poor permeabilities are associated with high water saturation. This can be visualized from the following observations:

1. Low permeability rocks may contain a relatively large number of small pores having smaller pore throats.

2. Formation water being the wetting fluid, it was not displaced from the smaller pores when oil migrated into the reservoir during geologic times, leading to high connate water saturation. (The wettability of rock is discussed later.)

Permeability anisotropy

Geologic formations that do not exhibit uniform rock properties in all directions are referred to as anisotropic. Of particular significance is the permeability anisotropy that may exist in a petroleum reservoir. Many reservoirs exhibit a definite permeability trend in one direction, as depicted in Figure 2–8. For example, the horizontal permeability in the east-west direction could be greater than that in the north-south direction.

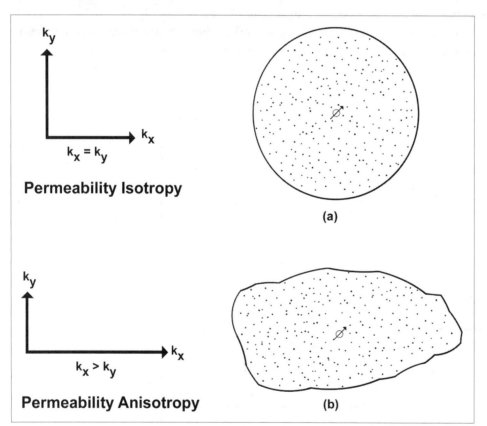

Fig. 2–8. Effects of permeability anisotropy on the shape of the fluid bank around an injection well

Information regarding directional permeability plays a critical role in designing and engineering a secondary recovery or enhanced recovery operation. In the case of water injection in a reservoir that provides additional energy for oil production, injection and production wells are aligned transverse to the directional permeability trend of the reservoir. This scheme avoids premature breakthrough of water and attains better sweep efficiency in the reservoir. Sweep efficiency is a measure of the area or volume of the reservoir contacted and swept by the injected fluid with a view to displacing petroleum fluid. The injected fluid could be water, gas, steam, or a chemical substance, among others. Improved and enhanced recovery methods, including waterflooding, are discussed in chapters 16 and 17.

Furthermore, it is frequently observed that the vertical permeability in a geologic formation is significantly different, and usually less, than its horizontal permeability. Due to the very nature of the settling and compaction process during deposition over millions of years, the elongated side of deposited grains is aligned horizontally rather than vertically. Consequently, the horizontal permeability is frequently found to be greater than the vertical permeability in a geologic formation. In many instances, the contrast between vertical and horizontal permeability is one order of magnitude or more.

High permeability streaks

Many heterogeneous reservoirs are associated with thin stratigraphic sequences of very high permeability, while the rest of the formation is of poor permeability. This usually leads to premature production of water or gas during improved oil recovery by fluid injection. As a result, oil production may diminish significantly in wells experiencing water or gas breakthrough. In the presence of highly conductive channels, water may be produced unexpectedly in oil wells soon after water injection is initiated in the nearby injectors. A similar phenomenon may be observed when the reservoir experiences water influx from an adjacent aquifer. High permeability streaks are the results of variations in the depositional environment during geologic times, among other factors. These channels may or may not be continuous throughout the reservoir. Figure 2–9 shows the occurrence of a high permeability streak in a formation where permeability measurements were obtained by a formation tester tool at close intervals of reservoir depth. In the absence of intensive data collection, existence of high permeability streaks could remain undetected until an unexpected breakthrough occurs at the producing well. In general, the existence of high permeability streaks in a reservoir may pose a significant challenge to reservoir engineers in relation to optimizing oil and gas recovery. Strategies leading to better reservoir management include conformance control, which is discussed in chapter 19.

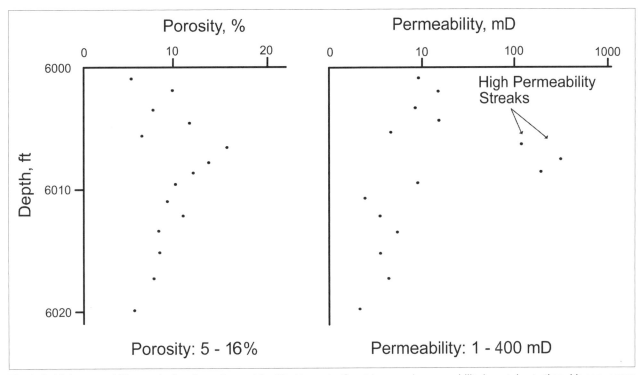

Fig. 2–9. High permeability streaks in a formation as identified by a significant increase in permeability in certain stratigraphic sequences

Alteration of rock permeability

No discussion of rock permeability is complete without mentioning the phenomenon of permeability degradation around the wellbore, frequently affecting well performance. The ability of a porous medium to transmit fluid in the immediate vicinity of the wellbore may be hindered due to a number of reasons. Two of the notable reasons include the following:

1. The introduction of an external fluid in the reservoir during drilling, hydraulic fracturing, or an enhanced recovery operation can lead to the swelling of clay material in the formation, thereby restricting fluid flow through the microscopic pore spaces.

2. The migration of fines dislodged from the relatively less-compacted formation may clog the flow channels.

The resulting phenomenon is frequently referred to as permeability damage, and the wellbore is visualized as having a "skin" at its outer perimeter. In unconsolidated sand formations, migration of fines and blockage of pore channels may pose a formidable challenge to producing oil and gas efficiently. The net result of permeability damage, also known as skin damage, is a reduction in well productivity. Remedial measures include acidizing the damaged formation or creating a fracture artificially. These lead to permeability enhancement and a subsequent increase in oil and gas production rates. Again, an optimum production rate is maintained below a threshold value in order to avoid the migration of fine particles into the producing stream. The extent of permeability damage or enhancement near the wellbore is assessed by a transient well test analysis, which is described in chapter 5. A schematic of a zone of altered permeability around a wellbore is also shown in chapter 5.

Damage in rock permeability can be induced by stress. Formations having high compressibility are subjected to pore compaction, reduction in pore throat size, and consequent decrease in permeability as reservoir pressure declines during production.

Sources of permeability information and data integration

In reservoirs having an extensive data acquisition program, information regarding rock permeability may be obtained from several sources, including the following:

1. Laboratory studies of core samples

2. Nuclear magnetic resonance (NMR) logs

3. Drillstem and other pressure transient tests

4. Formation testing following openhole logging

5. Mathematical models or correlations that attempt to predict permeability from porosity, lithology, and other data obtained from core and log studies

6. Stochastic or geostatistical methods to generate permeability models for the reservoir

Well testing and formation tester tools are described in chapter 5. It is interesting to note that the sources mentioned above provide average values of formation permeability at various scales, usually ranging from several inches to several hundred feet. Core and log studies represent permeability data on a much smaller scale and may highlight the heterogeneities present within the formation at different locations and depths, including fractures and high permeability streaks. By contrast, permeability data obtained from transient flow tests represents the average effective permeability. This permeability is based on the radial flow of fluid toward the well over a large area when multiple fluid phases are present, such as oil and water. Both types of information are integrated to characterize a reservoir and effectively manage it.

When unknown heterogeneities are present near the well, permeability values obtained by well test and laboratory core analysis could be very different. This is readily apparent in reservoirs with high permeability streaks present in certain intervals. The average permeability of the formation obtained by well tests may differ significantly from the permeability data obtained in laboratory studies of cores obtained from intervals where such streaks are not present. A similar phenomenon may be encountered in naturally fractured reservoirs. However, it must be kept in mind that the permeability value obtained by well testing is the effective permeability of the reservoir in the area of investigation, while core studies usually point to the absolute permeability of the rock sample.

Key points—rock permeability

To recapitulate the role of rock permeability in a reservoir and its relationship to other rock properties, the following points should be noted:

1. Rock permeability is a measure of the ability of a porous medium to transmit fluid under a certain driving force. It is primarily influenced by shape, size, pattern, and connectivity of microscopic pore channels that develop during deposition of the rock.

2. In most instances, permeability is instrumental in determining the production performance of a petroleum reservoir. Formulation of an effective strategy to develop and manage a reservoir depends largely on rock permeability and associated reservoir heterogeneities.

3. In specific circumstances, such as with near-homogeneous sandstone formations, reservoir permeability trends may correlate to porosity in a straightforward manner. The log of permeability is plotted against corresponding porosity values to obtain a straight line with a reasonable degree of accuracy. In contrast, limestones exhibiting a high degree of heterogeneity may demonstrate significant variations in permeability even within a narrow range of porosity.

4. Relatively low porosity and high connate water saturation obtained from log studies may point to a low range of rock permeability.

5. Factors like localized degradation of permeability, the presence of high permeability channels, and permeability anisotropy, to name a few, may significantly affect oil and gas production. Detrimental effects may include unsatisfactory well productivity, unwanted production of water or gas, and relatively low recovery from the reservoir.

6. According to Darcy's law, the permeability of a porous medium can be calculated from the observed fluid flow rate, and vice versa. To do so, certain information related to the reservoir and fluid characteristics must be known. In fact, Darcy's law serves as a cornerstone in all reservoir fluid flow analyses.

7. The range of permeability as observed in a typical reservoir is usually much greater than that of porosity in the same reservoir. Correlations between porosity and permeability are usually reservoir-specific and could be difficult to model or predict when significant heterogeneities are present.

8. Information regarding absolute and effective rock permeabilities can be obtained and validated from a number of sources that include core studies, well tests, production history matching, and applicable correlations. All available information is subsequently integrated in various reservoir studies.

9. Rock permeability may be altered in the vicinity of the wellbore due to various processes. These include the migration of fine particles, invasion of drilling fluid, and swelling of clay when an incompatible fluid is injected into the formation. Permeability damage may lead to a significant loss in well productivity, requiring remedial measures such as hydraulic fracturing or acidization of the damaged formation.

10. Sources of rock permeability data include core studies, NMR logs, formation tester studies, and well tests. Each source represents permeability values at different scales. The information, based on various sources, is integrated to characterize the reservoir. When unknown heterogeneities exist in the reservoir, permeability values obtained by various methods can differ significantly.

Formation compressibility

During primary production of oil and gas, rock and fluid compressibility contribute to driving energy. Sometimes a water or gascap drive is absent. In this case, the volumetric expansion of fluids and reduction in pore volume of rock are the primary mechanisms of initial production in an oil reservoir. Knowledge of rock and fluid compressibility is important in reservoir studies in order to evaluate flow characteristics accurately under changing reservoir pressure conditions.

The compressibility of a hydrocarbon-bearing formation under an isothermal condition is a function of the rate of change of pore volume with change of pressure. In mathematical form, formation compressibility can be expressed as in the following:[12]

$$c_f = -\frac{1}{V_p} \left(\frac{\delta V_p}{\delta p} \right)_T \tag{2.36}$$

where

c_f = compressibility of formation, $1/psi$ or psi^{-1},

V_p = pore volume of rock, ft^3, and

p = pressure exerted on formation, psi.

(Note that T is added as a subscript in the above expression to indicate an isothermal condition during compression or expansion of the rock.)

Following depletion in pressure in a reservoir due to oil and gas production, a slight decrease in rock porosity may be encountered, as the overburden pressure remains unchanged. However, certain reservoirs having highly compressible rock may experience a significant reduction in pore openings through which fluid must be transmitted to the wells. Consequently, microscopic pore channels in the rock become more restrictive. Fluid flow through these channels declines noticeably, as experienced in certain gas reservoirs. Examples of compaction-induced permeability reduction can be found in unconsolidated sandstone reservoirs in many petroleum regions. One such region includes the Diana Basin, located in the Gulf of Mexico (GOM), in which up to 80% reduction in flow capacity is encountered.[13]

Values of formation compressibility are usually in the order of 10^{-6} psi^{-1}. Formation compressibilities ranging between 3×10^{-6} and 15×10^{-6} psi^{-1} are encountered frequently. Formation compressibility, sometimes referred to as pore compressibility, is usually found to be in the same order of magnitude as reservoir oil and water compressibility. Hall proposed the following correlation to estimate formation compressibility as a function of porosity:[14]

$$c_f = 1.87 \times 10^{-6} \, \phi^{-0.415} \tag{2.37}$$

The following correlations are also available in the literature to estimate the formation compressibility when the reservoir lithology is also known:[15]

$$c_f = \frac{97.32 \times 10^{-6}}{(1 + 55.8721 \, \phi)^{1.42859}} \tag{2.38}$$

The above equation is valid for sandstone formations having porosity in the range of 0.02 to 0.23. A similar correlation exists for limestone formations, as in the following:

$$c_f = \frac{0.853531}{(1 + 2.47664 \times 10^6 \, \phi)^{0.9299}} \tag{2.39}$$

The above correlation is valid for porosity values ranging between 0.02 and 0.33. The average absolute error was found to be 2.6% in the case of a sandstone formation and 11.8% in the case of a limestone formation.

The change in porosity as a consequence of change in pressure can be estimated as in the following:

$$\phi = \phi_0 \exp \left[c_f (p - p_0) \right] \tag{2.40}$$

where

p_0 = original pressure, psi, and

ϕ_0 = original porosity, fraction.

The exponential term in the above equation becomes negative as pressure decreases, resulting in a reduction of porosity. Although the reduction in pore volume due to change in pressure is small at first glance, it contributes to the energy needed for primary oil production in an undersaturated reservoir. This is demonstrated in chapter 12, where the principles of material balance are utilized to predict reservoir performance.

Rock compressibility is composed of bulk and pore volume compressibility. Furthermore, total compressibility of a subsurface rock-fluid system is based on the pore (formation) compressibility and the compressibility of various fluids present in the pores of rock. Knowledge of compressibility is required in a number of reservoir studies, including well test analysis (chapter 5), material balance methods (chapter 12), and reservoir simulation (chapter 13). Total and effective compressibility are defined in chapter 3.

Key points—formation compressibility

The following points about formation compressibility are summarized here:

1. Formation compressibility is a measure of the rate of change in pore volume as a consequence of change in the prevailing pressure following the production of oil and gas.

2. Pore volume of the rock is reduced as reservoir pressure declines following production. However, except for highly compressible formations, the change is very small in most cases, as the typical value of formation compressibility is given in 10^{-6} psi^{-1}. Published correlations can be used to estimate formation compressibility based on the porosity of the rock.

3. Compressibility of the formation and fluids contained within the pores plays a significant role in primary oil production in the absence of a gas phase. Again, in certain gas reservoirs having significant formation compressibility, production may decrease markedly following a decline in reservoir pressure. This occurs as rock porosity and pore openings decrease, restricting the flow of reservoir fluid in porous media.

Dynamic properties of rocks

The simultaneous existence of more than one fluid in the pores under dynamic conditions of flow leads to the "dynamic" properties of rock that are influenced by rock-fluid interaction and interaction between immiscible fluids.The resulting properties include the wettability of the rock, interfacial tension, capillary pressure, and relative permeability, as described in the following discussion. Dynamic rock properties are strong functions of individual fluid saturation and usually vary with time and location as a reservoir is produced. Hence, the description of dynamic rock properties in this chapter is preceded by a detailed treatment of the saturation of fluids in porous media. Furthermore, it must be borne in mind that most of the dynamic properties of rock are interrelated.

Reservoir fluid saturation

Knowledge of fluid saturation is of prime importance in every phase of reservoir study. This includes the estimation of initial oil or gas in place at discovery and identification of reservoir zones where a large quantity of oil is left behind. It also involves evaluation of the enhanced oil recovery process. The relative amounts of oil, gas, or water that will flow when more than one phase is present in the porous medium are dependent on individual phase saturation. As a reservoir is produced, subsurface fluid saturations can alter significantly with time. This phenomenon can be observed due to the appearance of a new phase (such as free gas or condensate liquid). It also can occur following introduction of a driving fluid (such as water or gas) to augment recovery. Fluid phase changes that occur due to a decline in reservoir pressure are discussed in chapter 3. Moreover, as specific gravities of fluids are different, reservoirs exhibit dominance of a particular fluid saturation at various depths. For example, an oil zone having high oil saturation can be overlain by a gas cap and underlain by formation water.

Fluid saturation is usually expressed as a fraction of pore space or in percentages. For example, a reservoir having an oil saturation of 70% implies that 70% of the pore space in the rock is occupied by liquid petroleum. This saturation value represents an aggregate of hydrocarbon components present as liquid. (Hydrocarbons are discussed in chapter 3.) In the case of a reservoir where there are only two fluid phases, oil and water, the remaining 30% of the pore space is occupied by formation water, as expected. If a gas cap exists, overlying the oil zone in the reservoir, gas saturation would dominate in the top section.

In an oil reservoir, initial oil and water saturations add up to unity, or 100% of the pore space, when no free gas is present:

$$S_{oi} + S_{wi} = 1 \tag{2.41}$$

where

S_{oi} = initial oil saturation, fraction, and
S_{wi} = initial water saturation, fraction.

This equation is readily derived from the fact that the oil and water phases must occupy the entire pore space and add up to 100% of the pore volume. Following production, oil saturation would decrease, accompanied by the dissolution of gas or by an increase in water saturation due to the various reservoir drive mechanisms that may exist as described in chapter 8. Individual fluid saturations vary with time and location in a producing reservoir, and they control the relative flow of oil, gas and water towards the wellbore. As long as any dissolved gas remains in solution, the following is valid during the life of a reservoir:

$$S_o + S_w = 1$$
$$S_o, S_w = f \text{ (location in reservoir, time)}$$
$$q_o = f \text{ } (S_o, \text{ various other rock and fluid properties)}$$

where

q_o = average flow rate of oil in porous media.

As shown later in the chapter, the oil in place can be estimated based on knowledge of the reservoir's extent, average net thickness, porosity, and initial water saturation.

Initial water saturation is obtained from wireline log studies in a newly discovered reservoir. It is not expected to vary significantly in a lateral direction in a new reservoir unless certain geologic discontinuities exist. However, it does vary in a vertical direction in the transition zones between oil and water as a consequence of capillary effects. Due to the higher specific gravity of water, the saturation of water tends to increase with reservoir depth. The trend could be gradual or sharp, depending on rock and fluid characteristics. Topics on capillary pressure and transition zone are discussed later.

In a dry gas reservoir, where there is no liquid condensate present, initial gas saturation can be calculated in a manner similar to Equation 2.41:

$$S_{gi} + S_{wi} = 1 \tag{2.42}$$

where

S_{gi} = initial gas saturation, fraction.

Producing oil and gas reservoirs usually exhibit initial hydrocarbon saturations in excess of 70%. The rest of the pore space is filled with formation water that is not mobile in most instances. This is often referred to as connate water saturation (S_{wc}), which fills the pores during deposition of the rock.

When all three phases (oil, gas, and water) are present in a reservoir, such as an oil reservoir with an initial gas cap as shown in Figure 1–2, the saturation fractions of all three phases must add up to unity. At certain depths, however, the saturation of one fluid could be zero. Descending from the top of the reservoir to the bottom, the saturations of gas, oil, and formation water can be summarized as follows at any point during production:

Gas cap:

$$S_g + S_{wc} + S_{og} = 1 \tag{2.43}$$

Gas-oil transition zone:

$$S_g + S_o + S_{wc} = 1; S_o > S_{og} \tag{2.44}$$

Oil zone:

$$S_o + S_{wc} = 1; S_g = 0 \qquad (2.45)$$

Oil-water transition zone:

$$S_o + S_w = 1; S_w > S_{wc} \qquad (2.46)$$

Bottom water zone:

$$S_w = 1; S_o = 0 \qquad (2.47)$$

where

S_{og} = gas-cap interstitial-oil saturation, and

S_{wc} = connate water saturation

When the reservoir produces oil and gas simultaneously below the bubblepoint, knowledge of the saturation of any two phases is necessary in order to calculate the saturation of the third phase in the reservoir. The basis of calculation is in the following:

$$S_o + S_g + S_w = 1 \qquad (2.48)$$

Transition zones are treated later in the chapter. Important fluid properties, including the bubblepoint pressure of oil in a reservoir, are discussed in chapter 3.

The other example would be a gas condensate reservoir in which three fluid phases, namely gas, condensate liquid, and water, are present under certain reservoir pressures and temperatures.

Certain other dynamic rock properties, such as relative permeability and capillary pressure, are strong functions of fluid saturation as discussed in later sections of this chapter. Reservoir behavior due to evolution of fluid phases accompanying significant changes in saturation is discussed in chapter 3.

Changes in fluid saturation during the reservoir life cycle

This section describes fluid saturations that are "milestones" in a reservoir during production, along with their relationship to ultimate oil and gas recovery. These are:

- Irreducible saturation
- Critical saturation
- Movable saturation
- Residual saturation

Furthermore, knowledge of initial and residual oil saturations leads to the estimation of ultimate recovery. The following is a brief explanation of the fluid saturation terminology mentioned above.

Irreducible water saturation. Irreducible saturation is a certain saturation level below which the fluid will not flow through the microscopic pores and channels in the porous medium. The reservoir fluid will adhere to the surface of the pores due to the existence of certain forces. These forces include the surface tension between the fluid and the rock surface and the interfacial tension between two immiscible fluids present in the pores, among other factors. Connate water saturation, S_{wc}, as obtained from log and core studies, is the minimum water saturation that would remain adhered to the pores and not become mobile. During geologic times, rock pores were originally filled with subsurface water. Initial water saturation of 100% was later reduced to a considerably low value (typically 20%–30%) due to the migration of oil into the pore space and expulsion of pore water. This value is known as connate water saturation and is generally found to vary inversely with formation permeability in basin-wide studies. In contrast, interstitial water saturation is simply the water saturation found in the pores, which may or may not date back to the time of formation of the rock. At the time of the discovery of a reservoir, the initial hydrocarbon in place (IHCIP) is computed based on the value of connate water saturation, as shown in the following:

Initial hydrocarbon pore volume = Total pore volume − Volume of connate water

Critical gas saturation. Consider an oil reservoir in which no gas evolves out of solution within the reservoir as long as the reservoir produces above the bubblepoint. When the reservoir pressure declines below the bubblepoint, gas evolves out of solution but is not immediately mobile. Following the buildup of free gas saturation to a certain threshold value, referred to as critical gas saturation, the vapor phase begins to flow towards the wellbore. Critical saturation is a term used in conjunction with increasing fluid saturation. The effects of critical gas saturation on reservoir performance are shown in chapter 8.

Movable oil saturation. It is important to recognize that only a fraction of the oil in place is ultimately produced in most reservoirs. This poses a challenge to attain better recovery, requiring a thorough understanding of reservoir behavior. This necessitates the estimation of movable oil saturation, which represents the maximum volume of oil that can be moved or produced ultimately from a reservoir. Hence, movable oil saturation is defined as follows:

Movable oil saturation = Initial oil saturation − Residual oil saturation

$$\begin{aligned} S_{om} &= S_{oi} - S_{or} \\ &= 1 - S_{wc} - S_{or} \end{aligned}$$

(2.49)

Similarly, producible volumes in a gas reservoir can be estimated based on the initial and final gas formation volume factors discussed in Example 2.11.

Residual oil saturation. At the end of the productive life of the reservoir, the oil saturation that is left behind in the reservoir is referred to as residual oil saturation. This is illustrated in chapter 17, which presents an oil saturation profile with depth, as obtained by core studies in a heavy oil reservoir following thermal recovery. Some studies prefer the term remaining oil saturation (ROS) in quantifying the oil left behind following the primary or a subsequent recovery. Knowledge of residual oil saturation is of great interest to reservoir engineers as it points to the ultimate recoverable reserves.

Residual oil saturation, as can be obtained from laboratory displacement studies, points to a value below which oil is no longer mobile within porous media. The term residual saturation is used in connection with decreasing fluid saturation.

Oil recovery factor. Based on the knowledge of oil saturations, the recovery factor for an oil reservoir can be estimated as follows:

$$E_R = \frac{(S_{oi}/B_{oi} - S_{or}/B_{or})}{S_{oi}/B_{oi}} \times 100$$

(2.50)

where
E_R = recovery factor, %,
S_{or} = average value of residual oil saturation in reservoir, fraction,
B_{oi} = initial oil formation volume factor (FVF), rb/stb, and
B_{or} = residual oil formation volume factor, rb/stb.

The oil formation volume factor is defined in chapter 3. It relates to changes in oil volume that occur between the reservoir and the surface conditions due to the reduction in pressure. Gas recovery factor is treated in Example 2.11.

Remaining oil saturation. Following the primary depletion of a typical reservoir based on natural drive mechanisms, a substantial portion of movable oil is usually left behind in the reservoir. As a result, the remaining oil saturation encountered in the reservoir could be rather high. Reservoir engineers are interested to know the remaining oil saturation in a matured field, especially in the zones of bypassed oil for further recovery. The task requires multidisciplinary studies involving petrophysics, geology, geophysics, production, and reservoir simulation.

Improved and enhanced oil recovery. A reservoir asset team designs, simulates, pilot tests, and implements appropriate improved or enhanced oil recovery (IOR or EOR) operations, described in chapters 16 and 17. Following a successful enhanced oil recovery operation, incremental oil is produced, and the remaining oil saturation is reduced further in a reservoir. For example, study of a typical oil reservoir throughout its life cycle may lead to the following scenario, as shown in Table 2–1:

Table 2-1. Typical changes in oil saturation throughout the reservoir life cycle

Average Value of Oil Saturation in the Reservoir	Saturation, fraction	Source of Data
Initial	0.80	Petrophysical studies
Following primary production	0.68	Production history/reservoir studies
Following improved and enhanced recovery	0.55	Production history/reservoir studies
Irreducible/residual	0.22	Petrophysical studies

During primary production, a reduction in oil saturation in rock pores may be associated with an increase in gas or water saturation, or both. This is due to the evolution of gas from in-situ oil or water influx from adjacent aquifer, as discussed in chapters 3 and 4. At the end of improved oil recovery, further reduction in oil saturation is compensated for by an increase in the saturation of the injected water, gas, or other fluids. In essence, the saturation fractions of reservoir fluids, namely oil, gas, and water, add to unity.

It is apparent from Table 2–1 that not all of the movable oil can be produced even by implementing improved or enhanced recovery methods. Beyond a certain point, it is no longer feasible to produce a reservoir in an economic sense. The ultimate recovery from oil reservoirs is found to average around 35% worldwide (as illustrated in chapter 10). It is the primary goal of reservoir engineers to improve recovery based on technological innovation and new approaches in reservoir management, including detailed studies and intensive monitoring.

Furthermore, recovery factor calculations must take into consideration the lateral and vertical extent of a reservoir swept by an injected fluid in order to recover oil as discussed in chapters 16 and 17.

In a gas reservoir, however, the recovery factor is usually quite high (80% or more), as gas is much more mobile than oil and water. This implies that, in contrast to oil reservoirs, residual gas volume at the end of reservoir life is markedly low.

Key points—reservoir fluid saturation

To recapitulate, the following points can be noted about reservoir fluid saturation:

1. Either two or three fluid phases are encountered in a petroleum reservoir at discovery. These are: (i) oil and formation water; (ii) gas and formation water; or (iii) oil, gas, and formation water.

2. Experience has shown that petroleum reservoirs need to have high oil or gas saturation to be viable.

3. The saturation calculations of the individual phases require the knowledge of the saturation of at least one phase when two phases are present in porous media. However, a third fluid phase may appear (gas or condensate) as the reservoir pressure declines, requiring knowledge of the saturations of two phases to determine the saturation of the third phase.

4. Individual fluid saturation is expected to change significantly during the life of a typical reservoir. Knowledge of initial and residual oil saturations is of prime significance to reservoir engineers, as the information points to movable oil saturation and ultimate recovery from the reservoir.

5. Reservoir simulation studies need to estimate individual phase saturations at all locations based on available data in order to predict the relative flow of oil, gas, and water during the life of the reservoir.

Interfacial tension

Petroleum reservoirs contain at least two fluid phases in the porous rock: either oil and water or gas and water. Certain other reservoirs contain all three phases: oil or liquid condensate, gas, and interstitial water. When more than one immiscible fluid phase is present in the porous medium, a thin film develops at the boundary between the two fluid surfaces. At the interface of the fluids, the force exerted by two immiscible fluid phases is dissimilar, leading to

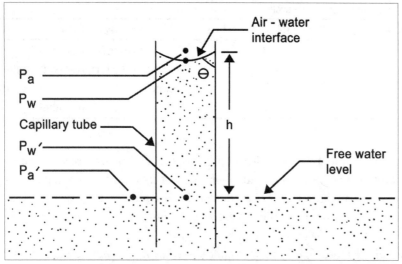

Fig. 2–10. Fluid column in thin tube due to interfacial tension between immiscible fluids. The symbols p_a and p_w denote pressure exerted by air and water, respectively, at the interface. The capillary pressure is equal to the difference: $p_a - p_w$.

the phenomenon of interfacial tension. This results in resisting miscibility between the two fluids, which generates a certain resistance to fluid flow in a porous medium. The resisting force would otherwise be nonexistent if the rock pores were entirely filled with a single phase of fluid.

Figure 2–10 is a schematic of a capillary tube experiment familiar to most students of science. When a thin glass tube of very small diameter, often referred to as a capillary tube, is immersed in a large water-filled container, the water level in the tube rises higher than that in the container due to the "capillary effect." In this case, the resulting surface tension is a function of radius of the tube, the height of water column in the tube, the densities of water and air, and the contact angle. Interfacial tension between two immiscible fluids can be expressed as in the following:

$$\sigma = \frac{r \, h \, g \, \Delta\rho}{2 \cos \theta}$$

(2.51)

where

σ = interfacial tension between two immiscible phases, dynes/cm,

r = radius of capillary tube, cm,

h = height of capillary column, cm,

$\Delta\rho$ = difference in densities between two fluids, g/cm^3 (or g/cc), and

θ = angle of contact.

In visualizing certain other dynamic properties of rock, the concept of interfacial tension and surface tension, as existing between (a) two reservoir fluid phases, and (b) between a fluid phase and a rock surface, is absolutely necessary. Such dynamic properties include wettability, capillary pressure, and relative permeability. The influence that these rock properties have on multiphase fluid flow and overall reservoir performance is discussed next.

Wettability of reservoir rock

Reservoir rocks can be either water-wet or oil-wet, depending on the tendency of one immiscible fluid phase over the other to adhere to, or "wet," the pore walls in the microscopic network. Wettability depends on the type of minerals in the rock matrix and on the composition of the fluids, oil, and water in the pores of rock. It is a function of the interfacial tension that exists between the oil phase and pore surface, between the water phase and pore surface, and between the two fluid phases as follows:

$$\sigma_{os} - \sigma_{ws} = \sigma_{ow} \cos \theta_c$$

(2.52)

where

σ_{os} = interfacial energy between the oil phase and the solid surface, dynes/cm,

σ_{ws} = interfacial energy between the water phase and the solid surface, dynes/cm,

σ_{ow} = interfacial tension between the oil and water phases, dynes/cm, and

θ_c = contact angle at the interface of oil, water, and the solid surface, degrees.

The contact angle is measured through the water phase as shown in Figure 2–11. It is observed that the angle of contact differs significantly depending on the wettability characteristic of the rock. When the surface of the rock pore is strongly water-wet, water droplets spread out at the periphery due to greater affinity between the water phase and the solid surface, leading to a contact angle significantly less than 90°. In contrast, a strongly oil-wet rock tends to have a contact angle significantly greater than 90°. Rock samples of intermediate wettability are also encountered, having a contact angle around 90°.

It is generally believed that reservoir rocks are more likely to be water-wet than oil-wet, since porous rocks are originally occupied by formation water through geologic times. During migration of the oil into the water-saturated formation, only a part of water is expelled from the rock's pores. The remaining portion is left behind, adhering to pore walls and overcoming the forces of oil migration. However, oil-wet or mixed wettability reservoirs are not uncommon.

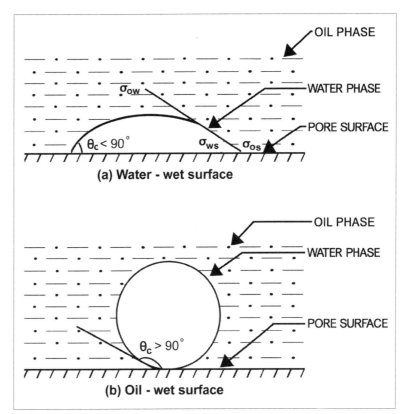

Fig. 2–11. Conceptual depiction of wettability. In an oil-wet porous medium, water droplets are less likely to adhere to the pore walls.

Water-wet reservoirs are considered to be better candidates for improved oil recovery by water injection into the reservoir. Oil is expected to flow through the pore channels with relative ease, as it has little or no tendency to adhere to the rock surface. In contrast, the oil phase exhibits a greater affinity for remaining in the pores in oil-wet reservoirs. Consequently, residual oil saturation following water injection is relatively high, and ultimate recovery is lower. Laboratory core studies indicate that the wetting fluid tends to occupy the smaller pores and does not flow through pore channels during drive by a nonwetting fluid.

Key points—wettability

In conclusion, the following points are summarized:

1. Rocks exhibit a characteristic known as wettability, where one fluid preferentially adheres to the pore surface over the other. The phenomenon occurs due to the existence of interfacial tension between immiscible fluids, as well as between fluids and the pore surface.

2. The wettability of a hydrocarbon-bearing rock has a profound impact on preferential flow of one fluid over another. The wetting fluid phase tends to occupy smaller pores in the rock and remains there under dynamic reservoir conditions.

3. All factors being the same, water-wet rock is expected to perform better when water is injected in the reservoir to displace in-situ oil, as the latter has little or no tendency to adhere to the rock surface.

4. Studies indicate that reservoir rocks may exhibit intermediate wettability characteristics, in addition to being either water-wet or oil-wet. Moreover, rock wettability may alter with time as certain liquids come in contact with the pore surface.

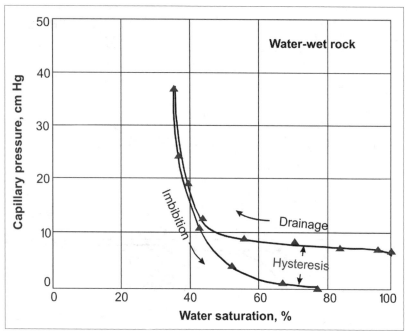

Fig. 2–12. Capillary pressure during drainage and imbibition in strongly water-wet rock. *Source: C. R. Killins, R. F. Nielsen, and J. C. Calhoun, Jr. 1953. Capillary desaturation and imbibition in rocks.* Producers Monthly. *Vol. 18 , no. 2: pp. 30–39.*

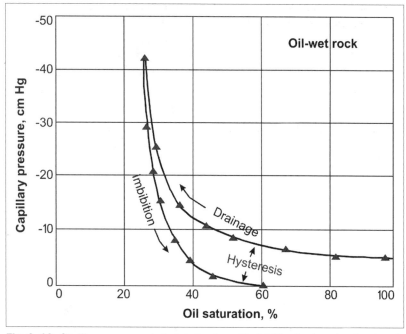

Fig. 2–13. Capillary pressure during drainage and imbibition in strongly oil-wet rock. *Source: C. R. Killins, R. F. Nielsen, and J. C. Calhoun, Jr. 1953. Capillary desaturation and imbibition in rocks.* Producers Monthly. *Vol. 18, no. 2: pp. 30–39.*

Capillary pressure

As discussed in the preceding section, when two immiscible fluid phases, such as oil and water, are present in a porous medium, one of the phases preferentially "wets" the pore surface over the other. As a result, a pressure differential is found to exist between the two phases that can be expressed as capillary pressure. Pressure exerted by the nonwetting phase is higher than that exerted by the wetting phase. This condition is necessary, as the nonwetting phase with a higher pressure, such as migratory oil, enters the pores of rock initially filled with the wetting phase at a lower pressure, such as formation water. The magnitude of the capillary pressure in a porous medium is influenced by fluid saturations, interfacial tension between the two fluid phases, and the radius of the pores, among other factors. As mentioned earlier, interfacial tension originates from the fluid behavior whereby the surface of one fluid phase in contact with another acts as a thin membrane and resists miscibility of the two phases.

A generalized expression for capillary pressure as it relates to fluid phases in porous media is as follows:

$$p_c = p_{nw} - p_w \qquad (2.53)$$

where

p_{nw} = pressure exerted by nonwetting phase, psi, and

p_w = pressure exerted by wetting phase, psi.

The oil-water capillary pressure can be expressed as follows:

$$p_{c,wo} = p_o - p_w \qquad (2.54)$$

where

$p_{c,wo}$ = capillary pressure at the water-oil interface, psi,

p_o = pressure exerted by the oil phase, psi, and

p_w = pressure exerted by the water phase, psi.

In a water-wet reservoir, the value of the capillary pressure is positive, as oil is the nonwetting phase. Similarly, gas-water and gas-oil capillary pressures are defined as in the following:

$$p_{c,go} = p_o - p_g \qquad (2.55)$$

$$p_{c,gw} = p_g - p_w \qquad (2.56)$$

where

$p_{c,go}$ = capillary pressure at the gas-oil interface, psi,

$p_{c,gw}$ = capillary pressure at the gas-water interface, psi, and

p_g = pressure exerted by the gas phase, psi.

Figures 2–12 and 2–13 plot the capillary pressure behavior of water-wet and oil-wet rock samples, respectively. In each plot, two sets of curves represent values of capillary pressure during the phenomena of drainage and imbibition. During drainage in a core, the wetting phase fluid is replaced by a flowing nonwetting phase. In water-wet rock (as

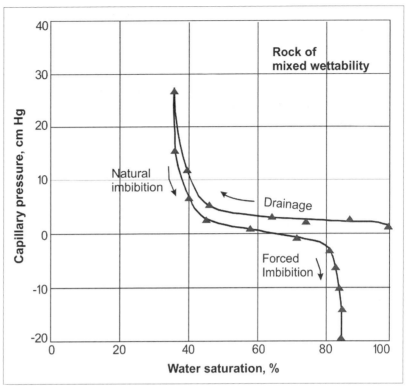

Fig. 2–14. Capillary pressure during drainage and imbibition in a rock of intermediate wettability. *Source: C. R. Killins, R. F. Nielsen, and J. C. Calhoun, Jr. 1953. Capillary desaturation and imbibition in rocks.* Producers Monthly. *Vol. 18, no. 2: pp. 30–39.*

shown in fig. 2–12), water saturation is reduced as a consequence of the drainage process, while the saturation of the nonwetting phase, oil, is increased. This process can be viewed as the desaturation of the wetting phase in porous media. In contrast, the imbibition process involves an increase in the saturation of the wetting phase. During imbibition, the wetting fluid phase is allowed to imbibe into the core, thereby increasing its saturation.

Figures 2–12 and 2–13 are strikingly similar in demonstrating capillary pressure behavior of rocks having strong water- or oil-wetting characteristics. In both cases, a certain threshold pressure is required to commence drainage of the oil or water phase. When the rock is either strongly water-wet or strongly oil-wet, the threshold pressure, also referred to as the displacement pressure, tends to be greater.

In contrast, Figure 2–14 demonstrates the drainage and capillary pressure behavior of a core of intermediate wettability. It is noted that the threshold pressure is comparatively lower. Moreover, at the end of drainage, water readily imbibed to a relatively low saturation value of 62%. However, saturation is more than 78% following imbibition in the case of the strongly water-wet rock, as shown in Figure 2–12. Additionally, the threshold pressure is a function of the interfacial tension between the wetting and nonwetting fluid phases and the pore diameter. Intuitively, small pore diameters in a rock lead to relatively high threshold pressure.

Effects of hysteresis

A review of Figures 2–12 and 2–13 reveals that the paths followed by capillary pressure during drainage and imbibition of the wetting phase are not same. This phenomenon, referred to as the hysteresis effect, points to the fact that the capillary pressure in a porous medium would be influenced by the history of saturation changes. While studying a petroleum reservoir, it is important to know whether fluid saturation is either decreasing or increasing to determine the capillary pressure, in addition to the value of fluid saturation.

For the same wetting phase saturation, capillary pressure is observed to be greater during drainage than during imbibition. In other words, a relatively large pressure differential between the nonwetting and wetting fluid phases would be required at a given saturation to expel or drain the wetting phase than to imbibe it. The relative permeability trends of the individual fluid phases are also dependent on whether the process is drainage or imbibition of the wetting phase. Relative permeability of reservoir fluids is discussed later.

The above is an important point to note, as the relative flow of fluids in porous media is subjected to the effects of hysteresis, among others.

Leverett J function

A dimensionless function, commonly referred to as the J function, is proposed by Leverett that accounts for capillary pressure as a function of water saturation. It is given in the following:[16]

$$J = \frac{p_c}{\sigma \cos \theta} (k/\phi)^{1/2}$$

(2.57)

where

J = Leverett J function, dimensionless,
σ = interfacial tension between the oil and water phases,
θ = wettability angle of the rock,
k = permeability of the formation, and
ϕ = porosity of the formation.

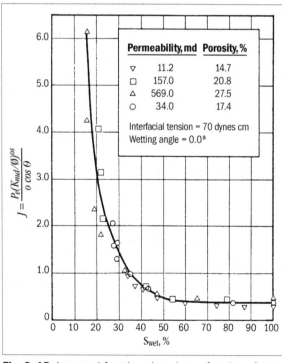

Fig. 2–15. Leverett J function plotted as a function of water saturation. *Source: H. C. Slider. 1976. Practical Petroleum Reservoir Engineering Methods. Tulsa: PennWell.*

Note that capillary pressure behavior as presented in Figures 2–12 and 2–13 typically varies from one location to the other in a formation. The Leverett J function offers an advantage in that a generalized curve can be developed as a function of saturation for the entire reservoir (fig. 2–15).

Example 2.6. Calculation of the J function. Calculate the J function for a core sample where observed capillary pressure is 4.5 psi at S_w = 40% in laboratory measurements. Porosity is 17%, and permeability is 70 mD. Assume interfacial tension is 50 dynes/cm and the angle of contact is 45°.

Solution: In oilfield units, Equation 2.57 can be written as:[17]

$$J = 0.21645 \frac{p_c}{\sigma \cos \theta} (k/\phi)^{1/2}$$

$$= 0.21645 \frac{4.5}{50 \cos (45°)} (70/0.17)^{1/2}$$

$$= 0.559 \text{ at } S_w = 40\%$$

The J function can be calculated for a series of saturation values based on laboratory measurements. Then a capillary pressure versus saturation plot can be constructed for the entire field based on studies conducted on multiple cores having different rock properties.

Capillary number

The capillary number is a ratio of viscous forces to forces arising out of interfacial tension. It is found that the measure can be correlated to residual oil saturation following enhanced recovery by water injection. Residual oil saturation represents the amount of oil that remains behind in the reservoir following production, and is treated in chapter 3. The capillary number is defined as the following:[18,19]

$$N_{ca} = C \frac{k_w \Delta p}{\phi \, \sigma_{ow} \, L} \qquad\qquad (2.58)$$

where

N_{ca} = capillary number, dimensionless,
C = a coefficient depending on the units used,
k_w = effective permeability to water,
ϕ = porosity of the reservoir, fraction, and
σ_{ow} = interfacial tension between oil and water.

The capillary number is of great significance in enhanced oil recovery operations that strive to reduce interfacial tension between fluids in the rock pores. (These processes are described in chapter 17.) Injection of surfactants is one example. During a successful oil recovery operation, forces arising out of interfacial tension are reduced, and viscous forces dominate the flow in porous media. Thus the capillary number is relatively large, and oil recovery is expected to be better. In contrast, capillary numbers are relatively lower when forces due to interfacial tension are significant, leading to large residual oil saturation. Capillary numbers may typically vary from 10^{-8} to 10^{-2}.

Transition zone in the reservoir

The phenomenon of capillary pressure exerted by one fluid over another in porous rock leads to a number of important observations. In most hydrocarbon-bearing formations, oil and water are not separated by a sharp boundary in the vertical direction. The porous network in the rock can be viewed as a bundle of capillary tubes arranged in a tortuous fashion. The water would rise to a certain height in the presence of oil in a manner similar to the rise of water observed in capillary tube experiments illustrated in Figure 2–11. It is observed that during the transition from the water zone to the oil zone, water saturation decreases gradually from 100% to a limiting value in an upward direction. The limiting saturation is usually referred to as the irreducible or connate water saturation. Within the transition zone, both oil and water phases are mobile.

Due to the effects of capillary pressure, formation water rises through the pores of the rock. This leads to the development of an oil-water transition zone in which water saturation decreases in an upward direction (fig. 2–16). In Figure 2–16, the oil/water contact is the height above which water saturation becomes less than 100%, and oil is found in the remaining pore space. At further depth, the free water level (FWL) is found, which would be the height of freestanding water in the absence of any capillary forces.

In certain cases, the reservoir rock at the lower part of the transition zone is water-wet. However, wettability alteration to an oil-wet system is encountered towards the top, where oil saturation is relatively high. Hence, transition zones could be of mixed wettability. Transition zones could be long due to low rock permeability, sometimes hundreds of feet, from which oil production can take place. However, the production of oil is accompanied by a water cut, as both phases are mobile in the transition zone. Characterization of the transition zone is required in correctly assessing the oil in place and corresponding reserves. Long transition zones are encountered in a large number of carbonate reservoirs worldwide.

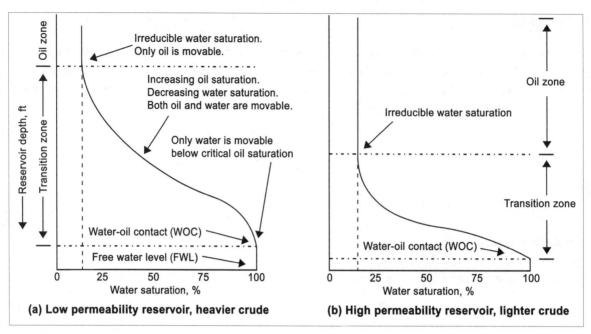

Fig. 2–16. Development of a transition zone in a reservoir as a function of capillary forces, rock characteristics, and density differences between immiscible fluids

Referring to Figure 2–10, a relationship between the height of the capillary rise above the free water level and capillary pressure can be found as in the following:[20]

$$p_c = p_{nw} - p_w \tag{2.59}$$

$$= (p_{nw}' - gh\rho_{nw}) - (p_w' - gh\rho_w) \tag{2.60}$$

In Equation 2.59, p_a is replaced by p_{nw}, the pressure exerted by the non-wetting phase. Pressure values at the capillary interface (p_{nw} and p_w) are evaluated by considering corresponding fluid phase pressure values (p_{nw}' and p_w') at a point located at the base of the capillary rise (h). The head of the fluids is subtracted from these values. However, $p_{nw}' = p_w'$, as these points are located in the same elevation. Hence Equation 2.60 reduces to the following:

$$p_c = gh(\rho_w - \rho_{nw}) \tag{2.61}$$

When p_c is in psi, h is in ft and ρ is in lb/ft³, the following can be obtained:

$$p_c = \frac{h}{144}(\rho_w - \rho_{nw}) \tag{2.62}$$

Equation 2.62 is a general relationship between capillary pressure and the height of capillary column that is valid for oil, gas, and water. The capillary height above the free water level can be computed when capillary pressure and fluid densities are known, as shown below. It is further noted that capillary height is an inverse function of the difference between fluid densities. In other words, two immiscible fluids may lead to a long capillary height if their densities are similar. In cases where the fluid densities are very dissimilar, such as at the gas/oil and gas/water contacts, the transition zone is rather short.

$$h = 144\frac{p_c}{\rho_w - \rho_{nw}} \tag{2.63}$$

Based on the above equation, the height of the oil/water contact as measured from the free water level can be expressed as the following:

$$h_{FWL-WOC} = 144\frac{p_d}{\Delta\rho} \tag{2.64}$$

where

p_d = displacement pressure, psi.

Displacement pressure can be obtained from capillary pressure versus saturation plots (see Figure 2–17).

In water-drive reservoirs, the oil/water contact is dynamic and rises with oil production as water encroaches into the reservoir from an aquifer located at the bottom or edge. It is interesting to note that in highly faulted and compartmentalized formations, an oil/water contact can be found "perched" at an anomalous depth due to the lack of connectivity between compartments. In naturally fractured formations, the transition zone between oil and water could be negligible due to the very high vertical conductivity of the fractures.

Key points—transition zones and fluid distribution

The following should be recalled regarding transition zones and the distribution of fluids in subsurface porous media:

1. In a broader sense, the three fluid phases, oil, gas, and water, are distributed in the reservoir according to an equilibrium attained between capillary and gravity forces when no viscous forces are present. The reservoir fluids are vertically separated due to the effects of gravity, but saturation transition zones do exist at the interface due to capillary forces.

2. The extent and shape of the transition zones are functions of fluid density, interfacial tension, and pore geometry, among other factors.

3. Transition zones are usually long in low permeability reservoirs, as the porous medium is associated with pore channels of smaller diameter.

4. The difference in densities between the wetting and nonwetting phases influences the length of a transition zone, as seen from Equation 2.63. A transition zone between heavier oil and water (where the contrast in fluid densities is comparatively small) would be longer than that between lighter oil and water, all other factors being equal. When a large density difference exists between the immiscible fluids, such as gas and water, the transition zone is relatively short.

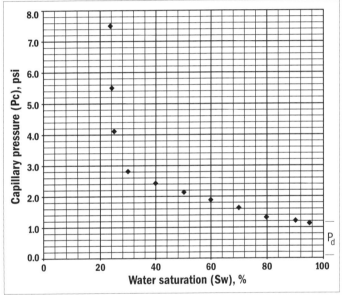

Fig. 2–17. Capillary pressure versus saturation data of an oil reservoir

5. Transition zones are identifiable from water saturation information varying between connate water saturation and 100%, as obtained by wireline logs. A transition zone between oil and water could be as large as 100 ft, or even more, due to certain fluid and rock characteristics.

6. Production from an oil/water or gas/water transition zone would lead to early water production, since water is mobile in this zone. Wells are completed above the transition zone as standard industry practice in order to maximize oil or gas production and reduce the possibility of high water cut.

7. The oil/water contact rises with production in a water-drive reservoir due to the water influx from an aquifer located at the bottom or edge of the reservoir. In highly faulted and compartmentalized formations, multiple oil/water contacts at anomalous depths can be encountered.

8. When a capillary pressure versus saturation curve is available for a reservoir as shown in Figure 2–17, a relationship between water or oil saturation versus depth is readily obtained by employing Equation 2.63. In fact, reservoir simulation studies perform this task to check the validity of the input data. The static saturation distribution as obtained by calculation is compared against the log results and other sources.

Example 2.7. Transition zone characteristics. The following information is obtained from a newly discovered oil reservoir. Capillary pressure versus saturation data based on core samples is plotted as shown in Figure 2–17. The objective is to calculate the following:

(a) The thickness of the transition zone

(b) The base of the oil zone

(c) The depth where water saturation is 40%

(d) The free water level

The necessary data is given in Table 2–2.

Solution:

(a) Examining Figure 2–17, it can be observed that $p_d = 1.1$ psi, approximately. It can be further noted that connate or irreducible water saturation of 24% is reached when p_c is about 6.5 psi. Based on Equation 2.63, the thickness of the transition zone can be obtained, as in following:

Table 2–2. Data and data sources for calculation of transition zone

Parameter	Value	Source
Oil/water contact, ft	7,050	Resistivity log
Density of oil, lb/ft³	42.8	Laboratory measurements
Density of water, lb/ft³	64.05	Laboratory measurements
Interfacial tension, dynes/cm	52	Literature review

$$h = \frac{144\,(6.5\text{-}1.1)}{(64.05\text{-}42.8)}$$

$$= 36.6 \text{ ft}$$

(b) The depth to the base of the oil zone is calculated as in the following:

$$\text{Depth } S_{wc} = 7{,}050 - 36.6 = 7{,}013.4 \text{ ft}$$

(c) From the plot, $p_c = 2.4$ psi where $S_w = 40\%$. The depth where $S_w = 40\%$ is calculated as in the following:

$$\text{Depth}_{Sw=40\%} = 7050 - \frac{144\,(2.4)}{(64.05\text{-}42.8)}$$

$$= 7{,}050 - 16.26$$

$$= 7{,}033.74 \text{ ft}$$

(d) The depth to the free water level can be estimated by Equation 2.64:

$$\text{Depth}_{FWL} = 7050 + \frac{144\,(1.1)}{(64.05\text{-}42.8)}$$

$$= 7{,}057.45$$

The results of the calculations are summarized in Table 2–3.

Table 2–3. Summary of results

Level	Depth, ft
Base of oil zone ($S_w = 24\%$)	7,013.4
$S_w = 40\%$	7,033.7
Oil/water contact ($S_w = 100\%$)	7,050.0
Transition zone	7,013.4–7,050.0
Free water level	7,057.5

Relative permeability

As described earlier, the absolute permeability of rock is a measure of its ability to transmit fluid where the fluid completely saturates the porous medium. This condition is simulated in core samples in laboratory studies for the measurement of absolute rock permeability. In reality, multiple fluid phases are commonly encountered in a petroleum reservoir. It is observed that one of the fluid phases, such as oil, usually experiences a decrease in flow when a second phase, such as water, is present in the microscopic pore network. In contrast to absolute permeability, the effective permeability of a rock to a fluid is determined by the relative saturation of all the fluid phases present in a porous medium, in addition to rock characteristics. Relative permeability is a non-linear function of individual fluid saturation.

Petroleum fluids diminish in saturation in subsurface porous media as the reservoir is produced following the water entry. Consequently, the effective permeability to oil becomes progressively less, until a limiting value of saturation is reached. The latter is referred to as residual oil saturation. On the other hand, effective permeability to water would increase as water saturation increases in the rock pores. This may occur as water enters the reservoir from an adjacent aquifer or is introduced from the surface through injection wells to augment oil recovery.

Relative permeability to a fluid phase is defined as the ratio of its effective permeability to the absolute permeability of rock. The relative permeabilities of the oil, gas, and water phase may be expressed as in the following:

$$k_{ro} = k_o / k \tag{2.65}$$

$$k_{rg} = k_g / k \tag{2.66}$$

$$k_{rw} = k_w / k \tag{2.67}$$

where
  k_{ro} = relative permeability of oil, dimensionless,
  k_{rg} = relative permeability of gas, dimensionless,
  k_{rw} = relative permeability of water, dimensionless,
  k_o = effective permeability of oil, mD,
  k_g = effective permeability of gas, mD, and
  k_w = effective permeability of water, mD.

Since relative permeability is dimensionless, it serves as a common standard in reservoir studies regardless of the magnitude of absolute or effective permeability encountered in a reservoir. Relative permeability of a fluid phase always varies between 0 and 1 in any instance. Knowledge of relative permeability is crucial in understanding multiphase fluid flow behavior in a reservoir and for predicting future reservoir performance. Relative permeability trends are of great significance when undesirable water or gas flow is anticipated in an oil reservoir. As noted previously, the relative permeability of a fluid phase is a function of the saturations of all of the fluid phases present in the rock. For example, in an oil reservoir where all three phases are present, the following is observed:

$$k_{ro} = f(S_o, S_g) \tag{2.68}$$

The absolute permeability of a rock sample is a unique number. In contrast, the relative permeability to different fluid phases constitutes a set of values that depends on fluid saturation, as illustrated in Figures 2–18 and 2–19. Several correlations are available in the industry to predict relative permeability as a function of the fluid phase saturation in a reservoir. Selected correlations are described later. Petrophysical studies are commonly performed on core samples to obtain relative permeability data. Well tests serve as an excellent source to obtain effective permeability to oil and gas under actual field conditions. Well test analysis is treated in chapter 5. Validation of relative permeability data can be accomplished by production history matching in simulation models, as described in chapter 13.

It is highly important to note that the relative permeability characteristics of reservoir fluids usually change from one location to another. Various rock facies in a reservoir may exhibit very different relative permeability trends. A large amount of relative permeability data as obtained from core samples is typically incorporated in reservoir models in order to make realistic predictions of recovery.

Examination of typical relative permeability curves for water-wet and oil-wet cores, as shown in Figure 2–18, reveals the following:

1. Values of relative permeability range between 0 and 1, as it is the ratio of effective permeability over absolute permeability. For a fluid phase, whether it is oil, gas, or water, a relative permeability value of 0 is encountered at a limiting saturation where the phase in question ceases to flow in a porous medium. In the case of the water phase, this point is referred to as the irreducible water saturation, $S_{w,irr}$. The limiting saturation for the oil phase is known as the residual oil saturation, S_{or}.

2. The relative permeability of a fluid phase increases with an increase in the saturation of the phase in porous media. The relationship between the two parameters is nonlinear. Apart from direct measurements in the laboratory, various correlations exist in the literature to estimate relative permeability values based on fluid saturation. Selected correlations are described later in the chapter.

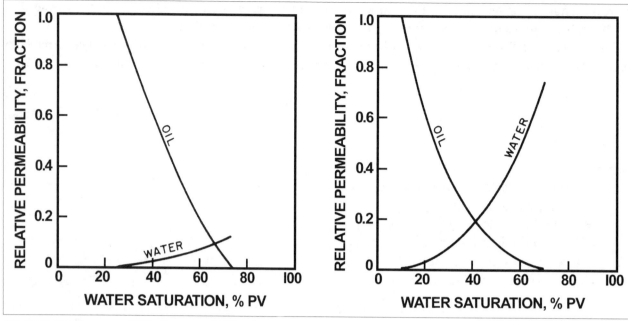

Fig. 2–18. Relative permeability of water and oil in (a) a water-wet reservoir and (b) an oil-wet reservoir. *Source: F. F. Craig. 1971. The reservoir engineering aspects of waterflooding. SPE Monograph 3. Dallas: Society of Petroleum Engineers. © Society of Petroleum Engineers. Reprinted with permission.*

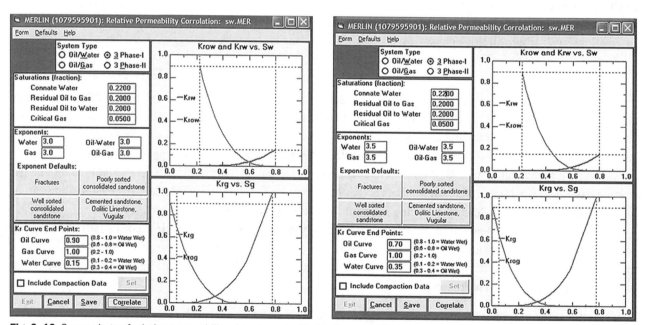

Fig. 2–19. Screenshots of relative permeability plots generated by Merlin reservoir simulation software for (a) water-wet and (b) oil-wet systems. Data requirements to generate the results include selection of system type (two-phase or three-phase, well-sorted or poorly sorted sandstone, etc.), residual fluid saturations, and relative permeability end points, among others. *Courtesy of Gemini Solutions, Inc.*

3. In the case of an oil-wet rock, both oil and water relative permeability curves are likely to shift to the left as shown in Figure 2–19. Oil is a wetting fluid with an affinity to adhere to the pore walls rather than to flow through larger channels. Thus it may become immobile and adhere to the pore walls at a relatively large saturation. Consequently, recovery from oil-wet reservoirs could be less optimistic. Since water is the nonwetting phase in an oil-wet formation, it tends to become mobile at relatively less saturation, with minimal affinity to adhere to the pore walls. It is mobile through relatively small pores. Again, reservoirs having mixed wettability characteristics will encourage the flow of the water phase when compared to strongly water-wet rock.

The following rules of thumb are found in the literature to distinguish between the wettability preferences of a rock unit, depending on connate water saturation and relative permeability characteristics:[21]

	Water-Wet	Oil-Wet
Connate water saturation	> 20%–25% PV, in the usual case	In general, < 15% PV, and frequently < 10% PV
Saturation at which $k_{ro} = k_{rw}$	> 50% PV water saturation	< 50% PV water saturation
k_{rw} at maximum water saturation, i.e., at floodout	In general, < 30%	> 50% and approaching 100%

Effect of imbibition and drainage processes

It has been observed that the relative permeability characteristics of fluid phases are not the same during the saturation and desaturation processes of a particular fluid in porous media. This phenomenon, known as hysteresis, was discussed in connection with capillary pressure earlier. In geologic times, petroleum reservoir rock pores were originally filled with subsurface water, and the porous media were considered to be water-wet prior to the migration of petroleum fluids. In later stages, oil migrated into the reservoir by displacing the connate water to a large degree. The saturation of water was reduced to a minimum value known as connate or irreducible water saturation, while oil filled up the rest of the pore space. This process in which a nonwetting fluid phase (migrating oil) displaces the wetting phase (interstitial water) is known as drainage.

In contrast, imbibition is a process whereby the wetting phase displaces the nonwetting phase in porous media. The process of imbibition is exemplified by waterflooding in a water-wet reservoir. Oil, being the nonwetting phase, is displaced by water introduced through injection wells. Oil is eventually produced, resulting in a decrease in the nonwetting phase saturation.

Oil and water relative permeability curves are found to shift to the left during imbibition when compared to the drainage curves. This implies that when water is the wetting phase, it may be mobile at certain low values of saturation during the imbibition process, while it is immobile during drainage at the same saturation. A similar hysteresis effect is observed for a nonwetting fluid phase during imbibition and drainage. Consequently, the correct relative permeability data and rock wettability trend must be used in all reservoir studies, including simulations leading to the design of waterflooding.

Laboratory measurements of relative permeability

Two-phase relative permeability data is usually obtained from laboratory tests by either of the following:
- Dynamic method
- Static method

In the dynamic method, oil is displaced from a core by gas or water, and relative permeabilities are calculated from the observed production data. In the static method, the flow of each phase is measured separately, when, at a particular saturation, capillary equilibrium between oil and gas or water is carefully maintained. Laboratory methods are treated in detail in the literature.[22–24]

Relative permeability correlations

Several correlations are proposed in the literature between the relative permeability of a fluid phase and its saturation. Wyllie and Gardner proposed the following correlations for oil-water or gas-oil systems during the drainage cycle:[25]

Well-sorted sand (unconsolidated).
I. Oil-water relative permeabilities:

$$k_{ro} = (1 - S^*)^3 \tag{2.69}$$

$$k_{rw} = (S^*)^3 \tag{2.70}$$

II. Gas-oil relative permeabilities:

$$k_{ro} = (S^*)^3 \tag{2.71}$$

$$k_{rg} = (1 - S^*)^3 \tag{2.72}$$

Poorly sorted sand (unconsolidated).

I. Oil-water relative permeabilities:

$$k_{ro} = (1 - S^*)^2 (1 - S^{*1.5}) \tag{2.73}$$

$$k_{rw} = (S^*)^{3.5} \tag{2.74}$$

II. Gas-oil relative permeabilities:

$$k_{ro} = (S^*)^{3.5} \tag{2.75}$$

$$k_{rg} = (1 - S^*)^2 (1 - S^{*1.5}) \tag{2.76}$$

Cemented sandstone, oolitic limestone, and vugular rocks.

I. Oil-water relative permeabilities:

$$k_{ro} = (1 - S^*)^2 (1 - S^{*2}) \tag{2.77}$$

$$k_{rw} = (S^*)^4 \tag{2.78}$$

II. Gas-oil relative permeabilities:

$$k_{ro} = (S^*)^4 \tag{2.79}$$

$$k_{rg} = (1 - S^*)^2 (1 - S^{*2}) \tag{2.80}$$

where

$$S^* = \frac{S_o}{1 - S_{wc}} \text{, for gas-oil system,} \tag{2.81}$$

$$S^* = \frac{S_w - S_{wc}}{1 - S_{wc}} \text{, for oil-water system, and} \tag{2.82}$$

S_{wc} = connate water or irreducible water saturation, fraction.

Correlations are also available to estimate relative permeability values based on capillary pressure data. Computer-based reservoir engineering applications are capable of producing relative permeability values of oil, gas, and water based on industry-accepted correlations, as shown in Figure 2–19. Subsequently, the results are utilized in various reservoir studies, including performance prediction of wells and estimation of reserves.

Example 2.8. Estimation of two-phase relative permeability based on correlation. Calculate the following:

(a) The phase relative permeabilities of an oil and gas system having a gas saturation of 34%. Consider the following porous media: (i) well-sorted unconsolidated sand; (ii) poorly sorted unconsolidated sand; and (iii) cemented sandstone. Assume the irreducible water saturation is 0.21 or 21%.

(b) The phase relative permeabilities of an oil-water system in an oolitic limestone using the correlations given above. Assume the following: the critical oil saturation is 0.2, and the connate water saturation is 0.15. Will the results be useful in designing a waterflood project in a water-wet reservoir? (Waterflooding a reservoir is described in chapter 16.)

Solution:

(a) First, the oil saturation is computed to obtain S^* based on Equation 2.81. The oil and gas relative permeabilities can then be estimated by available correlations:

$$S_o = 1 - S_g - S_{wc}$$
$$= 1 - 0.34 - 0.21$$
$$= 0.45$$
$$S^* = \frac{0.45}{1 - 0.21} = 0.5696$$

Well-sorted sand (unconsolidated):

$$k_{ro} = (S^*)^3 = (0.5696)^3 = 0.1848$$
$$k_{rg} = (1 - S^*)^3 = (1 - 0.5696)^3 = 0.0797$$

Poorly sorted sand (unconsolidated):

$$k_{ro} = (S^*)^{3.5} = (0.5696)^{3.5} = 0.1395$$
$$k_{rg} = (1 - S^*)^2 (1 - S^{*\,1.5}) = (1 - 0.5696)^2 (1 - 0.5696^{1.5}) = 0.1056$$

Cemented sandstone:

$$k_{ro} = (S^*)^4 = (0.5696)^4 = 0.1053$$
$$k_{rg} = (1 - S^*)^2 (1 - S^{*2}) = (1 - 0.5696)^2 (1 - 0.5696^2) = 0.1251$$

(b) Based on equations 2.77, 2.78, and 2.82, the values of the oil and water relative permeabilities are summarized in Table 2–4.

Table 2–4. K_{ro}, k_{rw} versus Sw

S_w	S^*	k_{ro}	k_{rw}
0.15	0.0000	1.0000	0.0000
0.20	0.0588	0.8827	0.0000
0.25	0.1176	0.7678	0.0002
0.30	0.1765	0.6571	0.0010
0.35	0.2353	0.5524	0.0031
0.40	0.2941	0.4552	0.0075
0.45	0.3529	0.3665	0.0155
0.50	0.4118	0.2874	0.0287
0.55	0.4706	0.2182	0.0490
0.60	0.5294	0.1594	0.0786
0.65	0.5882	0.1109	0.1197
0.70	0.6471	0.0724	0.1753
0.75	0.7059	0.0434	0.2483
0.80	0.7647	0.0230	0.3420

Relative permeability correlations used in the study are based on the drainage process. In water-wet reservoirs, the waterflood operation is essentially an imbibition process, as water, being the wetting phase, displaces oil. Estimation of changes in oil saturation based on drainage correlations could be misleading. Pirson proposed a correlation for relative permeability of water that is applicable in both drainage and imbibition processes:[26]

$$k_{rw} = (S_w^*)^{1/2} S_w^3$$

Oil displacement studies in the laboratory are routinely carried out in cores obtained from a specific reservoir before any waterflood project is implemented. Reservoir models are utilized to test the validity of the relative permeability data, as the cores studied in the laboratory may or may not represent the entire reservoir.

Three-phase relative permeability correlations

Three-phase relative permeability correlations are also available to estimate the phase relative permeabilities when oil, gas, and water are simultaneously present in porous media. Depending on individual saturation, the flow of one phase may dominate in various stages of production from a reservoir. Correlations for water-wet systems, based on the work of Corey, Rathjens, Henderson, and Wyllie, are listed by Ahmed as follows:[27]

I. Cemented sandstone, oolitic limestone, and vugular rocks:

$$k_{ro} = \frac{S_o^3(2S_w + S_o - 2S_{wc})}{(1-S_{wc})^4} \tag{2.83}$$

$$k_{rg} = \frac{S_g^2[(1-S_{wc})^2 - (S_w + S_o - S_{wc})^2]}{(1-S_{wc})^4} \tag{2.84}$$

$$k_{rw} = (S_w^*)^4 \tag{2.85}$$

II. Unconsolidated sandstone, well-sorted grains:

$$k_{ro} = \frac{S_o^3}{(1-S_{wc})^3} \tag{2.86}$$

$$k_{rg} = \frac{S_o^3(2S_w + S_o - 2S_{wc})^4}{(1-S_{wc})^4} \tag{2.87}$$

$$k_{rw} = (S_w^*)^3 \tag{2.88}$$

where

$$S_w^* = \frac{S_w - S_{wc}}{1 - S_{wc}} \tag{2.89}$$

Triangular diagrams can be generated to depict three-phase relative permeabilities for porous media. The three axes in the plot represent oil, gas, and water saturations in a scale of 0 to 100%, and relative permeability values are shown as contours. In addition to the above, familiar three-phase relative permeability correlations include the equations proposed by Stone as follows:[28]

$$k_{ro} = (k_{row} + k_{rw})(k_{rog} + k_{rg}) - (k_{rw} + k_{rg}) \tag{2.90}$$

where

k_{row} = relative permeability to oil in oil-water system ($S_g = 0$)
k_{rog} = relative permeability to oil in gas-oil system ($S_w = 0$)

Relative permeability ratio

The relative permeability ratio of a two-phase fluid system in porous media is defined as in the following:

I. Oil-water system:

$$\frac{k_{ro}}{k_{rw}} \tag{2.91}$$

II. Gas-oil system:

$$\frac{k_{rg}}{k_{ro}} \tag{2.92}$$

In a gas-oil system, the relative permeability ratio is low when the gas saturation is low. However, with an increase in gas saturation, the ratio may increase by several orders of magnitude, as gas is significantly more mobile. A semilog plot is used to show the significant increase in the gas-oil relative permeability ratio as the gas saturation increases. A similar trend is observed in an oil-water system where water displaces oil during waterflooding. The oil-water relative permeability ratio decreases markedly as water saturation increases. Values of the relative permeability ratio readily indicate the dominance of one fluid phase over another as a function of phase saturation in porous media. These values are utilized in multiphase fluid flow studies.

Pseudorelative permeability

Absolute permeability as well as relative permeability usually differ from layer to layer in a petroleum reservoir. Pseudorelative permeability values can be derived from layer-specific relative permeability values by considering permeability-thickness weighted averages.

I. Pseudorelative permeability of wetting phase:

$$\frac{\Sigma (kh)_i (k_{r,w})_i}{\Sigma (kh)_i} \tag{2.93}$$

II. Pseudorelative permeability of nonwetting phase:

$$\frac{\Sigma (kh)_i (k_{r,nw})_i}{\Sigma (kh)_i} \tag{2.94}$$

These equations for pseudorelative permeabilities under dynamic flow conditions find widespread application in reservoir simulation studies. In certain instances, a multilayered reservoir can be viewed as a single-layer system when pseudorelative permeability values are employed along with weighted averages of other layer properties. Such properties could include absolute permeability, porosity, and saturation. This aids in building a relatively simple reservoir model and increases computing efficiency without sacrificing the desired accuracy in results.

Effect of Rock Properties on Reservoir Performance

As mentioned earlier, rock properties and their variations from one location to another significantly influence reservoir performance and ultimate recovery. Due to the limitation of resources concerning reservoir data collection, not all the rock characteristics are known in a typical reservoir. The effects of rock heterogeneities on future reservoir performance cannot be predicted accurately. Due to the variations in rock properties both laterally and vertically, hydrocarbon-bearing formations are increasingly characterized as being comprised of multiple "flow units." Some flow units may perform better than others during primary production. The rest of the units may need particular attention to recover residual oil. Again, certain units must be monitored carefully during improved oil recovery operations involving fluid injection in order to achieve desired results. An outline of reservoir characterization is provided later.

Some of the rock characteristics and associated heterogeneities that reservoir engineers frequently deal with are discussed in the following sections.

Rock permeability. Reservoirs having relatively low permeability, frequently referred to as "tight" reservoirs, are difficult to produce, as more energy is needed to drive petroleum fluids towards the wellbore. The ultimate recovery from tight reservoirs could be quite low. However, a low permeability formation having a fracture network is producible in many instances. Development strategies for tight reservoirs are described in chapter 19. On the other end, reservoirs having high permeability values, in hundreds of millidarcies, are expected to produce substantially at the onset. However, the producing wells may experience early water production when the formation receives water from an external source. Design and implementation of various engineering schemes at the reservoir and well level would be needed to address the problem of undesirable breakthrough of water or gas.

Stratified reservoir. Petroleum reservoirs are frequently encountered where fluid flow may occur through more than one distinct porosity-permeability system. These include stratified reservoirs and fractured reservoirs.

In stratified reservoirs, multiple stratigraphic sequences are separated by relatively thin shale beds. The intervening shale layer acts as a complete or partial barrier between the producing sandstone or carbonate layers. When a layer exhibits a tendency for early water production, the well is completed selectively in better-performing layers so as to avoid the problem.

Two scenarios may be encountered in a stratified reservoir:

- **Commingled flow.** Stratigraphic sequences are separated by an impermeable shale barrier. Pressure communication between the layers is possible only through the wellbore, as the same well is completed in multiple productive layers. Relatively simple analytic and numerical models are available to evaluate fluid flow behavior and develop the reservoir optimally.

- **Crossflow.** Stratigraphic sequences are either fully or partially communicating, and communication (crossflow) among the layers occurs in the reservoir as well as in the wellbore. The degree of crossflow is a function of vertical permeability across the intervening shale beds, which usually varies from location to location within the reservoir. Experience has shown that reservoirs with partially communicating layers are rather difficult to understand, model, and develop in an efficient manner.

In general, reservoir performance is usually more complex to evaluate in a stratified reservoir. This is seen specifically in relation to early water or gas breakthrough from one or more layers and with uncertainties related to the degree of crossflow.

Fractured reservoir. The reservoir is visualized as being comprised of two systems: fracture and matrix. Fractures are thought to have infinite or near-infinite conductivity with little porosity. Rock matrix may have very little conductivity in relation to fractures. The matrix may contribute to production directly to the wellbore or through the fracture network. The existence of fractures may either aid or hinder reservoir performance, depending on the matrix-fracture interaction, the effects of capillary, gravity, and viscous forces, and well placement. In adverse situations such as production at a higher-than-optimum rate, fractured reservoirs may exhibit early decline. Premature breakthrough of injected gas or water can also be experienced in certain cases. Management of two naturally fractured reservoirs is discussed in chapter 19.

Lithology. Many giant reservoirs are hydrocarbon-bearing limestone and dolomite formations. Carbonate reservoirs, as opposed to sandstone reservoirs, may exhibit the presence of high permeability streaks, channels, and fractures. These are due to the various geologic and geochemical processes that can affect limestones and carbonates in a unique manner during deposition and at later stages. Following water or gas injection to augment oil recovery, the response of a carbonate reservoir could be quite different than that of a sandstone reservoir. Some of these reservoirs could have permeabilities in the range of darcies. Their performance in producing oil could be quite remarkable as long as premature breakthrough of water or gas is either avoided or kept under control. However, high permeability streaks may be present in a reservoir that could provide a rapid flow conduit for a less-desirable fluid phase (water or gas). This requires that a variety of measures be considered by the reservoir engineering team to abate the problem. Issues related to early breakthrough in producing wells and strategies to address them are discussed in chapters 16 and 19.

The presence of styolites in carbonates having negligible porosity may affect reservoir performance under water and gas injection in an unpredictable manner. Some carbonate reservoirs are of low to very low permeability, resulting in poor recovery.

Reservoir connectivity. Certain hydrocarbon-bearing formations are compartmental in nature due to a lack of areal connectivity or limited connectivity between various sections. This occurs as a result of complex faulting, among other geologic processes. Compartmental reservoirs often exhibit unexpected behavior, including anomalous reservoir pressure, multiple oil/water contacts, and rapid pressurization under fluid injection. In certain cases, unexplained loss of injected fluid has been encountered as certain barriers tend to become conductive under injection pressure. Unless detailed reservoir characterization efforts are undertaken, enabling appropriate well planning and reservoir management, compartmental reservoirs may lead to disappointing recovery. A case study of a time-lapse seismic survey conducted regularly in a compartmental reservoir in order to track the complex movement of fluids is reviewed in chapter 19.

Relative permeability. The relative permeability characteristics of the oil, gas, and water phases present in the reservoir may have a pronounced effect on reservoir performance during enhanced recovery operations. Certain relative permeability curves may exhibit high residual oil saturation or an abrupt increase in the relative permeability of the water phase when the latter becomes mobile. These reservoirs may not prove to be good candidates for enhanced recovery. Ultimate recovery from reservoirs with unfavorable relative permeability trends could be comparatively less. Results obtained from reservoir simulation studies depend heavily on the phase relative permeability data used.

Wettability. In any engineering study related to the future performance of a reservoir, knowledge of the wettability of the reservoir rock is of prime importance. When the reservoir is oil-wet as opposed to water-wet, ultimate recovery could be poor. The oil phase tends to adhere to the rock surface rather than flow towards the wellbore during waterflooding. Again, some of the reservoirs exhibit intermediate wettability. Recovery could be less than expected if the reservoir is initially thought to be water-wet, and the enhanced oil recovery operation was designed accordingly. It is even possible that the reservoir rock may experience a change in wettability with time or location, as observed by some analysts. The reservoir engineer must recognize all of the uncertainties associated with the wettability of a reservoir when assessing the performance of the reservoir.

Capillary pressure. When the reservoir rock exhibits relatively high oil-water capillary pressure, a long transition zone between the oil and underlying water is observed in the formation. This usually leads to completion of a well considerably above the oil/water contact in order to avoid premature water breakthrough.

Formation transmissibility and storavity

In reservoir engineering studies, two parameters are often sought to characterize the reservoir. These are the transmissibility and storavity of the formation, which are defined in the following:

I. Transmissibility:

kh/μ, mD-ft/cp

II. Storavity or storage capacity:

$\phi\, c_t\, h$, psi^{-1} ft

High rock permeability in a thick pay zone having less viscous petroleum fluid leads to a comparatively large value of transmissibility. When associated with large storage capacity, it is viewed as good reservoir quality. Sources of data to assess formation transmissibility and storavity include transient pressure tests, log analyses, and core studies, among others.

Measures of reservoir heterogeneity

In reservoir engineering studies, the degree of reservoir heterogeneity is used to characterize a formation and predict the performance of a reservoir. These include, but are not limited to, the following:
- Permeability variation factor
- Crossflow index
- Lorenz coefficient

The first two of these measures are described in chapter 16 in connection with improved oil recovery (IOR) from heterogeneous and stratified formations, respectively. The widely known permeability variation factor proposed by Dykstra and Parsons is obtained from permeability data of a number of core samples.[29] This factor is based on the assumption of lognormal distribution of rock permeability in a reservoir, as encountered frequently. The lognormal distribution of a rock property, along with other probability distribution functions, is illustrated in chapter 18. The permeability variation factor, abbreviated as V, typically varies between 0.5 and 0.9, indicating that a geologic formation is moderately to highly heterogeneous. For a perfectly homogeneous formation, V = 0. However, the factor approaches the upper limit in the case of severe reservoir heterogeneity, as evident from core data. Based on reservoir simulation, the effect of the permeability variation factor on reservoir performance is demonstrated in chapter 16.

The crossflow index is a measure of the degree of fluid communication anticipated between two adjacent layers in a stratified reservoir. The communication between the layers of a stratified reservoir may range between noncommunicating and fully communicating. Knowledge of the degree of communication between the layers is vital in understanding the flow dynamics during fluid injection in this type of reservoir.

The Lorenz coefficient is based on a plot of permeability-thickness product (kh) data against porosity-thickness product (øh) data obtained from the core samples in a reservoir.[30] The points are plotted in order of decreasing values of k/ø. In an ideal scenario of uniform rock properties, the points essentially fall on a diagonal drawn from the upper right to the lower left corner of the plot, suggesting that the porosity is a linear function of the rock permeability. However, in a more realistic case where the formation is heterogeneous and such a relationship does not exist, certain points are located away from the diagonal on the Lorenz plot. With the increasing degree of reservoir heterogeneity, the plotted points are shifted further away from the diagonal. The area between the diagonal and the deviated points serves as the basis for determining the value of the Lorenz coefficient as follows:

$$\text{Lorenz coefficient} = \frac{\text{Area enclosed by plot points and the diagonal}}{\text{Area enclosed by the diagonal and bottom right corner of the plot}}$$

In ideal circumstances, the value of the Lorenz coefficient is 0, as the value of the numerator is 0. With increasing heterogeneity, however, the Lorenz coefficient approaches a value closer to unity. One application of the Lorenz plot involves the characterization of various flow units in a geologic formation. The flow unit in a geologic formation is described in the following section.

Introduction to reservoir characterization

A major challenge for the integrated reservoir team involves the realistic characterization of the reservoir to a certain achievable degree. Reservoir engineering literature is replete with reservoir characterization studies that attempt to add value to the asset, i.e., producible petroleum reserves. Experience has shown that detailed reservoir characterization is vital in successfully conducting enhanced oil recovery operations, including waterflooding, in heterogeneous formations having faults, natural fractures, and high permeability streaks. It is estimated that more than 100 billion bbl of oil remain untapped in the United States alone. Reservoir characterization can be instrumental in augmenting ultimate recovery from the matured fields.

The preceding sections have described the major rock properties and their influence on reservoir performance. A myriad of factors related to the skeletal and dynamic properties of rock can affect oil and gas production in unpredictable manners that are not easily identified or understood. Reservoir characterization studies have a wide scope and attempt to identify the key elements in rock properties. Such properties include storage capacity (a function of porosity and thickness) and flow capacity (a function of permeability and thickness). A geologic formation can be viewed as being composed of a number of depositional facies that are usually arranged in a vertical sequence. In turn, several ranges of porosity and permeability values may be encountered within a single geologic sequence that may consist of multiple flow units. Each flow unit is characterized by its storage capacity, flow capacity, and other indicators of reservoir quality. Besides having similarity in porosity, permeability, and capillary pressure characteristics, flow units are viewed as being areally and vertically continuous. They are also viewed as having the same position in the sedimentary sequence. Correct recognition of flow units having good or poor reservoir quality holds the key to reservoir characterization.

Figure 2–20 illustrates a relatively simple exercise in reservoir characterization based on the primary rock properties of porosity, permeability, and formation thickness.[31] The pertinent information can be collected and integrated from a variety of sources, including laboratory core studies, well logs, and seismic studies, for example. A plot of cumulative flow capacity versus cumulative storage capacity, measured from top to bottom, is drawn. Two distinct flow units are observed that may characterize the reservoir. A break in slope marks the beginning of a new flow unit based on the dynamic characteristics of the formation. It is of significance to note that the boundary of a new flow unit may or may not coincide with a lithologic boundary. The flow unit located in the lower part of the reservoir has the greater slope and is expected to be more productive. Reservoir characterization studies also focus on various other rock properties, such as wettability, capillary pressure, pore geometry, and fracture orientation, that affect fluid flow.

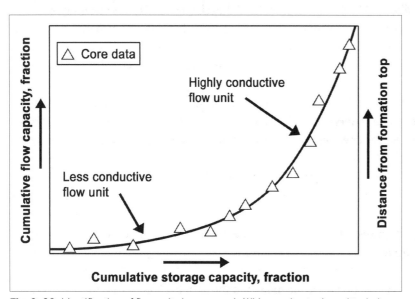

Fig. 2–20. Identification of flow units in a reservoir. Wide-ranging tools and techniques can be utilized to accomplish the above, including geological, geophysical, petrophysical, production, and simulation studies.

Some of the tasks involved in the reservoir characterization study are listed below:

- Preparation of reservoir maps showing the geologic structure and depositional facies
- Detailed description of the porosity, permeability, water saturation, net pay, and hydrocarbon pore volume (HCPV) data
- Identification and mapping of areal and vertical heterogeneities, including the variations in rock permeability

The tools and techniques utilized in conducting reservoir characterization studies include, but are not limited to, the following:

- Microscopic core studies (thin sections, CT scan, and others)
- Conventional and special core analyses (SCAL)
- In-situ determination of rock and fluid properties by a formation tester
- PVT analyses of fluids collected at different locations

- Wireline and cased-hole log studies
- 3-D and 4-D seismic studies
- Vertical seismic profiles (VSPs)
- Interwell tomography
- 3-D earth models based on geological, geophysical, and geostatistical studies
- Production data, including flow rate, pressure, gas/oil ratio, and water/oil ratio
- Production logging data
- Pressure transient test results, single well and multiwell
- Tracer injection survey
- Reservoir simulation
- Reservoir visualization
- Geochemical studies
- Rock compaction and subsidence studies
- Pilot waterflood or enhanced oil recovery projects

Some of these terminologies are briefly explained in the appendix. Examples of efficient reservoir management based on reservoir characterization are described in chapters 16 and 19.

Challenges in reservoir characterization include resource requirements in collecting, analyzing, and integrating multidisciplinary data, both static and dynamic. The data is collected by various tools at different locations and times and may not be in agreement in certain instances, requiring further investigation. Furthermore, workflow related to reservoir characterization changes significantly during the life of a field. In the early stages, 3-D geophysical studies play a major role, as well-related data is sparse. However, as the field is developed, dynamic data related to well production and injection, tracer studies, and pressure transient tests gains significance in characterizing a reservoir.

In conclusion, the ultimate goal of reservoir characterization is to augment petroleum production and add value to proved reserves at each phase of the reservoir life cycle. The task typically involves an integrated team approach, intensive data collection, and computer-aided analyses. It also requires an in-depth understanding of rock heterogeneities, on both a macroscopic and microscopic scale, which play a key role in reservoir performance.

Determination of oil and gas in place

This section illustrates how reservoir engineers utilize basic knowledge of rock properties to seek the most valuable information about a reservoir, i.e., estimated oil or gas in place and petroleum reserves. In addition, it briefly addresses the limitations of such methodology and the strategies adopted to improve the accuracy of the estimates. The basic philosophy behind this application applies to virtually all reservoir engineering tasks. The reservoir engineer must recognize the limitations associated with the study being conducted due to the assumptions made. The next step is to envisage and implement a suitable strategy to develop the reservoir.

The extent of a petroleum reservoir is assessed both laterally and vertically. Knowledge of the basic rock properties described earlier then leads to the estimates of reservoir pore volume and original oil in place, as given in the following:

$$OOIP = A \, h \, \phi \, S_{oi} \tag{2.95}$$

where

$OOIP$ = volume of original oil in place, ft^3,

A = areal extent of the reservoir, ft^2,

h = average net thickness of the reservoir, ft,

ϕ = average porosity of the formation, dimensionless, and

S_{oi} = initial oil saturation in pore space, dimensionless.

Saturation of reservoir fluids, a dynamic property, is discussed later. In oilfield units, the reservoir area is converted to acres, and original oil in place to barrels, using the given conversions:

1 acre = 43,560 ft^2

1 barrel (bbl) = 5.615 ft^3

Equation 2.95 thus can be rewritten as follows:

$$OOIP = (A, acre)\left(43,560 \frac{ft^2}{acre}\right)(h, ft)(\o, fraction)(S_{oi}, fraction)\left(\frac{1 \, bbl}{5.615 \, ft^3}\right)$$

$$= 7,758 \, A \, h \, \o \, S_{oi}$$

$$= 7,758 \, A \, h \, \o \, (1 - S_{wc}) \text{ reservoir barrels, or rb} \qquad (2.96)$$

Note that the initial oil saturation is determined based on the connate water saturation, S_{wc}, in the absence of free gas. Equation 2.96 reflects the volume of oil under reservoir conditions of pressure and temperature. However, the reservoir engineer is more interested in determining oil and gas volumes under stock-tank conditions above ground. In surface facilities, volatile hydrocarbon components evolve out of the reservoir fluid as the fluid pressure declines from thousands of pounds per square inch to levels that are equal to or close to atmospheric conditions. Consequently, crude oil undergoes "shrinkage" or a reduction in volume due to the dissolution of the gas phase, mainly comprised of lighter hydrocarbons. The formation volume factor (B_o) is a measure of the change in oil volume and is treated in chapter 3. For present purposes, it is sufficient to know that Equation 2.96 can further be extended to estimate the oil in place in terms of stock-tank barrels (stb).

$$OOIP = \frac{7,758 \, A \, h \, \o \, (1-S_{wc})}{B_{oi}} \qquad (2.97)$$

where

B_{oi} = oil formation volume factor at the initial reservoir pressure, rb/stb,

rb = reservoir barrels, implying oil volume under high pressure and temperature, possibly containing appreciable quantities of dissolved gas, and

stb = stock-tank barrels, implying altered oil volume at surface conditions due to shrinkage following the evolution of any dissolved gas.

Equation 2.97 is widely known as the equation for the volumetric estimate of original oil in place. It is often referred to as the volumetric equation, and this method is essentially related to a static description of the reservoir. Described in later chapters, other methods are routinely used by reservoir engineers that involve dynamic oil and gas production information.

The volume of hydrocarbons in the reservoir is also expressed in barrels per acre-feet (bbl/acre-ft) as follows:

$$OOIP' = 7,758 \, \o \, (1 - S_{wc}) \qquad (2.98)$$

where

OOIP' is the original oil in place given in reservoir bbl/acre-ft.

It should be further noted that hydrocarbon quantities, such as original oil in place and reserves, are sometimes reported in metric tons or tonnes, abbreviated as t. However, this unit requires knowledge of the hydrocarbon density. One ton of crude oil having an API gravity of 33° is approximately equivalent to 7.3 bbl of oil.

Sources of data and methodology

Sources of reservoir data used to estimate oil or gas in place include geophysical, geological, log, and core studies. Reservoir limit tests are also conducted to identify reservoir boundaries. The areal extent of a reservoir can be wide ranging, from very small fields that barely produce for a short time to giant fields capable of producing millions of barrels of oil a day for decades. The thickness of a reservoir can vary from less than a foot, sometimes referred to as a stringer, to hundreds of feet. Initial oil saturation is found from the known connate or interstitial water saturation as obtained from electric logs. Initial oil or gas saturations in commercial fields usually range between 70% and 90%.

Areal, vertical, and volumetric averaging techniques for reservoir rock properties described earlier can be utilized to improve the accuracy of the estimates of original oil and gas in place. However, geostatistical methods are frequently employed to model porosity and other rock properties throughout the reservoir, which attempt to take into account the inherent uncertainties in data beyond the well. In the process, multiple realizations of porosity distribution in the reservoir are generated. Consequently, a probabilistic approach is adopted to obtain the most likely scenario, including the upper and lower ranges of estimates of hydrocarbon in place (HCIP). The familiar volumetric equations described above serve as a basis for computing oil and gas volumes. However, instead of attempting to obtain a unique value of original oil in place or reserves, a probability distribution function (PDF) is generated based on a Monte Carlo simulation. The results of the simulation indicate what probability (e.g., 50%, 70%, etc.) is attached to a specific value of hydrocarbon in place or oil reserve in the field. The technique is demonstrated in chapter 18. Proved, probable, and possible oil and gas reserves are customarily defined in terms of probability of occurrences and are treated in chapter 15.

Example 2.9. Estimation of original oil in place (OOIP) in a reservoir based on minimum data. Calculate the original oil in place in an undersaturated oil reservoir in reservoir barrels per acre-feet and in stock-tank barrels per acre-feet. The interstitial water saturation (S_{wi}) is 21% PV, and the initial oil formation volume factor (B_{oi}) is 1.18 rb/stb. The porosity value for the reservoir from Example 2.2 will be used.

Solution: Original oil in place per acre-ft, OOIP', is obtained by simply dividing Equation 2.96 by (A, in acres)(h, in feet), as shown in the following:

$$OOIP' = \frac{OOIP}{(A, \text{acre})(h, \text{ft})} = \left(43{,}560 \frac{ft^2}{acre}\right)(\text{ø, fraction})(S_{oi}, \text{fraction})\left(\frac{1 \text{ bbl}}{5.615 \text{ ft}^3}\right)$$

$$OOIP' = 7{,}758 \text{ ø } S_{oi} \frac{bbl}{acre\text{-}ft}$$

Initial oil saturation is calculated as:

$$S_{oi} = 1 - S_{wi}$$

Original oil in place per acre-feet, OOIP':

$$OOIP' = 7{,}758 \ (0.1679)(1 - 0.21)$$

$$= 1{,}029 \text{ rb/acre-ft}$$

$$= \frac{1{,}029 \text{ rb/acre-ft}}{1.18 \text{ rb/stb}}$$

$$= 872 \text{ stb/acre-ft}$$

A similar equation can be used to estimate the original gas in place (OGIP) in a gas reservoir. In this case, Equation 2.96 takes the following form:

$$OGIP = \left(43{,}560 \frac{ft^2}{acre}\right)(A, \text{acre})(h, \text{ft})(\text{ø, fraction})(S_{gi}, \text{fraction}), \text{ ft}^3 \qquad \textbf{(2.99)}$$

where

    OGIP = initial or original gas in place, ft^3,

    A = reservoir area, acres,

    h = average net reservoir thickness, ft,

    ø = average porosity of the formation, dimensionless, and

    S_{gi} = initial gas saturation in the pore space, dimensionless.

Natural gas remains in the subsurface in a highly compressed state, as the prevailing pressure is much higher than atmospheric pressure. Once produced at the surface, it would expand significantly. Gas in place, in terms of standard cubic feet (scf), can be estimated when the gas formation volume factor is known. Hence, Equation 2.99 is modified to take the following form:

$$\text{OGIP} = \frac{\left(43,560 \, \frac{ft^2}{acre} \times \frac{1 \, rb}{5.615 \, cft}\right)(A, acre)(h, ft)(ø, fraction)(S_{gi}, fraction)}{\left(B_{gi}, \frac{rb}{scf}\right)} \quad (2.100)$$

$$= \frac{7,758 \, A \, h \, ø \, S_{gi}}{B_{gi}} \, scf$$

where

OGIP = initial or original gas in place, scf,

scf = volume of gas following expansion at the surface under standard conditions, and

B_{gi} = gas formation volume factor (initial), rb/scf.

Note that 1 bbl = 5.615 ft^3, hence Equation 2.100 is modified to conform to the unit of gas formation volume factor in reservoir barrels per cubic feet under standard conditions.

Example 2.10. Estimation of the original gas in place in a dry gas reservoir. Calculate the original gas in place in a dry gas reservoir. The following data is given:

Reservoir area (A, acres) = 5,056.

Net reservoir thickness (h, ft) = 34.

Average porosity (ø, %) = 15.

Interstitial water saturation (S_{wi}) = 0.20.

Gas formation volume factor (B_{gi}, rb/scf) = 0.00095 (= 0.95 rb/Mscf).

Several examples of volumetric estimation of oil, gascap gas, and solution gas are presented in chapter 9.

Solution: The original gas in place is estimated as shown in the following:

$$\text{OGIP} = \frac{(43,560/5.615)(5,056)(34)(0.15)(1-0.2)}{0.00095}$$

$$= 1.6845 \times 10^{11} \, scf \, (1.6845 \times 10^2 \, bscf)$$

Petroleum reserves

A distinction must be made at this point between original hydrocarbon in place and reserves in an oil or gas field. Two reservoirs may have the same oil in place but very different reserves. The actual volume of petroleum that can be optimally recovered from a reservoir is tied to the estimation of oil and gas reserves. Reserves could be quite high in the case of a dry gas reservoir where near-ideal conditions exist, ranging as high as 80% or more of the initial gas in place (IGIP). However, reserves could be much lower in the case of oil reservoirs. Oil reserves may vary from low single digits to 60% of the original oil in place, depending on oil and rock characteristics, natural drive mechanisms, and efficiency of improved or enhanced oil recovery. Fields with quite low reserves are categorized as marginal, and they would only be developed using resource-intensive technologies when market conditions indicate a high demand for crude. A case in point is certain enhanced oil recovery processes described in chapter 17, which become viable only at certain oil price levels.

Equation 2.97 serves as a basis to estimate oil reserves when the residual oil saturation and oil formation volume factors are known with some degree of confidence at the end of the producing life of the reservoir. The ultimate oil recovery and reserves can be estimated based on the following general equation:

$$\text{Reserve, oil reservoir} = 7,758 \, A \, h \, ø \left[\frac{S_{oi}}{B_{oi}} - \frac{S_{or}}{B_{or}}\right] \quad (2.101)$$

where

S_{or} = residual oil saturation, fraction, and

B_{or} = oil formation volume factor at S_{or}, rb/stb.

Residual oil saturation can be estimated from petrophysical studies of core samples obtained from the reservoir. In fact, reserve estimates are strongly influenced by a myriad of technical and nontechnical issues. These include geologic complexity, reservoir drive mechanism, technological innovation, petroleum economics, government policy, and environmental issues. In essence, petroleum reserves are defined as follows:

Reserves = (Petroleum initially in place) × (Estimated recovery efficiency, fraction)

Depending on the increasing probability of finding hydrocarbon accumulations and commercial viability for production, petroleum reserves can be classified and reported as possible, probable or proved, which are discussed in chapter 15.

New technologies employed in recent years include production enhancement by drilling of horizontal wells and reservoir monitoring based on 4-D seismic studies. With the emergence of these technologies, petroleum reserves have increased notably in many fields, although the initial hydrocarbon in place estimation has remained essentially unchanged. (Estimation of petroleum reserves is discussed in chapter 15, with an example. The definition of reserves, published by the Society of Petroleum Engineers, is also included.)

In the present-day literature, reservoir management is frequently referred to as asset management. An attempt to improve the ultimate recovery of oil and gas by implementing a reservoir engineering technology is viewed as adding value to the asset.

Recovery efficiency

The recovery efficiency is defined as the ratio of recoverable oil volume over original oil in place. An expression for recovery efficiency for oil reservoirs is provided in Equation 2.50. Based on a multitude of field studies, correlations for estimating recovery efficiency have been developed that may be used as a guide for newly discovered reservoirs where actual production data is either unavailable or insufficient. Chapter 8 explains estimation of recovery efficiencies as influenced by reservoir drive mechanisms. Chapter 10 presents empirical correlations to estimate recovery from a reservoir. The latter approach would require knowledge of various reservoir and fluid properties, as well as reservoir pressure at abandonment.

Example 2.11. Calculation of petroleum reserves—data requirements. In the gas reservoir described in Example 2.10, what would be the minimum information required in order to estimate the reserve? Make necessary assumptions.

Solution: In a dry gas reservoir where there is no water encroachment, gas and water saturations remain virtually the same as the reservoir is depleted. Due to production, the gas becomes less compressed in porous media, resulting in a significant increase in the gas formation volume factor. The remaining volume of gas in the reservoir at abandonment can be expressed as shown:

$$\frac{7{,}758\,A\,h\,\emptyset\,S_{gi}}{B_{ga}}$$

Since the gas reserve is defined as the difference between the initial gas in place and the remaining gas at the end of the producing life of the reservoir, it can be estimated as in the following:

$$\text{Reserve, gas reservoir} = 7{,}758\,A\,h\,\emptyset\,S_{gi}\left[\frac{1}{B_{gi}} - \frac{1}{B_{ga}}\right] \tag{2.102}$$

This equation indicates that in order to estimate the gas reserve, knowledge of abandonment pressure and gas formation volume factor at abandonment would be required as a minimum. In chapter 3, it is shown how easily the gas formation volume factor can be estimated at a particular reservoir pressure once the gas gravity or composition is known from sample analysis. Typical gas reservoir pressures at abandonment is quite low resulting in good recovery.

Lastly, it is noted that the recovery efficiency of a dry gas reservoir can be expressed in terms of gas formation volume factors alone:

$$E_R = \left[\left(\frac{1}{B_{gi}} - \frac{1}{B_{ga}} \right) / \frac{1}{B_{gi}} \right] \times 100 \qquad\qquad (2.103)$$

Key points—volumetric estimations and petroleum reserves

Before leaving the topic, the key aspects related to volumetric estimation of oil and gas in place, as well as petroleum reserves, can be summed up as follows:

1. The original oil and gas in place in a reservoir can be estimated from a basic knowledge of a few reservoir characteristics. These include reservoir area and average thickness, average porosity of the formation, and the saturation of petroleum fluid in the rock pores.

2. However, the volumetric estimate is only an approximation for a newly discovered field. Better accuracy is achieved when several wells are drilled, and the reservoir production history becomes available. Various techniques are employed to assess rock properties in locations of the reservoir where there are no wells.

3. The petroleum reserve concerns the portion of oil and gas in the reservoir that can be economically produced under government regulations. Reserve of an oil or gas field depends both on technical and nontechnical factors. These include rock and fluid properties, technological breakthroughs, demand for oil and gas, and environmental issues, among others.

Sources of Rock Properties Data—An Overview

In order to gain insight concerning a reservoir, information related to rock properties is collected from a variety of sources based on wide-ranging disciplines in science and engineering. As noted in the beginning of the chapter, the gathering of necessary information over the entire reservoir is resource intensive, as it often requires the drilling of new wells. Common reservoir engineering studies are associated with some uncertainties in rock and fluid data, as discussed briefly in the following sections.[32]

Original hydrocarbon in place (OHCIP) estimates. Sources of uncertainty are porosity, connate water saturation, and formation volume factor data. In poor quality rocks, uncertainties related to porosity may introduce an error between 10% and 20%. In volatile oil reservoirs, the formation volume factor may not be known with certainty.

Recovery efficiency (E_R) estimates. Sources of uncertainty are residual oil saturation, relative permeability data, and oil viscosity. In a reservoir, uncertainties in residual oil saturation may introduce an error up to 20%. In heavy oil reservoirs, significant uncertainties may exist with viscosity and relative permeability, and the resulting error in estimating recovery can be greater than 15%.

In order to address the uncertainties inherent in studies, reservoir engineers and earth scientists resort to alternate means, such as the following:

- Industry-recognized correlations
- Geostatistical models
- Production history matching by reservoir simulation based on "known" data
- Development of probable scenarios based on uncertainties

In well-managed fields, the reservoir engineering team not only gathers vital data from an array of sources, but also attempts to integrate all available information using current technology. Apparently conflicting data obtained from multiple sources often points to the need for further data collection and resolution of the apparent anomaly. "Unexpected" data could also indicate possible geologic heterogeneities that were not previously envisioned.

Common sources of reservoir data include the following:
- Laboratory core analysis
- Log analysis
- Well test analysis
- Earth science studies, including geology, geophysics, and geochemistry
- Emerging technologies

The methodology of data acquisition in a reservoir is discussed in chapter 6. Chapter 7 outlines how integrated reservoir models are developed to enhance and predict reservoir performance. Reservoir simulation can be used to evaluate rock and fluid properties through history matching of the past production performance. Chapter 13 presents reservoir simulation concepts, simulator types, and simulation process, application, and examples.

Detailed treatment of the tools and techniques used to collect data related to reservoir rock properties is outside the scope of this book. However, the following section highlights the sources of reservoir data based on a literature review.[33–45]

Core analysis. Coring is the backbone for investigating the depositional environment of a reservoir and its internal anatomy. Core samples of the reservoir rock are collected during drilling and recompletion of wells. The intensity of sample collection and subsequent analysis may depend on the degree of complexity of the reservoir, as well as on reservoir economics. Core analysis provides direct measurements of porosity, permeability, lithology, compressibility, and residual fluid saturations. Additionally, it provides information regarding wettability, capillary pressure, and relative permeabilities of fluid phases in the pores. The results of the core analysis reflect localized reservoir properties and do not generally represent the entire reservoir.

Log analysis. Each newly drilled well is routinely logged open hole prior to its completion as a producer or injector. Logging is used to identify rock types, such as porous and permeable sands and limestones, as well as nonpermeable shales. Several types of logging equipment have been developed in the industry to collect information regarding lithology, rock porosity, thickness, producing zone depth, and fluid saturation, among others. Common logging methodology involves propagation of an electrical current, radiation energy, or sound waves in the subsurface formation, and the subsequent detection of the results. Logging tools, when placed down hole, can gather data from a few inches to a few feet into the reservoir. Data obtained by logs run at various wells in a reservoir provide well-to-well correlation and aid in characterizing the reservoir. Compared to core studies, well logging is less resource intensive and provides continuous measurement of the rock properties against depth. Studies are commonplace where log data and core data are correlated and integrated to build a detailed view of the reservoir.

Although the details of well logging theory, methodology, and analysis of results are outside the scope of this book, widely known logging techniques are outlined in later sections.

Transient well test analysis. Virtually all wells are subjected to some sort of pressure transient analysis at some point in time. In well tests, fluid flow pattern into or out of the well is altered by design, leading to a change in the pressure response from the reservoir. The response is influenced by fluid and rock properties around the well and is subsequently analyzed to estimate various rock characteristics. The information obtained from transient tests covers a relatively larger area, typically hundreds of feet into the reservoir.

Chapter 5 provides transient fluid flow theory, well test types, analyses, and results, along with well test design.

Geosciences studies. Geosciences professionals, technologies, tools, and data play a key role in the exploration, discovery, development, production, and operation of oil and gas reservoirs. Geoscientists and engineers need feedback from each other throughout their work.

Geologists usually are involved in determining structures, stratigraphy, depositional environments, rock types, and mineralogy, which provide a geological model for the reservoirs. Since petroleum engineers are normally required to take some geology and petrophysics courses, it is not necessary to devote any time to geology. However, petrophysics will be briefly discussed. While a detailed study is not within the scope of this book, the fundamentals of seismology and geostatistics, with example illustrations, will be presented in later sections.

Emerging technologies. A number of emerging technologies are being employed to collect rock properties and characterize the reservoir at various stages, as with the following:

- During drilling of a new well, an array of sensors or imaging devices is placed in the core barrel above the drill bit assembly that can construct cross-sectional and 3-D images of the core. This leads to measurement of rock porosity and fluid saturation of the core before bringing it to the surface.

- Wireline formation testers are employed along with a logging tool in order to measure horizontal and vertical permeabilities of the rock in a newly drilled well.

- Borehole imaging technology, comprised of acoustic, visual, and other tools, is used to characterize fracture density and orientation, secondary porosity, and faults.

- Interwell seismic studies lead to the interpretation of in-situ rock properties.

- Fluid movement in the reservoir over time can be tracked accurately with 4-D seismic studies. This can greatly improve the quality of a reservoir simulation model. The underlying principle is that the seismic waves are a function of fluid composition, pressure, and temperature.

- Downhole video cameras may help in pinpointing the water entry point and the possible existence of a high permeability channel.

- Fiber-optic technology can be used, whereby laser light is passed through a fiber placed in the wellbore to obtain a well temperature versus depth profile. This is used to identify high permeability steam breakthrough channels during thermal recovery.

With this background, the traditional sources of reservoir data can now be addressed in greater detail.

Core Analysis

Laboratory core studies are based on core samples obtained during drilling of wells in oil and gas reservoirs. Core samples are routinely collected in order to measure reservoir properties that are utilized in all facets of reservoir development and management. Once again, it is emphasized that reservoir information related to rock properties obtained from one well is location specific. It cannot be used to represent the entire reservoir. Large petroleum reservoirs typically have a coring program in which cores are obtained from a number of key wells. Highly heterogeneous reservoirs may require an intensive coring program. The results of core analyses are then integrated with information obtained from other sources, such as wireline logging and seismic studies.

In addition to the routine tests, special core analyses are needed for detailed analysis of reservoir characteristics. This includes determination of capillary pressure, wettability, and relative permeability.

The array of information as obtained from core analysis may be classified into three broad categories:

- **Factual data.** Thickness, permeability, porosity, and their distributions, and the presence or absence of hydrocarbon fluids.

- **Interpretation data.** Fluid distribution, type of fluid production, water or gas coning possibilities, and injected water or gas breakthrough possibilities from high permeability streaks.

- **Evaluation data.** Initial hydrocarbon in place, recoveries, production rate, productivity index, injection rate, and injectivity index.

Coring practices

Obtaining a sample of the reservoir matrix can be of great aid in defining the reservoir. A sample must be of sufficient size for an analysis to be done. There are four types of coring:

- **Conventional coring.** The core bit is shaped like a donut, and as the hole is drilled, the core is extruded up into the core barrel.

- **Rubber sleeve coring.** This is much the same operation as a conventional core job but has a rubber sleeve insert in the core barrel that surrounds the core as it is taken. This type of operation is done in unconsolidated or highly friable zones where a conventional core would not work.

- **Sidewall coring.** The coring tool is sent into the hole on a wireline. The tool sends core cutters into the wall of the hole either hydraulically or with percussion caps. These cutters are attached to the tool by cable, and when the tool is removed from the hole, a sample of core is taken from each of the cutters. This method does not have a high rate of recovery.

- **Pressure coring.** The coring operation is the same as for the conventional coring, except the core is sealed in the barrel while it is still in the hole at reservoir conditions. This is achieved by hydraulically closing a valve above the bit and below the core. Once on the surface, special handling techniques are used to keep the core sample at reservoir conditions. This type of core generally offers the best information that can be obtained about the formation. However, limited success has been obtained with this tool in unconsolidated sands.

The core is subjected to changes in pressure coring due to the very process of cutting the core and bringing it to the surface. Every case would have its own characteristic changes, and the core analyst must be well aware of the effects of these changes on the properties to be measured.

The selected coring fluids must be compatible with the coring objective, as shown below:

Determinations	Suitable Core Fluids	
Porosity and permeability	Water-based mud	
Interstitial water	Oil-based mud	
	Water-Wet Formation	**Oil-Wet Formation**
Water-oil relative permeability, Water-oil capillary pressure	Water-based	Nonoxidized crude oil
Gas oil relative permeability	Water-based	Nonoxidized crude oil
Gas-oil capillary pressure	Any fluid	

Core preservation

Commonly used preservation techniques for cores include the following:

1. Submersion in deaerated water.

2. Submersion in nonoxidized crude oil or refined oil treated to remove polar compounds.

3. Enclosure in plastic bags (short term storage, two to three days only).

4. Enclosure in plastic wrap and aluminum foil coated with wax or strippable plastic.

5. Canning (however, cans rust and leak with time, and the core may dry and deteriorate).

6. Freezing or chilling with dry ice. (The core is in a CO_2 atmosphere as the ice sublimes. This is not recommended if permeability measurements are to be made.)

7. No preservation except insulation to prevent breakage.

8. Rubber sleeve cores may be preserved by capping and taping the sleeve ends.

Core sampling

Since the measured data can be no better than the sample from which it is obtained, it is imperative that core samples be representative of the zone of interest and that the samples be properly preserved until the analysis can be performed. Prior to sampling the core, pertinent field and reservoir information should be obtained. Mud data and coring conditions should also be recorded. Cores are subjected to invasion by drilling fluids. Furthermore, volatile fluids contained in pore spaces evaporate as the core is brought to the surface. Changes in wettability characteristics may also be encountered.

The frequency of sampling for conventional analysis is generally one sample per foot. Where the formation consists of laminated sand and shale, or shows other irregularities in lithology, more frequent sampling is desirable. About 200 g to 300 g of sample are ordinarily taken for analysis from the portion of each foot that appears to be most representative of the productive formation within that foot. For whole core analysis, the sample interval is determined by the lengths of the segments of full-diameter core available. The lengths of these samples can vary from about 8 cm to 60 cm (3 in.–24 in.) but generally average about 30 cm or 12 in. The entire sidewall sample recovered is used in the analysis if the sample is suitable.

Core analysis classification

Core analysis can be classified into several main types.

Conventional analysis. Conventional or plug analysis is the type most frequently employed. It involves the use of a relatively small sample of core to represent an interval of the formation to be tested. It is used where the uniformity of the reservoir formation permits such a sample to accurately represent the interval.

Whole core analysis. Whole core analysis is employed where fractures, vugs, or erratic porosity development exists. The volume of individual pore spaces may be relatively large in comparison to the plug samples used in the conventional analysis. It involves the use of essentially the entire core, in pieces as large as possible, recovered from the interval to be tested.

Sidewall core analysis. Sidewall core analysis is the analysis of cores recovered by any of the sidewall coring techniques. The small sample size and the conditions under which they are taken limit the value of the measured data in reservoir evaluation.

Conventional core analysis is the least expensive way to obtain reservoir matrix properties. Two procedures are used to determine the permeability of the core. The first is plug analysis. In this procedure, plugs taken horizontally are used. The alternative method is whole core analysis, done using the whole core.

The measurement of porosity can be generally the same by conventional and whole core analyses. The permeability, on the other hand, can be dramatically affected. The use of plugs to obtain permeability values is subject to bias, since plugging a shale lamina would not give useful results. Even when cores are taken from the same formation, the vertical permeabilities can be clearly different, due to the selective plugging. Permeability measurements done on a whole core can provide directional permeability at directions that are 90° apart.

Sidewall coring is a less-desirable alternative and is used when other means of obtaining reservoir samples are not available. The force of the sidewall cylinders can greatly affect any measurements obtained. In friable or unconsolidated zones, compaction may occur, and in hard rock, fractures are a problem.

Analysis of conventional cores does not provide quantitative fluid saturation values. However, there are situations in which the values of saturation obtained from analysis may be used in a qualitative way. One such case occurs with dead oil reservoirs, in which dissolution of the gas phase does not occur due to a decrease in prevailing pressure. However, great care should be exercised with this approach. The only relatively common method of obtaining realistic fluid saturations is by pressure coring. This is done by sealing the core barrel while on the bottom and retrieving the sample under pressure. This operation is relatively expensive and is less applicable to unconsolidated sands.

Core measurements

The minimum basic measurements made on cores are for porosity, permeability, and residual fluid saturations. In addition, various supplementary routine tests are made when required, such as the following:
- **Chloride content.** Indicates the degree of flushing of the core by the drilling mud fluid.
- **Oil gravity.** Indicates the quality of the producible oil and, indirectly, its viscosity.
- **Fluorescence.** Indicates the relative amount of flushing in the center and outside of the core due to the drilling mud filtrate.

- **Vertical permeability.** Indicates the extent of vertical flow or crossflow.
- **Permeability to water.** Indicates permeability in contact with fresh and salt water.

Well Log Analysis

Wireline logging of the wellbore provides valuable subsurface information for understanding and determining the reservoir properties. Every well is logged by lowering a sonde, which is attached to a conductor cable, into the wellbore. Typical measurements recorded on well logs include spontaneous potential, natural gamma radiation, induced radiation, resistivity, acoustic velocity, density, and caliper.

Logging is used to identify rock types, such as porous and permeable sandstones and limestones, as well as nonpermeable shales. Log analysis results of interest to reservoir engineers include, but are not limited to, the following:

- Producing zone depth
- Reservoir zone thickness and porosity
- Formation lithology and any heterogeneity such as fault or pinchout
- Fluid saturation distribution, including OWC and GWC
- Horizontal and vertical permeabilities (by wireline tester)

Rock properties are indirectly determined from the log responses by making certain assumptions about the interaction of the logging device with the rocks and fluids. Before the analysis can be done, the logs must be adjusted or "environmentally corrected" for the effects of the tool geometry and borehole fluids. Since any particular log may detect one reservoir condition and be insensitive to another, several types of logs are usually compared to determine the true nature of the reservoir.

Additional information is needed concerning the mud, such as its resistivity, temperature, density, and composition. Knowledge of borehole temperature and tool geometry also is required for proper formation evaluation. This data can be found on the log header. It is also necessary to know which logging company recorded the data, since the environmental corrections are different for each company's tools.

Types of logs

A large and growing number of logging tools are available to assist in the determination of reservoir properties. Understanding the details of their operation and the subtleties of interpreting the measurements is the domain of the log analysts.

Table 2–5 provides brief summary of basic logging systems, which include the following:

- Spontaneous potential (SP)
- Natural gamma
- Neutron (general)—CNL, SNP
- Density (gamma-gamma), FDC, densilog
- Sonic—velocity log, BHC
- Resistivity—normal, lateral, focused, etc.
- Induction
- Caliper
- Microlog, minilog, contact log (pad-type devices)
- Microresistivity (focused)—microlaterolog, F_oR_{xo}, minilog, proximity log, micro-SFL (pad-type devices)

A brief description follows of the basic logs used to determine the values of water saturation and porosity against the depth of the subsurface geologic formation.

Table 2–5. Brief summary of basic logging systems*

Logging System	Can Run in Csg.	Basic Borehole Measurement	Value Recorded on Log
Spontaneous Potential (S.P.)	No	Difference of potential between electrode in borehole and a distant electrode	+ or − deflection from arbitrary shale base lines
Natural Gamma	Yes	Natural radioactivity emitted from formation	Gamma ray intensity
Neutron (General) CNL SNP	All but SNP	Capture gammas or neutrons located some distance from a source emitting neutrons	Counting rate or porosity index
Density (Gamma-Gamma) FDC Densilog	No	Back scattered gamma rays reaching detector located some distance from a source of gamma rays	Bulk density
Sonic Velocity Log BHC	No	Acoustic travel time through given length of rock	Interval travel time − Δt
Resistivity (Normal, lateral focused, etc.)	No	Voltage difference between potential measuring electrodes	Resistivity
Introduction	No	Component of the voltage induced in receiver coil by the secondary magnetic field	Conductivity (also reciprocated to resistivity)
Caliper	Yes	Mechanical measure of hole size	Borehole diameter
Microlog Minilog Contact Log (Pad-type devices)	No	Voltage difference between potential measuring electrodes of two different short-spaced electrode arrangements. One lateral, one normal	Resistivity of both measurement
Microresistivity (focused) Microlaterolog $F_0 R_{xo}$ Minilog Proximity Log Micro SFL (Pad-type devices)	No	Voltage difference between potential measuring electrodes	Resistivity

*CNL = Compensated neutron log; SNP = Sidewall neutron porosity log; FDC = Compensated formation density log; SFL = Spherically focused log;

Water saturation determination

Resistivity logs. Valuable information about the reservoir water saturation can be deduced from measurements of electrical resistivity. Archie's classic equation gives the water saturation (S_w) as a function of porosity (ø), water resistivity (R_w), and true formation resistivity (R_t):

$$S_w = \left(\frac{aR_w}{\text{ø}^m R} \right)^{1/n} \tag{2.104}$$

Constants a, m, and n are experimentally determined. The parameters can vary but are often in the range of a = 1, m = 2, and n = 2. This relationship works well in clean formations but not as well in shaly formations or formations in which the connate water is fresh. Resistivity is used to determine water saturation, since interstitial water in the rocks is usually conductive. The conductivity is proportional to the chloride (salt) content of the water, the temperature, and the water volume in the pores.

Water saturation can be determined by using the more familiar Archie relationship:

$$S_w = (R_o / R_t)^{1/2} \tag{2.105}$$

Sometimes resistivity is used as the base log to pick out formation tops and thicknesses and to correlate with other wells. The formation resistivity factor is defined as:

$$F = R_o / R_w \tag{2.106}$$

Units of Log Recording	Quantitative Formation Parameter Interpreted from Recorded Info.	Other Uses
+ or – millivolts	R_w – Resistivity of interstitial water in formation	Lithology (distinguish sands and limestones from shales). Qualitative estimate of formation shaliness, correlation, indicated ionic permeability.
API units	Amount of natural radioactivity present in rocks	Lithology (distinguish shales from nonshales). Qualitative estimate of formation shaliness, correlation.
Counts/sec, API units, or % pore space	Porosity (Indicates total ø)	In combination with other logs to indicate lithology, gas, correlation.
g/cm^3	Porosity (Indicates total ø)	Identify lithology, in combination with other logs help locate gas, estimate secondary porosity, correlation.
Microseconds/ft	Porosity (Indicates primary ø)	Identify lithology, in combination with other logs help locate gas, estimate formation shaliness, estimate secondary porosity, assist in interpreting seismic data, correlation
Ohm-m	Formation resistivity depending on electrode spacing, R_t, R_o, R_{xo}, etc.	In combination with other logs can be used as indication of invasion (qualitative permeability). Correlation, basic parameter to derive S_w
Milliohms/meter (ohm-meter after reciprocation)	Formation resistivity, R_t	In combination with other logs can be used to give qualitative indication of invasion (permeability), correlation, basic parameter to derive S_w
Inches	None	Estimate mud cake thickness, qualitative indication of permeability
Ohm-meters	ø, R_{xo}, and mud cake thickness. (Recommend use as only qualitative)	Qualitative indication of permeability. Used to delineate feet of pay
Ohm-meters	R_{xo} (use qualitatively only)	Qualitative indication of porosity

R_{xo} = Resistivity of zone invaded by mud; R_t = Uninvaded zone resistivity; R_o = Resistivity of uninvaded zone 100% saturated with water.

where

R_o = the resistivity of the formation, and

R_w = the resistivity of the formation 100% saturated with water.

F, as related to porosity, can be expressed in the general form:

$$F = a / ø^m \tag{2.107}$$

where

a = an empirically derived constant, and

m = cementation factor, which is a function of rock type.

For general use, a = 1 and m = 2. Thus:

$$F = 1 / ø^2 \tag{2.108}$$

For unconsolidated Gulf Coast sediments, a = 0.62 and m = 2.15. Then the following applies:

$$F = 0.62 / ø^{2.15} \tag{2.109}$$

This is called the Humble formula.

Example 2.12. Estimation of water saturation in a reservoir based on resistivity data. Compute the water saturation at the following depth for unconsolidated sand in a Gulf Coast reservoir when the porosity of the formation is known from a density log. The depth, in feet, is 6,012. The porosity (fraction) is 0.21.

Based on the results of an induction log, the values of R_w and R_t are 0.028 ohm-m and 18 ohm-m, respectively. Make necessary assumptions.

Solution: Assuming n = 2, water saturation is computed from Equation 2.104, as in the following:

$$S_w = \left[\frac{(.62)(.028)}{(0.21)^{2.15}(18)} \right]^{1/2}$$

$$= 0.166 \text{ or } 16.6\%$$

Example 2.13. Calculation of the formation resistivity factor and porosity. Additionally, the following data is given: $R_0 = 0.8$ ohm-m.

Calculate the formation resistivity factor and porosity using the Humble formula. Compare with the porosity value obtained in Example 2.12.

Solution:

$$F = R_0/R_w = 0.8/0.028 = 28.57$$

$$\varnothing = (0.62/F)^{1/2.15} = 0.168 \text{ or } 16.8\%$$

Spontaneous potential (SP)

This tool measures naturally occurring electrical currents when fluids of different salinities are in contact. Since the salinity of the reservoir fluid generally is different than that of the drilling mud, SP currents are generated by the interaction of the two fluids. This tool is normally used with freshwater-based drilling muds.

The generated potential (in millivolts) is given by the following:

$$SP = -(60 + 0.133 \, T) \log (R_{mf}/R_w) \tag{2.110}$$

where
  T = temperature, °F,
  R_{mf} = mud filtrate resistivity, ohm-m, and
  R_w = water resistivity, ohm-m.

If all of the other parameters are known, this equation can be solved for R_w. Impermeable shale zones can be used to set a "baseline" from which to measure SP variations. The SP deflection is a qualitative indicator of permeability, but it can be reduced by shale or hydrocarbons.

Porosity determination

Density log. An estimate for porosity is also obtained from the measurement of the electron density of the formation using a chemical source of gamma radiation and two gamma detectors:

$$\varnothing = \frac{\rho_{matrix} - \rho_{bulk}}{\rho_{matrix} - \rho_{fluid}} \tag{2.111}$$

where
  ρ = density.

This log works in air-drilled holes or with any type of fluid. Its penetration is shallow, so ρ_{fluid} is usually taken to be the drilling fluid density (1.0 for fresh mud, 1.1 for salt-based mud). The density log results in a pessimistic value for S_w in the presence of shale unless corrected, but the porosity is not affected much by shale. On the other hand, the density log yields high porosity values in the presence of gas.

Acoustic velocity log (sonic log). The difference in travel time, Δt, is measured from a sound source at one end of the tool to two receivers at different distances along the tool. This difference represents the travel time of sound through the portion of the reservoir between the two receivers. Both the rock matrix and the fluid in the pores influence the result, which leads to an estimate of the porosity:

$$\emptyset = \frac{\Delta t_{\log} - \Delta t_{matrix}}{\Delta t_{fluid} - \Delta t_{matrix}} \qquad (2.112)$$

This log is good for primary porosity but does not indicate all secondary porosity. It yields high porosity values in shaly zones.

Combination porosity logs. Comparing the measurements from two or more porosity tools can resolve the uncertainties presented by the individual devices. This is helpful in differentiating liquids from gas, identifying lithology, and determining shale volume. Both acoustic velocity and density logs have a weak response to gas, while neutron logs indicate a much lower porosity for gas-filled rock than for fluid-filled rock. Density, acoustic velocity, and neutron logs all respond differently to various lithologies. The ratios of the density-derived porosity to acoustic velocity or neutron porosity will yield the shale volume.

Analysis procedure

The steps involved in performing the log analysis include:
1. Data entry. Loading raw log curves with other log information.
2. Log data preparation. Quality control, trace edit, depth shift, and borehole environmental corrections.
3. Well log preparation. Parameter selection and computation of reservoir properties.
4. Verification of results. Accomplished with cores, drill cuttings, and other wells.

The procedure is essentially the same (except for data loading) whether done with paper logs or by computer.

Example log analysis results

Figure 2–21 shows log analysis results including lithology, water saturation, and other parameters. Figure 2–22 shows log cross sections. Cross sections of the raw logs can be used to correlate geologic tops from well to well and to isolate the zone to be analyzed. Geologic layers can be defined on the cross sections. Then the summation process determines the representative value of each log property in each layer for each well.

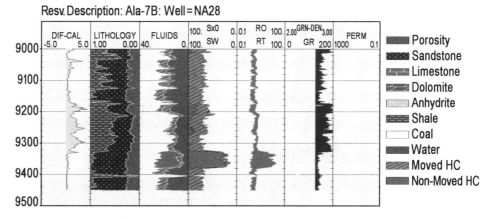

Fig. 2–21. Results of typical log analysis in a well. *Source: A. Satter, J. Baldwin, and R. Jespersen. 2000. Computer-Assisted Reservoir Management. Tulsa: PennWell Books.*

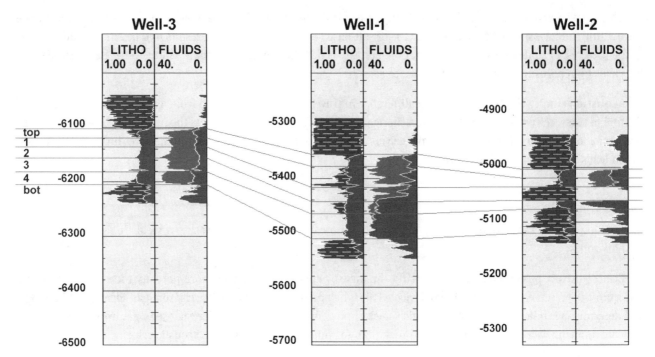

Fig. 2–22. Example of a cross-section study that correlates log data across wells. *Source: A. Satter, J. Baldwin, and R. Jespersen. 2000. Computer-Assisted Reservoir Management. Tulsa: PennWell.*

Geophysics

Seismology plays an important role in defining or characterizing reservoirs and locating a potential discovery well. Its application minimizes dry holes and poor producers; more importantly, it assists in the economic development of oil and gas fields.

Seismic survey results provide depth to reservoir, structural shape, faulting, boundaries, and 3-D survey results to give 3-D visualization of the reservoir. It may also be possible to identify a porous interval, a hydrocarbon reservoir, or an overpressure zone. It may further help in determining geologic history, connectivity between wells, movement of fluids, and fracture orientation.

The seismic process involves the following:

- Selection of location on land or offshore based upon geological information and any known discoveries in adjacent areas
- Acquisition of seismic data
- Data processing, interpretation, and results

While a detailed treatment is not within the scope of this book, it will be worthwhile to introduce some fundamentals of seismic measurements and present example results.

The following discussion briefly explains basic seismic concepts to provide a better understanding of the technology.

Seismic measurements and processing

The basic principle is based upon propagation of sound waves. As sound energy is introduced into the ground by means of an explosive or a vibrating source, energy starts to propagate. Seismic reflection occurs at an interface where rock properties change, as illustrated in Figure 2–23. This simplified example of a seismic recording shows the ray path of energy from shot 1 being partially reflected at the top of each rock layer and recorded at receiver 1.

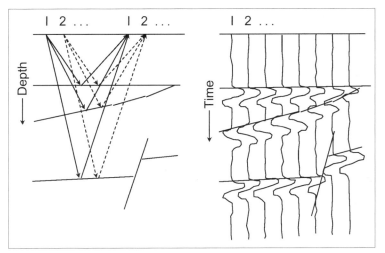

Fig. 2–23. The seismic record section in a reservoir. *Source: A. Satter, J. Baldwin, and R. Jespersen. 2000. Computer-Assisted Reservoir Management. Tulsa: PennWell.*

Normal Incidence Reflection

$$i\downarrow \quad \uparrow r = R_o i \qquad \rho_1 v_1 \quad \text{rock layer}$$
$$\downarrow t = (1+R_o)i \qquad \rho_2 v_2 \quad \text{interface}$$

$$R_o = \frac{\rho_2 v_2 - \rho_1 v_1}{\rho_2 v_2 + \rho_1 v_1}$$

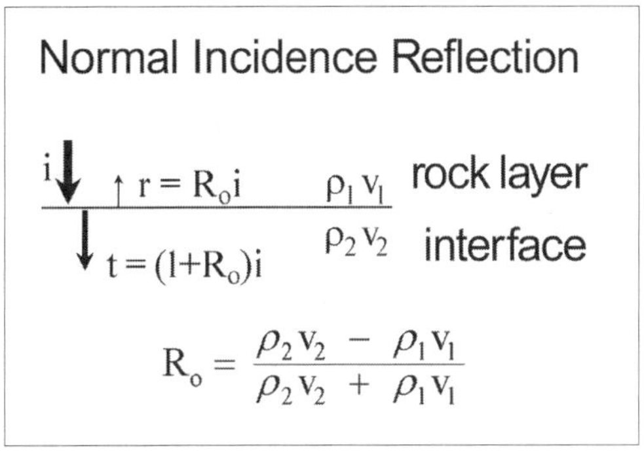

Fig. 2–24. Seismic reflection amplitude. *Source: A. Satter, J. Baldwin, and R. Jespersen. 2000. Computer-Assisted Reservoir Management. Tulsa: PennWell.*

The display of the resulting seismic trace number 1 is shown on the cross section at the adjacent figure. Each reflection appears as a wavelet at the point corresponding to its arrival time at the receiver. The procedure is repeated, with shot 2 and receiver 2 moved some distance down the survey line, and so on.

The size of the recorded wavelet depends on the reflection strength, which is not determined solely by the properties within one rock layer. It also is determined by the contrast in acoustic impedance (which equals velocity × density) of the two layers above and below the interface (fig. 2–24). Normal incidence reflection is given by the following:

$$R_0 = \frac{\rho_2 v_2 - \rho_1 v_1}{\rho_2 v_2 + \rho_1 v_1} \qquad (2.113)$$

where

ρ_1 = bulk density of the upper layer,

ρ_2 = bulk density of the lower layer,

v_1 = acoustic velocity in the upper layer,

v_2 = acoustic velocity in the lower layer, and

R_0 = the reflection strength at 0° angle of incidence, i.e., perpendicular to the interface.

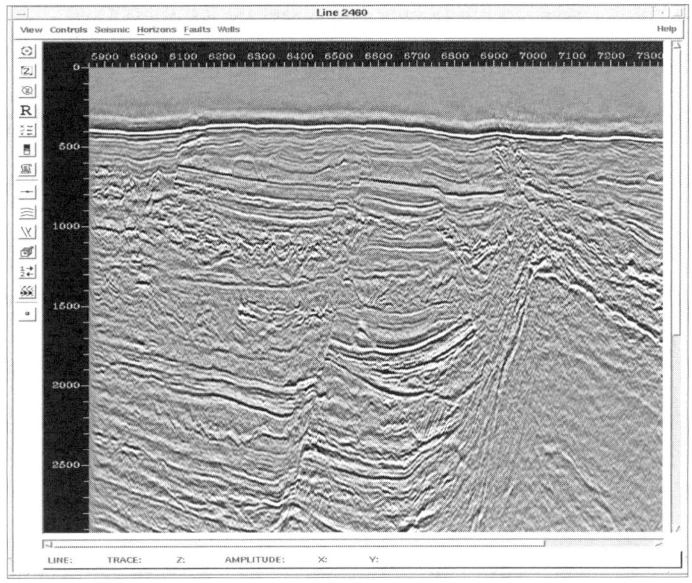

Fig. 2–25. A typical seismic cross section. *Source: A. Satter, J. Baldwin, and R. Jespersen. 2000. Computer-Assisted Reservoir Management, Tulsa: PennWell.*

Structural interpretation

Figure 2–25 shows a typical seismic cross section from an offshore area. The horizontal axis represents map position. The vertical axis is the two-way travel time for the seismic waves (sort of a nonlinear depth scale). The color or shade of gray indicates the strength of the reflected signal. A lot of information about the geology is apparent to the trained eye.

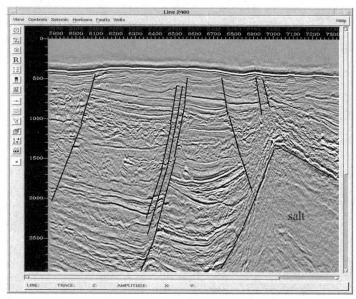

Fig. 2–26. Seismic interpretation. *Source: A. Satter, J. Baldwin, and R. Jespersen. 2000.* Computer-Assisted Reservoir Management. *Tulsa: PennWell.*

Figure 2–26 shows the same data with some faults interpreted and the outline of a salt dome indicated. By correlation with well logs (if available), the interpreter will determine which of the seismic events corresponds to the reservoir. The interpreter will then follow it throughout the 3-D volume, producing a structure map of the top of the reservoir.

Geostatistics

Geostatistics started gaining popularity in the mid-1980s. It provides a major tool for reservoir characterization, which accounts for vertical and horizontal variations of rock and fluid properties.

Computerized mapping is a simple extension of the maps drawn by hand contouring. Statistics provides an alternative procedure for measuring statistical variations of the data points and then creating maps having similar statistical properties throughout.

Like conventional mapping, geostatistics seeks to answer the question, "What are the reservoir property values between the wells?" (See fig. 2–27.) Analysis of the statistical distribution of the data values can lead to more detailed estimates of the map values between the measured points. It offers the following:

- Statistical variation of the data points and creation of maps having similar statistical properties throughout
- Better description of reservoir heterogeneity using statistical distribution of data values
- Means for displacing uncertainty of the interpolated values

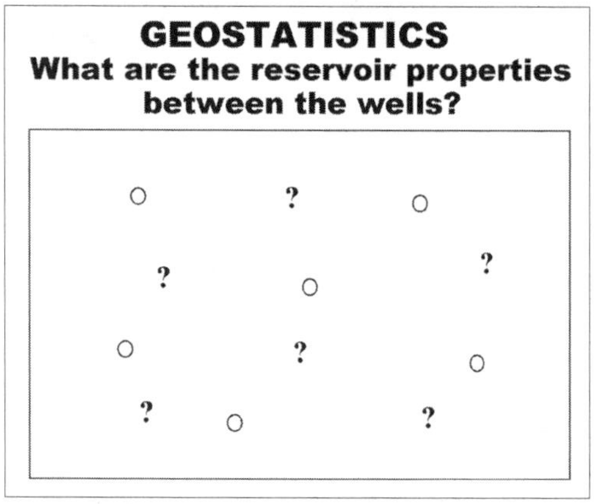

GEOSTATISTICS
What are the reservoir properties between the wells?

Fig. 2–27. The actual reservoir information is limited to drilled well locations only (shown as circles). *Source: A. Satter, J. Baldwin, and R. Jespersen. 2000.* Computer-Assisted Reservoir Management. *Tulsa: PennWell.*

The usual measure of this statistical variation is called a variogram. It is a mathematical expression involving the measured data points and represents how the data values change from one location on the map to another. The amount of change is measured as a function of the distance between data points, sometimes as a function of direction. The variogram serves as a tool for interpolating the map property between known points, while preserving the degree of variation seen in the data.

Some fundamentals of geostatistics, with example results, are presented in the following sections.

Conventional mapping

Well log analysis generally provides the measured data, which serves as the basis for mapping reservoir properties, such as gross thickness, net thickness, and porosity.

Figure 2–28 shows the map of net reservoir thickness for a Rocky Mountain field after analysis of all available logs. The task at hand is to estimate the net thickness at all locations between and surrounding the wells.

The conventional technique for estimating values is to contour the data, either by hand or by using a computer algorithm. If hand-contoured by a geologist, the resulting map can incorporate knowledge of the geologic trends and depositional patterns. However, maps made by 10 different people would likely give 10 somewhat different values of thickness at any point that is not fairly near a well.

Mathematical algorithms can provide a consistent means of contouring the data, with results like those of Figure 2–28. Such algorithms generally cannot incorporate geologic character unless provided with additional control points or contours, or both, by the geologist.

Statistical mapping. Geostatistics provides a means of interrogating the data to determine how a reservoir property varies with distance and direction from any given point. A map that includes these trends can then be created.

The common measure of variation with distance (and sometimes direction) is called a variogram, as mentioned previously. (See *Computer Assisted Reservoir Management* for detailed discussions.[46]) Figure 2–29 shows a variogram ellipse for the data in Figure 2–28, incorporating the variation in all directions.

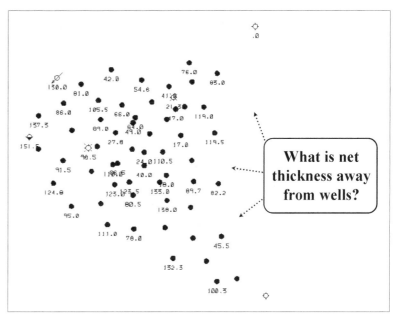

Fig. 2–28. Net reservoir thickness of a Rocky Mountain field. Well locations are shown as dots. *Source: A. Satter, J. Baldwin, and R. Jespersen. 2000. Computer-Assisted Reservoir Management. Tulsa: PennWell.*

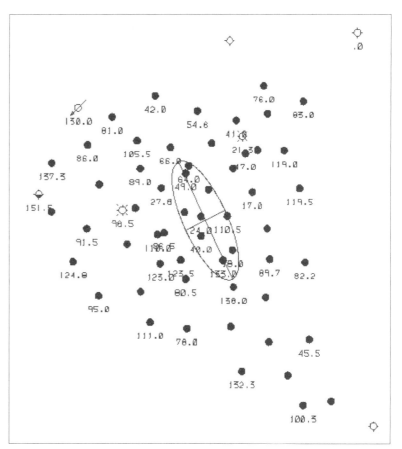

Fig. 2–29. Example of a variogram ellipse. *Source: A. Satter, J. Baldwin, and R. Jespersen. 2000. Computer-Assisted Reservoir Management. Tulsa: PennWell.*

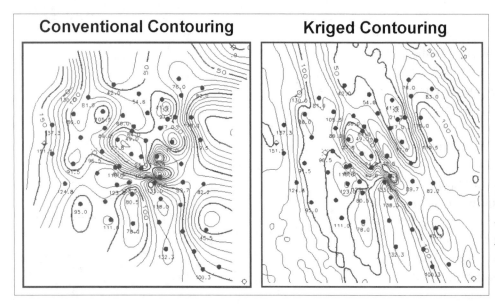

Fig. 2-30. Conventional versus kriged contouring of reservoir thickness. *Source: A. Satter, J. Baldwin, and R. Jespersen. 2000. Computer-Assisted Reservoir Management. Tulsa: PennWell.*

Kriging. Once the variogram ellipse has been calculated, a contouring algorithm can be used to guide the estimation of the values of net thickness throughout the map, as shown in Figure 2–30. This is called kriging. It results in a map that ties the measured values at the wells and also honors the statistical trends determined by the directional variograms. Note that contour shapes tend to mimic the variogram ellipse.

In Figure 2–30, a conventional contouring algorithm is compared to geostatistical kriging. Both maps tie the wells, but they differ in the areas away from the wells. In locations more distant from the wells, the conventional contouring does not indicate the geological trends measured by the variograms and incorporated in the kriged map.

Simulation

An alternative to the single map produced by kriging is to generate many maps spanning the range of possible estimates between the bounding lines. This is called simulation. Simulation creates a mathematical model of the mapped property that has the same spatial statistics as the actual data. In the usual case where the simulation is forced to honor known data points, the process is called conditional simulation.

Each individual simulation map, called a realization, is one of many possible estimates. Maps produced by simulation include more fluctuations than kriged maps, because they allow for the extremes in the estimated values. No single map represents the "most likely" case. Instead, the set of all realizations represents the range of possibilities. Each of the realizations is equally likely to represent the actual reservoir condition.

Measuring uncertainty

Geostatistics provides an opportunity to measure the uncertainty of the maps it produces. For an example, the kriged map can predict a relatively large net thickness in the area of a shallow well that did not penetrate this reservoir. It has been proposed to deepen the well and complete it here. But how reliable are the values of the net thickness shown on this map? This can be ascertained by generating the probability distribution function from the simulation for the grid cell containing the proposed well. It can also be evaluated by generating the plot for the cumulative probability distribution, also called cumulative distribution function (CDF).

Summing Up

Rock and fluid properties are the building blocks in any reservoir engineering study that lead to the formulation of a successful reservoir management strategy.

Reservoir rocks, mostly sedimentary in origin, are classified as given in the following:

- **Clastic rocks.** Formed from pre-existing rocks by erosion, transformation, and deposition. These are sands, sandstones, and conglomerates, and less importantly, siltstones and shales.
- **Carbonate rocks.** Formed from organic constituents and chemical precipitates. These are dolomites, reef rocks, limestones, and chalks.

Rock properties

The basic properties of rocks can be classified as skeletal, concerning the skeleton of the rocks, and dynamic, concerning interaction of the rocks and fluids contained in the pores.

- **Skeletal.** Properties including porosity, pore size distribution, compressibility, surface area, and absolute permeability.
- **Dynamic or interaction.** Properties including wettability, capillary pressure, saturation, and relative permeability.

Rock porosity

Rock porosity is a measure of the pore volume of the rock over its bulk volume. Reservoir engineers are interested in the knowledge of porosity in determining the following, among other:

- Pore volume of the rock
- Hydrocarbon volume and initial oil or gas in place
- Movable hydrocarbon volume and recovery

Porosity is classified as absolute porosity and effective porosity, as follows:

- Absolute porosity is the volume of connected and nonconnected pores as a fraction of the bulk volume of porous rocks.
- Effective porosity is the volume of interconnected pores as a fraction of the bulk volume of porous rocks. Nonconnected pores do not contribute to oil and gas recovery.

The porosity of a rock can be affected by a host of factors during deposition, as well as in the long periods following deposition. Shape, angularity, packing, and sorting of grains in a rock would dictate the volume of void space during the depositional environment. Porosity can be primary or secondary, as follows:

- The porosity that initially develops in a reservoir rock during its deposition in geologic times is known as primary porosity.
- Secondary porosity may develop following original deposition due to various geological and geochemical processes, leading to significant alteration in rock characteristics.

Examples of secondary porosity are vugs or cavities that are typically observed in limestone formations. Circulation of certain solutions, dolomitization of carbonate rocks, and development of fractures in the rock matrix may lead to secondary or induced porosity of the rock.

Porosity measurements. The absolute or total porosity of a core can be determined by comparing its volume before and after crushing core samples.

In contrast, the effective porosity can be determined by allowing a fluid of known density to enter the empty pores of a dry core. The volume of fluid that enters the core is readily known from the increase in weight of the saturated core and the density of the fluid. The fluid can enter only the interconnected pores of the rock and can be used to calculate effective porosity.

Core analysis provides direct measurement of porosity, permeability, lithology, and residual fluid saturations. It also provides wettability, capillary pressure, and gas-oil and water-oil relative permeabilities.

Logging is used to identify rock types, such as porous and permeable sands and limestones, as well as nonpermeable shales.

Well test analysis provides permeability and porosity, bottomhole flowing and reservoir pressure, wellbore conditions, well connectivity, heterogeneities, and reservoir boundaries. The information obtained from transient test analyses covers a relatively larger interwell area as compared to core and log data.

History matching of the reservoir performance and reservoir simulator provides a means of validating reservoir rock and fluid properties for accuracy and reliability.

Cutoff porosity and net thickness of reservoir. Many reservoirs are encountered where porosity is rather low in certain vertical sequences of a geologic formation. Shaliness is also observed in certain sequences requiring computation of net thickness. It is a common practice in the industry to use a cutoff value for porosity in reservoir studies. Depending on reservoir characteristics, typical porosity cutoff points around 5% are used in oil reservoirs. The concept of cutoff porosity leads to the introduction of net thickness as opposed to gross thickness of a reservoir in estimating oil and gas reserves. Net thickness represents the portion of the hydrocarbon-bearing formation that can be produced by conventional means where porosity is relatively high. Commonly encountered values for the net to gross thickness (NTG) ratio may range about 0.95 or less in reservoirs.

Rock permeability

Rock permeability is a measure of the capability of a porous medium to transmit fluid through the network of microscopic channels. Rock permeability can be classified as effective and absolute permeability. Absolute permeability of a rock is a skeletal property, while effective permeability of a rock to a specific fluid in the presence of multiple fluid phases in pores is a dynamic property.

Measure of rock permeability—Darcy's law. In 1856, French engineer Henry Darcy conducted an experiment in order to study fluid flow behavior through a bed of packed sand particles that might emulate a subsurface aquifer or petroleum reservoir. He observed that the volumetric flow rate of water through the packed bed is a function of (a) the dimension of the porous medium, and (b) the difference in hydraulic head.

Darcy's law is found to be valid for other fluids, such as petroleum, when his equation is modified to include the viscosity of the fluid. His empirical equation points to the fact that a porous medium would have a permeability of 1 darcy when a fluid having viscosity of 1 cp flows at a rate of 1 cm^3/sec under a pressure gradient of 1 atm/cm. Permeability of producing reservoirs is known to range from few microdarcies to several darcies. Darcy's law serves as the foundation of reservoir engineering.

Formation compressibility

During primary production of oil and gas, the phenomenon of pore space compression and fluid expansion due to a decline in reservoir pressure contributes to driving energy. The compressibility of the hydrocarbon-bearing formation under isothermal conditions is a function of the rate of change of pore volume with change of pressure.

Values of formation compressibility are given in units per pounds per square inch, and typically fall within the range of 10^{-6} to 10^{-5}. Following depletion in pressure in a reservoir due to oil and gas production, a slight decrease in rock porosity may be encountered, as the overburden pressure remains unchanged.

Dynamic properties of rocks

The simultaneous existence of more than one fluid under dynamic conditions of flow gives rise to rock-fluid interaction and interaction between immiscible fluids in the rock pores. The resulting properties include wettability, capillary pressure, and relative permeability. Dynamic rock properties are strong functions of individual fluid saturation and usually vary with time and location as the reservoir is produced.

Reservoir fluid saturation. Knowledge of fluid saturation is of prime importance in every phase of reservoir study. Saturation values are needed to estimate initial oil in place or initial gas in place at discovery. They are also needed to simulate multiphase fluid flow in order to visualize future performance. Fluid saturation is usually expressed as a fraction of pore space or in percentages.

When all three phases (oil, gas, and water) are present in a reservoir, such as an oil reservoir with an initial gas cap, the saturation fractions of all three phases must add up to unity.

Interfacial tension and wettability. Petroleum reservoirs usually contain at least two fluid phases, which are either oil and water or gas and water. At the interface of the fluids, the force exerted by two immiscible fluid phases is dissimilar, resulting in interfacial tension.

Reservoir rocks can be either water-wet or oil-wet, depending on the tendency of one immiscible fluid phase over the other to adhere to, or "wet," the pore walls in the microscopic network. Wettability of a reservoir is a function of the interfacial tension that exists between the oil phase and the pore surface, between the water phase and the pore surface, and between the two fluid phases. This is explained in the following sections.

Capillary pressure and transition zone. When two immiscible fluid phases are present in a porous medium, such as oil and water, one of the phases preferentially "wets" the pore surface over the other. As a result, a pressure differential is found to exist between the two phases that can be expressed as capillary pressure. The pressure exerted by the nonwetting phase is higher than that exerted by the wetting phase. Capillary pressure is equal to the pressure exerted by the nonwetting phase minus the pressure exerted by the wetting phase.

The paths followed by capillary pressure during drainage and imbibition of the wetting phase are not same. The phenomenon, referred to as the hysteresis effect, points to the fact that capillary pressure in a porous medium would be influenced by the history of saturation.

Leverett J function. A dimensionless function, commonly referred to as the J function, is proposed by Leverett that accounts for capillary pressure in a formation in terms of reservoir porosity and permeability.

The Leverett J function offers a distinct advantage. A generalized curve can be developed as a function of saturation alone for the entire reservoir, instead of an array of capillary pressure plots for each porosity and permeability value.

Transition zone in a reservoir

Transition zone indicates changes in oil, gas, or water saturations over a finite vertical interval as dictated by gravity and capillary forces. In most hydrocarbon-bearing formations, oil and water are not separated by a sharp boundary in the vertical direction. It is observed that during the transition from the water to oil zones, water saturation decreases gradually from 100% to a limiting value in an upward direction. The oil/water contact (OWC) is the height above which water saturation becomes less than 100%, and oil is found in the remaining pore space. At further depth, the free water level (FWL) is found, which would be the height of freestanding water in the absence of any capillary forces.

Relative permeability

The absolute permeability of the rock is a measure of its ability to transmit fluid that completely saturates the porous medium. However, when multiple fluid phases are encountered, one of the fluid phases, such as oil, may experience a decrease in flow when a second phase, such as water, is present in the pore network. Relative permeability to a fluid phase is defined as the ratio of its effective permeability to the absolute permeability of the rock.

Since relative permeability is dimensionless, it serves as a common standard in reservoir studies regardless of the magnitude of the effective permeability in a reservoir. Knowledge of relative permeability is crucial in understanding multiphase fluid flow behavior in a reservoir and predicting future reservoir performance. Again, relative permeability of a fluid phase is a function of the saturations of all the fluid phases present in the rock.

Transmissibility, storavity, and reservoir quality

In reservoir engineering studies, two parameters are often sought to characterize the petroleum reservoir. These are the transmissibility and storavity of the formation. Transmissibility is a function of formation permeability, thickness, and fluid viscosity, while the storavity or storage capacity depends on the formation thickness, porosity, and total compressibility of the system. Hydrocarbon-bearing formations having good storavity and transmissibility are viewed as having good reservoir quality, implying a relatively high recovery.

Reservoir characterization

A myriad of factors related to skeletal and dynamic properties of rock can affect oil and gas production in an unpredictable manner that is not easily identified or understood. Reservoir characterization studies attempt to identify the key elements in rock properties. Such elements include storage capacity (a function of porosity and thickness) and flow capacity (a function of permeability and thickness). The ultimate goal of reservoir characterization is to augment petroleum production and add value to proved reserves at each phase of reservoir life cycle. The task typically involves an integrated team approach, intensive data collection, and computer-aided analyses. It also involves in-depth understanding of rock heterogeneities, on both a macroscopic and microscopic scale, which play a key role reservoir performance.

Sources of rock properties data

Common sources of reservoir data include:

- Laboratory core analysis
- Log analysis
- Well test analysis
- Earth sciences study, including geology, geophysics, and geochemistry
- Emerging technologies

Reservoir simulation techniques can be used to test reservoir rock and fluid properties for accuracy and reliability.

Core analysis. Core analysis provides information regarding porosity, permeability, wettability, capillary pressure, and the relative permeabilities of fluid phases in pores. Results of core analysis reflect localized reservoir properties and do not generally represent the entire reservoir.

Log analysis. Virtually every new well is logged prior to completion as a routine procedure. Logging is used to identify rock types, such as porous and permeable sands and limestones, as well as nonpermeable shales. Several types of logging equipment have been developed in the industry to collect information. Input is gathered regarding lithology, rock porosity, thickness, producing zone depth, and fluid saturation, among others.

Data obtained by logs run at various wells in a reservoir provide well-to-well correlation and aid in characterizing the reservoir. Compared to core studies, well logging is less resource intensive and provides continuous measurement of the rock properties against depth. Studies are commonplace where log and core data are correlated and integrated to build a detailed view of the petroleum reservoir.

Transient well test analysis. Virtually all wells are subjected to some sort of pressure transient analysis at some point in time. The information obtained from transient tests covers rock permeability and porosity, reservoir pressure, wellbore storage, skin, and completion. It also helps in evaluating flow efficiency, reservoir boundaries, and faults.

Geosciences. Geosciences professionals, technologies, tools, and data play a key role in exploration, discovery, development, production, and operation of oil and gas reservoirs.

Geologists usually are involved in determining structures, stratigraphy, depositional environments, rock types, and mineralogy, which provide a geological model for the reservoirs.

Seismology plays an important role in defining and characterizing reservoirs and locating a potential well. Its applications minimize dry holes and poor producers, and much more importantly, assist in the economic development of oil and gas fields.

Geostatistics provides a major tool for reservoir characterization, which accounts for vertical and horizontal variations in rock and fluid properties.

Geoscientists and engineers need feedback from each other throughout their work.

Emerging technologies

A number of emerging technologies are being employed to collect rock properties and characterize the reservoir at various stages, such as the following:

- During the drilling of a new well, an array of sensors or imaging devices is placed in the core barrel above the drill bit assembly. Data from these sensors is used to construct cross-sectional and 3-D images of the core, leading to measurement of rock porosity and fluid saturation of the core before bringing it to the surface.
- Wireline formation testers are employed to measure horizontal and vertical permeabilities of the rock in a newly drilled well.
- Interwell seismic studies lead to the interpretation of in-situ rock properties.
- Results of 4-D seismic studies are used to accurately locate fluid movement over time in the reservoir, which can greatly improve a reservoir model.
- Downhole video cameras may help in pinpointing the water entry point and the possible existence of a high permeability channel.
- Data types used include the following:
  - *Interpretation data.* Fluid distribution, type of fluid production, water or gas coning possibilities, and injected water or gas breakthrough possibilities from high permeability streaks.
  - *Evaluation data.* Initial hydrocarbon in place, recoveries, production rate, productivity index, injection rate, and injectivity index.

Class Assignments

Questions

1. What is petrophysics? Name the principal objectives of collecting rock property information from a petroleum reservoir. What are the two principal classifications of rock properties, and how they are distinguished from each other? Describe at least three important rock properties from each category.

2. Consider the following stages during the course of exploration and development of oil and gas fields:

   (a) A 2-D geophysical survey has been performed in an unexplored region of a petroleum basin. Possible existence of a large field is indicated.

   (b) An exploratory well is drilled. Initial production is 1,200 barrels per day (bpd).

   (c) A second well is drilled at a distance of about 2,500 ft from the first well. Initial production is 300 bpd.

   (d) What rock properties are most likely to be obtained in (a), (b), and (c)? What information is least likely to be available in each case? Organize the answers in a table. What could be the reason for the wide variation in production rates? What would be likely recommendations for developing the reservoir?

3. Would the recommendations be any different if the expected field size is rather small? Why or why not?

4. Why do rock properties vary from one location to another within the reservoir? Why is the knowledge of reservoir heterogeneities important to reservoir engineers? To what degree are the variations in rock properties usually known or estimated? Based on what has been learned, what rock property may be varying spatially and having the greatest influence on reservoir performance?

5. Conduct a literature survey to briefly describe the following, with examples:

   (a) Naturally fractured reservoir

   (b) Reservoir with a fault boundary

   (c) Stratified reservoir with at least three zones

   (d) Channel-shaped reservoir

   (e) Tight gas reservoir

   (f) Reservoir at a steep inclination angle

   (g) Watered-out reservoir

   (h) Chalk formation characteristics

   (i) Vertical seismic profile

   (j) 4-D seismic study

   (k) Geostatistical modeling

6. Define the porosity and permeability of a rock. Is permeability related to porosity? Name at least five microscopic features of a rock that influence its permeability. Experience has shown that a trend is clearly identifiable when porosity and permeability values are plotted on a semilog scale in certain reservoirs. In other cases, however, no such trend is observed. Explain why two rock samples of same porosity can have significantly different permeabilities. Based on a literature review, provide a range for pore throat diameter in rock samples cored from producing oil reservoirs.

7. Distinguish between the following, with examples:

   (a) Sandstone and carbonate reservoirs

   (b) Grain volume and pore volume

   (c) Total porosity and effective porosity

   (d) Primary porosity and secondary porosity

   (e) Net thickness and gross thickness of a formation

   (f) Absolute permeability and effective permeability

   (g) Directional permeability and uniform permeability

   (h) Harmonic average and geometric mean

   (i) Core permeability and average layer permeability

   (j) Dynamic reservoir model and earth model

8. Describe Darcy's law in linear and radial form. What information can be obtained by applying Darcy's law? Provide two examples. What are the assumptions on which the law is based? Consider the following and explain where Darcy's law may not be valid in a high permeability reservoir:

   (a) Flow of oil or gas condensate near the wellbore

   (b) Flow of gas near the wellbore

   (c) Flow of gas far from a producing well

   What modifications of Darcy's law are necessary in such circumstances to correctly describe the flow?

9. Based on Darcy's law, discuss the interrelationships of fluid flow rate, permeability, fluid viscosity, and pressure gradient in porous media. What would be the objective or objectives in order to enhance oil and gas production based on this interrelationship? Why is knowledge of the permeability-thickness product (kh) important in layered reservoirs?

10. What is the unit of permeability, and how it is defined? Give approximate ranges of permeability in cases where an oil reservoir is considered to be of low, moderate, or high permeability. Provide examples of the three cases by conducting a literature review.

    Experience has shown that very tight gas reservoirs would produce, but oil reservoirs would not. However, tight gas reservoirs require hydraulically created fractures in order to produce commercially. Explain the above phenomenon on the basis of Darcy's law.

11. Why is air permeability routinely measured in laboratory studies instead of liquid permeability? What is the Klinkenberg effect?

12. Following months of initial production, it is observed that productivity is declining in several wells completed in the same formation. However, reservoir pressure has not declined significantly in the same relatively short period. List the probable causes for the drop in well productivity.

13. Several new wells are drilled in a limestone reservoir where water is injected to maintain reservoir pressure. An early breakthrough of water is observed in one of the wells following four months of production, while the other wells produced 100% oil. Discuss the probable causes of premature breakthrough in the affected well.

14. Are the rock permeability values generally expected to be the same in both the vertical and horizontal directions? Explain from a geological perspective. What are the sources of permeability data in a reservoir? Would the same or similar values of rock permeability be expected from each source? Why or why not?

15. There are plans to inject water in a reservoir that exhibits a significant directional trend in rock permeability. The trend is along a southeast/northwest direction. How should the placement of the injectors be planned in relation to the producing wells? Why?

16. Define formation compressibility and give its commonly observed range in petroleum reservoirs. How does the compressibility of a formation contribute to production? Does the porosity of a formation change as the reservoir is depleted? The properties of two reservoirs are being reviewed. The formation compressibilities are given as 5.8×10^{-6} psi^{-1} and 5.8×10^{-5} psi^{-1}, respectively. Which formation is more compressible?

17. Why are relative permeability, wettability, capillary pressure, and certain other properties of rock referred to as dynamic properties? Give at least two examples of the significance of the dynamic properties of rock in evaluating reservoir performance.

18. Define hydrocarbon fluid saturation in a rock in terms of pore volume. What are the typical ranges of hydrocarbon fluid saturation in petroleum reservoirs at discovery? Consider a newly discovered oil reservoir with a gas cap. Describe the changes in fluid saturation expected to occur with increasing depth in the reservoir.

19. How is water saturation obtained in a newly drilled well? What is irreducible water saturation? At discovery, why do the pores of the rock almost always have some degree of water saturation in oil and gas zones? Does the formation or interstitial water usually produce along with oil or gas?

20. Is the interface between various fluids expected to be sharp in the reservoir? Describe oil/water and gas/water contacts.

21. Define movable and residual oil saturations. Describe their significance in estimating oil recovery from a reservoir. How can the residual oil saturation be measured when a reservoir is discovered? What steps can be taken to increase movable oil saturation in a reservoir?

22. What is interfacial tension? How does it affect the movement of fluids in porous media?

23. Define the capillary pressure between the oil and water phases in porous media. In what circumstance can the capillary pressure be negative? Define drainage, imbibition, and threshold pressure.

24. Describe the role of capillary pressure in creating a transition zone in the reservoir. How might the extent of a transition zone in a reservoir affect development and production? Explain with examples.

25. Consider the following scenarios. Where would the longest and shortest transition zones likely be located?

    (a) Gas/water contact in reservoir having high vertical permeability

    (b) Oil/water contact in a volatile oil reservoir of moderate porosity

    (c) Gas/oil contact in a heavy oil reservoir of relatively low porosity

26. Define wettability of a formation. Distinguish between water-wet and oil-wet rocks. Do rocks of different wettability affect oil production? Describe the significance of wettability in designing a waterflood operation for an oil reservoir. What is the preferred wetting phase during water injection?

27. Does the wettability of a rock formation change during the life of a reservoir and from one location to another in the same formation? If all other factors are equal, where would the higher oil saturation likely be located when the field is abandoned: in a formation with strongly water-wet rock, or in a reservoir that exhibits mixed wettability characteristics?

28. Water is being injected in an oil reservoir in order to augment oil production. In what circumstances is it a drainage process for oil? In what other circumstances is it imbibition of water?

29. How do gravity and capillary forces influence the distribution of fluids in a reservoir? Explain, giving at least two examples.

30. Define the relative permeability of various fluid phases (gas, oil, and formation water). How does it differ from the effective permeability? Why are relative permeability values always reported between 0 and 1, when the absolute permeability of a formation can vary by several orders of magnitude?

31. Consider an oil reservoir with no free gas present. Are the following statements true or false? Explain.

    (a) The relative permeabilities of oil and water add up to 1 during waterflood operations, since the sum of the saturation fractions of oil and connate water is always 1.

    (b) Relative permeability values of oil and water are not expected to change from one location to another in the formation.

    (c) The relative permeability of water and oil will change continuously with time when water is injected into the reservoir over the life of reservoir.

    (d) Better recovery is expected when the relative permeability curves for oil and water shift to the left in Figure 2–18.

    (e) At the abandonment of an oil or gas field, the relative permeability of oil is always 0.

    (f) Based on oil-water relative permeability curves, the maximum amount of movable oil in a reservoir can be estimated.

32. Consider a reservoir where oil, gas, and water are present. Are the following statements true or false? Explain.

    (a) Below critical gas saturation, the relative permeability of gas is 0.

    (b) Relative permeability to gas is a function of gas saturation alone.

    (c) Three-phase relative permeability values cannot be determined in the laboratory and are obtained only by correlation.

33. How can representative values of effective permeability to oil be obtained from the field? Would it be the same as the effective permeability obtained from laboratory studies?

34. Define pseudorelative permeability. Describe its significance in reservoir studies.

35. What is reservoir characterization? What basic information would be required to characterize a reservoir? Discuss the concept of flow units in effectively producing a layered reservoir. How would a fractured reservoir be characterized?

36. Based on a literature survey, describe a reservoir characterization study. Explain how reservoir characterization was utilized to enhance reservoir performance.

37. Define formation transmissibility and storavity. Qualitatively compare the performance of the following layers in a reservoir:

    Lower layer: very high transmissibility and low storavity

    Upper layer: relatively low transmissibility and high storavity

    (a) Which layer is likely to be responsible for early water breakthrough? What remedial action can be taken?

    (b) Should any difference be anticipated in reservoir performance if good vertical communication exists between the two layers?

38. Distinguish between original hydrocarbon in place and the petroleum reserve. List the data necessary in order to make estimates of both of these. Consider the following stages in developing a large field, and explain in which circumstances the best estimate of the reserve can be given:

    (a) Following a geophysical survey of the unexplored field

    (b) Following a review of data from nearby fields

    (c) Following drilling of several new wells at the early stages of the field

    (d) Following performance prediction by simulating a reservoir model

    (e) Following attainment of peak production

    (f) Following well testing in a few wells that identified reservoir boundaries

    (g) Near the abandonment of the field as significant water cut is encountered in most of the producers

39. Name the rock properties and reservoir characteristics that would be necessary to estimate original hydrocarbon in place. Define the recovery factor of oil and gas. Name at least three sources of recovery factor the reservoir engineer may seek for an oil field where few wells are drilled and production has been ongoing for about a year.

40. Describe the sources and techniques of collecting information related to reservoir rock properties. Why it is necessary to develop the reservoir first in order to gain adequate knowledge of rock properties? Why are probabilistic models needed in reservoir engineering studies?

41. Describe core analysis and list the information, along with their categories, that can be obtained about the reservoir. As a well is drilled and cores are collected, what change or changes can a porous rock undergo? How can the changes affect the results of the core studies? Does every well need to be cored in a reservoir?

42. Distinguish between sidewall plug and sidewall whole-core analyses. What would be a better method in the case of a limestone formation having secondary porosity? Twenty core samples are obtained from a single well drilled in a newly discovered reservoir. The following information is obtained:

    (a) Effective porosity

    (b) Bulk permeability

    (c) Formation compressibility

    (d) Residual saturation

    Rearrange the above properties, from least likely to most likely, of their expected validity over the entire reservoir.

43. What is well logging? What information is usually obtained from a well log? Is every new well logged customarily? Briefly describe the principle of operation of the following:

    (a) Resistivity log

    (b) Acoustic log

    (c) Density log

44. Based on a literature survey, describe how information obtained from core and log studies can be integrated. Include in this description how the integrated data was utilized in enhancing reservoir performance.

45. What is the basic principle behind seismic measurements? How does a seismic survey aid in oil and gas exploration? Describe a 4-D seismic survey and its application in a reservoir during production. Conduct a literature survey, if necessary.

46. Describe a geostatistical reservoir model. How it is used in reservoir studies?

Exercises

2.1. Develop an expression for Darcy's law in the following cases:

    (a) Linear flow of gas in a core sample (Equation 2.15)

    (b) Radial flow in an oil reservoir

    (c) Radial flow in a gas reservoir

    (d) Inclined flow of oil in a dipping reservoir

Include all of the assumptions made in deriving the above.

Calculate the oil flow rate down dip at an inclination angle of 30°. The following data is available:

Length of the flow system, ft, = 200.
Width, ft, = 150.
Height, ft, = 10.
Pressure drop across the system, psi, = 125.
Fluid viscosity, cp, = 0.8.

2.2. (a) What gain in average rock permeability can be obtained in a tight gas reservoir by acid stimulation? Consider the following data:

Drainage radius of well, ft, = 1,320.
Radius of wellbore, ft, = 0.397.
Average permeability of the reservoir, millidarcies, = 0.8.
Average thickness of formation, ft, = 125.

Stimulated zone:

Approximate radius, ft, = 5.5.
Estimated permeability, darcies, = 1.8.

    (b) Calculate the steady-state gas flow rate before and after stimulation with the following additional information:

        Reservoir pressure at the external boundary, psi, = 1,850.
        Bottomhole flowing pressure, psi, = 1,078.
        Reservoir temperature, °F, = 138.

Assume an average gas viscosity and gas deviation factor of 0.0166 cp and 0.798, respectively. Make any other assumptions necessary to perform the analysis.

    (c) Estimate the well production rate in the following five years of the stimulation operation. Nearby wells are known to decline at rates between 12.5% and 15% each year.

    (d) Do you expect the wells drilled in a reservoir of significantly greater permeability having similar formation pressure and size to decline at similar rates? Why or why not?

2.3. The following permeability data is obtained from a reservoir (see table 2–6).

    (a) Calculate the geometric average permeability of the formation.

    (b) Calculate the fluid flow rate based on the following information:

        Length of the flow system, ft, = 400.
        Width, ft, = 200.
        Height, ft, = 12.5.
        Pressure drop across the system, psi, = 200.
        Fluid viscosity, cp, = 0.8.

    (c) Recalculate (b), ignoring the most permeable interval. Compare results in (b) and (c), and draw conclusions.

Table 2-6. Horizontal permeability at various depths

Sample	Formation Interval, ft	Permeability, mD
1	6,191.5–6,193.0	22
2	6,193.0–6,194.4	18
3	6,194.4–6,196.0	17
4	6,196.0–6,197.7	14
5	6,197.7–6,199.8	16
6	6,199.8–7,000.6	121

2.4. A sensitivity study is being performed related to the occurrence of transition zones in a reservoir.

(a) Plot the height of the transition zone between oil and water versus specific gravity of oil, which varies between 20° and 40° API. Make valid assumptions of rock and fluid characteristics necessary in computations.

(b) Repeat (a) at the gas/oil contact. Assume the same specific gravity of gas in the gas cap in all cases, while varying oil specific gravity.

(c) Plot the sensitivity of the J function to the porosity of the rock between 0.5 and 0.35. Assume that the permeability of the rock can be estimated by Equation 2.35. Again, make all other assumptions necessary.

(d) Clearly state all the assumptions made in (a), (b), and (c).

2.5. Based on the three-phase relative permeability correlations for an unconsolidated sandstone reservoir, prepare oil, gas, and water phase isopermeability contours in separate triangular diagrams. Assume a connate water saturation of 0.185. Example contours are shown as dotted lines in Figure 2–31. The three apexes of triangle represent 100% saturations of oil, gas and water. Draw conclusions from the study.

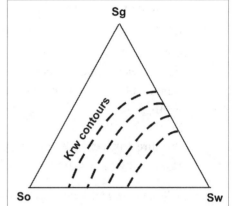

Fig. 2–31. Triangular diagram of three-phase relative permeability

2.6. The following well data shown in Table 2–7 is available for the oil reservoir shown in Figure 2–32.

Table 2–7. Well data for oil reservoir

Well	Porosity, fraction	Thickness, feet	Connate Water Saturation, fraction
D-1	0.32	46	0.19
D-2	0.34	47	0.21
D-3	0.31	42	0.21
D-4	0.23	40	0.24
D-5	0.18	32	0.28
D-6	0.28	45	0.22
D-7	0.28	35	0.24
D-8	0.25	36	0.22
D-9	0.17	32	0.25
D-10	0.2	35	0.24
D-11	0.3	35	0.22
D-12	0.15	28	0.24
D-14	0.14	26	0.26
D-15	0.14	29	0.28

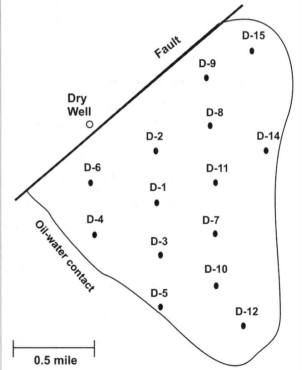

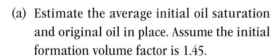

Fig. 2–32. Oil reservoir showing location of wells

(a) Estimate the average initial oil saturation and original oil in place. Assume the initial formation volume factor is 1.45.

(b) Draw approximate contours of hydrocarbon pore volume (HCPV) in barrels per acre-feet of reservoir.

2.7. Based on a literature review, describe the fracture-related properties of rock usually sought in (a) naturally fractured reservoirs and (b) hydraulically fractured wells in evaluating performance. Provide the ranges of fracture characteristics that were listed.

References

1. Darcy, H. 1856. *Les fontaines publiques de la ville de Dijon*. Paris: Victor Dalmont.

2. Amyx, J. W., D. M. Bass, Jr., and R. L. Whiting. 1960. *Petroleum Reservoir Engineering—Physical Properties*. New York: McGraw-Hill.

3. Craft, B. C., and M. Hawkins. 1990. *Applied Petroleum Reservoir Engineering*. 2nd ed., rev. by R. E. Terry. New Jersey: Prentice-Hall.

4. Amyx, J. W., D. M. Bass, Jr., and R. L. Whiting. 1960.

5. Ahmed, T. H. 2001. *Reservoir Engineering Handbook*. 2nd ed. Houston: Gulf Professional Publishing Co.

6. Warren, J. E., and H. S. Price. 1961. Flow in heterogeneous porous media. *Society of Petroleum Engineers Journal*. September.

7. Amyx, J. W., D. M. Bass, Jr., and R. L. Whiting. 1960.

8. Ahmed, T. H. 2001.

9. Craft, B. C., and M. Hawkins. 1990.

10. Klinkenberg, L. J. 1941. *The Permeability of Porous Media to Liquids and Gases*. API Drilling and Production Practice.

11. Tiab, D., and E. C. Donaldson. 1996. *Petrophysics: Theory and Practice of Measuring Rock and Fluid Transport Properties*. Houston: Gulf Publishing Co.

12. Towler, B. F. 2002. *Fundamental Principles of Reservoir Engineering*. Richardson, TX: Society of Petroleum Engineers.

13. Guenther, K., S. Perkins, B. Dale, R. Pakal, and P. Wylie. 2005. South Diana, Gulf of Mexico, U.S.A., a case study in reservoir management of a compacting gas reservoir. IPTC Paper #10900. International Petroleum Technology Conference, Doha, Qatar, November 21–23.

14. Hall, H. N. 1953. Compressibility of reservoir rocks. *Transactions*. AIME. Vol. 198: pp. 309–311.

15. Newman, G. H. 1973. Pore-volume compressibility. *Journal of Petroleum Technology*. February.

16. Leverett, M. C. 1941. Capillary behavior in porous solids. *Transactions*. AIME.

17. Ahmed, T. H. 2001.

18. Wilhite, G. P., 1986. *Waterflooding*, SPE Textbook Series Vol. 3, Richardson, Texas, Society of Petroleum Engineers.

19. Tiab, D., and E. C. Donaldson. 1996.

20. Leverett, M. C. 1941.

21. Craig, F. F. 1971. The reservoir engineering aspects of waterflooding. In *SPE Monograph 3*. Dallas: Society of Petroleum Engineers.

22. Amyx, J. W., D. M. Bass, Jr., and R. L. Whiting. 1960.

23. Tiab, D., and E. C. Donaldson. 1996.

24. Ahmed, T. H. 2001.

25. Wyllie, M. R. J., and G. H. F. Gardner. 1958. The generalized Kozney-Carmen equation—its applications to problems of multi-phase flow in porous media. *World Oil*.

26. Pirson, S. J., ed. 1958. *Oil Reservoir Engineering*. New York: McGraw-Hill.

27. Ahmed, T.H., 2001.

28. Stone, H. L. 1973. Estimation of three-phase relative permeability and residual oil data. *Journal of Canadian Petroleum Technology*, (Oct.), pp. 53–59.

29. Dykstra, H., and R. L. Parsons.1950. The prediction of oil recovery by waterflooding. In *Secondary Recovery of Oil in the United States*. 2nd ed. Washington, DC: American Petroleum Institute. pp. 160–174.

30. Schmalz, J. P., and H. D. Rahme. 1950. The variation of waterflood performance with variation in permeability profile. *Producers Monthly*. Vol. 18, no. 9: pp. 9–12.

31. Aminian, K., and B. Thomas. 2002. A new approach for reservoir characterization. SPE Paper #78710, prepared for 2002 SPE Regional Meeting, Lexington, KY, October 23–26.

32. Honarpour, M. M., N. R. Nagarajan, and K. Sampath. 2006. Rock/fluid characterization and their integration—implications on reservoir management. *Journal of Petroleum Technology*. September: 120–131.

33. Wardlow, N. C., and J. P. Cassan. 1989. Oil recovery efficiency and the rock-pore properties of some sandstone reservoirs. In *Reservoir Characterization*. Vol. I. SPE Reprint Series no. 27. Richardson, TX: Society of Petroleum Engineers. p. 214.

34. Lucia, F. J. 1989. Petrophysical parameters estimated from descriptions of carbonate rocks: a field classification of carbonate pore space. In *Reservoir Characterization*. Vol. I. SPE Reprint Series no. 27. Richardson, TX: Society of Petroleum Engineers. 89.

35. Craig, F. F. 1971.

36. Western Atlas. 1992. *Introduction to Wireline Log Analysis*.

37. Schlumberger. 1989. *Log Interpretation Principles and Applications*.

38. Texaco EPTD. 1983. *Open Hole Analysis and Formation Evaluation*.

39. Satter, A., J. Baldwin, and R. Jespersen. 2000. *Computer-Assisted Reservoir Management*. Tulsa: PennWell.

40. Brown, A. R. *Interpretation of 3-Dimensional Seismic Data*. AAPG Memoir 42.

41. Jenkins, S. D., M. W. Waite, and M. F. Bee. 1997. Timelapse monitoring of the Duri steamflood. *The Leading Edge*. Vol. 16, no. 9: 1,267–1,273.

42. Jespersen, R. Personal communication.

43. DeLage, J. Scientific Software-Intercomp, Houston, Texas, personal communication.

44. Hohn, M. E.1988. *Geostatistics and Petroleum Geology*. New York: Van Nostrand Reinhold.

45. Isaaks, E. H., and R. Mohan Srivastava. 1989. *An Introduction to Applied Geostatistics*. Oxford: Oxford University Press.

46. Satter, A., J. Baldwin, and R. Jespersen. 2000.

3 · Fundamentals of Reservoir Fluid Properties, Phase Behavior, and Applications

Introduction

Reservoir fluids consist of naturally occurring gas and liquid hydrocarbons, in addition to subsurface formation water. Hydrocarbons are chemical compounds and derive their name from their composition, which is based on two elements: hydrogen and carbon. Hydrocarbon fluids are often found to be associated with impurities such as carbon dioxide, nitrogen, and hydrogen sulfide in the porous network of a geologic formation. Reservoir fluid properties are primarily dependent upon prevailing pressure and temperature, in addition to fluid composition. Petroleum reservoirs usually contain a whole range of hydrocarbon components, which vary in important properties such as molecular weight, viscosity, and vaporization or condensation characteristics.

Initial pressure, temperature, and fluid composition determine the reservoir types, such as black oil, volatile oil, dry gas, and gas condensate reservoirs. The reservoir types are described later in the chapter. Furthermore, the combined influences of the rock characteristics and fluid properties need to be known with reasonable accuracy in order to develop, produce, and manage a petroleum reservoir. Reservoir engineers are interested in the pressure-volume-temperature (PVT) properties of petroleum fluids. Such properties are influenced by the prevailing pressure and temperature under reservoir conditions, as well as the volume or density of a fluid. A checklist of fluid properties needed in conducting various reservoir evaluations and in planning an optimal production strategy is included later in the chapter.

As reservoir pressure declines during production, certain changes pertaining to fluids occur in porous media, affecting reservoir performance significantly. The changes include the following:
- Changes in reservoir fluid volume and density
- Changes in fluid compressibility, viscosity, and mobility
- Changes in fluid composition due to the following:
  · Evaporation of lighter hydrocarbon components from the liquid
  · Condensation/precipitation of liquid droplets from the gas
- Reduction of gas solubility in the oil
- Transformation of rich gas to lean gas, meaning that the relatively heavy hydrocarbons condense out of the gas phase
- Changes in the gas/oil ratio (GOR) or gas/condensate ratio (GCR), as evident from well production

These affect the following:
- Well production rates
- Ultimate recovery from a reservoir
- Formulation of oil and gas recovery strategy
- Design of surface facilities

The learning objectives of this chapter are as follows:
- Significance of reservoir fluid properties in reservoir management
- Hydrocarbon accumulation and fluid properties
- The role of pressure and temperature in influencing dynamic reservoir behavior
- Estimation of reservoir pressure in oil and gas zones

- Hydrocarbon composition of reservoir fluids
- Classification of petroleum reservoirs based on reservoir fluid properties
- Characteristics of reservoir fluids in various reservoir types
- Pressure-volume-temperature (PVT)—dependent properties of reservoir fluids, including gas, gas condensate, oil, and formation water
- Sources of fluid properties data
- In-situ and laboratory measurements of fluid properties
- Industry-recognized correlations of important fluid properties
- Introduction to vapor/liquid equilibrium relationships
- Phase behavior of reservoir fluids illustrated by phase diagrams
- Dynamic response of reservoirs as influenced by fluid phase behavior
- Effects of various fluid properties on reservoir performance, including fluid gravity, viscosity, initial reservoir pressure, and bubblepoint pressure
- Fluid properties checklist in order to conduct various reservoir studies, including determination of oil and gas in place and field recovery

Hydrocarbon accumulations and fluid properties

Fluid compositions can differ significantly from one reservoir to another, depending on the nature of the petroleum source rock and the prevailing pressure and temperature in the subsurface geologic formation. The depth of the hydrocarbon accumulation may show certain trends reflecting prevailing conditions during the formation of oil in the geologic ages. For example, volatile oil and gas or gas condensate reservoirs are encountered at relatively greater depths in many petroleum regions. This occurs due to the intense heat encountered during the formation of petroleum at great depths. On the other hand, shallow accumulations of heavy oil at depths of a few hundred to a few thousand feet are commonplace.in certain regions across the continents. Studies have generally indicated that volatile oil, gas condensate, and gas reservoirs are discovered with increasing frequency as exploratory wells are drilled deeper in a petroleum horizon.

In many oil reservoirs, areal variations in fluid properties within a geologic layer are not significant. Hence, the properties can be customarily obtained with relative ease based on routine laboratory studies of fluid samples obtained from a single well or a handful of wells. However, in certain cases, such as in compartmentalized formations or in heavy oil reservoirs, areal variations in important oil and gas properties may be encountered. Such properties include specific gravity, dew point, and gas solubility.

Again, fluid characteristics may vary from one geologic layer to another within a reservoir when the layers are segregated by impermeable shale. This may be due to the differences in the origin and migration history of the oil present in the individual layers.

Fluid property measurements

Modern downhole sensors are capable of measuring a number of fluid properties in situ. When field data is unavailable, as in the case of exploratory wells or when fluid samples are not representative of reservoir conditions, suitable correlations may be utilized to estimate various fluid properties. Certain widely known PVT correlations are presented in later sections. With the advent of the digital age, petroleum software applications are capable of computing a myriad of fluid properties from these correlations in a fraction of a second when performing reservoir studies. These are used to predict future well rates associated with any fluid phase changes, to name only one application.

As a new well is drilled, vital properties of the reservoir fluid are sought immediately to evaluate the petroleum reservoir in terms of future performance. Traditionally, fluid samples are retrieved by conducting a drillstem test (DST), described in chapter 5. Modern tools that employ downhole sensors and analyzers in real time are capable of measuring various PVT properties and composition of the fluid. This information is supplemented by conventional laboratory studies of retrieved fluid samples. Most downhole tools interface directly with related computer applications and an integrated database system.

Reservoir management strategy

Initial assessment of fundamental fluid properties, including the behavior of reservoir fluids with changing pressure, determines whether an improved recovery process can produce the reservoir effectively. For example:

- In order to augment recovery from a highly viscous oil reservoir, a thermal recovery project may be necessary. A significant amount of energy is needed to drive oil towards the wellbore by lowering its viscosity.

- In volatile oil reservoirs, oil production can be hindered due to evolution of gas when the reservoir pressure declines. In this case, external water injection may be considered early in the life of field. This may be used in order to maintain the reservoir pressure at a relatively high level.

- When an oil reservoir is discovered with an overlying gas cap, as opposed to a reservoir where no free gas is present, the reservoir development strategy could be significantly different. This could include the reinjection of produced gas in order to maintain pressure and optimize the recovery of the liquid phase.

- In a gas condensate reservoir where lean gas is produced predominantly, leaving the enriched hydrocarbon components behind, a gas cycling operation would be required. This would be necessary to maintain the reservoir pressure above a certain level to improve recovery of enriched hydrocarbons.

Some of the techniques in developing petroleum reservoirs described above are treated in detail in later chapters.

Pressure and Temperature

The reservoir energy that drives petroleum fluids towards wells is directly related to the prevailing reservoir pressure. Since both reservoir pressure and temperature control fluid properties, these are discussed first. In fact, hydrocarbon fluid composition, pressure, and temperature determine whether the fluid would initially exist in a single phase (oil or gas) or in two phases (oil with a gas cap). It is common practice to determine the reservoir pressure and temperature at discovery, and to conduct pressure surveys periodically or even continuously at various wells during the life cycle of the reservoir. Most reservoir engineering studies, including reservoir simulation, require knowledge of the reservoir pressure response as a function of time and location during production and shutdown of the wells. A robust pressure monitoring program may readily point to reservoir drive mechanisms, effectiveness of fluid injection, and suspected heterogeneities in the rock, among other factors. Modern reservoir monitoring practices employ downhole sensors that continuously provide a wealth of pressure and rate data.

Estimation of reservoir pressure is primarily based on the following factors:
- Hydrostatic pressure gradient of fresh water
- Specific gravity of formation water present in the reservoir
- Depth of the reservoir

In oilfield units, reservoir pressure is commonly expressed in pounds per square inch (psi). Reservoir temperature is reported in degrees Fahrenheit (°F). The hydrostatic pressure gradient of fresh water can be obtained from its density, as shown in the following:

Hydrostatic gradient of fresh water:

$$= \left(62.4 \frac{\text{lb-m}}{\text{ft}^3}\right) \times \left(\frac{1 \text{ ft}^2}{144 \text{ in}^2}\right) \times \left(32.2 \frac{\text{ft}}{\text{sec}^2}\right) / \left(32.2 \frac{\text{lb-m ft}}{\text{lb-f sec}^2}\right) = 0.433 \frac{\text{psi}}{\text{ft}} \qquad \textbf{(3.1)}$$

Noted previously, psi stands for pounds per square inch, as abbreviated in most engineering studies. Pressure measurements are usually reported either as gauge pressure (psig) or absolute pressure (psia). Implicit in the conversion in Equation 3.1 is that 1 pound-mass (abbreviated as lb-m) exerts a force of 1 pound (lb) under standard acceleration of gravity.

Due to the presence of dissolved solids in significant quantities, subsurface formation water, also referred to as connate water or interstitial water, is usually heavier than fresh water. Connate water is distinguished from interstitial water by the fact that the former fills the rock pores during the formation of the rock, while the latter may enter the pores in the postdepositional period. The dissolved solids in formation water include salts of sodium and potassium, among others. Consequently, the density of the formation water is higher, and its specific gravity is usually greater than unity. The hydrostatic gradient of the formation water is calculated as shown in the following:

Hydrostatic gradient of formation or connate water:

$$= 0.433\ \gamma_w \quad \text{psi/ft} \tag{3.2}$$

where

γ_w = specific gravity of formation water, dimensionless.

The values of the formation water gradient typically range between 0.435 and 0.5 psi/ft. However, higher gradients are not uncommon in the field. When the depth of the reservoir and the specific gravity of the formation water are known, the reservoir pressure can be estimated from the following equation:

$$p_{res} = p_{atm} + 0.433\ \gamma_w D \quad \text{psia} \tag{3.3}$$

$$= p_{atm} + \left[\frac{dp}{dD}\right]_w D \quad \text{psia} \tag{3.4}$$

where

p_{res} = reservoir pressure, psia,

p_{atm} = atmospheric pressure, usually assumed to be 14.69 psi,

γ_w = specific gravity of formation water, dimensionless,

$\left[\dfrac{dp}{dD}\right]_w$ = formation water gradient, psi/ft, and

D = reservoir depth, ft.

However, pressure in certain reservoirs is found to be markedly higher than that calculated by Equation 3.3. These reservoirs are referred to as overpressured. Certain overpressured gas reservoirs have been discovered with an apparent initial pressure gradient of 0.8 psi/ft or even more. Figure 3–1 shows an overpressured formation along with the hydrostatic gradient and overburden pressure. In such cases, the above equations must be modified, as shown in the following:

$$p_{res} = p_{atm} + 0.433\ \gamma_w D + C \quad \text{psia} \tag{3.5}$$

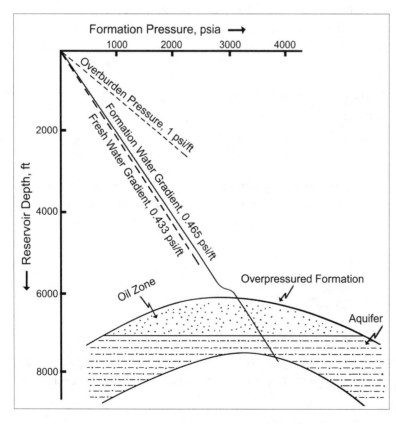

Fig. 3–1. Hydrostatic gradient of abnormally pressured reservoirs

The constant C in Equation 3.5, expressed in psi, is positive in the case of an overpressured reservoir, and negative when the geologic formation is underpressured. The geopressured zones are formed as a result of various processes that occur over the geologic time scale. These could include accelerated sedimentation and subsidence, incomplete expulsion of interstitial water from pores, faulting, uplifting, geologic unconformities, and others. Abnormally pressured geologic formations may pose operational challenges during drilling, resulting in well blowouts in certain cases.

Furthermore, it can be noted that the drilled well depth is customarily reported in either true vertical depth (TVD) or measured depth (MD). Measured depth is greater than true vertical depth in the case of deviated or horizontal wells, where the angle of deviation is needed to compute the true vertical depth of a reservoir. Worldwide statistics have indicated that the major accumulations of conventional resources occur between 4,000 and 12,000 ft, although the current trend is to drill and explore much deeper in both onshore and offshore locations.

Reservoir engineers frequently encounter various measures of pressure when conducting engineering studies, as given in the following description.

Reservoir pressure. This represents the reservoir fluid pressure in the rock pores and is sometimes referred to as the formation pressure or pore pressure. Reservoir pressure is measured by digital or analog devices placed downhole before and after completion of a well. Reservoir pressure usually increases with depth according to the specific gravity of fluid (oil, gas, or formation water) present in the pay zone or zones.

Initial reservoir pressure. This is the reservoir pressure at discovery before any production takes place. The major source of initial reservoir pressure data includes drill-stem testing described in chapter 5. Typical values of initial reservoir pressure range from less than 2,000 pounds per square inch to greater than 10,000 pounds per square inch. Certain abnormally pressured reservoirs located at great depths may exhibit significantly higher initial pressure than what is usually anticipated. In the absence of any external source of energy, such as water influx from the surrounding aquifer, reservoir pressure is expected to decline continuously with production.

Average reservoir pressure. During the life of a reservoir, the average reservoir pressure represents at any time the pressure that would be obtained when all of the flow in the porous media ceases. Under the circumstances, an equilibrium in reservoir fluid pressure is eventually established. For a well producing from its own drainage area, the average reservoir pressure in the drainage area can be obtained by conducting a well test (described in chapter 5).

Abandonment pressure. The abandonment pressure is the pressure at which commercial recovery of oil and gas is no longer feasible from a reservoir. The producing well is said to have reached its economic limit and is abandoned.

Flowing bottomhole pressure. This pressure is measured in a producing well under dynamic conditions of oil and gas flow. The term bottomhole usually refers to the depth where the well is completed in the producing zone. Similarly, bottomhole injection pressure is measured and reported in the case of injectors.

Static bottomhole pressure. This pressure is measured in a well when a static condition prevails. A static condition may be achieved by shutting in the well for a considerable period. In a producer, the static bottomhole pressure is higher than the flowing bottomhole pressure. During production, the reservoir pressure decreases continuously towards the wellbore but stabilizes to a higher level when production is ceased for a sufficient length of time. Certain wells in a reservoir are designated as observation wells that neither produce nor inject but are utilized to measure the reservoir pressure at the location.

Wellhead pressure. This pressure is measured at the wellhead. In a producing well, the wellhead pressure is less than the bottomhole pressure due to the hydrostatic column and frictional losses during flow.

Fracture pressure. The threshold pressure at which the subsurface formation is fractured by injecting fluid is referred to as the fracture pressure. Certain well tests are specifically designed to determine the fracture gradient of the formation. In tight reservoirs, wells are hydraulically fractured to boost productivity. Again, care is exercised so as not to exceed the fracture pressure during water injection in a reservoir undergoing waterflooding.

Pressure at datum. The reference depth at which the reservoir pressure (initial, average, etc.) is reported. The top of the reservoir or the depth of fluid contact within the reservoir, such as the oil/water contact, may be designated as the datum.

Overburden pressure. The overburden pressure refers to the combined pressure exerted by the formation rock and the fluid that exists in the pore spaces of the rock. A typical value of overburden pressure is 1 psi/ft.

Reservoir temperature. Reservoir temperature is a direct function of reservoir depth and is found to increase by 1°F for about 60 ft in many reservoirs. Reservoir temperature can be estimated based on the following equation:

$$T_{res} = T_s + (\text{Thermal gradient}) \ D/100°F \tag{3.6}$$

where

T_{res} = reservoir temperature, °F, and
T_s = surface temperature, °F.

Figure 3–2 is a graphical representation of the increase in subsurface temperature with depth. Temperature gradients from various oil and gas regions are included in the plot. Certain reservoirs exhibit temperature anomalies as a result of geothermally active processes.

Example 3.1. Calculate the pressure and temperature of a newly discovered reservoir at a depth of 7,000 ft. The specific gravity of the formation water from the same producing zone in the area is known to be about 1.048 (water = 1). Assume a geothermal gradient of 1.8°F per 100 ft. List all other assumptions necessary.

Solution: Referring to Equation 3.3, the estimated reservoir pressure is calculated as in the following:

$$p_{res} = 14.7 + 0.433 \ (1.048) \ (7,000) = 3,191.2 \ \text{psia}$$

Reservoir temperature can be found based on the assumption that the mean surface temperature is 60°F. Temperature gradient and mean surface temperature are generally available in oil regions.

$$T_{res} = 60 + (1.8)(7,000)/100 = 186°F$$

Reservoir pressure in oil and gas zones

Estimation of the reservoir pressure in oil and gas zones requires knowledge of the oil and gas gradients, respectively. Consider an oil reservoir with a gas cap. Gas, having the least density, segregates to the top portion of the reservoir, followed by an intermediate oil zone located between the gas and water zones (fig. 1–2). Reservoir fluids are distributed vertically in accordance with the equilibrium between gravity and capillary forces, as described in chapter 2. Transition zones are frequently observed at the gas/oil and oil/water contacts. Gas, water, and oil zones are usually characterized by openhole logs run in a newly drilled well. The following example illustrates the computation of reservoir pressure in an oil or gas zone.

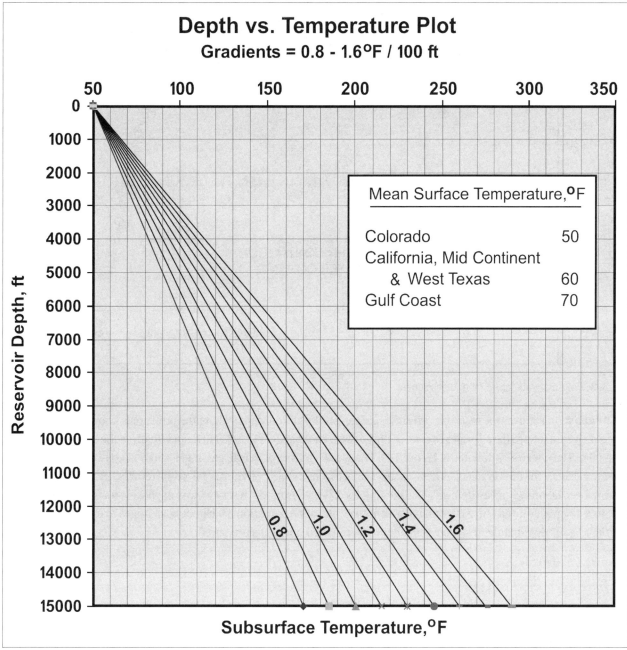

Fig. 3–2. Plot of temperature gradients in a subsurface formation as a function of depth. For surface temperatures other than 50°F, add or subtract the difference from the result. For example, when the surface temperature is 60°F, add 10°F to the result. Mean surface temperatures for various regions are provided in plot. *Source: Halliburton Co. 1985. Welex Log Interpretation Charts.*

Example 3.2. Consider a newly discovered gas reservoir where the hydrocarbon zone is located between 4,900 ft and 4,990 ft. The reservoir pressure, recorded in the underlying water zone, is 2,275 psia at 5,000 ft.

(a) Calculate the expected pressure at the top of the reservoir, considering an oil zone between 4,900 ft and 4,990 ft. The pressure gradient of the oil is 0.38 psi/ft.

(b) Recalculate the expected pressure when a gas zone, instead of an oil zone, exists between 4,900 ft and 4,990 ft.

Make necessary assumptions.

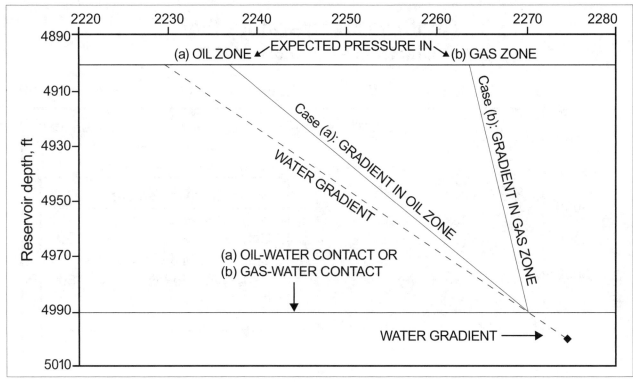

Fig. 3–3. Pressure gradients in oil and gas zones

Solution: To solve for the reservoir pressure, one can start from the depth where the pressure is known. This work will proceed upward toward the oil/water contact or the gas/water contact (GWC) at 4,990 ft, and finally to the top of the reservoir at 4,900 ft. The reservoir is depicted in Figure 3–3. The specific gravity of the oil, gas, and formation water must be known in order to calculate the fluid gradients. These can be obtained from a variety of sources, including recorded pressure versus depth information, laboratory measurements of fluid samples collected from the reservoir, and applicable correlations or charts.

Since the reservoir pressure is available within the water zone, Equation 3.4 can be rearranged to calculate the water gradient, assuming that the reservoir is neither overpressured nor underpressured.

$$\left[\frac{dp}{dD} \right]_w = \frac{p_{res} - p_{atm}}{D}$$
$$= (2{,}275 - 14.7) / 5{,}000$$
$$= 0.452 \text{ psi/ft}$$

Again, the equation can be rewritten to calculate the reservoir pressure at a certain depth with reference to measured pressure at a known depth in the reservoir as follows:

$$p = p_{ref} + \left[\frac{dp}{dD} \right]_f (D - D_{ref}) \tag{3.7}$$

The subscript "ref" in Equation 3.7 is associated with a known depth and pressure. Furthermore, the subscript "f" is associated with the gradient of the reservoir fluid that exists at the depth in question.

The expected reservoir pressure at the oil/water contact or gas/water contact of 4,990 ft is found as follows:

$$p_{res,\,4{,}990'} = 2{,}275 + 0.452\,(4{,}990 - 5{,}000)$$
$$= 2{,}270.5 \text{ psia}$$

(a) Above the oil/water contact, the oil gradient is used to estimate reservoir pressure at any point within the oil zone. Since the oil gradient is 0.38 psi/ft, reservoir pressure at 4,900 ft is found as in the following:

$$p_{res,\,4{,}990'} = 2{,}270.5 + 0.38(4{,}900 - 4{,}990)$$
$$= 2{,}236.3 \text{ psia}$$

(b) When a gas zone is present between 4,900 ft and 4,990 ft, the reservoir pressure is recalculated by using the gas gradient. By assuming a gas gradient of 0.075 psi/ft, reservoir pressure at 4,900 ft is found as in the following:

$$p_{res,\,4,990'} = 2{,}270.5 + 0.075(4{,}900 - 4{,}990)$$
$$= 2{,}263.8 \text{ psia}$$

Finally, the results obtained in (a) and (b) are compared to the pressure based on the gradient of formation water:

$$p_{res,\,4,990'} = 14.7 + 0.452(4{,}900)$$
$$= 2{,}229.5 \text{ psia}$$

As evident from the above, subsurface formations with oil or gas zones are expected to exhibit relatively high pressure compared to the pressure obtained by utilizing the gradient of the formation water alone. The deviation is more pronounced in the case of a gas zone due to the marked contrast between the water and gas gradients. Furthermore, the reservoir pressure could be significantly more where the gas zone extends hundreds of feet. This would require careful review and planning of drilling operations, as noted by Dake.[1]

Vertical and lateral discontinuities in reservoir pressure

Stratified reservoirs may exhibit vertical discontinuity in the measured pressure values over depth (fig. 3–4). Such data can be obtained by running a wireline formation tester in a newly drilled well. The above observation clearly demonstrates that any vertical communication between the adjacent layers is either limited or virtually nonexistent. In a stratified reservoir, knowledge pertaining to the degree of vertical communication is vital in designing well completions and waterflood operations.

Tools are also available to measure formation pressure along the wellbore trajectory of a horizontal well. Any anomalies in the observed reservoir pressure may indicate possible reservoir heterogeneities in the lateral direction, including the presence of faults, permeability degradation and reservoir compartmentalization.

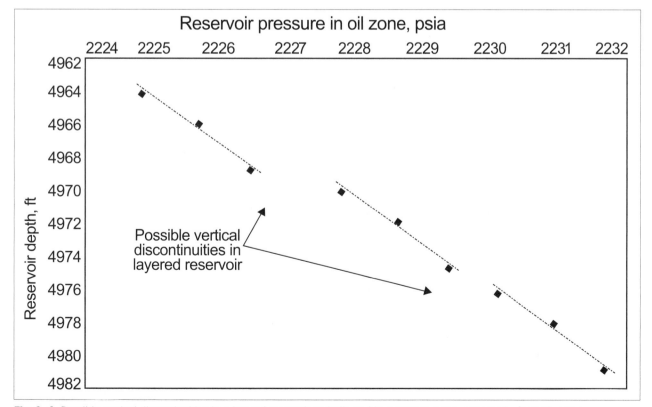

Fig. 3–4. Possible vertical discontinuities in a layered reservoir as indicated by pressure measurements by formation tester

Key points—fluid properties, reservoir pressure, and temperature

A summary of the points related to basic fluid properties, such as pressure, temperature and composition, is given in the following:

1. Petroleum reservoirs are primarily classified according to the composition of in-situ fluids (oil, gas, volatile oil, heavy oil, etc.).

2. Key factors that influence the production behavior of a petroleum reservoir depend on the following, among others:
   - Reservoir pressure and temperature controlling the fluid properties.
   - Composition of the reservoir fluid, i.e., fractions of light to heavy hydrocarbons.
   - Fluid viscosity determines how mobile a fluid would be in porous media given the reservoir permeability and driving pressure.
   - Saturation pressure marks the appearance of a new fluid phase. The pressure maintenance or a gas cycling operation is designed accordingly.
   - The gas/oil ratio influences the relative flow of oil and gas in porous media. Well completion and surface facilities are designed based on the expected volume of gas and liquid.
   - Specific gravity and compressibility play an important role in the ultimate recovery of the reservoir in many circumstances. Certain reservoirs are produced by gravity drainage.
   - Multiphase behavior of hydrocarbons may be encountered, including vaporization, condensation, and revaporization during production.

3. The fluid properties mentioned above are generally interrelated. Furthermore, the effects of various fluid properties on reservoir performance are highlighted later in the chapter as well as throughout the book.

4. Key fluid properties determine how a reservoir should be optimally developed and managed. The reservoir engineering team takes necessary steps, such as waterflooding, steam injection, and gas cycling, in order to augment the recovery of petroleum.

5. Reservoir pressure is primarily a function of depth of burial and density of the formation water. However, overpressured and underpressured reservoirs are not uncommon due to certain geologic processes that may occur. These include rapid subsidence, incomplete expulsion of the formation water, faulting, and other geologic events.

6. Monitoring of reservoir pressure is vital to any reservoir management strategy. Frequently encountered terms in reservoir engineering related to pressure measurements include initial reservoir pressure, static and flowing bottomhole pressure, and wellhead pressure.

7. Reservoir temperature is usually obtained from direct measurements. Moreover, charts are available to estimate reservoir temperature in various oil regions of the world. Again, a reservoir temperature that is higher or lower than expected could be encountered due to geothermal activities.

8. Oil and gas gradient information is required to determine the reservoir pressure in oil and gas zones, respectively, when direct measurement is not available. Reservoir pressure at the top of a thick gas zone may be considerably higher than the pressure calculated by using the formation water gradient alone.

9. Observed discontinuities in the pressure or pressure gradient between adjacent layers in a stratified reservoir usually suggest limited or nonexistent vertical communication.

10. Reservoir fluid properties are primarily obtained by the following methods:
    - Laboratory study of collected fluid samples
    - In-situ measurements by downhole sensors
    - Mathematical correlations applicable to the specific reservoir

    Most petroleum software applications allow the estimation of fluid properties when certain basic data about the reservoir fluid is available.

Classification of Reservoirs Based on Fluid Properties

Petroleum reservoirs can be broadly classified into five major categories based on fluid properties, as in the following:

Gas reservoirs

These reservoirs produce either dry or wet gas, as described below:

Dry gas reservoir. A dry gas reservoir is found initially with hydrocarbon components in the gas phase alone. No hydrocarbon component is found in a liquid form. During production of the reservoir, both the reservoir gas and produced gas remain essentially in single phase, i.e., in the gas phase. Besides conventional gas reservoirs, unconventional gas resources include coalbed methane, which is trapped in underground coal deposits, and fractured shales containing gas, among others. Certain extremely tight formations are also known to be the source of commercially producible gas and deemed as convential resources.

Wet gas reservoir. The reservoir is found initially with all the hydrocarbon components in the gas phase, as in a dry gas reservoir. When the reservoir pressure declines upon production, the gas remaining in the reservoir would be entirely in a single phase, without any condensation in the subsurface formation. However, a portion of gas produced through the wells condenses out due to the reduction in pressure and temperature at the surface. This occurs due to the presence of certain hydrocarbons in reservoir gas that condense under surface conditions. The compounds are heavier than those found in a dry gas reservoir.

Gas condensate reservoirs

Upon depletion of the reservoir, a portion of the gas dominated by heavier hydrocarbons condenses out and deposits in the subsurface porous network. This occurs as reservoir pressure declines below the dew point of the reservoir fluid. Condensation could be significant near the wellbore due to the relatively large drop in pressure. The phenomenon is referred to as retrograde condensation since pure substances evaporate, not condense under declining pressure. The prevailing reservoir temperature is below the cricondentherm defined as the limiting temperature above which a fluid can only exist as gas. Revaporization of the condensate may take place to a certain degree when the reservoir pressure becomes sufficiently low. However, revaporization is inhibited as the condensation and vaporization characteristics of the deposited hydrocarbons alter in an unfavorable manner. The adverse effects of retrograde condensation result in the production of lean gas only. Hence, dry gas is reinjected to maintain the reservoir pressure above the dew point leading to the efficient recovery of relatively rich components. Typical characteristics exhibited by gas condensate reservoirs are listed later in the chapter. Dew point and bubblepoint pressure are described later.

Oil reservoirs

Crude oil is often referred to as light, intermediate or heavy, which leads to the following classification of oil reservoirs:

Volatile oil reservoirs. Hydrocarbon components are initially found in the liquid phase. However, an oil reservoir may be discovered with an overlying gas cap when the initial reservoir pressure is lower than the bubblepoint of the crude oil. When the reservoir pressure declines, relatively light hydrocarbon components may evolve out of the liquid phase in large quantities. Typical API gravity of crude oil is relatively high, 38° or above. Crude oil with an API gravity of 45° or greater is referred to as near-critical oil, implying its location near the critical point on the phase diagram and the abundance of highly volatile hydrocarbons. A phase diagram is depicted in Figure 3-30. API gravity of oil is defined in a later section. Volatile crude typically shows light to dark amber color.

Black oil reservoirs. The term "black oil" is frequently cited in reservoir simulation studies as various volatile components are collectively considered to be single gas phase due to the relatively low volatility of crude encountered in many reservoirs. It is used here to categorize oil reservoirs that fall in-between the volatile and heavy oil reservoirs in terms of the specific gravity of crude. The typical range of crude gravity would be 38°–22.3° API (approximately). Abundance of intermediate hydrocarbons is a typical characteristic of black oil samples. Upon depletion of a black oil

Table 3–1. Composition and characteristics of reservoir fluids. *Source: C. H. Whitson and M. R. Brule. 2000. Phase Behavior. SPE Monograph. Vol. 20. Richardson, TX: Society of Petroleum Engineers.*

Component	Dry Gas	Wet Gas	Gas Condensate	Near-Critical (Highly Volatile) Oil	Volatile Oil	Black Oil
Typical Fluid Composition in Mole %						
Hydrocarbons						
Methane	86.12	92.46	73.19	69.44	58.77	34.62
Ethane	5.91	3.18	7.8	7.88	7.57	4.11
Propane	3.58	1.01	3.55	4.26	4.09	1.01
i-Butane	1.72	0.28	0.71	0.89	0.91	0.76
n-Butane		0.24	1.45	2.14	2.09	0.49
i-Pentane	0.5	0.13	0.64	0.9	0.77	0.43
n-Pentane		0.08	0.68	1.13	1.15	0.21
Hexanes		0.14	1.09	1.46	1.75	1.61
Heptanes & higher		0.82	8.21	10.04	21.76	56.4
Nonhydrocarbons						
CO_2	0.1	1.41	2.37	1.3	0.93	0.02
N_2	2.07	0.25	0.31	0.56	0.21	0.34
Total:	100	100	100	100	100	100
Reservoir Fluid Characteristics						
Mol wt., heptane +		130	184	219	228	274
GOR[a], scf/stb	—	105,000	5,450	3,650	1,490	300
Oil gravity[a], °API	—	57	49	45	38	24
Usual color of liquid phase		Translucent to clear	Straw	Light amber	Light to dark amber	Green to black

[a] Values are provided as a general guideline.

reservoir, the gas/oil ratio would typically range in the hundreds of scf/stb, as indicated in Table 3–1. Depending on the initial reservoir pressure, hydrocarbon components can either completely or partially be in liquid phase. The color of crude oil is green to black.

Heavy oil reservoirs. Heavy and extra-heavy oil reservoirs exhibit a dominance of heavier hydrocarbons in the composition of crude. According to API classification, reservoirs with crude oil having gravities less than 22.3° are considered to be heavy oil reservoirs. Typical oil viscosity is 10 cp or greater, exceeding 10,000 cp in certain instances. Heavy oil composition exhibits low hydrogen/carbon ratios and high asphaltene content. Ultra heavy oil having API gravity in the teens or less (bitumen, tar in oil sands, and others) is referred to as an unconventional resource. It cannot generally be produced by conventional means. A new era is emerging in the petroleum industry in which substantial recovery is expected from unconventional resources in the future based on innovative technology. Updated definitions of petroleum reserves reflect the significance of unconventional resources.

These categories of reservoirs are based on the relative dominance of light, intermediate, and heavy petroleum fractions, as shown in Table 3–1. (Compositional analysis of crude oil from a Nigerian field is shown in a later section.) Hydrocarbons mentioned in Table 3–1 are discussed in the following section. Liquid and gas phase behavior exhibited during production by the preceding categories of reservoirs also are discussed later in the chapter.

Note that with the progression from dry gas to black oil, the percentages and molecular weights of heavier components, heptanes and above (C_{7+}), increase significantly. Consequently, the specific gravity of the reservoir fluid is found to be progressively higher. Furthermore, the amount of dissolved gas in the petroleum fluid is markedly lower when heavier hydrocarbons are present.

Saturated and undersaturated oil reservoirs

In addition to the above, reservoir engineers frequently refer to oil reservoirs as either saturated or undersaturated. In undersaturated oil reservoirs, reservoir pressure is above the bubblepoint, and no free gas is present at discovery. However, in saturated oil reservoirs, initial reservoir pressure is at or below the bubblepoint. This is one of the most important properties of oil in relation to reservoir development strategy and is discussed later in the chapter. At the bubblepoint, dissolved gas starts to evolve out of the liquid phase.

Reservoir development and management

It must be borne in mind that each category of reservoir fluid mentioned above usually exhibits a unique behavior during production. This leads to reservoir-specific development strategies, including planning of future wells, optimization of production rates, and selection of a suitable pressure maintenance or enhanced recovery program. Additionally, the design of surface facilities to handle the produced liquid and gas depends on the hydrocarbon fluid properties in a given reservoir. Some of the possible development scenarios depending on fluid properties are mentioned in the beginning of this chapter.

Fluid mobility and recovery

Worldwide experience in petroleum production indicates that recovery from gas reservoirs is significantly higher than from reservoirs having heavier hydrocarbons, such as black oil or heavy oil. Natural gas, being significantly less viscous, overcomes resistance to flow in porous media with relative ease and is rapidly driven towards the producing wells. In general, the mobility of a hydrocarbon fluid relates directly to the extent of ultimate recovery, all other factors being the same. The major approach to the recovery of heavy oil is to reduce its viscosity and increase its mobility by adding thermal energy, as described in chapter 17. The effects of certain fluid properties on overall reservoir performance are illustrated with the aid of simulation in the final sections of this chapter.

Hydrocarbon composition of reservoir fluids

Before reviewing the properties of various petroleum fluids any further, a familiarity with the fundamentals of reservoir fluid composition will be necessary. Hydrocarbon compounds are chemical substances essentially made up of hydrogen and carbon. The molecular structure of hydrocarbons could range from relatively simple to highly complex as they occur in reservoirs. Naturally occurring petroleum fluids may contain a whole range of hydrocarbon components of varying characteristics such as molecular structure, gravity, and viscosity. Sulfur, oxygen, and nitrogen may also be present, besides hydrogen and carbon.

Typically, a reservoir fluid may contain hundreds or even thousands of petroleum compounds of varying complexity and molecular weight. The composition of the in-situ fluid depends on the source of organic matter that leads to the formation of petroleum. It also is affected by the depositional environment, geologic events, and the maturation process, among other factors. Consequently, a regional trend in fluid composition may be observed in oil and gas reservoirs. In the following sections, hydrocarbon compounds commonly found in reservoir fluids are outlined in order of their abundance.[2,3]

Paraffins. Occurring most frequently in petroleum fluids, paraffins are also referred to as alkanes or saturated hydrocarbons. Alkanes have the chemical formula of C_nH_{2n+2}. This implies that, for a certain number (n) of atoms of carbon present in a molecule of alkane, there would be $2n+2$ atoms of hydrogen. Familiar examples are methane (CH_4), ethane (C_2H_6), and propane (C_3H_8), as they dominate the composition of fluids found in a typical gas reservoir. With relatively low boiling points, these compounds appear mostly in gaseous form both inside the reservoir and at surface facilities. Heavier alkanes, with a large number of carbon atoms, such as pentanes (C_5), hexanes (C_6), and heptanes (C_7), appear in liquid form. Compounds that are heavier than heptanes, such as C_{16+}, can appear as highly viscous or waxy crude.

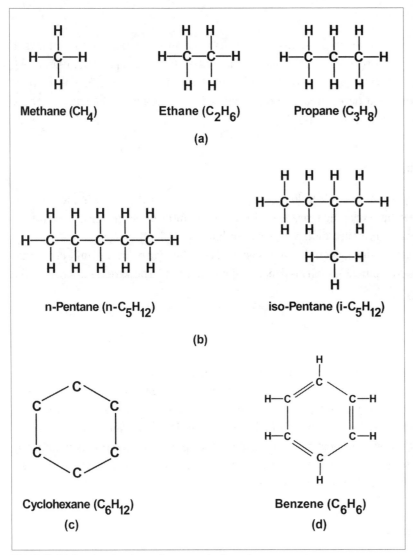

Fig. 3–5. Molecular structure of common hydrocarbons

Figure 3–5(a) shows the structure of the simplest of the alkanes, namely, methane and ethane. However, heavier paraffins may exhibit a branched chain configuration, in addition to the straight-chained molecular structure observed in relatively light components of the family. The prefixes n- and i- are used to distinguish between the "normal" configuration of the component and its isomer having the same number of carbon and hydrogen atoms but a different configuration. An example is $i\text{-}C_5H_{12}$ and $n\text{-}C_5H_{12}$, as illustrated in Figure 3–5(b). Although isomers have the same molecular weight as their normal counterparts, they have different properties, such as boiling point.

Cycloparaffins. As the name implies, this family of hydrocarbons is characterized by a cyclic structure of carbon atoms, in contrast to the chained configuration found in alkanes. The chemical formula is C_nH_{2n}. Cycloparaffins, also known as cycloalkanes, can be broadly classified as cyclopentanes and cyclohexanes. Cyclopentanes have five carbon atoms in the ring, and cyclohexanes are based on six carbon atoms, as shown in Figure 3–5(c). Due to their relatively high boiling point, cycloparaffins appear in a liquid phase under atmospheric conditions.

Aromatic compounds. These compounds are characterized by a molecular structure based on a ring of six carbon atoms. Common aromatic compounds found in crude oil include toluene, xylenes, and benzene. The aromatic ring is further characterized by the existence of a double chain between carbon atoms, as illustrated in Figure 3–5(d). Selected properties of various hydrocarbon components are summarized in Table 3–2.

Key points—reservoir classification and fluid properties

The following points are summarized in relation to reservoir classification and properties of fluids as encountered in the reservoir:

1. Petroleum reservoirs are generally classified into five broad categories. Given in order of increasing molecular weight of hydrocarbon components, these are: gas, gas condensate, volatile oil, black oil, and heavy oil. Relatively large quantities of heavier hydrocarbons are encountered in the latter.

Table 3-2. Properties of selected hydrocarbons and nonhydrocarbons encountered in petroleum reservoirs. *Source: W. D. McCain, Jr. 1990. Properties of Petroleum Fluids. 2nd ed. Tulsa: PennWell; and W. J. Lee and R. A. Wattenbarger. 2002. Gas Reservoir Engineering. Richardson, TX: Society of Petroleum Engineers.*

Component	Chemical Formula	Mol. Wt.	Boiling Point, °F	Liquid Sp. Gr. (Water =1.0)	Critical Pressure, psia	Critical Temperature, °R
Hydrocarbons						
Methane	CH_4	16.043	−258.7		666.4	343
Ethane	C_2H_6	30.07	−127.5		706.5	549.59
Propane	C_3H_8	44.097	−43.7	0.508	616	665.73
Butane	C_4H_{10}	58.123	31.1	0.584	550.6	765.29
Pentane	C_5H_{12}	72.15	96.9	0.631	488.6	845.47
Hexane	C_6H_{14}	86.177	155.7	0.664	436.9	913.27
Heptane	C_7H_{16}	100.204	209.2	0.688	396.8	972.37
Octane	C_8H_{18}	114.231	258.2	0.707	360.7	1,023.89
Nonane	C_9H_{20}	128.258	303.4	0.722	331.8	1,070.35
Decane	$C_{10}H_{22}$	142.285	345.5	0.734	305.2	1,111.67
Nonhydrocarbons						
Nitrogen	N_2	28.01			507.5	227.16
Carbon Dioxide	CO_2	44.01			1,071	547.58
Hydrogen Sulfide	H_2S	34.08			1,306	672.35

2. In dry and wet gas reservoirs, hydrocarbon fluids essentially remain in a gaseous state throughout production. However, in the latter case, the wet gas stream separates into gas and condensate at the surface facilities under atmospheric or near-atmospheric conditions. In a gas condensate reservoir, however, condensation and revaporization of the heavier hydrocarbons in porous media can occur with the decline of reservoir pressure.

3. Recovery from a gas reservoir is usually higher than that from an oil reservoir. This is due to the fact that light hydrocarbons, having low viscosity, require relatively less energy to flow in porous media. Heavier hydrocarbons, on the other side of spectrum, have significantly higher viscosity and less mobility in a porous medium.

4. Volatile oil reservoirs may exhibit better recovery than reservoirs having highly viscous crude. However, the former is usually associated with a high gas/oil ratio.

5. Natural gas is primarily composed of a handful of lighter hydrocarbons, methane being the dominant component. Crude oil, on the other hand, may contain hundreds or even thousands of heavier components, including heptanes and hydrocarbon compounds of higher molecular weight. Hydrocarbons commonly found in reservoir fluids include paraffins, cycloparaffins, and aromatic compounds.

6. As mentioned earlier, a reservoir development and management strategy depends significantly on the properties of the reservoir fluid encountered at discovery. Development of a petroleum reservoir also depends on rock characteristics and reservoir heterogeneities. The influence of rock properties on reservoir performance is described in chapter 2.

PVT Properties of Reservoir Fluids

In the following sections, the PVT properties of reservoir fluids, namely natural gas, condensate, crude oil, and formation water, are described. Their significance in reservoir engineering is also presented.

Properties of Natural Gas

The composition and properties of natural gas are important to the reservoir engineer. They can lead to the determination of the gas reserve at discovery, as well as evaluation of production performance during the life of the reservoir. Hydrocarbons that are typically present in natural gas are of low molecular weight and viscosity, including methane, ethane, propane, and butane. Methane is the dominant hydrocarbon compound, ranging between 70% and 98% in dry gas, and between 50% and 92% in wet gas.[4] The next major component is found to be ethane, followed by propane and butane. Heavier hydrocarbons (C_{5+}) and impurities such as carbon dioxide, nitrogen, and hydrogen sulfide are also found in smaller quantities. However, dry gases usually do not contain heptanes or heavier hydrocarbons. On the other hand, gas condensate reservoirs indicate relatively high concentrations of heptane and heavier hydrocarbons, as seen from Table 3–1.

Ideal gas law

An understanding of the behavior of natural gas in relation to pressure and temperature begins with the examination of the ideal gas law. In essence, the volume of an "ideal" gas is dictated by prevailing conditions of pressure and temperature. A gas is considered ideal when the following assumptions are met:[5]

- The volume of molecules that constitutes the ideal gas is negligible compared to the total volume occupied by the gas.
- The molecules in an ideal gas have neither attractive forces nor repulsive forces among them.
- The collision between the gas molecules is perfectly elastic and does not lead to dissipation of internal energy.

The ideal gas law is based on Boyle's law and Charles's law. Boyle's law states that the pressure exerted by the ideal gas is inversely proportional to its volume given that the temperature remains constant. For instance, one could consider a certain mass of gas undergoing isothermal compression or expansion from one state of pressure and volume to another. According to Boyle's law:

$$p_1 V_1 = p_2 V_2 \qquad\qquad (3.8)$$

where

p_1, p_2 = pressure of gas at state 1 and 2, respectively, psia, and
V_1, V_2 = volume of gas at state 1 and 2, respectively, ft^3.

During the entire process, the prevailing temperature remains constant at T.

Charles's law relates the volume and temperature of the ideal gas and states that the volume of the ideal gas is directly proportional to the temperature when the pressure is held constant. It is given as the following:

$$\frac{V_1}{T_1} = \frac{V_2}{T_2} \qquad\qquad (3.9)$$

where

T_1, T_2 = temperature of the gas at state 1 and 2, respectively, °R.

In addition, Avogadro's law states that equal volumes of all ideal gases contain the same number of molecules under the same pressure and temperature conditions. It is noted that the volume of one mole of any ideal gas, which represents its amount, is same and known at standard conditions. It is further noted that pound-mass of gas is dependent on its molecular weight and the number of moles. For example, a pound-mole of ethane contains 30.04 pound-mass (lbm) of gas, since the molecular weight of ethane is 30.04.

Hence, the ideal gas law relating pressure, volume, temperature, and the number of moles of gas can be expressed as the following:

$$pV = nRT \tag{3.10}$$

where
    p = prevailing pressure, psia,
    V = volume of gas under study, ft^3,
    n = number of pound-moles of gas, lbm-mol,
    R = gas-law constant (psia)(ft^3)/(°R)(lbm-mol), and
    T = prevailing absolute temperature, °R.

Equation 3.10 is also referred to as the equation of state (EOS) for an ideal gas. In relation to the units of pressure, temperature, volume, and moles of gas mentioned previously, the value of the gas law constant (R) is 10.73.

Properties of real gases

Under conditions of elevated pressure and temperature that are characteristic of subsurface petroleum reservoirs, gas properties deviate significantly from the ideal gas law. When the prevailing pressure is relatively low, the gas molecules are far apart, and any attractive forces between them are negligible. However, under high pressure, the molecules are close enough that attractive forces between the molecules must be considered. The assumptions that the volume of gas molecules is negligible and the molecules do not have any attractive or repulsive force among them would no longer be valid. Under the circumstances, gas properties computed from the straightforward relationships upon which the ideal gas law is based would not be representative. The margin of error could be as high as several hundred percent, according to a literature survey.

Gas deviation factor

The behavior of a real gas deviates significantly from that of an ideal gas, notably at high pressures and temperatures. Consequently, the ideal gas law is modified to develop an equation of state for real gases, as shown in the following:

$$pV = z\,n\,RT \tag{3.11}$$

where
    z = gas deviation factor, a function of prevailing pressure and temperature.

The gas deviation factor, also referred to as the gas compressibility factor, or simply the z factor, must be determined experimentally in order to utilize Equation 3.11 in determining the properties of real gases. Once the gas deviation factor is known, vital fluid properties, such as density, viscosity, and compressibility are readily obtained. This, in turn, leads to the computation of the gas flow rate through porous media, among other characteristics. Various plots and correlations are available in the literature. These can facilitate the estimation of the gas deviation factor depending on pressure, temperature, and impurities present in a typical gas sample obtained from a petroleum reservoir. A host of software applications have been developed in the petroleum industry to calculate petroleum fluid properties under a multitude of reservoir conditions.

Pseudoreduced pressure and temperature

One of the significant aspects of the gas deviation factor lies in the fact that similar gases can relate to the same gas deviation factor relationship when their pseudoreduced pressure and temperature are known. Thus it is not necessary to develop individual sets of z factor charts or correlations for gases or gas mixtures found in different reservoirs in order to determine the required gas properties. It is sufficient to know the values of the pseudoreduced pressure and temperature of a hydrocarbon component or a mixture of hydrocarbons.

Noting that the z factor is a function of pseudoreduced pressure and temperature alone in the absence of significant impurities, a general expression can be written, as in the following:

$$z = f(p_{pr}, T_{pr}) \qquad (3.12)$$

where
p_{pr} = pseudoreduced pressure of a gas mixture, dimensionless, and
T_{pr} = pseudoreduced temperature of a gas mixture, dimensionless.

Values of pseudoreduced pressure and temperature can be calculated when the pressure data and temperature data of a hydrocarbon component at the critical point are available. The critical point represents a state whereby the properties of the vapor and liquid phases of a pure substance are indistinguishable:

$$p_{pr} = \frac{p}{p_{pc}} \qquad (3.13)$$

$$T_{pr} = \frac{T}{T_{pc}} \qquad (3.14)$$

where
p, T = prevailing pressure (psia) and temperature (°R), respectively, and
p_{pc}, T_{pc} = critical pressure (psia) and temperature (°R) of the component, respectively.

For a multicomponent hydrocarbon mixture in a gas phase, the equation can be extended by summing the product of the mole fractions of the individual hydrocarbon components, y_i, as in the following:

$$p_{pc} = \sum y_i \, p_{c,i} \qquad (3.15)$$

$$T_{pc} = \sum y_i \, T_{c,i} \qquad (3.16)$$

It is noted that the critical pressure and temperature values based on the above equations do not represent the actual critical values for the mixture. Rather, these are used in the estimation of the gas deviation factor.

Determination of z factor

One of the most familiar tools utilized in the analysis of gas properties is the z factor chart in Figure 3–6.[6] The gas deviation factor is plotted against pseudoreduced pressure and temperature values commonly encountered in petroleum reservoirs. Developed several decades ago, the chart has been the most widely accepted tool in the industry for estimation of z factors. The values are found to be sufficiently accurate in case of "sweet" natural gases, where sulfur compounds such as hydrogen sulfide are not detected. A suitable correction needs to be applied for "sour" gases that contain sulfur, and is treated later.

On the z-factor chart, the pseudoreduced pressure is plotted along the x-axis, while the gas deviation factor is plotted along the y-axis. A number of curved lines of pseudoreduced temperature are plotted as the third parameter in the plot. Note that the z factor deviates significantly from ideality (z = 1.0) under certain pressure and temperature conditions as indicated in the chart. The more pronounced the deviation of the curved lines from unity; the more the gas phase departs

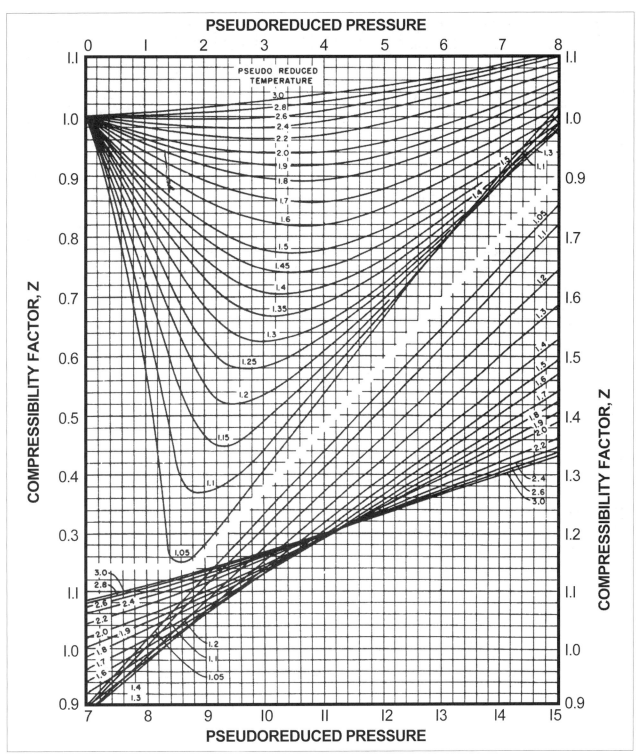

Fig. 3–6. Z factor as function of pseudoreduced pressure and temperature of natural gas. *Source: M. B. Standing and D. L. Katz. 1942. Density of natural gases. Transactions. AIME. Vol. 146. © Society of Petroleum Engineers. Reprinted with permission.*

from the ideal behavior. As indicated in Figure 3–6, one would find the gas deviation factor to be as low as 0.25 and as high as 1.75. Both values represent extremes in nonideal behavior under the pressure and temperature ranges represented by the plot. On the other hand, there would be a number of cases where the z factor is close to unity, indicating near-ideal behavior of the real gas mixture under certain other conditions.

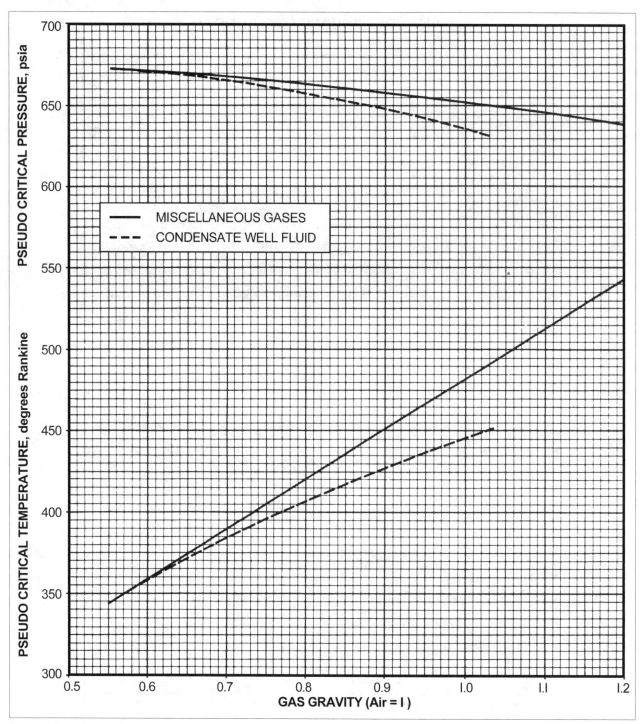

Fig. 3–7. Pseudocritical pressure and temperature as a function of the specific gravity of a hydrocarbon fluid. *Source: Brown, G. G., D. L. Katz, G. B. Oberfell, and R. C. Alden. 1948.* Natural Gasoline and Volatile Hydrocarbons. *Tulsa, OK: NGAA. 44. Reproduced courtesy of Gas Processors Suppliers Association.*

In some cases, the composition of a hydrocarbon gas mixture is not known with certainty, but the specific gravity is known. In these circumstances, certain charts and correlations are available to estimate the pseudocritical pressure and temperature. Figure 3–7 is a plot of pseudoreduced pressure and temperature as a function of fluid specific gravity.[7] Direct measurement of the gas deviation factor in the laboratory involves pressurization of a known volume of gas in a closed container. The altered volume at higher levels of pressure is then noted. (The temperature of the container is maintained at the temperature of the reservoir.) The gas deviation factor can be calculated by simply applying the equation of state for real gases, as in the following:

$$z_{(p,T)} = \frac{pV \text{ (at elevated pressure)}}{p_{sc}V_{sc}}$$ (3.17)

where

p_{sc} = standard pressure, psia,
V_{sc} = volume of gas sample at standard pressure, ft^3, and
T = reservoir temperature, °R.

Last, but not least, several correlations are available to determine the gas deviation factor as a function of gas composition, specific gravity, and reservoir pressure and temperature. These are integrated in various petroleum software applications. Based on the Starling-Carnahan equation of state, Hall and Yarborough proposed the following correlation to compute the z factor based on pseudoreduced properties of natural gas:[8]

$$z = \frac{C\, p_{pr}\, t \exp\{-1.2(1-t)^2\}}{y}$$ (3.18)

where

$C = 6.125 \times 10^{-2,}$
p_{pr} = pseudoreduced pressure, psia,
t = the reciprocal of pseudoreduced temperature, T_{pc}/T, and
y = reduced density factor.

The value of y can be determined by iteration and subsequent convergence from the following nonlinear equation:

$$-C\, p_{pr}\, t \exp\{-1.2(1-t)^2\} + y' - t'\, y^2 + t''\, y^{(2.18+2.82t)} = F$$ (3.19)

where

$y' = \dfrac{y + y^2 + y^3 + y^4}{(1-y)^3}$,
$t' = 14.76\, t - 9.76\, t^2 + 4.58\, t^3$, and
$t'' = 90.7\, t - 242.2\, t^2 + 42.4\, t^3$.

Equation 3.19 is solved by assuming an initial value of y and performing iterations until the correct value of y is obtained. Note that:

- If $F \neq 0$, the assumed value of y is incorrect; perform more iterations.
- If $F = 0$, the assumed value of y is correct; calculate the z factor.

Based on the recursive procedure, a new value of y is estimated as in the following:

$$y_{new} = y_{current} - \frac{F}{dF/dy}$$ (3.20)

where

$dF/dy = y'' - t'''y + (2.18 + 2.82t)\, t''\, y^{(1.18 + 2.82t)}$,
$y'' = (1 + 4y + 4y^2 - 4y^3 + y^4) / (1 - y)^4$, and (3.21)
$t''' = 29.52\, t - 19.52\, t^2 + 9.16\, t^3$.

Following one iteration, y_{new} becomes $y_{current}$, and the next value of y_{new} is calculated by Equation 3.20. After performing several iterations, the value of F becomes progressively small. The recursive process is continued until F approaches zero and a convergence for y is achieved up to the desired decimal places:

$$y_{new} = y_{current}$$

Finally, the converged value of y is utilized in Equation 3.18 for computing the z factor.

Table 3–3. Pseudoreduced pressure, temperature, and apparent molecular weight of a naturally occurring gas mixture

Component	Mole Fract., i	p_c	T_c	p_c (i)	T_c (i)	Mol. Wt.	M.W. (i)
	(1)	(2)	(3)	(4)=(1)×(2)	(5)=(1)×(3)	(6)	(7)=(1)×(6)
Methane	0.9	666.4	343.33	599.760	308.997	16.043	14.439
Ethane	0.075	706.5	549.92	52.988	41.244	30.070	2.255
Propane	0.015	616.4	666.06	9.246	9.991	44.097	0.661
n-Butane	0.01	527.9	734.46	5.279	7.345	58.123	0.581
Total:	1.00			667.27	367.58		17.94

Example 3.3. Calculate the gas deviation factor of a natural gas sample under reservoir pressure and temperature of 2,000 psia and 130°F, respectively. The composition of the natural gas, as measured in the laboratory, is the following: methane, 90%; ethane, 7.5%; propane, 1.5%; and n-butane, 1%.

Solution: As a first step, the pseudocritical pressure and temperature of the gas mixture are computed by multiplying the individual mole fractions with the pseudocritical pressure and temperature data found in Table 3–2. Calculations are shown in Table 3–3.

Next, the pseudoreduced pressure and temperature of the gas mixture are calculated based on Equations 3.13 and 3.14:

$$p_{pr} = \frac{2,000}{667.27}$$
$$= 2.997$$
$$T_{pr} = \frac{(130+460)}{367.58}$$
$$= 1.605$$

Finally, the z factor value is read from Figure 3–6:

$z = 0.828$ at $p_{pr} = 2.997$ and $T_{pr} = 1.605$

Molecular weight, density, and specific gravity of natural gases

The gas phase in a petroleum reservoir commonly occurs as a multicomponent mixture of relatively light hydrocarbon compounds such as methane, ethane, propane, and butane, among others. The molecular weight of the multicomponent gas phase is referred to as the apparent molecular weight. It can be based on the mole fractions of the individual components and their respective molecular weights, as in the following:

$$M = \Sigma\, y_i\, M_i \tag{3.22}$$
$$M = y_1 M_1 + y_2 M_2 + y_3 M_3 \ldots + y_n M_n \tag{3.23}$$

where
   M = apparent molecular weight of gas mixture having n components,
   y_i = mole fraction of the ith component in gas mixture, and
   M_i = molecular weight of the ith component.

The number of moles of a gas (n) can be expressed in terms of its mass over molecular weight:

$$n = \frac{m}{M} \tag{3.24}$$

Moles represent the amount of a substance.

Density is defined as its mass per unit volume:

$$\rho = \frac{m}{V} \qquad (3.25)$$

Familiar units of density of reservoir and drilling fluids include pounds mass per cubic feet (lb_m/cft, or lb_m/ft^3), grams per cubic centimeter (g/cc, or g/cm^3), and pounds per gallon (ppg).

Thus Equation 3.11 can be recast to obtain an expression for gas density:

$$\rho_g = \frac{pM}{zRT} \qquad (3.26)$$

At standard conditions of pressure and temperature, the gas deviation factor is unity, and the above equation reduces to the following:

$$\rho_{g,sc} = \frac{p_{sc}M}{RT_{sc}} \qquad (3.27)$$

where

the subscript sc denotes standard conditions.

At reservoir conditions of pressure and temperature, Equation 3.26 can be utilized to calculate the density of a compressed gas. Alternately, the following equation can be used when the gas specific gravity or density at standard conditions is available:

$$\rho_g = \frac{35.35\,\rho_{g,sc}\,p}{zT} \qquad (3.28)$$

The specific gravity of a gas is defined as the ratio of the density of the gas over that of air, both measured under the same prevailing conditions of pressure and temperature. Since the molecular weight of air is 28.966, an expression for specific gravity can be obtained based on Equation 3.26:

$$\gamma_g = \frac{\rho_g}{\rho_{air}} \qquad (3.29)$$

$$= \frac{(pM)_{gas}/(pM)_{air}}{(RT)_{gas}/(RT)_{air}}$$

Since the pressure and temperature are the same for both gas and air at standard conditions, and the gas-law constant (R) cancels out, the above equation takes the final form:

$$\gamma_g = \frac{M_{gas}}{28.966} \qquad (3.30)$$

Example 3.4. Using the natural gas sample in Example 3.3:

(a) Calculate the specific gravity of the natural gas sample under atmospheric conditions.

(b) Calculate the density of the same gas both at standard and reservoir conditions. Assume standard conditions to be 14.69 psia and 60°F.

Solution:

(a) First, it is necessary to calculate the apparent molecular weight of the gas mixture based on Equation 3.23, followed by computation of the specific gravity by Equation 3.30. From Table 3–3, it is noted that M = 17.94. Hence, the specific gravity of the gas mixture is calculated as the following:

$$\gamma_g = \frac{17.94}{28.966}$$

$$= 0.619$$

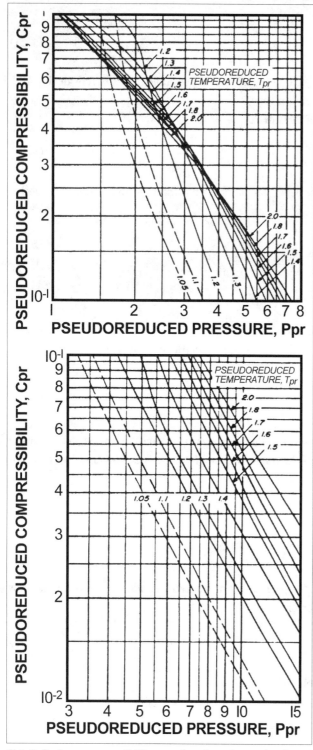

Fig. 3–8. Gas compressibility as a function of pseudoreduced pressure and temperature *Source: A. S. Trube. 1957. Compressibility of natural gases. Transactions. AIME. Vol. 210.* © *Society of Petroleum Engineers. Reprinted with permission.*

(b) The gas density at standard conditions is calculated by Equation 3. 27:

$$\rho_{g,sc} = \frac{(14.69)(17.94)}{(10.732)(60+460)} \qquad (3.30)$$

$$= 0.0472 \text{ lb/ft}^3$$

Similarly, the gas density at reservoir pressure and temperature, 2,000 psia and 130°F, respectively, can be computed by Equation 3.27. The gas deviation factor must be considered due to the departure from ideal gas behavior at reservoir conditions:

$$\rho_{g,res} = \frac{(2,000)(17.94)}{(0.828)(10.732)(130+460)}$$

$$= 6.844 \text{ lb / ft}^3$$

This example indicates that natural gas is significantly denser in the subsurface reservoir conditions than under the standard conditions encountered at the surface.

Example 3.5. Recompute the gas deviation factor of the natural gas sample based on the specific gravity calculated in Example 3.4, and compare it with the result obtained in Example 3.3.

Solution: The pseudocritical properties of the gas mixture are estimated based on Figure 3–5:

$$T_{pc} = 369°R, p_{pc} = 669.5 \text{ psia}$$

Hence,

$$P_{pr} = \frac{2,000}{669.5}$$

$$= 2.987$$

$$T_{pr} = \frac{(130+460)}{369}$$

$$= 1.6$$

Finally, the z factor is calculated from Figure 3–4:

$$z = 0.83$$

It can then be noted that the values for the gas compressibility factor calculated by both methods are relatively close. In this case, the deviation in results is 0.24%.

Isothermal gas compressibility

Gas compressibility is a measure of change in the volume of a gas with respect to a change in prevailing pressure. Of interest is the isothermal gas compressibility, since the prevailing temperature generally remains unchanged, as in a reservoir during production. Expressed mathematically, the definition of compressibility takes the following form:

$$c = -\frac{1}{V}\left(\frac{\partial v}{\partial p}\right)_T \tag{3.31}$$

Figure 3–8 shows the plots of pseudoreduced compressibility as a function of pseudoreduced pressure and temperature. Once pseudoreduced compressibility is obtained from the chart, the value of the gas compressibility can be calculated as in the following:

$$c_g = \frac{c_{pr}}{p_{pc}} \tag{3.32}$$

Compressibility is given in units over pounds per square inch (psi^{-1}). The compressibility of natural gas is significantly higher than the reservoir oil, interstitial water, and formation compressibility in a reservoir, typically by two orders of magnitude.

Viscosity of natural gases

The viscosity of a fluid is a measure of its resistance to flow. Gas, having significantly lower viscosity than oil and water, tends to dominate multiphase flow in the reservoir. The oilfield unit of viscosity is centipoise, abbreviated as cp. The viscosity of the natural gas can be estimated from correlations or plots available in the industry. Figure 3–9 shows a plot of gas viscosity under atmospheric conditions when the molecular weight of the gas and the prevailing temperature are known. Additionally, the chart provides corrections for viscosity values for any impurities in the natural gas, namely, nitrogen (N_2), carbon dioxide (CO_2), and hydrogen sulfide (H_2S). In the following example, Figure 3–10 is used to estimate gas viscosity under elevated pressure with the knowledge of pseudoreduced pressure and temperature.

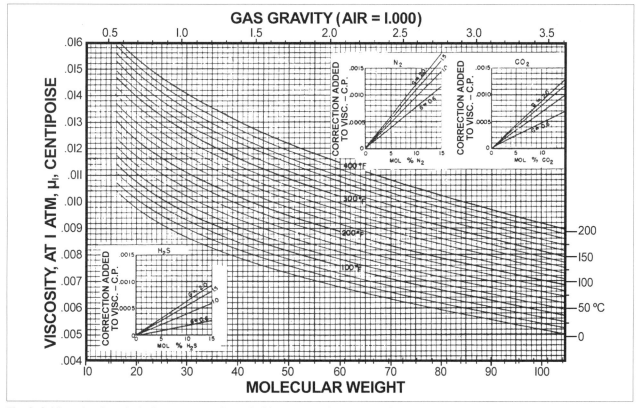

Fig. 3–9. Viscosity chart, including correction for H_2S, CO_2, and N_2. *Source: N. L. Carr, R. Kobayashi, and D. B. Burrows. 1954. Viscosity of hydrocarbon gases under pressure. Transactions. AIME., 264. © Society of Petroleum Engineers. Reprinted with permission.*

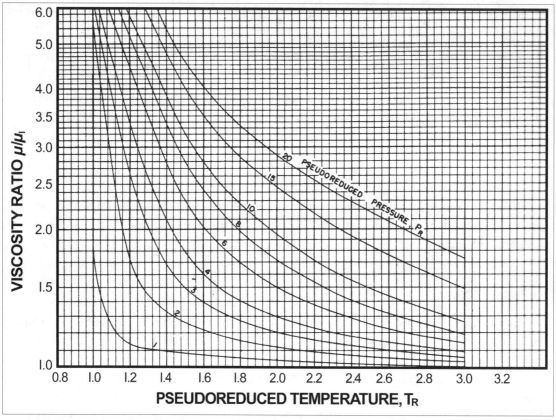

Fig. 3–10. Viscosity chart at elevated pressure. *Source: N. L. Carr, R. Kobayashi, and D. B. Burrows. 1954. Viscosity of hydrocarbon gases under pressure. Transactions. AIME., 264. © Society of Petroleum Engineers. Reprinted with permission.*

Example 3.6. Calculate the viscosity and compressibility of the gas mixture described in Example 3.3 at reservoir conditions.

Solution: First, the gas viscosity at 1 atm is found from Figure 3–9, followed by reading the viscosity ratio at reservoir pressure and temperature from Figure 3–10. Since the apparent molecular weight of the gas mixture is calculated to be 17.94, the following is obtained:

$\mu_{1\ atm} = 0.0119$ cp at reservoir temperature of 130°F

$\mu / \mu_{1\ atm} = 1.395$ at pseudoreduced pressure and temperature of 2.997 and 1.605, respectively.

Hence:

$\mu = 1.395 \times 0.0119 = 0.0166$ cp

The pseudoreduced compressibility of the gas mixture is found from Figure 3–8 based on the pseudoreduced pressure and temperature.

$c_{pr} = 0.37$

Hence:

$$c_g = \frac{c_{pr}}{P_{pc}}$$

$$= \frac{0.37}{667.27}$$

$$= 5.545 \times 10^{-4} \text{ psi}^{-1}$$

Formation volume factor of natural gases

The gas formation volume factor is the volume of gas under elevated pressure and temperature in a subsurface reservoir over the volume that would be occupied by the same mass of gas under standard conditions. The values for standard pressure and temperature are typically represented by 14.65 psia and 60°F, respectively. When expressed in terms of an equation, the gas formation volume factor takes the following form:

$$B_g = \frac{V_{res}}{V_{sc}} \tag{3.33}$$

where

B_g = gas formation volume factor, rb/scf,
V_{res} = volume of gas under reservoir conditions, rb, and
V_{sc} = volume of gas under standard conditions, scf.

Since natural gas undergoes significant expansion at surface conditions following production from the reservoir, the value of the gas formation volume factor is typically small. Note that in the numerator of Equation 3.33, gas volume is expressed in terms of reservoir barrels (rb) in order to conform to the oil volume formation factor. The latter is commonly expressed as reservoir barrels per stock-tank barrels (rb/stb). This is a necessity, as both oil and gas volume factors are used jointly in equations relating to multiphase hydrocarbon volume calculations in the reservoir.

Equations 3.11 and 3.33 can be combined as shown in the following:

$$B_g = \frac{(znRT/p)_{res}}{(znRT/p)_{sc}} \frac{cft}{scf} \tag{3.34}$$

The gas constant and the number of moles of gas are the same at both reservoir and standard conditions, and the z factor can be assumed to be unity under standard conditions. Thus Equation 3.34 reduces to the following:

$$B_g = \frac{(zT/p)_{res}}{(T/p)_{sc}} \frac{cft}{scf} \tag{3.35}$$

The values of pressure, 14.65 psia, and temperature, 60°F or 519.67°R, at standard conditions can be used in Equation 3.35. Noting that 1 bbl of fluid is equivalent to 5.615 cft, an expression for the gas volume factor in oilfield units can finally be obtained:

$$B_g = \frac{(zT/p)_{res}}{(519.67/14.65)} \frac{cft}{scf} \times \frac{1}{5.615} \frac{rb}{cft} \tag{3.36}$$

$$= 5.021 \times 10^{-3} \left(\frac{zT}{p}\right)_{res} \frac{rb}{scf} \tag{3.37}$$

The range of the gas formation volume factor is usually a very small number. Hence, it is often expressed in terms of reservoir barrels per thousand standard cubic feet (rb/Mscf), as follows:

$$= 5.021 \times 10^{-3} \left(\frac{zT}{p}\right)_{res} \frac{rb}{scf} \times \frac{1000\ scf}{1\ Mscf} \tag{3.38}$$

$$= 5.021 \left(\frac{zT}{p}\right)_{res} \frac{rb}{Mscf} \tag{3.39}$$

In case a standard pressure of 14.69 psia is assumed, the coefficient in Equation 3.39 is 5.035. An alternate measure used in the gas industry is the gas expansion factor, which is expressed as scf/cft.[9] It is a measure of the volume (in standard cubic feet) that 1 scf of reservoir gas would occupy when brought to the surface:

$$E = \frac{V_{sc}}{V_{res}} \frac{scf}{cft} \tag{3.40}$$

where

E = gas expansion factor, scf/cft.

Inspecting the units of E and B_g, the following can be readily shown:

$$E = \frac{1}{B_g} \frac{scf}{rb} \times \frac{1}{5.615} \frac{rb}{cft} \qquad (3.41)$$

$$= \frac{1}{5.615\, B_g} \frac{scf}{cft} \qquad (3.42)$$

Example 3.7. Calculate the formation volume factor and expansion factor of the natural gas sample in Example 3.3.

Solution: Since the gas deviation factor at reservoir pressure and temperature has already been calculated, the formation volume factor for the gas mixture is found by Equation 3.37:

$$B_g = 5.021 \times 10^{-3} \left[\frac{(0.828)(130+460)}{2,000} \right]_{res} \frac{rb}{scf}$$

$$= 0.001226 \text{ rb/scf or } 1.226 \text{ rb/Mscf}$$

The expansion factor is computed based on Equation 3.42:

$$E = \frac{1}{(5.615)(0.001226)} \text{ scf/cft}$$

$$= 145.26 \text{ scf/cft}$$

Estimation of gas deviation factor in the presence of common impurities

Natural gas typically consists of more than four components as given in the preceding examples. However, the methodology to calculate the gas deviation factor and other properties, as illustrated in earlier sections, is the same, as long as no impurities or contaminants are present in the gas mixture. Common nonhydrocarbon compounds in natural gases are hydrogen sulfide, carbon dioxide, and nitrogen, as mentioned previously. Natural gas is referred to as sour or sweet, depending on whether it contains hydrogen sulfide (H_2S) or not. A correction factor is commonly applied to pseudocritical pressure and temperature values to improve the accuracy of the gas deviation factor. A widely used correction factor, proposed by Wichert and Aziz, is given in the following:[10]

$$T_{pc} = T_{pc} - \varepsilon \qquad (3.43)$$

$$p_{pc} = \frac{p_{pc}\, T_{pc}}{T_{pc} + y_{H_2S}\left(1 - y_{H_2S}\right)\varepsilon} \qquad (3.44)$$

where

p_{pc} = pseudocritical pressure corrected for H_2S, CO_2, and N_2, psia,
T_{pc} = pseudocritical temperature corrected for H_2S, CO_2, and N_2, °R,
$\varepsilon = 120(A^{0.9} - A^{1.6}) + 15(y_{H_2S}^{0.5} - y_{H_2S}^{4})$, °R, $\qquad$ (3.45)
$A = y_{CO_2} + y_{H_2S}$,
y_{N_2} = mole fraction of nitrogen in natural gas,
y_{CO_2} = mole fraction of carbon dioxide, and
y_{H_2S} = mole fraction of hydrogen sulfide.

In Equations 3.42 and 3.43, the value of ε can either be calculated or read from Figure 3–11.

The methodology to compute the pseudocritical properties of natural gas having impurities is illustrated in the literature.[11–13] A procedure to compute the gas deviation factor in the presence of impurities is outlined in the following:

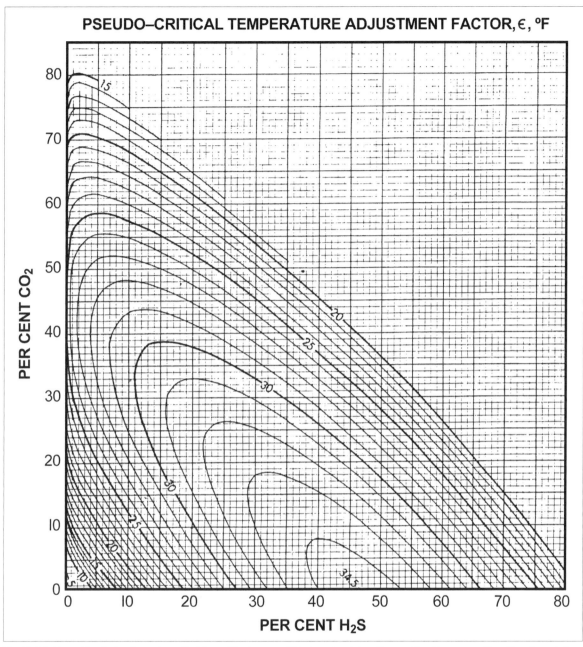

Fig. 3–11. Correction factor for carbon dioxide and hydrogen sulfide in natural gas. *Source: Gas Processors Suppliers Association. 1978. Engineering Data Book. 10th ed. Tulsa: GPSA. Reproduced courtesy of Gas Processors Suppliers Association.*

Step I. Determination of the pseudocritical properties of the hydrocarbon components in natural gas. The following correlations, proposed by Sutton, can be used when the specific gravity of hydrocarbons is known:[14]

$$P_{pc,\,HC} = 756.8 - 131.0\,\gamma_h - 3.6\,\gamma_h^2 \tag{3.46}$$

$$T_{pc,\,HC} = 169.2 + 349.5\,\gamma_h - 74.0\,\gamma_h^2 \tag{3.47}$$

where

$P_{pc,\,HC}$ = pseudocritical pressure of the hydrocarbon components, psia,

$T_{pc,\,HC}$ = pseudocritical temperature of the hydrocarbon components, °R, and

γ_h = specific gravity of the hydrocarbons in the gas mixture.

If no impurities are present in the natural gas, and no condensate is formed, the value of the separator gas gravity, γ_g, can be used in Equations 3.46 and 3.47 to compute the pseudocritical properties. However, γ_h is not the same as γ_g when one of three conditions apply, as in the following:

1. $y_{H_2S} > 0.0$
2. $y_{N_2} > 0.03$
3. $y_{CO_2} > 0.12$

Under the above circumstances, the specific gravity of the hydrocarbons in the natural gas is computed as in the following:

$$\gamma_h = \frac{(\gamma_g - 1.1767\, y_{H_2S} - 1.5196\, y_{CO_2} - 0.9672\, y_{N_2} - 0.622\, y_{H_2O})}{A'} \qquad (3.48)$$

where

y_{H_2O} = mole fraction of water vapor,
γ_h = specific gravity of the hydrocarbons in the mixture, and
$A' = 1 - (y_{H_2S} + y_{CO_2} + y_{N_2} + y_{H_2O})$.

When the natural gas contains water vapor, Equation 3.48 is modified accordingly to account for moisture content.

Step II. Determination of the uncorrected pseudocritical properties of natural gas with impurities. As previously shown, the pseudocritical properties of the hydrocarbons in the natural gas are computed by Equations 3.46 and 3.47. Next, the values of the pseudocritical pressure and temperature of the entire gas mixture can be calculated, as in the following:[15]

$$p_{pc} = A'(p_{pc,\,HC}) + 1{,}306\, y_{H_2S} + 1{,}071\, y_{CO_2} + 493.1\, y_{N_2} + 3{,}200.1\, y_{H_2O} \qquad (3.49)$$

$$T_{pc} = A'(T_{pc,\,HC}) + 672.35\, y_{H_2S} + 547.58\, y_{CO_2} + 227.16\, y_{N_2} + 1{,}164.9\, y_{H_2O} \qquad (3.50)$$

Equations 3.49 and 3.50 have been modified from their original form in order to account for the water vapor content in the gas mixture.[16]

Step III. Determination of the Wichert and Aziz correction factor. The Wichert and Aziz correction factor is either read from Figure 3–11 or calculated by Equation 3.45.

Step IV. Determination of the corrected pseudocritical pressure and temperature of the gas mixture. Corrected values of the pseudocritical pressure and temperature are obtained by Equations 3.43 and 3.44.

Step V. Determination of the gas deviation factor. The pseudoreduced pressure and temperature are then calculated by the usual methods. The final step involves the determination of the gas deviation factor from Figure 3–6. The methodology is illustrated in the following example.

Example 3.8. A natural gas sample contains hydrogen sulfide, carbon dioxide, and nitrogen in the following mole fractions: $H_2S = 0.08$; $CO_2 = 0.02$; and $N_2 = 0.015$.

The rest are relatively light hydrocarbon compounds. Laboratory measurements of a wellstream fluid sample indicate that the specific gravity of the gas mixture is 0.71. Calculate the gas deviation factor and viscosity of the natural gas at 3,191 psia and 186°F, as obtained in Example 3.3.

Solution:

Step I. Since the sample contains hydrogen sulfide, the specific gravity of the hydrocarbons is calculated by Equation 3.48 as in the following:

$$A' = 1 - (0.08 + 0.02 + 0.015)$$
$$= 0.885$$
$$\gamma_h = \frac{0.71 - 1.1767(.08) - 1.5196(.02) - 0.9672(.015)}{0.885}$$
$$= 0.645$$

Once the specific gravity is calculated, the pseudocritical properties of the hydrocarbons in the mixture are calculated as follows (Equations 3.46 and 3.47):

$$p_{pc, HC} = 756.8 - 131.0\,(0.645) - 3.6\,(0.645)^2$$
$$= 670.81 \text{ psia}$$
$$T_{pc, HC} = 169.2 + 349.5\,(0.645) - 74.0\,(0.645)^2$$
$$= 363.84°R$$

Step II. The uncorrected pseudocritical properties of the gas mixture, including impurities, are computed by Equations 3.49 and 3.50 as in the following:

$$p_{pc} = (0.885)(670.81) + 1,306\,(0.08) + 1,071\,(0.02) + 493.1\,(0.015)$$
$$= 726.96 \text{ psia}$$
$$T_{pc} = (0.885)(363.84) + 672.35\,(0.08) + 547.58\,(0.02) + 227.16\,(0.015)$$
$$= 390.15°R$$

Step III. The Wichert and Aziz correction factor for the pseudoreduced temperature is calculated next (Equation 3.45).

$$A = 0.08 + 0.02 = 0.1$$
$$\varepsilon = 120(0.1^{0.9} - 0.1^{1.6}) + 15(0.08^{0.5} - 0.08^4)$$
$$= 16.33$$

Step IV. The corrected pseudocritical properties are calculated by Equations 3.43 and 3.44 as in the following:

$$T_{pc}' = 390.15 - 16.33$$
$$= 373.82°R$$
$$p_{pc}' = \frac{(726.96)(373.82)}{390.15 + (0.08)(1 - 0.08)(16.33)}$$
$$= 694.39°R$$

Step V. Following computation of the pseudoreduced pressure and temperature of a natural gas with impurities, the gas deviation factor is obtained from Figure 3–6:

$$p_{pr} = 3,191/694.39 = 4.6$$
$$T_{pr} = (186 + 460)/373.82 = 1.73$$
$$z = 0.882$$

Key points—natural gas properties

1. Natural gas is typically composed of light hydrocarbons such as methane, ethane, propane, and butane. Methane is the dominant hydrocarbon compound, followed by ethane, propane, and butane in progressively smaller quantities. Heavier hydrocarbons (C_{5+}) and impurities such as carbon dioxide, nitrogen, and hydrogen sulfide are also found.

2. Properties of natural gas that are important in classical reservoir engineering studies include, but are not limited to, the composition of the gas mixture, viscosity, specific gravity, compressibility, and gas formation volume factor.

3. The pressure-volume-temperature (PVT) relationship of the natural gas is based on the equation of state for real gases that incorporates a gas deviation factor, or z factor. This factor can be determined when either the composition or the specific gravity of the natural gas is known.

4. In most cases, available correlations and charts provide sufficiently accurate results, as either gas composition or specific gravity is known from samples obtained from the field. Certain correction factors need to be applied when the reservoir gas contains impurities and relatively heavy hydrocarbon components.

Properties of Gas Condensate

As is evident from Table 3–1, gas condensate reservoirs contain certain hydrocarbons, namely heptanes and heavier compounds. In general, C_{7+} fractions are virtually nonexistent in dry gas reservoirs and may occur in very small amounts in wet gas reservoirs. However, fluids from gas condensate reservoirs exhibit relatively high quantities of these hydrocarbons. When a gas reservoir produces below the dew point of the fluid system, condensation of heavier hydrocarbons (C_{7+}) may occur in porous media. As observed in the surface facilities, the producing wellstream consists of both gas and liquid phases. Furthermore, gas condensate reservoirs usually exhibit the following characteristics:[17]

- The gas/oil ratio is relatively high at the surface (6,000 cft/bbl or greater).
- The tank oil is lightly colored, with low gravity (45°API or higher).
- The methane content of the reservoir fluid is 65% or higher.

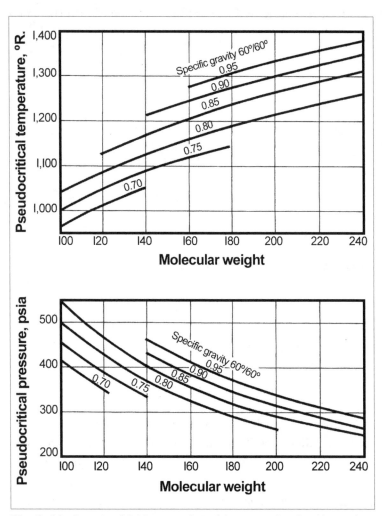

Fig. 3–12. Pseudocritical properties of heptane-plus fractions when molecular weight and specific gravity are known. *Source: Brown, G. G., D. L. Katz, G. B. Oberfell, and R. C. Alden. 1948.* Natural Gasoline and Volatile Hydrocarbons. *Tulsa, OK: NGAA., 44. Reproduced courtesy of Gas Processors Suppliers Association.*

Dew point

The dew point of a hydrocarbon fluid system is defined as the pressure at which certain hydrocarbon components begin to drop out of the gas to form a liquid phase. When reservoir pressure declines, dews, or droplets of liquid, condense from the gas phase in the pore channels of the rock as the fluids are driven to the wellbore. The definition of the dew point of a gas or vapor phase mirrors that of the bubblepoint of a liquid phase. The bubblepoint of a fluid is defined in this chapter in the discussion of oil properties. Certain hydrocarbon components in a condensed liquid phase may be left behind in the reservoir, as these are relatively viscous compared to the gas phase.

Fluid density in gas condensate reservoir

Density and certain other properties of a gas condensate system can be calculated with reasonable accuracy. This can be accomplished by utilizing the equation of state for real gases in the same manner shown for a dry gas when the volume of the liquid phase is limited to 10%.[18] However, representative values of pseudocritical pressure and temperature of the C_{7+} fractions (heptanes and heavier) are required, as these hydrocarbons exist in the reservoir fluid. Correlations and charts are available to provide the required information.[19] A chart showing the pseudoreduced properties of a gas condensate as a function of fluid gravity and molecular weight is shown in Figure 3–12.

Formation volume factor of gas condensate

The producing stream from a gas condensate reservoir consists of two phases: gas and liquid condensate. Heavier hydrocarbons condense out of the gas phase when the reservoir produces below the dew point. The formation volume factor of a gas condensate (B_{gc}) is defined as in the following:

$$B_{gc} = \frac{\text{Volume of gas + condensate vapor in gas in rb}}{\text{Volume of liquid condensate produced in stb}}$$

(3.51)

where

B_{gc} is given in rb/stb.

In essence, it is the volume of gas, including any condensate in vapor form under reservoir pressure and temperature, that would produce 1 stock-tank barrel (stb) of condensate liquid at the surface. The volume of the gas is measured in reservoir barrels in order to conform to the oil formation volume factor defined under oil properties in this chapter.

Various methods used to estimate the formation volume factor are compiled by McCain.[20] These require knowledge of one or more of the following:

- Producing gas/oil ratio
- Composition of produced gas and condensate
- Specific gravities of fluids
- Two-phase formation volume factor

The two-phase formation volume factor is defined in the oil properties section. Standing developed a correlation between the formation volume factor, gas/oil ratio, total gas gravity, tank oil gravity, and reservoir pressure and temperature based on a large number of experimental data.[21] The results of the study are shown in Figure 3–13.

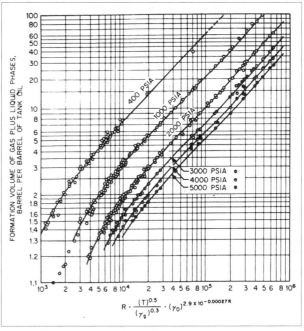

Fig. 3–13. Formation volume factor for gas condensate and dissolved gas systems. *Source: M. B. Standing. 1947. A pressure-volume-temperature correlation for mixtures of California oils and gases. In Drilling and Production Practice. Washington, DC: American Petroleum Institute, 285–287. Reproduced courtesy of American Petroleum Institute.*

Properties of Reservoir Oil

The reservoir engineer is primarily interested in oil properties, as such knowledge is critical in performing most reservoir engineering studies. These include, but are not limited to, studies that allow the following:

- Estimation of the hydrocarbon in place and petroleum reserves
- Analysis of the oil phase behavior, as evolution of a gas phase may occur
- Development of a reservoir model to predict ultimate recovery
- Placement and design of future wells
- Evaluation of the reservoir performance in terms of well rates and reservoir pressure
- Design of enhanced oil recovery operations, whereby oil is displaced by an injected fluid
- Study of the effects of fluid viscosity and gravity on ultimate recovery in a particular reservoir setting
- Investigation of early breakthrough of water or gas
- Design and analysis of pressure transient tests
- Evaluation of the effects of phase change on fluid sample collection from the reservoir
- Study of the effects of asphaltene precipitation and gas hydrate formation

The list is by no means comprehensive. It is presented here to emphasize the importance of fluid characteristics in virtually every reservoir study. Knowledge of vital oil properties is needed for classical reservoir studies. Such properties include, but are not limited to, the following:

- Specific gravity
- Viscosity
- Compressibility
- Gas solubility
- Saturation pressure
- Degree of oil volume shrinkage, over the entire range of pressure that may be encountered in the reservoir under study

Factors affecting reservoir fluid properties

Besides the initial reservoir pressure, oil properties are influenced by the prevailing temperature and fluid composition, i.e., the dominance of either lighter or heavier hydrocarbons. As can be observed in the following sections, oil properties are generally a strong function of reservoir conditions, such as changing pressure, as the oil is produced. At sufficiently high reservoir pressure and in the absence of a gas phase, oil properties are controlled by compressibility, among other factors, as the reservoir pressure declines. Substantial changes in the various oil properties may be observed as pressure declines further and evolution of gas ensues. Under these circumstances, oil properties are predominantly controlled by changes in composition. The fluid is progressively replete with heavier hydrocarbons as the relatively volatile components come out of solution. Finally, at the surface under stock-tank conditions, the produced oil is referred to as dead oil, as it does not contain any volatile hydrocarbon compounds of significance.

All of the volatile components separate from the liquid phase under standard conditions. Thus common laboratory studies of oil properties involve the recombination of oil and gas samples collected from the separators to simulate reservoir conditions.

Oil density, specific gravity, and API gravity

The density of crude oil, as in any substance, is defined as its mass per unit volume. In oilfield units, crude oil density is expressed as pound-mass per cubic feet. The density of crude oil is associated with reference pressure and temperature values, which are 14.7 psia and 60°F, respectively. Oil density calculations at reservoir pressure and temperature require knowledge of the volume and density of any dissolved gas. In addition, the change in the volume of oil when it is brought to the surface must be known. This is treated later in the section.

The specific gravity of crude oil is defined as the ratio of the oil density over the density of the water, both measured at the same reference temperature and pressure. (Specific gravity is thus dimensionless.) Specific gravity measurements are usually based on a temperature of 60°F. The specific gravity of crude oil usually varies between 0.8 and 0.97 in most instances.

A more familiar term in the petroleum industry is the API gravity, which is expressed in degrees. (API is the acronym for the American Petroleum Institute.) The API gravity of a crude oil is defined as follows:

$$^{\circ}API = \frac{141.5}{\gamma_o} - 131.5 \tag{3.52}$$

where

γ_o = specific gravity of oil, dimensionless.

Equation 3.52 can be rearranged to compute the specific gravity of a crude oil when the API gravity is known:

$$\gamma_o = \frac{141.5}{131.5 + {}^{\circ}API} \tag{3.53}$$

Note that the API gravity of the crude oil is inversely correlated to specific gravity. Heavier crude, having higher specific gravity, would lead to a lower API gravity. Light to intermediate crude oil having API gravity ranging from $46°$ to $29°$ is encountered frequently across oil regions. The specific gravity of the crude is an important property in predicting the reservoir performance where forces due to gravity control the fluid flow behavior in porous media.

Oil viscosity

Viscosity is a measure of the internal resistance to flow. A viscous crude would require more energy to flow towards the wellbore than low viscosity oil, provided other factors are the same. Bitumen, having very high viscosity, would not flow at all in porous media unless thermal recovery methods are engineered to reduce viscosity, as described in chapter 17. Laboratory methods are available to measure the viscosity of a crude sample obtained from the field. Determination of viscosity is a routine part of the tests conducted by laboratories when reporting fluid properties. In oilfield units, the dynamic viscosity of reservoir fluid is commonly expressed in centipoise, abbreviated as cp. Note that 1 mPa·s = 1 cp.

Oil viscosity under reservoir conditions is found to range from less than 1 cp to thousands of cp. It would depend on several factors, including prevailing temperature, pressure, composition, and the presence of dissolved gas. When reservoir pressure declines due to production, the viscosity of the liquid phase decreases due to expansion, as long as the pressure remains above the bubblepoint. Once the reservoir pressure declines further, accompanied by the dissolution of volatile hydrocarbons, the viscosity of the remaining liquid phase increases as the heavier components are left behind. Dead oil at surface conditions, virtually devoid of volatile components, is found to have relatively high viscosity.

Crude oil viscosity is one of the major factors dictating how mobile the oil phase will be in the reservoir during production, as evident from Darcy's law, as discussed in chapter 2. Oil recovery tends to be lower in reservoirs having highly viscous crude. Effects of oil properties on reservoir performance are discussed later in the chapter.

Isothermal compressibility

Oil compressibility is a function of the rate of change in the volume of crude oil per unit change in pressure. The measure is commonly based on isothermal conditions, whereby the prevailing temperature is kept constant. Mathematically, compressibility can be expressed in the form of a differential equation, as in the following:

$$c = -\frac{1}{V}\left(\frac{\partial V}{\partial p}\right)_T \tag{3.54}$$

where
  c = fluid compressibility,
  V = fluid volume,
  p = fluid pressure, and
  T = fluid temperature.

The subscript T denotes that the change in fluid volume as a consequence of change in pressure takes place under isothermal conditions. The crude oil density and formation volume factor are directly proportional to the oil volume, as in the above equation. Compressibility can also be defined as the rate of change in the oil density or formation volume factor per unit change in pressure. The unit of oil compressibility is given in units per pounds per square inch (psi^{-1}), and typical values of oil compressibility range from 5×10^{-6} to $15 \times 10^{-6}\,psi^{-1}$. One concrete example can be examined in which the reservoir contains 1 million bbl of oil having a compressibility of $10 \times 10^{-6}\,psi^{-1}$ at 3,000 psia. If the average reservoir pressure drops by 1 psi to 2,999 psia, a volumetric expansion of oil under reservoir conditions would be expected, as shown in the following:

$$(10^6\,rb)\,(10 \times 10^{-6}\,psi^{-1})\,(1\,psi) = 10\,rb$$

In the above computation, it is assumed that the reservoir pressure is maintained above the bubblepoint, and no free gas is present in the reservoir.

Total and effective compressibility

The total compressibility of the system takes into account the compressibilities of all the fluids present in the system, plus the compressibility of the rock. Total compressibility can be expressed as in the following:

$$c_t = c_f + c_o S_o + c_g S_g + c_w S_w \qquad \text{(3.55)}$$

where

subscripts t, f, o, g, and w denote total, formation, oil, gas, and water, respectively.

In the case of an undersaturated oil reservoir producing above the bubblepoint pressure, there is no gas present in the subsurface system, and the gas compressibility term drops out of the equation. Similarly, the oil compressibility term is neglected when computing the total compressibility term in a gas reservoir. In many undersaturated oil reservoirs, the value of total compressibility ranges between 15×10^{-6} and 20×10^{-6} psi^{-1}.

The effective compressibility of a particular fluid phase is obtained by dividing the total compressibility, as in Equation 3.55, by the saturation of that phase in the porous media. Hence, in an undersaturated oil reservoir, the effective compressibility of the oil phase can be expressed as follows:

$$c_e = \frac{c_f + c_o S_o + c_w S_w}{1 - S_{wi}} \qquad \text{(3.56)}$$

Bubblepoint pressure

The bubblepoint of a reservoir fluid is defined as the pressure above which the fluid essentially remains in a liquid phase, and all of the volatile components are dissolved in the liquid. Due to the elevated pressure levels normally found in subsurface formations, hydrocarbon fluid may initially be discovered only in a liquid phase (oil) throughout the reservoir. However, reservoir pressure declines once oil production starts, and at a certain point, the gas will begin to evolve out of solution. The point at which these bubbles of gas first appear is referred to as the bubblepoint for the fluid system. At the bubblepoint pressure, the liquid is in equilibrium with an infinitesimally small volume of evolved gas. The bubblepoint pressure of a liquid phase is also referred to as the saturation pressure, since the liquid is completely saturated with dissolved gas.

With continued production, the reservoir pressure would decline further, producing appreciable quantities of gas that may eventually dominate the multiphase flow of fluids in the reservoir.

For example, consider an oil reservoir with a bubblepoint known to be at 1,750 psia. This implies that, as long as the reservoir pressure is maintained above 1,750 psia, no free gas, mobile or immobile, is expected in the reservoir. Fluid flow towards the wellbore is essentially single phase. However, if the reservoir pressure declines below 1,750 psia, both oil and gas phases are expected. Once enough gas is produced, a high gas/oil ratio at the producing wells is anticipated.

Major decisions in reservoir engineering require knowledge of the bubblepoint pressure. Initiation of an early pressure maintenance scheme may be necessary to maintain reservoir pressure above the bubblepoint and circumvent gas evolution and its eventual dominance in production. If the initial reservoir pressure is below the bubblepoint with a gas cap present, reinjection of the produced gas may be necessary to maintain reservoir pressure at an optimum level. This results in the production of the nonvolatile (heavier) hydrocarbons as much as possible.

Solution gas/oil ratio

An oil reservoir typically contains a liquid phase with a certain quantity of gas dissolved in it. When the reservoir pressure declines due to production, lighter hydrocarbons begin to evolve out of solution and form a gas phase. The solution gas/oil ratio is an important parameter indicating the volitility of crude. It is a measure of evolved gas volume per unit volume of oil and is reported in scf/stb. This ratio indicates the volume of gas that would dissolve into 1 stb of oil when both oil and gas are subjected to the elevated pressure and temperature conditions encountered in the subsurface formation. The solution gas/oil ratio is defined as in the following:

$$R_S = \frac{\text{Volume of gas evolved from oil in scf}}{\text{Volume of produced oil following gas evolution in stb}} \qquad (3.57)$$

where

R_S = the solution gas/oil ratio, scf/stb.

For a concrete example, one could consider a case in which the solution gas/oil ratio of a volatile crude oil sample is evaluated to be 900 scf/stb. This indicates that at the starting point, there are 900 cft of evolved gas and 1 stb of produced crude, measured under standard conditions. If both fluids are transported back down the reservoir, the gas would be expected to dissolve completely in the oil due to the elevated pressure and temperature.

The solution gas/oil ratio is a function of the reservoir pressure, temperature, and specific gravities of the fluids. One could consider the case of a petroleum reservoir in which the oil production is accompanied by a continuous evolution of dissolved gas as the reservoir pressure declines. Consequently, the solution gas/oil ratio of the in-situ oil becomes progressively less. At sufficiently low reservoir pressure, when virtually all the dissolved gases have evolved and been produced, the value of the solution gas/oil ratio may become negligible. The solution gas/oil ratio is sometimes referred to as gas solubility.

Producing and cumulative gas/oil ratio

A distinction must be made between the solution gas/oil ratio and the producing gas/oil ratio. Oil reservoirs are encountered with or without a gas cap at discovery. A gas cap may exist on top of the oil zone, as shown in reservoir B of Figure 1–2 in chapter 1. In this case, the producing gas volume is based on two components: free gas flow from the gas cap and evolution of lighter hydrocarbons from the crude oil. Hence, the producing gas/oil ratio is expected to be greater than the solution gas/oil ratio. Furthermore, the cumulative gas/oil ratio is defined as the cumulative gas production over the cumulative oil production from a well or a reservoir up to a certain period during production.

Oil formation volume factor

The oil formation volume factor is a measure of the shrinkage or reduction in the volume of crude oil as it is produced. Accurate evaluation of the oil formation volume factor is of prime importance as it relates directly to the calculation of the petroleum reserve and oil in place under stock-tank conditions. It is the ratio of reservoir barrels of oil plus the volume of dissolved gas under reservoir pressure and temperature over stock-tank barrels of dead oil at the surface. Thus the oil formation volume factor may be defined as follows:

$$B_0 = \frac{\text{Volume of oil + dissolved volatiles: both under reservoir pressure and temperature}}{\text{Volume of produced oil under stock tank conditions following gas liberation}} \qquad (3.58)$$

where

B_0 = oil formation volume factor, rb/stb.

As oil is produced, its volume is reduced as lighter components evolve out of the liquid phase due to a reduction in pressure. As a concrete example, one could start with 1.25 bbl of oil in a reservoir under elevated pressure and temperature of 2,000 psia and 150°F, respectively. As oil is produced and stored in stock tanks at 14.7 psia and 60°F, the volume of oil is reduced to 1.0 bbl. This occurs as volatile hydrocarbons evolve out of solution due to the change in pressure. Hence, the formation volume factor of oil would be:

$$B_0 = \frac{1.25}{1.0} = 1.25 \; \frac{\text{rb}}{\text{stb}}$$

It can be readily inferred that the volume of dissolved gas that is liberated in the process is the difference in oil volumes between the reservoir and the surface.

Volatile crude tends to exhibit a relatively high oil formation volume factor, as significant quantities of gas come out of solution upon pressure depletion. Typical values of the oil formation volume factor could range between 1.5 rb/stb and 2.7 rb/stb in volatile oil reservoirs. It is closer to unity for low shrinkage crude and may typically vary between 1.02 rb/stb and 1.2 rb/stb.

Two distinct trends are observed as the reservoir pressure declines first above and then below the bubblepoint. Above the bubblepoint, the oil formation volume factor tends to increase slightly with declining pressure due to the expansion of the fluid. At the bubblepoint, the oil formation volume factor reaches its maximum. Any further decline in reservoir pressure below the bubblepoint triggers the evolution of dissolved gas from the oil, and a marked decrease in oil formation volume factor is observed. Above the bubblepoint, the formation volume factor is a function of oil compressibility and can be estimated as follows:

$$B_o = B_{ob} \exp[-c_o(p - p_b)] \tag{3.59}$$

where

B_{ob} = formation volume factor at bubblepoint, rb/stb, and

p_b = bubblepoint pressure, psia.

Below the bubblepoint, however, the effect of liquid expansion becomes relatively small compared to shrinkage as the lighter hydrocarbons evolve.

Two-phase formation volume factor

The two-phase formation volume factor is the sum of the individual formation volume factors of oil and evolved gas. With a decline in reservoir pressure, the volume of gas that comes out of solution per stock-tank barrel of oil can be determined as in the following:

$$R_{si} - R_s \text{ scf/stb}$$

where

R_{si} = initial solution gas/oil ratio, scf/stb, and

R_s = solution gas/oil ratio at the pressure where the two-phase formation factor is sought, scf/stb.

If the gas formation volume factor (B_g) is known, as defined in Equation 3.33, the equivalent reservoir barrels of the evolved gas per stock-tank barrel of oil can be computed as follows:

$$(R_{si} - R_s) B_g \text{ rb/stb}$$

Hence, the two-phase formation volume factor can be expressed as in the following:

$$B_t = B_o + (R_{si} - R_s) B_g \text{ rb/stb} \tag{3.60}$$

where

B_t = two-phase formation volume factor, rb/stb.

Above the bubblepoint pressure, $R_s = R_{si}$ and $B_t = B_o$. However, once the reservoir pressure declines below the bubblepoint, the two-phase volume factor may increase significantly due to the evolution of the dissolved gas and its expansion.

Oil density at reservoir conditions

The density of oil under reservoir conditions can be calculated when the following information is available from laboratory studies:

- Density of the oil under standard conditions
- Amount of dissolved gas in the oil under reservoir conditions
- Density of the gas under standard conditions
- Change in volume of the oil when produced

Based on material balance, the oil density can be expressed as follows:[22]

$$\rho_o = \frac{62.4\,\gamma_o + 0.0136\,R_s\,\gamma_g}{B_o} \tag{3.61}$$

where

ρ_o = density of the oil in the reservoir at elevated pressure and temperature, lb/cft,

γ_o = oil specific gravity in stock tanks at the surface, dimensionless, and

γ_g = gas specific gravity, dimensionless.

Alternately, Equation 3.61 can be used to compute the solution gas ratio or oil formation volume factor when other parameters are known.

Above the bubblepoint, however, the density of oil can be estimated as follows:

$$\rho_o = \rho_{ob}\,exp\,[c_o(p_R - p_b)] \tag{3.62}$$

where

ρ_{ob} = density of the oil at the bubblepoint pressure, lb/cft, and

p_R = reservoir pressure above the bubblepoint, psi

Key points—oil properties

1. Reservoir-specific knowledge of oil properties, as functions of pressure and temperature, is vital in all reservoir studies. Such properties include, but are not limited to, specific gravity, viscosity, isothermal compressibility, solution gas/oil ratio, bubblepoint pressure, and oil formation volume factor over the entire range of reservoir pressures encountered. This extends from the initial pressure to conditions at abandonment.

2. Throughout the productive life of a reservoir, individual oil properties may alter significantly as reservoir pressure and in-situ fluid composition change. During the decline in reservoir pressure, oil properties are controlled by compositional changes as gas evolves out of the liquid phase.

3. Most oil properties exhibit two distinct trends depending on whether the reservoir pressure is decreasing above or below the bubblepoint. Interestingly, the two trends are usually in opposite directions. With a decline in pressure, the following processes occur: (i) above the bubblepoint, oil expands in volume, and no gas evolution is observed; and (ii) oil shrinks in volume below the bubblepoint as gas evolves out of the liquid phase.

4. Bubblepoint pressure serves as a milestone in the life of an undersaturated reservoir, since gas comes out of solution at and below this pressure. Evolved gas eventually drives toward the wellbore with much higher mobility than liquid. Consequently, oil composition and properties change significantly as the liquid phase becomes progressively lean in lighter hydrocarbons. Ultimate recovery tends to be higher if the reservoir is produced at a pressure higher than the bubblepoint of the fluid. This is accomplished by fluid injection.

5. When the reservoir pressure declines due to production, the viscosity of the oil is initially reduced above the bubblepoint due to the volumetric expansion of the liquid. However, when the reservoir pressure declines below the bubblepoint, viscosity increases as the lighter (and less viscous) hydrocarbon components separate to form a free gas phase.

6. The solution gas/oil ratio, at a certain reservoir pressure, is a measure of how much gas is dissolved in the oil. The volumes of gas and oil are measured in standard cubic feet (scf) and stock-tank barrels (stb), respectively. The solution gas/oil ratio remains unchanged above the bubblepoint pressure, as no dissolved gas can evolve. Once the bubblepoint is reached, it decreases continuously as free gas is formed in the porous media. The solution gas/oil ratio eventually becomes insignificant as the oil is produced and stored in stock tanks.

7. The oil formation volume factor is indicative of the reduction or shrinkage in liquid volume as lighter hydrocarbons come out of solution due to a decline in pressure. This is required to convert barrels of oil in the reservoir, including any dissolved gas under elevated pressure, to stock-tank barrels of oil containing no such dissolved gas.

8. The oil formation volume factor increases above the bubblepoint pressure as the reservoir pressure declines. This factor then reaches its maximum value at the bubblepoint and decreases as more and more volatile components are liberated. When no further gas evolution or shrinkage takes place, the formation volume factor is equal to unity. The oil formation volume factor is relatively high in the case of volatile oil reservoirs.

Properties of Formation Water

Knowledge of formation water properties is needed in various reservoir studies along with oil and gas properties. Of interest to reservoir engineers are the salinity, viscosity, density, and compressibility of the formation water. Since hydrocarbon gas is soluble in water under elevated pressures and temperatures, the solution gas/water ratios and water formation volume factors are necessary information in conducting reservoir studies. These studies include, but are not limited to, well test analysis and multiphase fluid flow simulation, among others. Properties of formation water are generally dependent on reservoir pressure, temperature, and concentration of salt compounds. Water found in subsurface geologic formations usually has a high salt content due to the dissolution of minerals and related geochemical processes that occur throughout geologic times. Last, but not least, subsurface water is found to be incompatible with surface water that is injected into the formation during waterflood operation. The loss of injectivity is encountered due to the swelling of clay materials and precipitation of solids in porous media.

Various charts and correlations are widely available to estimate the important water properties. Selected charts that aid in the determination of viscosity, compressibility, gas solubility, and the formation volume factor of the formation water are presented in this chapter. Correlations predicting the properties of formation water are integrated into petroleum software, much in the same manner as the properties of hydrocarbon fluids.

Sources of Fluid Properties Data

When a new reservoir is discovered, fluid samples are collected during the initial studies conducted in the subsurface formations.[23] Downhole monitoring tools are employed, which include drillstem test tools, wireline formation testers, and optical fluid analyzers, among others. Various properties of the reservoir fluid are evaluated early on in order to further develop the reservoir. Such properties include reserve estimation, planning of future wells, formulation of recovery enhancement strategy, and economic evaluation.

Fluid properties are usually determined by conducting industry-standard laboratory studies. Results from one such study and a compositional analysis of the crude sample used in the study are presented later. Additionally, certain charts and correlations are employed to estimate the properties of reservoir fluids if the relevant reservoir information is not available from the reservoir. Before the widespread use of computers in reservoir studies, engineers frequently resorted to printed charts. These were used to estimate a specific fluid property, with the relationship between the fluid property and controlling factors laid out in graphical form. However, with the advent of the digital age, suitable correlations in mathematical equations are routinely integrated into petroleum software applications. Some of the equations may require an iterative procedure until convergence is attained for the result sought. Figure 3–14 presents one such example. Various properties of oil, gas, and formation water as a function of pressure, along with user-selectable options, are shown. The plots are generated by Merlin Software of Gemini Solutions.[24] Many of these are based on correlations proposed by Standing, as described in the following section.

As indicated earlier, information related to reservoir fluid properties is frequently referred to as pressure-volume-temperature (PVT) data in the petroleum industry. PVT data is obtained from the several sources, as described in the following sections.

Applicable correlations. Reservoir engineers employ a host of computer-aided tools to understand, develop, and evaluate a reservoir. This invariably requires the computation of a large array of fluid properties. (Fluid properties needed to accomplish various studies are listed in a table later in this section.) In many cases, extensive field data is not available, and PVT correlations are utilized instead. Correlations that estimate fluid properties with an acceptable degree of confidence are integrated into most petroleum software applications as necessary. However, it must be recognized that correlations provide estimated values of fluid properties and are region-specific in most instances. Reservoir-specific correlations are frequently developed by the oil industry, which may or may not be available for general use.

Laboratory studies. The classical method of obtaining reservoir fluid properties involves a variety of laboratory studies of fluid samples retrieved from the subsurface formation or the produced fluid phases recombined at the surface. The studies include routine laboratory analyses as well as special procedures. The former typically involves determination of the fluid composition, density, viscosity, compressibility, gas/oil ratio, bubblepoint pressure, and formation volume factor. Special laboratory procedures may involve fluid displacement studies when the feasibility of an enhanced recovery project is evaluated, for example. Descriptions of laboratory studies can be found in the literature.[25–28]

In-situ measurements. As fluid samples are brought to the surface, the prevailing pressures and temperatures change, leading to alteration of the fluid characteristics. This can lead to the separation of the gas phase. Modern techniques focus on in-situ determination of various fluid properties.[29,30] In-situ measurement tools include a fluid analyzer based on the optical properties of the reservoir fluid. Response from the optical sensor is used to determine fluid type (oil, gas, mud, or water), fluid composition (C_1, C_2–C_5, and C_{6+}), gas/oil ratio, and other factors. This has many advantages, including the enhancement of sampling efficiency, real-time monitoring of the reservoir fluid under reservoir conditions, and identification of any contrast in oil properties between two layers.

This section begins with selected correlations to estimate the various fluid properties, as these are incorporated extensively in modern software applications utilized in virtually all reservoir studies. Next, flash and differential vaporization processes are highlighted for the determination of fluid properties in the laboratory, along with a field example. The concept of vapor-liquid equilibria and their use in computation of fluid properties are also discussed. The next section describes an emerging technology of in-situ fluid characterization, with a field application. Finally, charts are provided to estimate the various properties of formation water.

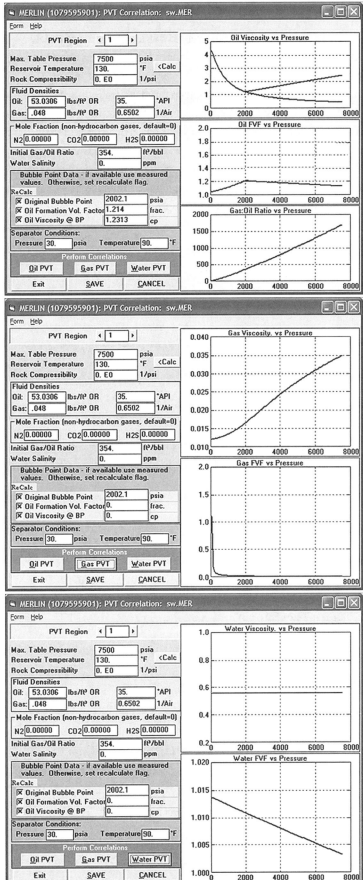

Fig. 3–14. PVT properties calculated by Merlin Software for (a) oil, (b) gas, and (c) water phases. These include solution gas/oil ratio, fluid viscosity, and formation volume factor for each phase as a function of reservoir pressure. *Courtesy of Gemini Solutions, Inc. 2005.*

Selected Correlations to Estimate Oil Properties

Oil property correlations presented in the following are widely recognized throughout the industry. The studies include, but are not limited to, a large number of fluid samples obtained from reservoirs located in California, the North Sea, the Middle East, and the Gulf of Mexico. Correlations of Standing, Glaso, Marhoun, Vasquez and Beggs, Petrosky and Farshad, and others are compiled in various references, including Mian, Ahmed, and Towler.[31–37]

De Ghetto, Paone, and Villa evaluated the accuracy of a large number of PVT correlations introduced in the industry between 1947 and 1993.[38] The study was based on approximately 200 samples of crude oil obtained from various petroleum regions worldwide, including the North Sea, Persian Gulf, Mediterranean Basin, and Africa. The authors classified crude oil into light, medium, heavy, and extra heavy crude according to API gravity. They proposed certain modifications of the available correlations in order to predict the bubblepoint, solution gas/oil ratio, isothermal compressibility, and viscosity in each category of crude oil with improved accuracy.

Solution gas/oil ratio

Standing (based on California reservoirs):

$$R_s = \gamma_g \left(\frac{p \times 10^x}{18} \right)^{1.2048}$$

(3.63)

where

R_s = gas solubility, scf/stb,

γ_g = solution gas specific gravity, dimensionless,

$x = 0.0125 \, (\gamma_0, \, °API) - 0.00091(T - 460)$.

p, T = reservoir pressure (psia) and temperature (°R), respectively, and

Hence, for a 33° API crude having a bubblepoint pressure of 3,000 psia and solution gas gravity of 0.78, the solution GOR at bubblepoint is estimated to be 703 scf/stb, the reservoir temperature being 200° F.

Petrosky and Farshad (based on Gulf of Mexico reservoirs):

$$R_s = \gamma_g + \left[\left(\frac{p}{112.727} + 12.34 \right) \gamma_g^{0.8439} \, 10^x \right]^{1.73184}$$

where

$x = 7.916 \times 10^{-4} \, (\gamma_0, \, °API)^{1.514} - 4.561 \times 10^{-5} \, (T - 460)^{1.3911}$

Glaso (based on North Sea reservoirs):

$$R_s = \gamma_g + \left[\frac{API^{0.989}}{(T-460)^{0.172}} \, (p_b^*) \right]^{1.2255}$$

(3.64)

where

$p_b^* = 10^x$

$x = 2.8869 - [14.1811 - 3.3093 \log (p)]^{0.5}$

Marhoun (based on Middle East reservoirs):

$$R_s = [185.843208 \, \gamma_g^{1.87784} \, \gamma_0^{-3.1437} \, T^{-1.32657} \, p]^{1.398441}$$

(3.65)

where

γ_g = specific gravity of gas, dimensionless,

γ_0 = specific gravity of stock-tank oil, dimensionless,

T = temperature, °R,

The correlations above are valid for reservoir fluids at or below the bubblepoint pressure. When the reservoir produces above the bubblepoint, fluid compressibility must be taken into account to compute R_s at various pressures.

Bubblepoint

In general, correlations for the bubblepoint of a reservoir fluid indicate that it is a nonlinear function of several other fluid properties, namely, oil and gas gravity, temperature, and gas solubility.

Standing*:*

$$p_b = 18.2[(R_s / \gamma_g)^{0.83} (10)^a - 1.4] \tag{3.66}$$

where

R_s = solubility of gas at the bubblepoint, scf/stb,

γ_g = specific gravity of gas under surface conditions, and

$a = 0.00091(T - 460) - 0.0125 (\gamma_o, °API)$.

The correlation above has limitations when certain impurities are present.

Petrosky and Farshad*:*

$$p_b = 112.727 \left[\frac{R_s^{0.5774}}{\gamma_g^{0.8439}} 10^x - 12.34 \right]$$

where

$x = 4.561 \times 10^{-5} (T - 460)^{1.3911} - 7.916 \times 10^{-4} (\gamma_o, °API)^{1.514}$

Glaso*:*

$$\log (p_b) = 1.7669 + 1.7447 \log (p_b^*) - 0.30218 [\log (p_b^*)]^2 \tag{3.67}$$

where

$p_b^* = (R_s/\gamma_g)^{0.816} (T - 460)^{0.172} (\gamma_o, °API)^{-0.989}$

The exponent for temperature is changed from 0.172 to 0.130 in the case of a volatile crude.

Marhoun*:*

$$p_b = 5.38088 \times 10^{-3} R_s^{0.715082} \gamma_g^{-1.87784} \gamma_o^{3.1437} T^{1.32657} \tag{3.68}$$

Example 3.9. Consider the reservoir in Example 3.1. Would a gas cap be expected to be present in the reservoir at discovery? Data from two nearby fields indicating the following average values of fluid properties is given:

Field A: oil API gravity = 35°, gas specific gravity = 0.72, and gas solubility = 750 scf/stb.

Field B: oil API gravity = 37°, gas specific gravity = 0.71, and gas solubility = 850 scf/stb.

Make all necessary assumptions.

Solution: The bubblepoint pressure of the reservoir fluid needs to be estimated in order to predict whether a gas cap will be present. In Example 3.1, the original pressure of the reservoir is found to be 3,191 psia. If the bubblepoint pressure is estimated to be less than that, a gas cap would be present on top of the oil zone. Since actual bubblepoint pressure data from nearby fields is not available, a suitable correlation can be used to estimate the bubblepoints. In this case, Standing's correlation in Equation 3.66 is used to determine the bubblepoints.

Field A:
$a = -0.26824$; $p_b = 3,112$ psia

Field B:
$a = -0.29324$; $p_b = 3,299$ psia

Results of the analysis are inconclusive, as the original reservoir pressure of 3,191 psia is in between the two values of the bubblepoints as estimated. In reality, the existence of a gas zone is identified during drilling, and bubblepoint data is generally available from nearby fields. Nevertheless, the preceding example highlights the uncertainties a reservoir engineer typically encounters when actual field data is not available.

Oil formation volume factor

Several correlations are available in the literature to estimate the oil formation volume factor given certain other fluid properties. These properties include the gas solubility in crude oil, oil and gas gravities, and reservoir temperature, in most cases. Selected correlations are outlined in the following discussion.

Standing:

$$B_o = 0.9759 + 12 \times 10^{-5} [R_s (\gamma_g/\gamma_o)^{0.5} + 1.25(T - 460)]^{1.2} \tag{3.69}$$

where

R_s = solution gas/oil ratio, scf/stb,
γ_g = specific gravity of gas, dimensionless,
γ_o = specific gravity of oil, dimensionless, and
T = reservoir temperature, °R.

Petrosky and Farshad:

$$B_o = 1.0113 + 7.2046(10^{-5})[R_s^{0.3738}(\gamma_g^{0.2914}/\gamma_o^{0.6265}) + 0.24626(T - 460)^{0.5371}]^{3.0936} \tag{3.70}$$

Glaso:

$$B_o = 1.0 + 10^A \tag{3.71}$$

where

$A = -6.58511 + 2.91329 \log B_{ob}^* - 0.27683 (\log B_{ob}^*)^2$, and
$B_{ob}^* = R_s (\gamma_g/\gamma_o)^{0.526} + 0.968(T - 460)$.

Marhoun:

$$B_o = 0.497069 + 0.862963 \times 10^{-3} T + 0.182594 \times 10^{-2} F + 0.318099 \times 10^{-5} F^2 \tag{3.72}$$

where

$F = R_s^a \gamma_g^b \gamma_o^c$,
$a = 0.742390$,
$b = 0.323294$, and
$c = -1.202040$.

Various authors, including Glaso and Marhoun, also proposed correlations for two-phase FVF.

Example 3.10. Using the preceding correlations, calculate and compare the formation volume factors at the bubblepoint of the fluid samples in Example 3.9.

Using Equations 3.54 and 3.69 through 3.72, the following is obtained:

Field A: specific gravity of oil = 141.5/(35°API + 131.5) = 0.8498.
Standing: B_o = 1.4338 rb/stb.
Glaso: B_{ob}^* = 867.4078; a = −0.41513; and B_o = 1.3845 rb/stb.
Marhoun: F = 149.0139; and B_o = 1.3973 rb/stb.
Petrosky: B_o = 1.3959 rb/stb.

Field B: specific gravity of oil = 141.5/(37°API + 131.5) = 0.8398.
Standing: B_o = 1.4862 rb/stb.
Glaso: B_{ob}^* = 958.218; a = −0.36002; and B_o = 1.4365 rb/stb.
Marhoun: F = 165.1395; and B_o = 1.4428 rb/stb.
Petrosky: B_o = 1.4434 rb/stb.

This example highlights certain points that must be borne in mind when using fluid property correlations in reservoir studies. First, all correlations based on various petroleum regions worldwide are not applicable with good accuracy to a particular field. These can differ significantly among each other. Second, the formation volume factor changes noticeably between the two samples, underscoring the importance of the quality of data used in correlations. A small deviation in the estimated value of B_o may lead to noticeable error in estimating oil volume in a large field. Uncertainties inherent in such analyses must be recognized when predicting well and reservoir performance. Reservoir engineers resort to a sensitivity analysis that examines how a small variation in a correlated fluid property may affect the outcome of the reservoir study being performed.

It is worth noting that above the bubblepoint pressure, B_o is computed based on oil compressibility (c_o) and the oil formation volume factor evaluated at the bubblepoint (B_{ob}) as follows:

$$B_o = B_{ob} \exp[-c_o(p-p_b)], \quad p > p_b \text{ and } B_o < B_{ob}$$

Density of oil

The density of oil under reservoir conditions can be calculated by Equations 3.61 and 3.62 when relevant PVT data is available.

Viscosity of oil

Charts and correlations are available to estimate the viscosity of the crude oil in the absence of laboratory measurements. Besides hydrocarbon composition, the viscosity of the liquid phase under reservoir conditions would depend on several factors, including prevailing temperature, pressure, and the amount of dissolved gas. This is in contrast to the viscosity of the dead oil at the surface, where all the volatile components have evolved out of solution. Again, the liquid phase in the reservoir is either saturated or undersaturated. As reservoir pressure declines below the bubblepoint, the viscosity can increase significantly due to the evolution of dissolved gas. Hence, estimation of the oil viscosity under reservoir conditions is a two-step or three-step process, depending on whether the fluid is above or below the bubblepoint pressure.

Estimation of dead-oil viscosity. Beal proposed the following correlation in order to estimate the dead oil when the API gravity of crude oil and reservoir temperature are known.[39] Originally presented in graphical form, the correlation was later expressed in mathematical form by Standing:

$$\mu_{od} = [0.32 + (1.8 \times 10^7)/(\gamma_o, °API)^{4.53}][360/(T-260)]^a \tag{3.73}$$

where
$$a = 10^{(0.43 + 8.33/\gamma_o, °API)}.$$

Glaso proposed the following correlation to estimate the viscosity of dead oil. Again, the correlation is based on the API gravity of the oil and on the reservoir temperature:

$$\mu_{od} = C(T - 460)^{-3.444}[\log(\gamma_o, °API)]^a$$

where
$$C = 3.141 \times 10^{10}, \text{ and}$$
$$a = 10.313[\log(T - 460)] - 36.447.$$

Estimation of reservoir oil or live-oil viscosity. Since crude oil would contain dissolved gas under reservoir conditions, a correction for dead-oil viscosity is necessary. Chew and Connally proposed the following correlation that requires knowledge of the solution gas/oil ratio.[40] The correlation was developed for saturated oil at or below the bubblepoint. It is valid over the usual range of reservoir pressure and temperature, with dead-oil viscosity of 0.377 cp to 50 cp and gas solubility of 51 scf/stb to 3,544 scf/stb:

$$\mu_{ob} = (10)^a (\mu_{od})^b$$

where

μ_{ob} = viscosity of oil at the bubblepoint, cp,
$a = R_s [2.2(10^{-7})R_s - 7.4(10^{-4})]$,
$b = 0.68/10^c + 0.25/10^d + 0.062/10^e$,
$c = 8.62(10^{-5})R_s$,
$d = 1.1(10^{-3})R_s$, and
$e = 3.74(10^{-3})R_s$.

Viscosity of undersaturated oil. If the reservoir pressure is above the bubblepoint, another correction would be necessary for the undersaturated oil as it is compressed further. Vasquez and Beggs proposed the following correlation when the viscosity of oil at the bubblepoint pressure is available.[41] The correlation is valid under a wide range of reservoir conditions, including API gravity of 15.3° to 59.5°, oil viscosity of 0.117 cp to 148 cp, gas gravity of 0.511 to 1.351, and gas solubility of 9.3 scf/stb to 2,199 scf/stb.

$$\mu_o = 1 + \mu_{ob} (p/p_b)^m \tag{3.74}$$

where

p_b = bubblepoint pressure, psia,
$m = 2.6 \, P^{1.187} \exp(a)$, and
$a = -11.513 - 8.98 \times 10^{-5} \, p$.

Compressibility of oil

McCain proposed the following correlation to calculate oil compressibility below the bubble point.[42] Reservoir pressure, temperature, bubblepoint, oil API gravity, and gas solubility at the bubblepoint are needed to estimate the compressibility of saturated oil.

$$\ln(c_o) = -7.573 - 1.450 \ln(p) - 0.383 \ln(p_b) + 1.402 \ln(T) + 0.256 \ln(\gamma_o, °API) + 0.449 \ln(R_{sb})$$

where

γ_o = °API gravity of oil, and
R_{sb} = solution gas/oil ratio at bubblepoint pressure, scf/stb.

Vasquez and Beggs proposed a correlation to estimate the compressibility of oil above the bubblepoint pressure, as in the following:[43]

$$c_o = [5 \, R_{sb} + 17.2 \, (T - 460) - 1,189 \, \gamma_{gs} + 12.61 \, (\gamma_o, °API) - 1,433]/(p \times 10^5) \tag{3.75}$$

where

$$\gamma_{gs} = \gamma_g \left[1 + 5.912 \times 10^{-5} (\gamma_o, °API)(T_{sep} - 460)\left(\log \frac{p_{sep}}{114.70}\right) \right]$$

Note that γ_{gs} is the corrected gas gravity depending on separator pressure and temperature, P_{sep} and T_{sep}, respectively.

Laboratory Measurement of Reservoir Fluid Properties

The properties of reservoir fluids are routinely determined from field samples in order to conduct important reservoir studies. Typical fluid property measurements include fluid composition, specific gravity, viscosity, compressibility, bubblepoint pressure, and gas solubility, among others. In addition to direct retrieval of fluid samples from the subsurface, separated liquid and gas phases obtained from surface facilities are recombined in correct quantities to reproduce fluid samples under simulated reservoir conditions. The recombination of fluid samples is guided by the initial solution gas/oil ratio of the crude oil. Detailed descriptions of laboratory tests on reservoir fluid samples are provided by McCain and Ahmed, and others.[44,45]

Flash and differential vaporization of a reservoir fluid in a PVT cell are two familiar studies conducted in the laboratory to measure important fluid properties. Flash vaporization is also referred to as constant composition expansion. The pressure of the sample fluid is gradually lowered in a confined cell, causing expansion in the fluid volume; no change in overall fluid composition is allowed. Any gas evolved during the process remains in contact with the liquid phase until equilibrium between the two phases is reached. Objectives of flash vaporization include the determination of the following:

- Bubblepoint or saturation pressure
- Specific volume at saturation pressure
- Coefficient of thermal expansion
- Isothermal compressibility of the liquid above the bubblepoint

A differential vaporization test involves the immediate removal of the vapor phase following its evolution from the crude oil under declining pressure. This is viewed to emulate the processes occurring in porous media more closely. Any gas evolved from the liquid phase is highly mobile in porous media and is driven rapidly towards the wellbore, while the oil phase lags behind. Results of the study include the following:

- Solution gas/oil ratio
- Relative oil volume
- Relative total volume
- Density of the oil
- Gas deviation factor
- Gas formation volume factor
- Incremental gas gravity
- Fluid viscosity as a function of pressure

Besides the preceding tests, a multistage separator test can be conducted in the laboratory in order to study fluid volume changes due to a reduction in pressure and temperature. The study simulates surface operation. The fluid sample is maintained at saturation pressure and reservoir temperature in a PVT cell, followed by flash separation into liquid and gas phases, usually in two or three stages. The objective is to maximize the liquid recovery by determining the optimum separator conditions as the hydrocarbon feed is received from the producing wells.

A typical information sheet for an oil sample is shown in Table 3–4. The information is usually stored in a database for further reservoir studies.

Table 3–4. Fluid sampling information summary. *Courtesy of Core Laboratories, Inc.*

General Reservoir Information	
Producing formation	Cretaceous limestone
Date first well completed	10/12/1990
Original reservoir pressure	3,660 psi @ 7,902 ft
Initial produced gas/oil ratio	690 scf/bbl
Initial production rate	540 bbl/d
Separator pressure and temperature	210 psig, 75°F
Original gas cap	None

Well Information	
Well number	5-B
Producing interval	7,895–7,988 ft MD
Tubing size and depth	2¾ in.; 7,750 ft MD
Well productivity index	1.27 bbl/d/psi @ 440 bbl/d
Last reservoir pressure	3,408 psi @ 7,980 ft MD
Date recorded	11/5/1993
Reservoir temperature	212°F at 7,980 ft MD
Status of well	Shut-in for 96 hrs
Pressure gauge	Amerada
Average production rate	390 bbl/d
Gas/oil ratio	750 scf/bbl
Water cut	0%

Sampling Information	
Sampling depth	7,950 ft MD
Status of well	Shut-in for 96 hrs
Tubing pressure	1,309 psig
Sampler type	Wofford
Sampled by	John Doe

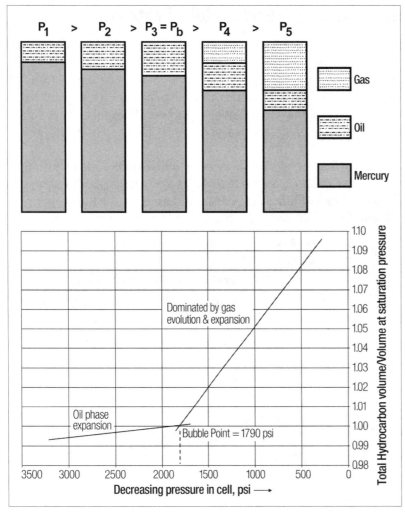

Fig. 3–15. Flash vaporization (constant composition expansion) of hydrocarbons in cells

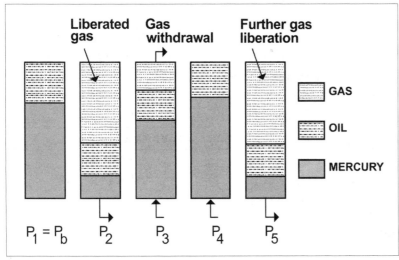

Fig. 3–16. Differential vaporization of hydrocarbons in cells

Constant composition expansion measurements

In a constant composition expansion study, a reservoir fluid sample or a recombined sample obtained from a surface separator or stock tank is used. This sample is introduced into a cell where the pressure is initially higher than the reservoir pressure (fig. 3–15). Any changes occurring in the cell can be monitored visually. The temperature of the cell is maintained to reflect the reservoir conditions. The pressure acting on the confined fluid is gradually lowered in a series of steps by withdrawing mercury from the cell. The increment in the fluid volume is noted with the decrease in prevailing pressure. At the end of each step, the vapor and liquid phases are allowed to reach equilibrium. When the bubblepoint of the fluid system is reached, a sharp change in slope is observed. The fluid volume begins to expand markedly with pressure decline at and below the bubblepoint due to the liberation of the vapor phase. No hydrocarbon component is withdrawn from the experimental cell. Several recombined samples with varying amounts of dissolved gas are usually employed to study the effect of the gas/oil ratio on the bubblepoint. Vital information that can be obtained from flash vaporization measurements includes the values of the formation volume factor and the gas/oil ratio as a function of pressure.

An example of the results obtained from a typical constant composition expansion study is presented in Table 3–5.

Table 3–5. Results of constant composition expansion of reservoir fluid. *Courtesy of Core Laboratories, Inc.*

Reservoir Fluid Data

a. Saturation pressure (bubblepoint)	2,090 psi
b. Specific volume at bubblepoint	0.021 cft/lb
c. Thermal expansion @ 4,500 psi	1.078 (210°F/75°F)
d. Average compressibility above the bubblepoint	
Pressure: 4,500–3,500 psi	$12.88 \times 10^{-6}\,psi^{-1}$
3,500–2,500 psi	$15.64 \times 10^{-6}\,psi^{-1}$
2,500–2,090 psi	$18.29 \times 10^{-6}\,psi^{-1}$

Pressure-Volume Data at 210°F

Cell pressure	Relative volume
Psi	V/V_{sat}
4,500	0.9570
4,000	0.9603
3,500	0.9655
3,000	0.9768
2,500	0.9871
2,200	0.9964
p_{sat}: 2,090	1.0000
2,075	1.0018
2,054	1.0046
2,008	1.0195
1,855	1.1462
1,634	1.2255
1,366	1.4357
1,066	1.8858
705	2.6690
442	3.6703

Table 3–6. Composition of reservoir fluid sample (by flash/extended chromatography) Courtesy of Core Laboratories, Inc.

Component	Mol.%	Wt.%	Liquid Density (gm/cc)	M.W.
Hydrogen sulfide	0.00	0.00	0.8006	34.080
Carbon dioxide	0.91	0.43	0.8172	44.010
Nitrogen	0.16	0.05	0.8086	28.013
Methane	36.47	6.25	0.2997	16.043
Ethane	9.67	3.10	0.3558	30.090
Propane	6.95	3.27	0.5065	44.097
iso-Butane	1.44	0.89	0.5623	58.123
n-Butane	3.93	2.44	0.5834	58.123
iso-Pentane	1.44	1.11	0.6241	72.15
n-Pentane	1.41	1.09	0.6305	72.15
Hexanes	4.33	3.88	0.6850	84
Heptanes	2.88	2.95	0.7220	96
Octanes	3.15	3.60	0.7450	107
Nonanes	2.16	2.79	0.7640	121
Decanes	2.12	3.03	0.7780	134
Undecanes	1.85	2.90	0.7890	147
Dodecanes	1.61	2.77	0.8000	161
Tridecanes	1.40	2.62	0.8110	175
Tetradecanes	1.23	2.50	0.8220	190
Pentadecanes	1.08	2.38	0.8320	206
Hexadecanes	0.95	2.25	0.8390	222
Heptadecanes	0.85	2.15	0.8470	237
Octadecanes	0.76	2.04	0.8520	251
Nonadecanes	0.70	1.97	0.8570	263
Eicosanes plus	12.55	43.54	0.9004	325
Total:	100.00	100.00	—	—
Total sample:	—	—	0.7015	93.66

Differential vaporization measurements

As previously discussed, with flash vaporization, liquid and vapor phases are in contact with each other at all times during the study. In contrast, differential vaporization involves the removal of the vapor phase from the cell as soon as it is liberated at each step of reduction in pressure (fig. 3–16). As a result, the

Table 3–7. Overall composition of plus (heavier) fractions

Plus Fractions	Mol. %	Wt. %	Density	M.W.
Heptanes plus	33.29	77.49	0.8515	218
Undecanes plus	22.98	65.12	0.8736	265
Pentadecanes plus	16.89	54.33	0.8887	301
Eicosanes plus	12.55	43.54	0.9004	325

composition of the hydrocarbon fluid components in the experimental cells changes continuously due to the withdrawal of the gas phase. Differential vaporization studies involve measurements of liberated gas volume as well as oil volume undergoing shrinkage as the cell pressure is reduced. In turn, this leads to the determination of gas and oil formation volume factors and solution gas volumes as a function of decreasing pressure below the bubblepoint. Combined with knowledge of the oil formation volume factor, the two-phase formation volume factors can be obtained as well.

The results of differential vaporization tests performed in laboratory on a 40.5° crude at 220°F are shown in Figures 3–17 through 3–22. The composition of the crude is shown in Tables 3–6 and 3–7. The bubblepoint is 2,620 psig, and the solution gas/oil ratio is 854 scf/bbl.

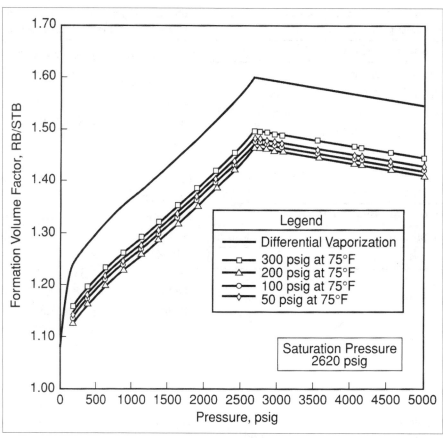

Fig. 3–17. Plot of formation volume factor versus pressure. *Courtesy of Core Laboratories, Inc.*

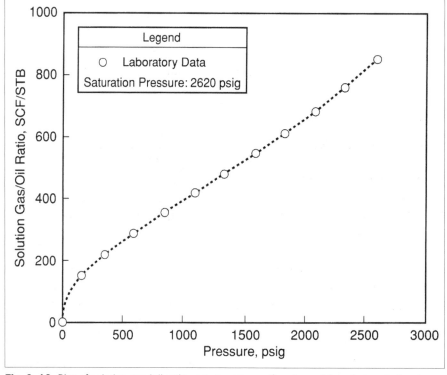

Fig. 3–18. Plot of solution gas/oil ratio versus pressure. *Courtesy of Core Laboratories, Inc.*

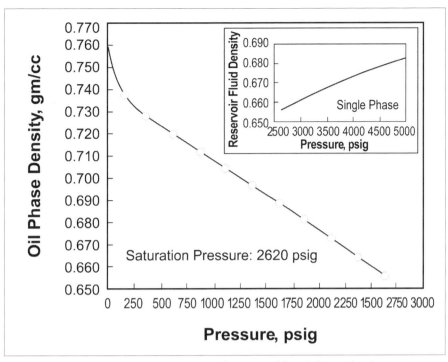

Fig. 3–19. Plot of oil density versus pressure. *Courtesy of Core Laboratories, Inc.*

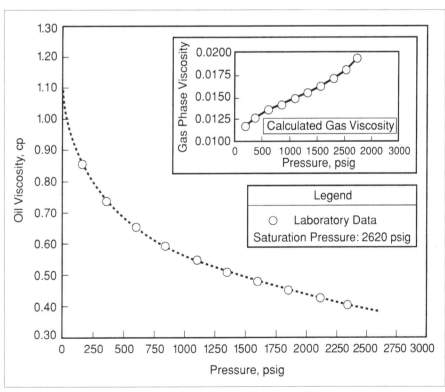

Fig. 3–20. Plot of oil viscosity versus pressure. *Courtesy of Core Laboratories, Inc.*

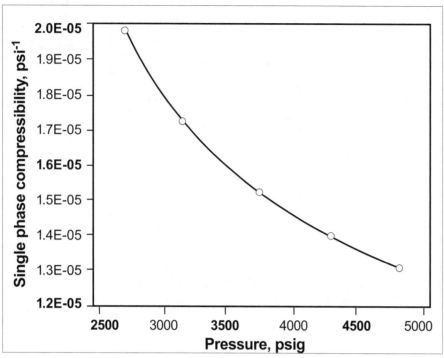

Fig. 3–21. Plot of average single-phase oil compressibility versus pressure. *Courtesy of Core Laboratories, Inc.*

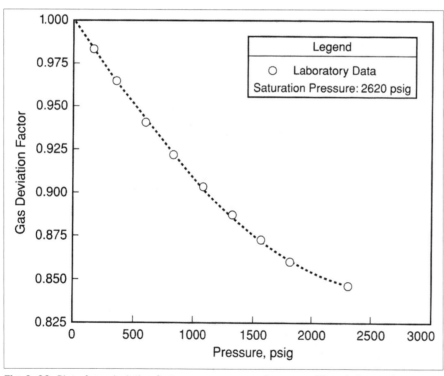

Fig. 3–22. Plot of gas deviation factor versus pressure. *Courtesy of Core Laboratories, Inc.*

Analysis of gas condensate

Fluid samples obtained from gas condensate reservoirs are subjected to constant composition expansion (CCE) and constant volume depletion (CVD) tests. The former is similar to a flash test conducted for oil samples. Pressure in a closed PVT chamber containing a fluid sample in the gas phase is gradually lowered until the dew point is reached, and the heavier components condense out. This allows determination of the dewpoint pressure. Additionally, z factors at or above the dew point are determined.

A constant volume depletion test determines important condensate properties, such as two-phase z factors, needed to predict reservoir performance. A predetermined volume of gas (including condensate in vapor form) is introduced in a PVT cell maintained at dewpoint pressure and reservoir temperature. Next, the cell pressure is reduced below the dew point by withdrawal of mercury, resulting in expansion of the hydrocarbon fluid volume and condensation of a liquid phase. Mercury is then reinjected into the cell in order to expel a certain amount of gas so that the initial volume of hydrocarbons in the cell is restored. Condensed liquid remains in the cell. This procedure simulates the production of gas from a reservoir, while the liquid condensate is left behind. Equation 3.11 can be used to calculate the two-phase z factor at each stage of depletion. The number of moles of fluid remaining in the cell can be found by subtracting the moles of cumulative gas expansion from the number of moles present initially in the cell.

Vapor/Liquid Equilibrium Relationships

The study of the vapor/liquid equilibrium relationships for reservoir fluids is important from the standpoint of the phase changes that occur during production. The changes may take place both in the subsurface reservoir and at the production facilities. Vapor/liquid equilibrium calculations can lead to the determination of the volume and composition of the gas and liquid phases when the crude oil is separated or flashed in surface facilities under reduced pressure and temperature. The objectives of the analysis include attainment of optimized operating conditions at surface facilities and calculation of various fluid properties under reservoir conditions.

Under equilibrium conditions, the relationship between the vapor and liquid phases is given as follows:

$$y_i/x_i = K_i \tag{3.76}$$

where

x_i = mole fraction of the ith component in the liquid phase,
y_i = mole fraction of the ith component in the gas phase, and
K_i = the gas/liquid equilibrium ratio of the ith component at a certain pressure, p, and temperature, T.

This relationship is based upon Dalton's and Raoult's laws for ideal gases. According to Dalton's law, the partial vapor pressure of any component in a two-phase fluid system is the product of its mole fraction in the vapor phase and total pressure:

$$p_i = y_i\, p \tag{3.77}$$

where

p_i = partial pressure of the component i in the vapor phase, psia, and
p = total pressure of the system, psia.

According to Raoult's law, the partial pressure exerted by a component of a liquid phase is equal to the product of its mole fraction in the liquid phase and vapor pressure:

$$p_i = x_i\, p_{vi} \tag{3.78}$$

p_{vi} = vapor pressure of component i in the liquid phase, psia.

When the vapor and liquid phases are in equilibrium, the partial pressure exerted by a specific component in the vapor phase is equal to the partial pressure exerted by the same component in the liquid phase. Hence, equating the partial pressure terms for a component in Equations 3.77 and 3.78, the following can be obtained:

$$p_i = y_i\, p = x_i\, p_{vi} \tag{3.79}$$

$$y_i/x_i = K_i = p_{vi}/p \tag{3.80}$$

However, real gases depart from ideal gas behavior at higher pressures and temperatures, and the gas/liquid equilibrium ratio must be determined experimentally. Furthermore, several equation of state correlations are also available to estimate the equilibrium ratio.

The compositions of the equilibrium ratios can be correlated through convergence pressure. This is the critical pressure of the system's composition with a critical temperature equivalent to the system temperature. K values approach the ideal behavior at low pressures, i.e., the slope of the line of a plot of log K versus log P tends to be $-45°$. The equilibrium ratios of all of the components converge to unity at the convergence pressure.

One could first consider a mole of a hydrocarbon stream having a certain number (n) of hydrocarbon components. This sample is flashed or separated into certain mole quantities of gas (V) and liquid (L) in a surface separator at a predetermined pressure and temperature. The hydrocarbon stream is from a producing well. The operating pressure and temperature of the flash separator determine the moles of gas and liquid produced. The separated liquid phase can go either to stock tanks for sales or to a second-stage separator operating at a lower pressure. Similarly, the evolved gas phase is either transported via pipeline for sales or sent to a gasoline plant for further recovery of liquid. The process of separation into liquid and gas phases leads to the following:

$$\Sigma \, z_i = L + V = 1.0 \qquad (i = 1...n) \tag{3.81}$$

$$\Sigma \, x_i = 1.0 \qquad (i = 1...n) \tag{3.82}$$

$$\Sigma \, y_i = 1.0 \qquad (i = 1...n) \tag{3.83}$$

where

z_i = mole fraction of the ith component in a hydrocarbon feed (liquid + gas),

L = total moles of liquid, and

V = total moles of vapor or gas.

Equations 3.81 to 3.83 simply state the following:

1. Since the basis is 1 mole of feed, the sum of liquid and vapor moles (L and V, respectively) is unity.
2. The mole fractions of all of the components in the liquid phase, L, must add up to 1.
3. The mole fractions of all of the components in the vapor phase, V, must be equal to unity by the same reasoning.

A material balance for an individual component, i, in 1 mole of feed, V moles of vapor, and L moles of liquid would lead to the following:

$$z_i = y_i \, V + x_i \, L \tag{3.84}$$

Combining the above with Equation 3.76, the following can be obtained:

$$z_i = x_i \, K_i \, V + x_i \, L \tag{3.85}$$

Equation 3.85 can be solved for x_i, the mole fraction of ith component in the liquid phase, as in the following:

$$x_i = \frac{z_i}{L\left[(V/L)\,K_i + 1\right]} \tag{3.86}$$

Based on Equations 3.82 and 3.86, the following can be written:

$$\Sigma \, x_i = \Sigma \, \frac{z_i}{L\left[(V/L)\,K_i + 1\right]} = 1 \tag{3.87}$$

Rearranging the above equation, an expression of the moles of liquid, L, can be derived:

$$L = \Sigma \, \frac{z_i}{\left[(V/L)\,K_i + 1\right]} \tag{3.88}$$

Similarly, based on Equations 3.76 and 3.84, the following can be deduced:

$$y_i = \frac{z_i \, K_i}{V\left[(L/V) + K_i\right]} \tag{3.89}$$

$$\Sigma \, y_i = \Sigma \, \frac{z_i \, K_i}{V\left[(L/V) + K_i\right]} = 1 \tag{3.90}$$

$$V = \Sigma \, \frac{z_i \, K_i}{\left[(L/V) + K_i\right]} \tag{3.91}$$

The vapor/liquid relationships derived above lead to the following applications, among others.

Determination of liquid and gas phase compositions. Solving Equations 3.88 and 3.91 requires an iterative process. The steps are:

1. Assume initial values of L and V.
2. Obtain corresponding K values.
3. Iterate until both sides of the equation become equal.

In each iteration, $V = 1 - L$, based on Equation 3.81. Once L and V are obtained, the composition of the liquid and gas phases can be found by Equations 3.86 and 3.89. [The values of the mole fraction of each component in the feed (z_i) and the corresponding equilibrium ratio (K_i) are known.]

Determination of the bubblepoint and dew point. A bubblepoint pressure for the hydrocarbon mixture in question is assumed. At the bubblepoint, virtually all components are in a liquid phase. An infinitesimal amount of gas begins to evolve out of the solution, which can be assumed to be negligible for all practical purposes. Hence, at the bubblepoint pressure and reservoir temperature, the following condition must satisfy:

$$\Sigma\, x_i\, K_i = A \qquad\qquad (3.92)$$

where

   $A = 1$.

If the bubblepoint pressure is assumed correctly, $A = 1$. However, if $A < 1$, then the values of K_i are too low, and the assumed bubblepoint pressure is too high. If $A > 1$, the values of K_i are too high, and the assumed pressure is too low.

Similarly, the dew point for a hydrocarbon mixture is obtained by satisfying the following equation:

$$\Sigma\, z_i\, /\, K_i = A \qquad\qquad (3.93)$$

If the dew point is assumed correctly, the value of A becomes unity. If A is greater than unity, one must choose the values of K_i at a lower pressure. Conversely, if A is less than one, a higher pressure must be assumed to calculate the dew point.

Example 3.11. Calculate the amount of liquid recovery from the following feed in a single-stage separator: ethane (C_2) = 0.33, n-butane (n-C_4) = 0.33, and hexane (C_6) = 0.34.

The separator operates at 15 psia and 60°F. Use Figure 3–23 to obtain the gas/liquid equilibrium ratios of the hydrocarbon components.

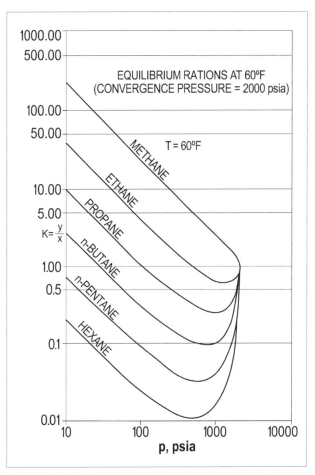

Fig. 3–23. Equilibrium ratio versus pressure chart of various hydrocarbon components. Equilibrium curves converge at 2,000 psia in the case considered. *Courtesy of Gas Processors Suppliers Association.*

Solution: This calculation requires iterative steps, as follows:

1. Read the values of K_i for the ith component from the figure. In the first iteration, assume L = 0.5, i.e., V/L = 1.0.
2. Calculate (V/L) K_i + 1.
3. Using Equation 3.86, calculate $L_i = Lx_i = \dfrac{z_i}{(1+K_i (V/L))}$.
4. Calculate sum as $\Sigma L_i = L$.
5. Compare the value of L with the value assumed in step 1.
6. Repeat steps 1 through 5 until convergence is achieved (see Table 3–8).

Table 3–8. Flash calculations for single-stage separator

Single-Stage @ 15 psi, 60°F			Iteration #1		Iteration #2		Iteration #3	
		Assume:	L = 0.500 (V/L = 1)		L=0.333 (V/L = 2)		L = 0.367 (V/L = 1.725)	
	z_i	K_i	$1 + K_i{*}V/L$	L_i	$1 + K_i{*}V/L$	L_i	$1 + K_i{*}V/L$	L_i
		Fig. 3–23	Calc.	Eq. 3.86	Calc.	Eq. 3.86	Calc.	Eq. 3.86
Component								
Ethane	0.33	24.600	25.600	0.01289	50.200	0.00657	43.430	0.00760
n-Butane	0.33	1.720	2.720	0.12132	4.440	0.07432	3.967	0.08319
Hexane	0.34	0.135	1.135	0.29075	1.270	0.26772	1.233	0.27578
Sum ($L = \Sigma L_i$)				0.425		0.349		0.367

Example 3.12. Repeat the calculations in Example 3.11 by considering a two-stage separator as follows:

- First stage operating at 100 psia and 60°F
- Second stage operating at 15 psia and 60°F

To arrive at the solution, it is necessary to follow the iterative procedure as illustrated in the previous example for a first-stage separation. Next, the liquid fractions obtained from the first stage are used as feed for the second stage. The iterative procedure is the same for the second or any subsequent stage. Table 3–9 shows the results of these calculations.

As noted earlier, various equations of state are available to determine the equilibrium ratio of the individual hydrocarbon components, the dew point, and the bubblepoint pressures. The most familiar equation of state correlations include the following:

- Redlich-Kwong (1949) and subsequent modifications by Soave
- Peng-Robinson (1976)
- Martin (1979)

The correlations are based on the van der Waals equation that relates pressure, temperature, and molar volume, as follows:

$$p = \frac{RT}{v-b} - \frac{a}{v^2} \qquad (3.94)$$

where

a = parameter for molecular attraction, and
b = parameter for repulsion.

Table 3-9. Flash calculations for two-stage separator

First Stage			Iteration #1		Iteration #2		Iteration #3	
100 psi, 60°F		Assume	L = 0.600 (V/L = 0.667)		L = 0.75 (V/L = 0.333)		L = 0.8 (V/L = 0.25)	
	z_i	K_i	$1 + K_i{*}V/L$	L_i	$1 + K_i{*}V/L$	L_i	$1 + K_i{*}V/L$	L_i
Component								
Ethane	0.33	4.500	4.002	0.08247	2.500	0.13200	2.125	0.15529
n-Butane	0.33	0.300	1.300	0.25385	1.100	0.30000	1.075	0.30698
Hexane	0.34	0.029	1.029	0.32070	1.010	0.33674	1.007	0.33755
Sum ($L = \Sigma L_i$)				0.657		0.769		0.7998
Second Stage			Iteration #1		Iteration #2		Iteration #3	
15 psi, 60°F		Assume	L = 0.500 (V/L = 1)		L = 0.400 (V/L = 1.5)		L = 0.343 (V/L = 1.915)	
	z_i	K_i	$1 + K_i{*}V/L$	L_i	$1 + K_i{*}V/L$	L_i	$1 + K_i{*}V/L$	L_i
From first stage								
Ethane	0.15529	24.600	25.600	0.00607	37.900	0.00410	48.120	0.00323
n-Butane	0.30698	1.720	2.720	0.11286	3.580	0.08575	4.295	0.07148
Hexane	0.33755	0.135	1.135	0.13682	1.203	0.28071	1.259	0.26820
Sum ($L = \Sigma L_i$)	0.7998			0.256		0.371		0.343

The van der Waals equation can be written in terms of z factor, reduced pressure, and temperature, as follows:

$$z^3 - (B + 1)z^2 + Az - AB = 0 \qquad\qquad (3.95)$$

where

$A = (27\, p_r) / (64\, T_r^2)$
$B = p_r / (8\, T_r)$

Equation 3.95 serves as the basis of the many cubic equations of state proposed, such as those by Soave-Redlich-Kwong (SRK) and Peng-Robinson (PR). Detailed treatment of various equations of state can be found in the literature, including Whitson and Ahmed.[46,47]

Equation of state correlations find wide use in compositional simulators. These allow the changing composition of a volatile fluid or gas condensate to be simulated during production, in addition to reservoir pressure and fluid saturation. However, the correlations used in simulations are tuned by laboratory findings to improve accuracy in the results of the study.

In-Situ Measurements of Fluid Properties

In-situ assessment of reservoir fluid properties by downhole fluid analyzer tools is an emerging technology. It offers many advantages, including monitoring and decision making in real time, and more efficient fluid sampling. In certain cases, optical sensors are employed that can correlate certain fluid properties with the optical response received from the tool, such as the index of refraction and fluorescence characteristics.[48] A few example field applications of the downhole measurement technology are described here.

Investigation of vertical flow barrier. The downhole fluid analyzer assembly consisted of two probes, one positioned above the suspected flow barrier in the formation and the other below. The bottom probe was pumped for a finite period, followed by pumping of the top probe. The gas/oil ratios of the reservoir fluids in the two intervals, as measured by the fluid analyzers, were observed to be dissimilar. Moreover, the reservoir fluid was heavier, with a relatively low gas/oil ratio in the upper interval. This clearly indicated that the two intervals were not in vertical communication. In another instance, fluid samples in the two probes appeared to be very similar in properties, suggesting that the formation intervals were in communication.

In-situ determination of endpoint relative permeabilities. The success of an enhanced recovery operation, as predicted by reservoir simulation, is found to depend on accurate knowledge of relative permeability information. This information indicates the dominance of one fluid phase over others in the reservoir during multiphase flow. Of particular interest are the endpoint relative permeabilities, which point to the limiting saturation, when the flow of one fluid will commence or cease in the presence of the others. This information is directly tied to residual oil saturation following water injection, for example. The relative permeabilities of oil, gas, and water under changing reservoir fluid saturations are traditionally determined by laboratory studies or are based on correlations. However, state-of-the-art tools may aid in determining the endpoint relative permeability values in subsurface reservoirs under actual conditions. In one application, flow of both oil and water was detected by an optical analyzer from the same depth without moving the downhole sensor tool. By manipulating the pressure and rate, it was possible to flow virtually 100% oil or water in separate runs. This led to the determination of the endpoint relative permeability of the oil and water in situ.

Sampling quality. Downhole monitoring tools such as optical fluid analyzers detect the general composition of sampled fluid in real time. This ensures minimization of contaminants, such as mud filtrate, in fluid samples collected for laboratory studies.

Properties of Formation Water

Knowledge of the formation water properties is needed in various reservoir studies, along with knowledge of the oil and gas properties. Of interest to reservoir engineers are the salinity, viscosity, density, and compressibility of the formation water. Since hydrocarbon gas is soluble in water under elevated pressures and temperatures, the solution gas/water ratios and water formation volume factors are necessary information in conducting reservoir studies. These studies include, but are not limited to, well test analysis and multiphase fluid flow simulation. Properties of the formation water are generally dependent on the reservoir pressure, temperature, and concentration of salt compounds. Water found in subsurface geologic formations usually has a high salt content due to the dissolution of minerals and related geochemical processes that occur throughout geologic times. Last, but not least, subsurface water is found to be incompatible with surface water that is injected into the formation during waterflood operation. Loss of injectivity is encountered due to the precipitation of solids in porous media.

Various charts and correlations are widely available to estimate the important water properties. Selected charts that aid in determination of the viscosity, compressibility, gas solubility, and formation volume factor of the formation water are presented in this chapter. Correlations predicting the properties of the formation water are integrated into the petroleum software, much in the same manner as the properties of the hydrocarbon fluids.

Viscosity

The viscosity of the formation water primarily depends on the reservoir temperature and pressure, and the salinity of the water. Based on Figure 3–24, calculation of the formation water viscosity is a two-step process.[49] The first step is to read the value of viscosity at elevated reservoir temperature but under atmospheric pressure if the chloride salt content of the formation water is known. Next, a correction factor is applied to estimate the viscosity at elevated reservoir pressure. The viscosity of the formation water is a strong function of the reservoir temperature and salinity.

Isothermal compressibility

The compressibility of water depends on the reservoir pressure, temperature, and dissolved gas/water ratio. For undersaturated water, compressibility can be estimated based on the method by Dodson and Standing.[50] Again, it is a two-step process whereby the value of compressibility is first read from Figure 3–25 when reservoir pressure and temperature are known. A correction factor, based on Figure 3–26, is then applied to account for the dissolved gas in water.

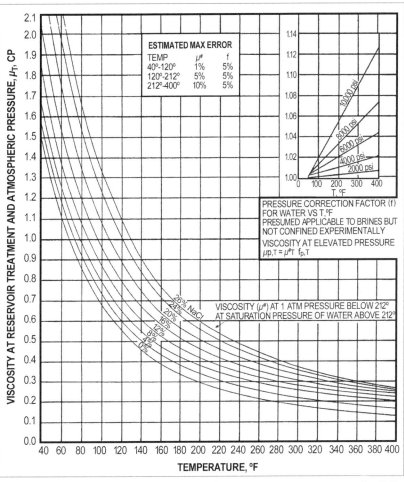

Fig. 3–24. Viscosity of formation water at various temperature, pressure, and salinity. *Source: C. S. Matthews and D. G. Russell. 1967. Pressure Buildup and Flow Tests in Wells. SPE Monograph. Vol. 1. Dallas: Society of Petroleum Engineers. © Society of Petroleum Engineers. Reprinted with permission.*

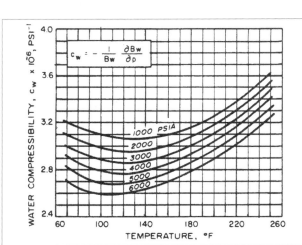

Fig. 3–25. Compressibility of water as a function of pressure and temperature. *Source: C. R. Dodson and M. B. Standing. 1944. Pressure-volume-temperature and solubility relations for natural gas–water mixtures. In Drilling and Production Practice. Washington, DC: American Petroleum Institute, 173–179. Reproduced courtesy of the American Petroleum Institute.*

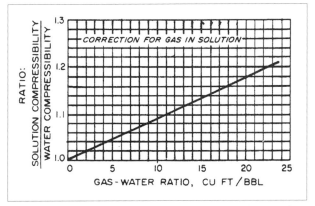

Fig. 3–26. Correction for dissolved gas in water. *Source: C. R. Dodson and M. B. Standing. 1944. Pressure-volume-temperature and solubility relations for natural gas–water mixtures. In Drilling and Production Practice. Washington, DC: American Petroleum Institute, 173–179. Reproduced courtesy of the American Petroleum Institute.*

The compressibility of formation water is computed as shown in the following:

Solution gas/water ratio

Natural gas exhibits limited solubility in formation water. The solution gas/water ratio for pure water is computed first, followed by correction for the salt content of the formation water. Figures 3–27 and 3–28, based on the correlations by Dodson and Standing, can be used to estimate the solution gas/water ratio once reservoir pressure, temperature, and salinity are known.[51]

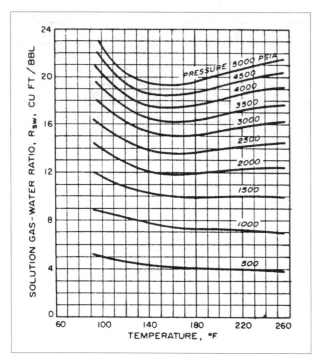

Fig. 3–27. Solubility of natural gas in water as a function of pressure and temperature. *Source: C. R. Dodson and M. B. Standing. 1944. Pressure-volume-temperature and solubility relations for natural gas–water mixtures. In* Drilling and Production Practice. *Washington, DC: American Petroleum Institute, 173–179. Reproduced courtesy of the American Petroleum Institute.*

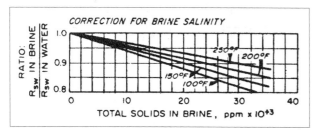

Fig. 3–28. Correction of gas solubility for solids content. *Source: C. R. Dodson and M. B. Standing. 1944. Pressure-volume-temperature and solubility relations for natural gas–water mixtures. In* Drilling and Production Practice. *Washington, DC: American Petroleum Institute, 173–179. Reproduced courtesy of the American Petroleum Institute.*

Water formation volume factor

Due to the solubility of hydrocarbon gases in formation water, it is necessary to consider the formation volume factor of the water in reservoir engineering calculations. It is defined as the volume of water under elevated pressure and temperature in the reservoir, including any dissolved gas, over 1 stb of water under standard conditions. The formation volume factors for pure water and gas-saturated water, proposed by Dodson and Standing, are presented in Figure 3–29.[52]

Key points—fluid properties, data sources, and laboratory measurements

1. Knowledge of the hydrocarbon fluid properties is vital in developing a reservoir and optimizing production performance. Such properties include, but are not limited to, fluid viscosity, compressibility, saturation (bubblepoint and dewpoint) pressure, and solution gas/oil ratio.

2. Sources of fluid properties data include: (a) laboratory studies, (b) in-situ measurements, and (c) applicable correlations. Determination of the actual fluid properties either in the laboratory or in situ is preferred over using correlations, especially with oil reservoirs.

3. Important natural gas properties can be determined with a fair degree of accuracy based on the methods illustrated in the chapter. Gas condensates can be treated in a similar manner as long as the occurrences of heavier hydrocarbons are not significant.

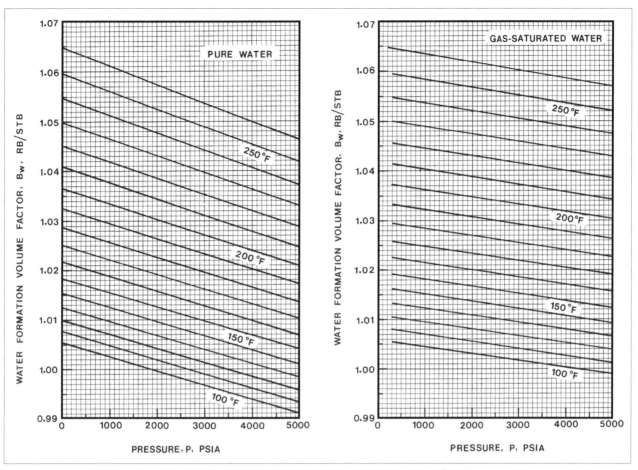

Fig. 3–29. Formation volume factor of water. *Source: C. R. Dodson and M. B. Standing. 1944. Pressure-volume-temperature and solubility relations for natural gas–water mixtures. In* Drilling and Production Practice. *Washington, DC: American Petroleum Institute,. 173–179. Reproduced courtesy of the American Petroleum Institute.*

4. Notable laboratory studies involve flash and differential vaporization of lighter hydrocarbons from reservoir oil samples. In flash vaporization, the evolved gas is kept in contact with the liquid phase at all times in a closed chamber. The process is similar to what takes place in surface separators. Flash tests are also referred to as constant composition expansion. In contrast, the procedure involving differential vaporization removes the gaseous hydrocarbons as soon as they evolve from the solution. This emulates the rapid gas movement towards the wells in porous media immediately following the gas dissolution.

5. Flash liberation studies determine the bubblepoint pressure of the reservoir fluid and oil compressibility above the bubblepoint, among other factors. Differential liberation tests determine the solution gas/oil ratio and the oil formation volume factor of a fluid as a function of declining cell pressure. This decline is accompanied by gas evolution and its subsequent removal from the cell.

6. Constant composition expansion and constant volume depletion tests are usually performed on gas condensate samples. The former is similar to constant composition expansion or flash vaporization of oil. The dew point of a gas condensate and the z factor above the dew point are obtained by the test. A constant volume test involves lowering of the cell pressure in stages, resulting in fluid volume expansion and condensation of heavier components. Mercury is reinjected into the cell to expel a certain quantity of gas and restore the original volume. A two-phase z factor is obtained by the test. This input is required in predicting gas condensate reservoir performance.

Phase Behavior of Hydrocarbon Fluids in the Reservoir

The phase behavior of hydrocarbons is essential in characterizing the following:

- Petroleum reservoir types
- Producing mechanisms of a reservoir
- Expected recovery efficiency

The classification of petroleum reservoirs, such as black oil, volatile oil, dry gas, and gas condensate, is dictated by the prevailing pressure, temperature, and fluid composition at discovery. Figure 3–30 illustrates the typical phase behavior of a multicomponent fluid system, with the fluid temperature plotted on the x-axis and pressure on the y-axis. This plot, or phase diagram, leads to prediction of the reservoir fluid behavior based on changes in pressure and temperature during production in a reservoir.

Depending upon the composition of the hydrocarbons, in-situ fluid in each reservoir will have its own phase diagram. Again, in a given reservoir, when a second phase (such as evolved gas) appears due to a decline in pressure, the original phase diagram of the liquid phase changes. This occurs as the composition of the liquid is altered. In other words, phase diagrams are dynamic, as fluid composition changes in a reservoir.

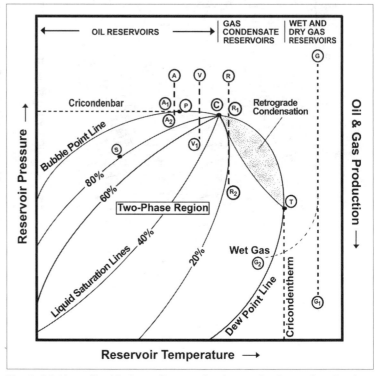

Fig. 3–30. Vapor/liquid phase diagram showing production paths of various reservoir fluids

Fluid phase behavior in the reservoir is best described with the help of phase diagrams. There are many important features of phase diagrams, as described in the following.

- **Single- and two-phase regions.** A phase diagram consists of two major regions: the single-phase region that lies outside the envelope (fig. 3–30), and the two-phase region that is enclosed by the envelope. The curved lines forming the two-phase region within the phase envelope are referred to as liquid saturation lines. They represent the amount of liquid volume as a fraction of the total volume of liquid and gas. For example, point S on the 80% saturation line would represent a reservoir fluid system comprised of 80% liquid and 20% vapor.

- **Critical point, bubblepoint, and dewpoint curves.** The envelope consists of the bubblepoint and dewpoint curves, which converge at the critical point. Liquid and vapor phases are indistinguishable and are in equilibrium at the critical point. Thus the physical properties of the two phases of the reservoir fluid at the critical point are identical.

- **Cricondentherm and cricondenbar.** The highest temperature above which hydrocarbon fluid can exist solely as gas is referred to as the cricondentherm. The farthest point to the right on the phase envelope indicates the cricondentherm for the fluid system, as illustrated by point T. It represents the highest temperature at which a liquid and vapor can coexist in equilibrium. Similarly, the highest pressure above which hydrocarbons exist in a liquid phase alone is known as the cricondenbar. In Figure 3–30, the cricondenbar is located on the highest point of the two-phase envelope illustrated by point P. It indicates the upper limit of pressure at which liquid and vapor can coexist in equilibrium.

Again, it is emphasized that the critical point, cricondentherm, cricondenbar, and other phase behavior characteristics vary from one hydrocarbon fluid system to another.

The various types of reservoirs where hydrocarbons initially occur can be located on the phase diagram as in the following:

- **Undersaturated oil reservoirs.** Occur to the left of the critical point (C) in the single-phase region above the phase envelope, as in point A shown in Figure 3–30. Highly volatile reservoirs are located closer to the critical point, as in point V. Undersaturated oil reservoirs essentially occur above the bubblepoint pressure. No free gas is present in the reservoir at discovery.

- **Gas condensate reservoirs.** Occur between the critical point and the cricondentherm over the phase envelope, as in point R.

- **Wet and dry gas reservoirs.** Occur farther to the right from the cricondentherm, as in point G. Again, the initial location is outside the phase envelope.

- **Two-phase reservoirs.** These reservoirs are essentially located within the phase envelope. Saturated oil reservoirs having a gas cap over the oil zone occur within the two-phase region. An example would be point S in the phase diagram. When the initial reservoir temperature and pressure are within the envelope, a gas cap is expected to exist in the reservoir at discovery. If the reservoir pressure equals the bubblepoint pressure, the point would be located on the bubblepoint line of the phase envelope, as in point P.

Reservoir Production and Phase Behavior

During the production of oil and gas, reservoir fluid behavior is often characterized by phase separation under certain conditions. Fluid composition, pressure, and temperature play a crucial role in shaping the performance of a petroleum reservoir. The characteristic trends of oil and gas reservoirs under production in relation to fluid phase behavior are described in the following sections.

Undersaturated black oil reservoirs

Undersaturated black oil reservoirs are encountered when the initial reservoir pressure and temperature are to the left of the critical point, and the pressure is above the bubblepoint. Figure 3–30 shows an undersaturated black oil reservoir with pressure and temperature initially at point A. All of the available gas is dissolved in the oil at the initial conditions. The vertical line from point A to point A_1 on the bubblepoint curve represents isothermal production, resulting in rock and liquid expansion. Reservoir pressure drops rapidly and continuously until the bubblepoint is reached. Hydrocarbons are essentially in the liquid phase as long as the reservoir pressure remains above the bubblepoint. Evolution of the gas occurs at surface facilities due to the reduction in pressure, and the resulting gas/oil ratio remains low and constant. The oil recovery mechanism is dominated by the volumetric expansion of the reservoir fluids and rock above the bubblepoint when no other external driving mechanism is present. Recovery efficiency is relatively less, and typically varies from 1% to 5%, with an average of 3%.

When the reservoir pressure declines below the bubblepoint due to production, dissolved gas starts to come out of solution, and a free gas phase is formed. Depletion below the bubblepoint causes the gas phase to increase rapidly in the reservoir. The dominant recovery mechanism is known as solution gas or depletion drive. Since the viscosity of the gas is much lower than the oil, the gas phase is significantly more mobile than the liquid phase in the reservoir. The gas/oil ratio is initially low, then rises to a maximum, and finally drops as most of the liberated gas is produced. Typical oil recovery due to solution gas drive could be from 10% to 25%, with an average of about 16%.

Saturated black oil reservoirs

When the initial reservoir pressure and temperature are within the two-phase region (as at point S in fig. 3–30), reservoirs with gas caps are encountered. Gas being lighter than oil, it rises above the oil zone due to gravity segregation. Reservoir pressure falls slowly and continuously. Initially, gas is produced because of the free gas saturation in the reservoir. The gas/oil ratio is low initially, and then rises to a maximum and finally declines. Production from the gas cap reservoir is due to the driving energy imparted by both solution and free gases, resulting in higher oil recovery than the solution gas drive alone. Oil recovery due to gas cap drive could be 15% to 35%, with an average of 25%.

Volatile oil reservoirs—undersaturated and saturated

In the case of an undersaturated volatile reservoir, the initial reservoir pressure and temperature is at point V in Figure 3–30, close to the critical point but above the bubblepoint. Compared to the black oil reservoir, volatile reservoirs have higher API gravity, in the range of 38° or more. The dissolved gas/oil ratio could be 1,500 scf/bbl or greater. As in black oil reservoirs, production mechanisms for volatile reservoirs, undersaturated and saturated, are due to rock and fluid expansion, and solution gas or depletion drive, respectively.

Like the black oil reservoir, a gas cap is encountered for the volatile reservoir when the initial reservoir pressure and temperature are within the two-phase region. Production from the gascap reservoir is due to both solution and free gas drives, resulting in higher oil recovery.

Because of the lighter oil with lower viscosity, recoveries from the volatile oil reservoirs could be more than from the black oil reservoirs.

Gas condensate reservoirs with retrograde condensation

In gas condensate reservoirs, the initial reservoir temperature is higher than the critical temperature of the fluid system but lower than the cricondentherm. Reservoir pressure is found to be above the dewpoint pressure (point R in fig. 3–30). The vertical line from point R to point R_1 located on the dewpoint curve represents isothermal production. As the reservoir pressure declines below the dew point, the heavier hydrocarbons condense out in the reservoir. The process of condensation increases to a maximum value, and then decreases again until the abandonment pressure is reached (point R_2). Isothermal retrograde condensation may result in the loss of certain intermediate to heavy hydrocarbon components in the reservoir due to poor mobility in comparison to the gases. The API gravity of the condensates may range between 45° and 120°.

Wet and dry gas reservoirs

Gas reservoirs exist in the single-phase region with the initial temperature exceeding the cricondentherm as in point G in Figure 3–30. When the gas phase undergoes isothermal depletion inside the reservoir without any condensation, it traces a path as shown from point G to point G_1 that lies in the single-phase region as well. When a portion of produced gas condenses in the surface separators under reduced pressure and temperature, the produced gas is referred to as wet gas. The path traced by the wet gas from reservoir conditions to surface facilities is illustrated by the curve from point G to point G_2. Greater condensate recovery could be realized by operating the separators at lower temperatures.

If the gas is sufficiently lean, the produced gas remains in a single phase under conditions of reduced pressure and temperature at the surface. The reservoir fluid would remain as a single phase in the reservoir due to isothermal depletion along the vertical line from point G to point G_1 (fig. 3–30). Dry gas reservoirs contain mostly lighter hydrocarbons with gas/oil ratios of more than 100,000 scf/stb of condensate. Gas recoveries could be as high as 80% or more at relatively low separator pressures. Gas pumps are installed to raise gas pressure to the pipeline delivery pressure.

Figure 3–31 shows the path of the dry gas from the reservoir to the surface under decreasing pressure and temperature. The phase diagram for each specific type of reservoir is significantly different from the general representation as in Figure 3–30.

It must be mentioned that the recovery factors cited in this book are meant for comparison purposes only. Recoveries from oil reservoirs in particular are usually reservoir-specific or region-specific, depending on the local geologic setting, among a host of other factors. Moreover, the recent advent of state-of-the-art tools and techniques in well drilling and operation, along with reservoir surveillance, visualization, and simulation, have resulted in increasing oil and gas recoveries in many cases.

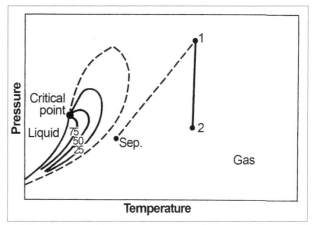

Fig. 3–31. Production path of a dry gas reservoir. A phase diagram is dependent on the composition of in-situ oil or gas. *Source: W. D. McCain, Jr. 1990. Properties of Petroleum Fluids. 2nd ed. Tulsa: PennWell.*

Key points—phase behavior of petroleum fluids

Knowledge of the phase behavior of petroleum fluids is critical in reservoir studies. Many factors are dictated by the phase behavior of the reservoir fluid. These include petroleum reservoir type (such as volatile oil, heavy oil, or gas condensate) and producing mechanisms (such as volumetric or solution gas drive). In addition, the expected recovery efficiency (higher for volatile oil than heavy oil, etc.) is also influenced strongly.

Phase behavior is best explained with the aid of a phase diagram, an example of which was presented earlier in Figure 3–30. Important features of a phase diagram are outlined as follows:

- **Single- and two-phase regions.** A two-phase envelope determines whether the fluid is in one phase (either liquid or gas) or in two phases (oil and gas, or gas and condensate). Single-phase regions are above and to the right of the envelope, indicating the state of relatively high pressure or temperature, or both, in the reservoir.
- **Liquid saturation lines.** These lines within the two-phase envelope indicate the relative amounts of oil and gas phases under the prevailing conditions.
- **Critical point, bubblepoint, and dewpoint curves.** The two-phase envelope mentioned above consists of a bubblepoint curve and a dewpoint curve. These converge at the critical point, where the liquid and vapor phases are indistinguishable and in equilibrium.
- **Cricondentherm and cricondenbar.** The highest temperature above which hydrocarbon fluids can exist solely as a gas is referred to as the cricondentherm. Similarly, the highest pressure beyond which fluids are completely in a liquid phase is known as the cricondenbar.

Effects of Fluid Properties on Reservoir Performance

This section highlights how reservoir performance is influenced by certain fluid properties. Selected fluid properties are the viscosity, bubblepoint pressure, solution gas ratio, and specific gravity of the reservoir fluid. The effects of reservoir pressure, which controls the fluid properties, are included in the following presentation. Several sensitivity studies will be considered using Merlin reservoir simulation software from Gemini Solutions, Inc. This software allows a specific fluid property to be varied in order to predict reservoir performance, while all other fluid and rock properties are kept the same. Reservoir performance is examined in terms of production rate, gas/oil ratio, and recovery efficiency, among other factors.

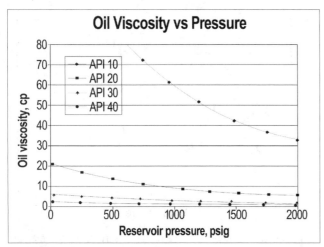

Fig. 3–32. Viscosity of 10°, 20°, 30°, and 40° API gravity oil below the bubblepoint used in the sensitivity study. *Courtesy of Gemini Solutions, Inc.*

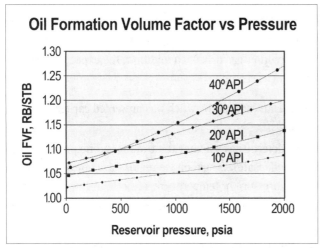

Fig. 3–33. Formation volume factor of 10°, 20°, 30°, and 40° API gravity oil used in the study. *Courtesy of Gemini Solutions, Inc.*

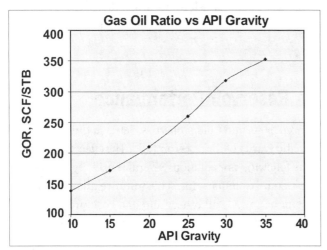

Fig. 3–34. Solution gas/oil ratio as a function of oil gravity at a specific saturation pressure. *Courtesy of Gemini Solutions, Inc.*

High- versus low-API gravity oil

In the first sensitivity study, four cases will be considered for a saturated oil reservoir. The API gravity of the liquid phase is varied in a wide range of specific gravities representing typical oil reservoirs worldwide. Starting with highly volatile oil having an API gravity of 40°, a gradual progression will be considered to heavy oil having an API gravity of 10°. The bubblepoint of the liquid is assumed to be 2,000 psia in all cases. The reservoir is simulated to produce from an initial reservoir pressure of 3,000 psia. Suitable correlations integrated in the software are used to estimate the corresponding oil viscosity, formation volume factor, and solution gas/oil ratio. These are presented in Figures 3–32 through 3–34. The following discussion examines the PVT behavior of the light to heavy oil used in the study.

The viscosity of oil increases with higher specific gravity (lower API gravity), as heavier hydrocarbons are relatively viscous. The viscosity of oil having a specific gravity of 10°API is significantly higher than that of 20°API oil. Moreover, as the reservoir pressure declines below the bubblepoint, the viscosity of the liquid phase increases due to the dissolution of lighter hydrocarbons.

The formation volume factor of 40°API oil at the bubblepoint pressure is the highest among the four cases studied. This is expected, as the light oil contains relatively more volatile hydrocarbons. On the other end of the spectrum, the heavy oil having 10°API gravity exhibits a relatively low formation volume factor value. The formation volume factor decreases with the reservoir producing below the bubblepoint pressure; the solution gas/oil ratio decreases due to the evolution of the dissolved gas.

The solution gas/oil ratio tends to be greater in the case of light oil having API gravity of 30° or more. This scenario represents the dominance of volatiles in crude oil, where the liquid phase contains lighter hydrocarbons in substantial amounts. These lighter hydrocarbons are ready to evolve in order to form a gas phase below the bubblepoint. In the case of 10°API oil, the amount of dissolved gas is minimal, as the fluid composition is dominated by heavier and nonvolatile components.

The formation volume factor of 40°API oil at bubblepoint pressure is the highest among the four cases studied. This is expected, as the light oil contains relatively more volatile hydrocarbons. On the other end of the spectrum, heavy oil having 10°API gravity exhibits a relatively low formation volume factor value. The formation volume factor decreases with the reservoir producing below the bubblepoint pressure; the solution gas/oil ratio decreases due to the evolution of dissolved gas.

Results from the sensitivity study are presented in Figures 3–35 to 3–38. These include cumulative oil production, oil and gas production rates, and average reservoir pressure over a period of 5,000 days. It is observed that the cumulative oil production (and the ultimate recovery) from the reservoir having 40°API gravity oil is the highest out of the four cases simulated. About 1,000,000 bbl of oil are produced within 1,000 days of production. In contrast, the recovery of heavy oil having 10°API gravity is the least, producing only 400,000 bbl in 5,000 days. The study further showed that the ultimate recovery for 10°, 20°, 30°, and 40° oil is 4.9%, 8.4%, 11.2%, and 13.5%, respectively. Heavier oil tends to have relatively low recovery.

The mechanism of oil recovery is based upon solution gas drive discussed in chapter 8. The light oil having an API gravity of 40° contains the maximum amount of dissolved gas initially. Upon production, reservoir pressure declines below the bubblepoint, accompanied by significant gas evolution and expansion, which provides the necessary energy to drive reservoir fluids towards the wellbore. Moreover, the liquid phase, with the lowest viscosity (0.7 cp) of the four cases, requires the least energy to be produced. In simulation studies, a limiting bottomhole pressure of 1,444 psi was used, and most of the production occurred at this pressure. Consequently, the producing gas/oil ratio remained steady in each case. However, when the gas/oil ratio becomes quite high, the flow of gas dominates, as can be seen from the gas/oil ratio relative permeability behavior.

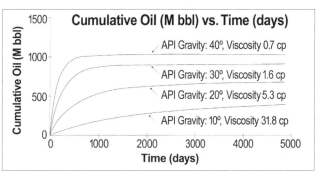

Fig. 3–35. Cumulative oil recovery from the reservoir simulated for oil having API gravity of 10°, 20°, 30°, and 40°. *Courtesy of Gemini Solutions, Inc.*

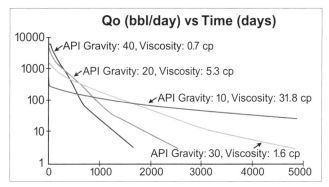

Fig. 3–36. Expected oil production rates with the four cases of 10°, 20°, 30°, and 40°API gravity oil. *Courtesy of Gemini Solutions, Inc.*

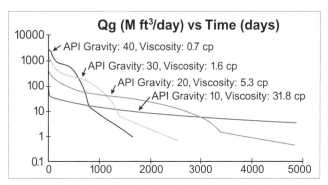

Fig. 3–37. Expected gas production rates simulated for the four cases of specific gravity and viscosity. *Courtesy of Gemini Solutions, Inc.*

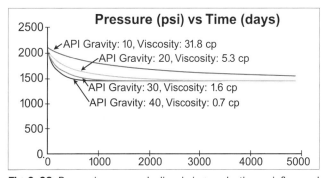

Fig. 3–38. Reservoir pressure decline during production as influenced by oil specific gravity. *Courtesy of Gemini Solutions, Inc.*

The results of the simulation are in agreement with what is generally observed in the field. Table 3–10 compares the oil viscosity encountered in various reservoirs to the amount of oil left behind at abandonment. Not unexpectedly, oil of relatively high viscosity is found to be associated with high residual saturation and poor recovery. A suitable enhanced recovery operation is implemented to improve the ultimate recovery in most cases.

Table 3–10. Effect of reservoir oil viscosity on residual oil saturation. *Source: R. C. Craze and S. E. Buckley. 1945. A factual analysis of the effect of well spacing on oil recovery. In Drilling and Production Practice. Washington, D.C.: American Petroleum Institute.*

Viscosity of Reservoir Fluid, cp	Residual Oil Saturation, %
0.2	30
0.5	32
1.0	34.5
2.0	37
5.0	40.5
10.0	43.5
20.0	64.5

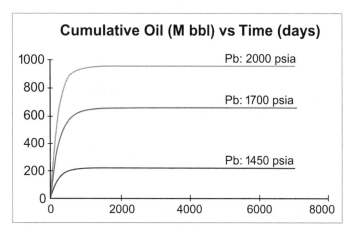

Fig. 3–39. Cumulative oil production for the three cases studied: bubblepoint at 2,000 psia, 1,700 psia, and 1,450 psia. *Courtesy of Gemini Solutions, Inc.*

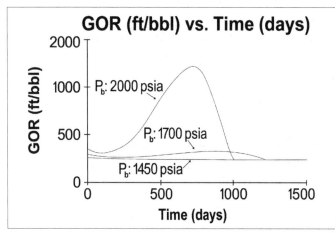

Fig. 3–40. Predicted gas/oil ratio for various bubblepoint pressures. *Courtesy of Gemini Solutions, Inc.*

The sensitivity study further indicates that the initial oil and gas production rates are the highest in the case of light oil having API gravity of 40°. The production rates are the least in the case of heavy oil of 10° API gravity. Although gas evolution does not take place in the reservoir above the bubblepoint pressure, the producing wellstream separates into oil and gas phases under reduced pressure. The contrasts between the initial production rates, as well as the primary recovery periods, are quite significant in the cases studied. Furthermore, it is observed that reservoir pressure declines at a significantly slow rate in the case of heavy oil having 10°API gravity. This is expected, as the least amount of hydrocarbons is produced in a given production period.

During secondary or tertiary recovery operations that strive to produce additional oil by injecting liquid or gas into the reservoir, the gravity of the in-situ fluid may play a crucial role. In the case of a marked contrast between displacing and displaced fluids, gravity override or underride may be encountered, which is detrimental to reservoir performance. Improved and enhanced recoveries of oil, including waterflooding, are discussed in chapters 16 and 17.

Effect of bubblepoint pressure

Since the bubblepoint pressure marks the appearance of a new fluid phase in the reservoir, it usually has a pronounced effect on reservoir performance. In this study, three cases will be considered in which the bubblepoints of the liquid are 1,450 psia, 1,700 psia, and 2,000 psia. The initial reservoir pressure is 3,000 psia, as in the preceding study. Results of the sensitivity study are presented in Figures 3–39 and 3–40. It is observed that the cumulative oil production is the highest when the reservoir fluid has a high bubblepoint (2,000 psia). As the bubblepoint of the reservoir fluid is set to lower values, such as 1,700 or 1,450 psia, oil recovery becomes relatively less. The phenomenon is due to the different reservoir drive mechanisms at work. When the bubblepoint is at a relatively high level, evolution of the gas phase commences early in the life of reservoir. Reservoir fluids are subjected to solution gas drive, which provides significant energy due to the expansion of the evolved gas. In contrast, when the bubblepoint is at 1,450 psia, much of the production takes place above the bubblepoint. The recovery mechanism is based on the energy provided by the compressibility of the oil and rock alone. Gas being highly compressible, solution gas drive provides much more energy to drive reservoir fluids towards the wellbore.

Furthermore, it is interesting to review the gas/oil ratios in the three cases studied (fig. 3–40). When the bubblepoint is at 2,000 psia, a significantly high gas/oil ratio is observed early. Once a sufficient amount of gas is produced, the gas/oil ratio is found to decrease as the liquid phase becomes deficient in volatile hydrocarbons. On the other hand, no surge in the gas/oil ratio can be observed when the bubblepoint is set at a relatively low pressure of 1,450 psia. The observed gas/oil ratio remains essentially flat as the wellstream separates into oil and gas phases under much-reduced pressure. Again, the reason behind the behavior is that the reservoir has much of its production above the bubblepoint pressure, and no gas evolves out of the liquid phase in the reservoir. The case of the reservoir liquid having a bubblepoint pressure of 1,700 psia exhibits an intermediate behavior, in which a relatively small increase in the gas/oil ratio is observed.

Effect of initial reservoir pressure

As mentioned, the reservoir pressure provides the necessary driving energy to produce petroleum fluids. Thus it can be inferred that a relatively high initial reservoir pressure would lead to substantially higher primary production under favorable conditions. Figure 3–41 represents a simulation study in which three undersaturated reservoirs were considered. The initial reservoir pressures were assumed to be 3,000 psia, 2,000 psia, and 1,250 psia, respectively, in the cases studied. Bubblepoint pressure was assumed to be 500 psia in each case so that the primary recovery mechanism is essentially based on depletion drive. The recovery efficiency is predicted to be about 8%–9% when the initial

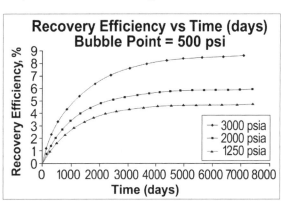

Fig. 3–41. Reservoir production performance with high, intermediate, and low initial pressure. *Courtesy of Gemini Solutions, Inc.*

reservoir pressure is set at 3,000 psia. However, at a low initial pressure of 1,250 psia, recovery efficiency is rather dismal, about 4%. This study highlights the importance of pressure maintenance operations as practiced widely in the industry. Injection of an external fluid, such as water, is typically required to maintain reservoir pressure and augment petroleum recovery.

Key points—effects of fluid properties

Simulation studies were performed to analyze the sensitivity of reservoir performance to the various fluid properties described in this chapter. The following observations are made based on the results of the simulations:

1. The ultimate recovery of less viscous fluid, dominated by lighter hydrocarbons, is notably higher than in the case where the oil is more viscous and is heavier. Initial production rates are also higher for the relatively light crude.

2. Reservoir pressure declines at a greater rate in the case of volatile oil as opposed to heavy oil, since the production of volatiles is accompanied by higher production rates.

3. Oil with a relatively high bubblepoint pressure is expected to produce better than the case where the bubblepoint of the oil is relatively less. This is due to the fact that in the former case, the reservoir pressure declines early below the bubblepoint, leading the way for solution gas drive to produce oil. However, when the bubblepoint pressure of the reservoir fluid is low, the reservoir produces above the bubblepoint by the mechanisms of fluid expansion and pore volume contraction alone for a longer period.

4. In reality, an early pressure maintenance operation by water and/or gas injection is initiated in many reservoirs to ensure that the reservoir produces above the bubblepoint. Experience has shown that the ultimate oil recovery is better when the oil production is based on a pressure maintenance scheme.

5. The production of oil having a high bubblepoint pressure is accompanied by a markedly high gas/oil ratio, as observed from the results of the simulation. This condition is associated with volatile oil reservoirs.

Fluid Properties Required in Reservoir Studies

The following table provides a listing of the oil, gas, and water properties required in classical reservoir studies. However, the list is not comprehensive. Special reservoir studies would require additional information. The following points are emphasized again:

1. In-situ monitoring tools and conventional laboratory studies are the preferred methods in obtaining the various properties of subsurface fluids from a reservoir. All fluid properties determined by these procedures are subjected to rigorous quality assurance and quality control.

2. Correlations are used to estimate fluid properties if field data is not available or is not reliable. It must be borne in mind that the correlations may not provide the desired degree of accuracy in results.

3. If the depositional conditions or the source of oil varied during geologic times, the fluid properties may not be the same in the various layers that are not in communication.

4. In some situations, certain oil properties, including specific gravity, may vary areally within the same formation. Examples of areal variation in fluid properties are found in heavy oil reservoirs occurring at shallow depths.

5. Care must be taken to collect fluid samples that are representative of the subsurface reservoir conditions. With a decrease in pressure, evolution of lighter hydrocarbon components may occur in an oil reservoir. Similarly, heavier components may be lost due to condensation, especially near the wellbore, when fluid samples are collected from a gas condensate reservoir.

Table 3–11. Fluid properties checklist

Topic	Reservoir Fluid Properties	Source of Data
Classification of petroleum reservoirs	Initial pressure, temperature, and composition of fluid Phase behavior of reservoir fluid	Logs, well tests, laboratory studies of collected samples, and in-situ measurements
Estimate of original oil or gas in place by volumetric method	Oil and gas formation volume factors	Differential vaporization studies or applicable correlations
Primary recovery efficiency based on API correlations	Initial pressure, oil formation volume factor, specific gravity, viscosity, and bubblepoint	Laboratory studies, field measurements, and applicable correlations
Reservoir performance studies—material balance	Reservoir pressure, formation volume factor, gas/oil ratio, and compressibility at initial reservoir conditions and at subsequent stages of production	Downhole measurements, laboratory analyses, and applicable correlations
Reservoir performance studies—reservoir simulation (black oil model)	Reservoir pressure, formation volume factor, gas/oil ratio, permeability, relative permeability characteristics, viscosity, and compressibility at initial reservoir conditions and at subsequent stages of production	Downhole measurements, laboratory analyses, and applicable correlations
Reservoir monitoring and evaluation	Reservoir pressure and temperature Water composition during water injection	Downhole measurements and laboratory analyses of collected samples
Design of surface facilities	Fluid composition, gas/oil ratio, presence of impurities	Laboratory studies and in-situ measurements

Honarpour, Nagarajan, and Sampath listed the dynamic rock and fluid properties vital in evaluating improved and enhanced oil recovery processes, described in chapters 16 and 17, respectively, as follows:[53]

Immiscible displacement (e.g., waterflooding):

- Wettability and rock-fluid compatibility
- Imbibition/drainage capillary pressure
- Two-phase relative permeability
- Critical gas saturation

Miscible displacement (e.g., injection of CO_2 or hydrocarbons):
- Physical and thermodynamic properties of fluids
- Wettability and rock-fluid compatibility
- Interfacial tension (IFT)–dependent relative permeability
- Remaining/residual oil saturation

Thermal and solvent-assisted processes for heavy oil (e.g., steamflooding):
- Variation of physical and thermodynamic properties with depth and temperature
- Two- and three-phase relative permeability, effect of temperature, and critical gas saturation
- Potential for emulsion
- Stress sensitivity and potential for compaction
- Potential for formation damage

Gas recycling (gas condensate reservoir):
- Gas relative permeability and capillary pressure
- Composition-dependent fluid properties, dew point, viscosity, condensate/gas ratio (CGR), z factor, interfacial tension, etc.
- Gas condensate relative permeability and effect of capillary number
- Trapped gas saturation

Summing Up

Petroleum reservoir fluids are hydrocarbon compounds having their composition based on two key elements: hydrogen and carbon. The fluids are initially in either a liquid or gas phase in the reservoir at the time of discovery. The subsurface pressure and temperature, and the composition of the petroleum fluid, determine its initial state. Fluid properties usually vary from one reservoir to another. Furthermore, properties may vary within a reservoir from one zone to another or vary areally, depending on geologic setting. As a petroleum horizon is explored at progressively greater depths, accumulations of heavier to lighter crude, followed by gas condensate and gas, are generally encountered with increasing frequency.

Petroleum reservoirs are broadly classified as oil, gas, or gas condensate reservoirs. Crude oil is often referred to as light, intermediate, and heavy, leading to the classification of oil reservoirs. In case of gas condensate reservoirs, condensation of the gas into liquid droplets, followed by revaporization, occurs in the reservoir as the pressure declines. Again, some reservoirs are discovered with both liquid and gas phases present in the porous medium, the oil zone being overlain by a gas cap.

A new category, frequently referred to as unconventional sources, is emerging in the petroleum industry. Prime examples of unconventional sources are oil sands and coalbed methane.

Reservoir engineers frequently refer to oil reservoirs as either saturated or undersaturated. In undersaturated oil reservoirs, the reservoir pressure is above the bubblepoint, and no free gas is present at discovery. However, in saturated oil reservoirs, initial reservoir pressure is at or below the bubblepoint.

Oil reservoirs are further classified according to fluid gravity and volatility. Some reservoirs are referred to as heavy oil reservoirs (API gravity $< 22.3°$), while the others are known either as black oil reservoirs or as volatile oil reservoirs. The reservoir fluid is composed of relatively light hydrocarbons in the latter case. Depending on the types of reservoir fluids and their behavior during production, field development and ultimate fluid recovery vary significantly.

Reservoir oil may typically consist of hundreds of hydrocarbon components of varying molecular weights. Reservoir gas samples exhibit relatively fewer components, chiefly the light hydrocarbons.

Each category of reservoir fluid (namely gas, gas condensate, light crude, heavy oil, etc.) usually exhibits unique behavior during production. This behavior leads to reservoir-specific development strategies, including the planning of future wells and optimization of production rates. It also leads to the selection of a suitable pressure maintenance or enhanced recovery program.

Reservoir pressure and temperature control the fluid properties. Reservoir engineers are interested in the PVT properties of the fluid, i.e., the properties that are dependent on pressure, volume, and temperature. Key fluid properties that influence production behavior are as follows:

- **Composition of the reservoir fluid.** Fractions of light to heavy hydrocarbons.
- **Fluid viscosity.** Determines how mobile a fluid would be in porous media given the reservoir permeability and driving pressure.
- **Saturation pressure.** Marks the appearance of a new fluid phase. The pressure maintenance or gas cycling operation is designed accordingly.
- **Gas/oil ratio.** Influences relative flow of oil and gas in porous media. Well completion and surface facilities are designed based on the expected volume of gas and liquid.
- **Compressibility and specific gravity.** Play an important role in the ultimate recovery of the reservoir in many circumstances. Initial production of an undersaturated oil reservoir is driven by fluid and pore compressibility. Some inclined reservoirs produce by gravity drainage.
- **Multiphase behavior of hydrocarbons.** Includes vaporization, condensation, and revaporization during production.

These fluid properties are generally interrelated.

Reservoir fluid properties are usually obtained by laboratory studies, in-situ measurements, and mathematical correlations. Based on correlations, most petroleum software tools are capable of estimating the properties of reservoir fluids when certain basic information is available.

Pressure and temperature are the most basic properties of reservoir fluids. They are recorded regularly or even continuously in oil and gas wells to closely monitor reservoir performance. Reservoir engineers deal frequently with various measures of pressure, including, but not limited to, initial reservoir pressure, abandonment pressure, wellhead pressure, and bottomhole pressure. The latter may be recorded in two cases: when the well is active, or when it is shut-in.

Besides direct measurements, initial reservoir pressure and temperature are estimated based on their anticipated gradients in a specific region. Knowledge of the specific gravity of the formation water is required in calculating its gradient. Typical values of the formation water gradient range around 0.435–0.5 psi/ft. Estimation of reservoir pressure within oil and gas zones requires knowledge of the oil and gas gradients, respectively, as these are less than the formation water gradient. In a thick gas zone, the deviation from the reservoir pressure as determined by the formation water gradient alone is substantial.

Any discontinuity in fluid pressure between adjacent layers in a reservoir indicates very limited or nonexistent vertical communication between the two.

Petroleum reservoirs are broadly classified into five major categories based on the properties of the fluid. These are given in the following:

1. **Gas reservoir.** In a dry gas reservoir, no liquid phase is produced either in the reservoir or at the surface. However, in a wet gas reservoir, a liquid phase or condensate is produced in the surface facilities under reduced pressure and temperature conditions.
2. **Gas condensate reservoir.** Liquid droplets condense out of the gas phase in the rock pores during production as the reservoir pressure declines below the dew point. The Gas/condensate ratio is typically in the thousands of scf/stb.
3. **Volatile oil reservoir.** Oil production is accompanied by the evolution of a gas phase in significant quantities. Typical GOR is 1200 scf/stb or greater. API gravity of crude is 38° or greater.
4. **Black oil reservoir.** API gravity of crude oil typically ranges between 23° and 38°. Gas evolution from oil takes place in low to moderate quantities with GOR ranging in hundreds of scf/stb. Color of crude is green to black.
5. **Heavy oil reservoir.** Crude oil composition, dominated by heavier and complex hydrocarbons, has API gravity less than 22.3°. Oil having gravity of less than 10° is sometimes referred to as extra-heavy. Oil viscosity is greater than 10 cp and can exceed 10,000 cp.

The categories above are primarily rooted in the relative dominance of light, intermediate, and heavy hydrocarbons that constitute crude oil or natural gas. The light to intermediate components of petroleum tend to be more volatile.

Typically, a reservoir fluid may contain hundreds or even thousands of petroleum compounds of varying complexity and molecular weight. Fluid composition depends on the organic matter, depositional environment, and maturation process of the hydrocarbons, among other factors.

The hydrocarbon compounds commonly found in reservoir fluids are broadly classified as paraffins, cycloparaffins, and aromatics.

There are a number of properties of natural gas that are important in classical reservoir engineering studies. These include, but are not limited to, the composition of the gas mixture, viscosity, specific gravity, compressibility, and the gas formation volume factor. The PVT relationship of natural gas is based on the equation of state for real gases that incorporates a gas deviation factor, or z factor.

Natural gas is typically composed of light hydrocarbons such as methane, ethane, propane, and butane, and methane is the prevalent component. As the percentage of heavier components (C_5 and above) increases progressively, wet gas or gas condensate reservoirs are encountered.

Gas condensate fluid is typically characterized by the following:
- Relatively high gas/oil ratio at the surface, 6,000 cft/bbl or greater.
- Lightly colored tank oil with low gravity, 45°API or higher.
- The methane content of the reservoir fluid is 65% or higher.

Most reservoir studies require knowledge of the in-situ and injected fluid properties. These include, but are not limited to, the following:
- Estimation of the hydrocarbon in place and petroleum reserves
- Analysis of oil phase behavior, as evolution of a gas phase may occur
- Development of a reservoir model to predict ultimate recovery
- Placement and design of future wells
- Evaluation of the reservoir performance in terms of well rates and reservoir pressure
- Design of enhanced recovery operations
- Investigation of early breakthrough of water or gas
- Design and analysis of well tests

Oil properties of interest to reservoir engineers are:
- Viscosity
- API gravity
- Compressibility
- Bubblepoint pressure
- Solution gas/oil ratio
- Formation volume factor

Bubblepoint pressure serves as a milestone in the life of an undersaturated reservoir, since the gas comes out of solution at and below this pressure. Evolved gas eventually drives toward the wellbore with much higher mobility than liquid. Consequently, the oil is likely to become relatively viscous and less mobile.

Above or below the bubblepoint, certain oil properties exhibit distinctly different behaviors. As reservoir pressure declines with production, the viscosity of the oil decreases due to volume expansion above the bubblepoint. Once the bubblepoint is reached, lighter hydrocarbons evolve out of solution, and the oil shrinks in volume.

Major decisions in reservoir engineering require knowledge of the bubblepoint pressure. Initiation of an early pressure maintenance scheme may be necessary to maintain reservoir pressure above the bubblepoint, circumvent gas evolution, and attain maximum oil recovery. If the initial reservoir pressure is below the bubblepoint with a gas cap, gas reinjection may be necessary. It could be used to maintain reservoir pressure at an optimum level in order to produce the nonvolatile heavier hydrocarbons as much as possible.

The solution gas/oil ratio is a measure of the quantity of gas dissolved in the oil, with the volumes of gas and oil measured under standard conditions. The ratio remains unchanged above the bubblepoint pressure, as no dissolved gas can evolve. Once the bubblepoint is reached, the solution gas/oil ratio decreases continuously as free gas is formed in porous media.

The oil formation volume factor is a measure of the degree of reduction or shrinkage of liquid volume. This occurs when the reservoir pressure declines, and lighter hydrocarbons evolve out of solution and form a free gas phase. This factor is required to convert barrels of oil in the reservoir, including any dissolved gas under elevated pressure, to stock-tank barrels of oil at the surface. As long as the reservoir produces above the bubblepoint, fluid volume expansion occurs within pores due to the decline in pressure resulting in slight increase of the oil formation volume factor. It reaches its maximum value at the bubblepoint and then decreases continuously as the increasingly volatile hydrocarbons are liberated.

Throughout the productive life of a reservoir, the individual oil properties may alter significantly as the reservoir pressure declines and the in-situ fluid composition changes as a consequence. During the decline in reservoir pressure, the oil properties are controlled by compositional changes as gas evolves out of the liquid phase as the bubblepoint pressure is reached.

Knowledge of the formation water properties is needed in various reservoir studies, along with knowledge of the oil and gas properties. Hydrocarbon gases are soluble in formation water under reservoir pressure and temperature conditions. Properties of interest to reservoir engineers are the salinity, viscosity, density, and compressibility of the formation water. Various charts and correlations are widely available to estimate the important water properties.

Notable laboratory studies involve flash and differential vaporization of the lighter hydrocarbons from the reservoir oil samples. In flash vaporization (constant composition expansion), the evolved gas is kept in contact with the liquid phase at all times in a closed chamber. The process is similar to what takes place in surface separators.

In differential vaporization, the gaseous hydrocarbons are removed as soon as they evolve from the solution. This emulates the rapid gas movement towards the wells in porous media immediately following the dissolution of the gas.

Fluid samples obtained from gas condensate reservoirs are subjected to constant composition expansion and constant volume depletion tests. The former is similar to a flash test conducted on oil samples. A constant volume depletion test determines important condensate properties, such as a two-phase z factor, needed to predict reservoir performance.

The study of vapor/liquid equilibria calculations is important. It leads to the determination of the volume and composition of the gas and liquid phases when the crude oil is separated or flashed in surface facilities under reduced pressure and temperature. The objectives of the analysis include the attainment of optimized operating conditions at the surface facilities and calculation of the important fluid properties under reservoir conditions. Bubblepoint and dewpoint pressures of a hydrocarbon mixture can also be obtained.

Several equations of state are available to calculate equilibrium constants of hydrocarbon components. These are utilized extensively in the compositional simulation of volatile oil or gas condensate production from a reservoir. These correlations are usually tuned by laboratory findings to improve the accuracy in the results obtained by simulation study.

Knowledge of the phase behavior of petroleum fluids is critical in reservoir studies. Several factors are dictated by the phase behavior of the reservoir fluid. These include petroleum reservoir type (such as volatile oil, heavy oil, or gas condensate reservoir) and producing mechanisms (such as volumetric or solution gas drive). The expected recovery efficiency (higher for volatile oil than heavy oil, etc.) is also dictated by the phase behavior of the reservoir fluid. Phase diagrams have unique characteristics depending on the fluid system considered.

Important features of phase diagrams are given in the following:

- **Single- and two-phase regions.** A two-phase envelope determines whether the fluid is in one phase (either liquid or gas) or in two phases (oil and gas, or gas and condensate). Single-phase regions are above and to the right of the envelope, indicating the state of relatively high pressure or temperature, or both, in the reservoir.

- **Liquid saturation lines.** These lines within the two-phase envelope indicate the relative amounts of oil and gas phases under the prevailing conditions.

- **Critical point, bubblepoint, and dewpoint curves.** The two-phase envelope mentioned above consists of a bubblepoint curve and a dewpoint curve. They converge at the critical point (C), where the liquid and vapor phases are indistinguishable and are in equilibrium.

- **Cricondentherm and cricondenbar.** The highest temperature above which hydrocarbon fluids can exist solely as gas is referred to as the cricondentherm. Similarly, the highest pressure beyond which fluids are completely in a liquid phase is known as the cricondenbar.

Undersaturated oil reservoirs as well as gas reservoirs are located in the single-phase region of the phase diagram. Oil reservoirs are to the left of the critical point, and gas reservoirs are located to the right. Volatile oil reservoirs occur closer to the critical point.

Saturated oil reservoirs below the bubblepoint appear within the two-phase envelope. A gas cap usually overlies the oil zone. Saturated oil reservoirs, at the bubblepoint and with negligible free gas, are located on the two-phase line.

With a decline in reservoir pressure, undersaturated oil and gas condensate reservoirs enter the two-phase envelope. A second phase (gas or condensate) appears in the porous media. The net effect is the rapid transport of the less viscous hydrocarbons towards the wells, leaving the more viscous and heavier components behind. This phenomenon constitutes a perennial challenge in reservoir engineering for the optimization of recovery.

Sensitivity studies based on reservoir simulation indicate that the ultimate recovery of volatile oil, dominated by the lighter hydrocarbons, is likely to be much higher than heavy oil. The initial production rate of the less viscous crude (high API gravity) is found to be noticeably higher.

The reservoir pressure declines at a greater rate in the case of volatile oil as opposed to heavy oil, since the production of volatiles is accompanied by higher production rates.

Oil with a relatively high bubblepoint pressure is found to produce more by solution gas drive, as the reservoir pressure declines below the bubblepoint earlier. In the latter case, however, the drive mechanism is depletion for a relatively long production time. However, in many reservoirs, a pressure maintenance operation is initiated before the bubblepoint is reached in order to attain optimum oil recovery. Furthermore, the production of oil having a high bubblepoint pressure is accompanied by a markedly high gas/oil ratio, as observed from the results of simulation.

Class Assignments

Questions

1. Describe the general classification of petroleum reservoirs. What factors determine the type of reservoir? Describe any major changes in fluid characteristics that may accompany production in each type of reservoir. Name one large field from each category, along with its location and reserve, by conducting a literature review.

2. Why does the general strategy to develop a petroleum reservoir depend on fluid properties? Consider a company that has recently discovered two oil reservoirs located in two basins having crude oil specific gravities of 25°API and 45°API, respectively. The latter reservoir is found to have a gas cap. Based on this information alone, briefly outline a development strategy in each case.

3. As more information becomes available, it is found that the reservoir with 25°API oil in question 2 is naturally fractured. On the other hand, the reservoir with 45° oil is suspected to have rather low permeability, ranging between 3 mD and 10 mD. Would this additional information change the perception of the two reservoirs in terms of ultimate recovery, leading to a revised field development strategy? Why or why not?

4. A gas condensate reservoir in the North Slope, Alaska, is discovered at a depth of 10,780 ft. Reservoirs in the same basin are known to be significantly overpressured. What would be the minimum initial reservoir pressure that would be anticipated? Conduct a literature review to include pressure and temperature gradients associated with similar reservoirs in the region.

5. Describe the typical fluid characteristics that would be expected in the reservoir described in question 4. What are the important phenomena that may occur during production from a gas condensate reservoir? Discuss a development plan to attain optimum recovery from the reservoir.

6. Why are pressure measurements important in a reservoir? How would the initial reservoir pressure in a new field be measured? Describe various measures of pressure that reservoir engineers work with on a regular basis. Include the frequency of measurements for each category that would be required to effectively manage a reservoir. How do pressure measurements aid in characterizing a stratified reservoir?

7. Distinguish between static and flowing bottomhole pressures. Why is the latter relatively less in a producer? There are plans to shut-in a well to record the static bottomhole pressure. Make a qualitative comparison of the shut-in time necessary to achieve a stabilized bottomhole pressure in the following cases:

   (a) Reservoir with a single well. Average permeability of the formation is 5 mD.

   (b) Reservoir with a single well. Average permeability of the formation is 50 mD.

   (c) Reservoir with several active injectors and producers located nearby. Average permeability of the formation is 500 mD.

   In which case or cases can the recorded shut-in pressure be regarded as the average reservoir pressure? Assume the values of the formation thickness are about the same in all three cases.

8. What are the chief components of oil and gas? Briefly describe the composition and structure of the chemical compounds typically found in petroleum fluids, including any impurities. What is the range of specific gravities of common chemical compounds in the fluids? How many chemical components can be found in a crude oil sample?

9. What is the ideal gas law, and how it is modified to account for the behavior of real gases? Why and how are the pseudoreduced properties of natural gas calculated?

10. Why is knowledge of the dew point, gas/oil ratio, and gas condensate formation volume factor necessary to effectively develop and produce a gas condensate reservoir? In a gas condensate reservoir, where would condensation be expected to occur first—near the reservoir boundary or at the producing wells? What effects might naturally occurring fractures have on condensate recovery?

11. A company has made an oil discovery. The reservoir does not appear to have a gas cap. What fluid properties would be important to know, and why? List at least six tasks in which this data could be utilized, with detailed descriptions.

12. If the reservoir has multiple layers, in what circumstances would the fluid properties be expected to vary between layers? How might this information aid in effectively producing the reservoir?

13. Name the most significant phenomenon that can take place in an undersaturated oil reservoir as the pressure declines with production. Define the following fluid properties and discuss their significance in effectively producing the reservoir:

   (a) Bubblepoint pressure

   (b) Viscosity of the oil

   (c) Gas/oil ratio

   (d) Oil gravity

   (e) Compressibilities of the oil and gas

   The rate of change in pressure in an undersaturated reservoir usually accelerates once the pressure declines below the bubblepoint. Why or why not?

14. Describe any changes in the following fluid properties both above and below the bubblepoint pressure:

    (a) Liquid composition

    (b) Viscosities of the oil and gas

    (c) Oil formation volume factor

    (d) Two-phase formation volume factor

    (e) Liquid density

Are these properties interrelated?

15. Once the reservoir pressure decreases below the bubblepoint, would oil be produced in as much quantity as above the bubblepoint? Explain by identifying the fluid properties mentioned in question 13, which are responsible for any changes in fluid flow dynamics within the reservoir. How does critical gas saturation affect fluid flow below the bubblepoint?

16. Distinguish between the following, with clear illustrations:

    (a) Bubblepoint and dewpoint pressures

    (b) Formation volume factors of volatile and nonvolatile crude samples

    (c) Formation volume factors at and below the bubblepoint pressure

    (d) Gas formation volume factor and gas expansion factor

    (e) Single-phase and two-phase formation volume factor

    (f) Single-phase and two-phase z factor

    (g) Solution gas/oil ratio of light and heavy crude

    (h) Solution gas/oil ratio and producing gas/oil ratio

    (i) Cumulative gas/oil ratio and producing gas/oil ratio

    (j) Phase diagrams of wet gas and gas condensate

    (k) Gas/oil ratio and gas/condensate ratio

    (l) Reservoir pressure in oil and gas zones

    (m) Pore pressure and overburden pressure

    (n) Reservoir barrels and stock-tank barrels

    (o) Initial and average reservoir pressure

    (p) Static and flowing bottomhole pressure

    (q) kPa and MPa

    (r) Fluid gradient and fracture gradient

    (s) Critical point and critical saturation

    (t) Cricondentherm and cricondenbar

    (u) Single-phase region and two-phase region in a phase diagram

    (v) Black oil and volatile oil

    (w) Total compressibility and effective compressibility

    (x) Density of volatile oil in the reservoir and at surface facilities

    (y) Surface and in-situ measurement of fluid properties

    (z) API gravity and specific gravity (water = 1)

17. Why can well spacings in gas reservoirs be larger than in oil reservoirs, given the same rock characteristics? Explain.

18. Describe the usual sources of fluid properties data. In instances where limited or no data is available, what recourse does a reservoir engineer have to conduct a meaningful reservoir study?

19. Tabulate the data requirements and possible sources in estimating the following fluid properties based on various correlations:

    (a) Bubblepoint pressure

    (b) Solution gas/oil ratio

    (c) Oil formation volume factor

    (d) Viscosities of live and dead oil

    (e) Oil compressibility

20. Describe the processes related to constant composition expansion (CCE), differential vaporization, and constant volume depletion (CVD). Tabulate the parameters that can be obtained from each test, with brief comments explaining how the results are utilized in reservoir studies.

21. Define relative volume as reported in the results of a laboratory study. What test leads to the determination of relative volumes? Why does the relative volume increase monotonically above and below the bubblepoint with decreasing pressure? What information can be obtained from the relative volume data?

22. Why is the study of vapor/liquid equilibria important in reservoir engineering? Describe the methodology used in calculating various fluid properties based on vapor/liquid equilibrium relationships. Define convergence pressure and explain its role in these calculations.

Exercises

3.1. What are the gas deviation factor and the gas formation volume factor? Discuss their significance in calculating fluid properties and estimating reserves. Develop a computer-assisted method to calculate the gas deviation factor and the gas formation volume factor in rb/Mscf, given reservoir pressure and temperature and the specific gravity of the gas. Compare this work with published data.

3.2. A low permeability gas reservoir is discovered at a depth of 9,250 ft (TVD). The specific gravity of the natural gas is found to be 0.669. Based on this information alone, can the reserve be calculated in MMscf per acre-ft? If so, perform the calculation, showing all the steps. If not, clearly state any other assumptions necessary in the estimation of the reserves. In case of an overpressured formation, will the estimate change?

3.3. Based on the typical ranges of pressure, temperature, and gas composition, including impurities as encountered in dry gas reservoirs, calculate the most likely ranges of in-situ gas viscosity and compressibility. Illustrate the results with the aid of appropriate plots. How would the values be different in a gas condensate reservoir? Explain.

3.4. Develop a computer-aided application to estimate the essential PVT properties of the in-situ oil within the common ranges of pressure, temperature, oil, and gas gravities. Sought fluid properties are:

    (a) Bubblepoint pressure

    (b) Solution gas/oil ratio above and below the bubblepoint pressure computed in (a).

    (c) Oil formation volume factor

    (d) Two-phase oil formation volume factor

    (e) Viscosity of oil

    (f) Compressibility of oil

    Use at least one correlation for each. Compare this work with published results.

3.5. Using appropriate correlations described earlier in the chapter, plot the oil formation volume factor and viscosity of a 40°API crude as a function of declining pressure. Consider the following cases of gas solubility at the bubblepoint:

   (a) 500 scf/stb

   (b) 700 scf/stb

   (c) 900 scf/stb

The starting point would be the initial reservoir pressure of 4,010 psia. What is the estimated bubblepoint pressure of the reservoir fluid in each case? The specific gravity of the solution gas is 0.674. Assume the reservoir temperature to be 188°F.

3.6. The following data for a reservoir is given: Average porosity of the formation is 28.2%. Assume the following fluid saturations and properties for all cases: $s_o = 0.6$; $s_g = 0.15$; oil gravity $= 41°API$, and the specific gravity of the gas $= 0.768$. Make any other assumptions necessary.

   (a) Estimate the total compressibility for the following reservoir conditions:

      i.   1,500 psia, 160°F

      ii. 2,000 psia, 170°F

      iii. 3,000 psia, 180°F

   (b) Recalculate the total compressibility values in (a) by assuming that only two phases are present in the reservoir: oil and water. Compare the results and explain.

   (c) What is the effective compressibility of the oil in the three cases considered in (b)?

3.7. Calculate the effluent liquid composition from a two-stage separator. The operating pressure and temperature of the first stage are 125 psia and 60°F, respectively. The second stage operates at 15 psia. The composition of the feed is as follows:

   Ethane = 0.51.
   Propane = 0.32.
   n-Butane = 0.11.
   n-Pentane = 0.04.
   Hexane = 0.02.

3.8. Describe the construction, capabilities, installation, and integration of a permanent downhole gauge. Explain how an integrated asset team could benefit from the deployment of a permanent downhole gauge in the following:

   (a) Production well

   (b) Observation well

   (c) Reservoir producing below the bubblepoint

References

1. Dake, L. P. 1978. *Fundamentals of Reservoir Engineering*. Amsterdam: Elsevier Scientific Publishing Co.
2. Selley, R. C. 1997. *Elements of Petroleum Geology*. 2nd ed. Academic Press.
3. McCain, Jr., W. D. 1990. *Properties of Petroleum Fluids*. 2nd ed. Tulsa: PennWell.
4. McCain, Jr., W. D. 1990.
5. McCain, Jr., W. D. 1990.
6. Standing, M. B., and D. L. Katz. 1942. Density of natural gases. *Transactions*. AIME. Vol. 146.
7. Brown, G. G., D. L. Katz, G. B. Oberfell, and R. C. Alden. 1948. *Natural Gasoline and Volatile Hydrocarbons*. Tulsa, OK: NGAA. p. 44.
8. Yarborough, L., and K. R. Hall. 1974. How to solve equation of state for z-factor. *Oil & Gas Journal*. February 18.
9. Dake, L. P. 1978.
10. Wichert, E., and K. Aziz. 1972. Calculate Z's for sour gases. *Hydrocarbon Processing*. May. Vol. 51, no. 5:, 119–122.
11. Lee, W. J., and R. A. Wattenbarger. 2002. *Gas Reservoir Engineering*. 2nd ed. Richardson, TX: Society of Petroleum Engineers.

12. Mian, M. A. 1992. *Petroleum Engineering Handbook for the Practicing Engineer*. Tulsa: PennWell Books.

13. Ahmed, T. H. 2001. *Reservoir Engineering Handbook*. 2nd ed. Houston: Gulf Professional Publishing Co.

14. Sutton, R. P. 1985. Compressibility factors of high-molecular-weight reservoir gases. SPE Paper #14625. Society of Petroleum Engineers Annual Technical Meeting and Exhibition, Las Vegas, September 22–25.

15. Standing, M. B. 1977. *Volumetric and Phase Behavior of Oil Field Hydrocarbon Systems*. Richardson, TX: Society of Petroleum Engineers., 125–125.

16. Lee, W. J., and R. A. Wattenbarger. 2002.

17. Bradley, H. B., ed. 1992. *Petroleum Engineering Handbook*. Richardson, TX: Society of Petroleum Engineers.

18. Bradley, H. B., ed. 1992.

19. Matthews, T. A., C. H. Roland, and D. L. Katz. 1942. High pressure gas measurements. *Proceedings*. National Gas Association of America., 41–51.

20. McCain, Jr., W. D. 1990.

21. Standing, M. B. 1947. A pressure-volume-temperature correlation for mixtures of California oils and gases. In *Drilling and Production Practices*. Washington, DC: American Petroleum Institute., 285–287.

22. Ahmed, T. H., 2001.

23. Tiab, D., and E. C. Donaldson. 1996. *Petrophysics: Theory and Practice of Measuring Rock and Fluid Transport Properties*. Houston: Gulf Publishing Co.

24. Merlin Simulation Software. 2005. Houston: Gemini Solutions Inc.

25. Dake, L. P. 1978.

26. McCain, Jr., W. D. 1990.

27. Ahmed, T. H., 2001.

28. Amyx, J. W., D. M. Bass, Jr., and R. L. Whiting. 1960. *Petroleum Reservoir Engineering*. New York: McGraw-Hill.

29. Van Dusen, A., S. Williams, F. H. Fadness, and J. Irving-Fortesque. 2003. Determination of hydrocarbon properties by optical analysis during wireline fluid sampling. SPE Paper #85753. In *SPE Reservoir Evaluation and Engineering*. Richardson, TX: Society of Petroleum Engineers., 286–292.

30. Elshahawi, H., M. Hashem, C. Dong, P. Hegeman, O. C. Mullins, G. Fujisawa, and S. Betancourt. 2005. In-situ characterization of formation-fluid samples—case studies. *Journal of Petroleum Technology*. February.

31. Standing, M. B. 1977.

32. Glaso, O. 1980. Generalized pressure-volume-temperature correlations. *Journal of Petroleum Technology*. May:, 785–795.

33. Marhoun, M. A. 1988. PVT correlation for Middle East crude oils. *Journal of Petroleum Technology*. May:, 650–665.

34. Petrosky, Jr., G. E., and F. Farshad. 1993. Pressure-volume-temperature correlations for Gulf of Mexico crude oils. SPE Paper #26644. Presented at the 68th Annual SPE Technical Conference and Exhibition, Houston, TX, October, 3–6.

35. Mian, M. A. 1992.

36. Ahmed, T. H. 2001.

37. Towler, B. F. 2002. *Fundamental Principles of Reservoir Engineering*. SPE Textbook Series. Vol. 8. Richardson, TX: Society of Petroleum Engineers.

38. De Ghetto, G., F. Paone, and M. Villa. 1994. Reliability analysis on PVT correlations. SPE Paper #28904. European Petroleum Conference, London, U.K., October 25–27.

39. Beal, C. 1946. The viscosity of air, water, natural gas, crude oil and its associated gases at oil field temperatures and pressures. *Transactions*. AIME. Vol. 165:, 94–112.

40. Chew, J., and C. A. Connally, Jr. 1959. A viscosity correlation for gas saturated crude oils. Transactions of AIME. Vol. 216:, 270–275.

41. Vasquez, M., and D. Beggs. 1980. Correlations for fluid physical properties prediction. *Journal of Petroleum Technology*. June.

42. McCain, Jr., W. D., J. B. Rollins, and A. J. Villena-Lanzi. 1988. The coefficient of isothermal compressibility of black oils at pressures below the bubblepoint. *Society of Petroleum Engineers Formation Evaluation*. Sept.

43. Vasquez, M., and D. Beggs. 1980.

44. McCain, Jr., W. D. 1990.

45. Ahmed, T. H. 2001.

46. Whitson, C. H., and M. R. Brule. 2000. *Phase Behavior*. SPE Monograph. Vol. 20. Richardson, TX: Society of Petroleum Engineers.

47. Ahmed, T. H. 2001.

48. Elshahawi, H., M. Hashem, C. Dong, P. Hegeman, O. C. Mullins, G. Fujisawa, and S. Betancourt. 2005.

49. Matthews, C. S., and D. G. Russell. 1967. *Pressure Buildup and Flow Tests in Wells*. SPE Monograph Series. Vol. 1. Dallas: Society of Petroleum Engineers.

50. Dodson, C. R., and Standing, M. B. 1944. Pressure-volume-temperature and solubility relations for natural gas-water mixtures. In *Drilling and Production Practice*. Washington, DC: American Petroleum Institute., 173–179.

51. Dodson, C. R., and Standing, M. B. 1944.

52. Dodson, C. R., and Standing, M. B. 1944.

53. Honarpour, M. M., N. R. Nagarajan, and K. Sampath. 2006. Rock/fluid characterization and their integration—implications on reservoir management. *Journal of Petroleum Technology*. September., 120–131.

4 · Fundamentals of Fluid Flow in Petroleum Reservoirs and Applications

Introduction

The foundation of reservoir engineering is based upon the principles of fluid flow through porous media in rocks. The flow of fluids in subsurface reservoirs is characterized by the following:

- **Condition of flow.** Fluid flow in porous media can be steady state, unsteady state, or pseudosteady state. These relate to the changes in the pressure trend of the reservoir fluids during flow and are described later in the chapter. Another classification is whether the flow is laminar or turbulent. The flow of oil in porous media is laminar. In the vicinity of gas wells, however, turbulence may develop due to high fluid velocity.

- **Flow geometry.** In reservoir studies, including reservoir simulation, fluid flow is conceptualized and modeled as linear (x, y, or z), radial (r), spherical (r and z), and 3-D (x, y, and z) in direction. In certain circumstances, fluid flow in porous media is studied in a 2-D cross-sectional plane (horizontal or vertical). Most analytic models are developed on the basis of simplified flow geometry, such as radial and linear. However, robust numerical simulators are based on 3-D flow of fluids in reservoirs coupled with production or injection through the wellbore.

- **State of fluids.** Significant changes in reservoir pressure are encountered during production and injection, influencing the flow of fluids in porous media. Fluids can be described as compressible, slightly compressible, or incompressible. Oil is generally considered to be slightly compressible. Natural gas is treated as a compressible fluid.

- **Phases of fluid flowing through the porous media.** These include the following:
  - Single phase (oil or gas or water). An example would be oil production from an undersaturated reservoir without any water cut.
  - Two-phase (oil and gas, oil and water, or gas and water). Examples are an oil reservoir producing below the bubblepoint or a gas reservoir producing both gas and liquid condensate.
  - Three-phase (oil, gas, and water). Certain reservoirs produce oil, gas, and water concurrently. Three-phase flow can be encountered in a saturated oil reservoir under water injection, for instance.

In multiphase flow, individual fluid saturation is a strong function of effective permeability to the specific fluid phase.

The learning objectives in this chapter are the following:
- Fluid flow through porous media, involving:
  - Steady-, pseudosteady-, and unsteady-state flow
  - Compressible, slightly compressible, and incompressible fluids
  - Radial, pseudoradial, linear, and spherical flow
  - Flow through fractures
  - Multiphase fluid flow
- Mechanisms of fluid flow in reservoirs
- Well productivity index
  - Vertical and horizontal wells
- Water and gas coning in the vicinity of oil producers

- Immiscible displacement, such as water displacing oil during waterflooding
- Mobility ratio of fluids
- Water influx from aquifers
  - Steady and unsteady state
- Applications of fluid flow equations in well and reservoir management
- Sensitivity analyses of rock and fluid properties to reservoir or well performance
- Example problems and class exercises

The following examples are provided in the chapter in order to gain insight into the various processes related to the flow of fluids in porous media:

- Application of the diffusivity equation: reservoir characterization
- Prediction of rate decline and cumulative oil production for a well producing at a constant bottomhole pressure in a bounded reservoir
- Transient flow characteristics in formations having the same transmissivity but dissimilar rock permeability
- Analysis of fluid flow in a dipping formation: concept of fluid potential function
- Estimation of the pressure at the reservoir boundary: compressible and slightly compressible fluids
- Prediction of rate for a planned well in a gas reservoir
- Estimation of the average reservoir pressure under pseudosteady-state conditions
- Effects of skin factor, shape of drainage area, and well configuration on average reservoir pressure determination
- Performance prediction of a future well under uncertainty
- Calculation of the pseudopressure function used in the analyses of compressible fluid flow based on numerical integration
- Comparison between pseudopressure and pressure-squared approach
- Estimation of fracture permeability
- Evaluation of the injectivity index of a well for waterflood design
- Analysis of the productivity index and production rate of a horizontal well
- Comparison of the productivity indices between horizontal and vertical wells
- Horizontal well performance: sensitivity analysis
- Effects of permeability anisotropy and oil viscosity on water coning
- Calculation of water influx volume: small aquifer
- Water influx based on unsteady-state aquifer model

Moreover, many important concepts related to the flow of fluids in petroleum reservoirs, including analytic reservoir models, are discussed throughout the book, in addition to what is presented in this chapter. Some of these are listed in the following:

- Chapter 2: Darcy's law and the computation of rock permeability
- Chapter 5: the diffusivity equation and well test analysis
- Chapter 16: immiscible displacement and waterflood performance

This chapter primarily focuses on reservoir engineering applications of basic fluid flow equations. Detailed treatment, including derivations, of the fluid flow equations in porous media is widely available in the literature.

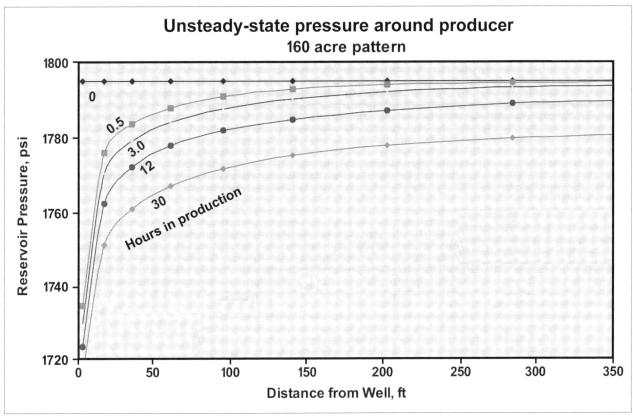

Fig. 4–1. Time-lapse study of a pressure profile in the reservoir as the well is produced. The flow characteristic is "infinite-acting" at first, until the effects of the reservoir boundary are felt. A pseudosteady-state flow condition develops eventually, and the rate of decline in pressure is the same everywhere in the drainage area. *Courtesy of Gemini Solutions, Inc.*

Important Concepts in Fluid Flow through Porous Media

Unsteady-state, steady-state, and pseudosteady-state flow

An important characterization of fluid flow relates to how the reservoir pressure changes in time and space during production, injection, and due to the boundary effects, including the influence by an aquifer. Figure 4–1 shows a time-lapse study of declining pressure in the vicinity of the wellbore, based on simulation of reservoir pressure around a well producing from a drainage area of 160 acres. Unsteady-state flow is encountered as soon as a well is opened for production in a reservoir where equilibrium in pressure prevailed prior to production. The rate of change in the reservoir pressure is the greatest in the immediate vicinity of the wellbore.

During the earliest time period, pressure at the outer boundary does not decline. In other words, no boundary influence is evident on the well, and it behaves in a manner as if it is producing from an infinite reservoir. Under these conditions, the reservoir is "infinite acting" and fluid flow in porous media is characterized as unsteady state or transient. As pressure and flow rate change in both time and location, unsteady-state flow can be described as follows:

$$\frac{\partial p(x,y)}{\partial t} = f(t) \tag{4.1}$$

where
    p is the fluid pressure at a location (x,y) in a 2-D flow geometry, and
    f(t) is a function of time.

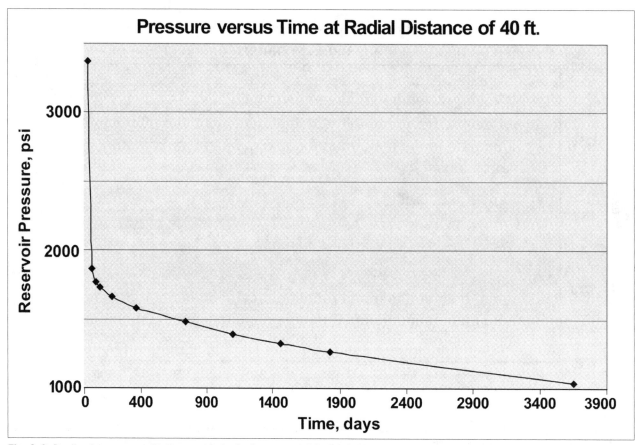

Fig. 4–2. Decline in pressure with time at a point in the reservoir following well production. A change in slope may indicate the transition from one flow characteristic to another. Changes in pressure trends can be identified and analyzed, as discussed in chapter 5. *Courtesy of Gemini Solutions, Inc.*

Depending on the pressure support by an aquifer at the outer boundary, a steady-state flow condition can eventually develop once the boundary effects are felt at sufficiently long times. Pressure support in a reservoir can be provided by fluid injection or the presence of a gas cap as well. In essence, a constant pressure boundary condition is a prerequisite for steady-state flow. The pressure profile of the reservoir fluid does not change with time once a steady-state flow condition is attained in the drainage area. Consequently, the reservoir pressure at a specific location does not change with time. Steady-state flow in a two-dimensional plane (x,y) can be characterized as follows:

$$\frac{\partial p(x,y)}{\partial t} = 0 \qquad\qquad (4.2)$$

However, for well production to occur, finite fluid pressure gradients directed towards the wellbore must exist:

$$\frac{\partial p}{\partial x} \neq 0 , \frac{\partial p}{\partial y} \neq 0 \qquad\qquad (4.3)$$

When a steady-state condition prevails, the pressure gradient between two points is constant over time. Last, but not least, it must be borne in mind that a transitory flow period is likely to be observed before pseudosteady-state or steady-state flow characteristics can develop fully. This is sometimes referred to as late transient flow. Attainment of true steady-state condition is not very common in reservoirs for a number of reasons. These include the individual scheduling of injectors and producers, unsteady well rates, and the uncertainties involved at the drainage area boundary. Such factors often lead to certain changes in the pressure with time at a given location.

In other scenarios where the reservoir is bounded, the effects of the "no-flow" boundary become evident as the transients in pressure propagate outward from the well to the boundary. Consequently, the rate at which the reservoir pressure declines due to production becomes the same everywhere within the reservoir. This is referred to as pseudosteady-state flow, which can be characterized by the following:

$$\frac{\partial p(x,y)}{\partial t} = C \qquad\qquad (4.4)$$

where

C is a constant. In oilfield units, the unit of C is psi/d.

It should be borne in mind that although the *rate* of change in pressure is constant, the *pressure* of the flowing fluid is by no means steady and does decline with time. The time to reach pseudosteady-state flow following the start of well production depends on the rock characteristics, fluid properties, and well drainage geometry. The equation describing the decline in pressure during pseudosteady-state flow, along with the equation to estimate the time to reach pseudosteady-state flow condition from an unsteady-state condition, is provided later. Figure 4–2 shows the numerical simulation of the usual decline characteristics of reservoir pressure in a bounded reservoir with time at a distance of 40 ft from the wellbore following production. A noticeable change in slope occurs at the transition from one flow characteristic to another within a few hundred days, as the effects of the boundary are felt. In a developed field, where each well produces from an individual drainage area, pseudosteady-state flow conditions can be evident.

The three distinct conditions (or regimes) of fluid flow are identifiable and interpretable in pressure response data obtained during well tests, as illustrated in chapter 5. Well tests are designed to gather valuable information about the well and various reservoir characteristics based on the fluid flow regimes. The basis of interpretation is the mathematical models pertaining to fluid flow under unsteady-state, pseudosteady-state, and other conditions. The fundamentals of classical well test theory, based on the radial diffusivity equation, are discussed in this chapter and in chapter 5.

Incompressible, Slightly Compressible, and Compressible Fluids

In order to discuss the flow characteristics of incompressible, compressible, and slightly compressible fluids, the following general explanations are given.

Incompressible fluid. The volume and density of incompressible fluids remain unchanged following any increase or decrease in fluid pressure (or temperature) encountered in a reservoir following production and injection.

Slightly compressible fluid. The fluid volume and density change with a change in pressure according to the following approximation:[1]

$$V = V_{Ref} [1 + c(p_{Ref} - p)] \qquad\qquad (4.5)$$
$$\rho = \rho_{Ref} [1 - c(p_{Ref} - p)] \qquad\qquad (4.6)$$

where

V_{Ref} = fluid volume at a reference pressure, pRef,

p_{Ref} = reference or initial pressure,

ρ_{Ref} = density at the reference pressure, and

c = the compressibility of the reservoir fluid as defined in chapter 3.

The value of c is very small for slightly compressible fluids, usually in the order of 10^{-6} psi^{-1}.

Compressible fluid. Natural gas being a compressible fluid, the following definition of compressibility is used in deriving flow equations of the gas phase in porous media:

$$c_g = \frac{1}{p} - \frac{1}{z}\left(\frac{dz}{dp}\right)_T \qquad\qquad (4.7)$$

where

z = the gas deviation factor.

Equation 4.7 is based on the definition of compressibility in terms of fluid density as a function of pressure. The density, in turn, is a function of pressure, temperature, and the gas deviation factor for real gases. The compressibility of natural gases is discussed in chapter 3.

During compression or expansion, the physical properties of a compressible fluid are observed to vary significantly, which must be taken into account in performing the relevant studies. A stabilized pressure profile of compressible and incompressible fluids during production is presented later.

Radial, Pseudoradial, Spherical, and Linear Flow

Fluids in oil and gas reservoirs exhibit various flow patterns that must be identified in order to conduct a meaningful study of reservoir dynamics. Major fluid flow patterns observed in specific circumstances are described in the following sections.

Radial flow

Whether for a single well producing in a newly discovered reservoir, or for several wells producing from their own drainage area in a developed field, the flow pattern around a well is usually visualized as radial. The presence of geologic heterogeneities, such as a sealing fault or a pinchout located nearby, or severe imbalances in production and injection, tend to distort the radial flow pattern. Other types of fluid flow patterns in the porous media are not uncommon. These are described briefly in the following.

Pseudoradial flow

At the early stages of production from a hydraulically fractured well, the fluid flow pattern is not radial due to the geometry of the fracture. However, pseudoradial flow is found to develop eventually, as depicted in chapter 5. Similarly, a late radial flow pattern emerges around a horizontal wellbore after sufficient passage of time. (Other flow patterns that may be observed in a horizontal well are illustrated in chapter 5.) The development of the radial flow pattern in hydraulically fractured and horizontal wells can be identified from the interpretation of well tests.

Linear flow

During transient testing of a hydraulically fractured well, linear flow is encountered in the early stage, as the fracture geometry is linear. Horizontal well testing usually exhibits a linear flow regime at some stage of the test due to the long well trajectory. Linear flow may also be observed at late stages of a well test in channel-shaped reservoirs due to the geometry of the reservoir boundaries.

Spherical flow

Spherical flow typically develops in partially completed wells where fluid is drawn to the well across the entire thickness of the permeable formation. Spherical flow is also observed where a pinchout boundary is present.

Cone-shaped flow

Referred to as water coning in oil wells, cone-shaped flow may arise in the immediate vicinity of the well due to the presence of movable water. Similarly, gas coning may also be observed if a gas cap is present. The phenomenon of coning, triggered by a high rate of production, is illustrated later in the chapter.

It is again noted that the various flow patterns (linear, radial, spherical, and others) leave distinct signatures in pressure response plots obtained during well tests. These are identified and analyzed subsequently in order to gather valuable information about the well, rock characteristics, and reservoir boundaries. Well test interpretation is treated in detail in chapter 5.

Driving Forces and Mechanisms of Flow

The following primarily influence fluid flow in a reservoir:

- Viscous forces
- Gravity drainage
- Capillary effects

In most circumstances, viscous forces dominate as various fluids are produced from (or injected into) a petroleum reservoir. Under viscous flow condition, the flow rate of the fluid is laminar and is proportional to the imposed pressure gradient that exists in the reservoir. However, some reservoirs produce predominantly by the mechanism of gravity drainage. Examples include dipping hydrocarbon-bearing formations and cases where gas is injected in updip wells to augment recovery. Capillary forces arise due the surface tension that exists between the fluid phase and the surface of the pore walls, as described in chapter 2. These effects are most evident in reservoirs indicating long oil-water transition zones. Both gravity and capillary forces are instrumental in determining the initial distribution of the fluid phases (oil, gas, and water) in the porous network in the absence of viscous effects. Capillary and gravity forces act in opposite directions.

The driving forces and mechanisms related to the fluid flow in petroleum reservoirs are discussed in detail in chapters 8, 16, and 17. These chapters describe the primary, secondary, and enhanced recovery of hydrocarbons, respectively. In essence, the movement of the reservoir fluids may occur due to the following:

- Depletion (i.e., a decline in reservoir pressure)
- Compressibility of the rock/fluid system
- Dissolution of the gas phase from the liquid
- Formation dip
- Capillary rise through microscopic pores
- Additional energy provided by an aquifer or overlying gas cap
- External fluid injection
- Thermal, miscible, and other processes engineered to augment recovery which alter certain properties of reservoir fluids, resulting in greater mobility of fluids

In many reservoirs, more than one factor is responsible for the flow of fluids. For example, the chief driving forces in a saturated reservoir could be the dissolution of gas combined with water encroachment from an adjacent aquifer. In a dipping heavy oil reservoir, the mechanisms of recovery may involve displacement by injected steam, viscosity reduction by thermal processes, and the drainage of oil through downdip wells. In gas reservoirs, the principle mechanisms of fluid flow include depletion and water influx.

Key points—important concepts in reservoir fluid flow

1. Fluid flow in porous media can be characterized as unsteady-state, steady-state, and pseudosteady-state. The characterization is based on how the fluid pressure changes in a reservoir in response to production, injection, and reservoir boundary conditions, including any water encroachment from the aquifer, among others.

2. Unsteady-state or transient flow of the reservoir fluid develops following the start (or end) of well production or injection, where stabilized conditions existed prior to the change in well status. Under unsteady-state flow conditions, the reservoir pressure is a function of both time and space. The rate at which any change in pressure occurs would vary with time at a point where transient conditions prevail.

3. The transient condition generated at the well in the form of perturbations in the fluid pressure eventually propagates to the reservoir boundary. At the reservoir boundary, any decline in pressure may be compensated for by strong water influx from an adjacent aquifer. A constant pressure boundary can be encountered due to water injection or the presence of a gas cap. Under such conditions, a steady-state flow can develop eventually in the reservoir. When steady-state conditions are attained in a reservoir, the observed pressure does not change with time and location. Fluid flow from the well is essentially constant.

4. In other circumstances, a no-flow boundary condition can be encountered in a reservoir due to the presence of a geologic barrier. Such reservoirs do not receive any pressure support from an adjacent aquifer. Typical examples of flow barriers are sealing faults and facies changes in rocks. Again, a no-flow boundary is created artificially in the case of multiple wells producing from their individual drainage areas in a field. Under such conditions, flow towards the well may take place under a pseudosteady-state condition, meaning that the rate of change in pressure is constant. Reservoir pressure is observed to decline at the same pace at all locations. The rate of change in the pressure is inversely proportional to the drainage pore volume.

5. The three flow characteristics encountered in porous media can be summarized as follows:

Fluid Flow Characteristic	Prevailing Condition in the Reservoir	Change in Reservoir Pressure with Time
Unsteady-state	Infinite-acting reservoir, no boundary effects	Not constant
Steady-state	Constant pressure at the reservoir boundary	None
Pseudosteady-state	No flow at the reservoir or drainage boundary	Constant

6. The transient and other fluid flow conditions in the reservoir are identifiable and interpretable in the pressure response data obtained during well tests, which are described in detail in chapter 5. A period of transition is likely to be observed when fluid flow conditions change from one state to another, as evident from well test results.

7. Fluid flow in porous media is affected by the compressibility of the fluids, in addition to other physical properties. Oil is treated as slightly compressible. In contrast, the analysis of natural gas flow takes into account the significant variations of the physical properties of the fluid with changes in the reservoir pressure. Oil and gas compressibility are described in chapter 3.

8. In reservoir studies, including reservoir simulation, fluid flow in porous media is conceptualized and modeled as linear (x, y, or z), radial (r), spherical (r and z), 2-D, and 3-D (x, y, and z) in direction.

9. Various fluid flow patterns may develop in a reservoir that are deemed vital in understanding the dynamics of flow in the reservoir. By and large, the fluid flow pattern is thought to be radial around an active well. In a hydraulically fractured well where a long fracture is present, the flow pattern eventually becomes pseudoradial. Similarly, a late radial flow pattern is encountered in a horizontal well. Other notable patterns include the following:
   - Linear and bilinear flow, as commonly evident at early times following the hydraulic fracturing of a well
   - Spherical flow in a partially completed formation
   - Cone-shaped flow of water and gas in the immediate vicinity of a well, referred to as water and gas coning, respectively

10. Fluid flow in reservoirs can be single phase, as in a dry gas reservoir, or two-phase, as in a saturated oil reservoir where water is immobile. It can also be three-phase, as in a saturated oil reservoir where water breakthrough is experienced.

11. The driving forces behind fluid flow in petroleum reservoirs include viscous forces, gravity drainage, and capillary effects. Viscous forces prevail in most scenarios where a reservoir is under production (and injection). The flow rate of the fluid in porous media is proportional to the imposed pressure gradient that exists between two points along the flow path. In a dipping reservoir, the gravity drainage of the oil can be significant. The process can be accentuated by gas injection in updip wells.

12. In petroleum reservoirs, the driving mechanisms for the flow of fluids include the following:
   - Depletion
   - Fluid expansion and pore compression
   - Gas dissolution
   - Effects of aquifer influx or gas cap
   - Existence of formation dip
   - External fluid and gas injection
   - Various thermal, miscible, and other processes.

   In many circumstances, oil and gas reservoirs are produced by multiple mechanisms of fluid displacement rather than by a single mechanism.

Basic Fluid Flow Equations

Fluid flow models in porous media are primarily based on the mathematical equations that attempt to predict fluid pressure, flow rate, and phase saturation over time and at various location within the reservoir. The propagation of fluid pressure in porous media relates to the phenomenon of diffusion and is analogous to the conduction of heat in solids. Various solutions of the diffusivity equation, based of the equation of continuity, Darcy's law, and the equation of state, dominate the fluid flow analyses in the course of reservoir studies. The diffusivity equation, shown later, is a nonlinear partial differential equation, second-order in space and first-order in time. It is nonlinear, since certain parameters appearing explicitly in the equation are not constant and vary with fluid pressure (Equation 4.25).

Hence, certain assumptions and approximations are introduced in order to linearize the equation and provide an analytic solution. Derivation of the diffusivity equation is presented in the following section.

Depending on the complexity of the reservoir model, the diffusivity equation is solved either analytically or numerically for various boundary conditions, such as no flow or constant pressure. A no-flow boundary is encountered when the reservoir is isolated from any influence of an aquifer, for instance. This occurs due to the presence of an impermeable barrier. Again, strong aquifer support may lead to a constant pressure boundary. The inner boundary is located at the wellbore, while the outer boundary is at the reservoir limit. At the inner boundary, a constant rate or constant pressure condition can be imposed. The diffusivity equation is solved for compressible, incompressible, and slightly compressible fluids. Analytic solutions in radial coordinates are commonplace, as the flow is usually visualized to be radial towards the wellbore during production.

The analytic solutions are mostly based on various simplifying assumptions involving the idealized depiction of the reservoir and the production (or injection) scenario. These are rarely encountered in the real world, and the assumptions are described later. Nevertheless, the solutions are instrumental in visualizing the dynamics of flow in the porous media and the overall reservoir behavior. Reservoir engineers routinely design and interpret well tests, and classical well test theory is based on analytical solutions of the diffusivity equation. Notable aquifer influx models are based on the analytical solutions. Benchmark tests are performed to evaluate the accuracy of numerical simulators against analytical solutions. In relatively straightforward cases where certain simplifying assumptions about the reservoir can be made, an analytic approach can contribute significantly in reservoir engineering studies.

Analytic solutions for fluid flow

Some of the notable analytic solutions of the diffusivity equation are in the following:

- Radial flow in an "infinite-acting" reservoir. The well is viewed as a line source of zero radius in order to simplify the derived solution.

- Radial flow in a bounded reservoir.

- Production at a constant rate, including various well tests.

- Production at a constant bottomhole pressure.

- Flow of slightly compressible and compressible fluids, i.e., oil and gas, respectively. A pseudopressure function, described later, is introduced to account for the variability of fluid properties in the latter case.

- Flow in various drainage shapes other than radial.

- Modified solution incorporating the effects of formation damage and wellbore storage.

- Flow through natural and hydraulic fractures.

- Multiphase flow of oil, gas, and water.

- Water influx from aquifers: steady state and unsteady state.

- Injection of fluids into the reservoir.

The solutions to the diffusivity equations are frequently derived in terms of dimensionless pressure, time, and distance. This leads to the depiction of the solutions in a convenient generic form, regardless of the wide-ranging values of rock and fluid properties. For instance, a standard set of type curves relating dimensionless pressure drop over dimensionless time and distance can be generated for the entire range of rock and fluid properties. The sought result, such as the actual pressure decline over time at any location within the reservoir, can be obtained from these curves with relative ease once the dimensionless quantities are evaluated for the reservoir under study. (The evaluation of well performance based on dimensionless rate and volume over dimensionless time is illustrated in Example 4.2.)

Analytic treatment of fluid flow equations under various reservoir scenarios is widely available in the literature, including Lee and Wattenbarger, Collins, and Matthews and Russell.[2–4] Numerical simulation methods and applications are also available in the literature, including those presented by Aziz and Settari, and Thomas.[5,6]

Development of a Mathematical Model—An Overview

The diffusivity equation is based on the following:

- The law of conservation of mass
- Darcy's law, an empirical relationship between flow rate and the imposed pressure gradient in a porous medium. (Darcy's law is introduced in chapter 2.)
- The equation of state (EOS), describing fluid volume as a function of pressure and temperature.

Assumptions

It is important to recognize the following assumptions associated with various analytic solutions of the diffusivity equation:

1. A homogeneous and isotropic reservoir, implying that rock porosity, permeability, and other properties are uniform.
2. Horizontal fluid flow, with negligible effects of gravity.
3. A single fluid phase of small and constant compressibility.
4. The viscosity of the fluid is independent of the reservoir pressure.
5. Isothermal and laminar flow of the fluid.

Mass balance of fluid in motion over an elemental volume

The derivation of the radial diffusivity equation begins by performing a mass balance of fluid flowing in a radial geometry through a volume element in the porous media towards the well during production. This is depicted in Figure 4–3.

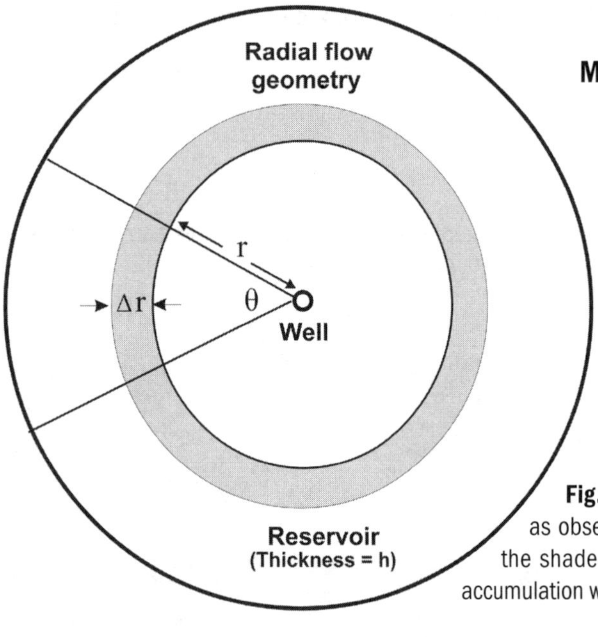

Fig. 4–3. Flow of fluid through a volume element of a porous medium as observed from the top. A mass balance for the fluid entering and leaving the shaded volume over a specific time interval, along with the rate of mass accumulation within the element, lead to the development of the equation of continuity.

The inner edge of the element is located at a distance r from the center of the producing well. The outer edge of the volume element is at r+Δr. The porous medium has a thickness of h. Since mass can neither be created nor be destroyed, the following relationship must be valid:

Mass of fluid flowing into the element during a specific time period
= Mass of fluid flowing out of the element + Rate of accumulation of mass within the element in the same period

Lee and Wattenbarger, among others, provide the detailed derivation of the equation of continuity.[7] In terms of fluid density, velocity, and flow volume, mass flow rates of fluid into and out of an element can be obtained as follows:

$$w_{in} = -\rho\, u_r\, (r + \Delta r)\, \theta\, h \tag{4.8}$$

where

w_{in} = mass inflow rate, m/t,
ρ = fluid density, m/L^3,
u = fluid velocity, L/t, and
$(r + \Delta r)\, \theta$ is the length of an arc, L.

A negative sign appears in the above expression when the flow of fluid takes place in the direction of decreasing radial distance from the wellbore.

Similarly,

$$w_{out} = -[\rho\, u_r - \Delta(\rho\, u_r)]\, r\, \theta\, h \tag{4.9}$$

where

w_{out} = mass outflow rate, m/t,
$\Delta(\rho\, u_r)$ = change in mass flux inside the element, $m/L^2 t$.

The mass of fluid within the pore volume of the element is obtained by multiplying its bulk volume with porosity and fluid density, as in the following:

$$m = r\, \theta\, \Delta r\, h\, \varnothing\, \rho \tag{4.10}$$

where

m = mass of fluid, and
$\varnothing$ = porosity of the porous medium, fraction.

The rate of accumulation of mass in the element between time t and t + Δt is calculated to obtain the change of mass of the fluid in the interval Δt, as follows:

$$q_m = \frac{(r\, \theta\, \Delta r\, h\, \varnothing\, \rho\,)_{t + \Delta t} - (r\, \theta\, \Delta r\, h\, \varnothing\, \rho)_t}{\Delta t} \tag{4.11}$$

where

q_m = rate of accumulation of mass, m/t.

In Equations 4.8 through 4.11, consistent units in mass (m), length (L), and time (t) are used. Combining Equations 4.8, 4.9, and 4.11, and performing necessary simplifications, the following can be obtained:

$$\frac{1}{r}\left[\rho u_r + r\frac{\Delta(\rho u_r)}{\Delta r}\right] = -\frac{\Delta(\varnothing\, \rho)}{\Delta t} \tag{4.12}$$

The differential form of the equation is formulated by considering infinitesimally small radial distance and time:

$$\Delta r \rightarrow 0\,,\, \Delta t \rightarrow 0 \tag{4.13}$$

Hence, Equation 4.12 is transformed as in the following:

$$\frac{1}{r}\left[\rho u_r + r\frac{\partial(\rho u_r)}{\partial r}\right] = -\frac{\partial(\emptyset\,\rho)}{\partial t} \tag{4.14}$$

However, the left side of Equation 4.14 can be written in the condensed form by noting the chain rule of differentiation:

$$\frac{\partial(r\,\rho\,u_r)}{\partial r} = \rho\,u_r\frac{\partial r}{\partial r} + r\frac{\partial(\rho\,u_r)}{\partial r} \tag{4.15}$$

Equation 4.14 is finally written in the following form, which is referred to as the equation of continuity:

$$\frac{1}{r}\left[\frac{\partial(r\,\rho\,u_r)}{\partial r}\right] = -\frac{\partial(\emptyset\,\rho)}{\partial t} \tag{4.16}$$

Application of Darcy's law

Darcy's law correlates fluid flow rate (and apparent velocity) in porous media to the existing pressure gradient that causes fluid to flow. Accordingly, the following expression for fluid velocity can be written in terms of pressure gradient, rock permeability, and fluid viscosity in radial geometry:

$$u_r = -\frac{k}{\mu}\frac{\partial p}{\partial r} \tag{4.17}$$

where
  k = permeability of porous medium, and
  μ = fluid viscosity

In Equation 4.17, the flow of fluid is assumed to occur in the horizontal direction. For the flow of fluids that takes place between points located at different elevations in porous media, i.e., inclined or vertical flow, Darcy's law is based on fluid potential instead of pressure. Fluid potential is defined later.

Substituting the value of u_r from Equation 4.17 into the equation of continuity, the following can be obtained:

$$\frac{1}{r}\frac{\partial}{\partial r}\left(r\rho\frac{k}{\mu}\frac{\partial p}{\partial r}\right) = \frac{\partial(\emptyset\,\rho)}{\partial t} \tag{4.18}$$

As the fluid viscosity and rock permeability are assumed to be uniform over the radial distance r, the equation of continuity can be rewritten as follows:

$$\frac{1}{r}\frac{\partial}{\partial r}\left(r\rho\frac{\partial p}{\partial r}\right) = \frac{\mu}{k}\frac{\partial(\emptyset\,\rho)}{\partial t} \tag{4.19}$$

Equation of state—slightly compressible fluids

The definitions of rock and fluid compressibilities are provided in chapters 2 and 3. Since the total compressibility of the system can be obtained by combining the compressibilities of the rock and fluid, the following can be written:

$$c_t = c + c_f \tag{4.20}$$

$$= \frac{1}{\rho}\left(\frac{\partial\rho}{\partial p}\right)_T + \frac{1}{\emptyset}\left(\frac{\partial\emptyset}{\partial p}\right) \tag{4.21}$$

where

    c_t = total compressibility,

    c = fluid compressibility, and

    c_f = compressibility of the formation

Furthermore, for slightly compressible fluids, the following approximation can be made based on Equation 4.6:

$$\frac{\partial \rho}{\partial p} = c\rho \tag{4.22}$$

Equation 4.19 can be recast by applying the chain rule of differentiation as follows:

$$\frac{\rho}{r} \frac{\partial}{\partial r}\left(r\frac{\partial p}{\partial r}\right) + \frac{\partial p}{\partial r}\frac{\partial \rho}{\partial p}\frac{\partial p}{\partial r} = \frac{\mu}{k}\,\emptyset\,\rho \left(\frac{1}{\rho}\frac{\partial \rho}{\partial p}\frac{\partial p}{\partial t} + \frac{1}{\emptyset}\frac{\partial \emptyset}{\partial p}\frac{\partial p}{\partial t}\right) \tag{4.23}$$

Next, the expressions for compressibility as obtained in Equations 4.21 and 4.22 are introduced in Equation 4.23:

$$\frac{\rho}{r} \frac{\partial}{\partial r}\left(r\frac{\partial p}{\partial r}\right) + c\rho\left(\frac{\partial p}{\partial r}\right)^2 = \frac{\emptyset \mu c_t}{k}\,\rho \left(\frac{\partial p}{\partial t}\right) \tag{4.24}$$

The term $c\,(\partial p\,/\,\partial r)^2$ is negligible compared to the other terms in Equation 4.24, as the value of fluid compressibility is small. Noting this fact and dividing by ρ, the equation can finally be obtained in the following form:

$$\frac{\partial^2 p}{\partial r^2} + \frac{1}{r}\frac{\partial p}{\partial r} = \frac{\emptyset \mu c_t}{k}\left(\frac{\partial p}{\partial t}\right) \tag{4.25}$$

In oilfield units, the diffusivity equation can be expressed as follows:

$$\frac{\partial^2 p}{\partial r^2} + \frac{1}{r}\frac{\partial p}{\partial r} = \frac{1}{2.637 \times 10^{-4}}\frac{\emptyset \mu c_t}{k}\left(\frac{\partial p}{\partial t}\right) \tag{4.26}$$

where k is given in mD,

    μ is in cp,

    r is in ft,

    c_t is in psi^{-1}, and

    t is in hrs.

It is to be noted that the following is defined as the hydraulic diffusivity or diffusivity coefficient for the rock-fluid system:

$$\eta = \frac{2.637 \times 10^{-4}\,k}{\emptyset \mu c_t} \tag{4.27}$$

where

    η is the diffusivity coefficient.

The rate of propagation of fluid pressure in a porous medium is proportional to the diffusivity coefficient. The derivation of Equation 4.26 is left as an exercise at the end of the chapter.

In 1-D and 2–D linear geometry, the diffusivity equation takes the following forms, respectively:

$$\frac{\partial^2 p}{\partial x^2} = \frac{1}{\eta}\frac{\partial p}{\partial t} \tag{4.28}$$

$$\frac{\partial^2 p}{\partial x^2} + \frac{\partial^2 p}{\partial y^2} = \frac{1}{\eta}\frac{\partial p}{\partial t} \tag{4.29}$$

Key points—equations describing fluid flow in porous media

1. Equations for fluid flow attempt to predict the fluid pressure, flow rate, and phase saturation in terms of time and location within the reservoir. The propagation of fluid pressure in porous media is a diffusive process, and the governing equation is referred to as the diffusivity equation.

2. Depending on the complexity of the reservoir model, the diffusivity equation is solved either analytically or numerically for various boundary conditions, such as no flow or constant pressure. In most classical solutions, a uniform initial pressure throughout the reservoir or drainage area is assumed.

3. The diffusivity equation is solved for compressible, incompressible, and slightly compressible fluids. Analytic solutions in radial coordinates are commonplace, as the flow is usually visualized to be radial towards the wellbore during production.

4. Major assumptions in developing the analytic solutions include the following:
   - Homogeneous and isotropic rock properties.
   - Laminar flow (implying that Darcy's law is valid).
   - Horizontal flow of single fluid.
   - Fluid viscosity and compressibility are independent of fluid pressure.
   - No thermal effects on fluid properties.

5. The analytic solutions, although based on the assumptions of homogeneous rock properties and ideal fluid flow conditions, provide invaluable insight into the reservoir dynamics. Some of the applications of the analytic solutions include the following:
   - Well test interpretation
   - Estimation of the flow rate and reservoir pressure in circumstances where simplified assumptions are acceptable
   - Modeling of the aquifer influx
   - Benchmarking of the reservoir simulation tools based on numerical solution methods

6. Some of the notable analytic solutions to fluid flow and their applications described in this book include the following:
   - Line-source solution of unsteady-state flow in an "infinite-acting" reservoir. This predicts the changes in reservoir pressure under the conditions of uniform initial pressure and a constant well rate. Applications include reservoir characterization under dynamic conditions, classical well test interpretation, and others.
   - Estimation of average reservoir pressure and drainage pore volume under pseudosteady-state flow conditions.
   - Well production at a constant bottomhole pressure in a bounded reservoir.
   - Prediction of flow rate and cumulative volume produced with time.
   - Estimation of the production rate of a planned well.
   - Fractional flow of water and oil during waterflooding.
   - Water influx from aquifers: steady-state and unsteady-state.

7. The derivation of the diffusivity equation is based upon the following:
   - The equation of continuity, developed on the basis of mass balance of the fluid flowing across an elemental volume in a porous medium.
   - Darcy's law, which correlates the apparent fluid velocity in terms of fluid pressure gradient, viscosity, and the permeability of the porous medium.
   - The equation of state, which describes the changes in the volume or density of a fluid with pressure. Two distinct approaches are adopted in this regard. This first is for oil, which is treated as slightly compressible. The second is for natural gas, which is considered compressible, with significant variations in physical properties.

8. It is noted that in the diffusivity equation, the term $\frac{k}{\phi \mu c_t}$ is referred to as the hydraulic diffusivity or the diffusivity coefficient. In porous media, the propagation of the fluid pressure is proportional to the diffusivity coefficient.

Solutions to the Diffusivity Equation—Unsteady-State or Transient Flow

Case I. Well producing at a constant rate in an infinite-acting reservoir

A highly familiar analytic solution of the diffusivity equation assumes that the well production originates from a line source and the radius of the well is zero. Referred to as the line-source solution of the diffusivity equation, it constitutes the foundation of classical well test interpretation, as described in chapter 5. The solution further assumes that the well flow rate is constant, and the flow is unaffected by the reservoir boundary, i.e., the flow of fluid occurs in an infinite porous medium. During the "infinite-acting" flow period, the reservoir pressure response at time t and distance r from the well can be estimated by the following:[8,9]

$$p_D = -\frac{1}{2}\, Ei\left[\frac{-948\,\phi\mu c_t\, r^2}{kt}\right] \qquad (4.30)$$

where

$$p_D = \frac{kh(p_i - p)}{141.2\, qB\mu}\;, \quad \text{dimensionless,} \qquad (4.31)$$

h = average net thickness of reservoir, ft,
p_i = initial uniform reservoir pressure, psi,
q = well flow rate measured in surface conditions, stb/d, and
B = oil formation volume factor, bbl/stb.

Conversely, if both pressure and flow rate can be monitored in a well over time, the above solution leads to the estimation of the formation transmissibility, kh/μ. This constitutes the basis of classical well test interpretation and is treated in chapter 5. The term Ei in Equation 4.30 stands for the exponential integral function and is described later.

Initial and boundary conditions for solution

The solution to the radial diffusivity equation is based on the following initial and boundary conditions in a cylindrical reservoir:[10]

- Initial condition. The reservoir pressure is at its initial uniform value:

$$p(r,t) = p_i\,, \quad t = 0 \qquad (4.32)$$

- Outer boundary condition. The pressure remains unaffected at the reservoir boundary during the infinite-acting flow period:

$$\frac{\partial p}{\partial t} = 0 \quad \text{at } r = \infty \qquad (4.33)$$

- Inner boundary condition. A constant flow rate at the wellbore is assumed based on Darcy's law:

$$q = -1.127 \times 10^{-3}\,\frac{kh}{B\mu}\left(2\pi r\right)\left(\frac{\partial p}{\partial r}\right) \quad \text{at} \;\; r = r_w \qquad (4.34)$$

During the infinite-acting flow period, the flow pattern is assumed to be radial in most cases. However, other types of flow pattern may develop as well.[11] In such cases, the transient flow equations need to be modified to account for nonradial geometry.

Table 4-1. Values of -Ei(-x). *Source: J. Lee and R. A. Wattenbarger. 1996. Gas Reservoir Engineering. SPE Textbook Series. Vol. 5. Richardson, TX: Society of Petroleum Engineers.*

x	−Ei(−x)	x	−Ei(−x)
0.001	6.332	0.6	0.454
0.002	5.639	0.7	0.374
0.003	5.235	0.8	0.311
0.004	4.948	0.9	0.26
0.005	4.726	1.0	0.219
0.006	4.545	1.1	0.186
0.007	4.392	1.2	0.158
0.008	4.259	1.3	0.135
0.009	4.142	1.4	0.116
0.01	4.038	1.5	0.1
0.012	3.858	1.6	0.0863
0.014	3.705	1.7	0.0747
0.016	3.574	1.8	0.0647
0.018	3.458	1.9	0.0562
0.02	3.355	2.0	0.0489
0.03	2.959	2.5	0.0249
0.04	2.681	3.0	0.013
0.05	2.468	3.5	0.00687
0.06	2.295	4.0	0.00378
0.07	2.151	4.5	0.00207
0.08	2.027	5.0	0.00115
0.09	1.919	5.5	6.41×10^{-4}
0.1	1.823	6.0	3.6×10^{-4}
0.11	1.737	6.5	2.03×10^{-4}
0.12	1.66	7.0	1.15×10^{-4}
0.14	1.524	7.5	6.58×10^{-5}
0.16	1.409	8.0	3.77×10^{-5}
0.18	1.31	8.5	2.16×10^{-5}
0.2	1.223	9.0	1.24×10^{-5}
0.3	0.906	9.5	7.18×10^{-6}
0.4	0.702	10.0	4.15×10^{-6}
0.5	0.56	10.5	2.41×10^{-6}

Line-source solution in dimensionless pressure and time

In the literature, the line-source solution of the diffusivity equation shown in Equation 4.30 is frequently expressed in terms of both dimensionless pressure and time, as follows:

$$p_D = \frac{1}{2}\left[-Ei\left(-\frac{1}{4t_D} \right) \right] \tag{4.35}$$

where

$$t_D = \frac{2.637 \times 10^{-4}\, k\, t}{\phi\, \mu\, c_t\, r^2}, \text{dimensionless} \tag{4.36}$$

Again, various solutions to the radial diffusivity equation incorporate the dimensionless radial distance as follows:

$$r_D = r/r_w \tag{4.37}$$

$$r_{eD} = r_e/r_w \tag{4.38}$$

Evaluation of the exponential integral function

Based on a series solution, the exponential integral function can be evaluated as follows:

$$Ei(-x) = \left[\ln x - \frac{x}{1!} + \frac{x^2}{2(2!)} - \frac{x^3}{3(3!)} + \dots \right] \tag{4.39}$$

Values of the exponential integral function at selected intervals are listed in Table 4–1.

Approximations for line-source solutions

In reservoir engineering calculations, the following approximations are fairly common:

(i) For x < 0.02:

$$Ei(-x) \approx \ln(1.781x) \tag{4.40}$$

(ii) For $t_D/r_D^2 > 100$:

$$p_D = \frac{1}{2}\left(\ln \frac{t_D}{r_D^2} + 0.80907 \right) \tag{4.41}$$

where

$$r_D = r/r_w .$$

(iii) For x ≥ 10.9:

$$Ei(-x) \approx 0 \tag{4.42}$$

The equations suggest the following characteristics of transient flow:

- **Equation 4.40.** Beyond a certain flow period, t is large enough so that x < 0.02, and the exponential integral function can be approximated by a natural logarithm function.
- **Equation 4.41.** At the flowing well, $r_D = 1$, and $t_D > 100$, except at very early times typically lasting from a few seconds to a few minutes. Hence, the pressure response at the well during the infinite-acting flow period can be approximated as a logarithmic function of time, which can be plotted as a straight line on semilog paper (p_D versus log t_D or p versus log t). This is the basis of well test analysis for infinite-acting radial flow, described in detail in chapter 5.
- **Equation 4.42.** At sufficiently long times, the response in pressure, as suggested by the line-source solution, becomes negligible.

Based on nonlinear regression, straightforward equations are available in the literature to evaluate the exponential integral function as a function of x within certain ranges. The results can be conveniently incorporated in spreadsheet calculations.[12]

Validity of line-source solutions

The line-source solution given in Equation 4.30 is valid under a specific time interval as determined by the reservoir and fluid properties. The solution is valid only after the early time, calculated in the following:[13]

$$t > \frac{3.79 \times 10^5 \, \varnothing \mu c_t r_w^2}{k} \tag{4.43}$$

At earlier times than what is indicated by Equation 4.43, the assumption of a zero well radius is not appropriate. Again, the equation is valid during the infinite-acting flow period. Once the boundary effects are felt, the pressure response is no longer infinite acting, and a steady-state or pseudosteady-state condition may emerge eventually. Hence, an upper limit of the validity of the line-source solution can be obtained as follows:

$$t < \frac{948 \, \varnothing \mu c_t r_e^2}{k} \tag{4.44}$$

where

r_e is the distance to the boundary of the reservoir (or drainage area) from the producing well, ft.

The duration of the infinite-acting flow period is a strong function of the external drainage radius and rock permeability. Beyond this limit, once steady-state or pseudosteady-state condition is attained following a transitory period, the diffusivity equation is solved with other boundary conditions to predict the appropriate fluid flow behavior.

Prediction of bottomhole pressure—the effects of skin and wellbore storage

In order to predict the changes in bottomhole flowing pressure at the well, Equation 4.30 needs to be modified to account for the permeability alteration, which is commonly encountered around the wellbore. The extent of permeability damage (or improvement following well stimulation) is expressed in terms of skin factor, as described in chapters 2 and 5. The modified equation is given in chapter 5.

Furthermore, deviations from predicted pressure changes at very early times following the commencement of flow are commonplace. This variation is due to the wellbore storage effects, as discussed in detail in chapter 5. Simply stated, the initial production characteristics are dominated by the flow of fluid "stored" in the wellbore, not by the fluid entering the wellbore from the formation. Hence, the line-source solution is not applicable as long as this condition persists, as the boundary condition is different.

It must be noted that the wellbore and near-wellbore phenomena as described in this section do not affect the prediction of changes in pressure obtained by Equation 4.30 at locations beyond the altered zone of permeability as encountered near the wellbore.

Example 4.1: Application of the diffusivity equation—reservoir characterization. Calculate the minimum pressure drop expected to occur in an observation well after 24 hrs of production from a nearby well. The latter is located at a distance of 300 ft and is producing at a steady rate of 500 stb/d. Core data obtained from the observation well indicates an average formation permeability of 60 mD. However, certain impermeable or semipermeable barriers are known to exist in nearby reservoirs in the same basin. Why is the study important? Assume that the reservoir pressure is uniform before the commencement of production. Use the following reservoir and fluid data:

Porosity of the formation, fraction, $= 0.25$
Formation thickness, ft, $= 30$
Viscosity of the oil, cp, $= 0.68$
Oil formation volume factor, rb/stb, $= 1.2$
Total compressibility, psi^{-1}, $= 17.9 \times 10^{-6}$
External radius of the reservoir, ft, $= 2,640$.
Radius of the wellbore, ft, $= 0.32$.

Solution: Based on Equations 4.43 and 4.44, the minimum and maximum time periods are computed for which the line-source solution is valid:

$$t > \frac{3.79 \times 10^5 \,(0.25)(0.68)(17.9\times10^{-6})(0.32)^2}{60}$$
$$> 0.00197 \text{ hrs}$$

$$t < \frac{948\,(0.25)(0.68)(17.9\times10^{-6})(2,640)^2}{60}$$
$$< 335 \text{ hrs}$$

The stipulated time period (24 hrs) is within the range calculated above. Thus the dimensionless pressure drop at a distance of 300 ft from the producing well is estimated by evaluating the Ei function as follows:

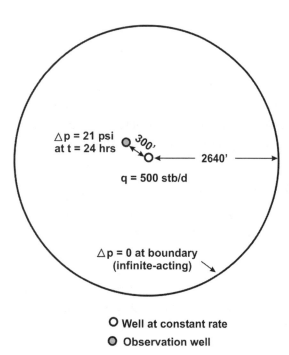

O Well at constant rate
◎ Observation well

Fig. 4–4. Investigation of reservoir heterogeneities between producing and observation wells based on the study of pressure transients (figure not to scale)

$$p_D = -\frac{1}{2} \text{Ei}\left[\frac{-(948)(0.25)(0.68)(17.9 \times 10^{-6})(300)^2}{(60)(24)}\right]$$
$$= -\frac{1}{2} \text{Ei}\,[-0.18]$$
$$= 0.655 \quad \text{(table 4–1)}$$

Finally, the expected pressure drop is calculated by Equation 4.31:

$$p_i - p = \frac{141.2(500)(1.2)(0.68)(0.655)}{(60)(30)}$$
$$\approx 21 \text{ psi}$$

A sketch of the reservoir under study showing the location of producing and observation wells is shown in Figure 4–4.

The presence of any geologic barrier between the producer and the observation well, including permeability degradation, would not lead to the expected pressure drop estimated here. Geologic discontinuities are often responsible for areas of high residual oil saturation and inefficient sweep during waterflooding or other improved recovery operations. The relevant issues are further discussed in chapters 2, 16, and 19. A well test is an effective tool in characterizing geologic discontinuities, provided a well is located in the area. This topic is treated in chapter 5. The major limitations of this type of study in the field include the following:

- It is not a common practice to shut in the well (or wells) for an extended period in order to attain stabilized reservoir pressure at the expense of oil and gas production.
- A constant well rate is rather difficult to maintain for an extended period of time.
- A nearby observation well may not be available to conduct the study.

Application in multiwell systems

An important aspect to note is that the solutions to the radial diffusivity equation are not limited to a single well producing or injecting in the middle of a reservoir. Based on the principle of superposition, and under appropriate conditions, the solutions can be extended to multiwell systems. (The principle of superposition is treated in chapter 5, with illustrations.) Based on the principle of superposition, any pressure changes in a reservoir where more than one well is producing can be determined as long as the initial and boundary conditions are met.

The core concept behind the principle of superposition is straightforward. The characterization of a complex scenario of fluid flow involving multiple wells, multiple rates, etc., can be accomplished by combining (or superposing) the elementary solutions of the flow of fluid involving a single well or single rate. In this case, the solution for the pressure drop at the observation point when multiple wells are producing is obtained by simply summing the solutions for the individual wells as follows:

$$\Delta p_{obs} = \Sigma \, \Delta p_k \, (q_k, r_k) \tag{4.45}$$

where

Δp_{obs} = total changes in pressure recorded at the observation well,

Δp_k = changes in pressure at the observation point due to production in the kth well,

q_k = production rate at the kth well, and

r_k = radial distance from the kth well to the observation well.

The individual pressure drops appearing under the summation sign are evaluated by Equation 4.30. In order to respect the initial condition of uniform reservoir pressure required in the analysis, stabilization in pressure must be achieved prior to the study. All producing wells must become active at the same time. In a heterogeneous and anisotropic medium, the anticipated pressure drop is also a function of average formation transmissibility and storavity between an individual well and the observation well. However, such real-world scenarios run counter to the assumption of a homogenous and isotropic medium as stipulated. Figure 4–5 shows five wells located in a typical 5-spot pattern in the field. Similar well patterns are common in fields under water injection and are described in chapter 16. The distances from the center well to the corner wells are not exactly the same.

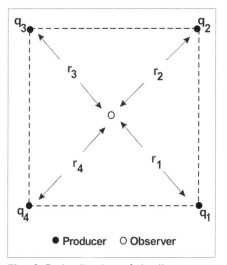

Fig. 4–5. Application of the line-source solution in predicting the pressure drop in a 5-spot pattern. Such patterns are commonly found in fields undergoing water injection.

A stabilized pressure is obtained in the reservoir following shutting-in of the wells for a necessary length of time. If the wells located at the four corners start producing at the same time but at different rates, the pressure drop experienced at the center well is obtained by the extension of Equation 4.30:

$$\Sigma \Delta p = \Delta p_1 + \Delta p_2 + \Delta p_3 + \Delta p_4 \tag{4.46}$$

$$= -\frac{70.6 \, q_1 \, B\mu}{kh} Ei\left[\frac{-948 \, \o\mu c_t r_1^2}{kt}\right] - \frac{70.6 \, q_2 \, B\mu}{kh} Ei\left[\frac{-948 \o\mu c_t r_2^2}{kt}\right] - \frac{70.6 \, q_3 \, B\mu}{kh} Ei\left[\frac{-948 \, \o\mu c_t r_3^2}{kt}\right] - \frac{70.6 \, q_4 \, B\mu}{kh} Ei\left[\frac{-948 \, \o\mu c_t r_4^2}{kt}\right]$$

It must be recognized that the applicability of Equation 4.45 is rather limited in a large field. This is because stabilized initial pressure is difficult to achieve when the nearby producers and injectors are opened and shut in periodically. Reservoir simulation based on a numerical approach is usually employed in large fields having many production and injection wells. Moreover, any known heterogeneities between the wells under study can be incorporated conveniently in simulation. Nevertheless, the analytical approach provides a way to gain insight into the system with significantly less effort.

Other important applications of the principle of superposition include the following:

- The well is located near an impermeable boundary, such as a sealing fault, rather than in a drainage area of large external radius.
- The rate at which the well produces is not constant. However, any changes in rate over time must be known.

One or more imaginary wells, referred to as image wells, are used in addition to the real well in order to perform the fluid flow analysis. The methodology is described in chapter 5.

Key points—line-source solution to the diffusivity equation

1. The line-source solution is based upon the following initial and boundary conditions:
   - Uniform initial reservoir pressure
   - The reservoir is "infinite-acting," implying that the prevailing condition at the reservoir boundary (constant pressure or no flow) does not influence the fluid flow characteristics
   - A constant well flow rate at the inner boundary

2. The line-source solution of the diffusivity equation predicts the changes in the reservoir pressure as a function of time and distance from the flowing well under the transient conditions. The solution requires evaluation of the exponential integral (Ei) function.

3. The reservoir pressure, time, and distance appearing in the solution are conveniently expressed in dimensionless forms. This allows a generic solution of dimensionless pressure drop versus dimensionless time, which can be displayed in a plot or a table, for example. This has the advantage that the exponential integral function need not be evaluated in individual cases of varying reservoir and fluid properties.

4. The line-source solution is not valid for the very early time, as the assumption of a line source for a well is not appropriate. However, the time period is usually very small, lasting only few seconds or minutes. Again, the solution also is not valid at late times when the flow is no longer infinite-acting, and the conditions at the reservoir boundary affect the pressure response.

5. Except at very early times, the dimensionless pressure drop is a linear function of the natural log of dimensionless time. This serves as the basis for classical well test interpretation, where the pressure response over time is drawn on a semilog plot. The resulting straight line is analyzed to determine various reservoir characteristics.

6. In reality, the transient pressure response at early times is affected by the phenomena of wellbore storage and skin. The response predicted by the line-source solution cannot be observed. At the start of flow, fluid "stored" in the wellbore is produced first, not the fluid from the formation. Again, skin refers to the permeability damage in the formation in the immediate vicinity of the well, which is encountered in virtually all cases. These topics are discussed in chapter 5.

7. The line-source solution can be extended to reservoirs where multiple wells are present. The pressure drop experienced at the observation well is the sum of the individual pressure drops caused by each well located in the vicinity of the producer. However, a uniform initial pressure is difficult to achieve in such reservoirs, as the wells are opened and shut in periodically. The principle of superposition, described in chapter 5, is applied to derive the combined solution.

Case II. Well producing at a constant bottomhole pressure in a closed drainage area

In arriving at the line-source solution described earlier, a constant well rate was assumed throughout the production period. However, a more likely scenario relates to the production of the well under stabilized bottomhole pressure in a bounded reservoir. In this case, the well rate is initially high, followed by a continuous decline until the abandonment of the well. The analysis of the decline in the well rate is simply referred to as decline curve analysis. It is one of the most widely used techniques in the industry to predict well and reservoir performance wherever applicable. Decline curve analysis is illustrated in chapter 11.

In the analytic solution of the diffusivity equation for wells producing under constant bottomhole pressure, the following initial and boundary conditions are assumed. Moreover, the reservoir geometry is assumed to be cylindrical.

(a) Initial condition

$$p(r,t) = p_i, t = 0 \tag{4.47}$$

(b) Inner boundary condition:

$$p = p_{wf}, \text{ constant} \tag{4.48}$$

(c) Outer boundary condition:

$$\frac{\partial p}{\partial r} = 0, r = r_e \tag{4.49}$$

The solution to the diffusivity equation predicting the well production rate and cumulative production during infinite-acting and finite-acting flow is given in terms of the Bessel function. This function is rather tedious to implement in manual computations. Thus an alternate solution predicting the flow rate during the infinite-acting period is proposed as follows:[14]

$$q_D = \frac{2.02623\, t_{Dw} (\ln t_{Dw} - 1) + 3.90086}{t_{Dw} (\ln t_{Dw})^2}, \ 200 < t_{Dw} < 1/4\, r_{eD}^2 \tag{4.50}$$

where the dimensionless quantities q_D and t_{Dw} are related to the well rate in stb/d and time in hrs, respectively, as follows:

$$q_D = \frac{qB\mu}{7.08 \times 10^{-3}\, kh\, (p_i - p_{wf})}, \quad \text{dimensionless, and} \tag{4.51}$$

$$t_{Dw} = \frac{2.637 \times 10^{-4}\, k\, t}{\varnothing\, \mu\, c_t\, r_w^2}, \quad \text{dimensionless.} \tag{4.52}$$

At late times, when $t_{Dw} > 1/4\, r_{eD}^2$, the flow is no longer in the infinite-acting period. In order to compute the well rate following the infinite-acting period, values of the dimensionless flow rate for a given t_{Dw} and r_{eD} are provided in Table 4–2.[15,16]

Table 4-2. Dimensionless production rate for a well producing at constant bottomhole pressure in a closed reservoir at selected values of r_{eD}. Source: B. F. Towler. 2002. Fundamental Principles of Reservoir Engineering. SPE Textbook Series. Vol. 8. Richardson, TX: Society of Petroleum Engineers; and J. Lee and R. A. Wattenbarger. 1996. Gas Reservoir Engineering. SPE Textbook Series. Vol. 5. Richardson, TX: Society of Petroleum Engineers.

$r_{eD} = 2,000$		$r_{eD} = 4,000$		$r_{eD} = 10,000$	
t_D	q_D	t_D	q_D	t_D	q_D
1×10^5	0.1604	9×10^5	0.1366	3×10^6	0.1263
2×10^5	0.1520	1×10^6	0.1356	4×10^6	0.1240
3×10^5	0.1475	1.3×10^6	0.1333	5×10^6	0.1222
4×10^5	0.1445	1.6×10^6	0.1315	6×10^6	0.1210
5×10^5	0.1422	2×10^6	0.1296	8×10^6	0.1188
6×10^5	0.1404	2.4×10^6	0.1280	1×10^7	0.1174
7×10^5	0.1389	3×10^6	0.1262	1.2×10^7	0.1162
8×10^5	0.1375	4×10^6	0.1237	1.4×10^7	0.1152
9×10^5	0.1363	5×10^6	0.1215	1.6×10^7	0.1143
1×10^6	0.1352	6×10^6	0.1194	1.8×10^7	0.1135
1.3×10^6	0.1320	8×10^6	0.1155	2×10^7	0.1128
1.6×10^6	0.1291	1×10^7	0.1118	2.4×10^7	0.1115
2×10^6	0.1254	1.2×10^7	0.1081	3×10^7	0.1098
2.4×10^6	0.1216	1.4×10^7	0.1046	4×10^7	0.1071
3×10^6	0.1166	1.6×10^7	0.1012	5×10^7	0.1050
4×10^6	0.1084	1.8×10^7	0.0979	7×10^7	0.0998
5×10^6	0.1008	2×10^7	0.0948	8×10^7	0.0975
7×10^6	0.0872	2.3×10^7	0.0902	9×10^7	0.0952
1×10^7	0.0701	3×10^7	0.0803	1×10^8	0.0930
1.3×10^7	0.0563	4×10^7	0.0681	1.2×10^8	0.0887
1.7×10^7	0.0421	5×10^7	0.0577	1.4×10^8	0.0846
2×10^7	0.0339	7×10^7	0.0415	1.7×10^8	0.0788
2.4×10^7	0.0253	8×10^7	0.0352	2×10^8	0.0734
3×10^7	0.0164	9×10^7	0.0298	2.4×10^8	0.0668
4×10^7	0.0079	1×10^8	0.0252	3×10^8	0.0580
5×10^7	0.0038	1.2×10^8	0.0181	4×10^8	0.0458
7×10^7	0.0009	1.4×10^8	0.0130	5×10^8	0.0362
1×10^8	0.0001	1.7×10^8	0.0079	6×10^8	0.0286
1.3×10^8	0.0000	2×10^8	0.0048	7×10^8	0.0226
		2.3×10^8	0.0029	8×10^8	0.0178
		2.6×10^8	0.0018	1×10^9	0.0111
		3×10^8	0.0009	1.4×10^9	0.0043
		4×10^8	0.0002	2×10^9	0.0011
		5×10^8	0.0000	3×10^9	0.0001

Table 4–3. Dimensionless production rate for a well producing at constant bottomhole pressure in a closed reservoir at selected values of reD. *Source: B. F. Towler. 2002. Fundamental Principles of Reservoir Engineering. SPE Textbook Series. Vol. 8. Richardson, TX: Society of Petroleum Engineers; and J. Lee and R. A. Wattenbarger. 1996. Gas Reservoir Engineering. SPE Textbook Series. Vol. 5. Richardson, TX: Society of Petroleum Engineers.*

$r_{eD} = 2{,}000$		$r_{eD} = 4{,}000$		$r_{eD} = 10{,}000$	
t_D	Q_{pD}	t_D	Q_{pD}	t_D	Q_{pD}
4×10^5	6.27×10^4	1.6×10^6	2.26×10^5	3×10^6	4.06×10^5
5×10^5	7.70×10^4	2×10^6	2.78×10^5	4×10^6	5.31×10^5
6×10^5	9.11×10^4	2.4×10^6	3.30×10^5	5×10^6	6.76×10^5
7×10^5	1.05×10^5	3×10^6	4.06×10^5	6×10^6	7.76×10^5
8×10^5	1.19×10^5	4×10^6	5.31×10^5	8×10^6	1.02×10^6
9×10^5	1.33×10^5	5×10^6	6.54×10^5	1×10^7	1.25×10^6
1×10^6	1.46×10^5	6×10^6	7.74×10^5	1.2×10^7	1.49×10^6
1.3×10^6	1.86×10^5	8×10^6	1.01×10^6	1.4×10^7	1.72×10^6
1.6×10^6	2.25×10^5	1×10^7	1.24×10^6	1.6×10^7	1.95×10^6
2×10^6	2.76×10^5	1.2×10^7	1.46×10^6	1.8×10^7	2.17×10^6
2.4×10^6	3.26×10^5	1.4×10^7	1.67×10^6	2×10^7	2.40×10^6
3×10^6	3.97×10^5	1.6×10^7	1.87×10^6	2.4×10^7	2.85×10^6
4×10^6	5.10×10^5	1.8×10^7	2.07×10^6	3×10^7	3.51×10^6
5×10^6	6.14×10^5	2×10^7	2.27×10^6	4×10^7	4.60×10^6
7×10^6	8.02×10^5	2.3×10^7	2.54×10^6	5×10^7	5.66×10^6
1×10^7	1.04×10^6	3×10^7	3.14×10^6	7×10^7	7.70×10^6
1.3×10^7	1.23×10^6	4×10^7	3.88×10^6	8×10^7	8.69×10^6
1.7×10^7	1.42×10^6	5×10^7	4.51×10^6	9×10^7	9.65×10^6
2×10^7	1.53×10^6	7×10^7	5.49×10^6	1×10^8	1.06×10^7
2.4×10^7	1.65×10^6	8×10^7	5.87×10^6	1.2×10^8	1.24×10^7
3×10^7	1.78×10^6	9×10^7	6.20×10^6	1.4×10^8	1.41×10^7
4×10^7	1.89×10^6	1×10^8	6.47×10^6	1.7×10^8	1.66×10^7
5×10^7	1.95×10^6	1.2×10^8	6.90×10^6	2×10^8	1.89×10^7
7×10^7	1.99×10^6	1.4×10^8	7.21×10^6	2.4×10^8	2.17×10^7
1×10^8	2.00×10^6	1.7×10^8	7.52×10^6	3×10^8	2.54×10^7
1.3×10^8	2.00×10^6	2×10^8	7.71×10^6	4×10^8	3.06×10^7
		2.3×10^8	7.82×10^6	5×10^8	3.47×10^7
		2.6×10^8	7.89×10^6	6×10^8	3.79×10^7
		3×10^8	7.94×10^6	7×10^8	4.04×10^7
		4×10^8	7.99×10^6	8×10^8	4.24×10^7
		5×10^8	8.00×10^6	1×10^9	4.53×10^7
				1.4×10^9	4.82×10^7
				2×10^9	4.96×10^7
				3×10^9	5.00×10^7

Additionally, values of the dimensionless cumulative volume produced, Q_{pD}, are given in Table 4–3.

Based on the above, the cumulative production is calculated as follows:[17]

$$Q = 1.119\, \varnothing\, c_t\, h\, r_w^2\, (p_i - p_{wf})\, \frac{Q_{pD}}{B} \qquad (4.53)$$

where

   Q = cumulative volume of the oil produced, stb, and

   Q_{pD} = dimensionless cumulative volume of oil produced.

After a sufficiently long time, $p_i \rightarrow p_{wf}$, and Tables 4–2 and 4–3 suggest that there is no further increase in the cumulative volume of oil, as there is no driving force available to produce the well any further. It must be noted that the solution of the diffusivity equation in this specific case is based on the single-phase flow of the fluid until the abandonment of the well. However, gas evolution takes place when the reservoir pressure declines below the bubblepoint, and the flow is no longer single phase. The recovery mechanism is dominated by solution gas drive, and an increase in production is likely to be observed.

In certain other cases, a constant pressure is maintained at the boundary due to the pressure support provided by an aquifer. A similar solution for a well producing under a constant rate in a reservoir where a constant pressure condition prevails at the boundary is available in the literature.[18]

Example 4.2: Prediction of rate decline and cumulative oil production in a bounded reservoir. A newly drilled well can be considered that produces at a constant bottomhole pressure of 1,250 psia in a reservoir with a closed boundary.

(a) Based on Equation 4.50, plot the declining rate of the well over time during the infinite-acting period.

(b) Plot the well performance following the infinite-acting period as the effects of the boundary are felt.

(c) Estimate the cumulative oil production at the end of the first year.

   Use rock and fluid data as in the following:

   Initial reservoir pressure, psia, = 3,300.

   Porosity of the formation, fraction, = 0.21.

   Average rock permeability, mD, = 45.

   Formation thickness, ft, = 18.

   Viscosity of the oil, cp, = 1.1.

   Oil formation volume factor, rb/stb, = 1.06.

   Total compressibility, psi^{-1}, = 13.2 × 10^{-6}.

   External radius of the reservoir, ft, = 1,300.

   Radius of the wellbore, ft, = 0.32.

Table 4-4. Rate of well producing under constant bottomhole pressure in infinite-acting period

t (hrs)	t_{Dw}	q_D	q (stb/d)
	Eq. 4.52	Eq. 4.50	Eq. 4.51
0.01	3.800×10^2	0.2840	2,864.0
0.1	3.800×10^3	0.2160	2,178.6
0.25	9.501×10^3	0.1971	1,987.6
0.5	1.900×10^4	0.1848	1,863.7
1	3.800×10^4	0.1739	1,754.1
2	7.601×10^4	0.1643	1,656.5
4	1.520×10^5	0.1556	1,569.2
6	2.280×10^5	0.1509	1,522.1
8	3.040×10^5	0.1478	1,490.2
12	4.561×10^5	0.1436	1,447.9
24	9.121×10^5	0.1369	1,380.6
48	1.824×10^6	0.1308	1,319.2
72	2.736×10^6	0.1275	1,285.7
96	3.648×10^6	0.1252	1,263.0
108	4.105×10^6	0.1243	1,253.9

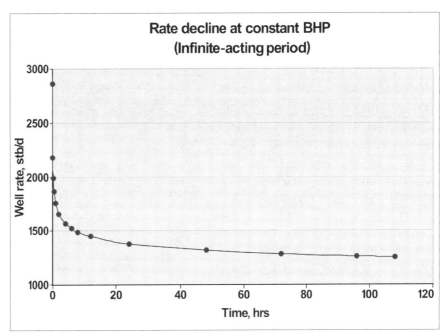

Fig. 4-6. Plot of rate decline of a well producing under constant bottomhole pressure during infinite-acting period

Solution:

(a) First, the time interval in which the solution is valid must be obtained. Based on the conditions stipulated in Equation 4.50, the lower limit of the time interval is calculated as follows:

$$200 = \frac{2.637 \times 10^{-4} \, (45) \, t}{(0.21)(1.1)(13.2 \times 10^{-6})(0.32)^2}$$

Hence,

t = 0.0053 hrs

The upper limit is found as per the following:

$$r_{eD}^2 = (1300/0.32)^2$$
$$= 1.65 \times 10^7$$

$$\frac{1}{4}(1.65 \times 10^7) = \frac{2.637 \times 10^{-4} \, (45)t}{(0.21)(1.1)(13.2 \times 10^{-6})(0.32)^2}$$
$$t = 108.5 \text{ hrs}$$

The results of the well rate over time during the infinite acting period are presented in Table 4–4 and Figure 4–6. Example calculations (t = 8 hrs)

$$t_{Dw} = \frac{2.637 \times 10^{-4} \, (45)(8)}{(0.21)(1.1)(13.2 \times 10^{-6})(0.32)^2}$$
$$= 3.04 \times 10^5$$

$$\ln(t_{Dw}) = 12.6248$$

$$q_D = \frac{2.02623 \, (3.04 \times 10^5)(12.6248 - 1) + 3.90086}{(3.04 \times 10^5)(12.6248)^2}$$
$$= 0.1478$$

$$0.1478 = \frac{q(1.06)(1.1)}{7.08 \times 10^{-3} \, (45)(18)(3300 - 1250)}$$
$$q = 1,490.2 \text{ stb/d}$$

Table 4–5. Rate of well producing under constant bottomhole pressure influenced by reservoir boundary effects

t_{Dw}	t, hrs	t, days	q_D	q
Table 4-2	Eq. 4.52		Table 4-2	Eq. 4.51
4.00×10^6	105.3	4.39	0.1237	1,247.6
6.00×10^6	157.9	6.58	0.1194	1,204.2
8.00×10^6	210.5	8.77	0.1155	1,164.9
1.00×10^7	263.1	10.96	0.1118	1,127.6
2.00×10^7	526.3	21.93	0.0948	956.1
3.00×10^7	789.4	32.89	0.0803	809.9
4.00×10^7	1,052.5	43.85	0.0681	686.8
5.00×10^7	1,315.6	54.82	0.0577	581.9
8.00×10^7	2,105.0	87.71	0.0352	355.0
1.00×10^8	2,631.3	109.64	0.0252	254.2
1.40×10^8	3,683.8	153.49	0.0130	131.1
2.00×10^8	5,262.5	219.27	0.0048	48.4
3.00×10^8	7,893.8	328.91	0.0009	9.1

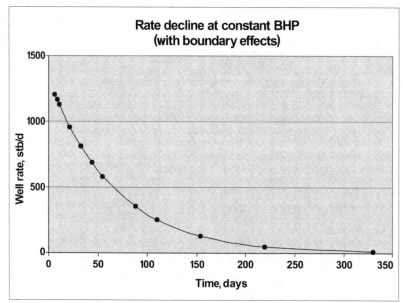

Fig. 4–7. Plot of rate decline of a well producing under constant bottomhole pressure influenced by reservoir boundary effects

(b) Based on Table 4–2, and assuming that reD ≈ 4,000, the values of the dimensionless production rate are read at selected values of dimensionless time, as shown in Table 4–5:

The results are plotted in Figure 4–7.

The advantages of solving the diffusivity equation in dimensionless quantities are obvious from the analysis shown in Table 4–5. If the well performance needs to be evaluated in a different reservoir having dissimilar porosity and permeability, values of t_{Dw} and Q_D need not be recomputed. Equations 4.51 and 4.52 can be employed in a straightforward manner to convert the dimensionless quantities into actual time and rate.

(c) In order to calculate the oil production at the end of the first year, the dimensionless time is calculated as follows:

$$t_{Dw} = \frac{2.637 \times 10^{-4}\,(45)(365)(24)}{(0.21)(1.1)(13.2 \times 10^{-6})(0.32)^2}$$
$$= 3.329 \times 10^8$$

The dimensionless cumulative volume oil produced at t_{Dw} (or t_w) = 3.329×10^8 can be obtained by interpolation from Table 4–3 as follows, assuming r_{eD} = 4000:

$$Q_{pD} = 7.957 \times 10^6$$
$$Q = 1.119\,(0.21)(13.2 \times 10^{-6})(18)(0.32)^2\,\frac{(3,300-1,250)\,7.957 \times 10^6}{1.06}$$
$$= 87,982\ \text{stb}$$

In about 11 months, it is calculated that the well rate will decline to less than 10 stb/d, indicating abandonment. This is expected to occur as the well produces from a finite oil reservoir of relatively small dimensions. Some of the major uncertainties in implementing the analysis in the field include the following:

- Unknown reservoir size and geologic heterogeneities
- Assumption of single-phase flow
- Drive mechanism for oil production
- Reservoir boundary conditions

Key points—solution for well producing under a constant bottomhole pressure

1. A common scenario in petroleum reservoirs involves the production of a well under a constant bottomhole pressure in a closed region. As a result, the production rate declines continuously until the abandonment of the well. Reservoir engineers routinely perform analysis of the decline rate of a well to predict future well and reservoir performance over the entire life cycle wherever applicable.

2. The solution to the diffusivity equation is based upon the following initial and boundary conditions:
    - Uniform initial reservoir pressure
    - Constant pressure at the inner boundary
    - No-flow condition at the outer boundary

3. The analytic solution involves the evaluation of the Bessel function, and subsequent manual computations could be tedious. Hence the values of the dimensionless flow rate and cumulative volume over dimensionless time are either provided in tabulated form in the literature or are approximated by simpler functions.

4. The solution is based on single-phase flow of fluid from the initial state to abandonment. Recovery by other mechanisms, such as solution drive, is not considered in this analysis.

Derivation of the diffusivity equation for compressible fluids

Natural gases in the reservoir are subjected to significant variations in physical properties, including the density, viscosity, and gas deviation factor, with changes in reservoir pressure. Various equations and charts were presented in chapter 3 to facilitate the computation of these properties.

In order to treat compressible flow, the relationship between the density of real gas with pressure, temperature, and the gas deviation factor is recalled from chapter 3:

$$\rho = \frac{pM}{zRT} \tag{4.54}$$

By replacing the density term in Equation 4.18, the equation of continuity is modified as in the following:[19]

$$\frac{1}{r} \frac{\partial}{\partial r} \left(r \frac{p}{\mu z} \frac{\partial p}{\partial r} \right) = \frac{1}{k} \frac{\partial}{\partial t} \left(\varnothing \frac{p}{z} \right) \tag{4.55}$$

The three basic methods are commonly adopted to account for the dependency of the fluid properties, namely, μ and z, on fluid pressure for compressible flow in porous media. These are listed in the following:
- Pressure approach, valid for relatively high reservoir pressure
- Pressure-squared approach, valid for relatively low reservoir pressure
- Pseudopressure approach, valid for all ranges of pressure

The various methods of analysis have their own advantages. Depending on reservoir pressure, the first two methods have limited applicability. However, they are relatively easy to implement in manual computations. The third method involves the evaluation of pseudopressure as a function of μ and z over the entire range of pressure, which is relatively involved. Pseudopressure is an integral function defined later. With the advent of digital machines, the pseudopressure approach is used routinely.

Formulation of the diffusivity equation based on the aforementioned methods is outlined in the following discussion.

Pressure approach. At relatively high pressure (>3,000 psia), the term p/μz is assumed to be constant, and the diffusivity equation can be presented in a manner very similar to the case of a slightly compressible fluid. The derivative term in the right side of Equation 4.55 can be expanded as follows:

$$\frac{\partial}{\partial t}\left(\phi\frac{p}{z}\right) = \phi\frac{p}{z}\frac{\partial p}{\partial t}\left[\frac{1}{\phi}\frac{\partial\phi}{\partial p} + \frac{z}{p}\frac{\partial}{\partial p}\left(\frac{p}{z}\right)\right] \tag{4.56}$$

Formation and gas compressibilities are defined in chapters 2 and 3, respectively.

$$c_f = \frac{1}{\phi}\frac{\partial\phi}{\partial p} \tag{4.57}$$

$$c_g = \frac{1}{\rho}\frac{\partial\rho}{\partial p}$$

$$= \frac{z}{p}\frac{\partial}{\partial p}\left(\frac{p}{z}\right) \tag{4.58}$$

As indicated earlier, the total compressibility is obtained by summation:

$$c_t = c_f + c_g \tag{4.59}$$

In case of single-phase gas flow in a porous medium, the following can be assumed:

$$c_t \approx c_g \tag{4.60}$$

Combining Equations 4.54, 4.55, and 4.59, and since the term p/μz is assumed to be constant, the following form of the diffusivity equation for compressible fluids can be obtained:

$$\frac{1}{r}\frac{\partial}{\partial r}\left(r\frac{\partial p}{\partial r}\right) = \frac{\phi\mu c_t}{k}\frac{\partial p}{\partial t} \tag{4.61}$$

Pressure-squared approach. At relatively low pressure (<2,000 psia), any variation in the product of gas viscosity and gas deviation factor (μz) with pressure is not significant. Under such assumption, Equation 4.55 can be written as follows:

$$\frac{1}{r}\frac{\partial}{\partial r}\left(r\,p\right)\frac{\partial p}{\partial r} = \frac{\phi\,\mu\,z}{k}\frac{\partial}{\partial t}\left(\frac{p}{z}\right) \tag{4.62}$$

The following can be noted:

$$\frac{\partial\,(p)^2}{\partial\,r} = 2p\frac{\partial p}{\partial t} \tag{4.63}$$

Thus, a simpler form of the diffusivity equation can be obtained as follows:

$$\frac{1}{r}\frac{\partial}{\partial r}\left[r\frac{\partial\,(p)^2}{\partial r}\right] = \frac{\phi\mu c_t}{k}\frac{\partial\,(p)^2}{\partial t} \tag{4.64}$$

Lee and Wattenbarger presented a series of plots investigating the values of the p/μz and μz terms over wide ranges of pressure.[20] In certain instances, significant deviations from linearity were observed, particularly for gases of relatively high specific gravity and in certain temperature ranges. Hence, the pseudopressure approach is adopted in the analyses in order to linearize the diffusivity equation for compressible gases.

Pseudopressure approach. This approach introduces the pseudopressure function for compressible fluids as follows:[21,22]

$$m(p) = 2 \int_{p_{ref}}^{p} \frac{p}{\mu_g(p)z(p)} \, \partial p \tag{4.65}$$

In addition, a pseudotime function is introduced in the analysis as a further enhancement of the method.[23]

$$t_{pseudo} = \int_{0}^{t} \frac{1}{\mu_g z} \, \partial t \tag{4.66}$$

where

p_{ref} = reference pressure, psi,

$m(p)$ = pseudopressure, $psia^2/cp$, and

t_{pseudo} = pseudotime.

The following can be noted:

$$\frac{\partial m(p)}{\partial r} = \frac{2p}{\mu z} \frac{\partial p}{\partial r} \tag{4.67}$$

and

$$\frac{\partial t_{pseudo}}{\partial t} \approx \frac{1}{\mu c_t} \tag{4.68}$$

The diffusivity equation can thus be obtained as follows:

$$\frac{\partial^2 m(p)}{\partial r^2} + \frac{1}{r} \frac{\partial m(p)}{\partial r} = \frac{1}{2.637 \times 10^{-4}} \frac{\phi \mu c_t}{k} \left[\frac{\partial m(p)}{\partial t_{pseudo}} \right] \tag{4.69}$$

The outstanding advantage of this approach is the improved accuracy. However, since the pseudopressure values are quite large and both pseudopressure and pseudotime functions are not reported in familiar units, these have been normalized and implemented in various studies.

Equation 4.69 is nonlinear, as the gas properties appearing explicitly in the equation are functions of pressure (and hence, pseudopressure). Based on numerical methods, an approximate solution is proposed by Al-Hussainy and Ramey. The following solution is used to predict the changes in bottomhole flowing pressure in a gas well producing at a constant rate under unsteady-state conditions as follows:[24]

$$m(p_{wf}) = m(p_i) - \frac{1.637 \times 10^3 \, q_{sc} \, T}{kh} \left[\log \left(\frac{kt}{\phi \, \mu_i \, c_{ti} \, r_w^2} \right) - 3.23 \right] \tag{4.70}$$

where

$m(p_{wf})$ = pseudopressure function evaluated at bottomhole flowing pressure, p_{wf}, psia,

$m(p_i)$ = pseudopressure function evaluated at initial reservoir pressure, p_i, psia,

q_{sc} = well rate, Mscf/d, and

T = reservoir temperature in °R.

Furthermore, μ_i and c_{ti} are gas viscosity and total compressibility, respectively, at the initial reservoir pressure of p_i.

Table 4–6. Evaluation of the pseudopressure function of a compressible fluid

PVT Data				Calculations			Results
p (psia)	μ (cp)	z	Δp (psia)	p/μz (psia/cp)	p/μz$_{avg}$ (psia/cp)	2Δp [p/μz$_{avg}$] (psia²/cp)	m(p) (psia²/cp)
(1)	(2)	(3)	(4)	(1)/(2)×(3)=(5)	(6)	2×(4)×(6)=(7)	Σ (7)
14.7	0.01180	0.9983	—	1.2479×10^3	—	—	—
200	0.01228	0.9831	185.3	1.6567×10^4	8.9073×10^3	3.3010×10^6	3.3010×10^6
400	0.01268	0.9682	200	3.2582×10^4	2.4574×10^4	9.8297×10^6	1.3131×10^7
600	0.01306	0.9541	200	4.8152×10^4	4.0367×10^4	1.6147×10^7	2.9277×10^7
800	0.01351	0.9409	200	6.2935×10^4	5.5543×10^4	2.2217×10^7	5.1495×10^7
1,000	0.01397	0.9291	200	7.7044×10^4	6.9990×10^4	2.7996×10^7	7.9491×10^7
1,200	0.01439	0.9188	200	9.0761×10^4	8.3903×10^4	3.3561×10^7	1.1305×10^8
1,400	0.01485	0.9107	200	1.0352×10^5	9.7141×10^4	3.8856×10^7	1.5191×10^8
1,600	0.01533	0.9032	200	1.1556×10^5	1.0954×10^5	4.3815×10^7	1.9572×10^8
1,800	0.01577	0.8984	200	1.2705×10^5	1.2130×10^5	4.8521×10^7	2.4424×10^8
2,000	0.01624	0.8949	200	1.3762×10^5	1.3233×10^5	5.2933×10^7	2.9718×10^8
2,200	0.01671	0.8919	200	1.4761×10^5	1.4262×10^5	5.7046×10^7	3.5422×10^8
2,400	0.01716	0.8891	200	1.5731×10^5	1.5246×10^5	6.0984×10^7	4.1521×10^8

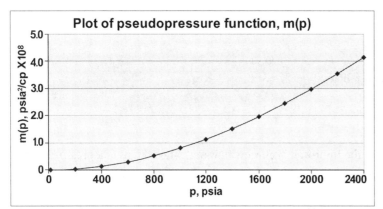

Fig. 4–8. Plot of pseudopressure function versus pressure for a compressible fluid. The function is unique for each gas sample having dissimilar values in viscosity and gas deviation factor. A similar plot presented in chapter 5 shows the pseudopressure function up to 6,000 psia.

Evaluation of the pseudopressure function

In order to compute the pseudopressure function, knowledge of the gas viscosity and gas deviation factor at various pressures is necessary. A suitable algorithm, such as the trapezoidal rule, can be utilized in a spreadsheet program or application software to evaluate the integral numerically as follows:[25]

$$m(p) = 2 \sum \frac{1}{2}\left[\left(\frac{p}{\mu z}\right)_{i-1} + \left(\frac{p}{\mu z}\right)_{i-1}\right](p_i - p_{i-1}) \quad \textbf{(4.71)}$$

The summation in Equation 4.71 is performed from 2 to n times, where there are n data points. Table 4–6 illustrates the computation methodology, which can be conveniently incorporated in digital computation.

Example 4.3: Calculation of the pseudopressure function based on PVT data. Calculate the values of the pseudopressure function of a reservoir gas sample. The PVT data of the fluid, including gas viscosity and gas compressibility factor, is provided in Table 4–6. Plot the results.

Example calculation (p = 400 psia)

Δp = 400 − 200 = 200 psia

p/μz @ 200 psia = 200/(0.01228)(0.9831) = 16,566.6 psia/cp

p/μz @ 400 psia = 400/(0.01268)(0.9682) = 32,581.8 psia/cp

p/μz$_{avg}$ = 0.5(16,566.6 + 32,581.8) = 24,574.2 psia/cp

2Δp (p/μz$_{avg}$) = 2 × 200 × 24,574.2 = 9.82968 × 10⁶ psia²/cp

m(p) = Σ 2Δp (p/μz$_{avg}$)

= 2Δp (p/μz$_{avg}$) @ 200 psia + 2Δp (p/μz$_{avg}$) @ 400 psia

= 3.301 × 10⁶ + 9.82968 × 10⁶ = 1.3131 × 10⁷ psia²/cp

The values of the pseudopressure function are plotted against pressure in Figure 4–8.

Example 4.4: Transient flow characteristics in formations of the same transmissibility but dissimilar rock permeability. A gas well is producing at a constant rate of 2,500 Mscf/d under transient conditions. Compare the decline in flow rate from the initial conditions up to 48 hrs in two different cases in which the rock permeabilities and formation thicknesses are dissimilar. All other properties, including the transmissibility, are the same. The following data is available:

Case I:
Rock permeability, mD, = 100.
Net formation thickness, ft, = 10.

Case II:
Rock permeability, mD, = 25.
Net formation thickness, ft, = 40.

Other data is as follows:
Porosity of the formation = 0.22.
Total compressibility, psi^{-1}, = 3.2×10^{-4}.
Initial reservoir pressure, psia, = 2,400.
Reservoir temperature, °R, = 595.
Radius of the wellbore, ft, = 0.29.

Assume that the reservoir is sufficiently large so that the infinite-acting flow period extends beyond the time period under study. The PVT properties of the gas are given in Table 4–6.

Solution:
Case I:

$$m(p_{wf}) = 4.1521 \times 10^8 - \frac{1.637 \times 10^3 (2500)(595)}{(100)(10)} \left[\log \frac{(100)t}{(0.22)(0.01716)(3.2 \times 10^{-4})(0.29)^2} - 3.23 \right]$$

$$= 4.1521 \times 10^8 - 2.43504 \times 10^6 \left[\log (9.8427 \times 10^8\ t) - 3.23 \right]$$

Case II:

$$m(p_{wf}) = 4.1521 \times 10^8 - \frac{1.637 \times 10^3 (2500)(595)}{(25)(40)} \left[\log \frac{(25)t}{(0.22)(0.01716)(3.2 \times 10^{-4})(0.29)^2} - 3.23 \right]$$

$$= 4.1521 \times 10^8 - 2.43504 \times 10^6 \left[\log (2.4607 \times 10^8\ t) - 3.23 \right]$$

The computed values of the pseudopressure function and corresponding flowing bottomhole pressures as obtained by interpolation are listed in Table 4–7.

From the analysis, it is evident that the rate of decline is greater during transient flow when the rock permeability is greater, although the same transmissibility is assumed in both cases. However, under steady-state conditions of flow where Darcy's law applies, two systems having the same transmissibility produce at the same volumetric flow rate, as illustrated in chapter 2.

Table 4–7. Decline in pressure under constant well rate: sensitivity to rock permeability

t	Case I k = 100 mD, h = 10 ft		Case II k = 25 mD, h = 40 ft	
hours	m(p), psia²/cp×10⁸	BHFP, psia	m(p), psia²/cp×10⁸	BHFP, psia
0.1	4.036	2,362.0	4.051	2,366.8
0.2	4.029	2,359.6	4.043	2,364.4
0.5	4.019	2,356.4	4.034	2,361.2
1	4.012	2,354.0	4.026	2,358.8
2	4.004	2,351.6	4.019	2,356.4
6	3.993	2,347.8	4.007	2,352.6
12	3.985	2,345.4	4.000	2,350.2
24	3.978	2,343.0	3.993	2,347.8
48	3.971	2,340.6	3.985	2,345.4

Key points—transient flow of compressible fluids

1. In characterizing the flow of compressible fluids in porous media, the density of real gases is incorporated in the equation of continuity as follows:

$$\rho = \frac{pM}{zRT}$$

   In the case of a slightly compressible fluid, the diffusivity equation is linearized by the assumption that fluid viscosity and compressibility are constant. In contrast, the solution for a compressible fluid requires special treatment, as gas properties such as viscosity and the gas deviation factor are strong functions of pressure.

2. There are three basic approaches to treat the nonlinearity of incompressible fluid properties, namely viscosity and gas deviation factor, as shown below:

Method for Solving Diffusivity Equation	Assumption	Customary Application	Comment
Pressure, p	Variation in $p/\mu z$ with pressure is negligible at relatively high pressure range	Reservoir pressure > 3,000 psia	
Pressure-squared, p^2	Variation in μz with pressure is negligible at relatively low pressure range	Reservoir pressure < 2,000 psia	Widely used in analyzing fluid flow in low pressure gas reservoirs
Pseudopressure, $m(p)$	Any variations in μ and z with pressure are accounted for by a pseudopressure function, which is used in lieu of pressure	Valid for all pressure ranges	Methodology requires numerical integration of the pseudopressure function over the entire pressure range

3. The pseudopressure function can be evaluated conveniently by implementing the trapezoidal rule or Simpson's rule, among others, in a spreadsheet or software application tool.

Steady-State Flow—Darcy's Law

Since steady-state flow is not a function of time, the diffusivity equation takes the following simplified form:

$$\frac{\partial^2 p}{\partial r^2} + \frac{1}{r}\frac{\partial p}{\partial r} = 0 \qquad (4.72)$$

Darcy's law provides the basis for characterizing the flow through a porous medium under a steady-state condition in a straightforward manner. Darcy's experiment was related to the flow of water through a permeable sand bed, as introduced earlier in chapter 2 in connection with the determination of rock permeability. It can be shown that Darcy's law for radial flow is a solution to Equation 4.72 under the appropriate initial and boundary conditions.[26]

Darcy's law is routinely utilized in evaluating one of the three unknowns when the values of the other two variables are known under a steady-state condition of flow in porous media:

- The average permeability of the porous medium
- The flow rate and apparent velocity of the fluid based on the bulk volume of the porous medium
- The pressure or potential gradient that exists along the flow path of the reservoir fluid

The assumptions of Darcy's law are discussed in chapter 2.

For horizontal flow in linear geometry, Darcy's law takes the following form:

$$q = -\frac{kA}{\mu}\frac{\partial p}{\partial s} \qquad (4.73)$$

where

q = fluid flow rate, cm^3/sec,

k = permeability of a porous medium, darcies,

A = the cross-sectional area of the porous medium in the direction of flow, cm^2,

p = reservoir pressure, atm,

$\frac{\partial p}{\partial s}$ = pressure gradient along the direction of flow, atm/cm,

s = distance along which flow occurs, cm, and

μ = viscosity of flowing fluid, cp.

The negative sign in Equation 4.73 appears as the fluid pressure decreases along the flow path, and the distance traversed by the fluid increases. It is noted that the fluid flow rate in Equation 4.73 is based on the bulk volume of the porous medium as opposed to the pore volume through which the actual movement of fluid takes place. Hence, the velocity of the fluid obtained on the basis of bulk volume is referred to as the apparent velocity:

$$v = \frac{q}{A} \tag{4.74}$$

A more general from of Equation 4.73 considers the effect of gravity during vertical or inclined flow by replacing the fluid pressure term with fluid potential. The fluid potential is defined as follows:

$$\Phi = p - \rho\, g\, \Delta z = p - 0.433\, \gamma\, \Delta z \tag{4.75}$$

where

γ = specific gravity of the fluid (water = 1), and

Δz = vertical distance between the point and an arbitrary datum level in the reservoir. In many instances, the datum is set at the oil/water contact (OWC).

By convention, z is considered to be positive downward. In evaluating the potential of a point, the following is observed based on Equation 4.75:

- The point is located at the same elevation as that of the datum, $\Delta z = 0$, and $\Phi = p$.
- The point is located below the datum level, Δz is positive, and $\Phi < p$.
- The point is located above the datum level, Δz is negative, and $\Phi > p$.

Based on Equations 4.73 and 4.75, the flow rate between two points can be obtained by integrating between the limits of potential and distance. The limits of the integral are as follows:

- Upstream: $\Phi = \Phi_1$, $s = 0$
- Downstream: $\Phi = \Phi_2$, $s = L$

In oilfield units, Darcy's law describing fluid flow in linear geometry (horizontal, vertical, or inclined) can be presented as follows:[27,28]

$$q = \frac{1.127 \times 10^{-3}\, k\, A\, (\Phi_1 - \Phi_2)}{\mu\, L} \tag{4.76}$$

where

k is given in mD,

A is in ft^2,

Φ is in psia, and

L is in ft.

When the two points are located at the same elevation, Equation 4.76 reduces to a simpler form, as the potential terms can be replaced by pressure:

$\Phi_1 = p_1$

$\Phi_2 = p_2$

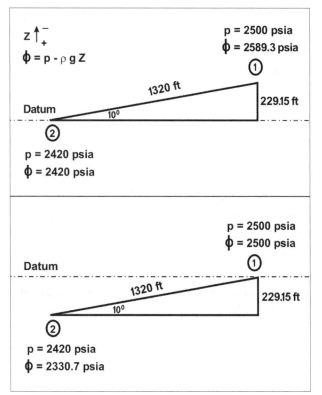

Fig. 4–9. Linear flow of fluid in an inclined plane from higher to lower potential. Two cases are shown where the datum is set at the level of either a downstream or upstream point. In both cases, the potential gradient between the two points is the same.

Example 4.5: Analysis of fluid flow in a dipping reservoir. Calculate the flow rate of oil in a reservoir between two points located at different elevations due to dip, as shown in Figure 4–9. The following data is available:

Length of the linear flow geometry, ft, = 1,320.
Cross-sectional flow area, ft², = 1,200.
Angle of dip, degrees, = 10.
Upstream pressure at point 1, psia, = 2,500.
Downstream pressure at point 2, psia, = 2,420.
Average rock permeability, mD, = 50.
Specific gravity of oil = 0.9.
Viscosity of oil, cp, =1.8.

Compare the results if the flow of fluid in the reservoir is assumed to be horizontal.

Solution: The vertical distance between the two points located at different elevations is determined as follows:

$$\Delta z = (L)\,(\sin 10°) = (1{,}320)\,(0.1736) = 229.15 \text{ ft}$$

Arbitrarily assuming the datum to be at point 2 located downstream, and recalling that Δz is considered positive downward (and hence, negative upward), the following values of potential are computed:

$$\Phi_1 = 2{,}500 - 0.433\,(0.9)\,(-229.15)$$
$$= 2{,}589.3 \text{ psia}$$

$$\Phi_2 = 2{,}420 - 0.433\,(0.9)\,(0)$$
$$= 2{,}420 \text{ psia}$$

Alternately, the datum can be assumed to be at point 1 located upstream, which leads to the following:

$$\Phi_1 = 2{,}500 - 0.433\,(0.9)\,(0)$$
$$= 2{,}500 \text{ psia}$$

$$\Phi_2 = 2{,}420 - 0.433\,(0.9)\,(+229.15)$$
$$= 2{,}330.7 \text{ psia}$$

Two points are evident from the above calculations of potential function:
- The value of fluid potential is dependent on the arbitrary selection of the datum.
- Regardless of the location of the datum, the difference of the potential values between two points essentially remains the same. In this case:

$$\Phi_1 - \Phi_2 = 2{,}589.3 - 2{,}420 = 2{,}500 - 2{,}330.7 = 169.3 \text{ psia}$$

Thus the datum can be arbitrarily set at any other level besides at the elevation at which the fluid pressure is measured. Finally, the flow rate between the two points is calculated based on Equation 4.76:

$$q = \frac{1.127 \times 10^{-3}\,(50)(1200)(169.3)}{(1.8)(1320)}$$
$$= 4.8 \text{ bbl/d}$$

If the flow is assumed to take place in a horizontal plane, $\Delta z = 0$ for points 1 and 2, and the fluid potentials are equal to the measured pressure values. Hence:

$$\Phi_1 - \Phi_2 = p_1 - p_2 = 2,500 - 2,420 = 80 \text{ psia}$$

The length of the horizontal flow path is calculated as follows:

$$L = (1,320)(\cos 10°) \approx 1,300 \text{ ft}$$

$$q = \frac{1.127 \times 10^{-3} \, (50)(1,200)(80)}{(1.8)(1,300)}$$

$$= 2.3 \text{ bbl/d}$$

In the former case, the gravitational effect contributes to the flow of oil, which is evidently substantial.

An alternate form of Equation 4.76, presented earlier in chapter 2, may be used to calculate the horizontal, inclined, or vertical flow in linear geometry without evaluating the potential function explicitly:

$$q = -1.127 \times 10^{-3} \frac{kA}{\mu} \left[\frac{\partial p}{\partial L} - 0.433 \, \gamma \, \cos \alpha \right] \tag{4.77}$$

where

α = angle of dip measured counterclockwise between the vertical direction downward and the inclined plane of fluid flow.

For flow in the vertical direction downward, the calculated flow rate would be significantly greater than what would be obtained for horizontal flow.

The various forms of Darcy's equation for single-phase flow are given in the following. Noting that the flow of fluid in porous medium is controlled by the imposed pressure gradient as well as gravity in the vertical direction, expressions for Darcy's law along the lateral (x or y) and vertical (z) axes are as follows:

$$q_x = -\frac{1.127 \times 10^{-3} \, k_x A_x}{\mu} \left(\frac{dp}{dx} \right) \tag{4.78}$$

$$q_y = -\frac{1.127 \times 10^{-3} \, k_y A_y}{\mu} \left(\frac{dp}{dy} \right) \tag{4.79}$$

$$q_z = -\frac{1.127 \times 10^{-3} \, k_z A_z}{\mu} \left(\frac{d\Phi}{dz} \right) \tag{4.80}$$

where

k_x, k_y, and k_z are the values of rock permeability in x, y, and z directions, respectively.

Field experience indicates that the vertical permeability is usually less than the permeability in the horizontal direction by one or two orders of magnitude. The existence of directional permeability is not uncommon in geologic formations, implying that $k_x \neq k_y$.

Steady-state linear flow—compressible fluids

Under steady-state conditions, the linear flow of compressible fluids can be described as follows:[29]

$$q_{sc} = \frac{3.164 \times 10^{-3} \, T_{sc} \, A \, k}{p_{sc} \, T \, z \, L \, \mu} \, (p_1^2 - p_2^2) \tag{4.81}$$

where

q_{sc} = gas flow rate in Mscf/d,

T = reservoir temperature in °R,

z = gas deviation factor (average),

μ = gas viscosity (average), cp,

p_{sc} = pressure at standard condition, psia, and

T_{sc} = temperature at standard condition, °R.

The previous equation is valid at a relatively low pressure range (p < 2,000 psia). Gas properties, such as viscosity and compressibility factor, are evaluated as follows:

$$p_m = \sqrt{0.5(p_1^2 + p_2^2)} \tag{4.82}$$

where

p_m = mean pressure at which the gas properties are evaluated, psia.

At relatively high reservoir pressures (p > 3,000 psia), the following correlation is more appropriate based on the assumption that the product of $\mu z/p$ is constant over the pressure range:

$$q_{sc} = \frac{6.328 \times 10^{-3} \, k \, T_{sc} \, A \, (P_1 - P_2)}{P_{sc} \, T \, (z \, \mu/p)} \tag{4.83}$$

In the above equation, the product of the gas deviation factor and viscosity over pressure is evaluated at the mean pressure:

$$p_m = \frac{P_1 + P_2}{2} \tag{4.84}$$

Steady-state radial flow—incompressible fluids

When the fluid flow to a well reaches steady state, the following equation may apply:[30]

$$q = \frac{7.08 \times 10^{-3} \, kh \, (P_e - P_w)}{\mu \, B_o \, \ln(r_e/r_w)} \tag{4.85}$$

In reality, it is difficult to attain steady-state flow in a well, as the production rate can change frequently due to operational constraints, including maintenance and downtime. Moreover, instabilities in the surrounding wells may influence the well under study.

Steady-state radial flow—slightly compressible fluids

The radial flow of a slightly compressible fluid in steady state is given by the following:[31]

$$q = \frac{7.08 \times 10^{-3} \, kh}{\mu \, B_o \, c_o \, \ln(r_e/r_w)} \, \ln[1 + c_o(p_e - p_w)] \tag{4.86}$$

where

q = flow rate evaluated at the bottomhole reference pressure, stb/d.

For the typical range of oil compressibility, the assumption of oil as an incompressible fluid may lead to underestimation of the reservoir pressure, as shown in the following example. However, the deviation may not be large.

Example 4.6: Estimation of pressure at the reservoir boundary—compressible and slightly compressible fluids. An oil well is producing at a stabilized rate of 300 stb/d from a circular drainage area estimated to have a radius of 2,640 ft. The reservoir is believed to receive strong pressure support from the aquifer. Assuming steady-state flow conditions, compare the reservoir pressures at the drainage boundary, considering the oil to be the following:

(a) An incompressible fluid

(b) A slightly compressible fluid

The following data is available:

Flowing bottomhole pressure, psia, = 1,463.
Oil viscosity, cp, = 0.58.
Oil formation volume factor, rb/stb, = 1.35.
Compressibility, psi^{-1}, = 28.5×10^{-6}.
Formation thickness, ft, = 13.
Average permeability, mD, = 29.
Radius of the wellbore, ft, = 0.29.

Assume that the well is located at the center of a circular drainage area.

Solution:

(a) Incompressible fluid:
By rearranging Equation 4.85, p_e can be computed as follows:

$$p_e = p_w + \frac{q_o\, \mu_o\, B_o\, \ln\,(r_e/r_w)}{7.08 \times 10^{-3}\, k\, h}$$

$$= 1,463 + \frac{(300)(0.58)(1.35)\ln\,(2,640/0.29)}{(7.08 \times 10^{-3})(29)(13)}$$

$$= 2,265.3 \text{ psia}$$

(b) Slightly compressible fluid:
Based on Equation 4.86, the following can be obtained:

$$p_e = p_w + (1/c_o)\, \exp\left[\frac{q\mu B_o\, c_o\, \ln\,(r_e/r_w)}{7.08 \times 10^{-3}\, k\, h}\right] - 1/c_o$$

$$p_e = 1463 + \frac{1}{28.5 \times 10^{-6}}\, \exp\left[\frac{(300)(0.58)(1.35)(28.5 \times 10^{-6})\ln(2,640/0.29)}{7.08 \times 10^{-3}\,(29)(13)}\right] - \frac{1}{28.5 \times 10^{-6}}$$

$$= 2,274.5 \text{ psia}$$

The deviation in results is about 0.4%.

Steady-state radial flow—compressible fluids

The following equation can be used to calculate the steady-state flow rate in a gas well at a pressure below 2,000 psi:[32,33]

$$q_{sc} = \frac{19.88 \times 10^{-6}\, T_{sc}\, k\, h\, (p_e^2 - p_w^2)}{p_{sc}\, T\, (z\, \mu)\, \ln\,(r_e/r_w)} \tag{4.87}$$

Assuming p_{sc} = 14.7 psia and T_{sc} = 520°R, Equation 4.87 can be rewritten as in the following:

$$q_{sc} = \frac{kh\, (p_e^2 - p_w^2)}{1422\, T\, z\, \mu\, \ln\,(r_e/r_w)} \tag{4.88}$$

where

q_{sc} is given in Mscf/d,
T is given in °R, and
z and μ are evaluated at average pressure, as defined in Equation 4.82.

The equation for flow based on the pseudopressure function is given below:

$$q_{sc} = \frac{kh\,[m(p)_1 - m(p)_2]}{1{,}422\,T\,[\ln{(r_e/r_w)}]} \tag{4.89}$$

where

$m(p)_1$ and $m(p)_2$ are evaluated at the drainage boundary and at the wellbore, respectively.

Example 4.7: Rate prediction for a planned well in a gas reservoir. A low permeability gas field is under development. Estimate the flow rate of a future gas well under steady-state conditions when the following data is available (the sources of the data are listed in parentheses):

> External radius of well drainage area, ft, = 660 (well spacing).
> Reservoir pressure at the external boundary of well drainage, psia, = 1,870 (field pressure monitoring program).
> Bottomhole flowing pressure, psia, = 1,145 (transient test in adjacent wells).
> Net formation thickness, ft, = 30 (openhole logs).
> Average rock permeability, mD, = 2.5 (multidisciplinary studies).
> Wellbore radius, ft, = 0.29 (completion records).
> Gas sp. gr. = 0.69 (laboratory studies).
> Reservoir temperature, °F, = 120 (downhole measurement).

Solution: Since the reservoir pressure is below 2,000 psia, the pressure-squared approach as shown in Equation 4.88 is used to estimate the flow rate of the planned well.

The average pressure to estimate the PVT properties is calculated as follows:

$$p_m = \sqrt{(1{,}870^2 + 1{,}145^2)/2}$$

$$= 1{,}550 \text{ psia (105 atm, approximately)}$$

> The pseudocritical pressure, psia, = 669 (from fig. 3–7, chapter 3).
> The pseudocritical temperature, °R, = 387 (from fig. 3–7, chapter 3).
> The pseudoreduced pressure = 1,550/669 = 2.32.
> The pseudoreduced temperature = (120 + 460)/387 = 1.5.
> The gas deviation factor, z, = 0.805 (from fig. 3–6, chapter 3).
> The gas viscosity at 1 atm and 120°F = 0.0113 (from fig. 3–9, chapter 3).
> The viscosity ratio = 1.33 (from fig. 3–10, chapter 3).
> The gas viscosity at 1,550 psia and 120°F, cp, = 0.0113 × 1.33 = 0.015.

$$q_{sc} = \frac{(2.5)(30)\,(1{,}870^2 - 1{,}145^2)}{1{,}422\,(580)\,(0.805)(0.015)\,\ln{(660/0.29)}}$$

$$= 2{,}130 \text{ Mscf/d}$$

Effects of bottomhole pressure and well radius on production rate

Figures 4–10 and 4–11 show the effects of bottomhole flowing pressure and well radius on the production rate under steady-state flow conditions. When all other factors are the same, a relatively low bottomhole pressure creates a relatively large drawdown and greater flow rate through the wellbore.

Comparison between compressible and incompressible fluids under steady-state conditions

Figure 4–12 presents a comparison between the steady-state pressure profiles that develop during compressible and incompressible flow of fluids towards the wellbore. The external radius, r_e, of the well is 660 ft, and the well radius is 0.25 ft.

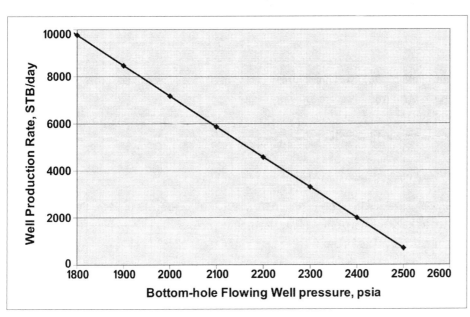

Fig. 4–10. Plot of oil production rate versus bottomhole flowing pressure. The inverse relationship between the two leads to the definition of well productivity index illustrated in Fig 4–16

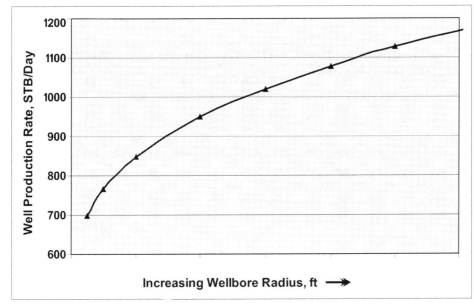

Fig. 4–11. Influence of wellbore radius on production rate. Depending on well productivity, the size of the wellbore is optimized.

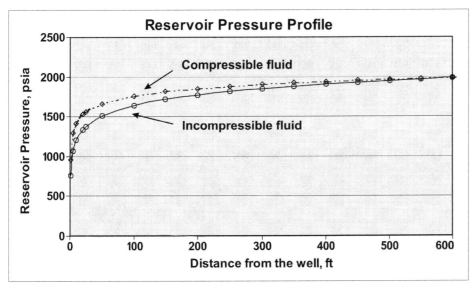

Fig. 4–12. Reservoir pressure during radial flow—comparison between incompressible and compressible fluids

Key points—steady-state flow

1. Fluid flow behavior under steady-state conditions is not a function of time. Darcy's law can be utilized to determine one of three parameters when the remaining two are known:
   - Average rock permeability of the flow medium
   - Stabilized pressure gradient along the flow path
   - Stabilized well rate

The assumptions inherent in Darcy's law are presented in chapter 2.

2. In generalized form, Darcy's law is expressed in terms of potential gradient in order to account for the flow of fluid between two points not located at the same horizontal plane, leading to inclined flow or vertical flow. In oilfield units, the fluid potential can be calculated as follows:

$$\Phi = p - 0.433 \, \gamma \, \Delta z$$

where

Φ = fluid potential, psia,

γ = specific gravity of the flowing fluid, and

Δz = difference in elevation between an arbitrary datum level and the point where the fluid potential is determined.

3. By convention, z is taken to be positive downward. Hence, the following values of z and Φ are observed when determining the potential at a point:

Location of Point	Δz, ft	Φ, psia
Same elevation as datum	0	$\Phi = p$
At lower elevation than datum	> 0	$\Phi < p$
At higher elevation than datum	< 0	$\Phi > p$

4. The potential values of individual points shift by an equal amount when the arbitrary datum level is changed. However, the *difference* in potential between any two points remains the same regardless.

5. Familiar equations describing fluid flow under steady-state conditions include the following:
   - Incompressible, slightly compressible and compressible fluids
   - Linear and radial flow

Solutions to the Diffusivity Equation—Pseudosteady-State Flow

A well steadily producing from a bounded reservoir leads to the development of pseudosteady-state flow once any pressure drop due to production reaches the reservoir boundary from the wellbore. Flow from multiple wells producing from their individual drainage areas is usually characterized by a pseudosteady-state flow condition. Under this condition, reservoir pressure continues to change with time; however, the *rate* at which the change occurs is constant. A no-flow condition is assumed at the outer boundary, as follows:

$$\frac{\partial p}{\partial t} = 0, \qquad r = r_e \tag{4.90}$$

Hence, the diffusivity equation as shown in Equation 4.30 can be solved to predict the fluid characteristics under pseudosteady-state condition as follows:[34]

$$p_D = \frac{2t_D}{r^2_{eD}} + \ln r_{eD} - \frac{3}{4} \tag{4.91}$$

where

$$p_D = \frac{kh(p_i - p_{wf})}{141.2qB\mu} \tag{4.92}$$

The equation is valid at long times given by the following:

$$t > \frac{948 \, \text{ø} \mu c_t r_e^2}{k}$$

(4.93)

In dimensional form, the flowing bottomhole pressure is given by the following:

$$p_{wf} = p_i - \frac{141.2 \, qB\mu}{k \, h} \left[\frac{5.274 \times 10^{-4} \, k \, t}{\text{ø} \, \mu \, c_t \, r_e^2} + \ln \, (r_e / r_w) - \tfrac{3}{4} \right]$$

(4.94)

Determination of pore volume

It can be further shown that during pseudosteady-state flow, the rate of change in the bottomhole flowing pressure is related to the reservoir (or drainage) volume as per the following:

$$\frac{\partial p_{wf}}{\partial t} = -\frac{0.234 \, qB}{c_t V_p}$$

(4.95)

where

$$V_p = \frac{A \, h \, \text{ø}}{5.615}$$

(4.96)

V_p = pore volume of the reservoir or well drainage region in bbl, and
A = area of the reservoir or well drainage area in ft^2.

The rate of decline in pressure, $\frac{\partial p_{wf}}{\partial t}$, is based on material balance and is inversely proportional to reservoir pore volume. In fact, Equation 4.96 is used to assess the extent of the reservoir (or individual well drainage area) in a reservoir limit test, as described in chapter 5. During the test, the bottomhole flowing pressure is recorded with time to determine the rate of decline. Values of other parameters used in Equation 4.95 are usually available.

Determination of average reservoir pressure

Under pseudosteady-state flow, the decrease in reservoir pressure is proportional to the volume produced and the total compressibility of the system as follows:

$$p_i - p_{av} = \frac{\Delta V}{c_t V_p}$$

(4.97)

where

p_{av} = volumetric average reservoir pressure, psia,
ΔV = volume of fluid produced, bbl, and
V_p = pore volume, bbl.

Writing the volume of fluid produced in terms of the production rate and time, and expressing the pore volume in terms of rock porosity and reservoir dimensions, the following can be obtained:

$$p_{av} = p_i - \frac{0.234 \, q \, B \, t}{\text{ø} \, c_t \, h(\pi \, r_e^2)}$$

(4.98)

The initial reservoir pressure can be eliminated from Equations 4.94 and 4.98. Then an expression for the average reservoir pressure under pseudosteady-state condition can be obtained in terms of the stabilized production rate and bottomhole flowing pressure:

$$p_{av} = p_{wf} + \frac{141.2 \, q \, B \, \mu}{k \, h} \, [\ln \, (r_e / r_w) - \tfrac{3}{4} + s)]$$

(4.99)

where

s = skin factor, positive or negative depending on permeability damage or stimulation

It is to be borne in mind that, under the pseudosteady-state condition, both the volumetric average reservoir pressure and the bottomhole pressure decline at the same pace as the well continues to produce at a stabilized rate. The rate of change in reservoir pressure is constant throughout. When plotted against time, the pressure trends are parallel straight lines with a negative slope.

Equation 4.99 is a snapshot in a specific point in time that relates the average reservoir pressure in the drainage area to the flowing bottomhole pressure of a well if the latter is available at that time. Determination of the average pressure in the well drainage area constitutes one of the major objectives in well testing and is treated in chapter 5 with an illustrative example. A reservoir team routinely seeks the average reservoir pressure from the well test results in a large field in order to evaluate the reservoir performance. Areas of unexpectedly higher or lower reservoir pressure often require appropriate remedial measures. These measures include, but are not limited to, manipulation of well rates, well recompletion, and drilling of new producers or injectors. The aforementioned measures can ensure optimal reservoir performance under the given conditions.

If the well drainage area is not circular, a shape factor is introduced in the equation to calculate the average reservoir pressure, as follows:[35]

$$p_{av} - p_{wf} = 141.2 \frac{q\,B\,\mu}{kh}\left[0.5\ln\left(\frac{10.06\,A}{C_A\,r_w^2}\right) - 0.75 + s\right] \qquad (4.100)$$

where

A = drainage area, ft², and

C_A = shape factor.

The values of C_A, which depend on the drainage geometry and relative location of the well, are available in the literature. Some common values of the shape factor are provided in Table 4–8. In certain instances, the well is not located at the center. A few of these are shown in Figure 4–13.

Table 4-8. Shape factor for various drainage geometries. *Source: R. C. Earlougher, Jr. 1977. Advances in Well Test Analysis.* SPE Monograph Series. Vol. 5. Richardson, TX: Society of Petroleum Engineers.

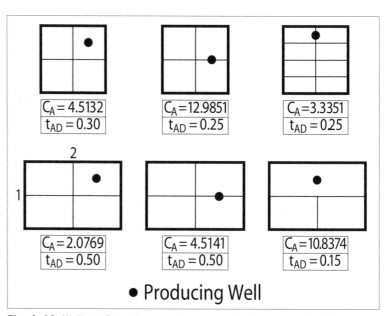

Fig. 4-13. Well configurations and corresponding shape factors for certain off-centered wells *Source: R. C. Earlougher, Jr. 1977. Advances in Well Test Analysis. SPE Monograph Series. Vol. 5. Richardson, TX: Society of Petroleum Engineers.*

Drainage Geometry	C_A	t_{DA}
Circular	31.62	0.06
Square	30.88	0.05
Hexagonal	31.6	0.06
Rectangular (2:1)	21.83	0.15
Rectangular (4:1)	5.38	0.30
Triangular	27.6	0.07

Note: (a) The well is located at the center of the drainage area in all cases.

(b) t_{DA} = dimensionless time, following which a pseudosteady-state solution can be used with < 1% error in results. In some literature, this term is referred to as t_{AD}. Dimensionless time can be converted to the time measured in hours as follows:

$$t > \frac{\o\,\mu\,c_t\,A\,t_{DA}}{2.637\times10^{-4}\,k}$$

Example 4.8: Estimation of average reservoir pressure under pseudosteady-state conditions. Estimate the average reservoir pressure in the circular drainage area where an oil well is producing under pseudosteady-state flow conditions at a rate of 250 stb/d. The following data is available:

Net formation thickness, ft, = 40.
Average permeability in the drainage region, mD, = 100.
Bottomhole flowing pressure, psi, = 2,000.
Oil viscosity, cp, = 0.9.
Oil formation volume factor, rb/stb, = 1.32.
Radius of drainage area, ft, = 2,979.
Wellbore radius, ft, = 0.328.
Skin factor (dimensionless) = 3.0.

Solution: First, the drainage radius is calculated, followed by estimation of the average reservoir pressure in the drainage area.

$$P_{av} = 2,000 + 141.2 \frac{(250)(1.32)(0.9)}{(100)(40)} [\ln(2,979/0.328) - 0.75 + 3]$$

$$= 2,119.1 \text{ psi}$$

Example 4.9. Effects of skin factor and well location on reservoir pressure determination. Redo Example 4.3 by the following:

(a) Neglect the skin factor.

(b) Assume that the well is located in a quadrant of a square drainage area shown in Figure 4–13. The area of the square is equal to the circular area used in Example 4.8.

Compare the results.

Solution:

(a) In this case, s = 0, and p_{av} = 2,087.7 psi, as obtained by Equation 4.100. The results indicate that the reservoir pressure is underestimated by 31.3 psi. This highlights the importance of taking into account the skin factor in determining the average reservoir pressure.

(b) Again, sound knowledge of drainage area and shape is necessary to make reliable estimates of reservoir pressure, as demonstrated below:

$$p_{av} = 2000 + 141.2 \frac{(250)(1.32)(0.9)}{(100)(40)} \left[0.5 \ln \frac{10.06(27,879,855)}{(4.5132)(0.328)^2} - 0.75 + 3 \right]$$

$$= 2,129.3 \text{ psia}$$

The average reservoir pressure is found to be 10.2 psia greater than the base case.

Example 4.10: Performance of a future well under uncertainty. A likely range of production rates needs to be estimated for a planned well in a reservoir under development. A pseudosteady-state flow condition is assumed for the wells once well production is stabilized following the initial high rate. The following data, some in ranges pertaining to the reservoir, is available:

Average reservoir pressure prior to drilling, psia, = 3,250.
Flowing bottom hole pressure in nearby wells, psia, = 2,080.
Drainage radius of the well, ft, = 1,320.
Net thickness of the formation, ft, = 25 – 29.
Permeability range from core studies, mD, = 15–25.
Viscosity of oil, cp, = 0.85–0.88.
Oil formation volume factor, rb/stb, = 1.05.
Wellbore radius, ft, = 0.32.
Skin factor (dimensionless) = –3.0 to 3.0.

Solution: Rearranging Equation 4.100, an expression for the well rate under pseudosteady-state condition is obtained as in the following:

$$q = \frac{(p_{av} - p_{wf})\,kh}{141.2\,\mu\,B_o\,[\ln\,(r_e/r_w) - \tfrac{3}{4} + s]}$$

The upper and lower limits are related to formation transmissibility (kh/μ), skin, and the extent of drawdown pressure. Based on the ranges of data given above, the limiting rates are computed below:

(a) Upper limit

$$q = \frac{(3,250 - 2,080)(25)(29)}{(141.2)(0.85)(1.05)\,[\ln\,(1,320/0.32) - \tfrac{3}{4} + (-3)]}$$

$$= 1,471\ \text{stb/d}$$

(b) Lower limit

$$q = \frac{(3,250 - 2,080)(15)(25)}{(141.2)(0.88)(1.05)\,[\ln\,(1,320/0.32) - \tfrac{3}{4} + 3]}$$

$$= 318\ \text{stb/d}$$

Pseudosteady-state flow—compressible fluids

For relatively low pressure (< 2,000 psi), an expression for pseudosteady-state flow can be obtained for compressible fluids as follows:[36]

$$q_{sc} = \frac{kh\,(p_{av}^2 - p_{wf}^2)}{1,422\,T\,z\,\mu\,[\ln\,(r_e/r_w) - 0.75 + s]} \qquad (4.101)$$

In the case of turbulent flow encountered in the vicinity of gas wells, the skin factor, s, can be replaced by apparent skin, s, as follows:

$$s = s + Dq \qquad (4.102)$$

where

$$D = F\,\frac{kh}{1,422T} \qquad (4.103)$$

(D is referred to as the turbulent flow factor; F is the non-Darcy flow coefficient. T is the reservoir temperature.)

The net effect of turbulence is the development of additional pressure drop near the wellbore. During a gas well test, the value of the turbulent flow factor (D) is estimated by flowing the well at different rates and then plotting apparent skin versus the flow rate. An example of the effects of the turbulent flow factor on transient pressure is shown in chapter 5. The following equation based on the pseudopressure function can be used for compressible fluids:

$$q_{sc} = \frac{kh\,[m(p_{av}) - m(p_{wf})]}{1,422\,T\,[\ln\,(r_e/r_w) - 0.75 + s]} \qquad (4.104)$$

At relatively high pressure, a simpler form of the solution of the radial diffusivity equation can be used, as in the following:[37]

$$q_{sc} = \frac{kh\,(p_{av} - p_{wf})}{1,422\,\mu B_g\,[\ln\,(r_e/r_w) - 0.75 + s]} \qquad (4.105)$$

where

$$B_g = \frac{5.04 \times 10^{-3}\, z\, T}{p}, \text{ and} \qquad\qquad (4.106)$$

$$p_m = \frac{p_{av} + p_{wf}}{2}$$

In the above equation, μ and B_g are evaluated at the mean pressure pm.

Example 4.11: Comparison between the pseudopressure and pressure-squared approach. A gas well is producing under a pseudosteady-state condition in a reservoir having an average permeability of 20 mD. However, the permeability of the formation in the immediate vicinity of the well is severely damaged, as indicated by a skin factor of 6. (Skin factor is treated in detail in chapter 5). Estimate the well production rate by the pseudopressure approach and compare the result with that of the pressure-squared approach. Fluid PVT data is given in Table 4–6. Additionally, the following data is available:

      Average reservoir pressure in the drainage area, psia, = 2,000.
      Bottomhole flowing pressure, psia, = 1,000.
      Ratio of external drainage radius over wellbore radius = 2,000.
      Reservoir temperature, °F, = 130.

Based on the pseudopressure approach shown in Equation 4.104, the well flow rate is calculated as follows:

$$q_{sc} = \frac{(20)(10)\ (2.9718 \times 10^8 - 7.9491 \times 10^7)}{1{,}422\ (590)\ [\ln (2{,}000) - 0.75 + 6]}$$

$$= 4{,}038 \text{ Mscf/d}$$

The values of pseudopressure at 2,000 and 1,000 psia were read from Table 4–6. In the pressure-squared approach, the mean pressure is determined first for evaluating the viscosity and the gas deviation factor:

$$p = [(2000^2 + 1000^2)/2]^{0.5}$$

$$= 1{,}581.1 \text{ psia}$$

From Table 4–6, the values of gas viscosity and deviation factor are obtained by interpolation, as in the following:

$$\mu = 0.01528 \text{ cp}$$

$$z = 0.9039$$

$$q_{sc} = \frac{(20)(10)\ [(2{,}000)^2 - (1{,}000)^2]}{1422\ (590)\ (0.9039)\ (0.01528)\ [\ln (2{,}000) - 0.75 + 6]}$$

$$= 4{,}029 \text{ Mscf/d}$$

The deviation between the results is about 0.2%.

Key points—pseudosteady-state flow

1. Pseudosteady-state flow is encountered in a reservoir with a no-flow boundary once the pressure transients originated at the well due to flow propagate to the boundary. Pseudosteady-state condition is also encountered when multiple wells produce from a field, and a no-flow boundary is created between the individual drainage areas. The diffusivity equation is solved with the following boundary condition:

$$\frac{\partial p}{\partial t} = 0, \qquad r = r_e$$

2. During pseudosteady-state flow, the pressure declines at the same rate everywhere in the closed system. The rate of decline is inversely proportional to the drainage volume and total compressibility.

3. Interpretation of the pseudosteady-state flow, as performed in the course of well test analysis, has the following important applications, among others:
   - Determination of reservoir or drainage pore volume
   - Determination of average pressure in the reservoir or drainage area

Table 4–9. Summary of selected fluid flow equations

Fluid Flow Model	Equation Relating Fluid Pressure, Rate, and Time
Unsteady-state flow, well at the center of circular drainage, constant flow rate, slightly compressible fluid, infinite-acting reservoir (line-source solution)	$p_D = \frac{1}{2}\left[-\operatorname{Ei}\left(-\frac{1}{4t_D}\right)\right]$ where $p_D = \frac{kh(p_i - p)}{141.2\,q\,B\,\mu}$ and $t_D = \frac{2.637 \times 10^{-4}\,k\,t}{\emptyset\,\mu\,c_t\,r^2}$
Unsteady-state flow, compressible fluid, well at the center of circular drainage, constant flow rate	$m(p_{wf}) = m(p_i) - \frac{1.637 \times 10^3\,q_{sc}\,T}{kh}\left[\log\left(\frac{kt}{\emptyset\,\mu_i\,c_{ti}\,r_w^2}\right) - 3.23\right]$
Steady-state 1-D flow in horizontal, inclined, or vertical plane, incompressible fluid	$q = \frac{1.127 \times 10^{-3}\,k\,A\,(\Phi_1 - \Phi_2)}{\mu L}$, where $\Phi = p - 0.433\,\gamma\,\Delta z$
Steady-state 1-D linear flow, compressible fluid, low pressure ($< 2{,}000$ psia)	$q_{sc} = \frac{3.164 \times 10^{-3}\,T_{sc}\,A\,k}{p_{sc}\,T\,z\,L\,\mu}\,(p_1^2 - p_2^2)$, where z and μ are evaluated at $p_m = \sqrt{0.5(p_1^2 + p_2^2)}$
Steady-state 1-D linear flow, compressible fluid, high pressure ($> 3{,}000$ psia)	$q_{sc} = \frac{6.328 \times 10^{-3}\,k\,T_{sc}\,A\,(p_1 - p_2)}{p_{sc}\,T\,(z\,\mu/p)}$ where z and μ are evaluated at $p_m = (p_1 + p_2)/2$
Steady-state radial flow, incompressible fluid	$q = \frac{7.08 \times 10^{-3}\,kh\,(P_e - P_w)}{\mu\,B_o\,\ln(r_e/r_w)}$
Steady-state radial flow, slightly compressible fluid	$q = \frac{7.08 \times 10^{-3}\,kh}{\mu\,B_o\,c_o\,\ln(r_e/r_w)}\,\ln[1 + c_o(P_e - P_w)]$
Steady-state radial flow, compressible fluid, low pressure ($<2{,}000$ psia)	$q_{sc} = \frac{kh\,(p_e^2 - p_{wf}^2)}{1{,}422\,T\,z\,\mu\,\ln\,(r_e/r_w)}$
Steady-state radial flow, compressible fluid, all pressure ranges	$q_{sc} = \frac{kh\,[m(p)_1 - m(p)_2]}{1{,}422\,T\,[\ln\,(r_e/r_w)]}$
Pseudosteady-state flow, constant rate, no-flow boundary, cylindrical drainage geometry, slightly compressible fluid	$p_{wf} = p_i - \frac{141.2\,q\,B\,\mu}{k\,h}\left[\frac{5.274 \times 10^{-4}\,k\,t}{\emptyset\,\mu\,c_t\,r_e^2} + \ln\,(r_e/r_w) - \tfrac{3}{4}\right]$
Pseudosteady-state flow, constant rate, no-flow boundary, various drainage geometries, slightly compressible fluid	$p_{av} - p_{wf} = 141.2\,\frac{q\,B_o\,\mu}{kh}\left[0.5\,\ln\left(\frac{10.06\,A}{C_A\,r_w^2}\right) - 0.75 + s\right]$
Pseudosteady-state flow, constant rate, no-flow boundary, slightly compressible fluid	$\frac{\partial P_{wf}}{\partial t} = \frac{0.234\,qB}{c_t\,V_p}$
Pseudosteady-state flow, constant rate, no-flow boundary, compressible fluid, all pressure ranges	$q_{sc} = \frac{kh\,[m(p_{av}) - m(p_{wf})]}{1{,}422\,T\,[\ln\,(r_e/r_w) - 0.75 + s]}$

Note: Under steady-state conditions, fluid flow characteristics are independent of time.

Flow through fractures

Under steady-state conditions, an expression for steady-state flow of incompressible fluids through a fracture under pressure gradient can be obtained as follows:[38]

$$q = 8.7 \times 10^9 \frac{W^2 A (p_1 - p_2)}{B \mu L} \tag{4.107}$$

where
    W = width of the fracture, ft, and
    A = cross-sectional flow area of the fracture.

Furthermore, the fracture permeability can be approximated as follows:

$$k = 7.7(10)^{12} W^2 \tag{4.108}$$

Example 4.12: Estimation of fracture permeability. Estimate the permeability of a fracture that has a width of 10^{-4} in.

Solution: Based on Equation 4.108, the fracture permeability is calculated as follows:

$$k = 7.7(10)^{12}(10^{-4}/12)^2$$
$$= 535 \text{ millidarcies}$$

If the fracture has a width of 10^{-3} in., $k = 53.5$ darcies. These calculations indicate that fractures can be highly transmissive in an otherwise tight rock matrix.

Multiphase fluid flow

When oil, gas and water phases are flowing simultaneously in porous medium, the equations for a single homogeneous fluid are modified to account for the individual phase pressures, relative permeabilities and viscosities as demonstrated in chapter 13. The flowrate of oil, gas and water are expressed in terms of their saturations. Furthermore, capillary pressure as function of saturations is used to correlate oil, gas and water phase pressures. In essence, the following can be written for horizontal flow, and the resulting equations are solved simultaneously for phase pressure and saturation in a simulator:

$$q_o = f\left(\frac{\partial p_o}{\partial x}, S_o, B_o, \mu_o, K_{ro}\right) \tag{4.109}$$

$$q_w = f\left(\frac{\partial p_w}{\partial x}, S_w, B_w, \mu_w, K_{rw}\right) \tag{4.110}$$

$$q_g = f\left(\frac{\partial p_g}{\partial x}, S_g, B_g, \mu_g, K_{rg}, R_s, \frac{\partial p_o}{\partial x}, S_o, B_o, \mu_o, K_{ro}\right) \tag{4.111}$$

The auxiliary equations are:

$$S_o + S_g + S_w = 1$$
$$p_{c,ow} = p_o - p_w$$
$$p_{c,go} = p_g - p_o$$

The subscripts o, w, and g denote oil, water, and gas, respectively. In eq. (4.111), Rs is the solution gas oil ratio, which takes into account the gas dissolved in oil phase. Oil-water and gas-oil capillary pressures are denoted by $p_{c,ow}$ and $p_{c,go}$, respectively. If the capillary pressure is negligible at a specific saturation, the same pressure drop exists between the phases. Lee and Wattenbarger, among others, have treated the equations describing multiphase flow in porous media under unsteady-state conditions in detail.[39]

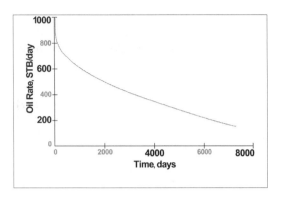

Fig. 4–14. Oil production rate in a reservoir where two-phase flow is encountered below the bubblepoint following a period of single-phase production above the bubblepoint. *Courtesy of Gemini Solutions, Inc.*

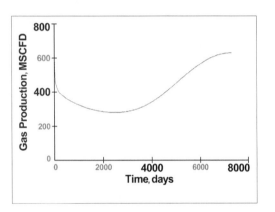

Fig. 4–15. Changes in gas production as the reservoir pressure declines below the bubblepoint. Above the bubblepoint, oil and gas production decline as the reservoir depletes. Once gas evolution occurs below the bubblepoint, a marked increase in gas production is encountered, while the oil rate continues to decline as shown previously. *Courtesy of Gemini Solutions, Inc.*

Figures 4–14 and 4–15 present the typical flow behavior of the individual fluid phases (oil and gas) during depletion. Above the bubblepoint, the oil rate declines with time, and so does the gas production at the separator. However, when the reservoir pressure declines below the bubblepoint, gas evolution from the liquid phase takes place in the reservoir, leading to a significant increase in gas production and in the gas/oil ratio. As the reservoir is depleted further, most of the dissolved gas comes out of solution, and a decrease in the gas/oil ratio is eventually observed.

Multiphase, Multidimensional Simulation of Fluid Flow

Equations describing fluid flow in porous media are the backbone of reservoir simulation models, which can either be analytical or numerical. As described in preceding sections, analytic models generally represent simplified flow conditions and reservoir geometry, and serve well in conceptualizing the dynamic processes occurring in the reservoir during its life cycle. However, rigorous modeling of fluid flow in reservoirs involves numerical simulation of the three fluid phases (oil, gas, and water) in three dimensions (x, y, and z). This allows consideration of changes in the fluid phase due to reservoir pressure changes. A suite of correlations for simultaneous flow of oil, gas and water in one dimension is outlined in Equations 4.109 through 4.111 followed by the auxiliary equations.

Reservoir simulation is discussed in detail in chapters 13 and 14. Reservoir simulation tools can range from simple to robust. They are employed to determine the dynamic changes in the saturation and pressure of each fluid phase in various locations in the reservoir due to production and injection. Compositional simulators attempt to track the composition of the fluid phases when simulating production of a near-critical oil or gas condensate undergoing significant phase changes. Influences of various geologic heterogeneities on fluid flow, including fractures and stratification, are also studied based on reservoir simulation models. Again, water influx from an adjacent aquifer and external water injection influence the fluid flow characteristics and reservoir performance. Simulations of thermal and miscible processes involved in enhanced oil recovery (EOR) processes are also performed to optimally design recovery operations in the fields.

The objectives of modeling fluid flow in porous media include, but are not limited to, the following:

- Conceptualize and characterize the reservoir
- Understand fluid dynamics, including phase changes
- Match production history of the individual wells and the field
- Validate the reservoir description from geosciences studies
- Design the horizontal well trajectory
- Build field development strategies and optimize production of oil and gas
- Evaluate a variety of what-if scenarios
- Predict reservoir performance in the future
- Ensure better overall management of the reservoir

Reservoir data acquisition and integrated models are described in chapters 6 and 7, with case studies.

In reality, the flow of fluids is likely to occur under transient conditions in a field. This is because the wells produce or undergo injection under conditions in which strict stabilization of rates is difficult to achieve. Moreover, certain wells are shut down periodically for routine maintenance, and a number of new wells may be drilled. The changes are rapid initially when a well either starts or ceases production (or injection). However, with the passage of time, the changes are observed to take place at a diminished pace. Multiphase, multidimensional reservoir simulators are based on unsteady-state fluid flow models. However, the simulators are capable of predicting reservoir performance under a variety of conditions, including pseudosteady-state or steady-state flow.

In the following sections, important applications of fluid flow models pertaining to well and reservoir performance are discussed, including the following:
- Vertical and horizontal well productivity indices
- Water and gas coning in an oil well
- Influx of water into the reservoir

The mathematical models discussed in the following are highly valuable in evaluating well, reservoir, and aquifer performance in reservoir engineering studies.

Productivity Index of a Well

Vertical oil well

The productivity index, simply abbreviated PI, is a measure of how efficiently an oil or gas well can produce under a specific pressure drawdown at the sandface. The productivity index of an oil well is defined as follows:

$$J = \frac{q_o}{\Delta p} \tag{4.112}$$

where

    J = productivity index, stb/d/psi,

    q_o = oil production rate, stb/d, and

    Δp = pressure drawdown experienced by the well, psi.

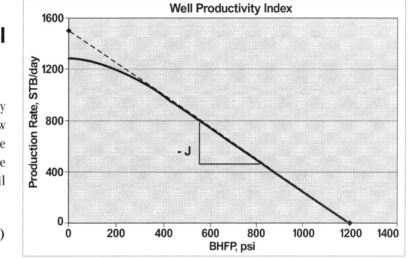

Fig. 4–16. Productivity index (J) is obtained from a plot of flow rate versus bottomhole pressure. At small bottomhole pressures (larger drawdown), the well production rate may exhibit an asymptotic behavior.

Considering pseudosteady-state flow, the productivity index can be defined for radial flow geometry and other drainage shapes as follows:

$$J = \frac{q_o}{p_{av} - p_{wf}} \tag{4.113}$$

$$= \frac{7.081 \times 10^{-3} k_o h}{[\mu_o B_o \ln(r_e/r_w) - \tfrac{3}{4} + s]} \tag{4.114}$$

$$= \frac{7.081 \times 10^{-3} k_o h}{\mu_o B_o [0.5\ln(10.06A/C_A r_w^2) - \tfrac{3}{4} + s]} \tag{4.115}$$

In Equation 4.113, p_{av} is the average reservoir pressure in the well drainage region. As seen from these equations, well productivity is a function of rock and fluid properties, formation damage, and reservoir geometry. Hence, wells from different fields cannot be compared solely based on productivity index. A representative value of a well productivity index can be obtained by producing the well at successively lower bottomhole pressures and noting the increment in production rate (fig. 4–16).

Equations 4.114 and 4.115 indicate that the productivity of a well is greater when the effective permeability to oil is high and the viscosity of the oil is low. A frequently encountered condition for reduction in permeability is formation damage near the wellbore, or skin. The net effect of skin is the apparent reduction in well productivity, which is detrimental to well performance and is discussed in the next chapter. Transient well tests are routinely performed to evaluate skin factor surrounding the wellbore.

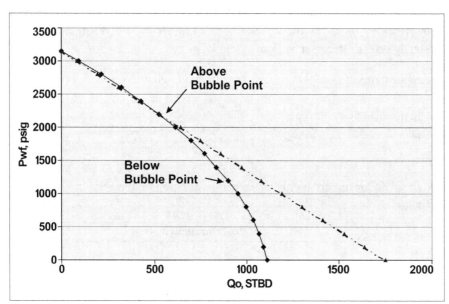

Fig. 4–17. Inflow performance relationships for a well producing above and below the bubblepoint pressure

The specific productivity is defined in terms of reservoir thickness as follows:

$$J_s = \frac{J}{h} \qquad (4.116)$$

In a manner analogous to productivity index, the injectivity index of an injector is defined as follows:

$$I = \frac{q_{inj}}{P_{wf} - P_{av}} \qquad (4.117)$$

The injectivity index is an important parameter used in the evaluation of waterflood performance. Waterflooding is described in chapter 16.

Inflow performance relationship

In the case of an undersaturated reservoir producing above the bubblepoint, the straight-line relationship is applicable:

$$q_o = J(p_{av} - p_{wf}) \qquad (4.118)$$

The linear correlation between drawdown and the productivity index adequately describes flow rate as a function of well flowing pressure (or drawdown) above the bubblepoint. However, a departure from the trend is observed when the well produces

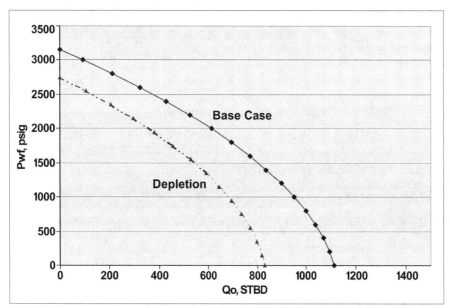

Fig. 4–18. The dynamic characteristic of the inflow performance relationship as the reservoir is depleted to relatively low pressure

below the bubblepoint in a solution gas drive reservoir. The inflow performance relationship (IPR) correlates expected flow rate with bottomhole flowing pressure for a producing well as in the following:[40]

$$\frac{q_o}{q_{o\,max}} = 1 - 0.2 \left(\frac{p_{wf}}{p_{av}} \right) - 0.8 \left(\frac{p_{wf}}{p_{av}} \right)^2 \qquad (4.119)$$

Figure 4–17 presents typical inflow performance relationships for a well producing above or below the bubblepoint. It must be noted that the inflow performance relationship for a well is not constant throughout the life of the reservoir. As the reservoir depletes, the inflow performance relationship curve shifts to the left, as shown in the figure. Figure 4–18 demonstrates the inflow performance relationship during production.

Example 4.13: Evaluation of the injectivity index of a well for waterflooding. The injectivity indices of the wells are known to range between 0.7 bbl/d/psi and 0.9 bbl/d/psi in a reservoir. In order to inject 500 bbl/d in a waterflood operation, what would be the maximum bottomhole injection pressure to be attained? Assume that the average reservoir pressure is 1,750 psi.

Solution: Rearranging Equation 4.117, and noting that q is negative in the case of injection, the following is obtained, using a conservative value of injectivity index:

$$p_{wf} = \frac{q}{J} + p_{av}$$
$$= 500/0.7 + 1,750$$
$$= 2,464 \text{ psi}$$

In certain shallow formations, the optimum injection pressure gradient may exceed the fracture gradient. Hence, the target injection rate may not be achieved in order to avoid the creation of fracture pathways in the formation.

Horizontal well

In recent times, an increasing number of horizontal wells have been drilled to achieve better reservoir performance. The advantages of developing a reservoir with horizontal wells are discussed in chapter 19, with case studies and illustrations of single-lateral and multilateral horizontal wells. The following equation can be used to estimate the productivity index:[41,42]

$$J_h = \frac{7.081 \times 10^{-3} \, h \, k_h}{\mu_o \, B_o \, [\ln(R) + (Bh/L)\ln\{(Bh/r_w(B+1))\}]} \tag{4.120}$$

where

$$R = \frac{a + [a^2 - (0.5L)^2]^{0.5}}{0.5L}, \tag{4.121}$$

$$a = \frac{L}{2}\left[0.5 + \sqrt{0.25 + (2r_{eh}/L)^4}\right]^{0.5}, \tag{4.122}$$

$$B = \sqrt{k_h/k_v}, \text{ and} \tag{4.123}$$

r_{eh} = drainage radius of the horizontal well, ft.

Example 4.14: Productivity index and production rate of a horizontal well. Calculate the productivity index and production rate of a horizontal well with the following data:

Length of lateral, ft, = 3,500.
Bottomhole flowing pressure, psi, = 1,350.
Formation permeability, mD, = 20.
Estimated drainage area, acres, = 160.
Average pressure of the drainage area, psia, = 2,250.
Radius of wellbore, ft, = 0.42.
Average formation thickness, ft, = 60.
Viscosity of oil, cp, = 1.2.
Oil formation volume factor, rb/stb, = 1.24.

Assume the porous medium in which the well is drilled is uniform and isotropic.

Solution: Since the formation is isotropic, B = 1.

$$r_{eh} = (160 \times 43{,}560/3.14)^{1/2}$$
$$= 1{,}490 \text{ ft}$$

$$a = 0.5L[0.5 + (0.25 + (2r_{eh}/L)^4)^{0.5}]^{0.5}$$
$$= 0.5(3{,}500)[0.5 + (0.25 + (2 \times 1{,}490/3{,}500)^4)^{0.5}]^{0.5}$$
$$= 2{,}056.3$$

$$R = \frac{2056.3 + [2056.3^2 - (0.5 \times 3{,}500)^2]^{0.5}}{(0.5)(3{,}500)}$$
$$= 1.792$$

$$J_h = \frac{7.081 \times 10^{-3}(60)(20)}{(1.2)(1.24)[\ln(1.792) + (60/3{,}500)\ln\{60/(0.42 \times (1+1))\}]}$$
$$= \frac{8.4972}{(1.2)(1.24)(0.5833 + 0.0732)}$$
$$= 8.7 \text{ stb/d/psi}$$

Hence, the production rate is:

$$q_o = 8.7 \ (2{,}250 - 1{,}350)$$
$$= 7{,}830 \text{ stb/d}$$

Further accuracy in calculations can be obtained by considering the effective length of the horizontal wells instead of the drilled length. The effective length of a horizontal wellbore, as determined by well test interpretation, could be less than the drilled length due to various factors. Such factors include permeability degradation across the well trajectory, high water saturation at certain sections of the wellbore, and sand control issues. An example of a horizontal well test interpretation is provided in chapter 5.

For horizontal wells producing under pseudosteady-state condition, the following inflow performance relationship can be used to estimate the flow rate when the flowing well pressure is known:[43]

$$\frac{q_h}{q_{h \, max}} = 1 + 0.2055 \left(\frac{p_{wf}}{p_{av}}\right) - 1.1818 \left(\frac{p_{wf}}{p_{av}}\right)^2 \tag{4.124}$$

Example 4.15: Compare the productivity indices between horizontal and vertical wells. Compare the horizontal well productivity obtained above with that of a vertical well drilled in the same reservoir. Assume $k_o \approx k_h$ and $s = 0$ in estimating the productivity index.

Solution: Assuming pseudosteady-state flow, and by using Equation 4.114, one can obtain the following:

$$J = \frac{7.081 \times 10^{-3}(20)(60)}{(1.2)(1.24)[\ln(1490/0.42) - \tfrac{3}{4}]}$$
$$= 0.77 \text{ bbl/d/psi}$$

It is noted that the productivity of a horizontal well computed in Example 4.14 is one order of magnitude greater than that of a traditional vertical well.

Example 4.16. Horizontal well performance—sensitivity analysis. Based on Equation 4.120, analyze the sensitivity of a horizontal well productivity index to formation permeability, anisotropy, formation thickness, and length of lateral.

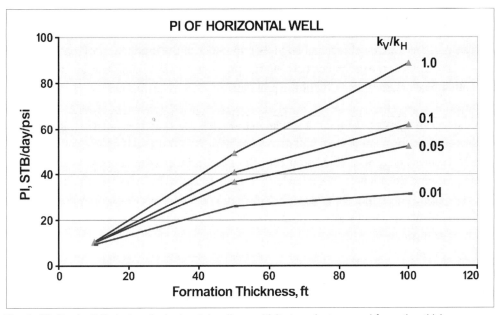

Fig. 4–19. Productivity index of a horizontal well—sensitivity to anisotropy and formation thickness

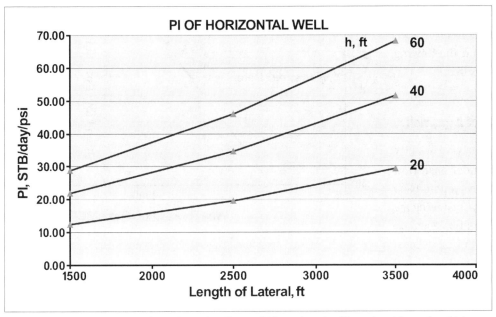

Fig. 4–20. Productivity index of a horizontal well—sensitivity to length of lateral and formation thickness

Solution: Figures 4–19 and 4–20 highlight the influences of the above parameters on well productivity. The following observations are made:

1. Inspection of the equation for productivity index indicates that horizontal well productivity has a direct correlation with formation permeability. The plot is not shown here.

2. Long horizontal bores tend to increase productivity. In chapter 19, an illustration is given showing oil production rate versus length of laterals in a thermal recovery operation. This is in agreement with the above conclusion.

3. Relatively thick net pay leads to better productivity.

4. Permeability anisotropy reduces productivity when the formation is relatively thick, all other factors being the same.

Key points—well productivity index

The productivity index of a well, abbreviated as PI, is a measure of the production rate achievable under a given drawdown pressure. The latter is the difference between the pressure at the external boundary of the drainage region and the flowing bottomhole pressure. The unit of the productivity index is barrels per day per pounds per square inch of pressure (bbl/d/psi).

The injectivity index of an injector is defined in a similar manner. It is a key parameter in designing waterflood operations.

From a reservoir engineering point of view, well management primarily depends on the well productivity index. The productivity index is dependent on rock and fluid properties. One of the principal causes of the reduction in well productivity is formation damage around the wellbore. Wells having the same productivity index in two different reservoirs cannot be equated. The index also changes in a reservoir with depletion. The productivity index of a horizontal lateral of a well is usually significantly higher due to the long exposure of the wellbore to the formation and relatively low drawdown.

The inflow performance relationship attempts to predict the well production rate at various flowing bottomhole pressures. In solution gas drive reservoirs, the inflow performance relationship is no longer a straight line below the bubblepoint. It is expressed by a nonlinear relationship between the flow rate and the flowing bottomhole pressure.

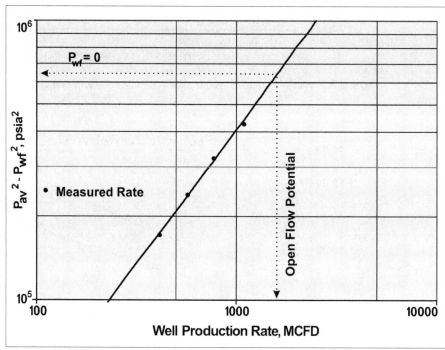

Fig. 4–21. Absolute open flow potential (AOFP) of a gas well

Flow capacity of a gas well

The productive capacity of a gas well under open flow conditions (i.e., open flow potential) is calculated using the steady-state radial flow equation for compressible fluid, as follows:

$$q_{sc} = c\,(p_f^2 - p_w^2)^n \tag{4.125}$$

where

$$c = \frac{703\,k\,h}{\mu_o\,zT\,\ln(r_e/r_w)} \tag{4.126}$$

q_{sc} = gas flow rate, scf/d,

k_g = effective permeability to gas, darcies,

μ = gas viscosity, cp,

p_f = formation pressure, psia, and

n = 1 for completely laminar steady-state flow; 0.5 for completely turbulent steady-state flow.

Then:

$$\ln q_{sc} = n \log (p_f^2 - p_w^2) + \log c \tag{4.127}$$

According to Equation 4.127, a plot of $(p_f^2 - p_w^2)$ versus q_{sc} on log-log paper should yield a straight line. For n = 0.971 and c = 500, typical open flow test data is shown in Figure 4–21. The open flow potential is the value at which $p_w = 0$.

In order to obtain meaningful results, the well must be produced at a stabilized rate for a sufficient period leading to the attainment of pseudosteady-state flow.

Water and gas coning

Coning of gas and water in oil wells is a common issue in reservoir performance that must be addressed. Figure 4–22 illustrates the coning effects in an oil well producing at a higher-than-optimum rate. Coning occurs due to imbalances between viscous and gravity forces that result from very high drawdown created in the vertical direction near the wellbore. Consequently, the in-situ fluids (both oil and water, although the latter has higher gravity) are driven towards the well perforations.

Consider the position of an oil/water interface in a reservoir. At a sufficient distance away from the well, equilibrium between the capillary, gravity, and viscous forces is achieved at a level below the perforation of the well. Care is taken during completion so that the wells are perforated above the oil/water contact. However, due to significantly high viscous forces that develop near the well, movable water is drawn up towards the perforations. The net result is formation of a water cone

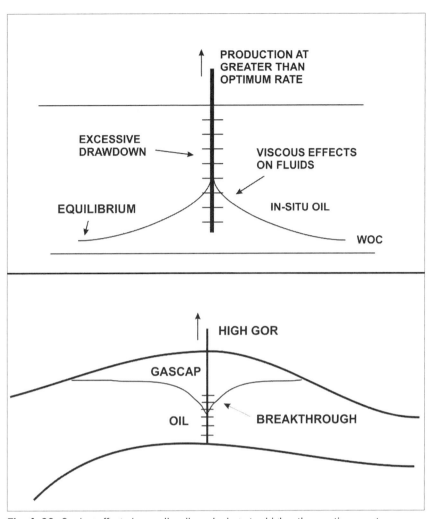

Fig. 4–22. Coning effects in an oil well producing at a higher-than-optimum rate

and a high water/oil ratio in the producing stream. A similar phenomenon is observed where a gas cap overlying the oil column exists, as free gas is coned downward due to viscous effects and is eventually produced through the wellbore, resulting in a high gas/oil ratio. In certain cases, both water and gas coning are observed in the same well.

Reservoir management practices to reduce the detrimental effects of coning are discussed in chapter 19, along with a case study. One approach to coning problems involves recompletion of the affected well as a horizontal well, where sufficient distance is kept between the new well trajectory and the oil/water contact or gas/oil contact. Furthermore, drawdown of horizontal wells could be significantly less, reducing the effects of coning. In other cases, the well is choked down or reperforated in a different position in order to reduce the flow of gas or water.

Various analytical methods used to analyze coning behavior in vertical and horizontal wells are compiled by Ahmed.[44] The equations can be utilized to predict the following:

- Critical rate for water and gas coning
- Optimum placement of well perforations
- Breakthrough time of gas or water in an oil well due to coning

The phenomenon of coning is a strong function of the vertical permeability of the formation. Chaperon proposed the following correlation that considers permeability anisotropy in order to estimate the critical rate for coning to occur:[45]

$$q_{oc} = C\, k_h\, [(h - h_p)^2 / \mu_o\, B_o]\Delta\rho\, q_c^* \tag{4.128}$$

where

q_{oc} = critical oil rate in stb/d, defined as the rate above which coning will lead to gas or water breakthrough,
$C = 7.83 \times 10^{-6}$,
$q_c^* = 0.7311 + 1.943/\alpha''$,
$\alpha'' = (r_e/h)\,(k_v/k_h)^{0.5}$,
k_h = horizontal permeability, mD,
$\Delta\rho$ = density difference between fluids, lb/ft³,
h = oil column thickness, ft, and
h_p = perforated interval, ft.

In horizontal wells, the following equations are applicable to evaluate critical rates for water and gas coning, respectively:

$$q_{oc} = a - b(\rho_w - \rho_o)\frac{k_h[h - (h - D_b)]^2}{\mu_o B_o} \tag{4.129}$$

$$q_{oc} = a - b(\rho_o - \rho_g)\frac{k_h[h - (h - D_t)]^2}{\mu_o B_o} \tag{4.130}$$

where

$a = 7.83 \times 10^{-2}$,
$b = 10^{-4}\,(L\, q_c^*/y_e)$,
$q_c^* = 3.9624955 + 0.0616438\,\alpha'' - 0.000504(\alpha'')^2$,
$\alpha'' = (y_e/h)\,(k_v/k_h)^{0.5}$,
D_b = distance between the water/oil contact and the horizontal wellbore,
D_t = distance between the gas/oil contact and the horizontal wellbore,
h = oil column thickness,
y_e = one-half of the distance between two lines of horizontal wells, and
L = horizontal well length.

Example 4.17. Effects of permeability anisotropy and oil viscosity on water coning. Evaluate the effects of the following on a well located in a drainage radius of 1,320 ft:

(a) Permeability anisotropy

(b) Water coning

The following data is available:

Formation thickness, ft, = 80.
Horizontal permeability, mD, = 50.
Oil density, lb/ft³, = 48.5.
Density of water, lb/ft³, = 63.5.
Viscosity of oil, cp, = 0.56.
Oil formation volume factor, rb/stb, = 1.2.
Perforated interval, ft, = 28.

Solution: The problem is analyzed by assuming three values of k_v/k_h as follows: (i) 0.01, (ii) 0.1, and (iii) 1.0. When $k_v/k_h = 0.1$, $\alpha'' = 5.2178$.

$q_c^* = 0.7311 + 1.943/5.2178$
$\quad = 1.1035$
$q_{oc} = 7.83 \times 10^{-6} \times 50 \times [(80 - 28)^2/(0.56)(1.2)](63.5 - 48.5)(1.1035)$
$\quad = 26.1 \text{ stb/d}$

Using the same methodology, the critical rates are computed for other values of anisotropy. The results are tabulated at the right:

k_v/k_h	q_{oc}, stb/d	
	($\mu_o = 0.56$ cp)	($\mu_o = 0.67$ cp)
0.01	45.1	37.6
0.1	26.1	21.7
1.0	20.1	16.7

The analysis suggests the following:

(a) Coning can develop at relatively low oil rates, except for low values of vertical permeability or permeability anisotropy.

(b) Viscous oil is less mobile, which aggravates water coning.

Key points—water and gas coning

Water and gas coning occur when oil wells produce over a critical rate, leading to severe imbalances between the viscous and gravity forces near the wellbore. Coning results in an excessive water/oil ratio or gas/oil ratio in the producing stream and a loss in oil production. The equilibrium attained between gravity and viscous forces in the reservoir is lost near the wellbore. As a result, water (having a higher density than oil) cones upward and is driven into the well perforations. Similarly, gas in the gas cap is coned downward into the wellbore.

Various equations are available in the literature to compute the optimum well rate beyond which coning is likely to occur. Remedial measures include recompletion of the well further away from the oil/water contact and producing the well at the optimum rate. To mitigate the effects of oil and gas coning, horizontal wells are drilled in order to reduce large drawdown and place the wellbore in precise interval.

Immiscible Displacement of Fluid

Displacement of oil from a porous medium by immiscible fluids, water or gas, can be described by the fractional flow equation, and the frontal advance theory, proposed by Buckley and Leverett.[46] The fractional flow equation for water displacing oil, as encountered during waterflooding a reservoir is as follows:

$$f_w = q_w / (q_w + q_o)$$

Expressing the fluid flow rates in terms of Darcy's law and noting that capillary pressure is the difference between non-wetting and wetting phase pressures, one can obtain:

$$f_w = \frac{1 + 0.001127 \dfrac{k\, k_{ro}}{\mu_o} \dfrac{A}{q_t} \left[\dfrac{\partial p_c}{\partial L} - 0.433\, \Delta\rho \sin \alpha_d \right]}{1 + \dfrac{\mu_w}{\mu_o} \dfrac{k_{ro}}{k_{rw}}} \qquad (4.131)$$

where:

A = area, ft^2,
f_w = fraction of water flowing,
k = absolute permeability, mD,
k_{ro} = relative permeability to oil,
k_{rw} = relative permeability to water,
μ_o = oil viscosity, cp,
μ_w = water viscosity, cp,
L = distance along direction of flow, ft,
p_c = capillary pressure = $p_o - p_w$, psi,
q_t = total flow rate = $q_o + q_w$, stb/d,
$\Delta\rho$ = water-oil density difference = $\rho_w - \rho_o$, g/cm^3, and
α_d = angle of formation dip to the horizon, degrees.

The fractional flow of water for given rock and fluid properties and flooding conditions is a function of water saturation. This is true because relative permeability and capillary pressure are functions of saturation only. Application of the frontal advance theory is illustrated in chapter 16.

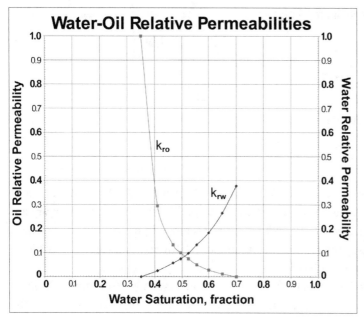

Fig. 4–23. Water-oil relative permeability curves

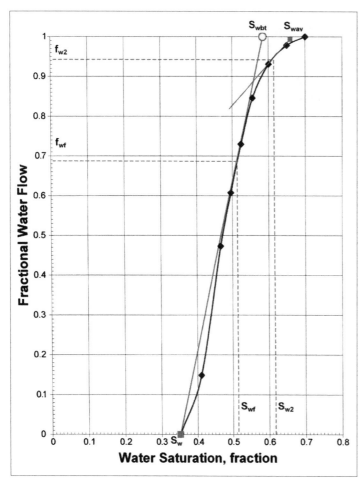

Fig. 4–24. Fractional flow curve

Neglecting gravity and capillary effects, the above fractional flow equation is reduced to the following:

$$f_w = \frac{1}{1 + \frac{\mu_w}{\mu_o}\frac{k_{ro}}{k_w}} \qquad (4.132)$$

Plots of water-oil relative permeabilities and fractional flow curves are shown in Figures 4–23 and 4–24, respectively. Immiscible displacement, including the Buckley and Leverett frontal advance theory, are discussed in more detail in chapter 16.

The linear frontal advance equation for water, based upon the conservation of mass and assuming incompressible fluids, is given by the following:

$$\left(\frac{\partial x}{\partial t}\right) = \frac{q_t}{A\emptyset}\left(\frac{\partial f_w}{\partial S_w}\right)_t \qquad (4.133)$$

This equation states that the rate of advance of a plane of a fixed water saturation, Sw, at a certain time, t, is equal to the total fluid velocity multiplied by the change in composition of the flowing stream caused by a small change in the saturation of the displacing fluid.

The frontal advance equation can be used to derive the expressions for average water saturation as follows:

At breakthrough:

$$\bar{S}_{wbt} - S_{wc} = \left(\frac{\partial S_w}{\partial f_w}\right)_f = \frac{S_{wf} - S_{wc}}{f_{wf}} \qquad (4.134)$$

After breakthrough:

$$\bar{S}_w - S_{w2} = \frac{1 - f_{w2}}{\left(\frac{\partial f_w}{\partial S_w}\right)_{S_{w2}}} \qquad (4.135)$$

where

f_{wf} = fraction of water flowing at the flood front,

f_{w2} = fraction of water flowing at the producing end of the system,

$\bar{S}_w$ = average water saturation after breakthrough, fraction,

S_{wf} = water saturation at the flood front, fraction,

$\bar{S}_{wbt}$ = average water saturation at breakthrough, fraction,

S_{wc} = connate water saturation, fraction, and

S_{w2} = water saturation at the producing end of the system, fraction.

Mobility ratio of fluid phases

According to Darcy's law, the flow rate is proportional to the mobility of the fluid, which is the ratio of the permeability of the flowing fluid to the fluid's viscosity. In the case of water displacing oil, the ratio of the water mobility to oil mobility is a governing factor in the immiscible displacement process.

In waterflooding, the permeabilities are for two different and separate regions in the reservoir. Craig suggested calculating the mobility ratio prior to water breakthrough, i.e., k_{rw} at the average water saturation in the swept region, and k_{ro} in the unswept zone, as given below.[47]

The mobility ratio is defined as follows:

$$M = \frac{\lambda_w \text{ in the water contacted portion}}{\lambda_o \text{ in the oil bank}} \tag{4.136}$$

$$= \frac{\dfrac{k_{rw}}{\mu_w}}{\dfrac{k_{ro}}{\mu_o}}$$

where

λ = mobility, effective permeability/viscosity,

k_r = relative permeability,

μ = viscosity, cp, and

w, o = subscripts denoting oil and water, respectively.

A mobility ratio of less than unity leads to the favorable displacement of oil by water. Displacement efficiency is further discussed in chapter 16.

Water/oil ratio and gas/oil ratio

The water/oil ratio (WOR) and gas/oil ratio (GOR) at reservoir conditions can be calculated according to the following:

$$WOR_{res} = \left(\frac{q_w}{q_o}\right) = \left(\frac{\mu_o}{\mu_w}\right)\left(\frac{k_w}{k_o}\right) \tag{4.137}$$

$$GOR_{res} = \left(\frac{q_g}{q_o}\right) = \left(\frac{\mu_o}{\mu_g}\right)\left(\frac{k_g}{k_o}\right) \tag{4.138}$$

These equations can be developed for surface conditions as well, provided the respective PVT data (i.e., formation volume factors, viscosities, and gas solubility values) is available as follows:

$$WOR_{surf.} = \left(\frac{B_o}{B_w}\right)\left(\frac{\mu_o}{\mu_w}\right)\left(\frac{k_w}{k_o}\right) \tag{4.139}$$

$$GOR_{surf.} = \frac{\dfrac{q_g}{B_g} + R_s\left(\dfrac{q_o}{B_o}\right)}{\dfrac{q_o}{B_o}} \tag{4.140}$$

Therefore, based on Equations 4.138 and 4.140, the gas/oil ratio is expressed as follows:

$$R = R_s + \left(\frac{B_o}{B_g}\right)\left(\frac{\mu_o}{\mu_g}\right)\left(\frac{k_g}{k_o}\right) \tag{4.141}$$

Water Influx from an Aquifer

Many petroleum reservoirs are influenced by water influx and pressure support from adjacent aquifers. The following scenarios are commonplace:

- **Bottom water drive.** Characterized by vertical flow of water. Bottom water drive in petroleum reservoirs is illustrated in Figures 1–2 and 1–3 in chapter 1.

- **Edge water drive.** Water influx occurs only from the edges of the reservoir, such as at the downdip edges of an anticlinal structure. The flow of water in a vertical direction is not considered to be significant.

- **Linear water drive.** Water influx occurs in a linear plane from one direction.

In the case of a radial aquifer, it must be noted that water encroachment may not occur throughout 360°, due to the presence of various geologic features, including faults.

The general equation for water influx from an aquifer is expressed as follows:

$$W_e = U \, S(p,t) \tag{4.142}$$

where

U = an aquifer constant, and

$S(p, t)$ = an aquifer function that is defined separately for different aquifer types.

The quantitative evaluation of the cumulative water encroachment, W_e, into a reservoir is one of the very important problems of primary production analysis. Primary recovery is generally better when pressure support is provided by the aquifer during depletion of a petroleum reservoir. Since aquifer influence is not amenable to direct measurement, its evaluation must necessarily be deduced from indirect estimates. In fact, to calculate water influx from an aquifer, the engineer confronts what is inherently the greatest uncertainty in the whole subject of reservoir engineering. Calculations of W_e require a mathematical model that in turn relies on the properties of the aquifer, i.e., fluid properties, permeability, thickness, and geometrical configuration, etc. However, these aquifer characteristics are seldom known, since wells are not intentionally drilled into the aquifer to obtain this data.

There are many uncertainties in the model (i.e., steady state or unsteady state, and the geometry, dimensions, and properties of the aquifer). Thus direct calculation of water influx, even though it is possible, is unreliable. For the best accuracy, water influx calculations are made in conjunction with the overall material balance of the reservoir. Examples are provided in chapter 12, where various aquifer models are employed to match the production history of a reservoir. The model that provides the best match is generally accepted as a valid aquifer model for the reservoir.

Reservoir simulators are routinely used to characterize aquifer support based on history matching with past reservoir performance. (Reservoir simulation is discussed in chapters 13 and 14.) Well test analyses can provide information related to dynamic conditions at the outer boundary of the reservoir, such as no flow or constant pressure. This aids in characterizing aquifer support. Well test interpretation is discussed in chapter 5.

Aquifer models

There are various aquifer models commonly used in reservoir engineering studies. Steady-state models are relatively simple and are based on the assumption that the rate of water influx is a function of pressure drop alone at the oil/water contact. However, in many circumstances, robust treatment of aquifer behavior is necessary. Hence, aquifer influx is considered to be time-dependent, leading to unsteady-state models. For example, a steady-state model can be based on Darcy's law, while an unsteady-state model employs a diffusivity equation to predict the aquifer performance. A relatively simple pseudosteady-state model is also proposed, which assumes that water influx is due to pseudosteady-state behavior of the aquifer. Such a model is found to be adequate in many cases. Aquifer models include the following:

Steady-state and pseudosteady-state models	Unsteady-state aquifer models
• Small (pot) aquifer	• Infinite linear
• Schilthuis	• Carter-Tracy
• Hurst (simplified)	• van Everdingen–Hurst
• Fetkovich	• Finite/infinite radial
	• Finite/infinite bottom water drive

Wang and Teasdale presented a list of theoretical functions and constants for small, Schilthuis steady-state, Hurst simplified steady-state, van Everdingen and Hurst infinite linear, and radial unsteady-state aquifers.[48–51] Two of the above-mentioned aquifer models, with various degrees of complexity, are described briefly in the following sections.

Small aquifer. A small or pot aquifer can be modeled as follows:

$$S(p,t) = p_i - p \tag{4.143}$$

$$U = (c_w + c_f) V_{aq} \tag{4.144}$$

where

p_i = initial reservoir pressure, psi,

p_n = reservoir pressure at any time following the initial condition, psi,

c_w = compressibility of the water, psi^{-1},

c_f = compressibility of the rock, psi^{-1}, and

V_{aq} = initial pore volume of the water in the aquifer, bbl.

If water influx does not occur from all sides into the reservoir, Equation 4.144 is modified as in the following:

$$U = (c_w + c_f) V_{aq} (\alpha / 360) \tag{4.145}$$

where

α = the angle of encroachment in degrees.

Example 4.18. Calculation of water influx volume—small aquifer. Consider a reservoir that is cylindrical in shape, with a radius of 5,000 ft. Calculate the volume of water influx into the reservoir for a pressure drop of 500 psi. Porosity of the formation is 0.2. Assume the following aquifer properties:

Formation compressibility, psi^{-1}, = 4.8×10^{-6}.

Compressibility of water, psi^{-1}, = 4.05×10^{-6}.

Aquifer thickness, ft, = 50.

Porosity = 0.14.

Estimated radius, ft, = 12,000.

Solution: A volumetric estimate of the water initially present in the pot aquifer can be made as follows:

$$V_{aq} = 3.14 \,(12{,}000^2 - 5{,}000^2)\,(50)\,(0.14) / 5.615$$
$$= 465.8 \times 10^6 \text{ bbl}$$
$$S(p,t) = p_i - p = 500 \text{ psi}$$
$$U = (c_w + c_f) V_{aq}$$
$$= (4.8 + 4.05) \times 10^{-6} \times 465.8 \times 10^6$$
$$= 4{,}122.3 \text{ bbl psi}^{-1}$$
$$W_e = 4{,}122.3 \times 500$$
$$= 2.06 \text{ MMbbl}$$

Assuming any change in compressibility is negligible, the water influx is directly proportional to the pressure decline in the reservoir.

Van Everdingen–Hurst unsteady-state aquifer model. The edge water aquifer model proposed by van Everdingen and Hurst is based on the radial diffusivity equation in a manner analogous to the prediction of fluid flow through a wellbore from the reservoir.[52] The radial diffusivity equation is derived earlier.

In the case of oil production, the inner boundary is at the wellbore. However, in modeling the edge water drive from an external aquifer, the inner boundary is at the reservoir/aquifer interface. The radial diffusivity equation is solved with a constant terminal pressure case. In the following steps, an outline of the solution methodology is presented.

Step 1. Calculation of dimensionless time and radius. In order to calculate water influx at t days, the dimensionless time and radius are calculated first as follows:

$$t_D = 6.328 \times 10^{-3} \frac{kt}{\varnothing \, \mu_w \, c_t \, r_e^2} \tag{4.146}$$

$$r_D = r_a / r_e$$

where

k = estimated permeability of the aquifer, mD,

$\varnothing$ = porosity of aquifer, fraction,

c_t = total compressibility $(c_f + c_w)$, psi^{-1},

r_e = radius of the reservoir, ft, and

r_a = radius of the aquifer, ft.

Step 2. Determination of dimensionless water influx from a chart or table. Once t_D and r_D are calculated, the next step is to read the corresponding value of the dimensionless water influx, W_{eD}, from relevant tables or charts.[53] These are reproduced in various publications.[54,55] Figure 4–25 presents values of W_{eD} for r_D between 5 and infinity, and t_D between 1 and 1,000.

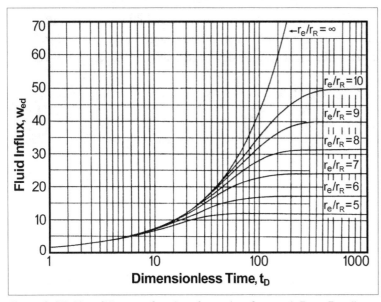

Figure 4–25. Plot of W_{eD} as a function of t_D and r_D. *Source: A. F. van Everdingen and W. Hurst. 1949. The application of the Laplace transformation to flow problems in reservoirs. Transactions. AIME. Vol. 186, no. 12:, 305–324. © Society of Petroleum Engineers. Reprinted with permission.*

Step 3. Calculation of water influx in barrels. The water influx in barrels is estimated by the following equation:

$$W_e = B \, \Delta p \, W_{eD} \tag{4.147}$$

where

$$B = 1.119 \, \varnothing \, c_t \, r_e^2 \, h \, (\theta / 360°), \tag{4.148}$$

h = aquifer thickness, ft,

θ = angle of encroachment, degrees, and

Δp = pressure drop at the reservoir/aquifer interface, psi.

In Equations 4.147 and 4.148, B is the water influx constant. It is expressed in barrels per pounds per square inch (bbl/psi).

Step 4. Repetition of steps 1–3 to calculate the influx for subsequent pressure drops. Water influx due to the first pressure drop at various times during the life of the reservoir is computed by using steps 1–3. Subsequent pressure drops at later times are calculated by the same methodology. Hence, the total water influx at any point in time is the sum of the individual water influx calculations performed for each pressure drop.

For example, assume that two pressure drops occur in the aquifer, one initially and the following drop at 100 days. The total water influx after 250 days due to the above pressure drops is computed as follows:

Total water influx, bbl = Water influx throughout 250 days due to the first pressure drop (occurring at t = 0 days)

+ Water influx for 150 days due to the second pressure drop (occurring at t = 100 days)

During the life of a reservoir, hundreds or thousands of instances of similar pressure drops in the aquifer need to be considered, which certainly requires high-speed digital computation. However, in order to demonstrate the methodology, example calculations are shown below.

Example 4.19. Water influx based on an unsteady-state aquifer model. Estimate the total water influx in barrels at 250 and 500 days due to the following two pressure drops at the reservoir boundary at two points in time:

At t = 0 days, $\Delta p_1 = 12$ psi
At t = 100 days, $\Delta p_2 = 15$ psi

Hence, the total pressure drop at the end of 100 days is $\Delta p_1 + \Delta p_2 = 27$ psi.

The following data is available:
Radius of reservoir, ft, = 5,000.
Porosity, fraction, = 0.25.
Permeability, mD, = 30.
Radius of aquifer, ft, = 40,000.
Thickness, ft, = 120.
Total compressibility, psi^{-1}, = 7.8×10^{-6}.
Viscosity of formation water, cp, = 0.617.

Solution: First, the water influx volumes are calculated at 250 days and 500 days due to the first pressure drop of 12 psi at t = 0 days. Next, the water influxes at 150 days and 400 days are computed due to the second pressure drop of 15 psi, which occurred 100 days later.

Calculations for the water influx at 250 days due to a pressure drop of 12 psi are given in the following:

$r_D = 40,000/5,000 = 8$

$$t_D = \frac{(6.328 \times 10^{-3})(30)(250)}{(0.25)(0.617)(7.8 \times 10^{-6})(5,000)^2}$$

$$= 1.58$$

From Figure 4–25, $W_{eD} = 1.89$. Next, values of B and W_e are computed based on Equations 4.148 and 4.147, respectively:

$B = 1.119 (0.25) (7.8 \times 10^{-6}) (5,000)^2 (120) (360° / 360°)$
$= 6.55 \times 10^3$ bbl/psi
$W_e = 6.55 \times 10^3 (12)(1.89)$
$= 1.48 \times 10^5$ bbl

Following the same methodology, Table 4–10 can be prepared in order to estimate total water influx at 250 days and 500 days.

Table 4-10. Aquifer influx summary

Time of Pressure Drop, days	Δp, psi	Elapsed Time, days	t_D	W_{eD}	Water Influx, 10^5 bbl	Elapsed Time, days	t_D	W_{eD}	Water Influx, 10^5 bbl
0	12	250	1.58	1.89	1.48	500	3.16	2.55	2.00
100	15	150	0.95	1.85	1.82	400	2.52	2.45	2.41
Total Influx, 10^5 bbl					3.30[a]				4.41[b]

[a] (at 250 days)   [b] (at 500 days)

Summing Up

Fluid flow in porous media is mathematically modeled to conceptualize, characterize, and predict the flow of fluids. Fluid flow is basically modeled in terms of fluid pressure, saturation, flow rate, and composition during production from the reservoir. Fluid can be treated as compressible or slightly compressible, as in the cases of natural gas and oil, respectively.

Various fluid flow patterns may develop in a reservoir, including radial, pseudoradial, linear, bilinear, spherical, and cone-shaped flow. Other considerations concern whether the fluid system is homogeneous (only one fluid) or heterogeneous (more than one fluid.)

Fluid flow in porous media may occur under steady-state, pseudosteady-state or unsteady-state conditions. In reality, well injection and production rates are not usually constant and wells are periodically shut-in; hence the flow of fluid is more likely to occur under unsteady-state conditions in a reservoir. The changes in pressure are rather rapid with time initially in the vicinity of a well as it is shut-in or put to production. However, the changes in pressure occur more slowly later. Pseudosteady-state or steady-state flow may eventually develop under certain circumstances. Under pseudosteady-state conditions, the rate of change in pressure is constant everywhere within a bounded reservoir. Steady-state condition prevails when reservoir pressure does not change with time anywhere due to the presence of a gascap or aquifer providing adequate support.

Viscous, gravity, and capillary forces are responsible for the flow of fluids in porous media. Under the circumstances of the production and injection processes that are usually encountered in a reservoir, viscous forces predominate. This is due to the imposed pressure gradient. In certain cases, including dipping reservoirs, gravity drainage may play a significant role. In long transition zones, the effects of capillary forces are evident.

Analysis of fluid flow through porous media is based on the diffusivity equation that models the unsteady-state, pseudosteady-state, and steady-state flow of fluids under various boundary conditions. The equation is derived from the following:

- Law of conservation of mass
- Darcy's law
- Equation of state

A well-known solution to the diffusivity equation is widely referred to as the line-source solution. It is used to predict the changes in pressure at a given location in the reservoir under unsteady-state conditions for a slightly compressible fluid under a constant rate of a well in an infinite-acting medium. The solution serves as the basis for well test interpretation. By applying the principle of superposition, the solutions to fluid flow can be extended to multiwell systems.

Other solutions to diffusivity equations are available for various other analyses of fluid flow characteristics. These include a no-flow or constant pressure condition at the reservoir boundary and a constant bottomhole flowing pressure at the well. Solutions include the flow of incompressible, slightly compressible, and compressible fluids in linear and radial flow geometry.

Solutions to various fluid flow models are often presented in the literature in a dimensionless form. This approach enables the reservoir engineer to obtain the sought parameters (pressure or rate) by using the dimensionless values for a wide range of rock and fluid properties. Regardless of the reservoir, this can be accomplished in a straightforward manner.

Darcy's law provides the fundamental equation for steady-state flow of an incompressible fluid through a uniform porous medium. The flow rate is proportional to the cross section and pressure gradient in the direction of flow and is inversely proportional to fluid viscosity. The proportionality constant is named after Henry Darcy, proponent of the law, and is called the permeability of porous media.

Analytic solutions to fluid flow equations are valuable in conceptualizing the reservoir dynamics. However, due to the complexity of multiphase, multidirectional flow, phase changes, and reservoir heterogeneities, numerical simulators are generally employed in detailed reservoir studies.

The productivity index of a well is a measure of the production rate achievable in an oil or gas well under a given drawdown, which is a function of fluid and rock properties. The unit of productivity index is barrels per day per pounds per square inch of pressure (bbl/d/psi). The injectivity index of an injector is defined in a similar manner.

The productivity index changes in a reservoir as the reservoir is depleted. The productivity index is also dependent on the skin factor of a well. Wells having the same productivity index in two different reservoirs cannot be equated, as the reservoir and fluid properties are not the same.

The inflow performance relationship attempts to predict the well production rate at various flowing bottomhole pressures. In solution gas drive reservoirs, the inflow performance relationship is no longer a straight line below the bubblepoint and is expressed by a nonlinear relationship between flow rate and flowing bottomhole pressure.

Water and gas coning occur due to severe imbalances between viscous and gravity forces near the wellbore in certain circumstances. The net effect of coning is an excessive water/oil ratio or gas/oil ratio in the producing stream, and a loss in oil production. The equilibrium attained between gravity and viscous forces in the reservoir is lost near the wellbore as oil is produced at higher-than-optimum rates. As a result, water (having higher density than oil) cones upward and is driven into the well perforations.

The Buckley and Leverett frontal advance equation can be used for immiscible displacement of oil by water. The waterflood displacement efficiency can be calculated using these equations. The topic of immiscible displacement of fluids is discussed in detail in chapter 16.

Steady-state, pseudosteady-state, and unsteady-state aquifer models are available to simulate the flow of water from an aquifer into a depleting petroleum reservoir. Due to the considerable uncertainties that exist in characterizing the aquifers, numerical simulators are employed to build various influx scenarios and obtain the best match with an aquifer model.

Class Assignments

Questions

1. Why is the study of fluid flow through porous media important to reservoir engineers? Explain with at least three major applications of fluid flow theory in reservoir engineering. What are the two principal methods of solving fluid flow equations in porous media? Discuss their relative advantages and disadvantages.

2. Discuss the effects of viscous, gravity, and capillary forces on the flow of fluid in porous media.
   True or false?
   (a) During fluid injection into a reservoir to recover oil, gravity forces do not play any role.
   (b) In highly volatile oil reservoirs where oil has relatively less viscosity, viscous forces play minimal role during production.
   (c) Evolution of gas from oil as reservoir pressure crosses the bubblepoint is due to gravity effects.
   (d) Viscous forces may arise in a reservoir due to the presence of a gas cap.
   (e) Primary production from a reservoir with bottom water drive is due to the capillary forces.
   (f) Dry gas reservoirs are usually produced by the forces related to earth's gravity.
   (g) In geologic times, oil migration into a reservoir occurred due to the prevailing viscous forces at that time.
   (h) During gas coning in a well the gravity forces predominate, and are responsible in downward movement of gas into the wellbore.

3. Distinguish between steady-state and pseudosteady-state flow in porous media. Consider two oil wells in different reservoirs having similar rock and fluid properties, one producing under a pseudosteady-state condition and the other under steady-state condition. Sketch a diagram showing the approximate shape of the pressure profile over

distance between the well and reservoir boundary under pseudosteady-state and steady-state conditions following transient production. If all other factors are the same, including the oil in place, is it possible to predict which well would produce more during the first half of its life cycle?

4. Describe the fluid flow patterns that may emerge in the following:

    (a) A partially completed well producing from a shoestring sand.

    (b) A horizontal well completed in three different geologic intervals in an inclined trajectory.

    (c) A producer surrounded by four injection wells in a pattern. The injectors located at northeast and southwest corners inject at twice the rate of others.

    (d) A gas reservoir where the wells are hydraulically fractured routinely for production. The rock permeability is in microdarcies.

    (e) A deviated well located in a circular reservoir.

5. Discuss at least four major limitations of the analytic approach in solving the diffusivity equation. In an oilfield where a number of injection and production wells are active, how can the limitations be overcome in studying the reservoir?

6. What is the hydraulic diffusivity coefficient? What rock and fluid properties affect the coefficient, and how?

7. Consider the production of the following newly drilled wells. Which well is expected to take the longest time to reach pseudosteady-state flow, if all other factors are the same? Demonstrate the answer by using separate sets of reservoir and fluid data. Compare the results.

    (a) Well in a gas reservoir with moderate to good permeability

    (b) Well in a gas reservoir with low permeability

    (c) Well in a volatile oil reservoir with moderate permeability

    (d) Well in a heavy oil reservoir with moderate permeability

8. In the following cases, which well is most likely to produce under steady-state conditions? Explain why.

    (a) Well producing in a volatile oil reservoir with a gas cap

    (b) Well producing in an undersaturated heavy oil reservoir with weak aquifer support

    (c) Well receiving pressure support from nearby injectors

9. Two new wells, one horizontal and the other vertical, are drilled in two different reservoirs having similar rock and fluid properties. Which well is expected to attain pseudosteady-state flow first? Which well will have a better productivity index? Explain.

10. Address the following concerns:

    (a) Describe the possible effects of an excessive production rate on steady-state flow in a reservoir having aquifer support.

    (b) When a new well is drilled, is the well expected to initially have a better production rate under unsteady-state flow conditions than under pseudosteady-state conditions? Explain.

    (c) Can Darcy's law be used to predict the earliest production rate from a new well? If not, what method or methods can be utilized to predict the initial well performance?

    (d) Why is a pseudopressure function used to characterize compressible flow in porous media? Demonstrate a method to calculate the pseudopressure function.

    (e) What is the net effect of turbulent flow in porous media? Describe how the effect of turbulence is incorporated in fluid flow equations.

11. Distinguish between the following, with examples:

    (a) Pseudosteady-state and steady-state flow

    (b) Slightly compressible and compressible fluid

    (c) Radial and spherical flow

    (d) Infinite-acting and finite-acting system

    (e) Unsteady-state flow and pseudosteady-state flow

    (f) Actual and apparent skin factor

    (g) Viscous and gravity effects on reservoir fluids

    (h) Productivity index and inflow performance relationship

    (i) Critical and subcritical rates in coning

    (j) Flow through fractures and matrix

    (k) Injectivity and productivity indices

    (l) Laminar and turbulent flow

    (m) Turbulent flow factor and non-Darcy flow coefficient

    (n) Positive and negative skin factor

    (o) Relative permeability and fractional flow curves

    (p) Fractional flow at and after breakthrough

    (q) Mobility and transmissibility

    (r) Unsteady-state and steady-state aquifer model

    (s) Pseudopressure function and pseudosteady-state flow

    (t) Constant pressure and no-flow boundary

    (u) Fracture permeability and matrix permeability

    (v) Unsteady-state aquifer and steady-state aquifer model

    (w) Edge water drive and bottom water drive

    (x) Permeability and diffusivity

12. Describe water and gas coning. How do they affect well performance? Describe the influence of the following factors on coning:

    (a) Thin formations

    (b) Viscous, gravity, and capillary forces

    (c) Viscosity of in-situ oil

    (d) Bottom water drive

    (e) Reservoirs with a gas cap

    (f) Horizontal and vertical permeability

    (g) Location of horizontal wellbore in the producing formation

    (h) Critical production rate

    (i) Choking back production

    (j) Sidetracking of an existing well

    (k) Undersaturated heavy oil reservoir

13. (a) Why is knowledge of the well productivity index necessary in reservoir management? Illustrate with two examples. Is well PI dependent on (a) formation transmissibility, (b) stratification, and (c) storavity?

    (b) Two exploratory wells in two reservoirs have the same productivity index of 1.0 stb/d/psi. Are these wells producing at the same rate? Explain.

    (c) Define inflow performance relationship. Why does the inflow performance relationship curve inward in a solution gas drive reservoir?

14. On the same plot, draw sketches of the inflow performance relationship for the following cases, giving explanations:

    (a) A vertical well with skin damage in an undersaturated oil reservoir

    (b) The same well when the problem is addressed by stimulation

    (c) An offset well where the reservoir quality is poor

    (d) The first horizontal well drilled in the same reservoir with a 2,500-ft-long wellbore

    (e) A vertical well in a nearby reservoir where the permeability is similar but the formation is only one-half as thick

15 Define the mobility ratio of two immiscible fluids and describe the significance of the mobility ratio in immiscible displacement of one fluid by another. Give an example of how knowledge of the mobility ratio may aid in understanding the reservoir performance.

16. List the key ingredients of 3-D, three-phase fluid flow simulation in petroleum reservoirs. How are these simulation studies utilized in reservoir engineering and management? Describe an application of reservoir simulation augmenting ultimate recovery based on a literature review.

Exercises

4.1. What role does the diffusivity equation play in characterizing the fluid flow in porous media? Derive the diffusivity equation in oilfield units. List all the assumptions necessary in deriving the equation. List and explain all of the boundary conditions employed in deriving the common solutions to fluid flow in petroleum reservoirs.

4.2. Perform the following analysis:

    (a) Based on the reservoir and fluid data provided in Example 4.1, generate pressure profiles between the two wells at time intervals of 0.24, 2.4, 24 and 240 hrs. Plot the profiles on the same graph and draw conclusions about the infinite-acting fluid flow characteristics. Plot dimensionless pressure (p_D) versus time (t_D) as expected in the observation well.

    (b) If the observed pressure drop is 9 psi after 24 hrs at the observation well, recalculate the average permeability of the rock. In order to produce at the same rate under the above circumstances, would the bottomhole pressure of the well increase, decrease, or remain unchanged? Explain with the help of illustrative calculations.

    (c) Discuss the probable causes of the apparent discrepancy between core data and results of this study, as the observed pressure drop is much less than expected.

4.3. In a multiwell system similar to that in Figure 4–5, calculate the pressure drop observed at the center well after 12 hours. The following data is given:

        Porosity of the formation, fraction, $= 0.32$
        Average rock permeability, mD, $= 90$
        Gross formation thickness, ft, $= 25$
        Net to gross thickness ratio $= 0.65$
        Viscosity of the oil, cp, $= 0.88$
        Oil formation volume factor, rb/stb, $= 1.18$
        Compressibility of the oil, psi^{-1}, $= 5.9 \times 10^{-6}$.
        Formation compressibility, psi^{-1}, $= 8.9 \times 10^{-5}$.
        Radius of the wellbore, ft, $= 0.32$.

4.4. Discuss the applicability of the line-source solution in the multiwell reservoir under the following scenarios:

    (a) One of the corner wells stopped producing after 4 hrs due to an equipment malfunction. Can Equation 4.28 still be used to estimate the pressure drop at the observation well following 12 hrs of production of the remaining three wells?

    (b) All four wells did not start production at the same moment. One of the wells started flowing 1 hr after the other three wells started producing.

Well	Distance from the Observation Well, ft	Well Flow Rate, stb/d
1	650	450
2	720	500
3	690	550
4	600	475

    (c) The recorded pressure at the observation well indicated a small pressure drop much earlier than anticipated.

4.5. What are the advantages of casting the solutions to the diffusivity equation in dimensionless variables? Redo Example 4.2 illustrated previously by assuming that the average rock permeability is twice the original value of 45 mD. Further assume that the initial reservoir pressure is 2,850 psia. Plot production rate and cumulative volume until the well reaches abandonment. With reduced initial reservoir pressure and enhanced formation permeability, would the cumulative volume of oil produced increase, decrease, or remain virtually the same? Explain.

4.6. Based on the data given in Example 4.2, provide an estimate for the percent of the oil volume recovered, i.e., the recovery factor. State all the assumptions made in the analysis. Does the anticipated recovery appear to be too high or low? Justify your answer. How would a relatively high bubblepoint pressure affect the recovery?

4.7. Calculate the oil production rate, assuming steady-state flow and an incompressible fluid. The following data is available:

        Average rock permeability, mD, $= 100$.
        Porosity of the formation $= 0.29$.

> Average formation thickness, ft, = 50.
> Oil viscosity, cp, = 0.846.
> Oil formation volume factor, rb/stb, = 1.252.
> Oil compressibility, psi^{-1}, = 4.5 × 10^{-6}.
> External drainage radius, ft, = 660.
> Reservoir pressure at drainage boundary, psig, = 2,554.
> Bottomhole flowing pressure, psig, = 1,226.
> Well radius, ft, = 0.328.

4.8. Redo Exercise 4.7 by taking into account the fact that oil is slightly compressible. Compare the results.

4.9. Assuming a slightly compressible fluid, evaluate the effects of formation permeability, external drainage radius, and pressure at the drainage boundary on the steady-state flow rate. Prepare plots for the following ranges of data:

   (a) Average formation permeability, mD, = 50–150.

   (b) Drainage radius, ft, = 660–1,320.

   (c) Reservoir pressure at drainage boundary, psig, = 2,450–2,650.

4.10. Based on the pseudopressure approach, estimate the gas production rate from a planned well by assuming steady-state flow. The following data is available:

> Reservoir pressure, psig, = 2,690.
> Reservoir temperature, °F, = 130.
> Estimated bottomhole pressure, psig, = 1,125.
> Gas gravity = 0.68.
> External drainage radius, ft, = 2,980.
> Radius of wellbore, ft, = 0.328.
> Permeability of formation, mD, = 50–60.
> Net thickness of formation, ft, = 120.

Based on available correlations, estimate fluid properties not reported here. The reservoir is believed to have good aquifer support. Make any other assumptions necessary.

4.11. If the formation is very tight (0.005–0.1 mD), would the assumption of steady-state flow still be valid? Why or why not? Make a conservative estimate of the production rate in this case.

4.12. In Examples 4.8 and 4.9, determine and compare the time taken by the following to reach the pseudosteady-state condition:

   (a) Well in the center of a circular drainage area, estimated to be 640 acres

   (b) Well in a quadrant of a drainage square, the area of which is the same as above.

   Explain the physical significance of the results obtained.

4.13. Using the data of Example 4.8, recalculate the average reservoir pressure when the drainage area is 40 acres. Compare results. How would the reservoir pressure be determined under steady-state flow?

4.14. Consider a newly drilled oil in the center of a rectangle shaped reservoir having sides in an approximately 4:1 ratio. The nearest boundaries are not expected to receive any aquifer support and are located 1,250 ft from the well on either side. Assuming that the well produces at a steady rate of 400 stb/d, plot the flowing bottomhole pressure of the well over time during transient and pseudosteady-state flow periods. What is the duration of the transient flow period? The following rock and fluid data is available:

Permeability, mD, = 55.
Net formation thickness, ft, = 26.
Oil viscosity, cp, = 0.74.
Oil formation volume factor, rb/stb, = 1.19.
Total compressibility, psi^{-1}, = 11.5 × 10^{-6}.
Initial reservoir pressure, psia, = 3,400.
Well diameter, ft, = 0.54.

4.15. Calculate the productivity index of an oil well producing under pseudosteady-state conditions for the following cases:

(a) Well in a developed field with 40-acre spacing. Assume a circular drainage area.

(b) Well located in the center of a rectangular drainage area 8,350 ft long and 2,100 ft wide.

(c) Well located at the upper right quadrant of the rectangle described in (b). Assume $C_A = 0.1155$.

(d) Estimate the time needed to reach a pseudosteady-state flow condition in the case described in (a) once the well starts production.

The following reservoir and fluid data is available:

Average permeability, mD, = 550.
Porosity (fraction) = 0.34.
Net formation thickness, ft, = 21.
Oil formation volume factor, rb/stb = 1.22.
Oil viscosity, cp, = 0.85.
Radius of wellbore, ft, = 0.39.
Skin factor = 2.6.
Total compressibility, psi^{-1}, = 12 × 10^{-6}.

Make any other assumptions necessary. Include comments based on the results of this study.

4.16. Estimate the increase in productivity that can be attained by drilling a 2,000-ft long horizontal well instead of a vertical well in a thin oil column. The following data is available:

Horizontal permeability, mD, = 60.
Ratio of vertical to horizontal permeability = 0.1.
Drainage area, acres, = 160.
Thickness of oil column, ft, = 10.
Radius of wellbore, ft, = 0.39.
Viscosity of oil, cp, = 0.9.
Oil formation volume factor, rb/stb, = 1.28.

(a) Upon completion of the horizontal well and its initial production, a well test was conducted that indicated an effective horizontal wellbore length of 1,400 ft. Update the productivity index of the horizontal well. Make any necessary assumptions.

(b) The primary recovery mechanism in the above reservoir is bottom water drive. What would be the best strategy to complete the horizontal well to delay water coning? What would be the reservoir performance if a vertical well is drilled instead?

4.17. The data given in Table 4–11 is obtained from a gas well test at four different flow rates. During each flow period, a pseudosteady-state condition was attained by flowing the well for a sufficiently long period of time.

(a) Calculate the absolute open flow potential of the well.

(b) Calculate the values of c and n as in Equation 4.50.

(c) Do the values of c and n remain unchanged throughout the life of the reservoir? Explain.

Assume a stabilized shut-in bottomhole pressure of 1,650 psia prior to the test.

4.18. By conducting a literature survey, briefly discuss the mathematical models and the limitations, if any, for the following:

(a) Flow through naturally occurring fractures with matrix-fracture interaction

(b) Stratified flow with or without crossflow between layers

(c) Productivity index of a multilateral horizontal well

Include any assumptions made in deriving the analytical or numerical models.

Table 4–11. Well test results

Flow Period	p_{wf}, psia	q, MMscfd
1	1,645	2.68
2	1,630	5.79
3	1,586	9.72
4	1,580	12.65

References

1. Ahmed, T. H. 2001. *Reservoir Engineering Handbook*. 2nd ed. Houston: Gulf Professional Publishing Co.

2. Lee, J., and R. A. Wattenbarger. 1996. *Gas Reservoir Engineering*. SPE Textbook Series. Vol. 5. Richardson, TX: Society of Petroleum Engineers.

3. Collins, R. E. 1961. *Flow of Fluids through Porous Materials*. New York: Reinhold.

4. Matthews, C. S., and D. G. Russell. 1967. Pressure buildup and flow tests in wells. SPE Monograph. Vol. 1. Dallas: Society of Petroleum Engineers.

5. Aziz, K., and A. Settari. 1979. *Petroleum Reservoir Simulation*. New York: Applied Science Publishers, Ltd. Reprinted in 1990 and available through the Society of Petroleum Engineers.

6. Thomas, G. W. 1982. *Principles of Reservoir Hydrocarbon Simulation*. Boston: International Human Resources Development Corp.

7. Lee, J., and R. A. Wattenbarger. 1996.

8. van Everdingen, A. F., and W. Hurst. 1949. The application of the Laplace transformation to flow problems in reservoirs. *Transactions*. AIME. Vol. 186:, 305.

9. Matthews, C. S., and D. G. Russell. 1967.

10. Craft, B. C., and M. Hawkins. 1990. Applied Petroleum Reservoir Engineering. 2nd ed. Revised by R. E. Terry. Englewood Cliffs, NJ: PTR Prentice-Hall.

11. Horne, R. N. 1995. *Modern Well Test Analysis*. 2nd ed. Palo Alto, CA: Petroway, Inc.

12. Ahmed, T. H. 2001.

13. Lee, J., and R. A. Wattenbarger. 1996.

14. Edwardson, M. J., H. M. Girner, H. R. Parkison, C. D. Williams, and C. S. Matthews. 1962. Calculation of formation temperature disturbances caused by mud circulation. *Journal of Petroleum Technology*. April: p. 416.

15. Towler, B. F. 2002. *Fundamental Principles of Reservoir Engineering*. SPE Textbook Series. Vol. 8. Richardson, TX: Society of Petroleum Engineers.

16. Lee, J., and R. A. Wattenbarger. 1996.

17. Lee, J., and R. A. Wattenbarger. 1996.

18. Lee, J., and R. A. Wattenbarger. 1996.

19. Lee, J., and R. A. Wattenbarger. 1996.

20. Lee, J., and R. A. Wattenbarger. 1996.

21. al-Hussainy, R., and H. J. Ramey, Jr. 1966. Application of real gas flow theory to well testing and deliverability forecasting. *Journal of Petroleum Technology*. May:, 637–642.

22. al-Hussainy, R., H. J. Ramey, Jr., and P. B. Crawford. 1966. The flow of real gases through porous media. *Journal of Petroleum Technology*. May:, 624–636.

23. Agarwal, R. G. 1979. Real gas pseudotime—a new function for pressure buildup analysis of gas wells. SPE Paper #8279. Presented at the 1979 SPE Annual Technical Conference and Exhibition, Las Vegas, NV, September 23–26.

24. al-Hussainy, R., and H. J. Ramey, Jr. 1966.

25. Horne, R. N. 1995.

26. Ahmed, T. H. 2001.

27. Ahmed, T. H. 2001.

28. Amyx, J. W., D. M. Bass, Jr., and R. L. Whiting. 1960. *Petroleum Reservoir Engineering—Physical Properties*. New York: McGraw-Hill.

29. Craft, B. C., and M. Hawkins. 1990.

30. Craft, B. C., and M. Hawkins. 1990.

31. Ahmed, T. H. 2001.

32. Craft, B. C., and M. Hawkins. 1990.

33. Ahmed, T. H. 2001.

34. Lee, J., and R. A. Wattenbarger. 1996.

35. Lee, J. 1982. *Well Testing*. SPE Textbook Series. Vol. 1. Richardson, TX: Society of Petroleum Engineers.

36. Ahmed, T. H. 2001.

37. Ahmed, T. H. 2001.

38. Craft, B. C., and M. Hawkins. 1990.

39. Lee, J., and R. A. Wattenbarger. 1996.

40. Vogel, J. V. 1968. Inflow performance relationships for solution-gas drive wells. *Journal of Petroleum Technology*. Vol. 20, no. 1:, 83–92.

41. Economides, M. J., A. D. Hill, and C. Ehlig-Economides. 1994. *Petroleum Production Systems*. Englewood Cliffs, NJ: PTR Prentice-Hall.

42. Joshi, S. D. 1988. Augmentation of well productivity with slant and horizontal wells. *Journal of Petroleum Technology*. June., 729–739.

43. Cheng, A. M. 1990. IPR for solution gas-drive horizontal wells. SPE Paper #20720. Presented at the 65th Annual Meeting of the Society of Petroleum Engineers, New Orleans, September 23–26.

44. Ahmed, T. H. 2001.

45. Chaperon, I. 1986. Theoretical study of coning toward horizontal and vertical wells in anisotropic formations: subcritical and critical rates. SPE Paper #15377. Presented at the Society of Petroleum Engineers 61st Annual Technical Conference and Exhibition, New Orleans, LA, October 5–8.

46. Buckley, S. E., and M. C. Leverett. 1942. Mechanism of fluid displacement in sands. *Transactions*. AIME. Vol. 146:, 107–116.

47. Craig, Jr., F. F. 1971. *The Reservoir Engineering Aspects of Waterflooding*. SPE Monograph Vol. 3. Richardson, TX: Society of Petroleum Engineers.

48. Wang, B., and T. S. Teasdale. 1987. GASWAT-PC. A microcomputer program for gas material balance with water influx. SPE Paper #16484. Presented at the Petroleum Industry Applications of Microcomputers meeting, Society of Petroleum Engineers, Del Lago on Lake Conroe, TX, June 23–26.

49. Schilthuis, R. J. 1936. Active oil and reservoir energy. *Transactions*. AIME. Vol. 118:, 33–37.

50. Hurst, W. 1958. Simplification of the material balance formulas by the Laplace transformation. *Transactions*. AIME. Vol. 213:, 292–303.

51. van Everdingen, A. F., and W. Hurst. 1949. The application of the Laplace transformation to flow problems in reservoirs. *Transactions*. AIME. Vol. 186, no. 12:, 305–324.

52. van Everdingen, A. F., and W. Hurst. 1949.

53. van Everdingen, A. F., and W. Hurst. 1949.

54. Craft, B. C., and M. Hawkins. 1990.

55. Ahmed, T. H. 2001.

5 · Transient Well Pressure Analysis and Applications

Introduction

Analysis of the transient pressure response due to a change in the well rate serves as a powerful tool in evaluating and enhancing well performance. Equally important is the fact that transient test analyses aid in characterizing a reservoir. This information can help to identify faults, fractures, reservoir boundaries, fluid communication between layers, and tracking of fluid fronts, to name a few applications. Commonly referred to as well testing, virtually all wells undergo some kind of pressure transient testing during various stages of their productive lives. Any newly drilled well is subjected to testing in order to assess the future potential of the well. In some fields, reservoir performance is intensively monitored due to the existence of complex geology, an ongoing enhanced recovery operation, or monitoring of specific production issues. In these situations, certain key wells could be selected for routine testing at relatively short intervals. Emerging technologies employ digital sensors to continuously monitor bottomhole pressure and flow rate during well production. Robust computer models attempt to interpret reservoir dynamics based on the information obtained.

Reservoir engineers place a high degree of confidence on well test interpretation in characterizing a reservoir and understanding well performance. In fact, well tests are regarded as "reality checks" in conceptualizing petroleum reservoirs. Conclusions based on well test results are utilized extensively as a valuable guide in the exploration, production, and development of petroleum reservoirs. A carefully designed and analyzed well test may provide vital information about the reservoir under dynamic conditions at distances of hundreds of feet or all the way to the reservoir boundary. This ability makes it a truly unique tool. It is rather difficult to quantify the inherent variations of rock properties from one point to another. Nevertheless, well tests may point to the combined effect of existing geologic heterogeneities on the flow of reservoir fluids, at least in the vicinity of the well being tested. Information obtained from well test analysis and various other sources is combined to build an integrated reservoir model that is updated on a regular basis.

This chapter is devoted to learning about the following:

- The role and objectives of well testing in field exploration, development, and production
- Basic principles and methodology of transient well testing
- Types of transient well tests as practiced in the industry
- Emerging technologies in well and formation testing
- Interpretation methodology—analytical and digital
- Factors affecting well tests
- Well test response in various geologic settings
- Interpretation of transient pressure tests
  - Important concepts in well test analysis
  - Development of fluid flow patterns
  - Well test interpretation methodology
  - Computer-aided qualitative and quantitative analyses
  - Type curve methodology
- Fundamentals of well test theory
- Design and scheduling of well tests

- Capabilities of well test software tools
- Examples of well test applications in reservoir studies in a variety of settings
- Class problems

The following cases are illustrated with the aid of well test software tools in this chapter, covering a wide range of applications in reservoir engineering:

- Pressure buildup test of a well producing from an area with constant pressure boundaries
- Drawdown test in a reservoir having no-flow boundaries
- Evaluation of a hydraulically fractured well performance
- Investigation and characterization of a pinchout boundary of a reservoir
- Interpretation of horizontal well testing (drawdown test)
- Interwell reservoir characterization based on computer-assisted design of an interference test
- Effect of non-Darcy flow on gas well transient testing
- Investigation of interlayer communication in a stratified reservoir
- Well test analysis in a tight gas reservoir based on a long drawdown test

Well Test Interpretation in the Digital Age

Since the emergence of digital computation and software-based analysis, workflow related to well tests used in reservoir characterization has significantly transformed the industry. In modern practice, a reservoir team evaluates the well test results in conjunction with information obtained from multiple sources. Integrated results of various tests, conducted at different locations throughout the life of reservoir, contribute significantly to understanding fluid flow behavior and reservoir performance as a whole. Test results are usually stored in a database having seamless integration with various applications in reservoir engineering. Such applications include geophysical and geological data, petrophysical studies, and reservoir simulation. These are evaluated along with well test results, leading to the development of robust reservoir models. This contrasts with early practices, when the well test response was hand drawn on a piece of paper, and test interpretation, performed by calculators, was mostly viewed as stand-alone information.

Objectives of pressure transient testing

The vital role played by the well testing program may be summarized in three key statements:

1. Know the well.
2. Know the reservoir.
3. Integrate the results of the well test interpretation with information obtained from other sources in order to manage the well and the reservoir.

All well test-related efforts, planned and executed by a reservoir team, lead to the development, production, and management of a reservoir on an individual well basis. Virtually all reservoirs have a well testing program to meet the overall management objectives.

The following section is a description of important objectives that transient pressure tests can accomplish in a reservoir.

Assessment of well productivity and reservoir viability. As soon as a new well is drilled, a pressure transient test is conducted to estimate the production potential of the well. Whether the well will be completed or abandoned largely depends on the results of the well test. In the case of an exploratory well drilled in a newly discovered reservoir, a transient test could indicate whether the reservoir would be technically or commercially viable.

Well and reservoir management plan. In closely monitored oil and gas fields, well tests are conducted at regular intervals to evaluate the performance of producers and injectors. A transient pressure test may readily identify any loss or gain in well productivity due to alteration of the rock permeability around the wellbore. The results may point to certain corrective actions, such as well stimulation, recompletion, or horizontal drilling of problematic wells. Additionally, a time-lapse study may be conducted based on regularly scheduled well tests. This could be used to track the advancing front of an injected fluid during an enhanced recovery project, leading to efficient waterflood management.

Evaluation of stimulated wells. Wells associated with relatively low productivity as a result of tight permeability or skin damage are frequently subjected to a hydraulic fracturing operation. The objective is to create a fracture intersecting the wellbore that facilitates unhindered fluid flow towards the well. Significant enhancement in oil or gas production is usually attained by this method. Following the hydraulic fracturing operation, a well test is conducted in order to estimate the increase in well productivity. The length and conductivity of the induced fracture are also determined from the test. Acidized wells are evaluated in a similar manner.

Determination of formation fracture pressure. In certain injection wells, step-rate tests are conducted to ascertain the fracture threshold pressure of the formation. This information is valuable in designing waterflood or enhanced recovery operations. In such operations, the injection pressure is maintained below the threshold value in order to ensure that the injected fluid is not lost through artificially created fractures during injection.

Identification of reservoir characteristics in the area of investigation. Common well tests provide a host of valuable information, including rock permeability and transmissibility. Furthermore, transient test results may provide definitive indications of reservoir heterogeneity, such as the existence of a sealing boundary, which may have a profound impact on reservoir performance. With the introduction of digital monitoring tools and diagnostic analysis, more and more well tests are designed to characterize the reservoir.

Investigation of reservoir boundary. Well tests can be utilized to make a definitive assessment of oil and gas in place. Certain well tests, when conducted for a sufficient length of time, aid in delineating the boundaries of a reservoir, leading to the estimation of hydrocarbon fluid volume. Geologic boundaries in the form of faults or facies changes can be identified and located by well tests. Additionally, the nature of a geologic boundary, either no flow or constant pressure, is identified. Existence of strong aquifer effects may be determined in the process.

Interwell characterization. Transient pressure tests may be based on either one well or multiple wells. In a multiple-well scenario, a perturbation in flow or a pulse is created in one well, while the subsequent pressure response is monitored in an adjacent well or wells. Multiwell tests are designed in a manner that identifies the flow behavior between wells influenced by existing geologic heterogeneities. For example, the reservoir may exhibit a directional flow characteristic due to permeability anisotropy, which may be detected by a distinct pressure response between two or more wells. Again, a test could be designed to identify and assess interlayer communication between wells completed in different layers. Zonal communication or crossflow may have a profound impact on well completion, production, and unwanted water breakthrough, among other factors. This category of well tests may aid significantly in understanding poor reservoir performance in a complex geologic setting. A computer-assisted model usually simulates the response from multiple wells obtained during the transient test, leading to realistic conceptualization of the reservoir.

Estimation of reservoir pressure. The initial reservoir pressure and subsequent changes in pressure are obtained by conducting well tests at various stages of reservoir development. Initial reservoir pressure is directly related to the plan for future field development. Estimation of average reservoir pressure in the drainage area of the well provides a vital piece of information when the reservoir pressure is declining under production. Optimum production could require additional energy in the form of external fluid injection. For example, in a reservoir where water injection provides additional energy, such tests may identify regions where the reservoir pressure is inadequate. This may lead to optimization of the injection rates in nearby wells.

Estimation of gas well deliverability. Gas wells are routinely subjected to a deliverability test to estimate the absolute open flow potential (AOFP) of the well. (The absolute open flow potential is the maximum rate at which the well would flow when the sandface pressure is equal to atmospheric pressure.) In the most common practice, a gas well is flowed at four different flow rates, and the resulting changes in flowing bottomhole pressure are noted. The subsequent interpretation of the test results, based on a log-log plot of data, leads to the determination of the absolute open flow potential.

Reservoir simulation model development. Carefully designed well tests provide valuable information about reservoir dynamics in the region surrounding a well. With the aid of a computer, this information is used to build an individual well model to investigate probable geologic heterogeneities that may affect well performance. Again, a model can be developed to estimate the future potential of an existing well in the case of redesign and recompletion, such as horizontal drilling. Turning to a larger scenario, well tests conducted in several locations at various stages of production contribute substantially towards building a complete reservoir model. Information obtained from a geosciences study, such as the existence of faults or pinchouts, can be verified by conducting carefully designed well tests.

Formulation of field or basin exploration strategy. Well tests conducted in a newly discovered field provide highly valuable information that may aid in formulating a strategy for future exploration in the petroleum basin. Based on optimistic results obtained from an initial well test, more resources could be allocated to explore the petroleum horizon in the future.

Key points—well testing applications

The following key points are reiterated in relation to well testing applications:

- Well testing is an indispensable tool for reservoir engineers in collecting and analyzing vital information about the well and the reservoir under dynamic conditions. Well tests contribute significantly to activities related to exploration, development, and production.
- The uniqueness of transient well testing lies in the fact that a properly designed test is capable of investigating all the way out to the reservoir boundary under dynamic conditions.
- Transient pressure testing techniques include the generation and measurement of variations in pressure response with time while the well is flowing or shut in. Well test data is used to determine the following:
  - **Reservoir rock properties.** Properties include such factors as permeability, alteration of permeability, and storage capacity, among others.
  - **Reservoir geometry and characterization.** This includes areal extent, fault, pinchout, compartmentalization, effect of an aquifer, vertical communication between layers, and existence of high-permeability streaks and fractures, among others.
  - **Pressure.** Pressure data encompasses the average reservoir pressure, the flowing bottomhole pressure, and the fracture gradient.
  - **Well characteristics.** These include the productivity index, partial completion, wellbore storage coefficient, and the presence of multiphase fluids, among others.
  - **Hydraulic fractures.** The length and conductivity of the fractures are determined.
  - **Interference from other wells.**
  - **Fluid flow pattern.** The extent of the injected fluid bank is measured.
  - **Petroleum basin exploration.** The viability of newly discovered fields is determined based on initial tests.

- All wells undergo some type of testing during the life of a reservoir. Virtually all reservoirs have a well testing plan that is carried out to gather information on a regular basis. Results of well tests are then integrated with other sources of data to build robust reservoir models leading to efficient management of the reservoir. The role of well testing in an integrated study of a reservoir generally involves the following workflow:
  - Identification of geologic features, including faults or boundaries, based on geophysical or geologic studies.
  - Confirmation of the above by conducting carefully designed well tests in one or more wells.
  - Simulation of an updated reservoir model in order to obtain a satisfactory match with the historical reservoir performance.
- With the advent of the digital age, test interpretations are almost exclusively performed by computers with well test software. In recent times, well tests are increasingly designed to aid reservoir characterization. Test results are no longer viewed as a single-well phenomenon. Information related to transient tests is stored in an integrated database, interacting seamlessly with data obtained from other sources and related studies.

Overview of Well Test Methodology

The basic principle behind a transient pressure test is relatively straightforward and is outlined in the following:

1. **Planned change in well rate.** A carefully designed perturbation or disturbance is created in the reservoir by conducting a step change in the production or injection rate in a test well.
2. **Effect on well pressure response.** The change in the fluid flow pattern triggers a new trend in the pressure response at the wellbore and into the reservoir. The nature of the response is transient.
3. **Monitoring and analysis.** The ensuing pressure response is then monitored over a length of time for subsequent interpretation. Depending on the well, reservoir, and fluid characteristics, the resulting response in the fluid pressure imprints one or more signatures that can be identified and analyzed. The interpretation is based on the theory of transient fluid flow in porous media under certain simplifying assumptions.

The basic workflow of well test methodology and analysis is presented in Figure 5–1. Data collection during the well test requires accurate monitoring of the well production or injection rate and bottomhole pressure with time. State-of-the-art electronic gauges placed downhole are capable of intensive data collection by recording minute changes in pressure at a high frequency. Recorded information is subsequently processed by well test software and stored in an integrated database.

Figure 5–2 illustrates the step changes in flow rate and resulting pressure response in several important types of well tests practiced in the industry. A general outline of test sequence and consequent response in pressure for each type of transient test is presented in the following section.

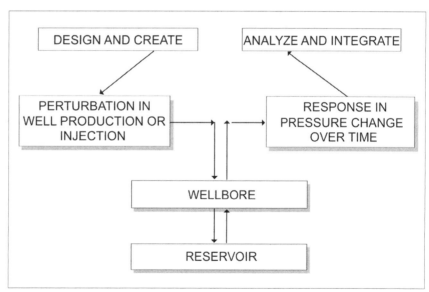

Fig. 5–1. Schematic flow diagram of well test methodology

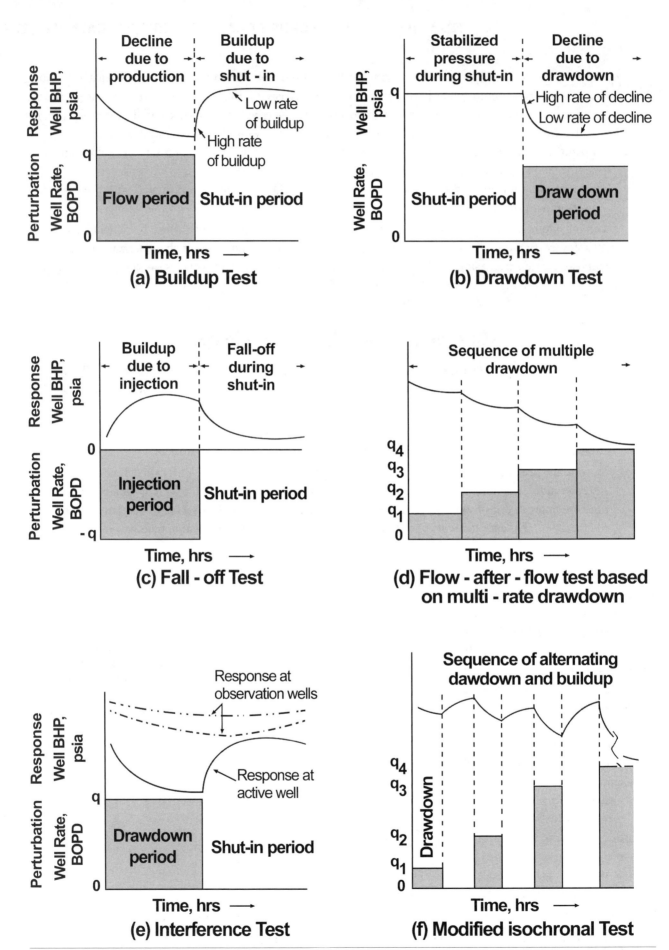

Fig. 5–2. Flow rate and pressure changes during common well tests

Types of Transient Well Tests

As noted earlier, well tests are based on the underlying principle that a step change in the well flow rate would trigger transient pressure responses in the reservoir. Depending on the characteristics of the reservoir and well, the response exhibits recognizable trends during the test. Certain tests are designed to involve multiple wells and several changes in flow rate.

Pressure buildup test

Pressure buildup tests are one of the most common well tests in the petroleum industry. The well is produced at a constant rate for a sufficient length of time (usually a few days to several months), followed by shutting in of the producing well to record the buildup of pressure at the well with time. The duration of the buildup period could be a few hours to a few days, depending on the reservoir characteristics and test objectives.

Figure 5–2(a) shows a typical pressure response in a well during a buildup test. The observed rate of pressure buildup is the highest at the initial stage of shut-in, followed by a gradual slowdown until some sort of stabilized state is attained. With the passage of time during the test, the observed pressure response is progressively influenced by rock and fluid properties further away from the well. Reservoirs having relatively low transmissibility characteristics usually require longer test periods, as pressure response from the formation to the well is transmitted rather slowly.

One limitation of a buildup test is the resulting loss in production and revenue during the long buildup period required in certain cases. Furthermore, ensuring a constant rate for well production prior to the test may prove to be difficult under practical operating conditions.

Drawdown test

The well is initially shut in for a sufficient length of time to allow static pressure to be reached in the reservoir, which is followed by production (drawdown) at a constant rate. Consequently, pressure declines at the well in a distinct trend that is monitored and analyzed. The rate of decline is most pronounced in the beginning as the drawdown commences, followed by a slowdown until some stabilization is attained. A typical pressure response from a drawdown test is shown in Figure 5–2(b).

A drawdown test may have an advantage over a buildup test in that the well is not required to be shut in for an extended period. However, attainment of static reservoir pressure prior to the drawdown test or maintaining a constant drawdown rate during the test may pose a challenge in certain circumstances. A drawdown test is ideally suited for newly discovered reservoirs where the initial reservoir pressure at discovery is static.

Falloff test

Falloff tests are conducted in fluid injection wells that are usually part of a pressure maintenance or enhanced recovery program in a petroleum reservoir. The well is first injected at a constant rate for a sufficient period to achieve stabilization in injection pressure, followed by shutting in of the injector. As a result, the bottomhole pressure at the well begins to decline (fall off), which is recorded and analyzed. Conceptually, it is a mirror image of a buildup test, as depicted in Figure 5–2(c). In water injection wells, a falloff test can indicate the leading edge of the injected fluid bank when the test is run for a sufficiently long period. A distinct change in pressure response is observed at the fluid phase boundary between the injected water phase and the in-situ oil phase.

Step-rate test

In order to determine the fracture pressure and fracture gradient of a formation, an injection well is subjected to series of injection rates while the injection pressure is recorded. The fracture pressure of the formation is the threshold pressure at which the subsurface formation is fractured. The injection rate involves a series of step changes in increasing order, the steps usually being of short duration. The observed pressure shows a distinctive change in trend as the fracture threshold is crossed, and a fracture is created in the formation. The rate of increase in pressure due to the increase in injection rate becomes markedly less in the presence of an artificially created fracture. In certain instances, rate-time-pressure data can be analyzed under appropriate assumptions to obtain reservoir parameters, such as formation transmissibility and skin.

Multirate test

In a multirate test, the well is flowed at multiple rates for definite time intervals, and the pressure response is recorded. Multirate tests are widely used in gas reservoirs in order to assess well potential and reservoir performance. Typical multirate tests are outlined in the following discussion.

Flow-after-flow test. The well is flowed successively at different but stabilized flow rates, and the bottomhole pressure is recorded. During the test, a step change in flow rate is made after the flowing bottomhole pressure is found to have stabilized. Gas well deliverability tests usually include four different flow rates in increasing order, as shown in Figure 5–2(d). The test is also known as a gas well deliverability test or a four-point test, with the objective being to estimate the absolute open flow potential of the well.

Isochronal test. Alternating sequences of drawdown and shut-in periods are implemented, with monitoring of the pressure response. Drawdown rates are constant within a sequence but vary from one sequence to another. In each sequence, shut-in of the test well continues until the bottomhole flowing pressure stabilizes. However, in tight reservoirs, the time needed to attain stabilization could be long. Hence, modified isochronal tests are designed to minimize the loss in production where the drawdown and shut-in periods are of equal duration, as shown in Figure 5–2(f). Although stabilized bottomhole pressure is not attained, the test can still provide meaningful results when analyzed appropriately.

Interference and pulse tests

These tests involve one active well and one or more observation wells located at a distance in neighboring locations in the reservoir. A source of interference is created at the active well in the form of predetermined step changes in the flow rate. This interference leads to alteration of the pressure response in both active and observation wells, as shown in Figure 5–2(e). In a variation of the above, an alternating sequence of flow and shut-in periods, or pulses, is generated at the active well, leading to measurement of the pressure response in multiple wells.

Interwell reservoir characteristics, among other factors, influence the nature and degree of the pressure response at the observation wells. The time lag between the initiation of interference in the active well and the ensuing response in the observation wells also depends on the reservoir and fluid properties. These tests may require two or more wells to be shut in for a relatively long period. In addition, sensitive monitoring equipment is deployed to record and interpret the slightest change in pressure at the observation wells. Reservoir simulators are frequently utilized to aid in test interpretation. Hence, multiwell tests are usually implemented only when complex issues arise in a reservoir, such as the possible existence of unidentified heterogeneities or premature water production.

Drillstem test

A drillstem test (DST) is routinely conducted in a new well prior to completion in order to assess the feasibility and potential of the new well in an unknown environment. Results of the drillstem test serve as the principal guide in assessing the feasibility of a newly discovered petroleum reservoir. The test is usually comprised of short sequences of multiple flow and shut-in periods, as follows:

1. A short flow period for 5 to 10 minutes (min.), followed by a buildup period of about 1 hr in order to determine the initial reservoir pressure.

2. A flow period of 4 to 24 hrs in order to establish stabilized flow to the surface.

3. The well is shut in again to conduct a buildup test, leading to the determination of the permeability-thickness (kh) product and flow potential.

The tool consists of a packer and valve assembly that can be operated from the surface. Following the opening or closure of the tool, fluid pressure is observed to fall or build up accordingly. The resulting response is analyzed before a decision is made to complete the well. A fluid sample is also collected during the test.

Wireline formation tester

This tool employs transient tests of short duration in order to estimate horizontal and vertical permeabilities at various depth intervals within the formation when a new well is drilled. The test is comprised of a drawdown of a predetermined amount of fluid in the tool chamber, followed by pressure buildup until a stabilized value is reached. One or more sensors closely monitor the entire sequence of drawdown and buildup. This information is used to estimate horizontal permeability (in the case of a single sensor) or horizontal and vertical permeabilities (in the case of multiple sensors). The formation tester tool embodies intelligent application of classical well test principles under the limitations of an openhole environment and relatively short test duration. The area investigated by a formation tester tool is relatively small. Mathematical models used in the analysis do not tend to idealize flow and shut-in conditions, as is usually encountered in a conventional test interpretation. Widely known formation testing tools are the Repeat Formation Tester (RFT) and the Modular Formation Dynamics Tester (MDT).[1]

Key points—types of transient well tests

Transient pressure tests in oil and gas reservoirs usually consist of the following:
- Drawdown and buildup tests on a production well
- Falloff test on an injection well
- Interference tests involving two or more wells, e.g., an injector and one or more surrounding producing wells
- Drillstem tests of new wells
- Tests performed by a formation tester in a new well in order to evaluate various formation characteristics following openhole logging

A transient test of a well is accomplished by initiating a step change in the well rate. This typically involves either (a) the shut-in of an active well or (b) flow at predetermined rate from a well that was inactive prior to drawdown. The change in rate creates a new response pattern of reservoir pressure that is carefully monitored by electronic recording devices, followed by detailed computer analysis.

Interpretation of Transient Pressure Tests

Topics in well test interpretation are organized in the following sections:
- Common factors affecting well response during transient tests
- Important concepts in well testing that help reservoir engineers to visualize and interpret a test
- Overview of well test interpretation methodology, including analytic and numerical methods
- Introduction to well test theory and its application in classical interpretation
- Flow regimes that are frequently encountered during well test interpretation
- Outline of specific techniques in well test interpretation, along with qualitative and quantitative interpretation of test data
- Identification of reservoir models based on diagnostic analysis
- Checklist for design, execution, and analysis of a test
- Pitfalls in well test interpretation
- Examples in computer-aided interpretation of well tests performed with a variety of reservoir engineering objectives

Factors Affecting Well Tests

Before embarking on well test interpretation, a well test analyst needs to be familiar with the factors typically influencing well response. A thorough understanding of these factors is instrumental in designing, conducting, and analyzing a test. The factors primarily include wellbore storage and skin effects, as introduced earlier in the book.

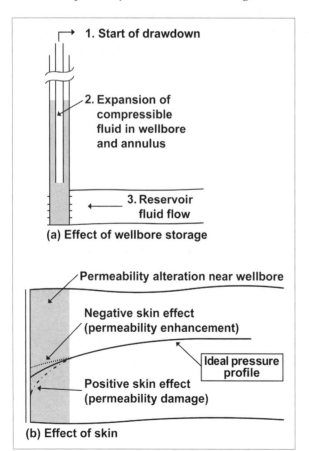

Fig. 5–3. Wellbore storage and permeability alteration (skin) during a well test

Storage effects of fluid in the wellbore

Wellbore storage effects invariably influence the transient pressure response at early stages of conventional well tests. When a step change in production or injection is made during the test, the fluid present in the wellbore dominates the initial response. For example, fluid initially flows from the wellbore and not from the formation as drawdown (flow) is initiated. Consequently, the initial pressure response is influenced by wellbore storage effects rather than by reservoir properties. Moreover, fluid from the annulus may contribute to the initial phase of drawdown when the well is completed open hole. Presence of fluid in wellbore is shown in Figure 5–3(a).

The coefficient of wellbore storage is defined as in the following:

$$C = \frac{V}{\Delta p} \tag{5.1}$$

where
- C = coefficient of wellbore storage, bbl/psi,
- V = volume of fluid produced, bbl, and
- Δp = pressure drop due to drawdown, psi.

With passage of sufficient time, virtually all of the fluid initially present in the wellbore is produced. Well production is essentially from the formation. The effects of wellbore storage disappear, identified by a change in trend in response. In certain cases, the effects of wellbore storage can last for a long period of time, masking the true behavior of the reservoir. Valuable information regarding subtle changes in flow pattern, particularly at the early stages of test, may not be visible due to the above-mentioned phenomena. Any reservoir heterogeneities that may exist in the immediate vicinity of the test well may remain undetected. One solution to the problem is deployment of a downhole shut-in valve during the well test. However, its use is limited by certain practical considerations, including well configuration and available resources.

Permeability alteration around the wellbore

The initial pressure response from the formation is usually affected by the phenomenon of permeability alteration that may take place in the immediate vicinity of a well during production or injection. The measure of permeability alteration is called skin factor, and it may vary during the life of the well. Skin factor can be either positive or negative. A positive value of skin signifies that the permeability of the formation is diminished near the wellbore. In contrast, a negative skin points to the fact that the permeability of the formation is enhanced by some means.

Permeability damage may occur due to the migration of fines through the pores of the rock during production. Another source of permeability reduction is rooted in swelling of clayey material present in the rock due to contact with an incompatible fluid during drilling or external fluid injection. The net result of positive skin is an increased pressure drop in the vicinity of the test well during drawdown, as shown in Figure 5–3(b). Hydraulic fracturing or acidizing is routinely conducted in a well to augment productivity when it is located in a tight formation or when a test indicates significant skin damage. If stimulation is successful, the skin factor changes from positive to negative. In fact, well tests are routinely conducted to determine the success of a stimulation operation.

It is to be noted that mechanical skin can also be encountered during well tests. Mechanical skin arises out of the transport of debris that may pose an obstruction to the flow of fluid through the well perforations or rock matrix. Again, a partially completed well experiences a greater drop in pressure than a well that is completed through the entire hydrocarbon-producing interval. The resulting skin is referred to as partial penetration skin. The evaluation of skin factor from well test data is described later.

Key points—well test data requirement and factors affecting well tests

- Well test data is typically comprised of the results of monitoring well pressure response over time, changes in well rates, fluid properties, rock characteristics, and information related to well configuration. Modern monitoring devices have the capability of interfacing directly with computers for subsequent interpretation of the test data.

- Most well tests are adversely affected by wellbore storage and altered permeability near the wellbore. These phenomena mask the true response in the early stages of the well test. Wellbore storage effects arise due to the fact that the earliest response from a well during drawdown or shut-in is dominated by fluid present in the wellbore.

- Wells are found to exhibit reduced permeability near the wellbore, leading to a positive skin factor. Well stimulation operations are routinely conducted in which formation permeability is enhanced either by acidizing the damaged formation or by creating a hydraulic fracture intersecting the well. These efforts result in negative skin. Both positive and negative skin factors exhibit distinct signatures on well test plots.

Important Concepts in Well Test Analysis

Having discussed the factors that affect well tests most frequently, certain concepts associated with transient test analysis are now introduced. These are essential in designing and interpreting well tests in general.

Characterization of fluid flow during the tests

In reviewing the response in pressure generated by well tests, including drawdown tests, three distinct types of fluid flow characteristics are usually observed. These are unsteady-state or transient, pseudosteady-state, and steady-state flow, which are discussed in detail in chapter 4. The flow of fluid is characterized by changes in the pressure trends with time and location, caused by injection, production, and boundary effects.

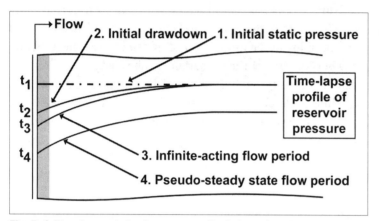

Fig. 5–4. Time-lapse study of pressure profile during a drawdown test

When a drawdown test is conducted in a well located in a newly discovered reservoir, the initial reservoir pressure is assumed to be uniform. As drawdown commences, the fluid pressure drops significantly close to the wellbore. No boundary effects are evident on the flow characteristics, as the perturbations in pressure caused by the drawdown require a certain period of time to travel to the reservoir boundary (fig. 5–4). It is frequently referred to as the infinite-acting flow period, since the reservoir acts like an infinite medium. The fluid flow is transient, implying that the reservoir pressure varies with time and location.

However, once the well flows for a sufficient length of time, boundary effects are observable. In a bounded reservoir with no external influence on pressure, the rate of change in the reservoir pressure eventually becomes constant everywhere due to the constant drawdown. This is referred to as pseudosteady-state flow. It must be emphasized that the reservoir pressure does change at all locations with time; however, the *rate of change* in the pressure is constant. In certain other circumstances, external pressure support is readily provided by an adjacent aquifer, gas cap, or injection well during drawdown. Consequently, reservoir pressure approaches a constant value, leading to the flow of fluid under steady-state conditions.

In a typical well test, the boundary effects are observable at late times, following an infinite-acting flow period. The fluid flow regimes commonly encountered in conventional well tests and their signatures on well test plots are described later in the chapter. It must be kept in mind that many reservoirs have complex geologic features, such as a localized barrier or a partially communicating aquifer at one edge. This may lead to fluid flow patterns that are not readily interpretable. Other sources of data, such as geophysical studies, may go a long way in interpreting such tests.

Radius of investigation

It is important to know the extent of the area investigated in a well test, or in other words, the area a test can "see" in the reservoir. When a change in flow rate is created, the resulting perturbation travels further and further into the reservoir with the passage of time. Consequently, the radius of investigation of the well test becomes larger. In well test design, it is important to determine the minimum duration of the test period required to investigate a reservoir boundary or heterogeneity located at a certain distance from the well. With all other factors remaining the same, a test of longer duration is required in the case of a reservoir of relatively low permeability. The equation to calculate the radius of investigation is presented later in the chapter.

Principle of superposition

The principle of superposition is a simple, yet powerful, concept in well test analysis. Well test theory is primarily based on a single ideal well, which operates at a constant rate in the middle of a reservoir without any effects of a geologic or drainage boundary. In reality, well rates are seldom constant, and certain boundary effects are observable. Multiple rates in wells and the existence of various drainage boundaries (such as no flow or constant pressure) can be easily simulated by applying the principle of superposition.

Simply stated, the anticipated pressure response from a well in a complex situation can be modeled by combining responses from two or more wells in simple circumstances for which straightforward solutions are available. Complex situations are encountered when a geologic boundary is encountered or when the well produces at multiple rates. The principle of superposition can be applied both in space and time as described in the following discussion.

Superposition in space. A well is considered that produces at a rate of q stb/d. A sealing fault is located nearby, and the distance between the fault and well is L, as shown in Figure 5–5(a). The effect of the fault on the pressure transient analysis is simulated by considering an image well located on the other side of the fault at the same distance from the fault (L) but in the opposite direction. The distance between the real well and the image well is 2L. Furthermore, the image well is considered to be producing at the

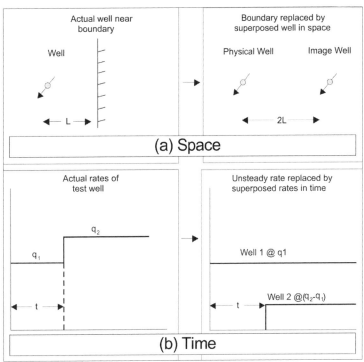

Fig. 5–5. Superposition in (a) space and (b) time

same rate (q) as the real well. This arrangement between the two wells simulates an impermeable boundary across which fluid flow cannot occur. The principle of superposition as applied in this case is summarized as in the following:

Pressure response of the real well in proximity of a sealing fault

= Response from the real well in absence of the fault

+ Response from the image well located equidistant from the fault on the other side to simulate no flow across the fault

The possibilities are virtually endless with implementation of multiple image wells when the well is located near more than one boundary.

Superposition in time. Next, a well will be considered that produced at two different rates during a drawdown test. The test initially started at a rate of q_1 stb/d. Following a certain time (t), the well rate increased to q_2 stb/d. The principle of superposition is applied in this case by considering two wells that produce at steady rates as shown in Figure 5–5(b). The first well produces at q_1 stb/d throughout the test. However, a second well is superposed to account for the increase in the flow rate at time t. The rate of the second well is expressed as $q_2 - q_1$. The combined production from the two wells is q_2 stb/d after time t. This can be summarized as in the following:

Pressure response from a real well having a rate change (from q_1 to q_2 stb/d at time t)

= Response from the well due to the original rate throughout the test (q_1 stb/d)

+ Response from a superposed well producing at the differential rate ($q_2 - q_1$ stb/d), commencing at the time of rate change (t)

The same logic can be applied when a decrease in the well rate occurs from q_1 to q_2 stb/d. A superposed well can be introduced that injects at the differential rate of $q_2 - q_1$ stb/d.

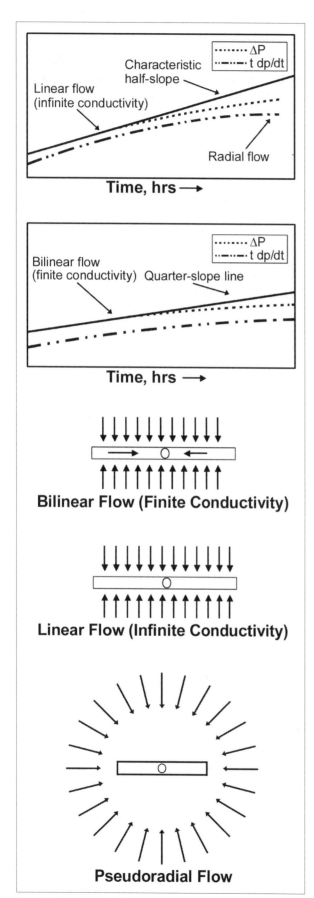

Fig. 5–6. Schematic of pseudoradial, linear, and bilinear flow

Development of fluid flow patterns

Various fluid flow patterns that may develop in porous media in the course of production are discussed briefly in chapter 4. It is interesting to note that the major fluid flow patterns encountered during well tests usually imprint their signatures on the pressure response. The signatures, often in the form of a straight line having a distinct slope, among other visible trends, are readily identifiable by appropriate interpretation techniques. This topic is discussed later in the chapter, with illustrations from field test examples.

An understanding of the various flow patterns that may be encountered during transient tests is essential for the following:

- Visualization of the test response in relation to a specific geologic setting.
- Recognition of reservoir model. No interpretation of a well test is valid unless the reservoir model is identified correctly.

For example, the initial flow pattern is significantly different in a hydraulically fractured well than in a well where no such fracture exists. Then again, the flow pattern that develops in a long, channel-shaped reservoir at later times is not evident in a reservoir of circular geometry. Common fluid flow patterns that may develop in a reservoir are discussed in chapter 4. These are restated here with their significance in well tests.

Radial flow. A well producing in the middle of a reservoir or its drainage area is envisaged to develop a radial flow pattern. During the infinite-acting flow period, a radial flow towards the test well is generally assumed, although nonradial flow patterns are quite possible. The initial flow patterns are observed to be nonradial in the case of horizontal or hydraulically fractured wells. However, radial or pseudoradial flow develops eventually with sufficient passage of time. Pseudoradial flow is illustrated in the lower diagram in Figure 5–6. In addition to radial flow in a horizontal direction, it may develop briefly in a vertical direction at the very early stages of a horizontal well test, as illustrated in the "Horizontal well model" section. Departures from ideally radial flow may be encountered in a reservoir due to the existence of significant heterogeneities (such as directional permeability). They may also occur due to a severe imbalance in injection and production rates among neighboring wells, among other factors.

Linear flow. A hydraulically fractured well having a fracture of infinite conductivity typically experiences linear flow in the early stages of a well test once the effects of wellbore storage diminish. In an infinitely conductive conduit, fluid does not experience any resistance during flow towards the wellbore. Linear flow in porous media is found to occur towards the fracture, as illustrated in Figure 5–6. Linear flow is also encountered in well tests performed in a channel-shaped or shoestring reservoir. The width of the flow channel is significantly small in comparison to its length, and a radial flow pattern would not develop as the reservoir produces. Horizontal well tests typically demonstrate a period of linear flow in the early stages of testing, as illustrated in the "Horizontal well model" section.

Bilinear flow. A typical example of bilinear flow is also shown in Figure 5–6, where the hydraulically created fracture is of finite conductivity and poses a certain resistance to flow. Hence, the flow is found to be linear but bidirectional. Bilinear and linear flow patterns are evident during the early stages of a test in a hydraulically fractured well. A pseudoradial flow pattern eventually develops if the test is carried out for a sufficient length of time. During intermediate times, an elliptical flow pattern may emerge, with the major axis of the ellipse aligned to the hydraulic fracture.

Spherical flow. When a well is partially completed in the formation, a spherical flow pattern develops. This occurs as fluid from the uncompleted portions of the formation flows towards the well perforations.

Overview of Test Interpretation Methodology

Traditional well test interpretation involved preparation of pressure versus time plots, followed by estimation of the sought parameters. This was aided by a limited number of charts and equations available to the analyst. The whole process was rather tedious, as virtually all of the required tasks were performed manually. With the widespread use of digital computation, however, the presentation of data and the underlying computations are completely automated. Because well test software can now incorporate multiple models and techniques, reservoir engineers are able to spend more time on the verification, validation, and integration of the test results.

Methods in interpreting well tests range from relatively simple to highly complex. Analytic models with certain simplifying assumptions are used in traditional well test interpretation. These assumptions include, but are not limited to, homogeneous rock properties, uniform reservoir pressure, and constant well flow rates. Modern computer-assisted methods, however, are capable of handling relatively complex scenarios in relation to geologic features and realistic flow behavior. In summary, well test interpretation may be categorized into the following:

- Classical interpretation
  - Computations are based on log-log and semilog plots of pressure response over the duration of the test. Conceptual reservoir models are relatively simple.
  - In type curve analysis, the well response as obtained in a test is matched with one of the available type curves showing a generic response.
- Computer-assisted methods
  - The recognition of the reservoir model and flow regimes is based on the diagnostic plot.
  - Nonlinear regression of the conceptual reservoir model is employed.
  - Numerical simulation of the well response is used in complex situations.

The computer-based methods are capable of performing all of the traditional analyses.

Well test interpretation—qualitative and quantitative

The task of interpreting transient well pressure response involves both qualitative and quantitative aspects as outlined in the following discussion.

Qualitative interpretation. In a qualitative interpretation, the distinctive flow regimes are diagnosed primarily by visual inspection of the carefully constructed plots. These log-log plots are constructed of pressure differential (Δp) versus time, in addition to pressure derivative data ($t \, \partial p / \partial t$) versus time ($\Delta t$). This is referred to as a diagnostic plot in modern well testing. Based on the diagnostic plot, the analyst is able to review the quality of the test and assess the wellbore and near-wellbore conditions. The analyst is also able to gain insight into the reservoir characteristics, including any boundary effects. Examples of qualitative interpretation include the identification of various flow pattern signatures during the test and assessment of the adequacy of the test duration. It could also include confirmation of a sealing fault or fractures, among other factors, and observation of the boundary conditions.

Quantitative interpretation. Quantitative tasks, on the other hand, employ deterministic models to gather valuable information. Such data can include formation transmissibility, average reservoir pressure, reservoir volume, and extent of the permeability alteration near the well (skin effects). It can also encompass the well productivity index, distance to a sealing fault, and the effective length of a horizontal well or a hydraulic fracture. In traditional practice, a simplified analytic approach based on semilog and log-log plots of pressure versus time data is the first option to interpret test results. Additionally, certain test plots may be drawn in Cartesian or other scales. Rendition of the well test response in this manner readily points to the applicability of simple models to the particular analysis. In a relatively homogeneous formation, the use of simple models is an acceptable practice. However, the analyst needs to be aware of the limiting assumptions associated with conventional well test models when dealing with a complex reservoir, where unknown heterogeneities and multiphase flow of fluids exist.

Before any further discussion of the classical well test interpretation, it is necessary to briefly describe the development of well test theory.

Fundamentals of Well Test Theory

The theory of pressure transient testing is based on the line-source solution of the diffusivity equation described in chapter 4. Well test models are treated in detail in the literature.[2–4] Technical papers are published on a regular basis that propose new or modified well test models under a variety of field and well settings. This book concentrates on the application of various well test applications from the reservoir engineering point of view.

It is important to recapitulate the assumptions inherent in the development of the diffusivity equation:
- Homogeneous and isotropic reservoir
- Horizontal fluid flow with negligible effects of gravity
- Single fluid phase of small and constant compressibility
- Isothermal and laminar flow

The first question raised about the assumptions relates to the inherent heterogeneities associated with any petroleum reservoir, on a microscopic as well as macroscopic scale. Microscopic variations in rock characteristics may not be discernible from the test response. However, interpretation of the well test allows the study of the overall reservoir behavior under dynamic conditions. Traditional well test interpretation is still valid in most circumstances, as the rock properties obtained from the analysis represent an operationally equivalent system extending throughout the well drainage area. Again, many heterogeneities existing on a macroscopic scale are detected in the well test as the flow pattern changes during the test, resulting in a different pressure response.

In the well test literature, familiar solutions to the radial diffusivity equation include the following:[5,6]
- The well as a line-source located in an infinite reservoir of cylindrical shape
- Modified solution that considers skin and wellbore storage
- Pseudosteady-state flow in a cylindrical reservoir with boundaries

- Generalized solution to account for various reservoir shapes
- Introduction of a pseudopressure approach to account for the highly compressible flow encountered in gas reservoirs

Most of the topics mentioned above are discussed in chapter 4. There are a number of common scenarios in which the fluid flow model requires modification. These include, but are not limited to: (i) multiphase flow, (ii) highly compressible fluid in gas and gas condensate reservoirs, and (iii) development of turbulence in gas wells.

The following outlines the solutions to the diffusivity equation that serve as the backbone of analytical as well as numerical interpretation methodology of transient tests.

Basic well test equations—drawdown test

The line-source solution to the diffusivity equation was introduced in the previous chapter with illustrative examples of its application. It is based on the assumption that fluid production originates from a line source, and the well has a zero radius. It is further assumed that the well produces at a constant rate from an infinite drainage area having uniform reservoir pressure prior to production. Attempts are made to implement the above mentioned initial and boundary conditions in drawdown and other well tests as follows:

- Prior to the drawdown test, the well is shut in for sufficient length of time to attain stabilized pressure in the drainage region.
- The well is flowed at a constant rate during the test.
- The pressure response is monitored at the well for a sufficient length of time to gather data from the infinite-acting flow period.

As mentioned in chapter 4, except for very early times, the following logarithmic approximation of the line-source solution to relate the dimensionless pressure with dimensionless time at the wellbore is employed in the interpretation of infinite-acting radial flow:

$$p_D = \frac{1}{2}\left(\ln t_D + 0.80907\right) + s \tag{5.2}$$

where

$$p_D = \frac{kh(p_i - p_{wf})}{141.2\, q\, B\, \mu}, \text{ dimensionless} \tag{5.3}$$

$$t_D = \frac{0.0002637kt}{\varnothing\mu c_t r_w^2}, \text{ dimensionless and} \tag{5.4}$$

s is the skin factor.

Combining Equations 5.2 through 5.4, the flowing bottomhole pressure can be expressed as fluid and rock characteristics as follows:[7]

$$p_{wf} = p_i - 162.6\frac{qB\mu}{kh}\left[\log t + \log\left(\frac{k}{\varnothing\,\mu\,c_t\,r_w^2}\right) - 3.2275 + 0.8686s\right] \tag{5.5}$$

Formation transmissibility

The plot of p_{wf} versus log t, referred to as the semilog plot in well test interpretation, is expected to exhibit a straight line of the following slope:

$$m = -\frac{162.6qB\mu}{kh} \tag{5.6}$$

The average transmissibility of the formation (kh/μ) within the radius of investigation of the well test is determined from Equation 5.6. Neither the formation transmissibility nor the skin factor can be estimated by classical analysis if the infinite-acting flow regime does not develop. Such tests are often considered to be inadequate.

Once the reservoir permeability is obtained, the skin factor is calculated by the following method.

Skin factor

It is noted that at t = 1, log t = 0. Hence the skin factor can be obtained by evaluating p_{wf} at 1 hr on the plot and substituting the value in Equation 5.5, as in the following:

$$s = 1.1513 \left[\frac{p_{1\,hr} - p_i}{m} - \log \left(\frac{k}{\phi \mu c_t r^2_w} \right) + 3.22275 \right] \tag{5.7}$$

It must be borne in mind that the recorded pressure response at 1 hr is usually distorted by wellbore storage effects, and $p_{wf(1\,hr)}$ is obtained by extrapolating the semilog straight line. It is also apparent that the skin factor is calculated only after the values of m and k are obtained from the interpretation.

The net effect of skin is to create an additional pressure drop around the wellbore. All factors being the same, a well with positive skin exhibits lower flowing bottomhole pressure than a well where no detrimental effects of skin are present. The following expressions of pressure drop due to skin effects are noted:

$$\Delta p_s = \frac{141.2\, q\, B\, \mu}{k\, h}\, s \tag{5.8}$$

$$\Delta p_{sD} = \frac{k\, h\, \Delta p_s}{141.2\, q\, B\, \mu} \tag{5.9}$$

$$\Delta p_{sD} = s \tag{5.10}$$

where

Δp_{sD} is the dimensionless pressure drop due to skin.

The skin factor is a function of the damaged zone characteristics, and the following can be shown:

$$s = \left(\frac{k}{k_s} - 1 \right) \ln \left(\frac{r_s}{r_w} \right) \tag{5.11}$$

where

k_s = rock permeability in the damaged or stimulated zone, mD, and

r_s = radius of the altered zone, ft.

Following a successful well stimulation operation, permeability around the wellbore is usually enhanced above the original value. This leads to a negative value of skin in Equation 5.11.

Effective wellbore radius

Due to the presence of skin, the effective radius of a wellbore is different than the actual radius, as in the following:

$$r_{wa} = r_w\, e^{-s} \tag{5.12}$$

where

r_{wa} is the effective wellbore radius.

In a damaged well where skin is positive, the flow rate is constricted, and the effective wellbore radius is less than the actual flow rate.

Flow efficiency

In a bounded reservoir, the flow efficiency of a well is defined as the actual productivity index over the ideal productivity index, evaluated at zero skin as follows:

$$FE = \frac{J}{J_{ideal}} = \frac{p_{av} - p_{wf} - \Delta p_s}{p_{av} - p_{wf}} \tag{5.13}$$

where

p_{av} = average reservoir pressure, psia, and

Δp_s = pressure drop at the well due to skin, psia.

Table 5–1 summarizes the range of values of skin-related terms as computed in well test analysis.

Table 5–1. Characteristics of damaged versus stimulated wells

	Damaged Well	Well neither Damaged nor Stimulated	Stimulated Well
Skin Factor	Positive	0	Negative
Pressure Drop due to Skin Effects	> 0	0	< 0
Effective Radius of the Wellbore	Less than actual	Unchanged	Greater than actual
Flow Efficiency	< 1.0	1.0	> 1.0
Well Productivity Index	$J < J_{s=0}$	$J_{s=0}$	$J > J_{s=0}$

Wellbore storage

In order to account for wellbore storage effects, the radial diffusivity equation is solved with the following boundary condition at the wellbore:[8]

$$\frac{q_{sf}}{q} = 1 - C_D \frac{dp_D}{dt_D} \tag{5.14}$$

where

$$C_D = \frac{0.8936\, C}{\phi c_t h r_w^2}\ , \text{ and} \tag{5.15}$$

q_{sf} = flow rate at the sandface, i.e., the rate of fluid entry into the wellbore from the reservoir, bbl/d.

During the early stages of the test, the flow from the reservoir, qsf, is either nonexistent or is negligible. Hence, Equation 5.14 reduces to the following:

$$0 = 1 - C_D \frac{dp_D}{dt_D} \tag{5.16}$$

It can be further shown that Equation 5.16 can be recast as in the following by taking logarithms:

$$\log p_D + \log C_D = \log t_D \tag{5.17}$$

The above equation has the following form representing a straight line with slope m and intercept c:

$$y = mx + c$$

Hence, inspection of Equation 5.17 suggests that a plot of p_D versus t_D drawn on a log-log scale would lead to a straight line of unit slope. This is evident on virtually all log-log plots where the phenomenon of wellbore storage is prevalent. Based on Equation 5.17, the coefficient of wellbore storage can be computed by selecting any point on the unit slope line, as in the following:

$$C_s = \frac{qB}{24} \frac{\Delta t}{\Delta p} \tag{5.18}$$

Additionally, the following approximation leads to the estimation of the wellbore storage coefficient based on the area of the wellbore and the fluid density:

$$C_S = \frac{25.65 \, A_{wb}}{\rho} \tag{5.19}$$

With the passage of time, the effects of wellbore storage diminish as fluid originally stored in the wellbore prior to the test is continuously produced. Following a transitory period, fluid flow is essentially from the formation, and a new trend emerges that reflects the infinite-acting flow behavior described earlier.

Radius of investigation

In any well test, it is vital to know the distance into the reservoir that is investigated. With the passage of time, any perturbation created at the wellbore travels further and further into the reservoir. The radius of investigation is defined as the distance from the well where a pressure transient response has a significant effect.[9] Based on the line-source solution of the diffusivity equation, it is possible to estimate the radius of investigation for a well flowing in a circular drainage area as follows:

$$r_{inv} = \left(\frac{kt}{948 \, \phi \, \mu \, c_t} \right)^{0.5} \tag{5.20}$$

Equation 5.20 suggests that a relatively long period is required to investigate the same drainage area in a reservoir of low permeability as opposed to one having high permeability, other factors being the same.

Effects of no-flow boundary—pseudosteady-state flow

When multiple wells are present in a reservoir, each well produces from its own drainage region, and a no-flow boundary is created at the edge of the drainage area. Under such circumstances, pseudosteady-state flow characteristics develop, and the pertinent solutions to the diffusivity equation as employed in well test interpretations are given in chapter 4. As noted earlier, pseudosteady-state flow has the following characteristics:

- The rate of change in the reservoir pressure is constant as the effects of the no-flow boundary are prevalent.

- Reservoir pressure declines at the same rate everywhere. When the bottomhole fluid pressure is known, the volumetric average reservoir pressure can be estimated. Correct estimation of the average reservoir pressure from a well test requires certain information. Such necessary input could include the shape of the drainage area (circular, rectangular, etc.) and the relative position of the well within the drainage area (centered or off-centered). Calculation of the average reservoir pressure is demonstrated in chapter 4 by manual computation, and later in this chapter with the aid of a well test software tool.

- The drop in reservoir pressure is inversely proportional to the total compressibility and the pore volume of the well drainage region. This can be very helpful in estimating the hydrocarbon pore volume of the reservoir or the drainage region. The estimation of the well drainage volume is illustrated in an example well test later.

Buildup test

The analysis for a buildup test is based on applying the principle of superposition in time. Prior to the shut-in period, the well produces at a steady rate (q). Once the well is shut in for the buildup test following a certain time period (t), an imaginary well producing at a rate of −q is superposed on the original well. Hence, the net production from the two-well system, one real and the other imaginary, is zero during the buildup period. Superposition in time with a view to representing changes in well rate was described previously.

The line-source solution of the radial diffusivity equation for fluid flow in an infinite reservoir during a buildup test can be utilized by considering a two-well system. In the following, t_p is the production period of the test well prior to shut-in, and Δt is the buildup period following shut-in, when both the real and imaginary wells are active:

- Test well flowing at a rate of +q for a combined period of $t_p + \Delta t$

- Superposed well flowing at a rate of –q since Δt

Hence, the shut-in bottomhole pressure during buildup has two components to consider. However, Horner proposed that the resulting model based on two Ei solutions could be approximated by a single Ei solution having appropriate producing rate and time.[10] Consequently, an expression for the well bottomhole pressure can be derived as in the following:

$$p_{ws} = p_i - 162.6 \frac{qB\mu}{kh} \log \frac{t_p + \Delta t}{\Delta t} \tag{5.21}$$

where

p_{ws} = shut-in bottomhole pressure, psia,

t_p = time the well produced prior to the buildup test, hrs, and

Δt = test interval corresponding to the value of p_{ws}, hrs.

The above equation is in the following form, suggesting the development of a straight line when plotted:

$y = m \log(x) + c$

Hence, a semilog plot of p_{ws} versus $(t_p + \Delta t)/\Delta t$ is expected to yield a straight line beyond any distortions caused by wellbore storage effects. It is to be noted that the log $(t_p + \Delta t)/\Delta t$ is usually referred to as the Horner time scale. Based on the slope of the line, the average permeability or transmissibility of the well drainage area can be calculated as in the following:

$$m = 162.6 \frac{qB\mu}{kh} \tag{5.22}$$

The semilog straight line appears as an ascending line, as the semilog plot is constructed in a manner such that the values of $(t_p + \Delta t)/\Delta t$ decrease from left to right. However, the configuration leads to the *increase* of actual test time from left to right. Equation 5.22 indicates an inverse relationship between the slope of a semilog straight line and the formation transmissibility. Other factors remaining the same, a small slope would indicate a highly permeable formation.

When the well produces for a significantly longer period prior to the buildup test ($t_p \gg \Delta t$), a semilog plot of p_{ws} versus the log of Δt can be utilized in interpretation. This plot, referred to as a Miller-Dyes-Hutchinson (MDH) plot, has gained popularity as it requires less computational work.[11]

Based on the same line-source solution, an expression for skin factor can be obtained by combining the following:

1. The equation for flowing bottomhole pressure at the instant the well is shut in

2. The equation for well pressure at a convenient time interval during the test, such as 1 hr following shut-in

The expression for skin takes the final form:

$$s = 1.1513 \left[\frac{p_{1\,hr} - p_{wf(\Delta t=0)}}{m} - \log \left(\frac{k}{\o\mu c_t r^2_w} \right) + 3.2275 \right] \tag{5.23}$$

where

p_{wf} = flowing bottomhole pressure at the time the well is shut in, psi, and

p_{1hr} = idealized static bottomhole pressure at 1 hr following shut-in, extrapolated from the infinite-acting line on a semilog plot, psi.

In calculating the skin factor, the bottomhole shut-in pressure at 1 hr into the test is found by extrapolating the semilog straight line, as the line represents the true infinite-acting response from the test. The apparent shut-in pressure recorded at early times, including at 1 hr, is often distorted by the effects of wellbore storage. Equation 5.23 suggests that

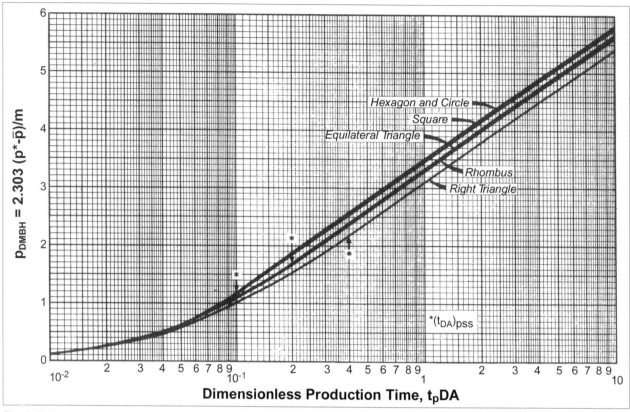

Fig. 5–7. Estimation of average reservoir pressure in commonly encountered drainage area configurations. *Source: C. S. Matthews, F. Brons, and P. Hazebroek. 1954. A method for determination of average pressure in a bounded reservoir.* Transactions. *AIME. Vol. 201, 182–191. © Society of Petroleum Engineers. Reprinted with permission.*

the magnitude of pressure buildup ($p_{1\,hr} - p_{wf}$) could be substantial for a severely damaged formation, other factors remaining the same.

The following related terms are familiar in the industry and are usually computed in well test interpretation to gauge the performance of a well.

Average reservoir pressure. Estimation of average reservoir pressure in the drainage area constitutes an important objective of any buildup test and is briefly discussed here. When the semilog straight line representing infinite-acting flow is extrapolated to infinite time [($t_p + \Delta t)/\Delta t = 1$], the resulting pressure is referred to as p*. The value of p* approaches that of the initial reservoir pressure only in the case of a newly discovered reservoir having negligible depletion. However, in a producing reservoir, the pressure is never restored to the original value unless there is an external source of energy.

Matthews, Brons, and Hazebroek provided a series of plots that enable the analyst to estimate the average reservoir pressure in a drainage area once the following parameters are either known or estimated:[12]

- Size and shape of the drainage area
- Location of the well in the drainage area
- Producing time prior to shut-in and time to reach steady state
- Fluid and rock properties, including formation permeability

The semilog plots are constructed with the following dimensionless groups:

y-axis: $p_{D\,MBH} = \dfrac{2.303\ (p^* - p_{avg})}{m}$ $\qquad$ (5.24)

x-axis: $t_{DA} = \dfrac{0.000264\ k t_p}{\varnothing \mu c_t A}$ $\qquad$ (5.25)

where

A = areal extent of drainage region, ft^2.

Figure 5–7 shows one such plot for the frequently encountered cases where the well is located in the center of a drainage area having a circular, square, hexagonal, or equilateral triangular shape. Other plots are available where the well is located off-center in a square or rectangular drainage area.

In order to estimate average reservoir pressure, p* is found by extrapolating the semilog straight line on the Horner plot to (tp + Δt)/Δt = 1. When a Miller-Dyes-Hutchinson plot is used, p* can be estimated as in the following:

$$p^* = p_{1\,hr} + m \log (t_p) \tag{5.26}$$

Next, t_{DA} is computed with the knowledge of formation permeability, flow period prior to shut-in, and drainage area, among other factors. Except in highly complex situations, the drainage area can be estimated with fair accuracy in a developed reservoir, as the producers and injectors are drilled in a regular pattern. Following computation of t_{DA}, the value of $P_{D\,MBH}$ is read from the plot. Finally, the average reservoir pressure in the drainage region is estimated by the following equation:

$$p_{avg} = p^* - \frac{m}{2.303} P_{D\,MBH} \tag{5.27}$$

In instances where the well is produced for a very long time prior to shut-in, the well rate may not remain steady. Even the shape of the drainage region may be altered due to the drilling of new wells in a reservoir, among other factors. In such cases, t_p could be significantly greater than t_{pss}, the time to reach steady state in the drainage area. Under such circumstances, the accuracy in calculation of the average reservoir pressure is improved by replacing t_p with t_{pss} in Equation 5.28. The time to reach steady-state flow can be estimated as in the following:

$$t_{pss} = \frac{\phi \mu c_t A}{0.000264k} (t_{DA})_{pss} \tag{5.28}$$

Values of $(t_{DA})_{pss}$ for various drainage shapes and well locations are available in the literature. If a well is located centrally in a circular or square-shaped drainage region, the value of $(t_{DA})_{pss}$ is 0.1.

Multirate tests. In many practical situations, the well flow rate during testing cannot be maintained at a constant level. Thus the assumption of single-rate flow in the interpretation may lead to a significant deviation in the test results. The same is true for buildup tests, where the well rate may vary noticeably prior to buildup. Hence, the principle of superposition in time is utilized to modify the fluid flow model. A complex pressure response, generally arising out of multiple well rates, is known as the phenomenon of convolution. Mathematical methods that attempt to interpret multirate tests are usually referred to as deconvolution. This is an active area of research in well testing, as new methods attempt to analyze varying well rates and the resulting response in well pressure recorded by permanent downhole gauges. Due to the complexity of the mathematical procedures, such models are almost exclusively computer based.

Traditional semilog analysis can still be employed to interpret a multirate test. A plot similar to a single-rate test is generated, where both the pressure response and test period are adjusted for varying well rates. Although manual computation of the rate-adjusted parameters could be rather tedious, the procedure poses little difficulty in computer-assisted interpretation. Digital analysis of multirate tests involves only the entry of varying rates over time in lieu of just one rate, and all the necessary calculations are performed automatically.

Equations for gas well testing

Well test theory as discussed so far is based on the assumption of a slightly compressible fluid, applicable to single-phase flow of oil in porous media. In contrast, natural gas is highly compressible, and the various physical properties of gas vary significantly during a typical transient pressure test. A widely accepted practice is to interpret gas well test results in terms of changes in pseudopressure instead of actual pressure versus time. The pseudopressure function, introduced in chapter 4, accounts for the variation in gas viscosity and the compressibility factor with pressure. Similarly, elapsed time during gas well testing can be modified in terms of variable gas properties.[13]

In gas reservoirs operating below 2,000 psia, the variation of the product of gas viscosity (μ) and the compressibility factor (z) with pressure is assumed to be negligible. This leads to the pressure-squared approach in analyzing gas well tests. This technique is practiced widely in low pressure gas reservoirs due to its obvious simplicity in computation.

A significant aspect of gas well testing is related to the observation of non-Darcy flow effects that develop near the wellbore. A component of the radial diffusivity equation is Darcy's law, which is based on the assumption of laminar flow of fluid through porous media. The assumption is quite satisfactory in oil well testing. However, in the case of gas flowing at substantially higher rates, turbulence develops in the vicinity of the wellbore, where the change in pressure is most pronounced. As a consequence, the pressure drop is higher than predicted by the models based on laminar flow. The phenomenon is addressed by defining a modified skin factor, as in the following:

$$s' = s + Dq_g \qquad (5.29)$$

where

    s' = apparent skin factor, dimensionless,

    s = skin due to formation damage or stimulation, dimensionless, and

    D = non-Darcy flow coefficient, d/Mscf.

An example of a gas well test with non-Darcy flow effects is presented later in the chapter.

Classical Interpretation of Pressure versus Time Plots

This is the most widely practiced approach, and it has been adopted by reservoir engineers since the inception of well testing.

Semilog or log-log plots. Pressure response versus time data is plotted in an appropriate scale, such as the semilog or log-log plots presented in Figure 5–8. Depending on rock, fluid, and well characteristics, the pressure response from the well may change trends several times during the test. The distinct signatures imprinted by the observed response are matched against simplified

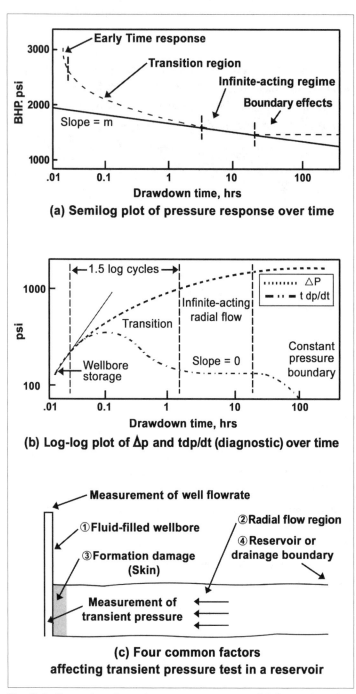

(a) Semilog plot of pressure response over time

(b) Log-log plot of Δp and tdp/dt (diagnostic) over time

(c) Four common factors affecting transient pressure test in a reservoir

Fig. 5–8. Illustration of transient pressure response: (a) flow regimes in a typical test; (b) diagnostic analysis to identify flow regimes and reservoir model; and (c) fluid flow and resulting pressure response affected by various factors during the test

analytic models, leading to the interpretation of the test results. For example, a typical test interpretation may involve the following:

- **Log-log plot.** A wellbore storage model is matched against the pressure response in the initial period. The effects of wellbore storage are identified as the unit slope line.

- **Semilog plot, middle time.** In the intermediate stage of a well test, an infinite-acting radial flow model is identified as a straight line with a measurable slope. It is assumed that this flow period starts one to one and one-half log cycles following the end of wellbore storage. The line-source solution of the diffusivity equation described in chapter 4 is employed for test interpretation.

- **Semilog plot, late time.** In the late stages of a test, further change in the pressure response trend may be encountered on the semilog plot. Appropriate models, including solutions to the diffusivity equation for pseudosteady-state flow, are utilized to gather information about the well drainage boundary.

Wellbore storage effects, infinite-acting radial flow, and pseudosteady-state flow were described previously.

Identification of Flow Regimes

Interpretation of transient well tests is accomplished by the application of the transient fluid flow theory outlined earlier. During a typical test, more than one flow state (unsteady, pseudosteady, and steady) as well as pattern (radial, linear, and others) may be observable. This section discusses the identification of major flow regimes as encountered in a test. Figure 5–8 illustrates the three flow regimes that are expected from a typical test:

1. The early time response, which is dominated by wellbore storage and skin. The true response from the reservoir is distorted by wellbore and near-wellbore phenomena.

2. The middle time region, in which the pressure response indicates an infinite-acting reservoir, i.e., the pressure disturbances created at the well have yet to reach the reservoir boundary. In conventional analysis, the fluid flow pattern is usually assumed to be either radial or pseudoradial, although other patterns are possible. During the early and middle times, the fluid flow is characterized as unsteady state or transient.

3. A departure from the infinite-acting trend at later stages of the test occurs as the effects of a no-flow or constant pressure reservoir boundary become perceptible. This leads to pseudosteady-state or steady-state flow.

Most well test plots presented in this chapter illustrate the early, middle, and late stage regions identified in different circumstances. A well test analyst primarily examines various flow regimes based on the following plots:

1. Semilog plot of the pressure response over the test period

2. Log-log plot of the pressure differential between static and flowing bottomhole pressures versus time

3. Pressure differential (Δp) and the derivative ($t\, \partial p/\partial t$) of the pressure response versus time plotted on a log-log scale.

The latter is generally referred to as a diagnostic plot. In computer-assisted interpretation, the flow regimes and reservoir model are primarily identified by this plot. Diagnostic plots and their applications are treated in detail later.

Early time region (ETR)

As mentioned earlier, the initial response from a well test is usually influenced by the expansion, contraction, and flow of fluid present in the wellbore or annulus at the time the test begins. The phenomenon, referred to as the wellbore storage effect, is readily identified by the characteristic unit slope on a log-log plot, as suggested by Equation 5.17. The early time region can last from a few minutes to several hours. The wellbore storage coefficient is computed from the unit slope portion of the pressure response.

The early time response, beyond any wellbore storage effects, can be influenced by a variety of reservoir or flow characteristics. Fortunately, many of the characteristics leave a distinct signature on the diagnostic plot. These include, but are not limited to, a partially penetrating well, phase separation in the wellbore, a naturally fractured reservoir, and flow through a hydraulic fracture. In stimulated wells, the initial flow is either linear or bilinear, and the resulting slope on the log-log or diagnostic plot is one-half or one-quarter, respectively.

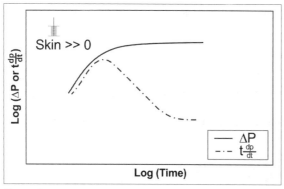

Fig. 5–9. Response of a damaged well during a buildup test

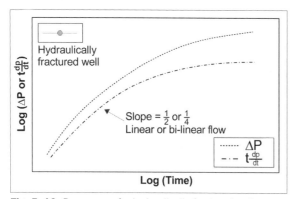

Fig. 5–10. Response of a hydraulically fractured well

Transition region

A transition region occurs as the effects of wellbore storage begin to diminish with time, and the well test response is progressively influenced by the reservoir characteristics. This region, located between the unit slope (characteristic of early time) and zero slope (characteristic of middle time as described later) sections of the plot, is identified with relative ease on the diagnostic plot. However, before the advent of computer-assisted diagnostic analysis, the true end of the transition regime was not always identified with certainty. This led to incorrect identification of the radial flow region, notably in cases of relatively short tests in which the true infinite-acting regime was not reached. Typically, analysts allow at least one and one-half log cycles on the time scale following the end of the unit slope line to analyze the infinite-acting flow regime.

In the case of severe skin damage, the transition zone may be characterized by a large "hump" over a sizeable portion on the diagnostic plot. An example is shown in Figure 5–9. In contrast, tests conducted in a hydraulically fractured or stimulated well having negative skin typically exhibit a long early time region of one-half or one-quarter slope, and the hump is not observed. Figure 5–10 shows a typical response from a hydraulically fractured well.

Middle time region (MTR)

Beyond early time and transition regions, a horizontal line is typically observed on the diagnostic plot. Known as the middle time region, the associated pressure response is infinite acting, as if the reservoir has no boundaries. At this stage, the test has been running long enough to "sense" the reservoir past the immediate vicinity of the wellbore, but not long enough to feel the effects of the reservoir boundary. In conventional interpretation, the flow pattern is assumed to be largely radial or pseudoradial in the middle time region.

The middle time region usually appears as a straight line having a finite slope on the semilog plot. The slope of a semilog straight line is customarily abbreviated as "m" on well test plots and in related computations. This yields information related to the transmissibility of the formation, one of the most important parameters to be obtained from a new well test. The information points to definitive assessment of the reservoir performance under actual flow conditions, even when all of the reservoir heterogeneities are not known with certainty. Skin factor is then computed based on the rock permeability and shut-in BHP.

A conventional test is designed to run for sufficient period in order to attain the infinite-acting radial flow regime as a minimum. Information obtained from a well test is rather limited when this region is not evident. However, with horizontal well testing, the time to attain radial flow regime could take several weeks or even months, as opposed to several hours or days in a vertical or deviated well.

Late time region (LTR)

At late times during the transient test, two types of boundaries are usually encountered: no flow and constant pressure. As described earlier, the aforementioned boundaries usually result in pseudo steady-state and steady-state flow, respectively. On the diagnostic plot of a drawdown test, a bounded reservoir is identified as an ascending line, while a continuously falling line indicates a constant pressure boundary. In a reservoir where several wells are producing from individual drainage areas, the effects of a drainage boundary are evident before the physical boundary of the reservoir is reached.

Interpretation of late time data may lead to the identification and characterization of a geologic heterogeneity. For example, a no-flow boundary can exist where certain geologic heterogeneities are present, including faults and pinchouts. Various trends are observable on the diagnostic plots at late times, depending on the type of geologic boundary as illustrated in the section on reservoir model diagnostics. It is interesting to note that a linear flow pattern can emerge at late times owing to the specific reservoir geometry. This can be diagnosed by a line with a slope of one-half, also called a half-slope line.

A horizontal well test usually exhibits additional flow regimes, which are discussed later.

Well interpretation based on type curves

Type curves are representations of various fluid flow models. They are used to predict the well response (transient pressure) under test conditions with the aid of certain relational curves, such as dimensionless pressure as a function of dimensionless time and storage. Dimensionless quantities are generic and are not limited to a specific reservoir. A pressure response plot, similar to a set of type curves, is generated from the well test. The test plot is then matched with one of the members of the type curve family. This leads to the determination of reservoir characteristics, as the values used to generate the type curve are known. Type curve matching, used in conjunction with semilog analysis, is a potent tool in identifying the reservoir model and flow regimes, and in determining the reservoir characteristics.[14]

Type curves are developed for a specific reservoir model, such as homogeneous, composite, bounded, or dual porosity, to name a few. One of the most familiar type curves was developed by Agarwal, al-Hussainy, and Ramey, and the technique is based on graphical depiction of the radial diffusivity equation solution.[15] A drawdown test with a constant rate at the surface was considered. Assumptions included: (a) flow of a slightly compressible fluid in an infinite-acting homogeneous reservoir; (b) uniform initial pressure in the drainage area; and (c) single-phase flow. These type curves constitute a family of log-log plots of pressure response versus time cast in a dimensionless form in order to make the curves generic. Each member of a type curve family is identified by a unique skin factor and a dimensionless storage coefficient as shown in Figure 5–11. The type curves are based on the following dimensionless quantities:

$$t_D = \frac{0.000264\,kt}{\varnothing\,\mu\,c_t\,r_w^2} \tag{5.30}$$

$$p_D = \frac{kh\,(p_i - p_{wf})}{141.2qB\mu} \tag{5.31}$$

$$C_{sD} = 0.894\,\frac{C_s}{\varnothing c_t h r_w^2} \tag{5.32}$$

In order to perform a type curve analysis, the actual well test response is plotted first. In the case of drawdown, $p_i - p_{wf}$ is plotted against the drawdown time, t, on a log-log scale. Before the advent of well test software, the plot was generated manually on a transparent sheet of paper. It was then matched against a specific curve by moving the transparent sheet horizontally and vertically over the set of curves. In modern day practice, the task of type curve matching is accomplished on the computer screen. Once the best fit is obtained, formation permeability is obtained from a pressure match as follows:

$$k = 141.2\,\frac{qB\mu}{h}\left(\frac{p_D}{p_i - p_{wf}}\right) \tag{5.33}$$

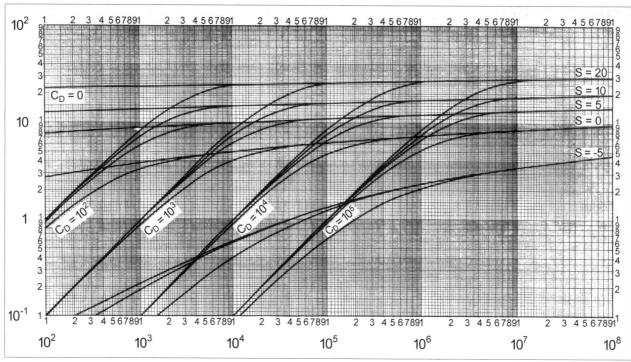

Fig. 5–11. Type curves that predict dimensionless pressure as a function of dimensionless time, wellbore storage, and skin in an infinite-acting homogeneous system. *Source: H. J. Ramey, Jr. 1970. Short-time well test data interpretation in the presence of skin effect and wellbore storage. Journal of Petroleum Technology. January: pp. 97–104; Transactions. AIME. Vol. 249. © Society of Petroleum Engineers. Reprinted with permission.*

In the preceding equation, p_D and $p_i - p_{wf}$ are evaluated at the match point. Similarly, one can obtain a porosity-thickness product from a time match based on the following equation:

$$\varnothing c_t = \frac{0.000264\,k}{\mu r_w^2}\left(\frac{t}{t_D}\right) \tag{5.34}$$

Again, t and t_D are evaluated at the match point.

In the above analysis, C_{sD} is estimated beforehand from the log-log plot of the actual test data by noting the following:

$$C_s = \frac{qB}{24}\left(\frac{t}{p_i - p_{wf}}\right) \tag{5.35}$$

In estimating C_s from the above equation, values of $p_i - p_{wf}$ and t are evaluated when the well pressure response is dominated by wellbore storage effects, as indicated by a straight line of unit slope. Prior knowledge of the wellbore storage coefficient enables the analyst to focus on a subset of type curves having the same value of C_{sD}.

There are situations in which type curves may offer an advantage over classical interpretation. In certain instances, the infinite-acting regime does not develop completely, or the boundary effects are found to dominate at relatively early times. Computer-based well test interpretation tools usually incorporate traditional analysis methods, including type curve–based solutions.

Computer-Assisted Methods in Well Test Interpretation

Reservoir model identification by diagnostic plot

Virtually all computer-assisted test interpretations begin with the diagnostic analysis of the test data. This is a relatively new tool that identifies flow signatures at various stages, leading to conceptualization of a reservoir model. The following are plotted together on a log-log scale to diagnose the reservoir model and resulting flow characteristics:

- Δp vs. Δt

- $t\frac{\partial p}{\partial t}$ vs. Δt

The basic idea is quite straightforward. As illustrated in Figure 5–8(b), subtle changes in the pressure response during a test can be much more perceptible on a pressure derivative plot (frequently referred to as a diagnostic plot). The *rate* of change in pressure provides more information than the change in pressure itself, and certain reservoir models are hardly recognized without a diagnostic plot. Unfortunately, traditional interpretations in the past concentrated on pressure versus time plots, and the powerful capabilities of pressure derivatives were largely unrecognized.

Regression analysis of a reservoir model

Computer-based well test applications incorporate the ability to review and confirm the results of test interpretation by regression analysis. The objective is to conceptualize a reservoir model and simulate the model within the likely range of reservoir properties. This is done to obtain the closest possible match with the actual response throughout the test, including the transition region. This may confirm the validity of the traditional analysis, which obtains results from each flow regime in isolation. Conversely, regression analysis could point to the possible inadequacy of the conceptualized reservoir model if a satisfactory match cannot be obtained.

During regression, reservoir parameters can be varied within reasonable limits to generate what-if scenarios. Horne points out that multiple pressure data are used in calculating the skin factor by this approach, as opposed to the single data as used in classical interpretation.[16] One limitation of the approach is rooted in the nonuniqueness of the solution. Multiple sets of reservoir parameters or more than one reservoir model may lead to a satisfactory match.

Numerical simulation

Certain reservoirs are highly heterogeneous in nature or have unknown heterogeneities that affect fluid flow behavior in an unpredictable manner. Well tests performed in such circumstances may not be amenable to conventional interpretation assuming simple reservoir geometry and homogeneous rock properties. In these cases, numerical models can be employed, which have the ability to overcome the limitations of the analytic approach. One example would be the development of a regional reservoir model based on the response obtained from multiple wells located in the target area following an interference or pulse test. Once a satisfactory match between reservoir simulation and actual well response is obtained, the model can be integrated into a larger reservoir model for better reservoir description.

Well test interpretation strategy

Results of a typical well test analysis are based on multiple interpretation methods, such as diagnostic, semilog, type curve, and regression analysis. Again, several reservoir models are investigated to achieve satisfactory results when reservoir heterogeneities are largely unknown or the initial well test analysis does not conform to the current perception of the reservoir. Further scrutiny may be warranted in the case of unsatisfactory or unexplainable well tests. This could include evaluation of data quality, review of well history, and planning of new tests to gain insight. Integration with other sources of data, including that obtained by geosciences studies or production history, is vital in resolving issues related to well test interpretation.

Resolution of test data

At the early stages of a test, wellbore storage and near-wellbore phenomena usually have profound effects on the test response. With the passage of time, regions located further and further away from the test well begin to influence the pressure response. In other words, the test is able to "see" deep into the reservoir. However, since transmission of the fluid pressure in porous media is a diffusive process, new trends developing due to reservoir heterogeneities could be rather muted. Hence, the resolution of any identifiable changes in trend could be poor at later times. In many circumstances, any subtle change is observable only on a diagnostic plot based on intensive data collection at high resolution.

Key points—well test interpretation methodology

The following key points are reiterated regarding well test interpretation methodology:

- Traditional methods of interpretation are based on the analysis of pressure-time data by plotting the test results in log-log and semilog scales. Various reservoir characteristics are determined based on line-source and other solutions to the diffusivity equation. Type curve analysis is also performed where the actual well response is matched against the type curve response. The amount of data collected during the test is limited.

- Current well test interpretation techniques implement computer-aided analysis. Modern electronic gauges are deployed to collect a vast amount of well test data at high resolution. For qualitative interpretation, a diagnostic plot is used that consists of log-log plots of the pressure differential and the pressure derivative over the time differential. Nonlinear regression and numerical model simulation also are employed, in addition to the traditional analysis.

- The task of well test interpretation involves both qualitative and quantitative aspects. As mentioned above, a pressure derivative or diagnostic plot is generated and is visually inspected to identify the correct reservoir model and underlying reservoir dynamics. These include the identification of radial or linear flow regimes, dual porosity behavior, reservoir boundaries and faults, and influence of aquifers, to name a few.

- Qualitative interpretation is followed by the utilization of appropriate mathematical models to quantify the sought parameters. Such parameters can include the estimation of wellbore storage, skin factor, formation transmissibility, reservoir pressure, and distance to a barrier, among others.

- Accurate well and production information is needed in order to perform a meaningful analysis. Appropriate values of rock and fluid properties are required as well. The analyst usually resorts to multiple interpretation techniques (diagnostic, semilog, type curve, regression, etc.) in order to verify and validate the results of a test.

- The success of well test interpretation hinges on the correct identification of a reservoir model. Multiple models are investigated based on test results when significant uncertainties exist in a reservoir. Last, but not least, a test could be repeated in case of inexplicable data.

In a well test, three flow regimes are usually observable, as in the following:

- **Wellbore storage–dominated flow in the early stages of the test.** The flow regime is identified by a unit slope line on a log-log pressure versus time plot. In a hydraulically fractured well, a half-slope or quarter-slope line is observed that points to linear or bilinear flow. In a fractured reservoir, a brief period of fracture flow is observable at early times. These are discussed in later sections.

- **Infinite-acting flow regime at an intermediate time period.** This regime, identified by a zero slope line on a pressure derivative plot, leads to the estimation of several important parameters sought in a test. Fluid flow, unsteady state in nature, is assumed to be radial or pseudoradial in nature. In a traditional analysis, the regime was identified as a straight line of constant slope on a semilog plot. The beginning of infinite-acting regime was estimated to be about one and one-half cycles later than the end of wellbore storage.

- **Steady-state or pseudosteady-state flow regime at late times.** The regime is identified by a distinct change in trend in the form of an ascending or descending line on the diagnostic plot having a different slope. For example, a unit slope line emerges during drawdown in the case of a bounded reservoir. Depending on the reservoir geometry and resulting fluid flow characteristics, one or more distinct signatures could be observed. However, any change in trend could be rather muted at late times due to the diffusive nature of pressure propagation in porous media.

Key findings of a test, including information about reservoir boundaries, are integrated with geosciences data to build better simulation models.

Reservoir Model Diagnostic

As mentioned earlier, pressure derivative plots serve as a powerful tool in recognizing fluid flow trends at various stages of transient tests. These plots thus aid in conceptualizing a reservoir model. Properly designed, executed, and interpreted well tests can provide much insight into the configuration and behavior of a reservoir. When uncertainties exist in characterizing a reservoir, the analyst can investigate the suitability of more than one mode with well test interpretation. This can be performed in conjunction with other sources of data until the reservoir team agrees upon the most likely model. Diagnostic analyses usually focus on the following:

- Identification of early time, middle time, and late time regions on the plot. Study of the transitions between the regions also is important.
- Analysis of distinctive signatures in well response in the form of changing line trends and slopes, among other factors.
- Verification of the reservoir description obtained from other sources.

Selected reservoir and well models identified by diagnostic analysis are described in the following sections.

Hydraulically fractured well

The transient pressure response from hydraulically fractured wells is dominated by linear flow or bilinear flow at early times. This occurs due to the linear flow towards the long vertical fracture from the reservoir, as depicted in Figure 5–9. Linear flow occurring through an infinite conductivity fracture is identified by a characteristic half-slope line in both log-log and pressure derivative plots. On the other hand, finite conductivity fractures involve bilinear flow, which leads to a quarter-slope line on the plots. Radial flow is eventually attained with the passage of time, as indicated by the appearance of a horizontal region on the diagnostic plot.

Well tests are usually conducted following a hydraulic fracturing operation in order to confirm its success. The length and conductivity of the fractures are determined based on the early time analysis. A negative value of skin and significant enhancement in productivity index serve as confirmation.

Bounded reservoir

Reservoirs are limited in their extent by a surrounding impermeable formation, frequently referred to as a no-flow boundary. During production, a reservoir with no-flow boundaries does not receive any pressure support from an aquifer. A diagnostic plot of a drawdown test shows the infinite-acting period, followed by an ascending line of unit slope at late times as the effect of the no-flow boundary is felt (fig. 5–12). However, there can be cases of a bounded reservoir of rectangular shape where two no-flow

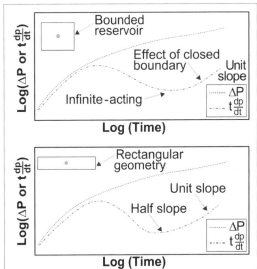

Fig. 5–12. Response from a well in a bounded reservoir during drawdown

boundaries are located significantly closer to the well than other boundaries. In this situation, a half-slope trend may be observed prior to the appearance of unit slope line. Well test interpretation of a well located in a closed system is illustrated later.

In contrast to drawdown tests, no-flow boundaries are difficult to characterize in a buildup test.[17] Since all buildup tests are preceded by well production (drawdown) for a considerable length of time, the boundary effects felt during the earlier drawdown influence the response at late times. A continuously falling line is observed on the diagnostic plot, which is the summation of response from both drawdown and buildup.

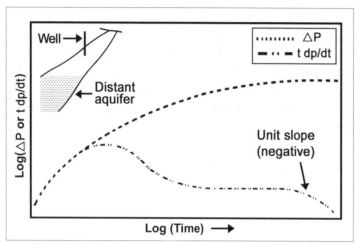

Fig. 5–13. Response from a well with constant pressure support from a distant aquifer

Reservoir with constant pressure boundary

Besides bounded reservoirs, certain reservoirs receive strong aquifer support and are characterized by a constant pressure boundary. Again, reservoirs with a gas cap fall in this category. A pressure derivative plot of a well test conducted in the presence of constant pressure boundaries on all sides exhibits a continuously falling line at late times, shown in Figure 5–7. When the source of the pressure support is located far from the well, a downward trending line with a negative unit slope can be observed. An example would be an inclined reservoir with bottom water drive, with the test well located far above the aquifer (fig. 5–13). Well test interpretation of a well located in a drainage area of constant pressure boundary is illustrated later.

Immiscible fluid boundary near a well

During a typical waterflood operation, an injected water bank advances towards the producers from an injection well in a radial or near-radial fashion. In highly heterogeneous formations, however, the injected fluid front does not flow uniformly in all directions and in all layers that may exist. Transient tests conducted in injection wells, when run for a sufficiently long period of time, may show two distinct horizontal lines on the diagnostic plot, as shown in Figure 5–14. A change in slope is also observed on a semilog plot. Two distinct regions are identified. The first one represents the injected fluid bank and is followed by a region beyond the injected fluid bank where the reservoir fluid is still not displaced. The transition region between the two horizontal lines marks the approximate location of the interface between the two immiscible fluids. When the transition region is short, a sharp edge of injected fluid bank may be envisioned, supported by computation of the radius of an injected fluid bank of known volume. However, a long transition between the two horizontal lines on the diagnostic plot may indicate diffused edges of the injected fluid bank, which may suggest a heterogeneous nature to the formation.

A time-lapse study can be conducted in which the injection well is tested at regular intervals in order to track the advance of the injected fluid front.

Reservoir with geologic faults

In some cases, a sealing fault exists on one side of a test well at some distance. In this situation, it has been long recognized from semilog plots that a doubling of the slope is identifiable once the effect of a no-flow boundary is felt (fig. 5–14). On the diagnostic plot, two horizontal lines separated by a transition are observed. The first horizontal line is established prior to the encounter with the sealing fault, and the transmissibility of the formation is estimated from this line. The ensuing transition regime on the plot marks the beginning of the influence by a linear impermeable boundary. Eventually, a second horizontal line develops on the diagnostic plot with passage of time. This trend continues until the effects of other boundaries are felt.

When two parallel sealing faults exist on two sides of a test well, the late time region on the diagnostic plot exhibits a positive half-slope line. This is due to the linear flow that develops at later times, influenced by the specific reservoir geometry. In the case of two faults intersecting at an angle, the diagnostic plots generally exhibit two horizontal lines separated by a transition period. Reservoir models with various fault configurations suggest specific signatures on the diagnostic plot.[18]

Single fault. A single fault is characterized by doubling of the slope on a semilog plot at late times during buildup or drawdown, as shown in Figure 5–15. Two horizontal lines on the diagnostic plot are observed, the second line at a level higher than the first. The horizontal lines are separated by a factor of two on a log-log scale. However, when the fault is located very close to the well, the first horizontal line may not be noticeable on the diagnostic plot.

Two faults at a right angle. This case is characterized by quadrupling of the slope on a semilog plot at late times. Two horizontal lines on the diagnostic plot are observed, the second line at a level higher than the first. The horizontal lines are separated by a factor of four on a log-log scale.

Two faults intersecting at 120°. In this case, the slope triples on a semilog plot at late times. Two horizontal lines on the diagnostic plot are observed, the second line at a level higher than the first. The horizontal lines are separated by a factor of three on a log-log scale.

Two faults in parallel. This case is characterized by a horizontal line in the middle time region followed by a second horizontal line and finally a half-slope line at late times on the diagnostic plot. The emergence of a half-slope line is typical of linear flow developing at late times.

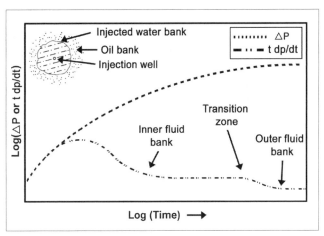

Fig. 5–14. Response from a well with an immiscible fluid boundary

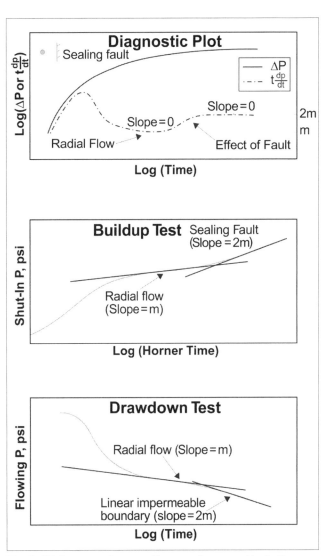

Fig. 5–15. Response from a well near an impermeable fault

Faults on three sides. In this case, a horizontal line in the middle time region is followed by a steeply ascending line that eventually trends into a half-slope line on the diagnostic plot.

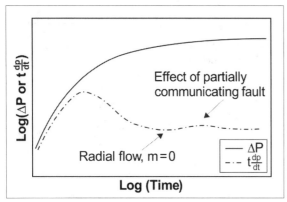

Fig. 5–16. Response from a well near a partially communicating fault

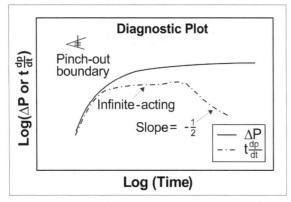

Fig. 5-17. Response from a well near a pinchout boundary

In the case of a partially communicating fault, the resulting signature on the diagnostic plot is relatively subtle (fig. 5–16). In the middle time region, a horizontal line usually is established before the fault is encountered by the test. Next, the effects of partial communication may be characterized by a modest hump. Finally, the line approaches approximately the same level of the horizontal line as was evident earlier.

In a certain field test, an interesting situation arose in which a fault was detected by geophysical study. A subsequent well test conducted in a nearby well produced a diagnostic plot that did not show the familiar signature of a sealing fault. The reservoir team concluded that the fault was nonsealing, and hence could not be detected from a transient pressure response. The above case highlights the importance of an integrated reservoir study.

Effect of pinchout boundary

Certain reservoirs are characterized by a decrease in formation thickness in one particular direction. Reservoir thickness eventually becomes zero, leading to a no-flow boundary. Such geologic characteristics are referred to as pinchouts. A diagnostic plot obtained from a drawdown test shows a characteristic horizontal line as long as the flow is infinite acting. However, when the effect of a pinchout boundary is felt, a hump in response may be observed, followed by the occurrence of a negative half-slope line (fig. 5–17).

Channel-shaped reservoir

Certain hydrocarbon reservoirs are shaped like long channels with limited width. These are frequently referred to as channel sand or shoestring reservoirs. Well tests conducted in the reservoirs are likely to exhibit a positive half-slope line in the late time region, as the late time flow pattern is dominated by linear flow through the long channel.

Dual porosity reservoir

The transient pressure response from naturally fractured reservoirs usually exhibits distinct flow behavior where a minimum or dip is observed on the diagnostic plot beyond the wellbore storage effects (fig. 5–16). A dual porosity flow model is generally utilized to explain the reservoir behavior. The model derives its name from the fact that the rock matrix and fracture network have distinctly different porosities. Naturally fractured reservoirs are modeled based on two distinct flow systems: one through the rock matrix and the other through the fracture network within the matrix. In a dual porosity model, the wellbore receives fluid from the fracture network alone. However, the rock matrix can contribute to production by fluid flow from the matrix to the fracture network first. In contrast, a dual permeability model envisions that the wellbore may receive fluid from both the matrix and the fracture network. It must be mentioned that apparent dual porosity behavior can be observed during a well test in other types of reservoirs, such as stratified reservoirs of highly contrasting transmissibility.

In the early stages of the test, fluid flow primarily originates from the network of fractures due to much greater transmissibility. This leads to establishment of an early flow regime as identified on diagnostic plots and other plots. With the passage of time, the fracture network is depleted progressively. It is consequently recharged by the flow of fluid from the rock matrix, creating a new trend. On a semilog plot, two straight lines are evident at earlier and later times, indicating a distinct flow characteristic from two porosity systems. The transition between the two flow regimes is diagnosed by a dip, the distinct signature of dual porosity (fig. 5–18).

Parameters estimated from well test analysis in a dual porosity system include the storativity ratio and transmissivity ratio. The storativity ratio is a function of the storage capacity of the fracture network to that of the entire system comprised of both matrix and fracture. A relatively high storativity ratio amplifies the separation between the two semilog lines on the plot. The transmissivity ratio, on the other hand, is representative of the ratio of the rock permeability to the fracture permeability, along with their geometries. In the case of a rock matrix having very low transmissivity, the value of the transmissivity ratio is significantly less.

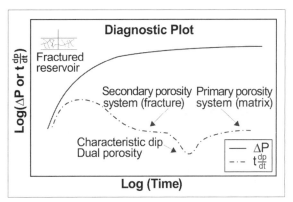

Fig. 5–18. Response from a dual porosity reservoir

Stratified reservoir

Wells completed in a stratified reservoir often exhibit transient responses similar to those of equivalent single-layer formations.[19] Two scenarios associated with multilayered formations are generally encountered. In the case of commingled production, various layers are connected at the wellbore. No vertical communication exists between the adjacent layers in the reservoir. The hydrocarbon-bearing layers are usually separated by impermeable shale. In the second scenario, vertical crossflow occurs between the layers through an intervening layer that is

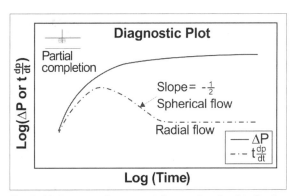

Fig. 5–19. Response from a partially completed well indicating spherical flow

semipermeable. In the case of extreme contrast between layer transmissibility and skin, the early time response can exhibit a dual porosity system. Stratified reservoirs are best characterized by running a formation tester tool following openhole logging when a new well is drilled. As outlined earlier, the tool can quantify the horizontal permeability of the individual layers and can be used to investigate the degree of communication that may exist between adjacent layers.

Partially completed well

Certain wells are partially completed where the fluid initially flows into the wellbore from the completed portion of the formation during drawdown. A straight line on a semilog plot may indicate this regime at the early stages of the test, provided that wellbore storage effects do not mask the flow regime. This is followed by a transition period during which fluid from the uncompleted part of the formation begins to enter the wellbore. The transition period involves the spherical flow of fluid into the wellbore and is characterized by a negative half-slope line on the diagnostic plot (fig. 5–19). A second horizontal line develops eventually, indicating the attainment of an infinite-acting regime in the entire formation. Reservoir properties are estimated from this regime. However, the computed value of skin incorporates the effects due to the partial completion of the well.

Phase segregation in tubing

Due to widely changing pressures during a test, phase segregation may take place in the tubing, indicated by a characteristic hump in the pressure versus time plot. The phases could be oil, gas, and gas condensate. The phenomenon of phase segregation during a transient test may render the test difficult to analyze. The problem can be circumvented by designing a two-rate test.

Horizontal well model

Horizontal well technology has made a profound impact on the petroleum industry in recent years. Horizontally drilled wells are capable of enhancing productivity by several fold as they contact more oil- or gas-bearing formation. Moreover, these wells can produce effectively in highly complex geologic settings.

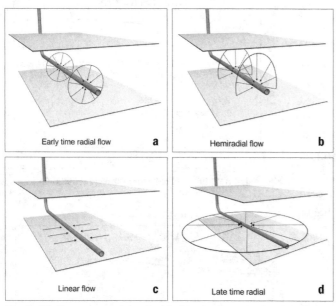

Fig. 5–20. Flow patterns associated with a horizontal well: (a) early time radial flow; (b) hemiradial flow when the horizontal section is not located at the midpoint in the vertical direction; (c) linear flow; and (d) late time radial flow. In certain cases, the latter may take weeks to develop.

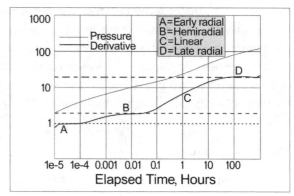

Fig. 5–21. Idealized depiction of diagnostic plot obtained from horizontal well testing. *Source: R. N. Horne. 1995. Modern Well Test Analysis. 2nd ed. Palo Alto, CA: Petroway, Inc.*

Table 5-2. Summary of diagnostics in well test interpretation. *Source: R. N. Horne. 1995. Modern Well Test Analysis. 2nd ed. Palo Alto, CA: Petroway Inc.; and Edinburgh Petroleum Services, Inc. 2006. PanSystem. Kingwood, TX: Edinburgh Petroleum Services, Inc. (a Weatherford Co).*

Signature on Plot	Location on Plot	Type of Plot/ Test
Unit slope line (ascending)	Early time	Log-log /Buildup and Drawdown
Half-slope line (ascending)	Early time	Derivative and Log-log/Buildup and Drawdown
Quarter-slope line (ascending)	Early time	Derivative and Log-log/Buildup and Drawdown
"Dip" with a zero slope at center	Early time	Derivative/ Drawdown
Half-slope line (descending)	Early time	Derivative/Buildup and Drawdown
"Hump" followed by a "spike"	Early time	Derivative/Buildup and Drawdown
Large "hump"	Transition zone	Derivative/Buildup and Drawdown
Zero slope line	Middle time	Derivative/Buildup and Drawdown
Doubling of slope	Between middle and late times	Semilog/Buildup and Drawdown
Continuously falling line	Late time	Derivative/Buildup[a] and Drawdown
Unit slope line (descending)	Late time	Derivative/Buildup[a] and Drawdown
Unit slope line (ascending)	Late time	Derivative/Buildup[b] and Drawdown
Half-slope line followed by unit slope line (ascending)	Late time	Derivative/Buildup[b] and Drawdown
Zero slope line followed by second zero slope line at a level higher than the first	First zero slope line in middle time, second zero slope line in late time	Derivative/ Drawdown
Zero slope line followed by a half-slope line (ascending)	First zero slope line in middle time, half-slope line in late time	Derivative/ Drawdown
Zero slope line followed by steeply ascending line, and then a half-slope line	First zero slope line in middle time, ascending line and half-slope in late time	Derivative/ Drawdown
Zero slope line followed by a modest "hump"	From middle to late time	Derivative/ Drawdown
Zero slope line followed by second zero slope line at a level below the first	From middle to late time	Derivative/ Drawdown
Two zero slope lines. The first line followed by appearance of the second at different level	From middle to late time	Derivative/ Drawdown

[a] Buildup without superposition or desuperposition

One prime example is a geologically faulted formation. A single horizontal well could replace multiple vertical wells that would be required to produce from different hydrocarbon-bearing sections separated by the faults. A horizontal well having multilateral branches is capable of producing oil and gas from multiple zones. The diagnostic plot derived from a horizontal well test may typically exhibit several distinct flow regimes that may develop during the early stages of the test. Figure 5–20 shows the distinct flow patterns that are associated with a horizontal well test under idealized circumstances. These patterns, as identified by diagnostic plot, are illustrated in Figure 5–21. These regimes are described in the following sections.[20–21]

Early time radial flow regime. The earliest response from a horizontal well test arises due to the development of radial flow around the horizontal wellbore as depicted in the Figure 5–20(a). This emulates an infinite-acting flow regime until one of the physical boundaries of the formation is reached in the vertical direction. This regime is identified by a typical horizontal line on a diagnostic plot (fig. 5–21). It is observed that wellbore storage effects often tend to obscure the early time response.

Hemiradial flow regime. A hemiradial flow regime develops when the vertical distances from the well to the upper boundary and lower boundary of the formation are not the same. Following the development of early radial flow in the vertical direction, one of the edges touches the vertical boundary of the formation, while the edge at the opposite side does not [fig. 5–20(b)].

Linear flow regime. In this phase of transient response, the horizontal wellbore acts like a long fracture, and fluid flow toward the wellbore is essentially linear [fig. 5–20(c)]. The linear flow regime is identified by a characteristic positive half-slope line on the diagnostic plot (fig. 5–21).

Late time radial flow regime. A radial flow or infinite-acting regime in a traditional sense may finally develop at a later time [fig. 5–20(d)]. In the case of long horizontal wells drilled in a tight formation, the time taken to reach the late radial flow regime could be rather long. Moreover, the regime may not be evident in the presence of certain reservoir complexities.

The transient analysis of horizontal well test data may provide valuable information. This could include the effective length of the horizontal wellbore and the ratio of vertical to horizontal permeability, among other factors. An example of transient test analysis is described later.

Depending on the horizontal well configuration and geologic complexities, one or more flow regimes as described above may or may not be discernible on a diagnostic plot. A numerical model may be required to study the flow behavior when traditional analysis is not adequate. Examples include horizontal wells drilled through several geologic strata or faults in a complex trajectory.

Table 5–2 summarizes the familiar signatures that certain reservoir models exhibit on a diagnostic plot. Caution must be used in developing a reservoir model based on diagnostic study alone.

Inferred Model/Condition	Comment
Wellbore storage effects	Common early time signature
Linear flow	Common in stimulated wells
Bilinear flow	Common in stimulated wells
Dual porosity	Common in naturally fractured reservoirs
Spherical flow	Common in partially completed wells
Phase redistribution in tubing	
Severe skin damage	
Infinite-acting radial flow	Usually expected in most tests
Single sealing fault	
Constant pressure boundary	Common in developed reservoir with pattern wells
Constant pressure boundary	Strong support from gas cap or aquifer in distance
Closed system: square shaped	No aquifer support or gas cap
Closed system: rectangular shaped	No aquifer support or gas cap
Intersecting faults (sealing)	Slope of line triples or more on semilog plot between middle and late time
Parallel faults (sealing)	
Faults on three sides (sealing)	
Partially communicating fault	Slope increases by less than a factor of two on semilog plot corresponding to the "hump" portion on derivative plot
Dual permeability	Vertical crossflow between uncompleted and completed layers
Radial composite system	Second line is at elevated level when the outer section is of low transmissibility and vice versa

ıp under desuperposition

The same data can be shown to fit or approximate more than one model if the test period is insufficient or the data quality is poor. The analysis could be further complicated by unforeseen reservoir heterogeneities and inappropriate interpretation methodology. Limitations of well test interpretation are described later.

Design and Scheduling of Well Tests

Selection of transient test type

The design of a well test begins with the detailed review of the objective or objectives of the test, followed by selection of a specific test type. As described earlier in the chapter, the objectives of the well test are wide ranging. They may encompass the determination of formation transmissibility, the evaluation of well productivity, and the identification of the type and location of a reservoir boundary. In the petroleum industry, the most frequently performed well tests include buildup and falloff tests for producing and injecting wells, respectively. Drawdown tests are commonly encountered as well. These tests are usually conducted to assess the effects of wellbore storage, the extent of skin damage or enhancement, and the formation permeability or transmissibility. These tests also may be used to determine the average reservoir pressure in the drainage area, the type of flow boundary, and the presence of geologic features in the vicinity of the test well. Other applications may include evaluation of a hydraulic fracturing operation, the location of an injected fluid bank, or a reservoir limit test, among others.

Less common are pulse or interference tests, designed specifically when certain geologic heterogeneities between two or more wells need to be identified, such as the presence of a geologic barrier or high permeability streaks.

Each newly drilled well undergoes evaluation by drillstem testing, and in many cases, by analyzing reservoir characteristics with the aid of a formation tester.

Duration of the well test

In general, a transient test of relatively long duration can identify reservoir conditions further into the reservoir, including the effects of flow or a geologic boundary. Practical considerations and resource constraints may limit the test period. This could include loss in production (in the case of a buildup test) or inability to maintain a stabilized flow (in the case of a drawdown test). Nevertheless, the analyst expects to find three distinct regions in a typical test: early time, middle time, and late time. The majority of transient tests will run anywhere from a few hours to a few days in an oil or gas reservoir. Formations having low permeability, usually 10 mD or less, require relatively long test periods in order to investigate the late time region. Hence, well tests in tight reservoirs are designed accordingly.

The end of an infinite-acting flow period or middle time region during a transient test, identified as a straight line on a semilog plot of pressure versus time, can be estimated from the following equations.[22] These equations may be used as a guide in designing the minimum duration of a test. In general, tests of longer duration are preferred, as information based on the late time region can be collected and analyzed.

Well centered in drainage area of radius r_e (buildup test):

$$\Delta t = \frac{237 \, \phi \, \mu \, c_t \, r_e^2}{k} \tag{5.36}$$

Well located at a distance L from a no-flow boundary (buildup test):

$$\Delta t = \frac{948 \, \phi \, \mu \, c_t \, L^2}{k} \tag{5.37}$$

In designing a drawdown test, an equation similar to (5.36) is used:

$$\Delta t = \frac{948 \, \phi \, \mu \, c_t \, r_e^2}{k}$$

Examination of the above equations points to the fact that certain parameters used in the equations, including formation permeability, are not known and may be part of the test objective. Unfortunately, the analyst confronts the challenge of the unknown in virtually every reservoir engineering study, including well test design. Conservative values of the key parameters, such as formation permeability and distance from the well to a boundary, need to be assumed. This could lead to a longer test period. It should be further noted that the radius of investigation of a designed test can be estimated by applying Equation 5.20.

Testing of horizontal wells usually requires a significantly long period to reach the pseudoradial flow regime. This may not always be feasible in view of the significant loss in production. Similarly, a much longer testing period is needed to determine reservoir limits (a reservoir limit test) or the degree of communication among several neighboring wells (an interference test). Tests could be of relatively short duration when the success of a hydraulic fracturing operation needs to be assessed. In the above, reservoir permeability is usually known from previous tests, and the sought parameters are negative skin factor and fracture length characteristics following a hydraulic fracturing operation. In essence, the duration of a specific well test is dictated by the test objective, in addition to reservoir characteristics.

Well rate

A literature review suggests that the complete range of changes in pressure response needs to be known in designing a test. In a reservoir producing above the bubblepoint, the drawdown rate should be determined optimally so as not to trigger the evolution of a gas phase. Virtually all well-testing software applications have the ability to design a well test pointing to an expected pressure response over different time periods when the best estimates of reservoir data are utilized. Well and fluid information are generally available during the design of a well test. However, most well test applications are capable of generating fluid properties based on industry-accepted correlations. Multiple scenarios in well test responses can be generated quickly when uncertainties exist, leading to a conservative estimate of buildup or drawdown time.

Selection of monitoring device

Accurate identification of various phenomena occurring at the time of a transient pressure test is only possible when minute changes in response can be obtained and plotted. The downhole analog gauge traditionally deployed to measure bottomhole pressure in the past has given way to more precise electronic devices. The latter, based on silicon-sapphire or quartz crystal transducers, are capable of recording a vast amount of time, pressure, and temperature data at very short intervals. In selecting a bottomhole pressure measurement device, the following considerations are necessary:

- Ranges of anticipated pressure and temperature, especially in a deep well environment
- Gauge accuracy: 0.02% of full range or better
- Gauge resolution: 0.01 psi or better
- Gauge drift: less than 1 psi per year
- Sampling rate: from 1 per second (or better) to several hours
- Frequency of data collection: variable at various stages of the well test
- Sampled data set: time, pressure, and temperature
- Memory capacity: hundreds of thousands of data points
- Interface: seamless integration with analytical software

An intensive data collection scheme with high-quality data is mandatory in many instances. These include, but are not limited to, the following:

- Analysis of an interference well test
- Confirmation of a reservoir heterogeneity or boundary having a subtle effect on the pressure response
- Identification of various flow regimes that develop during horizontal well testing
- Previous well test results based on analog data that does not have a high degree of confidence

Scheduling of well tests

Well tests are usually conducted in petroleum reservoirs at a certain frequency as part of an integrated reservoir management plan. For example, a reservoir management team may decide to test a producing well every few years. When unexpected issues emerge, such as loss in productivity, the well could be tested at a greater frequency to gain more insight into reservoir dynamics. Moreover, certain key wells may be identified in the reservoir to be tested at relatively short intervals. Possible scenarios include, but are not limited to, the following:

- Wells located in a strategic area during waterflooding or other enhanced recovery operations where early breakthrough may be encountered.
- Wells that aid in estimating average pressure in the drainage area during pressure maintenance.
- Wells requiring periodic stimulation to boost productivity.
- Well performance influenced by an unidentified geologic heterogeneity resulting in unpredictable performance. The heterogeneity may involve a fault, fracture network, or high permeability channel, among other factors.

Well test checklist

Gringarten recommends that specific questions be asked before starting a well test analysis, as detailed in the following list.[23]

Questions before analysis. The following checks are to be made before the analysis begins:

Reservoir questions
- What type of rock?
  - If limestone, carbonate, granite, basalt, or loose sand, expect possible double porosity behavior. (In a consolidated sandstone formation, however, double porosity behavior is not generally expected.)
  - If acidized carbonate formation, expect composite behavior (due to permeability enhancement near the wellbore following acidizing).
- Is this a layered reservoir?
- Any known boundary, producing or injecting well nearby, gas cap, or water contact?

Well questions
- Is the well vertical or horizontal?
- What was done to the well?
  - Acidized
  - Fractured
- Has the well been drilled through the entire formation?
- How long has the well been in production?

Fluid questions
- What is the dominant phase?
  - Oil, gas, or water
- How many phases?
  - In the wellbore
  - In the reservoir
- What is the bubblepoint pressure (oil) or dewpoint pressure (condensate gas)?
  - Expect composite behavior in low permeability reservoirs if the pressure falls below the bubblepoint (oil) or the dew point (condensate gas).

Checks during analysis. The following checks are to be made during the analysis:

Data validation
- Check accuracy of recording devices
- Check start and end of flow periods
- Check consistency in well flow rate

Model diagnostic
- Check time and pressure at the start of the flow period
- Check smoothing of the derivative (diagnostic) plot

Remember the checklist mentioned above and use common sense.

Pitfalls in well test analysis

Well test analysts need to be aware of the various issues involved in properly designing and interpreting a well test. The most common occurrences are listed in the discussion that follows.

Improper design of the test. Before conducting any well test, it must be ensured that the test design meets the desired objectives. Perhaps the most critical design criterion is the duration of a transient test. Many tests are not conducted for a sufficient period of time to ensure the emergence of infinite-acting radial flow region. Reasons include, but are not limited to, overestimation of reservoir transmissibility, operational constraints, and loss of revenue. Tests limited to early time and transition regions often fail to provide any meaningful results. A longer testing period is recommended as the test is repeated.

Data validation issues. An intensive data collection scheme based on state-of-the-art electronic devices should be employed in order to identify minute changes in the well response. Some tests fail due to malfunctions of the pressure recording device at some point during the test. Again, the test data may initially appear to be satisfactory but does not make a whole lot of sense once the diagnostic and certain other plots are generated. In this case, the well in question needs to be retested based on corrective measures.

Inability to identify flow regimes or reservoir model. In certain instances, the effects of wellbore storage may last for a long period. This could be followed by the appearance of reservoir boundary effects even before the infinite-acting radial flow regime develops. In such circumstances, the traditional interpretation of well test results may pose a challenge. Horne provides interesting field examples where more than one reservoir model (infinite acting, dual porosity, closed boundary, and fault) can be inferred from the same diagnostic plot.[24] The uncertainty is rooted in the following: (a) an insufficient test period in which the infinite-acting period was not reached, and (b) the middle time region was obscured by unidentified boundary effects.

Alternate methods of interpretation may be attempted to analyze a test in which the conceptual reservoir model is not established. In any case, reservoir information available from other sources should be incorporated in building the model. Again, the importance of an integrated reservoir study is emphasized.

Emerging technologies

As evident in many areas of reservoir engineering, well testing practice has undergone significant changes in recent years. Today, many reservoirs are monitored continuously in terms of rate and pressure information by utilizing permanent downhole sensors. Newly developed software applications analyze a vast amount of pressure and rate data associated with daily production by a number of techniques. When a well equipped with a downhole electronic monitoring tool is shut in, pressure buildup data is collected and analyzed. This is done to detect any possible changes in the well and reservoir performance. A significant benefit of this approach is that the well does not need to be shut in for days or weeks with the sole objective of conducting a transient test.

One study detailed the application of permanent sensors for continuous monitoring and subsequent well test analyses in a highly overpressured reservoir. This unconsolidated sandstone reservoir was located in Gulf Coast.[25] It was critical to evaluate the formation in terms of bottomhole pressure, temperature, rate, skin damage, and other properties. This needed to be done to ensure a high rate of production without creating potential sand-control issues. It was anticipated that sand production might ultimately lead to a significant loss in production, resulting in cost-intensive remedial measures. The other objectives of real-time monitoring included the realistic forecasting of the production rate and improved confidence in the reserve estimation.

Downhole data from the wells was obtained at intervals of a few seconds and was stored in a database for subsequent analysis. One such analysis involved identification of shut-in periods and buildup test interpretation of time-rate-pressure data. Transient test results obtained from the shut-in period were then reviewed in conjunction with downhole data obtained during the flowing period. Up-to-date values of skin factor and effective permeability were then used as a basis to determine an optimum production rate. If the resulting skin damage was observed to be greater than in previous instances, the well production rate was lowered to ensure operation under safe limits.

Well Test Examples

In this section, examples of transient tests are presented, highlighting their role in reservoir engineering, along with the illustration of basic interpretation methodology. These examples are based on computer-assisted interpretation by using PanSystem well test software.[26] Additionally, step-by-step computation of test results is included in certain instances in order to facilitate a thorough understanding of interpretation methodology. Well test data used in the illustrations was obtained from various sources.[27–30] Certain assumptions were made in performing interpretations where necessary information was not available.

The section is organized as follows:
- Capabilities of typical well test software
- Minimum data requirement to interpret a test and sources of data
- Computer-assisted interpretation based on various methods, including traditional semilog analysis, type curve matching, diagnostic analysis, nonlinear regression, and numerical simulation
- Verification of test results by manual computation in certain cases
- Design of a well test in characterizing a reservoir

Several class problems are provided at the end of the chapter for gaining hands-on experience.

Capabilities of well test interpretation software

Modern well test interpretation software is capable of handling an array of reservoir models and complex reservoir dynamics. Nonlinear regression and other methods may be utilized in conceptualizing a reservoir model that best fits the observed response. Notable features of well test applications, including various models, are given in the following lists.

Reservoir model
- Radial homogeneous
- Radial composite with different rock characteristics
- Vertical fracture—uniform flux, infinite or finite conductivity
- Dual porosity and dual permeability, including fractured formation
- Stratified formation with crossflow across layers
- Sealing and semisealing fault or faults
- Pinchout formation
- Partially penetrating well
- Gascap and aquifer support

Reservoir boundary type
- Infinitely acting
- Closed system
- Single, parallel, or intersecting boundaries by simulating image wells

Fluid phases
- Oil
- Gas, including non-Darcy flow
- Water
- Multiphase flow

Fluid flow rate
- Single rate
- Multirate

Well type
- Vertical
- Horizontal

Wellbore storage and skin
- Constant
- A function of flow rate

Sources of well test data

Successful interpretation of well test data, particularly in the case of complex systems, depends on wide-ranging sources of data. These include, but are not limited to, the following:
- Previous well tests performed in the reservoir, field, or basin
- Petrophysical, geological, and geophysical studies
- Integrated reservoir models
- Numerical simulation studies
- Region-specific correlations for estimating fluid and rock properties, when field data is not available
- Well completion record
- Well and reservoir production database

During the actual test, the following data-gathering scheme is implemented:
- Transient pressure response versus time as recorded by electronic monitoring devices installed bottomhole. Frequency of data collection is usually intensive and may involve hundreds or thousands of data points.
- Well flow rates prior to and during the test. Minute changes in the rate are recorded for accurate interpretation.

In modern-day well testing, monitors having direct interface capability with high-speed computers are deployed to collect rate, time, and pressure information during a test. Certain electronic devices are capable of transmitting data in real time for quick analysis. Once all the required information related to the test is obtained, interpretations involve data validation, reservoir diagnostics, and necessary computations.

Examples

Example 5.1. Buildup test of a well located in a drainage area of a constant pressure boundary. This example illustrates the methodology in interpreting a buildup test. Topics include:

- Overview of pressure buildup over the entire test period and identification of the early time region, middle time region, and late time region from log-log and diagnostic plots
- Steps involved in computer-aided interpretation, including analyses based on semilog plots (Horner and Miller-Dyes-Hutchinson)
- On-screen matching of type curves
- Manual verification of computer-generated results

The following data is available based on a 72-hr buildup test of a vertical well (Table 5–3).[31] Production records indicate that the well produced at a rate of 250 stb/d for 13,630 hrs prior to shut-in. Stabilized bottomhole flowing pressure prior to the shut-in was recorded as 3,534 psia. Other data includes the following:

Formation thickness, ft, = 69.
Porosity, %, = 3.9.
Total compressibility, psi^{-1}, = 17×10^{-6}.
Oil formation volume factor, rb/stb, = 1.136.
Oil viscosity, cp, = 0.8.
Well radius, ft, = 0.198.

Table 5–3. Shut-in bottomhole pressure versus time during buildup

Shut-in Period, hrs	Static BHP, psia	Shut-in Period, hrs	Static BHP, psia
0.15	3,680	8	4,350
0.2	3,723	12	4,364
0.3	3,800	16	4,373
0.4	3,866	20	4,379
0.5	3,920	24	4,384
1	4,103	30	4,393
2	4,250	40	4,398
4	4,320	50	4,402
6	4,340	60	4,405
7	4,344	72	4,407

The producer is located at the center of a square-shaped drainage region in a developed reservoir, where multiple producers are drilled in a predetermined pattern. The reservoir area from which the well produces oil is estimated to be 160 acres.

Analyze the test by making all necessary assumptions.

Solution: Steps taken to analyze the buildup test are outlined in the following discussion.

Step 1: Qualitative review of well test response. Figure 5–22 illustrates an overview of the buildup test in which the progressive buildup of bottomhole pressure is plotted against the test period in a Cartesian scale. Next, the following plots are generated in a log-log scale:

- Increase in shut-in bottomhole pressure (Δp) versus Horner time, defined earlier
- Pressure derivative (diagnostic) plot

A general inspection is made of the nature and shape of the plots shown in Figures 5–22 and 5–23. The pressure derivative of

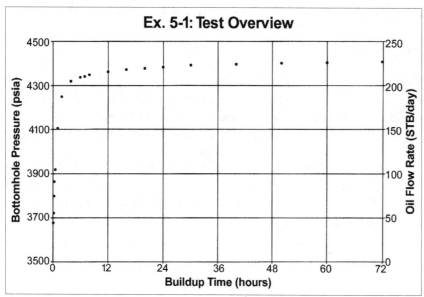

Fig. 5–22. Overview of well pressure against time. Significant buildup takes place at the very initial stages of test. *Courtesy of eP, a Weatherford Company.*

the diagnostic plot is subjected to detailed scrutiny in particular. The objectives include, but are not limited to, the following:

1. Review of data quality, including validity of the downhole measurement of pressure response.

2. Identification of various flow regimes dominated by wellbore storage, transition, infinite-acting radial flow, and late time boundary effects. In certain cases where the data appears to be highly scattered, smoothing of the derivative curve to a certain degree would be necessary in order to clearly identify the flow regimes and their duration.

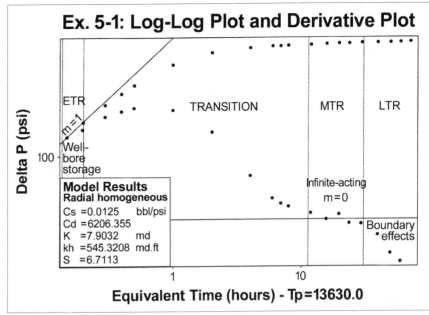

Fig. 5–23. Log-log plot of pressure buildup versus elapsed time. The rate of change of pressure buildup, widely known as the diagnostic, is plotted below. Most of the qualitative analysis in present-day well test interpretation is based on the diagnostic plot. *Courtesy of eP, a Weatherford Company.*

3. Conceptualization of a reservoir model based on observable trends on the plots.

In the absence of detailed information about the reservoir, the initial assumption is a relatively simple model: an oil well producing from a known drainage area in a developed reservoir. Furthermore, a constant pressure boundary is assumed. The well test model has three components:

- Wellbore storage: constant
- Fluid flow: radial
- Boundary condition: constant pressure

The buildup test illustrated here is based on data obtained by analog gauge. However, digital test data may require additional processing, as detailed in the following:

- **Data reduction.** The volume of data from a transient test obtained by modern electronic devices can easily run into thousands of items. Hence, an appropriate data reduction scheme is required. Algorithms are implemented by the well test software in order to reduce the volume of information. This could include data preservation only at longer intervals, or only where the deviation between two consecutive data values exceeds a certain percentage, among other tactics. Care must be taken during data reduction so as not to ignore any subtle changes in response that may lead to critical information.

- **Adjustment of the initial point on the pressure versus time curve.** At the commencement of buildup or drawdown, a time lag of a few moments may be noticeable. This lag can occur before the electronic gauge fully stabilizes to record the pressure response to a hundredth of a pound per square inch. Consequently, the derivative curve on the log-log scale may appear to be distorted. A shift in the initial time scale may be necessary to correctly represent the early time region on the plot.

Step 2: Identification of flow regimes. The following trends are identifiable from Figure 5–23:

- An early time region of wellbore storage indicated by a unit slope line through the first two data points. The wellbore storage coefficient is computed by well test application based on the unit slope line.

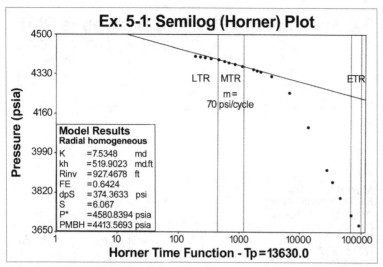

Fig. 5–24. Horner plot showing semilog plot of pressure versus Horner time. *Courtesy of eP, a Weatherford Company.*

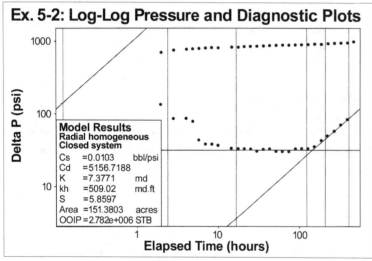

Fig. 5–25. Miller-Dyes-Hutchinson plot of pressure versus time. *Courtesy of eP, a Weatherford Company.*

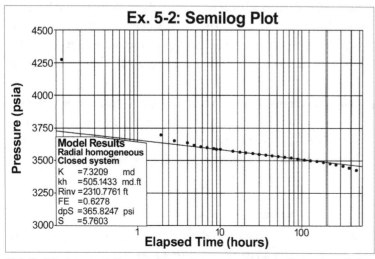

Fig. 5–26. Interpretation based on type curve match. *Courtesy of eP, a Weatherford Company.*

- A horizontal line drawn approximately between 12 and 50 hrs indicates the middle time region on the diagnostic plot. Computed values of permeability and skin based on the diagnostic plot are displayed.

- The late time region is identified as a continuously falling line beyond the middle time region. In a buildup test, reservoir pressure ultimately tends to be constant, hence the derivative of the pressure approaches zero, as observed during late times.

The early time region, middle time region, and late time region are selected on the diagnostic plot following detailed inspection as shown on the diagnostic plot.

Step 3: Classical analysis based on a Horner plot. A semilog plot of static bottomhole pressure versus Horner time, usually referred to as a Horner plot, is generated next, as shown in Figure 5–24. The software automatically generates a best-fitted line through the data points in the middle time region selected previously on the derivative plot. Alternately, the user has the complete freedom to select a different straight line through the points if considered more appropriate. Based on the above, several important parameters are computed. These include, but are not limited to, permeability, permeability-thickness product, radius of investigation, flow efficiency, skin, and pressure drop due to skin. Additionally, extrapolated pressure and average pressure in the drainage area are also estimated.

The stabilized production period prior to the test is significantly greater than the time to reach pseudosteady-state (infinite-acting radial flow regime). Thus, an analysis may be performed based on a Miller-Dyes-Hutchinson plot instead, as illustrated in Figure 5–25. The results obtained are very similar to the Horner analysis.

Step 4: Type curves. A type curve analysis is conducted to verify the results of the Horner analysis. Based on the entered information, a family of type curves is generated by the well test software to match the actual pressure response (shown in fig. 5–26). The derivative curve is also included in the matching process. A match can be obtained quickly between one of the curves and the test response by clicking and dragging the type curves on the plotted data. Only the early time region and middle time region are matched, as the type curves are based on the infinite-acting flow model described earlier. Apart from type curve analysis, nonlinear regression of pressure response data may be performed based on the reservoir model used in the exercise.

Step 5: Illustration of computational methodology. For purposes of illustration, some of the results obtained by computer-aided interpretation are manually computed in the following discussion.

First, the wellbore storage coefficient is computed from the log-log plot of the pressure increment, Δp, versus the elapsed time, Δt, based on Equation 5.17. The above plot is shown in Figure 5–23. Noting that the first point on the plot is part of the unit slope line, the following is obtained: $\Delta p = 146$ psia, and $\Delta t = 0.15$ hr.

Using these values, the wellbore storage coefficient is estimated as follows.

$$C_S = \frac{(250)(1.136)}{24} \frac{(0.15)}{(146)}$$

$$= 0.012 \text{ bbl/psi}$$

Next, the permeability of the formation is estimated based on Equation 5.22. This requires the calculation of the slope, m, of the straight line drawn through the infinite-acting flow regime on the Horner plot (fig. 5–24) or on the Miller-Dyes-Hutchinson plot (fig. 5–25). The unit of slope is a measure of the change in pressure over one log cycle in the time scale. A cycle represents the time period where the lower and upper limits differ by a factor of 10.

The equation to compute the slope of a semilog straight line is shown in the following:

$$m = \frac{p_2 - p_1}{\log(t_2/t_1)}$$

Referring to the Miller-Dyes-Hutchinson plot in Figure 5–25, the values of pressure are read on the straight line at $t_1 = 1$ hr and $t_2 = 10$ hrs. The computations are shown in the following:

$$m = \frac{4{,}358.6 - 4{,}288.6}{\log(10)}$$

$$= 70 \text{ psi/cycle}$$

$$kh = 162.6 \frac{(250)(1.136)(0.8)}{(70)}$$

$$= 527.8 \text{ mD-ft}$$

$$k = \frac{527.8}{69}$$

$$= 7.65 \text{ mD}$$

Next, Equation 5.23 is used to compute the skin factor. The value of the pressure at 1 hr is read from the straight line extrapolated from the infinite-acting flow period, not from the actual response curve distorted by wellbore storage.

$$s = 1.151 \left[\frac{4288.6 - 3534}{70} - \log \frac{7.65}{(0.039)(0.8)(17 \times 10^{-6})(0.198)^2} + 3.23 \right]$$

$$= 1.151 \left[10.8 - 8.566 + 3.23 \right]$$

$$= 6.29$$

The pressure drop due to skin is computed by Equation 5.8:

$$(\Delta P)_s = 141.2 \frac{(250)(1.136)(0.8)}{527.8} (6.29)$$

$$= 382.3 \text{ psi}$$

Due to a positive value of skin factor, as observed above, the well productivity is adversely affected, and the effective radius of the wellbore is less than the actual radius. The effective radius is computed by Equation 5.12:

$$r_{wa} = 0.189 \ e^{-6.29}$$

$$= 0.00035 \text{ ft}$$

To estimate the flow efficiency and average reservoir pressure in the well drainage area, the value of p* is calculated by extrapolating the semilog straight line on a Horner plot to $(t_p + \Delta t)/\Delta t = 1$. In order to extrapolate, the following is noted:

p = 4,368 psia at $(t_p + \Delta t)/\Delta t = 1,000$
m = −70 psi/cycle

The extrapolated value is computed as in the following:

$$p* = 4,368 + (-70) \log (1/1,000)$$

$$= 4,578 \text{ psia}$$

The computation of flow efficiency is based on Equation 5.13:

$$FE = \frac{4,578 - 3,534 - 382.3}{4,578 - 3,534}$$

$$= 0.63$$

As noted earlier, the horizontal portion on the diagnostic plot (fig. 5–23) is observed approximately between 10 and 50 hrs. The radius of investigation for the test is estimated by noting the time at the end of the infinite-acting flow regime:

$$r_{inv} = \left[\frac{(7.65)(50)}{948(0.039)(0.8)(17 \times 10^{-6})} \right]^{0.5}$$

$$= 872 \text{ ft}$$

The above calculation is based on Equation 5.20.
Estimation of average reservoir pressure in the well drainage area requires the following input:
- Computation of t_{DA} (or $t_{p\,DA}$) based on Equation 5.41
- Determination of the corresponding value of $p_{D\,MBH}$ from Figure 5–10
- Estimation of p_{avg} from Equation 5.40

First:

$$t_{DA} = \frac{(0.000264)(7.65)(13630)}{(0.039)(0.8)(17 \times 10^{-6})(2 \times 1320)^2}$$

$$= 7.45$$

Next, the value of $p_{D\,MBH}$ is read as 5.44 from Figure 5–7. Finally, the average reservoir pressure is estimated as follows:

$$5.44 = \frac{2.303 \ (4578 - p_{avg})}{70}$$

$$p_{avg} = 4,412.6 \text{ psia}$$

As noted earlier, the production period prior to buildup (t_p) is replaced by the time to reach pseudosteady state (t_{pss}) in the drainage region if t_{pss} is less than t_p.

Table 5–4 summarizes the results of the interpretation performed by various techniques.

The results in Table 5–4 are found to be in fair agreement given the overall uncertainties involved in reservoir characterization. It must be borne in mind that well test interpretation is a continuous process as further information becomes available with subsequent production, testing, and related-reservoir studies.

Example 5.2. Drawdown test of a well in a bounded reservoir.

This example illustrates the interpretation methodology of a drawdown test, which is very similar to a buildup test. Log-log, semilog, and diagnostic plots show the same signatures associated with wellbore storage and radial flow as in a buildup test. A unit slope line at late times clearly identifies a bounded system.

A drawdown test is performed in a well located in an estimated drainage area of 151 acres.[32] During the test, the well was produced at a stabilized rate of 250 stb/d for a period of 460 hrs. The initial reservoir pressure was 4,412 psia. All other available information, including recorded flowing bottomhole pressure versus time, is given in the following and in Table 5–5:

Formation thickness, ft, = 69.
Porosity, %, = 3.9.
Total compressibility, psi^{-1}, = 17×10^{-6}.
Oil formation volume factor, bbl/stb, = 1.136.
Oil viscosity, cp, = 0.8.
Wellbore radius, ft, = 0.198.

Interpret the drawdown test.

Solution: Workflow associated with the interpretation is very similar to the earlier example. To start with, the quality of the well test data was examined, followed by a detailed review of the diagnostic plot in order to identify a reservoir model. The diagnostic plot, shown in Figure 5–27, reveals the appearance of an infinite-acting radial flow regime having a zero slope during the middle time. This is followed by an ascending trend of unit slope at late times. The latter trend points to a closed system, such as a bounded reservoir or individual well drainage pattern. The model has the following assumptions:

- Wellbore storage: constant
- Fluid flow: radial
- Boundary condition: no flow

Table 5-4. Summary of interpretation results

	C_s, bbl/psi	k, mD	s	R_{inv}, ft	p^*, psia	p_{avg}, psia
Log-log Plot	0.0125	7.9	6.7	—	—	—
Semilog Plot	—	7.5	6.1	928	4,581	4,414
Type Curve Analysis	0.011	8.3	7.2	—	—	—
Verification[a]	0.012	7.7	6.3	872	4,578	4,413

[a] Performed manually to illustrate test interpretation methodology

Table 5-5. Flowing bottomhole pressure versus time during drawdown

Drawdown Period, hrs	Flowing BHP, psia	Drawdown Period, hrs	Flowing BHP, psia
0.12	3,812	43.0	3,537
1.94	3,699	51.5	3,532
2.79	3,653	61.8	3,526
4.01	3,636	74.2	3,521
4.82	3,616	89.1	3,515
5.78	3,607	107	3,509
6.94	3,600	128	3,503
8.32	3,593	154	3,497
9.99	3,586	185	3,490
14.4	3,573	222	3,481
17.3	3,567	266	3,472
20.7	3,561	319	3,460
24.9	3,555	383	3,446
29.8	3,549	460	3,429
35.8	3,544	—	—

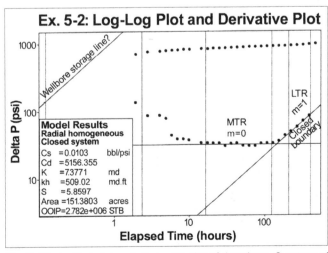

Fig. 5–27. Log-log plot and diagnostic plot of drawdown. *Courtesy of eP, a Weatherford Company.*

The reservoir model selected in this study can be validated by geophysical, geological, and reservoir simulation studies. The estimated drainage area and original oil in place is computed once the pseudosteady-state flow regime is identified beyond 200 hrs into the test.

Results of the drawdown test, including permeability, skin, flow efficiency, and the radius of investigation, are obtained by generating plots. These plots are the semilog of pressure response over drawdown time (fig. 5–28) and a plot showing the type curve matching technique (fig. 5–29).

Example 5.3. Evaluation of a hydraulically fractured well. Testing of stimulated wells, following either acidizing or hydraulic fracturing, is routine in the industry. The interpretation of transient tests based on an infinite-acting radial flow model was illustrated in Examples 5.1 and 5.2. In addition, the evaluation of a hydraulically fractured well involves linear or bilinear flow. Success of any fracturing operation is gauged by changes in skin factor, flow efficiency, and a long fracture path, among other factors. In certain cases, the well is tested prior to stimulation in order to compare this with the results of the follow-up test. For example, a change of skin factor from +4 to −6 is a definitive indicator of success.

In one case, a producing well was hydraulically fractured because its productivity had decreased. A drawdown test was conducted to measure the success of the operation. The pressure versus time data in Table 5–6 was obtained from the test.

Well, reservoir, and fluid data is given as follows:

Initial reservoir pressure, psia, = 5,000.
Drawdown rate, stb/d, = 2,000.
Formation thickness, ft, = 50.
Porosity = 0.24.
Total compressibility, psi^{-1}, = 14.8×10^{-6}.
Oil formation volume factor, bbl/stb, = 1.5.
Oil viscosity, cp, = 0.3.
Wellbore radius, ft, = 0.29.
Analyze the test.

Solution: The log-log pressure versus drawdown time plot and the diagnostic plot in Figure 5–30 indicate the following:

- A half-slope line develops at very early stages and extends to about 0.1 hrs. This is characteristic of linear flow through a fracture of infinite conductivity.

- The half-slope line is followed by a transition period, which eventually develops into a horizontal line between 20 and 50 hrs.

- At the late stages of the test, boundary effects are evident, as a line trending upward is observed.

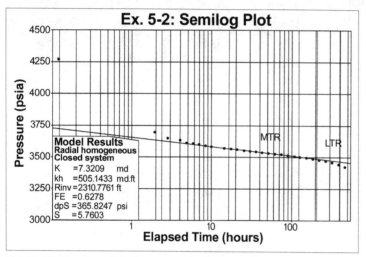

Fig. 5–28. Semilog plot of pressure decline over time during drawdown. Rate of drawdown is the largest at early stages of the test. *Courtesy of eP, a Weatherford Company.*

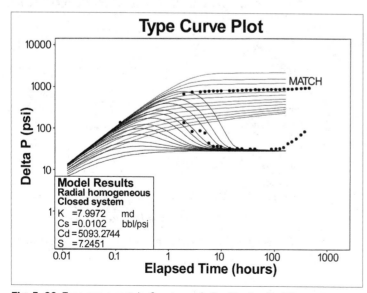

Fig. 5–29. Type curve match. *Courtesy of eP, a Weatherford Company.*

Table 5-6. Drawdown test data

Drawdown Period, hrs	Flowing BHP, psia	Drawdown Period, hrs	Flowing BHP, psia	Drawdown Period, hrs	Flowing BHP, psia
0.01002	4,980.61	0.276	4,907.1	7.58	4,678.74
0.01209	4,978.68	0.332	4,900	9.14	4,661.4
0.0142	4,976.57	0.396	4,892.37	11.02	4,643.13
0.0173	4,974.24	0.477	4,884.14	13.12	4,623.87
0.0209	4,971.98	0.568	4,875.29	15.81	4,607.71
0.0252	4,969.2	0.685	4,865.77	19.07	4,586.54
0.0308	4,966.14	0.836	4,855.52	21.99	4,568.78
0.0368	4,963.17	0.995	4,844.48	27.72	4,550.26
0.0436	4,959.52	1.2	4,832.6	32.99	4,530.93
0.0526	4,955.96	1.446	4,819.82	39.77	4,510.79
0.0634	4,951.59	1.74	4,806.06	47.94	4,495.11
0.0754	4,947.35	2.1	4,793.43	57.8	4,473.44
0.091	4,942.72	2.5	4,779.98	68.8	4,456.55
0.11	4,937.66	3.02	4,763.17	82.95	4,433.21
0.132	4,932.23	3.63	4,747.75	97.5	4,408.86
0.159	4,927.05	4.38	4,731.33	—	—
0.19	4,921.48	5.21	4,713.83	—	—
0.229	4,915.49	6.29	4,698.39	—	—

Based on the preceding information, the following model is selected:

- Wellbore storage: negligible as a unit slope line is not evident
- Fluid flow: linear in early stages, followed by pseudoradial flow
- Boundary condition: closed

The half-slope line is also evident on the diagnostic plot. Moreover, the two lines show a factor of separation by two. A half-slope line drawn through the log-log plot in the linear flow regime yields the fracture half-length. A horizontal line on the diagnostic plot reveals that the pseudoradial flow develops between 10 and 33 hrs. The transition period between linear and pseudoradial flow indicates that both types of flow pattern influence the pressure response.

As in other cases, reservoir permeability, skin, radius of investigation, and flow efficiency are computed from the semilog plot shown in Figure 5–31. A type curve analysis is performed to confirm the results obtained by semilog analysis

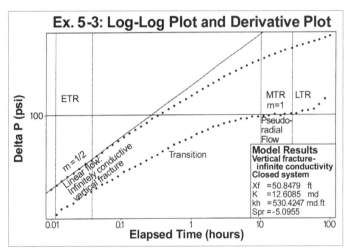

Fig. 5–30. Log-log plot. *Courtesy of eP, a Weatherford Company.*

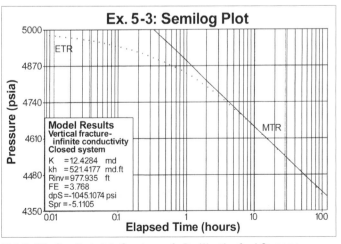

Fig. 5–31. Semilog plot. *Courtesy of eP, a Weatherford Company.*

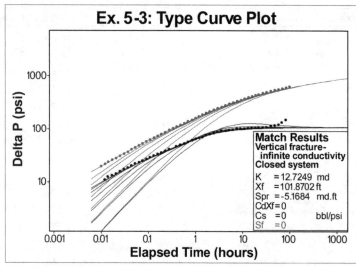

Fig. 5–32. Type curve. *Courtesy of eP, a Weatherford Company.*

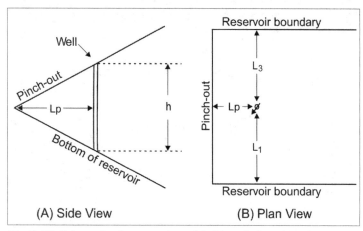

Fig. 5–33. Schematic of a reservoir model with a pinchout boundary

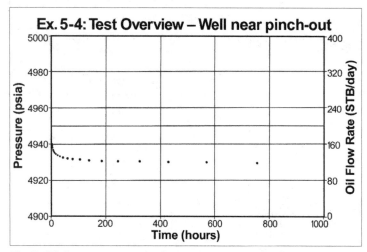

Fig. 5–34. Pressure response of a well located near a pinchout boundary. *Courtesy of eP, a Weatherford Company.*

(fig. 5–32). A high value of negative skin (–5.1) and good flow efficiency (2.77) are indicative of the success of the hydraulic fracturing operation. This can be compared with the results in Example 5.2, where the skin and flow efficiency of an unfractured well are found to be 5.7 and 0.76, respectively.

During the transient testing of a hydraulically fractured well, it is observed that a half-slope line (linear flow) appears in the early stages. In fact, a linear flow pattern can develop in the middle time region or late time region, depending on the reservoir geometry. Common examples include a channel-shaped reservoir or where the well drainage area is rectangular. Consequently, a half-slope line may emerge during the late time region.

Example 5.4. Investigation of a pinchout boundary. This example demonstrates features that are quite common in a computer-based interpretation of a heterogeneous reservoir. These features are given in the following:

- Confirmation of a pinchout boundary based on identification of a negative half-slope line on the diagnostic plot

- Nonlinear regression of a heterogeneous reservoir model to estimate the important parameters, including the location of the test well with respect to the reservoir boundaries

The widespread use of well test applications enables the review of the diagnostic plot and permits simulation of relatively complex reservoir models. Before this technology became available, however, such analyses were deemed to be uncertain or cumbersome at best.

In this case example, geologic studies indicate that a producing well may be located close to a boundary formed by a gradual decrease in formation thickness that eventually pinches out. Besides this geologic heterogeneity, the geologic studies indicate two parallel boundaries as shown in Figure 5–33. The objective of the reservoir team is to confirm the reservoir heterogeneity and estimate the distance of the well from the three boundaries. The test sequence consisted of a drawdown period of 1,000 hrs followed by 100 hrs of buildup (fig. 5–34). All available information is listed in the following:

Initial reservoir pressure, psia, = 5,000.

Well flow rate prior to buildup, stb/d, = 200.

Flowing bottomhole pressure prior to buildup, psia, = 4,929.39.

Formation thickness, ft, = 29.75.

Porosity = 0.36.

Total compressibility, psi^{-1}, = 8.2×10^{-6}.

Oil formation volume factor, bbl/stb, = 1.2.

Oil viscosity, cp, = 0.5.

Wellbore radius, ft, = 0.276.

Interpret the test.

Solution: Analysis of the drawdown portion of the pressure response is illustrated in the following discussion. However, the buildup portion can be analyzed in a similar manner.

The pressure response as observed from the test overview plot suggests a trend typical of any drawdown test. However, the diagnostic plot reveals a definitive signature of the existence of a pinchout boundary. The pinchout is identified by a negative half-slope trend following infinite-acting flow in the middle time (fig. 5–35). The semilog plot of the pressure response is presented in Figure 5–36. Based on reservoir model that incorporates a pinchout boundary, a regression analysis is performed that attempts to obtain the best fit of the resulting parameters with the actual pressure response. The results of the regression analysis are shown in Figure 5–37. In performing the analysis, there are many factors that were provided. These included the lower and upper ranges of probable skin, the wellbore storage coefficient, permeability, and the distance to the pinchout boundary, among other factors. The upper and lower limits of various parameters are obtained from geological studies and the results obtained from log-log and semilog analyses. The best match is obtained with the following values:

- Distance to the pinchout boundary is 288 ft.

- Distance to the parallel boundaries is 4,500 ft and 5,000 ft.

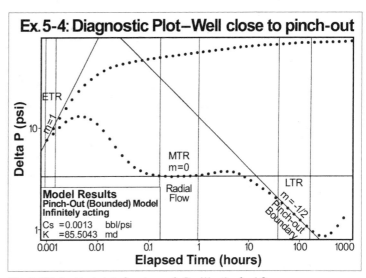

Fig. 5–35. Log-log plot. *Courtesy of eP, a Weatherford Company.*

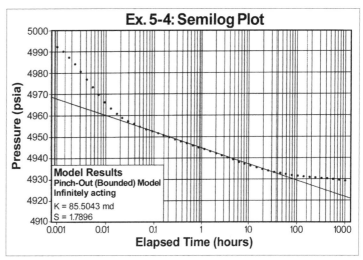

Fig. 5–36. Semilog plot. *Courtesy of eP, a Weatherford Company.*

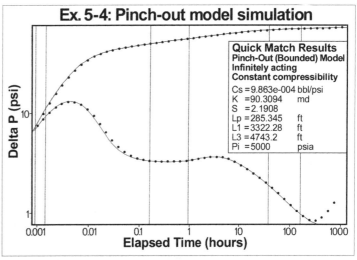

Fig. 5–37. Simulation of a reservoir model with a pinchout boundary. *Courtesy of eP, a Weatherford Company.*

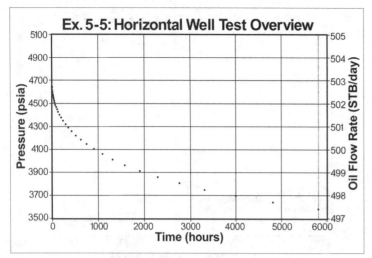

Fig. 5–38. Horizontal well test overview. *Courtesy of eP, a Weatherford Company.*

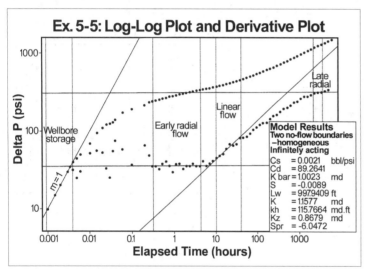

Fig. 5–39. Horizontal well test interpretation based on diagnostic plot. *Courtesy of eP, a Weatherford Company.*

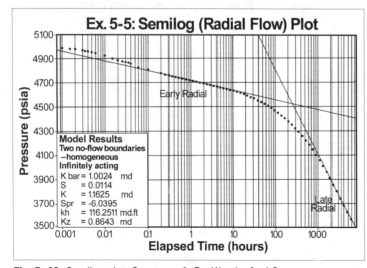

Fig. 5–40. Semilog plot. *Courtesy of eP, a Weatherford Company.*

Example 5.5. Horizontal well test interpretation. This example illustrates the various flow regimes observable in a typical transient test of a single-lateral horizontal well, followed by their interpretation. Due to the relative complexity of the test response from a horizontally drilled well, specifically at early times, well test software is used almost exclusively in this interpretation. Analytic models utilized may be inadequate in the case of a complex well trajectory or a reservoir heterogeneity, leading to grid-based numerical simulation.

A drawdown test is conducted in a horizontal well for 5,801.6 hrs. The horizontal section of the well is 1,000 ft long and is located midway between the upper and lower boundaries of the formation. The pressure response during the test period is plotted in Figure 5–38. The following information is available:

> Initial reservoir pressure, psia, = 5,000.
> Well flow rate prior to buildup, stb/d, = 200.
> Formation thickness, ft, = 100.
> Porosity = 0.23.
> Total compressibility, psi^{-1}, = 10×10^{-6}.
> Oil formation volume factor, bbl/stb, = 1.
> Oil viscosity, cp, = 1.
> Well offset (z_{wd}) = 0.5.

Analyze the test.

Solution: A horizontal well test reveals the development of additional flow regimes over those observed typically during the testing of a vertical well. (These flow regimes, arising out of the unique configuration of a horizontal well, were described previously.) Figure 5–39 shows the log-log pressure versus time plot and the diagnostic plot for the drawdown test analyzed here. Detailed review of this information indicates the following flow regimes:

- A unit slope line through the first few data points, indicating wellbore storage effects. The trend lasts for a very short period, up to about 0.004 hrs.

- A zero slope line between 0.4 hrs and 4 hrs, indicating an early radial flow regime. The radial flow pattern develops in the vertical direction and leads to estimation of the vertical permeability of the formation.

- A half-slope line between 10 hrs and 30 hrs, indicating a predominantly linear flow pattern between the horizontal wellbore and the adjacent formation. This is used in determining the effective length of the horizontal section of the well.

- A second zero slope line near the very end of the test period, indicating a late radial flow regime. The radial flow pattern is in a horizontal direction and leads to the calculation of average permeability.

Workflow related to the horizontal well test interpretation proceeds in the following sequence:

1. A typical horizontal test may involve thousands of data points, requiring certain data reduction algorithms prior to the generation of the plots.

2. Review of data quality based on test overview, log-log pressure versus time, and diagnostic plots.

3. Identification of various flow regimes based on slopes and trends in pressure response. As noted earlier, four distinct flow regimes are identified in the test.

4. Computation of test results as follows:

   - A unit slope line is drawn through the early time region, yielding the wellbore storage coefficient.

   - A horizontal line is drawn through the early radial flow regime. This leads to calculation of the vertical permeability of the formation in the vicinity of the horizontal section, once the horizontal permeability is available.

   - A half-slope line is drawn through the linear flow regime, leading to the estimation of the effective length of the horizontal section. In many cases, the effective length is found to be less than the actual length of the horizontal section.

   - A second horizontal line is drawn through the late radial flow regime, yielding the formation permeability in the horizontal direction. Furthermore, the pseudoradial skin factor is computed once this flow regime is established.

One observation from the test is that the hemiradial flow regime is not identified in the diagnostic plot. Hemiradial flow develops when one edge of the radial flow touches the upper or lower boundary of the formation, but the opposite edge does not. In this example, the horizontal section of the test well is located midway between the upper and lower boundaries of the formation. Hence, the above flow pattern is not expected. Again, a hemiradial flow regime leaves a signature of a zero slope line following the early radial flow regime and a transition period on the diagnostic plot.

Figure 5–40 shows the semilog plot of flowing well pressure versus drawdown time. Two straight lines having finite slope are drawn through the early radial and late radial flow regimes, leading to the estimated values of horizontal and vertical permeability, skin factor, and pseudoradial skin factor. The latter is indicative of the beneficial effects of horizontal drilling and the resulting productivity enhancement as a pseudoradial flow pattern develops.

In order to demonstrate the steps involved in interpreting a horizontal test with as much clarity as possible, an idealized example was selected. In this case, the well produces with a constant drawdown for a period of several months. In reality, a long uninterrupted flow of constant or near-constant rate is difficult to achieve in any well. Similarly, a long buildup test conducted in a horizontal well results in a significant loss in revenue. Consequently, most horizontal well tests are designed for a relatively short period, lasting from several hours to a few weeks. Consequently, the late radial flow regime is not identifiable unless the formation is highly permeable. Then again, vast numbers of horizontal wells are drilled in low permeability formations in order to enhance reservoir performance. Nevertheless, a short test may aid in estimating the effective length of the horizontal section, vertical permeability, and near-wellbore heterogeneities. A nonlinear regression technique is frequently used to interpret a horizontal test.

Last, but not least, it is noted that this horizontal well is a single-lateral well. As the technology matures, present-day horizontal wells are often designed to be multilateral, i.e., they have several branches drilled in various directions projecting from the vertical section. Moreover, the individual lateral sections have unique lengths and trajectories that are dictated by geology and oil saturation, among other factors. Hence the interpretation of multilateral well tests is quite involved and requires numerical model simulation.

Example 5.6. Interwell reservoir characterization. This example focuses on the investigation of reservoir characteristics between two wells using transient pressure tests as a tool. The first well is an active producer, and the other is an observer located 2,500 ft away. Based on core studies and single-well test results, the average permeability of the reservoir in the general area is assumed to be 125 mD. Due to the highly heterogeneous nature of the formation, the reservoir team has the following concerns:

- Is there any unknown heterogeneity in the area, such as a barrier or facies change, that will affect well productivity in the future?

- In contrast, does a high permeability zone exist that may lead to the premature breakthrough of injected water?

The answer may be rooted in the computer-assisted design of an interference test between the two wells based on all available information, including an assumed permeability of 125 mD. The pressure response at the observation well simulated by the reservoir model can be compared against actual data when the interference test is performed. Any significant departure from the predicted response would require further scrutiny of the reservoir. The following information is available:

Initial reservoir pressure, psia, = 4,000.
Formation thickness, ft, = 25.
Porosity = 0.2.
Total compressibility, psi $^{-1}$, = 12.5 × 10^{-6}.
Oil formation volume factor, bbl/stb, = 1.25.
Oil viscosity, cp, = 2.45.
Skin factor, active well, = 5.
Skin factor, observation well, = 0.
Wellbore radius, active well, ft, = 0.29.
Wellbore radius, observation well, ft, = 0.29.

Design an interference test that would predict the pressure response at the observation well due to predetermined changes in the production at the active well.

Solution: The interference test is designed as in the following sequence:

- **Active well.** Drawdown at a rate of 1,000 stb/d for 400 hrs; shut-in of the well for 100 hrs.

- **Observation well.** Monitoring of the well throughout the entire test period.

Based on a radial flow model in a homogeneous reservoir, the pressure response at the observation well is computed and plotted in Figure 5–41.

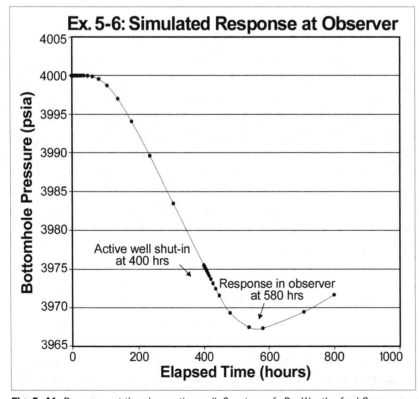

Fig. 5–41. Response at the observation well. *Courtesy of eP, a Weatherford Company.*

Example 5.7. Effect of non-Darcy flow on the transient test of a gas well. This example highlights an important difference between the interpretation of oil well and gas well tests. Due to the effects of turbulence, a pressure drop in the vicinity of gas wells is observed to be greater than that accounted for by permeability damage alone. Hence, the apparent skin observed in a gas well test is dependent on the magnitude of the well flow rate. The test, involving multiple flow rates, is analyzed by a non-Darcy or turbulent flow model that incorporates the following equation described earlier:

$$s' = s + Dq_g$$

Furthermore, the example illustrates the utilization of a pseudopressure function in analyzing gas well tests.

In this case, a flow-after-flow test is performed in a gas well where the flow rate was increased as per the following schedule:

Flow rate (MMscf/d)	Duration of flow (hrs)
2	100
4	100
6	100

The pressure response recorded during the test is presented in Figure 5–42. Related well test information, including the properties of the gas required to generate the pseudopressure function, is given below:

Initial reservoir pressure, psia, = 5,000.
Radius of the wellbore, ft, = 0.276.
Thickness of formation, ft, = 17.5.
Porosity = 0.36.
Reservoir temperature, °F, = 200.
Specific gravity of gas = 0.7.
Gas viscosity, cp, = 0.026.
Gas formation volume factor, cft/scf, = 0.0037.
Gas compressibility or z factor = 0.9899.
Gas compressibility, psi^{-1}, = 122.5 × 10^{-6}.
Water compressibility, psi^{-1}, = 2.952 × 10^{-6}.
Rock compressibility, psi^{-1}, = 2.857 × 10^{-6}.
Total compressibility, psi^{-1}, = 101.45 × 10^{-6}.
Water saturation = 0.2.

Analyze the test.

Solution: The first step in analyzing the test requires computation of the pseudopressure and pseudotime functions over the entire range of pressures observed. Knowledge of the viscosity and the gas compressibility factor over the range of pressures is necessary in order to evaluate the function. The trapezoidal rule can be used to integrate Equations 5.29 and 5.30 to calculate the pseudofunctions. Figure 5–43 presents a plot of pseudopressure function versus the actual pressure.

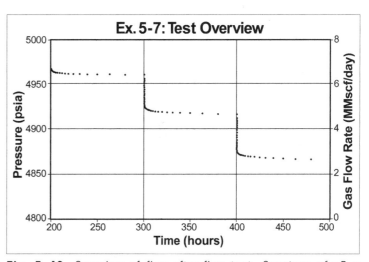

Fig. 5–42. Overview of flow-after-flow test. *Courtesy of eP, a Weatherford Company.*

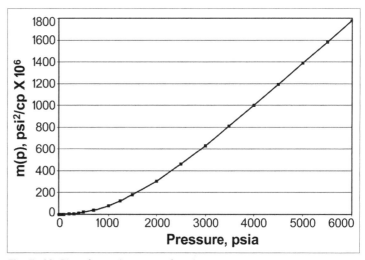

Fig. 5–43. Plot of pseudopressure function

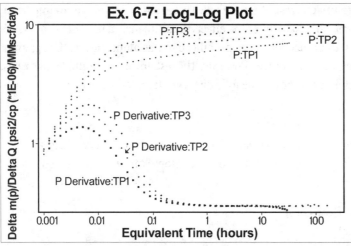

Fig. 5–44. Log-log plots of response at three different flow rates. Diagnostic plot shows variations in early time response, indicating changes in apparent skin. The largest "hump" is produced by the highest flow rate. The horizontal portion of all diagnostic curves converges in the middle time, indicating the same formation permeability. *Courtesy of eP, a Weatherford Company.*

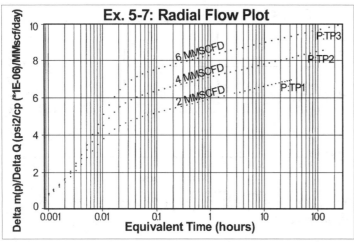

Fig. 5–45. Radial flow plot of transient pressure response at different flow rates. *Courtesy of eP, a Weatherford Company.*

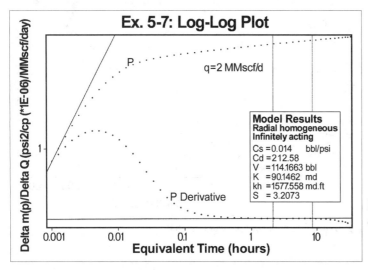

In the next step, log-log and semilog plots are generated (figs. 5–44 and 5–45) as in oil well test interpretation. However, in this case, the values of the observed pressure on the y-axis are replaced by the corresponding values of Δm(p) /Δq. Similarly, the time on the x-axis is replaced by adjusted time. In the plots, three sequences of drawdown at different well rates are compared, each lasting for 100 hrs. The striking aspect of the test is readily apparent in that the transient well response is dependent on the drawdown rate. From the diagnostic plot, it can be observed that the curves deviate from each other at early times, suggesting rate-dependent skin. However, the curves tend to converge at the middle time, leading to computation of very similar permeability values. This can be verified by analysis of the radial flow region for each individual drawdown.

Figures 5–46 and 5–47 present the test interpretations of two drawdown sequences: one performed at 2 MMscf/d and the other at 6 MMscf/d. While the calculated permeability is about 90 mD in both cases, the estimation of skin varies widely, from 3.2 to 7.2. Obviously, the conventional model used in interpretation is inadequate. A convenient method to interpret the test correctly is to perform a nonlinear regression analysis, in which a rate-dependent skin model is matched with the observed pressure over the entire test. By performing regression analysis, the true value of skin is estimated to be 2.3. A non-Darcy flow coefficient of 0.0005 (Mscf/day)$^{-1}$ matches the observed data. Results of the analysis are presented in Figure 5–48.

Fig. 5–46. Log-log plot. *Courtesy of eP, a Weatherford Company.*

Example 5.8. Investigation of interlayer communication by numerical method. This example highlights the role of numerical simulation in well testing. Reservoir engineers are frequently confronted by the possibility of vertical communication between two adjacent layers in a reservoir. Shale, which could be either impervious or semipervious, usually is found to separate the producing zones. Depending on the degree of communication, reservoir dynamics can change significantly, leading to a number of important issues. These can include well completion strategy, enhanced recovery program, identification of the source of water in the wells, and distribution of revenue, if the zones have different ownership.

The existence of crossflow between the layers in a two-layer system is investigated, with the well completed in the more permeable upper layer (fig. 5–49). Based on a numerical grid system, vertical crossflow between the two layers is simulated by using PanMesh software.[33] The following information is available:

Layer 1 (completed)
Initial pressure, psia, = 5,000.
Horizontal permeability, mD, = 91.
Vertical permeability, mD, = 91.
Porosity = 0.36.
Thickness, ft, = 17.5.
Mechanical skin = 3.
Radius of damaged zone, ft, = 1.

Layer 2 (uncompleted)
Initial pressure, psia, = 5,000.
Horizontal permeability, mD, = 10.
Vertical permeability, mD, = 0.5.
Porosity = 0.05.
Thickness, ft, = 17.5.

Other general data
Drawdown rate, stb/d, = 200.
Oil formation volume factor, rb/stb, = 1.2.
Oil viscosity, cp, = 0.5.
Total compressibility, psi^{-1}, = 8.2×10^{-6}.
Wellbore radius, ft, = 0.279.

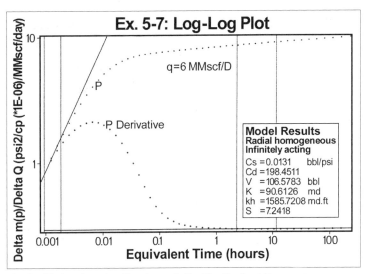

Fig. 5–47. Log-log plot. *Courtesy of eP, a Weatherford Company.*

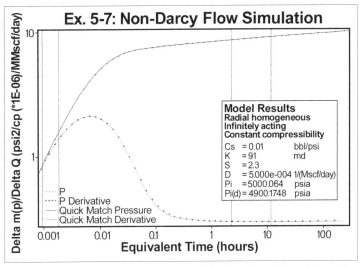

Fig. 5–48. Match of pressure response with simulation of non-Darcy flow. *Courtesy of eP, a Weatherford Company.*

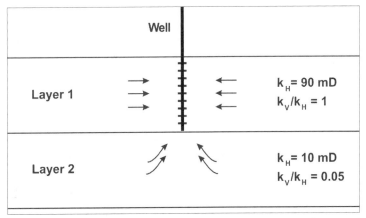

Fig. 5–49. Schematic showing well completion in upper layer and crossflow from lower layer due to vertical permeability. *Courtesy of eP, a Weatherford Company.*

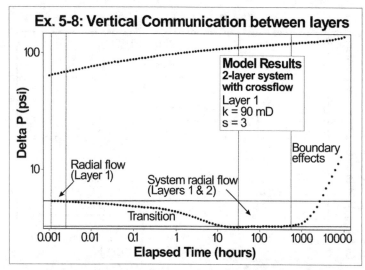

Fig. 5–50. Diagnostic plot showing vertical communication between two layers. *Courtesy of eP, a Weatherford Company.*

Table 5–7. Drawdown response of a hydraulically fractured well in a very tight gas reservoir. *Source: J. Lee and R. A. Wattenbarger. 1996. Gas Reservoir Engineering. Richardson, TX: Society of Petroleum Engineers.*

Time, hr	BHFP, psia	Time, hr	BHFP, psia
0.024	4,566	242.4	3,783
0.048	4,559	362.4	3,632
0.072	4,554	482.4	3,501
0.12	4,547	602.4	3,405
0.192	4,540	722.4	3,313
0.24	4,536	842.4	3,230
0.48	4,521	962.4	3,154
0.72	4,511	1,226.4	3,006
1.2	4,495	1,466.4	2,889
1.44	4,489	1,706.4	2,783
2.04	4,475	1,946.4	2,686
2.16	4,473	2,186.4	2,596
2.4	4,468	2,546.4	2,472
26.4	4,278	4,946.4	1,826
50.4	4,179	7,346.4	1,298
194.4	3,856	—	—

Figure 5–50 presents the results of a two-layer model simulation based on the preceding data. The following signatures are recognized from the diagnostic plot:

1. Initial radial flow from the completed layer at very early stages of the test, as observed from the horizontal line.
2. A transition region emerges within a minute due to the initiation of crossflow from the uncompleted layer.
3. A second radial flow regime representing the entire system is established beginning at about 20 hrs.
4. Boundary effects at the late stages of the test are represented by an ascending line.

Comparisons can now be made between the preceding information and the following scenarios:

- **Completed layer is not in communication with the uncompleted layer.** The diagnostic plot would not have indicated an initial radial flow region. Only one horizontal would be observed during the infinite-acting flow regime.

- **Vertical permeability of the uncompleted layer is significantly higher than 0.05 mD.** Crossflow would be more pronounced at the onset. The initial radial flow regime may not be identifiable. However, the calculated permeability-thickness product (kh) would be greater than what is expected from a single-layer system.

- **Effects of wellbore storage and skin.** In most wells, the early response during the test is dominated by wellbore storage and skin effects. The initial radial flow regime is likely to be obscured by a unit slope line on the log-log plot. Again, the calculated permeability-thickness product would be greater if interlayer communication exists.

This exercise points to the fact that well test interpretation can pose certain challenges to reservoir engineers. This is true even when state-of-the-art techniques in recording and analyzing the test are employed. In most cases, the answer is found in integrated reservoir studies, including reservoir simulation of multiple scenarios incorporating varying degrees of communication in the reservoir model.

Example 5.9. Evaluation of hydraulic fracture in an extremely tight gas reservoir. Reservoirs having ultra low permeability (0.1 mD or less) are increasingly developed in recent times due to the high demand for natural gas. Either the reservoirs are naturally fractured or the producers are hydraulically fractured in order to attain a viable production rate. This example, with data obtained from literature for a very long drawdown, highlights the methodology in interpreting tests conducted in low and ultra low permeability reservoirs. PanSystem well test software was used to interpret the test.

The pressure response obtained from a drawdown test lasting more than 300 days in a hydraulically fractured gas producer is shown in Table 5–7. The initial reservoir pressure was 4,598 psia. The well produced at a rate of 0.15 MMscfd. At the end of the drawdown period, the reservoir pressure decreased to 1,298 psia.

Other relevant data, including rock and fluid properties, are listed in the following:

Reservoir

Average porosity (fraction) = 0.076.

Average thickness, ft, = 30.

Formation compressibility, psi^{-1}, = 4.0×10^{-6}.

Temperature, °F, = 250.

Well

Radius of wellbore, ft, = 0.33.

Gas

Specific gravity = 0.744.

Formation volume factor, cft/scf, = 4.324×10^{-3}.

Average viscosity, cp, = 0.02364.

Average compressibility, psi^{-1}, = 1.53×10^{-4}.

Water

Saturation (fraction) = 0.33.

Compressibility, psi^{-1}, = 3.6×10^{-6}.

Solution: The first step of the interpretation is to review the diagnostic plot shown in Figure 5–51. The following observations are made:

- Even after 300 days of drawdown, the middle time region has not developed fully. The best guess is to draw a zero slope line slightly above the last recorded pressure located at the far right of the diagnostic plot. Reservoir permeability is found to be less than 0.1 mD.

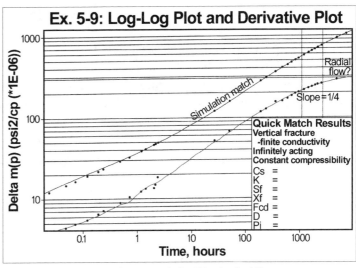

Fig. 5–51. Log-log plot. *Courtesy of eP, a Weatherford Company.*

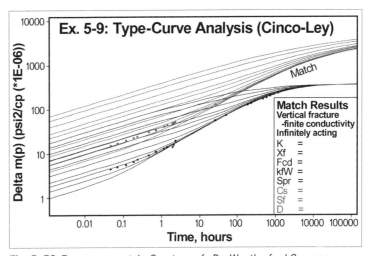

Fig. 5–52. Type curve match. *Courtesy of eP, a Weatherford Company.*

- A region following 1,000 hrs of drawdown shows a quarter-slope line on the diagnostic plot as indicated on the plot. The presence of a finite conductivity is confirmed. Although not shown here, the same quarter-slope line should be observed on the log-log line drawn above the derivative plot.

Based on the limited information obtained from the diagnostic plot, a type-curve match is attempted to estimate the half-length (x_f) and conductivity (k_{fw}) of the hydraulic fracture. The pseudoradial skin (s_{pr}) of the well is also estimated. Cinco-Ley type curves for finite conductivity fractures are utilized (fig. 5–52). A satisfactory match was obtained with one of the type curves in the family.

Finally, confirmation of type-curve analysis is obtained by simulation. A finite conductivity fracture model is simulated. The starting parameters are related to the fracture characteristics (x_f, k_{fw}, S_{pr}), and other reservoir properties were based on the results of the type-curve analysis. The aforementioned parameters are fine-tuned by nonlinear regression during simulation. A good match between the simulation results (continuous lines in fig. 5–51) and the field data (shown as dots) is obtained for the entire duration of the test period. However, the actual results obtained from this interpretation are not shown here and are left as an exercise. Nonetheless, the results of the simulation are quite close to the type-curve analysis, enhancing confidence in the interpretation. Looking at the practical aspects of the test, it must be borne in mind that it is very difficult to maintain a steady production for so long. However, drawdown tests with multiple rates are analyzed with relative ease by computer-based applications.

Summing Up

Well testing serves as an indispensable tool in developing, characterizing, and managing a reservoir, including surveillance of individual well performance. Well testing results point to reservoir and well behavior under dynamic conditions as in production. Therefore, these results carry a high degree of confidence and relevance to critical reservoir studies. Conclusions based on well test results are utilized extensively as a valuable guide in the exploration, production, and development of petroleum reservoirs.

Well testing is a unique tool, as it can "see" into the reservoir all the way to its boundary when planned diligently. In fact, well testing is referred to as a "reality check" in reservoir engineering by some. The important objectives of well testing are as follows:

- Assessment of well productivity and reservoir viability
- Well and reservoir management plan
- Evaluation of stimulated wells
- Reservoir description and conceptualization
- Interwell characterization
- Estimation of reservoir pressure
- Monitoring of improved oil recovery process performance
- Reservoir simulation model development

Well testing includes the following basic steps:

1. **Change in well rate.** A carefully designed perturbation or disturbance is created in the reservoir by conducting a step change in the production or injection rate.

2. **Pressure response.** The change in the fluid flow pattern triggers a new trend in the pressure response at the wellbore and into the reservoir. The nature of the response is transient.

3. **Monitoring and analysis.** The ensuing pressure response is then monitored over a length of time for subsequent interpretation. Depending on the well, reservoir, and fluid characteristics, the resulting response in the fluid pressure imprints one or more signatures that are identifiable and analyzable.

Virtually all wells undergo some type of testing during the life of a reservoir. Virtually all reservoirs have a well testing plan that is carried out to gather important information about the well and reservoir at a regular frequency. Results of the well tests are then integrated with other sources of data to build robust reservoir models.

With the emergence of the digital era, well testing has witnessed significant strides in hardware, software, and data integration. Traditional well test analyses often served as a stand-alone tool based on analog gauges and very limited data collection. Modern well test practices, however, involve digital gauges recording minute changes in pressure response and powerful software applications capable of analyzing the response. In addition, there are robust database systems that aid in integrating well test interpretation and reservoir information obtained from other disciplines.

Most of the familiar well test types as practiced in the industry are noted in the following:

Buildup test in the producer. The well is produced at a constant rate for a sufficient length of time, followed by the shut-in and recording of well response as the bottomhole pressure builds up. Buildup tests are very common in the oil industry. One drawback is the resulting loss in production during buildup when the duration of the test is long.

Drawdown test in the producer. The well is initially shut in for a sufficiently long period so that a static pressure is reached in the reservoir or the drainage area, followed by production (drawdown) at a constant rate. The resulting decline in bottomhole flowing pressure is recorded and analyzed. The test is ideal in newly discovered reservoirs. However, maintaining a steady drawdown rate could be difficult. Drawdown tests are also common.

Falloff test in the injector. The well is first injected at a constant rate for a sufficient period, followed by the injector being shut in. As a result, the bottomhole pressure at the well begins to decline (fall off), which is recorded and analyzed.

Multirate tests in the producer. In a multirate test, the well is flowed at multiple rates for definite time intervals, and the pressure response is recorded. Multirate tests are employed in gas reservoirs extensively where the well is flowed at predetermined rates for definite time intervals. Multirate tests include the following:

- Flow-after-flow test, in which the well is flowed successively at different but stabilized flow rates while the bottomhole pressure is recorded. The pressure needs to be stabilized before the well changes to a new flow rate. Gas well deliverability tests to evaluate the absolute open flow potential (AOFP) of the well fall in this category.

- Isochronal tests implement alternating sequences of drawdown and shut-in periods, along with monitoring of the pressure response. In each sequence, shut-in of the test well continues until the bottomhole flowing pressure stabilizes. In the case of tight gas reservoirs, modified isochronal tests are employed where the drawdown and shut-in periods in a sequence are the same.

Interference and pulse tests involving an active well and observation wells. These tests involve one active well and one or more observation wells in a neighboring location. These are designed to characterize the formation between two wells, such as directional permeability or confirmation of a geologic barrier. These tests may require two or more wells to be shut in for a relatively long period. They may also require deployment of sensitive monitoring equipment to record and interpret the slightest change in pressure at the observation well. Reservoir simulators are frequently utilized to aid in test interpretation.

Drillstem test. A drillstem test (DST) is routinely conducted in a new well prior to completion in order to assess the feasibility and potential of the new well in an unknown environment. Results of the drillstem test serve as the principal guide in assessing the feasibility of a newly discovered petroleum reservoir. The test is usually comprised of short sequences of multiple flow and shut-in periods. Fluid samples are also collected during the tests.

Wireline formation tester. This tool employs transient tests of short duration. These are conducted in order to evaluate reservoir pressures and horizontal and vertical permeabilities at various depth intervals within the formation when a new well is drilled. For evaluating vertical crossflow between layers, the tool is indispensable.

Common factors affecting the well test response include the wellbore storage effects and permeability alteration (skin) around the well. As a result, the early time data from a well test response is almost always distorted. The magnitude of the wellbore storage and skin can be evaluated in the well test response.

Wellbore storage effects are observed as a well is subjected to drawdown or is shut in during the test. The earliest response is from the fluid in the wellbore rather than from the fluid in the formation.

Skin damage may originate from drilling mud invasion, migration of fines through the pores, and clay swelling due to the injection of incompatible water. The net effect is an additional drop in pressure around the wellbore. When a well is hydraulically fractured, a determination of negative skin is usually based on well test interpretation.

Well test interpretation requires a clear understanding of the general flow characteristics that are encountered during a test. These are as follows:

- **Transient.** Reservoir pressure varies with both time and location in the reservoir following the start of drawdown or buildup as a perturbation is created. The well response is independent of any effects of reservoir boundary, as if it is producing in an infinite reservoir.

- **Pseudosteady-state.** Once the test is continued for a certain length of time, the rate at which the pressure declines becomes constant everywhere in the reservoir. Tests conducted in bounded reservoirs display this flow behavior.

- **Steady-state.** During the late stages of a test, reservoir pressure approaches a constant value, as any decline is readily compensated for by an external source of energy. The reservoir pressure profile no longer varies in time and location.

In a developed field where oil is produced from a number of wells, each well produces from its own drainage area. Well test results, including average reservoir pressure, are based on the individual drainage area.

The principle of superposition is applied in order to model the anticipated pressure response from a well in a complex situation arising out of the following:

- A geologic boundary is encountered.
- The well produces at multiple rates.

The principle of superposition can be applied both in space and time.

In well test design, the concept of radius of investigation is valuable. One important use is to determine the minimum duration of a test period required to investigate a reservoir boundary or heterogeneity located at a certain distance from the well. The radius of investigation is a direct function of the formation permeability and the test period.

Flow patterns that may develop during a test include the following:

- Radial flow
- Pseudoradial flow
- Linear flow
- Bilinear flow
- Spherical flow

Each flow pattern has a distinct signature on the pressure response and occurs in an expected sequence.

Well test interpretation methods can be categorized according to the general classification of either classical or computer-assisted.

Classical interpretation. These methods consist of the following:

- Computations are based on log-log and semilog plots. Conceptual reservoir models are relatively simple.
- Matching of well response with available type curves.

Computer-assisted methods. These methods consist of the following:

- Recognition of the reservoir model based on a diagnostic plot
- Nonlinear regression of a conceptual reservoir model
- Numerical simulation of the well response in complex situations

In the first method, the pressure response versus time data is plotted in an appropriate scale, such as semilog or log-log plots. Early time region, middle time region, and late time region on the plot are interpreted based on relatively simple analytic models. These include wellbore storage, infinite-acting, and boundary models, respectively.

Type curves are representative of the well responses under known reservoir conditions. These are based on dimensionless pressure, time, and storage, among other factors. When actual well response is matched against one of the curves, the corresponding values of well and reservoir properties can be determined.

Recognition of a reservoir model based on a diagnostic plot has become the first step in any well test interpretation since the introduction of computer-based analysis of well tests. The derivative of pressure change, in comparison to the change itself, provides more invaluable information for performing qualitative interpretation.

Well test software is capable of simulating the entire well test response based on a given range for the sought parameters (permeability, skin, etc.) and obtaining the best match with the field data. Furthermore, numerical methods are employed to simulate the well response in complex situations.

Multiple interpretation methods (semilog plot, type curves, simulation, etc.) are usually adopted in order to draw conclusions from a transient well test. Again, several reservoir models (homogeneous, layered, no-flow boundary, etc.) can be investigated.

Theory of well testing. The mathematical foundation of modeling the well test responses is built on unsteady-state fluid flow theory. This theory predicts the transient pressure behavior depending on the rock and fluid properties. Well test theory is based on the law of conservation of mass, Darcy's law, and the equation of state. The basic model is referred to as the hydraulic diffusivity equation. The equation is usually solved with appropriate boundary conditions in radial coordinates.

It is important to recognize the assumptions inherent in conventional well test theory:

1. The reservoir is homogeneous and isotropic.
2. Fluid flow is horizontal, with negligible effects of gravity.
3. There is a single fluid phase of small and constant compressibility.
4. Flow is isothermal and laminar.
5. The fluid does not react with the formation.

The hydraulic radial diffusivity equation as applied to oil well testing is based on the assumption of reservoir fluid, which is slightly compressible. Hence, the interpretation of gas well test results requires the introduction of a pseudopressure function, as gas is highly compressible. Hence, semilog plots to interpret the results are based on pseudopressure versus pseudotime, which are defined earlier.

In gas well testing where turbulent flow may develop, the apparent skin factor is dependent on the flow rate.

Flow regimes evident in well transient response. There are three flow regimes that are expected from a typical well test:

1. Early time response, dominated by the effects of wellbore storage and skin.
2. Middle time region, where the pressure response emulates an infinite-acting reservoir (i.e., the pressure disturbance created at the well has not reached the reservoir boundary). Flow is characterized as unsteady-state.
3. A departure from the infinite-acting trend at later stages of the test as boundary effects become perceptible

Most well test plots presented in this chapter illustrate the early time region, middle time region, and late time region as identified in different circumstances. The well test analyst primarily examines various flow regimes based on the following plots:

1. Semilog plot of the pressure response over the test period
2. Log-log plot of the pressure differential between static and flowing bottomhole pressures versus time
3. Derivative of the pressure response versus time overlain on the log-log plot (diagnostic plot)

Well test interpretation is both qualitative and quantitative. In qualitative analysis, the reservoir model is determined based on the diagnostic plot. Some of the reservoir models and well conditions that can be identified are as follows:

- Reservoir boundary (no flow, constant pressure, etc.)
- Faulted reservoir
- Dual porosity (fractured) reservoir
- Channel-shaped reservoir
- Pinchout formation
- Partially competed well
- Reservoir stratification

In quantitative analysis, examples of sought parameters include, but are not limited to, the following:

- Skin factor
- Wellbore storage coefficient
- Average transmissibility
- Average reservoir pressure
- Radius of investigation
- Reservoir limits
- Distance to a barrier
- Movement of an injected water bank

Design of a transient test involves selection of the test type best suited to meet the objectives, duration of test, flow rate, and measuring devices of desired accuracy. Certain wells need to be regularly tested for monitoring purposes, including the surveillance of an enhanced recovery operation.

There are pitfalls in well test design and interpretation. These include, but are not limited to: (a) issues related to the quality of data obtained from the test, (b) insufficient test duration, and (c) incorrect identification of the reservoir model.

Well test interpretation tools in the digital age are capable of analyzing a test based on wide-ranging techniques and reservoir models. Certain applications are capable of numerical simulation of the transient pressure response in a complex geologic setting. As a result, engineers are able to concentrate on analyzing well test data from different perspectives. They can integrate the results of the well test with relevant data obtained from other sources. This contrasts with classical practices, in which more time was spent in the preparation of plots and manual computations, viewing well test interpretation as a stand-alone study.

Class Assignments

Questions

1. What is transient well testing in a petroleum reservoir? Discuss the general principle on which a well test is based. Why it is called "transient"?

2. Name at least 10 applications of well testing. In what stage or stages in reservoir development do the well tests contribute? In what circumstances does a well need to be tested at regular intervals?

3. Name at least two aspects or capabilities that make well testing a unique tool. What degree of confidence is usually attached to well test results in characterizing a reservoir? Why?

4. What are the types of transient tests? Name at least six types of tests, including the two most common tests practiced in the industry for oil wells. In what aspects are the tests similar? Again, in what other aspects are they different?

5. What are the methods used in well test interpretation? Discuss the strengths and limitations of each method.

6. Discuss the basis and assumptions of well test theory. Name a few circumstances in which the conventional theory needs to be modified in order to correctly interpret a test.

7. Discuss the conceptual reservoir model or models used in traditional interpretation when the test is performed in a newly discovered reservoir.

8. How is pseudosteady-state flow distinguished from unsteady-state flow? Again, how does steady-state flow differ from pseudosteady-state flow? At what stages of a test would the three types of fluid flow be expected to be encountered?

9. What are the distinct flow regimes one could expect to identify from a properly designed test? How are the flow regimes recognized?

10. A well test is being designed in a newly discovered reservoir with very limited information. Develop a detailed workflow, including specification of downhole gauges, well test type, duration of test, interpretation methodology, and possible sources of data.

11. Regional studies indicate that many of the reservoirs in the same basin are faulted and have relatively low transmissibility. Would the well test design be any different where the formation is expected to have high permeability characteristics? Explain.

12. In the case of a low permeability reservoir, would a long transition zone usually be expected before the radial flow regime is observed? Why or why not?

13. Reservoir studies performed in the same basin reveal that the producing zones receive partial pressure support from the surrounding aquifer. In such a case, what kind of response would be expected from the transient test, and at what stage of the test? How would the response be different if the reservoir does not receive any aquifer support?

14. What is the principal tool one would plan to use in identifying any geologic heterogeneity from the test designed? Explain how the existence of the following would be recognized:
    • Natural fractures
    • Pinchout boundary
    • Sealing fault
    • Two parallel faults
    • Nonsealing fault

15. Is it necessary to know the bubblepoint of the reservoir fluid in designing the well test? Why or why not? What are the general methods used in estimating fluid properties in a new reservoir?

16. If a gas cap is present in the reservoir, what response in the transient pressure would be expected during the test?

17. What tool would be recommended for measuring the permeability anisotropy, i.e., the ratio of vertical to horizontal permeability of the formation? In what circumstances would this information be vital?

18. A well test is being planned in a tight gas reservoir in order to investigate any formation damage around the well. The estimated average permeability is 0.1 mD or less. What well test method would be recommended, and why?

19. The well tested above indicates severe skin damage. A hydraulic fracturing operation was performed to augment productivity. A postfracture test is designed to evaluate the success of the well stimulation. How might the response differ from the earlier test? How is the flow through the fractures characterized? Under what circumstances would transient pressure tests for this well be recommended at regular intervals?

20. Consider that the new reservoir is found to have moderate transmissibility following the first test. The reservoir is developed by drilling several new wells within a time frame of a few years. Unfortunately, the productivity has significantly declined in a couple of the wells. Describe a well test program that could be part of combating the problem.

21. A third well is showing traces of water in the production stream lately. The well is not originally thought to be located near any boundary. Can a well test be useful in this circumstance? What information might be obtained about the source of water by properly designing a well test? What sources of information, other than well testing, would significantly contribute to the study?

22. As part of a waterflood project, several producers are being converted to injectors in a reservoir. Would the permeability values obtained by well testing before and after the conversion of a producer be expected to remain the same? Explain.

23. In interpreting a transient well test, would one use the net formation thickness or the gross thickness? If the well is not completed throughout the vertical cross section, what type of flow might develop near the well? How is it identified?

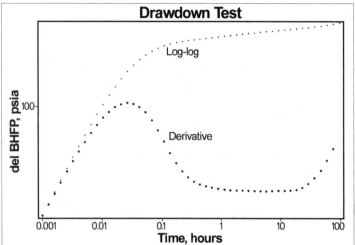

Fig. 5–53. Drawdown test diagnostic plot. *Courtesy of eP, a Weatherford Company.*

Table 5–8. Pressure response from a drawdown test

Drawdown, hrs	BHFP, Psia	Drawdown, hrs	BHFP, psia
0.001	4,192.65	0.34832	3,851.599
0.00126	4,189.44	0.4402	3,846.279
0.0016	4,185.47	0.55633	3,841.159
0.00202	4,180.52	0.70309	3,836.169
0.00255	4,174.439	0.88857	3,831.289
0.00322	4,166.989	1.12298	3,826.519
0.00407	4,157.929	1.41922	3,821.809
0.00515	4,147.019	1.79361	3,817.159
0.00651	4,134.019	2.26677	3,812.51
0.00822	4,118.729	2.86475	3,807.86
0.01039	4,101.059	3.62048	3,803.24
0.01314	4,080.999	4.57556	3,798.659
0.0166	4,058.819	5.78261	3,794.089
0.02098	4,034.909	7.30807	3,789.529
0.02652	4,010.019	9.23595	3,784.899
0.03351	3,985.029	11.67241	3,780.26
0.04235	3,961.009	14.75162	3,775.62
0.05352	3,938.909	18.64312	3,770.96
0.06764	3,919.529	23.56121	3,766.21
0.08549	3,903.219	29.7767	3,761.18
0.10804	3,889.979	37.63185	3,755.7
0.13654	3,879.349	47.5592	3,749.47
0.17256	3,870.759	60.1054	3,742.07
0.21808	3,863.569	75.96131	3,732.95
0.27561	3,857.309	96	3,721.5

24. Describe the distinct flow regimes that might develop during the transient testing of horizontal wells. Why are horizontal well tests not always run long enough to see the late time pseudoradial flow regime? A buildup test is conducted in a multilateral horizontal well having three branches drilled in three different layers. What method would one adopt to interpret the test?

Exercises

5.1. Based on a geophysical study, the existence of a boundary is indicated within 500 ft of a newly drilled well. A well test is being designed to confirm the existence of the boundary and characterize it (sealing, nonsealing, or partially sealing). What type of well test should be proposed? What would be the minimum test duration? How would the response between a sealing and partially sealing boundary be distinguished? The following data is available:

Average reservoir pressure, psia, = 3,250.
Reservoir permeability, mD, = 15–20.
Porosity of formation = 0.18–0.23.
Viscosity of oil, cp, = 0.515.
Oil saturation = 0.76.
Oil compressibility, psi^{-1}, = 8.22 × 10^{-6}.
Water compressibility, psi^{-1}, = 3.16 × 10^{-6}.

Make all necessary assumptions. Use appropriate correlations wherever needed.

5.2. A drawdown test is conducted by producing a well at a steady rate of 390 stb/d for 120 hrs. Relevant well, reservoir, and fluid data is provided as follows, including the diagnostic plot in Figure 5–53 and the data in Table 5–8.

Initial reservoir pressure in the drainage area, psia, = 4,205.
Average formation porosity (fraction) = 0.24.
Thickness of formation, ft, = 22.
Connate water saturation (fraction) = 0.195.
Oil formation volume factor, rb/stb, = 1.28.
Oil viscosity, cp, = 0.489.
Total compressibility, psi^{-1}, = 14.66 × 10^{-6}.
Radius of well, in., = 3.516.

(a) Identify the flow regions on the diagnostic plot and the duration of each regime.

(b) Calculate the average permeability, skin, flow efficiency, apparent wellbore radius, radius of investigation, drainage area, and original oil in place in the drainage region.

(c) Is the test period adequate?

(d) Would any strategy be recommended to enhance oil production from the well?

(e) One year following the test, the well is showing a water cut of 18% to 20%. The suspected source of water is an injector nearby. Would a similar response be expected if a new drawdown test is conducted? Explain.

5.3. A producing well is shut in for 168 hrs (one week), and the resulting pressure response is tabulated in Table 5–9.

Well, reservoir, and fluid data is given in the following:

Production period prior to buildup, days, = 20.
Stabilized rate prior to buildup, stb/d, = 660.
Flowing bottomhole pressure, psia, = 2,401.9.
Average porosity (fraction) = 0.31.
Average water saturation (fraction) = 0.21.
Thickness of formation, ft, = 15.
Oil formation volume factor, rb/stb, = 1.18.
Oil viscosity, cp, = 0.525.
Total compressibility, psi^{-1}, = 18.21 × 10^{-6}.
Radius of wellbore, ft, = 0.273.
Reservoir temperature, °F, = 176.

Table 5-9. Pressure response from buildup test

Buildup, hrs	Shut-in BHP, psia	Buildup, hrs	Shut-in BHP, psia
0.00101	2,427.623	0.46341	2,714.436
0.00128	2,434.331	0.59241	2,719.609
0.00165	2,442.817	0.75726	2,724.74
0.00208	2,452.157	0.96802	2,729.826
0.00266	2,463.918	1.23746	2,734.881
0.00342	2,478.006	1.58185	2,739.913
0.00436	2,493.631	2.02213	2,744.923
0.00558	2,511.216	2.5849	2,749.956
0.00714	2,530.227	3.30435	2,755.031
0.00912	2,550.111	4.224	2,760.257
0.01166	2,570.183	5.39963	2,765.684
0.01489	2,589.635	6.90244	2,771.428
0.01904	2,607.761	8.82349	2,777.521
0.02435	2,623.908	11.27924	2,784.02
0.03113	2,637.757	14.41846	2,790.898
0.03976	2,649.363	18.43137	2,798.144
0.05084	2,659.092	23.56116	2,805.692
0.065	2,667.342	30.11865	2,813.485
0.08307	2,674.537	38.50122	2,821.448
0.1062	2,681.053	49.2168	2,829.497
0.13577	2,687.117	62.91467	2,837.571
0.17355	2,692.887	80.42499	2,845.549
0.22186	2,698.452	102.8087	2,853.379
0.2836	2,703.88	131.4222	2,860.946
0.36252	2,709.193	168	2,868.258

(a) Qualitatively interpret the diagnostic plot (fig. 5–54) obtained from the buildup test. Can a reservoir heterogeneity be identified from the plot? If so, what would be its distance from the test well? (Hint: Note the vertical separation between the two zero-slope portions on the diagnostic plot.)

(b) Calculate the dimensionless wellbore storage, average reservoir transmissibility, and skin. What is p_{1hr}?

(c) In future buildup tests, what test duration would be optimal if further skin damage is suspected?

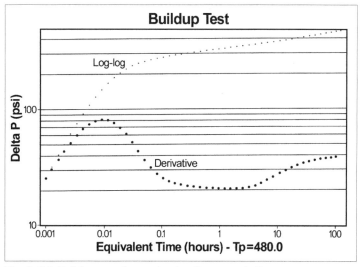

Fig. 5–54. Buildup test diagnostic plot. *Courtesy of eP, a Weatherford Company.*

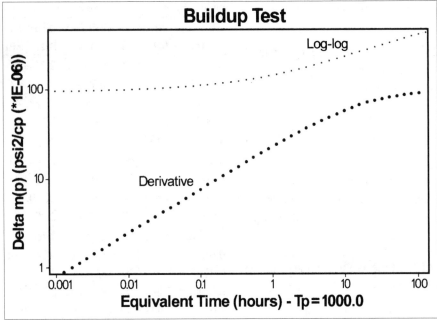

Table 5-10. Pseudopressure function

p, psia	m(p), psi^2/cp ($\times 10^6$)
100	0.733
200	2.938
300	6.62
400	11.782
500	18.42
700	36.1
1,000	73.446
1,300	123.3
1,500	163.128
2,000	283.687
2,500	430.134
3,000	597.12
3,500	779.688
4,000	973.69

Fig. 5–55. Buildup test diagnostic plot. *Courtesy of eP, a Weatherford Company.*

5.4. The diagnostic plot obtained from a gas well buildup test is presented in Figure 5–55.

(a) Qualitatively interpret the test based on the preceding information. Include all observations regarding the characteristics of the reservoir and well.

(b) Based on a Horner plot, estimate reservoir permeability and skin. Do the results validate the qualitative interpretation based on the diagnostics?

(c) What is the radius of investigation of the test?

(d) What is the extrapolated pressure (p*) from this interpretation? What additional information is needed to determine the average reservoir pressure?

All relevant data, including the pseudopressure function based on reservoir gas properties, is provided in the following, and in Table 5–10 and Table 5–11:

Reservoir
Average porosity (fraction) = 0.14.
Average thickness, ft, = 12.
Formation compressibility, psi^{-1}, = 4.23×10^{-6}.
Total compressibility, psi^{-1}, = 2.215×10^{-4}.
Temperature, °F, = 190.

Well
Radius of wellbore, ft, = 0.273.

Gas
Specific gravity = 0.578.
Formation volume factor, cft/scf, = 4.615×10^{-3}.
Average viscosity, cp, = 0.0205.
Initial compressibility, psi^{-1}, = 2.172×10^{-4}.

Table 5–11. Buildup test response

Buildup, hours	Shut-in BHP, psia	Buildup, hours	Shut-in BHP, psia
0.00098	2,552.594	0.34833	2,605.182
0.00128	2,553.029	0.44019	2,611.876
0.00159	2,553.416	0.55634	2,619.365
0.00201	2,553.898	0.70306	2,627.745
0.00256	2,554.449	0.88855	2,637.097
0.00323	2,555.046	1.12299	2,647.555
0.00409	2,555.722	1.41919	2,659.107
0.00513	2,556.453	1.79364	2,671.952
0.00653	2,557.333	2.26678	2,686.255
0.00824	2,558.284	2.86475	2,702.11
0.01038	2,559.341	3.62048	2,719.579
0.01312	2,560.552	4.57556	2,738.906
0.0166	2,561.913	5.78259	2,760.187
0.021	2,563.441	7.30804	2,783.542
0.02649	2,565.139	9.23596	2,809.115
0.03351	2,567.065	11.67242	2,837.022
0.04236	2,569.226	14.75165	2,867.283
0.05353	2,571.643	18.64313	2,900.097
0.06763	2,574.358	23.56122	2,935.236
0.08551	2,577.413	29.77673	2,972.765
0.10803	2,580.828	37.63184	3,012.58
0.13654	2,584.666	47.5592	3,054.304
0.17255	2,588.963	60.10535	3,097.825
0.21808	2,593.79	75.9613	3,142.561
0.27563	2,599.192	96	3,188.249

References

1. Schlumberger. Web literature. General Web page may be accessed at www.slb.com.
2. Matthews, C. S., and D. G. Russell. 1967. *Pressure Buildup and Flow Tests in Wells*. SPE Monograph Series. Vol. 1. Dallas: Society of Petroleum Engineers.
3. Lee, J. 1982. Well Testing. SPE Textbook Series. Vol. 1. Dallas: Society of Petroleum Engineers.
4. Earlougher, Jr., R. C. 1977. *Advances in Well Test Analysis*. SPE Monograph Series. Dallas: Society of Petroleum Engineers.
5. Matthews, C. S., and D. G. Russell. 1967.
6. Lee, J. 1982.
7. Earlougher, Jr., R. C. 1977.
8. Horne, R. N. 1995. *Modern Well Test Analysis*. 2nd ed. Palo Alto, CA: Petroway Inc.
9. Lee, W. J., and R. A. Wattenbarger. 2002. *Gas Reservoir Engineering*. 2nd ed. Richardson, TX: Society of Petroleum Engineers.
10. Horner, D. R. 1951. Pressure build-up in wells. Proceedings, Third World Petroleum Congress, The Hague. Sec. II, 503–523.
11. Miller, C. C., A. B. Dyes, and C. A. Hutchinson, Jr. 1950. The estimation of permeability and reservoir pressure from bottom hole pressure build-up characteristics. *Transactions*. AIME. Vol. 189, 91–104.
12. Matthews, C. S., F. Brons, and P. Hazebroek. 1954. A method for determination of average pressure in a bounded reservoir. Transactions. AIME. Vol. 201, 182–191.
13. Horne, R. N. 1995.
14. Lee, W. J., and R. A. Wattenbarger. 2002.
15. Agarwal, R. G., R. al-Hussainy, and H. J. Ramey, Jr. 1970. An investigation of wellbore storage and skin effect in unsteady liquid flow: I. analytical treatment. *Society of Petroleum Engineers Journal*. September, 291–297; *Transactions*. AIME. Vol. 249.
16. Horne, R. N. 1995.
17. Horne, R. N. 1995.
18. Edinburgh Petroleum Services, Inc. 2006. *PanSystem*. Kingwood, TX: Edinburgh Petroleum Services, Inc. (a Weatherford Co).
19. Horne, R. N. 1995.
20. Joshi, S. D. 1991. *Horizontal Well Technology*. Tulsa: PennWell.
21. Horne, R. N. 1995.
22. Lee, J., and R. A. Wattenbarger. 1996. *Gas Reservoir Engineering*. Richardson, TX: Society of Petroleum Engineers.
23. Gringarten, A. 1992. Well Test Interpretation Practice. Denver: Scientific Software-Intercomp Inc.
24. Horne, R. N. 1995.
25. Coludrovich, E. J., J. D. McFadden, M. R. Palke, W. R. Roberts, and L. J. Robson. 2005. Well management: use of permanent downhole flowmeters for pressure-transient analysis. *Journal of Petroleum Technology*. February.
26. Edinburgh Petroleum Services, Inc. 2006.
27. Lee, J. 1981.
28. Horne, R. N. 1995.
29. Edinburgh Petroleum Services, Inc. 2006.
30. Lee, J., and R. A. Wattenbarger. 1996.
31. Lee, J. 1981.
32. Lee, J. 1981.
33. Edinburgh Petroleum Services, Inc. 2006.

6 · Fundamentals of Data Acquisition, Analysis, Management, and Applications

Introduction

Present-day reservoir studies are seamlessly integrated by robust information management systems. An enormous amount of multidisciplinary data is acquired throughout the life of a reservoir. This process starts with exploration, continues through discovery, and is followed by delineation, development, production, and finally, abandonment. A realistic reservoir description, which is characterized by utilizing this data, is of vital importance for successful management of the reservoir. In recent years, the need for efficient management of all reservoir-related information has increasingly gained importance.

The fundamental classifications of reservoir data include the following:
- Microscopic. Microphotographs showing sand grains and pores from the cuttings or core slices.
- Macroscopic. Core data providing basic rock properties such as porosity, permeability, capillary pressure, and relative permeability, etc.
- Megascopic. Data from the drainage area around the well, such as porosity, and fluid saturation from well logs.
- Gigascopic. Data from the entire reservoir, such as seismic surveys, well pressure tests, and production and injection.

An efficient data management program consists of acquisition, analysis, validation, storage, and retrieval of data. All of these aspects are required to develop an integrated reservoir model (chapter 7) and to conceptualize the reservoir in its entirety. This data program serves as the foundation for evaluating reservoir performance (chapters 8 through 14), and ultimately managing the reservoir in the most efficient manner possible. An integrated approach involving all functions is necessary for planning, justifying, prioritizing, and timing the data collection program.

The objectives in this chapter are to learn about the following:
- Type and sources of data
- Data acquisition and analysis
- Validation of data
- Data storage and retrieval
- Data applications
- Data mining
- Example data
- Reservoir data management in the digital age
- Reservoir data acquisition and integration—a case study

Data Types and Sources

There are many types of multidisciplinary data collected, some of which are listed in the following discussion.

Geological. This data concerns depositional environments, diagenesis, lithology, structure, faults, and fractures. Exploration and development geologists acquire this data during exploration, discovery, and development of the field.

Seismic. This data can involve 2-D cross sections, 3-D visualization, and 4-D time-lapse studies. It can also include vertical seismic profiles and interwell tomography. Geophysicists and seismologists are responsible for acquiring this data during exploration and even during the production phases.

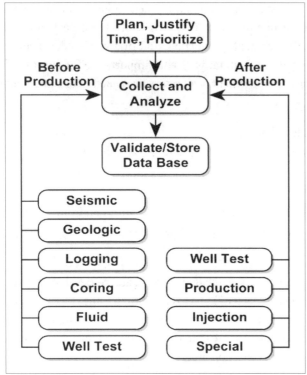

Fig. 6–1. Data collection before and after production. Special studies during production may include 4-D seismic, tracer injection and EOR pilot project, among others. *Source: A. Satter, J. E. Varnon, and M. T. Hoang. 1992. Reservoir management: technical perspective. SPE Paper #22350. SPE International Meeting on Petroleum Engineering, Beijing, China, March 24–27. © Society of Petroleum Engineers. Reprinted with permission.*

Petrophysical. Petrophysical data includes logging (open hole and cased hole) and coring (conventional whole core and side wall) data. Petrophysicists, geoscientists, engineers, and laboratory analysts are involved in collecting this data during discovery, delineation, development, and production.

Geochemical. Geochemical data concerns source rock chemistry and reservoir fluid composition.

Reservoir. Reservoir data relates to rock and fluid properties and pressure transient tests. Reservoir engineers and laboratory analysts acquire this data during discovery, delineation, development, and production.

Wells. Well data relates to well completion, configuration, and workover, and well history in general.

Production. Production data concerns oil, water, and gas production and injection profiles, including flow rate, pressure, and breakthrough of a new fluid phase. Modern electronic devices are capable of collecting vast amounts of downhole information at very short intervals. Production and reservoir engineers are responsible for collecting all relevant data during production and injection operations.

Chapters 2 and 3 provide descriptions of most of these data types. Various data types collected before and after production from the reservoir are presented in Figure 6–1.

Data Acquisition and Analysis[1]

Multidisciplinary groups of geophysicists, geologists, petrophysicists, and drilling, reservoir, production, and facilities engineers are involved in collecting various types of data. This process occurs throughout the life of a reservoir. Land, legal, and information technology professionals also contribute to the data collection process.

An effective data acquisition and analysis program requires careful planning and the ongoing, well-coordinated team efforts of geoscientists and engineers. A data acquisition program will require management approval due to the involvement of significant resources.

Planning, collecting, justifying, timing, prioritizing, and cost-effectiveness should be the guiding factors in the data acquisition and analysis program.

Data planning. Data planning needs to address the following:
- Type, quality, and quantity of data, and cost of acquisition.
- Who are the users?
- Is the data necessary?
- When will the data be used?
- What technical and financial benefits can be gained from the collected data?

Data collection. Data collection requires the following:
- Specification of what, when, and how much data will be gathered.
- Procedure and frequency to be followed during data collection.

Justification of data. Justification of the data must consider the following:
- Application of good practices for effective reservoir operation and management throughout the reservoir life cycle
- Justification to the management concerning the value of the data for cost-benefit effects. For example, why should the wells be shut in periodically for pressure tests?
- Answers to concerns: determine wellbore damage and also estimate reservoir pressures needed for performance analysis using classical material balance and reservoir simulation techniques.

Timing considerations. Timing considerations must address the following:
- *Coring before drilling the prospect zone.* Openhole sidewall coring may be damaged, and therefore the core may not be a good representative of the formation.
- *Fluid properties.* Fluid samples are obtained initially prior to production, particularly from condensate reservoirs. Compositions of the fluids would change below the dew point during production, and some of the condensate may be retained in the formation.
- *Initial reservoir pressure.* A well pressure test is conducted prior to production because the reservoir pressure changes with production. Again, the results of the initial well test data may point to certain geologic heterogeneities, such as faults or boundaries. This information would be very valuable for field development and reservoir performance analysis.
- *Initial oil/water and gas/oil contacts.* The preceding data is subject to changes due to the production of oil and gas.

Prioritizing. Considerations of priorities must address the following:
- The data selection is based upon its importance and on current and future needs.
- There must be a consideration of the vast amount of data available versus limited but vital data. Commonsense engineering judgments and cost-effectiveness are required in prioritizing the type and volume of data to be collected.

Data Validation

Validation of field data, geological maps, log and core data, reservoir rock and fluid properties, and reservoir performance data is essential to ensure quality and reliability.

Validation of different types of data

Field data. Data is subject to errors, e.g., sampling, systematic, and random, etc. Therefore, the data collected from various sources needs to be carefully reviewed and checked for accuracy, as well as for consistency.

Geological maps. Reasonable geological maps should be prepared by using knowledge of the depositional environment. The presence of faults and flow discontinuities, as evidenced in a geological study, needs to be investigated and validated by well pressure interference and pulse tests, as well as tracer tests.

Core and log data. In order to assess its validity, core and log analysis data should be carefully correlated, and frequency distributions should be made for identifying different geologic facies.

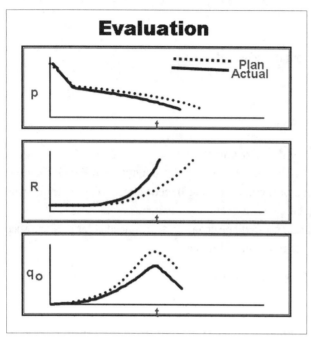

Fig. 6–2. Reservoir performance evaluation. *Source: A. Satter, J. Baldwin, and R. Jespersen. 2000. Computer-Assisted Reservoir Management. Tulsa: PennWell.*

Reservoir fluid properties. Data should be validated using equation of state calculations and empirical correlations.

Reservoir performance. The performance of a reservoir should be closely monitored while collecting routine well production, injection, and reservoir pressure data.

Figure 6–2 presents an example of pressure, gas/oil ratio, and oil production rate versus time data from a reservoir. The figure shows a discrepancy between the planned (expected) and actual data obtained from the field, raising the question as to whether the current reservoir management program needs to be reviewed and modified.

This situation relates to a case in which an integrated geoscience and reservoir engineering team determined that the reservoir was under depletion drive and was smaller than expected. That is why the actual pressure was lower, and there was a higher gas/oil ratio and lower production rate.

If past production and pressure data is available, classical material balance techniques and reservoir modeling can be very useful. They can help to validate the volumetric original hydrocarbons in place, along with aquifer size and strength.

Missing data

Laboratory rock properties, such as oil-water and gas-oil relative permeabilities, and fluid properties, such as PVT data, are not always available. Empirical correlations can be used to generate these data.

Data Storage and Retrieval

Development of an integrated computer database is required in order to store multidisciplinary data following reconciliation and validation. Stored data can be used to carry out multipurpose reservoir management functions, including monitoring and evaluating the reservoir performance.

Development of a seamless database is a major challenge, as the software and the data sets are rooted in different disciplines that may not be compatible. In addition, the communication between the multidisciplinary professionals may be inadequate.

In late 1990, several major domestic and foreign oil companies staged an integrated approach to solving the data management problem, forming Petrotechnical Open Software Corporation (POSC). POSC was created to establish industry standards and a common set of rules for applications and data systems within the industry. POSC's technical objective was to provide a common set of specifications for computing systems. This would allow data to flow smoothly between products from different organizations and allow users to move smoothly from one application to another.

Databases require updating on a regular basis, as new geoscience and engineering data becomes available during the various stages of the reservoir life cycle. Only authorized personnel should be allowed to update the database to ensure its integrity.

The database should be accessible to all interdisciplinary end users.

Data Application

Data is used for geological and geophysical maps, reservoir characterization to develop an integrated reservoir model for reservoir performance analysis, and for ultimate recovery evaluation

Geological maps. Geological maps, such as gross and net pay thickness, porosity, permeability, saturation, structure, and cross section, are prepared from seismic, core, and log analysis data. These maps, which also include faults, oil/water, gas/water, and gas/oil contacts, are used for reservoir delineation. They are also used for reservoir characterization, well locations, and estimates of original oil and gas in place.

Geophysics. A better representation of the reservoir is made from 3-D seismic information. The cross-well tomography also provides information on interwell heterogeneity.

Reservoir characterization. Data on the geology, geophysics, petrophysics, reservoir, and production engineering is needed for reservoir characterization. This is very important to build an integrated reservoir model for reservoir performance analysis and ultimate recovery efficiency calculations.

Integrated geosciences and engineering reservoir model. As mentioned previously, geosciences and engineering data is needed to build an integrated reservoir model for performance analysis and ultimate recovery efficiency calculations.

Well log data. This data provides the basic information needed for reservoir characterization, mapping, perforations, and estimates of the original oil and gas in place.

Production logs can be used to identify the remaining oil saturation in undeveloped zones in existing production and injection wells.

Time-lapse logs in observation wells can detect saturation changes and fluid contact movement. Also, log-inject-log data can be useful for measuring residual oil saturation.

Core analysis data. Unlike log analysis, core analysis results give direct measurement of the formation properties, and the core data is used for calibrating well log data. This data can have a major impact on estimates of the hydrocarbon in place, production rates, and ultimate recovery.

For wells with log porosities, core-derived porosity and permeability correlations can be utilized to determine permeability.

Reservoir rock and fluid properties. Rock and fluid properties are used for volumetric estimates of the original oil and gas in place and for reservoir performance analysis. Also, fluid properties can be used to determine reservoir type, i.e., oil, gas, or gas condensate reservoirs.

Well test data. Well test data is very useful for reservoir characterization and performance evaluation. Pressure buildup or falloff tests provide the best estimate of the effective permeability thickness of the reservoir, including reservoir pressure, stratification, and the presence of faults and fractures. Pressure interference and pulse tests provide reservoir continuity and barrier information.

Tracer tests. Multiwell tracer tests used in waterflood and enhanced oil recovery projects may indicate the preferred flow paths between the injectors and producers. Single-well tracer tests are used to determine residual oil saturation in waterflood reservoirs. Formation testers can measure pressure and rock permeability in stratified reservoirs, indicating varying degrees of depletion and communication in multiple zones.

Production and injection data. Production and injection data is needed for reservoir performance.

Basic Reservoir Performance Analysis Data

The data required for basic reservoir performance analysis using volumetric, decline curve, classical material balance, and reservoir simulation methods is listed below. The volumetric method is discussed in chapter 9, decline curve in chapter 11, material balance in chapter 12, and reservoir simulation in chapter 13.

Volumetric method

The volumetric method needs the following input data:
- Geometrical data: area, thickness
- Rock properties: porosity, oil or gas saturation
- Fluid properties: oil or gas formation volume factor corresponding to the reservoir pressure
- Production and injection data: not needed

This method is used for calculating the original oil or gas in place. The results are improved throughout the reservoir life cycle (exploration, discovery, delineation, development, and production) as the accuracy of the data is improved.

Decline curve method

The decline curve method needs the following input data:
- Production data: only oil and water production rates versus time for individual wells or field data

This method is used to calculate reserves, i.e., economically producible remaining oil at some economic rate, and the water/oil ratio. Having known the past oil production, the total recoverable oil can be calculated. Based upon the original hydrocarbon in place, the recovery efficiency can then be estimated.

Data can be readily available, and the technique is simple, but the method can be used only when production decline is established. The results are fair.

Material balance method

The material balance method needs the following input data:
- Geometrical data: same as volumetric
- Rock properties: porosity, saturation, rock compressibility, and absolute and relative permeabilities
- Fluid properties: PVT data involving oil, water, and gas compressibilities, formation volume factors, solubilities, and viscosities varying with pressure
- Production and injection data: oil, water, and gas production rates, cumulatives, pressure varying with time, and water injection

This method considers the reservoir to be homogeneous. It is used to calculate the original hydrocarbon in place, as well as natural drive mechanisms (i.e., solution gas drive, natural water influx, and gas drives). The results are improved, as more performance data is available with depletion of the reservoir.

Reservoir simulation method

The data needed is the similar to the classical material balance, but the technique is very comprehensive. This is due to the fact that the heterogeneities of the reservoir are taken into account. By history matching the past production and pressure performance, the future performance can be predicted. It can be used at any stage of the reservoir life cycle, but the results are more comprehensive and accurate with the depletion of the reservoir. Gathering data and analyzing it are time-consuming and expensive, but this is the preferred method.

Data Mining

According to Zangl and Hannerer, the traditional engineering approach is knowledge-driven, which depends on the degree of the practioner's knowledge.[2] They advanced the concept of data mining, which is an evolving discipline that uses hidden relationships and influences embedded in the data. Data collection and analyses (1960s), data access (1980s), and data queries (1990s) have evolved into data mining today. Readers are referred to Zangl and Hannerer's book for details of the process, which is summarized in the following discussion.

Data mining process can be divided into three main steps:
- **Preparing the data.** Time to complete is 75% of the process.
- **Surveying the data.** Time to complete is 18% of the process.
- **Modeling the data.** Time to complete is 7% of the process.

Not only is data preparation the most time-consuming step, it is also the most important step. Steps involved include data access, audit, and enrichment; looking for sampling bias; structure determination; and computer modeling. The first five steps are nonautomatic and require the users to make many decisions. The last step is automatic.

In the data preparation process, both the data and the multidisciplinary professionals working on it must be well-prepared. Although much of the data preparation can be automated, the manner in which the professionals interact with the data is essential for success.

After data preparation, surveying the data is needed, requiring about 18% of the completion time. It is also a very important aspect of data mining. Questions to answer include the following: What is in the data set? Can the posed questions be answered? Where are the danger areas? The data survey will yield a vast amount of insight into the general relationships and patterns in the data. The purpose is to build an overall map that users can depend on to make commitment to their activities.

After data preparation and surveying, modeling the data becomes a small part of the overall data mining process. Models are made only to compare the insights and discoveries, not to formulate them.

One of the most obvious applications in data mining is reservoir modeling, which can be divided into data gathering, history matching, and prediction steps. As pointed out, gathering input data is very time-consuming and expensive. History matching the past production and pressure performance is not a unique process. Reliability and accuracy depend upon the quantity and quality of the data used. The better the history match, the more reliable and accurate will be the future performance of a well or reservoir under existing operating conditions or some alternative development plans. Such plans could include infill drilling or waterflooding after primary recovery.

Case study

An example is presented in this section for the offshore Meren Field in Nigeria that involves study of the G-1, G-2, and G-3 reservoirs. The study data and results include basic reservoir, rock, and fluid properties, structure and cross-section maps, reservoir performance, and history matching.

The basic reservoir data is listed in Table 6–1. The multilayer sandstone reservoirs have good porosity and are

Table 6–1. Basic reservoir data of Meren Field

Basic Reservoir Data of Meren Field		
	Reservoir	
	G-1/G-2	**G-2/G-3**
Datum depth, ft subsea	−6,100	−6,000
Rock type	Sandstone	Sandstone
Average thickness, ft	138	126
Average porosity, %	27	32
Average permeability, mD	1,150	1,775
Average connate water saturation, %	24	14
Initial reservoir pressure (at datum), psig	2,660	2,560
Average bubblepoint pressure, psig	2,629	2,500
Initial oil volume factor, rb/stb	1.327	1.312
Initial solution gas/oil ratio, scf/stb	566	588
Initial oil viscosity, cp	0.575	0.46
Oil gravity, °API	34	33
Initial oil/water contact, ft subsea	−6,175	−6,197
Initial gas/oil contact, ft subsea	−6,000	−5,804
Original oil in place, MMstb	281.5	276.8
Original gas in place, bscf	205.6	176.7

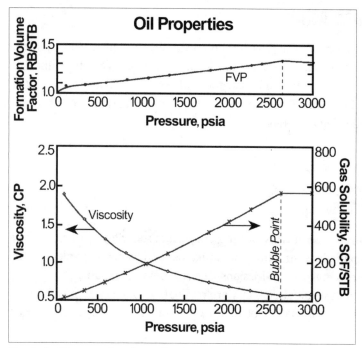

Fig. 6–3. Oil properties. *Source: A. Satter, J. Baldwin, and R. Jespersen. 2000. Computer-Assisted Reservoir Management. Tulsa: PennWell.*

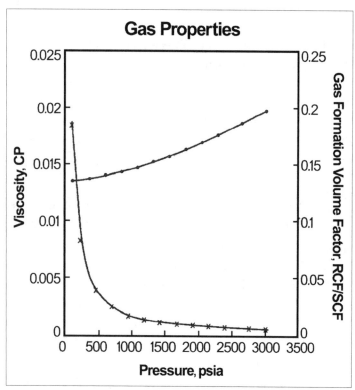

Fig. 6–4. Gas properties. *Source: A. Satter, J. Baldwin, and R. Jespersen. 2000. Computer-Assisted Reservoir Management. Tulsa: PennWell.*

highly permeable. Initial oil saturations and oil gravities are good. The reservoirs are undersaturated, which means the bubblepoint pressures are equal to the initial reservoir pressures.

The PVT data, including oil and gas properties, is given in Figure 6–3 and Figure 6–4, respectively. Figure 6–3 presents the oil formation volume factor, gas solubility, and viscosity varying with reservoir pressures. Figure 6–4 presents the gas formation volume factor and viscosity varying with reservoir pressures.

Gas-oil and water-oil relative permeability curves are presented in Figure 6–5.

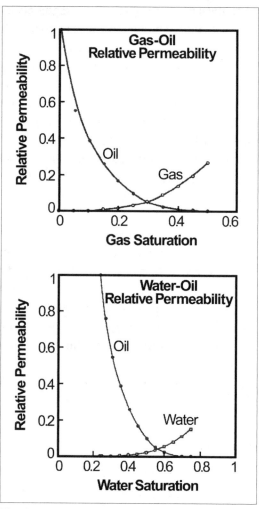

Fig. 6–5. Gas-oil and water-oil relative permeability curves. *Source: A. Satter, J. Baldwin, and R. Jespersen. 2000. Computer-Assisted Reservoir Management. Tulsa: PennWell.*

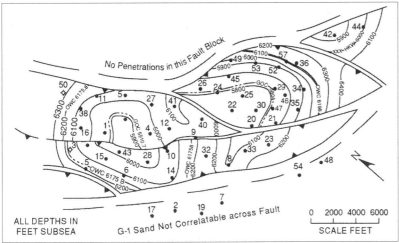

Structure maps of Sand G-1 and G-2 are shown in Figures 6–6 and 6–7, respectively. As is evident, the reservoirs are highly faulted.

Figure 6–8 shows Sand G in juxtaposition involving fault blocks A, B, C, and E.

A cross-section map involving wells MER-04, MER-01, MER-38, and MER-11 is shown in Figure 6–9.

Fig. 6–6. Structure map of Sand G-1. *Source: G. C. Thakur, R. B. Haulenbeek, A. Jain, W. P. Koza, S. D. Jurak, and S. W. Poston. 1982. Engineering studies of G-1, G-2, and G-3 reservoirs, Meren Field, Nigeria. Journal of Petroleum Technology. April, 721–732.* © Society of Petroleum Engineers. Reprinted with permission.

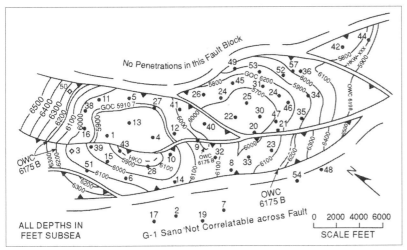

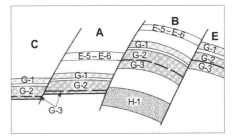

Fig. 6-8. Meren Field cross section showing Sand G juxtaposition. *Source: G. C. Thakur, R. B. Haulenbeek, A. Jain, W. P. Koza, S. D. Jurak, and S. W. Poston. 1982. Engineering studies of G-1, G-2, and G-3 reservoirs, Meren Field, Nigeria. Journal of Petroleum Technology. April, 721–732.* © Society of Petroleum Engineers. Reprinted with permission.

Fig. 6-7. Structure map of Sand G-2. *Source: G. C. Thakur, R. B. Haulenbeek, A. Jain, W. P. Koza, S. D. Jurak, and S. W. Poston. 1982. Engineering studies of G-1, G-2, and G-3 reservoirs, Meren Field, Nigeria. Journal of Petroleum Technology. April, 721–732.* © Society of Petroleum Engineers. Reprinted with permission.

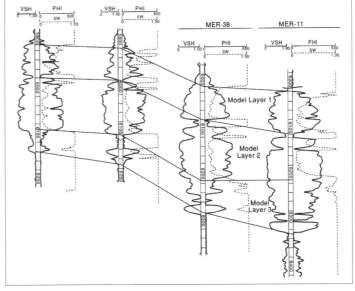

Fig. 6-9. Cross-section map. *Source: G. C. Thakur, R. B. Haulenbeek, A. Jain, W. P. Koza, S. D. Jurak, and S. W. Poston. 1982. Engineering studies of G-1, G-2, and G-3 reservoirs, Meren Field, Nigeria. Journal of Petroleum Technology. April, 721–732.* © Society of Petroleum Engineers. Reprinted with permission.

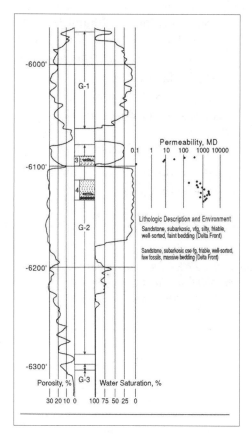

Fig. 6–10. Well logs. *Source: G. C. Thakur, R. B. Haulenbeek, A. Jain, W. P. Koza, S. D. Jurak, and S. W. Poston. 1982. Engineering studies of G-1, G-2, and G-3 reservoirs, Meren Field, Nigeria. Journal of Petroleum Technology. April, 721–732. © Society of Petroleum Engineers. Reprinted with permission.*

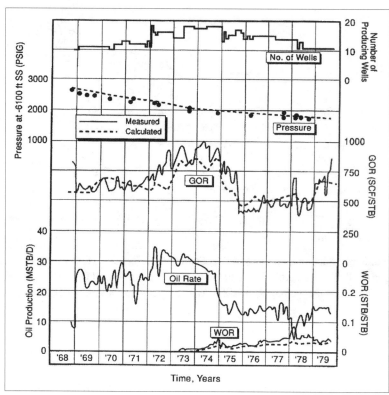

Fig. 6–11. Performances of G-1 and G-2 sands *Source: G. C. Thakur, R. B. Haulenbeek, A. Jain, W. P. Koza, S. D. Jurak, and S. W. Poston. 1982. Engineering studies of G-1, G-2, and G-3 reservoirs, Meren Field, Nigeria. Journal of Petroleum Technology. April, 721–732. © Society of Petroleum Engineers. Reprinted with permission.*

A type of well log showing the G-1, G-2, and G-3 formations and their depths is shown in Figure 6–10.

The historical reservoir performance of sands G-1/G-2 is shown in Figure 6–11. This includes the number of producing wells, measured reservoir pressures, well production rates, gas/oil ratios, and water/oil ratios. Included also are simulated reservoir pressures and gas/oil ratios. The pressure match is good, while the gas/oil ratio match is fair.

Performance of well 11 in the G-1/G-2 sands, including measured oil rate and gas/oil ratio, and simulated pressure and gas/oil ratio, are shown in Figure 6–12.

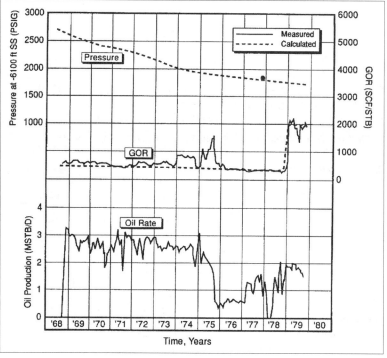

Fig. 6–12. Well 11 performance. *Source: G. C. Thakur, R. B. Haulenbeek, A. Jain, W. P. Koza, S. D. Jurak, and S. W. Poston. 1982. Engineering studies of G-1, G-2, and G-3 reservoirs, Meren Field, Nigeria. Journal of Petroleum Technology. April, 721–732. © Society of Petroleum Engineers. Reprinted with permission.*

Reservoir Data Management in the Digital Age

The deployment of electronic sensors allows the gathering of pressure, temperature, volume, composition, and other valuable data from a source well on continuous basis. With the use of these sensors, present-day petroleum fields are increasingly viewed as "digital fields," "smart fields," or "e-fields." The vision is to attain real-time or near–real-time control on the assets, including continuous optimization of production from the well to the point of sale.[3] The data obtained in real time is integrated with models that predict reservoir performance. The feedback is analyzed continually, leading to field-wide optimization.

Workflow related to the utilization of the data in the overall management of the reservoir is presented in Figure 6–13. In the transition process, where "data" turns into "information," and finally into "control action," four major phases can be identified as follows:

1. Surveillance of wells by deploying continuous downhole recording devices, wireless transmitters, and other tools.

2. Analysis, including data validation and visualization aided by software tools. Data becomes information at this stage.

3. Integration and optimization of the various processes related to the development and production of the reservoir. For instance, these may include fine-tuning of the injection and production rates in certain wells based on reservoir simulation.

4. Transformation of the overall field strategy, including the introduction of new processes and innovative technology. The structure and mode of operation of the reservoir team may also change. A decision to recomplete most of the producers in a field as multilateral horizontal wells may fall in this category.

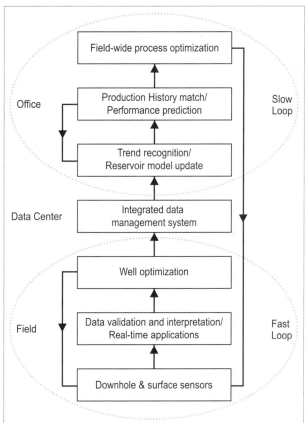

Fig. 6–13. Reservoir data management workflow

Again, actions resulting from newly acquired information from a field are associated with widely varying scales in time. For example, optimization of well production in real time by an operator or engineer belongs to a fast loop. In contrast, optimization related to total field performance, including enhanced recovery, is part of a slow loop, as the latter may involve years of planning and implementation. The entire process is viewed as a closed loop approach. In this approach, a subset of continuously collected data is utilized by reservoir prediction models in monitoring, trend recognition, history matching, and field-wide optimization.[4] A "smart field" may include, but is not limited to, the following:[5]

- Continuous monitoring of the downhole data and visualization of the fluid flow and reservoir characteristics in real time or near–real time.

- Installation of intelligent wells that selectively control injection and production in individual zones based on collected data without human intervention. Optimization of the field operation leads to maximizing the value from the existing wells and equipment.

- Installation of a permanent array of seismic receivers to conduct 4-D seismic surveying on a regular basis, which leads to tracking of the fluid fronts in real time and enhancement of recovery.

- Installation of optical fiber technology to conduct distributed temperature surveys at depth intervals in the well in order to monitor fluid injection and production.

- Data visualization in real time or near–real time, and collaborative efforts across multiple disciplines.

Topics related to intelligent well technology, 4-D seismic survey, and distributed temperature survey are covered in later chapters. An outstanding advantage of digital technology is that the surveillance data from a reservoir can be viewed and analyzed by one or several groups of experts from anywhere in the world in real time.

In conclusion, a robust database management system is indispensable in attaining both short-term and long-term objectives of the reservoir engineers. This includes optimization of reservoir performance and maximization of assets. In many instances, however, the full potential of an integrated database system, based on hundreds of thousands of pieces of data collected daily from a large field, is yet to be realized.

Online database search

In the age of the Internet, petroleum databases are available online that may aid reservoir engineers in accessing valuable information in wide-ranging areas. This may lead to the recognition of worldwide or regional trends in performing a task and providing solutions for field-related issues. Databases are available on worldwide exploration and production, enhanced recovery operations, and geological surveys, among others. Many oil companies have global operations spanning several continents, and the fields are located in wide-ranging geologic settings. Reservoir engineering staffs have access to a global knowledge base developed and maintained by the company. Moreover, technical publications are available online that focus on state-of-the-art technologies in petroleum engineering.

Reservoir data acquisition and integration—a case study

The following is a case study of data integration and reservoir management of a highly heterogeneous field located in offshore Canada. This example is selected because of the general applicability of the workflow in similar environments. The Hibernia field is located in Jeanne d'Arc Basin, Newfoundland, with estimated oil in place of approximately 3 billion bbl. The field is comprised of the Hibernia and Ben Nevis-Avalon reservoirs, the former being the major pay zone located at a depth of 12,100 ft. The lithology of the Lower Cretaceous reservoirs is sandstone. The geologic formation is an anticlinal structure with complex faults, compartments, and multiple stacked fluvial channels. Oil reserves are estimated to be about 615 million bbl, implying a recovery factor of slightly more than 20%. The relatively low value assigned to the ultimate recovery is due to the heterogeneous nature of the hydrocarbon-bearing formation. The average porosity of the formation is about 16%, and the rock permeability ranges from hundreds of millidarcies to 15 darcies or more.

Production from the reservoir began in 1997, with record initial production of 56,000 bopd. Water injection began in mid-1998, followed by commencement of gas injection about a year later. By late 1999, the combined production exceeded 150,000 bopd from seven wells.[6] In 2003, regulatory permission was received to increase production to 220,000 bopd.

Effective well planning and field development required detailed study in the following areas:[7]
- Identification of large- and small-scale faulting and their role in reservoir performance
- Nature of faults, i.e., sealing, partially communicating, etc.
- Potential occurrences of overpressured zones due to compartmentalization
- Increased water or gas saturation in certain areas
- Existence of relatively small net pay zones
- Existence of a perched oil/water contact at anomalous depths in contact with the oil column below
- Precise positioning of wells in the presence of highly faulted zones, multiple oil/water contacts, and reduced transmissibility across certain faults and other heterogeneities, which called for uncertainty analysis and contingency planning

A comprehensive data gathering program for the field included the following:

3-D seismic surveys. Geophysical studies included two 3-D seismic surveys conducted prior to reservoir development.

Geostatistical model. A geostatistically populated earth model incorporating several dozen faults was developed based on the 3-D seismic surveys. The model allowed the incorporation of new data (geological and petrophysical) as it became available during field development.

Reservoir visualization. Faults were validated by 3-D visualization of the seismic data.

Vertical seismic profile (VSP) data. Data from several vertical seismic profiles (VSPs) and dipole sonic studies was obtained at later stages. Vertical seismic profile data aided in obtaining better resolution of the structural model, including the imaging of fault planes.

Log, core, and formation testing data. Measurements-while-drilling and logging-while-drilling surveys were conducted during the drilling of new wells, which included resistivity, gamma ray, and density measurements. These also aided in the decision-making process during drilling in the fault zones.

Wireline logging tools were employed to collect relevant subsurface data, including that obtained by measurements-while-drilling and logging-while-drilling tools. Fluid samples were collected for PVT analyses.

Initial formation pressure data obtained from various wells played a vital role in interpreting communication or noncommunication across or within a fault block.

The oil/water contact and gas/water contact were also identified based on the preceding data.

Core samples were collected from each fault block in order to perform detailed reservoir characterization studies.

Integrated models. Integrated models involved the following:

- Geophysical studies contributed significantly in developing the structural model, stratigraphic analysis, reservoir simulation, and well planning.
- Log and core data were validated by results of the pressure transient testing based on measurements by permanent downhole gauges.
- Geological and reservoir models were updated as log and core analyses became available from a new well, thus leading to improvements in the oil in place (OIP) estimates and reservoir performance predictions.

Reservoir surveillance data. Permanent downhole gauges were installed in production wells to gather real-time production data, including pressure and flow rate. Reservoir monitoring included measurements of the following:

- Flowing bottomhole pressure
- Shut-in pressure
- Oil production volume
- Gas/oil ratio
- Densities of the fluid

Information management. Production information was stored in a robust database system for further analyses and model updates.

Reservoir management. Data gathering in real time led to effective development of reservoir management strategies, such as manipulation of the individual well rates and update of the 3-D earth model, leading to efficient well planning.

Summing Up

An enormous amount of multidisciplinary data is collected throughout the life of a reservoir, including geological, geophysical, petrophysical, geochemical, engineering, and production data. This data is needed to develop an integrated reservoir model, which provides the foundation for reservoir performance analysis and efficient management of the reservoir.

An integrated approach involving all of the functions is necessary for planning, justifying, prioritizing, and timing the data collection program. An efficient data management program consists of data acquisition, analysis, validation, storage, and retrieval.

Planning, collecting, justifying, timing, prioritizing, and cost-effectiveness should be the guiding factors in a data acquisition and analysis program.

Validation of field data, geological maps, log and core data, reservoir rock and fluid properties, and reservoir performance data are essential to ensure quality and reliability. Empirical correlations can be used to generate missing rock and fluid data.

An integrated computer database is required in order to store multidisciplinary data following reconciliation and validation. Close communication among multidisciplinary professionals as a team is crucial in achieving the goal. Stored information can be utilized to carry out multipurpose reservoir management functions, including monitoring and evaluating the reservoir performance.

The data gathered from the reservoir is used for reservoir characterization, with an objective of developing an integrated reservoir model. This model is utilized in history matching of the reservoir performance, future development strategy, and ultimate recovery evaluation.

The traditional engineering approach is knowledge-driven, while the recent concept of data mining uses hidden relationships and influences embedded in the data. The data mining process can be divided into preparing the data, surveying the data, and modeling the data. Preparing the data is time-consuming and expensive, and yet it is the most important step in the process.

This chapter includes an example from studies of the Meren field located offshore of Nigeria. The study data and results include basic reservoir, rock, and fluid properties, structure and cross-section maps, reservoir performance, and history matching.

Present-day petroleum fields are increasingly viewed as "digital fields," "smart fields," or "e-fields." The vision is to attain real-time or near–real-time control on the assets, including continuous optimization of the production from the well to the point of sale. Electronic sensors are deployed that are capable of gathering pressure, temperature, volume, composition, and other valuable data from a source well on a continuous basis. These sensors are part of the effort being made to evaluate reservoir data as part of overall reservoir management in the most efficient manner.

A smart field may include, but is not limited to, the following:

- Intelligent wells that continuously collect data in real time and control injection and production in various zones to maximize recovery
- A permanent array of seismic sensors to perform time-lapse seismic studies in order to track fluid movement
- A distributed temperature survey tool based on fiber optic technology to monitor zonal injection or production

Collected data can be visualized and analyzed in real time by a team of experts located anywhere in the world. The ultimate objective is to maximize the value of the asset based on the existing wells and equipment in the field.

Petroleum databases and technical journals are available online that may aid reservoir engineers in accessing valuable information in wide-ranging areas related to reservoir engineering and management.

A case study of an offshore field with complex faults and compartments illustrates that data integration across various disciplines is vital in characterizing and managing a reservoir. Such disciplines include geophysics, geology, geostatistics, petrophysics, and engineering.

Class Assignments

Questions

1. Why is reservoir data needed?

2. Professionals from which discipline(s) are involved in collecting reservoir data?

3. What are the essential elements in a data management program? When does the data collected from a field turn into information?

4. How would one go about earning management support and approval for a data collection program?

5. Why is the timing for data collection so important? Answer with field examples based on a literature search.

6. Why is data validation so important?

7. Give five examples for data application.

8. What opportunities could be recognized to apply knowledge from this chapter to one's work?

Exercises

6.1. Perform a literature review on permanent downhole gauges. Describe the capabilities of the gauges in collecting production data. Describe how the information can be utilized in a large field where the production level needs to be sustained.

6.2. Describe the data collection scheme of a large offshore field based on a literature review. Include the following in the presentation:
   • Type of data collected on a regular basis
   • Methodology of collecting the data
   • Previous and ongoing geological, geophysical, and laboratory studies
   • Computer simulation work based on the obtained data
   • Integration of the data with the overall reservoir management process

References

1. Satter, A., and G. C. Thakur. 1994. *Integrated Petroleum Reservoir Management*. Tulsa: PennWell.

2. Zangl, G., and J. Hannerer. 2003. *Data Mining*. Katy, TX: Round Oak Publishing Co.

3. Unneland, T., and M. Houser. 2005. Real-time asset management: from vision to engagement—an operator's experience. *Journal of Petroleum Technology*. December, 58–61.

4. Langley, D. 2006. Shaping the industry's approach to intelligent energy. *Journal of Petroleum Technology*. March, 40–46.

5. Reddick, C. 2006. Field of the future: making BP's vision a reality. SPE Paper #99777. 2006 SPE Intelligent Energy Conference and Exhibition, Amsterdam, April 11–13.

6. Sinclair, I. 2000. *Hibernia—Promise and Progress*. St. John's, Newfoundland, Canada: Hibernia Management and Development Co. Accessed at www.cseg.ca/conferences/2000/2000abstracts/1200.PDF.

7. Sydora, L. J. 2000. Integration of multiple data types for reservoir management. *World Oil*. January.

7 · Integration of Geosciences and Engineering Models

Introduction

Traditionally, data of different types has been processed separately to generate different models. These include the following models:

- Static (geological, geophysical, and geostatistical)
- Dynamic or flow (engineering)

The reservoir model is not just a geosciences model or an engineering model. Rather, it is an integration of static and dynamic models, which needs to be developed jointly by geoscientists and engineers. The integrated model is updated on a regular basis as further information about the reservoir is gathered by utilizing various tools and techniques during the life of a reservoir.

Working with an integrated team of multidisciplinary professionals, reservoir engineers collaborate to develop a reservoir characterization model, which can reasonably represent the behavior of an inherently heterogeneous reservoir. The integrated reservoir model can then be utilized with a good deal of confidence to predict the reservoir performance in terms of well rate, reservoir pressure, and ultimate recovery.

The economic viability of a petroleum recovery process is greatly influenced by the production performance of a reservoir under current and future operating conditions. The quality of the integrated reservoir model dictates the accuracy of the results of various reservoir studies, including reservoir performance analysis and estimation of the reserves.

An integrated reservoir model requires a thorough knowledge of the geology, geophysics, rock and fluid properties, and fluid flow and recovery mechanisms. It also requires knowledge of drilling and well completions, past production performance, and data obtained from routine reservoir surveillance.

This chapter is devoted to learning about the following:

- Data and its sources
- Contributions of geosciences and engineering
- Integrated reservoir models
- Synergy and teamwork
- Software tools used in reservoir studies
- Case studies in integrated model development

Data and Sources

Types of multidisciplinary data and their sources include geological, seismic, geochemical, petrochemical, reservoir, and production. These were described in the previous chapter, and the contributions of these disciplines are listed in the following section.

Contributions of geosciences and engineering

Geology. Geology provides information concerning the following:
- Origin of hydrocarbon deposits, migration, and accumulation
- Rock types
- Mineralogy
- Depositional environment
- Structures
- Stratigraphy
- Diagenesis
- Geological maps

Seismic. Seismic information includes the following:
- Depth to reservoir
- Structural shape, faulting, and salt boundaries
- Visualization of the reservoir
- Reservoir properties between wells
- Movement of fluids
- Fracture characterization

Furthermore, seismic surveys contribute to the following:
- Porous interval identification
- Hydrocarbon reservoir identification
- Geologic history
- Abnormally pressured zone identification

Geochemical. This information involves source rock chemistry and reservoir fluid composition.

Petrophysics. Typical measurements recorded on well logs include the following:
- Spontaneous potential
- Natural gamma radiation
- Induced radiation
- Resistivity
- Acoustic velocity
- Density
- Caliper

The results provided by petrophysical studies include the following:
- Producing zone depths
- Zone thicknesses
- Rock types
- Porosities
- Permeabilities
- Fluid saturations

Engineering. The engineering data includes the following:
- Rock properties, such as capillary pressure, wettabililty, and oil-water and gas-oil relative permeability
- Fluid properties, such as phase behavior and PVT data
- Well test data, such as reservoir pressure and temperature, well bore conditions, faults, effective permeability, and interwell continuity
- Well location and completion
- Injection and production data

- Material balance calculations to determine the original oil in place and natural producing mechanisms, including gascap size, aquifer size, and strength of the production mechanism
- Injection/production profiles to provide vertical fluid distribution

Geostatistics. Geostatistics provides the following:
- A means of estimating reservoir properties that incorporates measured trends
- Measurements of the uncertainty of the interpolated values
- A method for integrating independent measurements of reservoir properties

Integrated Reservoir Model[1,2]

An integrated reservoir model requires a thorough knowledge of the geology, geophysics, rock and fluid properties, and fluid flow and recovery mechanisms. It also requires knowledge of drilling and well completions, past production performance, and data obtained from routine reservoir surveillance.

Geoscientists play a vital role in developing a static reservoir model, sometimes referred to as an earth model. The distributions of the reservoir rock types and fluids determine the model geometry and the model type for reservoir characterization. A dynamic reservoir model evolves as production data becomes available following development of the field, together with results obtained from well testing, reservoir surveillance, and other studies. Full-field reservoir simulation is performed to match the production history in terms of the well flow rate, reservoir pressure, and relative quantities of various phases (oil, gas, and water) in the production stream.

The development and use of the reservoir model should be guided by both engineering and geological judgments. Geoscientists and engineers need feedback from each other throughout their work. For example, core analyses provide data for reservoir rock types, whereas well test analysis can confirm flow barriers and fractures recognized by the geoscientists. By discussing all the data as a team, each specialist can contribute available data and help other team members understand the significance of that data. In addition, each professional needs a basic understanding of the data provided by other professionals. For example, petrophysicists who provide rock and fluid properties data need to know how the reservoir engineers utilize their data. On the other hand, the engineers need to know how the petrophysical data is obtained, and the assumptions and limitations involved.

Furthermore, geoscientists and engineers should develop the reservoir model jointly as an integrated team for the following reasons:
- Interplay of effort results in a better description of the reservoir and minimizes the uncertainties of a model. The geoscientists' data assists engineering analysis, whereas the engineering data sheds new light on geoscientists' assumptions.
- The geoscientist-engineer team can resolve contradictions as they arise, preventing costly errors later in the field's life.
- In a fragmented effort, i.e., when engineers and geoscientists are not in communication with one another, each discipline may study only a fraction of the available data in isolation. Hence, the quality of the reservoir management may suffer as drilling decisions and depletion plans are adversely affected throughout the life of the reservoir.
- Multidisciplinary teams using the latest technology provide opportunities to tap unidentified reserves. For example, improved 3-D seismic data can aid in the surveillance of production operations in mature projects. It can also be used to identify the presence or lack of continuity between wells, thus improving the description of the reservoir model.
- One very important factor, which is often overlooked, is to discuss uncertainties related to the reservoir data. Should a number be taken as hard fact, or is there a range of possible values around the given value? Does hard-to-explain field data point to hitherto unknown reservoir heterogeneities? Utilizing reservoir models developed by multidisciplinary teams can provide practical techniques of accurate field description to achieve optimal production.

Integration & Alliance

Fig. 7–1. Integration and alliance

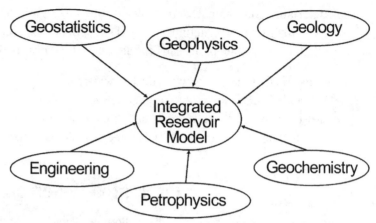

Fig. 7–2. Integrated reservoir model

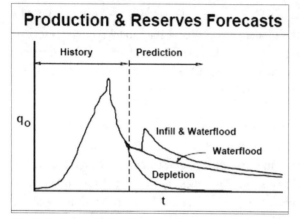

Fig. 7–3. Existing reservoir production rate and alternative development plans. *Source: A. Satter, J. E. Varnon, and M. T. Hoang. 1992. Reservoir management: technical perspective. SPE Paper #22350. Presented at the SPE International Meeting on Petroleum Engineering, Beijing, China, March 24-27. © Society of Petroleum Engineers. Reprinted with permission.*

Integration and alliance of data, professionals, tools, technology, and organizations provide the foundation of an integrated reservoir model (fig. 7–1).

Data required for an integrated model consists of geology, geophysics, petrophysics, geochemistry, geostatistics, and engineering sources. This data can be used to produce a realistic understanding of reservoir performance (fig. 7–2).

A case study of data collection and integration in a large reservoir located in offshore Canada is presented in chapter 6.

The integrated model can be used for realistic reservoir performance analysis and ultimate recovery forecasts, among others. It provides a means of understanding the current and future performance of a reservoir under various "what if" conditions, so that better reservoir management decisions can be made. Figure 7–3 shows the past oil production rate of an ongoing reservoir under depletion drive and projected future primary performance. Also shown are expected oil recoveries with water injection, and water injection with infill wells. The expected future performance results can be utilized to decide how to add more value to this asset.

Synergy and Teamwork

Synergy and team concepts are essential to build an integrated reservoir model based upon integration of geosciences and engineering data. It involves multidisciplinary professionals working together as an integrated team

Success in reservoir management requires synergy and team effort. It is increasingly recognized that reservoir management is not synonymous with reservoir engineering or reservoir geology. Success requires multidisciplinary, integrated team efforts. The players are all of those who have anything to do with the reservoir. The team members must work together to ensure development and execution of the management plan. Crossing the traditional boundaries and integrating their functions leads to better utilization of corporate resources and achievement of a common goal.

All development and operating decisions should be made by the reservoir management team, which recognizes the dependence of the entire system upon the nature and behavior of the reservoir. It is not necessary that all decisions be made by a reservoir engineer. In fact, a team member who considers the entire system, rather than just the reservoir aspect, will be a more effective decision maker. It will help tremendously if the person has background knowledge of reservoir engineering, geology, production, drilling engineering, well completion and performance, and surface facilities. Not many people in an organization have knowledge in all these areas. However, many persons develop an intuitive feel for the entire system and know when to ask for technical advice regarding various elements of the system.

The team effort in reservoir management cannot be emphasized enough. It is even more necessary when companies work with limited resources..

Furthermore, with the advent of technology and the complex nature of different subsystems, it is difficult for anyone to become an expert in all areas. Hence, it is obvious that the reduction of talent and the increasingly complex technologies must be offset by an increase in quality, productivity, and an emphasis on team effort.

The following efforts can enhance the team approach to better reservoir management:
- Overall understanding of the reservoir management process, technology, and tools through integrated training and integrated job assignments.
- Openness, flexibility, communication, and coordination.
- Facilitated communication among various engineering disciplines, geology, and operations staff by achieving the following:
  - Periodic meetings
  - Interdisciplinary cooperation in teaching each other's functional objectives
  - Development of trust and mutual respect

Each member of the team should learn to be a good teacher.
- The engineer, to some degree, must develop the geologist's knowledge of rock characteristics and depositional environment. The geologist also must cultivate knowledge concerning well completion and other engineering tasks relating to the project at hand.
- Each member should subordinate personal ambitions and egos to the goals of the reservoir management team.
- Each team member must maintain a high level of technical competence.
- The team members must work as a well coordinated "basketball" team rather than as a "relay" team. Reservoir engineers should not wait on geologists to complete their work and then start the reservoir engineering work. Rather, a constant interaction between the functional groups should take place. For example, it is better to know early if the isopach and cumulative oil/gas production maps do not agree. This is preferable to finalizing all the isopach maps and then finding out that cumulative production maps are indicating another interpretation of the reservoir. Using an integrated approach to reservoir management, along with the latest technological advances, will allow companies to extract the utmost economic recovery during the life of an oil field. It can prolong the economic life of the reservoir and add significant value to the asset.

Synergy is not a new concept. Michael Halbouty, chairman and CEO of Michael T. Halbouty Energy Co. in Houston, was a long-time advocate of synergy and the team approach. He recognized this concept as basic to future petroleum reserves and production. By and large, the concept is now supported across the industry. According to Robert Sneider, head of Sneider Exploration, "Synergy means that geoscientists, petroleum engineers, and others work together on a project more effectively and efficiently as a team than working as a group of individuals."[3]

Software tools used in reservoir engineering studies

A general categorization of reservoir engineering software available in the industry is provided in the following discussion. These interactive and user-friendly tools lead to the development of more realistic reservoir models based on teamwork. Users from different disciplines can work in collaboration with the software to form a simultaneously functioning team, much like a basketball team, rather than passing their individual results to each other like batons in a relay race.

Example applications of several types of software discussed below are used in the various chapters in this book.

PVT properties. Software tools can be used to estimate the PVT properties of reservoir fluids, namely oil, gas, condensate, and formation water. Applications are based on a multitude of published correlations and laboratory data, such as flash and differential vaporization. Certain packages are capable of computing the compositional properties of hydrocarbon mixtures, such as vapor/liquid equilibria, based on the equation of state.

Properties of fluids and their role in reservoir performance were described in chapter 3.

Hydrocarbon reserves. Estimates of oil and gas reserves are based on the volumetric calculation of hydrocarbon in place. Mapped volumes of planimeter data may be used as input. Also, software based upon techniques such as volumetric, decline curve analysis, classical material balance, and reservoir stimulation methods are used. The software may take into account the various drive mechanisms in estimating ultimate recovery and reserves. These methods are described briefly below and in more depth in the following chapters. Probability estimates of reserves can be accomplished by using Monte Carlo simulation techniques.

Volumetric method. Estimates of original hydrocarbon in place can be made based upon the bulk volume of the reservoir, porosity, initial fluid saturation, and formation volume factor. Reserves can be estimated with estimated recovery factor.

Volumetric methods are discussed in chapter 9.

Monte Carlo simulator. As indicated earlier, applications that estimate oil and gas reserves are likely to incorporate the Monte Carlo simulation technique. Risk and uncertainty are inherent in virtually any venture, including petroleum exploration, field development, and production. The Monte Carlo simulation method is extensively employed in the petroleum industry and in other industries to assess the likelihood of a certain outcome of a venture. Typical software may include a large number of probability distribution functions. From these, the most suitable ones can be selected in a given situation, such as a permeability distribution in a heterogeneous formation or the occurrence of a dry hole in a basin. Furthermore, the software may automatically perform a sensitivity analysis in order to identify the most critical factor or factors in a venture.

Monte Carlo simulation techniques as utilized in reservoir engineering are illustrated in chapters 9, 11, and 18.

Material balance analysis. Considering a homogeneous reservoir, the initial hydrocarbon in place can be computed when fluid properties and production and reservoir pressures are available. The strength of this method lies in recognizing the natural producing mechanisms, such as solution gas drive, gascap drive, and aquifer drive for oil reservoirs, and depletion and aquifer drive for gas reservoirs. If the abandonment pressure is known, gas reserves can be reliably estimated for a depletion drive reservoir.

Material balance methods are described in chapter 12.

Decline curve analysis. Estimates of the petroleum reserves, the future production profile, and the economic limit of the field can be made when sufficient production data is available and production is declining.

Decline curve analysis methods are discussed in chapter 11.

Well test analysis. Reservoir pressures and fluid properties based upon reservoir pressures are needed for material balance and reservoir simulator methods. In addition to reservoir pressure, well test analysis techniques provide effective permeability thickness of the reservoir, wellbore conditions, stratification, presence of faults and fractures, and reservoir continuity and barrier information.

The capabilities of well test software are illustrated in chapter 5.

Reservoir simulator. Perhaps the most important software application used by reservoir engineers is reservoir simulation that considers the inherent heterogeneities of the subsurface reservoir. Based on a conceptual reservoir model or models, a simulator predicts future reservoir performance and what-if scenarios as new wells or enhanced recovery programs are planned. Studies performed by other software tools mentioned earlier are used to complement the reservoir simulation model. Furthermore, present-day reservoir simulators are integrated with geological and geophysical models to varying degrees.

The topic of reservoir simulation is treated in detail in chapters 13 and 14.

Other tools. Besides the core applications related to reservoir development and management, reservoir engineers are likely to come across a variety of related petroleum software. Such software includes, but is not limited to, the following:

- Multiphase flow simulation in a wellbore
- Gas lift design
- Nodal analysis and well/flowline optimization
- Production logging interpretation tool
- Surface network simulation
- Economic evaluation
- Reservoir data management and integration
- Import of the geological model into a numerical simulation model
- Mapping and visualization
- Online information related to geology and production

Integrated software suite

Historically, reservoir-related software applications evolved in apparently incompatible platforms that made reservoir model integration and teamwork a formidable task. In recent times, however, one of the main efforts is geared toward achieving seamless integration between various software products. The objectives include the following:

- Better knowledge management
- Added value to assets, i.e., recoverable reserves
- Real-time or near–real-time decision making
- Shorter turnaround time in problem solving

A major breakthrough in reservoir modeling has occurred with the advent of integrated geosciences (reservoir description) and engineering (reservoir production performance) software. This software is designed to manage reservoirs more effectively and efficiently. For example, updates to a static model, including faults, boundaries, and other geologic features, are automatically reflected upon a dynamic model that adjusts the prediction of oil or gas production. This is accomplished by deployment of an open architecture across platforms, development of a robust database management system, and smart linking based on modern computer languages. Several service, software, and consulting companies are now developing and marketing integrated software installed in a common platform.

Components of an integrated software suite in the planning and development of oil and gas fields are outlined in the following discussion.

Database management system. An integrated software suite may be based on dozens of individual applications in various disciplines. These include geology, geophysics, geostatistics, petrophysics, modeling, visualization, simulation, production, and others. A robust database system provides the foundation for integration and connectivity between the various modules that can be accessed instantly from any of the modules in the integrated system. Besides serving as the ultimate repository of reservoir information, the database system can have several features, including the display of data hierarchy and workflow related to integrated study.

Seismic and structure modeling. Information obtained from a multitude of geosciences and well studies is integrated to build a structural model. The workflow involves seismic interpretation and stratigraphic correlation between wells, among other tasks. The static reservoir model is updated depending on information obtained from the dynamic behavior of the reservoir, including well test interpretation and production. Certain packages are capable of displaying both static and dynamic data, such as different realizations of a reservoir model and corresponding production performance scenarios on the same screen.

Simulation of reservoir performance. Following the development of a structural model, a software package performs simulation of the reservoir performance based on the simulation model. Connectivity between the two models may require intelligent coarsening of millions of grid cells in the structural model as it is imported into the other model. The simulator is capable of matching past performance of the reservoir and predicting future trends. During history matching, the structural model may need to be fine-tuned to achieve satisfactory results. Again, the whole process is interactive and integrated, leading to a high level of efficiency and accuracy.

Well planning. The integrated software suite aids significantly in determining the most optimum well trajectory. A 3-D visualization of static and dynamic models of the reservoir can be utilized seamlessly in designing a new well or in recompletion of an existing well. An integrated approach is critical in many complex scenarios, including the drilling of horizontal wells having several branches in a heterogeneous formation with faults and boundaries.

Production data monitoring. The final component of the software suite focuses on the collection, interpretation, and assimilation of daily production information in the integrated reservoir model. For example, a lower-than-expected reservoir pressure may require reassigning the strength of water influx in the simulation model or reviewing the reservoir boundaries in the geologic model, or both. Reservoir engineers and other professionals are able to access field data in real time. In addition, the software package facilitates analysis of historical production data using various methodologies.

Open connectivity. The modules in an integrated system not only can communicate between themselves but also are capable of interacting with other systems outside the system's domain. This is accomplished through open connectivity architecture. This capability enhances the versatility of the software suite, as reservoir studies are no longer limited to one particular system or platform.

The ultimate integration in software. Since studies in the various areas related to petroleum reservoirs were performed traditionally as stand-alone tasks, unintended consequences were not uncommon. Sometimes project delays and cost overruns resulted. The ultimate objective in software integration is to encompass virtually all aspects of reservoir management and thus reduce possible uncertainties. True integration can be achieved by establishing intelligent links among the various areas. Connected areas of study should include reservoir characterization, well data analysis, design of surface facilities, and reservoir economics in a rapidly changing market.

Integrated model workflow

Integrated reservoir modeling begins with the collection of multidisciplinary data, which is subject to established quality control processes. At the early stages of field development, the data on which the models are built chiefly consists of information obtained from geophysical surveys. This is combined with very limited petrophysical or production data, obtained from one or just a few wells. As the field is developed further, additional data pertaining to reservoir geology, logs, cores, formation tests, production rates, reservoir pressures, and well tests becomes available. This information is vital in updating and refining the models on a regular basis. Valuable information may be provided by 4-D seismic studies performed at specific time intervals, which may be used to update both static and dynamic models and reduce uncertainties. A robust database system seamlessly integrated with various modeling software applications is important in achieving this goal.

The reservoir models and related tasks may include, but are not limited to, the following:[4,5]

- Structural model, including faults, fractures, pinchouts, and other geologic features
- Lithofacies model
- Porosity, permeability, and net to gross thickness models
- Calibration of geophysical and other data to petrophysical data
- Oil/water and gas/water contacts
- Fluid saturation model
- PVT model
- Upscaling of the static model with fine resolution to a coarser grid for flow simulation
- Reservoir simulation model
- Reservoir characterization and visualization

Treatment of uncertainty

Due to significant uncertainties associated with rock properties throughout the reservoir, except at well locations, static models rely heavily on geostatistical methods, also referred to as stochastic modeling. Based on the above, rock properties are varied within certain bounds to generate multiple realizations (probable sets of reservoir description). Bounds for a parameter, such as porosity or net to gross thickness, are dictated by core, log, and seismic data. The following are some characteristics of uncertainty upon which stochastic modeling is based:

- Uncertainty is essentially zero at well locations where core and log studies are performed.
- The degree of uncertainty varies smoothly away from the wells.
- The variance depends on the quality of the seismic data and distance from the wells.

Integrated model studies based on stochastic techniques involve automatic iteration between static and dynamic models that can readily highlight uncertainties, sensitivities, and most-likely scenarios. In traditional modeling efforts, however, iterations are performed manually. Substantially more time and effort may be needed to gain insight concerning the reservoir under study.

Dynamic model simulation

Hundreds or thousands of realizations related to reservoir description may be utilized to produce a range of values of the original hydrocarbon in place or ultimate oil recovery. From the large set of results, values of the original hydrocarbon in place or ultimate oil recovery having various probabilities can be obtained (fig. 7–4). This information leads to the most (or least) likely scenarios. Static earth models are typically comprised of millions of cells. Upscaling the model to coarser grids is necessary prior to simulation of the corresponding dynamic model, where the number of cells is much less.

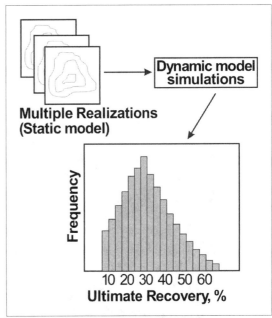

Fig. 7–4. Multiple realizations of static model lead to a large set of results obtained by a dynamic model. Each value in the results is attached to a particular probability distribution. Most likely scenarios are inferred from the analysis.

Development of a Simulation Model

Steps to build a model for a simulator are briefly outlined as follows:

1. Digitize paper copies of structure maps for various layers with well locations and water/oil and gas/oil contacts. The data sources are seismic, geological, and drilling information.

2. Build grid blocks on the digitized top structure map for a reservoir simulator and bring in the other maps, assigning the same grid configurations.

3. Import isopach data with thickness, porosity, and fluid saturations on the digitized maps with various layers, so that these parameters are defined. In addition, import a pressure survey map or calculate pressure at each block based upon pressure gradients.

4. Add PVT data for oil, water, and gas properties, and add oil-water and gas-oil relative permeability data to the model.

5. Initialize the map to verify initial saturations and pressure, and also to calculate the original hydrocarbon in place, which should be confirmed with the value from the volumetric method.

6. In the case of a history match study, validity of the model can be assessed from the history match results.

Figure 7–5 is a depiction of a reservoir model based on a numerical grid. Each grid is assigned a unique value of a reservoir property. Computer-assisted simulation of reservoir pressure and fluid saturation and composition are performed to analyze and predict performance.

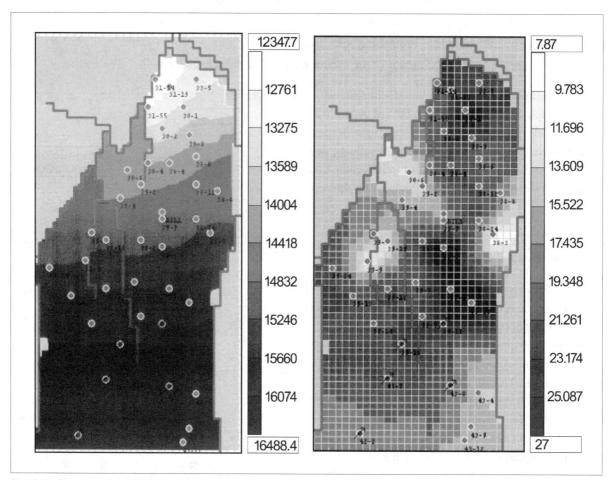

Fig. 7–5. Structure and thickness contour maps of a reservoir with well locations. Rectangular grids are placed on the maps as a step in reservoir modeling. *Courtesy Gemini Solutions, Inc.*

Case Studies in Integrated Model Development

Naturally fractured reservoir[6]

In one example from the literature, the objective of the study included the development of a dynamic flow model. This was needed in order to accurately predict well and reservoir performance in a giant field with naturally occurring fractures. The fracture network consisted of clusters trending in two major directions (N40E and N70E) and was known to influence well productivity and field reserves. Calculated fracture permeability ranged from a fraction of a millidarcy to more than 5 darcies, with significant permeability anisotropy among the three axes (x, y, and z). Fracture porosity was found to be very small, estimated at less than 0.38%, possibly as low as 1.5×10^{-5}%. Characterization of the fractures, including those under dynamic flow conditions, was based on industry standard tools and techniques, including the following:[7]

- Core studies
- Borehole imaging logs
- Seismic data, including seismic facies analysis
- Pressure transient tests
- Production logs
- Mud losses during drilling
- Well performance history, including water breakthrough

Three stochastic realizations of the fracture network at different scales of 500 m, 700 m, and 1,000 m were generated. Static modeling of the rock matrix was based on wireline log, core, and seismic data. Stochastically generated porosity and permeability data were made to honor the underlying geologic facies description. Realizations were ranked in accordance with (a) sweep efficiency of injected fluid, (b) original oil in place, and (c) the rock heterogeneity index. Sweep efficiency was evaluated by simulation of a streamline model.

Pressure buildup tests conducted at the wells provided estimates of permeability under dynamic conditions. Average matrix permeability was found to be more than 100 mD. A significant difference between the core-derived permeability data and the well test results indicated the presence of fractures dominating the flow dynamics around the well.

The earth model was integrated to a dynamic flow model. This was done by iteration between models in which the simulated reservoir flow capacity, oil production rate, and bottomhole pressure in the wells were matched with field data. In order to evaluate whether the well performance was dominated by a fracture system, comparative history match studies were made. One considered the matrix system only, while the other considered both the matrix and fracture systems. The integrated model was found to predict the performance of newly drilled wells successfully, and it enhanced the overall understanding of the fractured reservoir behavior. The net result was better reservoir management and addition to the reserve.

Evaluation of a geologically complex reservoir during the appraisal phase[8]

In the next example from the literature, the objective of the integrated study was the generation of probability distributions of critical elements such as original hydrocarbons in place, well rates, cumulative production, and net present value of the asset. Other probability distributions involved discounted cash flow during the early phase of a gas field in which a limited number of exploratory wells had been drilled. This information was required in deciding on the number of wells to be drilled and for scheduling of the wells in the face of significant uncertainties, as described below.

In this case, the reservoir consisted of several zones of varying reservoir quality and compartments with unknown connectivity. Considerable uncertainties existed in the rock and fluid properties, hydrocarbon in place (HCIP) volumes, ultimate recovery, and associated costs. Fluid PVT properties, such as the dew point and condensate volume, were not known with certainty from the entire reservoir. In addition, there was a lack of information (porosity, net to gross thickness, etc.) from unexplored areas of the reservoir.

The objectives of the integrated modeling included the following:
- Understanding the uncertainty existing between probable reserves and the required plateau production rate
- Improvement of the project by optimizing well counts and well group scheduling
- Investigation of the sensitivity of the project to uncertainties rooted in fluid PVT data

The workflow attempted to integrate information obtained from various sources noted below through a Monte Carlo simulation method. The probabilistic analysis method is described in chapter 18.
- Geology—hydrocarbon volume
- Engineering—well productivity
- Reservoir studies—production trend modeling related to plateau rate and decline
- Development planning—cost estimates and economic evaluation

Modeling components were based on these areas. Results from each component were provided as input to the next component of the model. A simulation run began with reservoir volume, petrophysical, and PVT data. Other pertinent information, including the number of wells, was utilized to generate a reservoir production profile. This was accomplished by the material balance method (as illustrated in chapter 12, with several examples). Next, capital and operating costs were estimated. All of the information was then used in an economic model to evaluate the project's economic value. (Economic indicators for oil and gas projects, such as net present value, discounted cash flow, and other factors are discussed in chapter 18.) The resulting histograms of the various parameters, similar to the one shown for recovery in Figure 7–4, led to vital information, including sensitivity analyses. Some are listed as follows:
- Likely range of gas reserve, upper and lower bounds
- Sensitivity of reserves to reservoir quality and connectivity
- Field development strategy, including optimum number of wells
- Probability of plateau production for x, a certain number of years before the onset of decline
- Probability attached to a negative net present value of the project

Summing Up

The reservoir model is not just a geosciences model or an engineering model. Rather, it is an integrated model that needs to be developed jointly by geoscientists and engineers. The integrated model provides the foundation for realistic reservoir performance analysis and ultimate recovery forecasts.

Integration of geosciences and engineering data collected throughout the reservoir life cycle is required to produce the reservoir model. Development of a sound reservoir model requires assembling geosciences and engineering data. It also requires geosciences and engineering professionals to work together as an integrated team. They must have an understanding of each other's data, a realization of measurement uncertainties, and access to a database that is updated as soon as new data is available.

Success in reservoir management requires synergy and team efforts, which are the essential elements for integration of geosciences and engineering, involving people, technology, tools, and data.

Overall understanding of the reservoir management process, technology, and tools, along with openness, flexibility, communication, and coordination, can enhance the team approach. The result is better reservoir management.

Several service, software, and consulting companies are now developing and marketing stand-alone and integrated software installed in a common platform.

Software tools used in reservoir studies and examples of case studies in integrated model development are presented.

An example reservoir simulator model building is presented, showing the numerical grid assignment of structure and thickness contour maps.

Class Assignments

Questions

1. Why is an integrated reservoir model so important?

2. Who are the contributors to the integrated model, and what are their contributions?

3. What are the procedures for developing an integrated model?

4. What is meant by the term "synergy"?

5. What should be the composition of an ideal reservoir management team?

6. How should the members of an integrated team work?

7. What can be done to improve on a team approach?

Exercises

7.1. Based on a literature review, describe an example of integrating a geological model with a reservoir simulation model for a large field with complex geology. Include the following:
   - Name, size, and location of the field
   - Types of data integrated
   - Professional disciplines involved in integration
   - Integration workflow
   - Economic analysis as part of the integration, if any
   - Goals achieved and lessons learned by developing an integrated reservoir model

References

1. Satter, A., and G. C. Thakur. 1994. *Integrated Petroleum Reservoir Management*. Tulsa: PennWell.

2. Satter, A., J. Baldwin, and R. Jespersen. 1994. *Computer-Assisted Reservoir Management*. Tulsa: PennWell.

3. Sneider, R. M. 1999. Geo-teams prove their E&P value. *AAPG Explorer Web site*. November. Accessed at www.aapg.org/explorer/1999/11nov/multi_teamexpl.cfm.

4. Gringarten, E. *Uncertainty Assessment in 3-D Reservoir Modeling: An Integrated Approach*. Earth Decision Sciences. Accessed at www.earthdecision.com/news/white_papers/AssessingUncertaintythru3DResModeling.pdf.

5. Statios LLC. *Some Practical Aspects of Reservoir Modeling—Reservoir Modeling with GSLIB*. San Francisco: Statios LLC. Accessed at www.statios.com/Training/.

6. Bahar, A., H. Ates, M. H. Al-Deeb, S. E. Salem, H. Badaam, S. Linthorst, and M. Kelkar. 2003. An innovative approach to integrate fracture, well-test, and production data into reservoir models. SPE Paper #84876. Presented at the 2003 SPE International Improved Oil Recovery Conference, Kuala Lumpur, October 20–21.

7. Bourbiaux, B., R. Basquet, M.-C. Cacas, J.-M. Daniel, and S. Sarda. 2002. An integrated workflow to account for multi-scale fractures in reservoir simulation models: implementation and benefits. SPE Paper #78489. Presented at the 2002 SPE Abu Dhabi International Petroleum and Exhibition Conference, Abu Dhabi, October 13–16.

8. Hegstad, B. K., S. Tollefsen, D. V. Arghir, A. S. Cullick, K. Narayanan, D. E. Heath, and J. C. Lever. 2004. Rapid scenario and risk analysis for a complex gas field with large uncertainties. SPE Paper #90961. Presented at the 2004 SPE Annual Technical Conference and Exhibition, Houston, TX, September 27–29.

8 · Evaluation of Primary Reservoir Performance

Introduction

Primary reservoir performance of oil and gas reservoirs is characterized by various field and individual well behavior as follows:

- Reservoir pressures with production period
- Oil, gas, and water production rates and cumulative volumes with time
- Gas/oil and water/oil ratios with time
- Water injection rates and cumulative volumes with time
- Gas injection rates and cumulative volumes with time

Natural factors governing the primary performance include the following:

- Geological and geophysical characteristics of the reservoir, including heterogeneities
- Rock and fluid properties on a microscopic and macroscopic scale
- Natural producing mechanisms
- Fluid flow dynamics

The operators of petroleum reservoirs have the capability of exercising some control on the reservoir performance through professional reservoir management, which is very important. For example, the same reservoir exploited by a different team of engineering and operating personnel with different strategies, equipment, and production facilities would differ in performance and ultimate recovery. Government policies and regulations would also be an influencing factor.

Reservoir engineers play a leading role in analyzing reservoir production performance under current and future operating conditions. Evaluation of past and present reservoir performance, followed by prediction of its future performance, is an essential aspect of the reservoir management process.

The contributions of geology, geophysics, and petrophysics to an integrated reservoir model are discussed in chapter 7. Rock properties, fluid properties, and fluid flow mechanisms are presented in chapters 2, 3, and 4, respectively. Natural producing mechanisms, heterogeneities, and reservoir performance as affected by variations in reservoir properties will be discussed in this chapter.

The objectives of this chapter are to learn about the following:

- The needs for reservoir performance analysis
- Natural producing mechanisms
- Reservoir heterogeneities
- Reservoir performance evaluation and prediction techniques
- Sensitivity of reservoir properties affecting production performance

Needs for Reservoir Performance Analysis

Successful reservoir management relies on the ability to generate reliable reservoir performance behavior. The primary questions that reservoir engineers are expected to answer are given in the following, in order of priority:

1. What are the expected quantities of original oil and gas in place (OOIP and OGIP)?

2. How much oil and gas can be economically recovered given the associated probabilities and risks?

3. How can a newly discovered field be developed, followed by implementation of the reservoir management plan and monitoring and evaluation of reservoir performance?

Natural Producing Mechanisms

There are natural sources of energy in oil reservoirs that control reservoir performance. These include the following:
- Liquid and rock compressibility drive
- Solution gas or depletion drive
- Gascap drive
- Aquifer water drive
- Gravity segregation drive
- Combinations of above[1]

Drive mechanisms in gas reservoirs are as follows:
- Gas expansion or depletion drive
- Aquifer water drive
- Combinations of above[1]

Figure 8–1 shows reservoir pressure versus recovery efficiency under various drive mechanisms in oil reservoirs under ideal conditions.[2]

Recovery of petroleum fluids due to liquid and rock expansion is relatively less, usually a few percent of total hydrocarbon in place. Solution gas or depletion drive is usually an efficient natural driving mechanism. Recovery from a gascap drive reservoir depends upon the size of the gas cap, movement of the gas cap through the entire segment of the reservoir, and effective gravitational segregation of oil and gas. Oil recovery due to gravity drainage reservoirs can be quite substantial. Recovery can depend upon the relief or dip of the reservoir, permeability in the direction of the dip, and the densities and viscosities of the oil and gas. Natural water drive is usually the most efficient driving force. Recovery can be very high under strong edge water drive through very porous and permeable reservoirs. On the other hand, recovery from bottom water reservoirs can be poor due to the phenomenon of water coning.

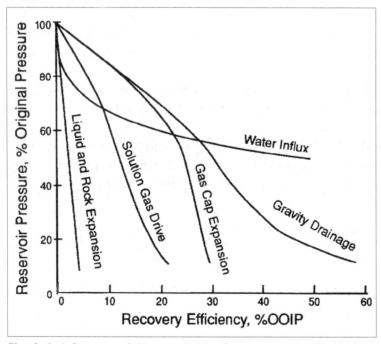

Fig. 8–1. Influences of drive mechanics for oil reservoirs. The highest recovery potential is associated with the reservoirs having water influx from an adjacent aquifer, followed by gravity drainage. *Source: A. Satter and G. C. Thakur. 1994. Integrated Petroleum Reservoir Management—A Team Approach. Tulsa: PennWell.*

Figure 8–2 presents the influence of drive mechanisms on recovery from gas reservoirs.

This figure shows reservoir pressure over gas compressibility, P/Z, for depletion and aquifer drive reservoirs.

Recovery efficiencies for the depletion drive gas reservoirs can be 80% to 90%. However, in the case of water drive, gas reservoir recovery efficiencies could be in the 50% to 60% range because of the bypassed gas and the high reservoir pressures.

Hydrocarbon phase behavior controls the formation of black oil, volatile oil, dry gas, and gas condensate reservoirs, which is discussed in chapter 3.

The drive mechanisms involved in these reservoirs will be now discussed.

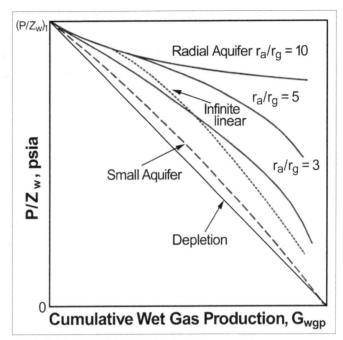

Fig. 8–2. Influence of drive mechanisms on recovery from gas reservoirs. The term shown in the figure, r_a/r_g, is defined as the ratio of the radius of the aquifer over the radius of the gas reservoir.

Liquid and rock compressibility drive —unsaturated black oil reservoirs

Unsaturated black oil reservoirs are encountered when the initial reservoir pressure and temperature are far from the critical point and above the bubblepoint. Figure 3–30 in chapter 3 shows an unsaturated black oil reservoir with pressure and temperature initially at point A, where all the available gas is dissolved in the oil. The vertical line from point A to point A_1 on the bubblepoint curve represents isothermal production as a result of liquid and rock expansion drive. Expansion of the rock due to a decline in reservoir pressure is accompanied by compaction of the pores in the rock. This phenomenon is significant in unconsolidated geologic formations. Hence the producing mechanism in such reservoirs is referred to as compaction drive.

Compressibilities of liquids and rocks are quite small and are usually in the order of 10^{-6} psi^{-1}. Under this mechanism of primary production, reservoir pressure drops rapidly and continuously until the bubblepoint is reached (fig. 8–1). Hydrocarbons are essentially in the liquid phase as long as the reservoir pressure remains above the bubblepoint (A_1 in fig. 3–30). Evolution of gas occurs at surface facilities due to the reduction in pressure and temperature. The producing gas/oil ratio remains low and constant. Oil recovery efficiency typically varies from 1% to 5%, with an average of 3%.

Solution gas drive—unsaturated black oil reservoirs

When the reservoir pressure declines below the bubblepoint due to production, dissolved gas starts to come out of solution (A_2 in fig. 3–30). The free gas is not produced until a critical gas saturation is developed when the gas phase is continuous through the porous media. Depletion below the bubblepoint causes the gas phase to increase rapidly in the reservoir. The mechanism of production of oil and gas is called solution gas drive or depletion drive.

Since the gas viscosity is much lower than the oil viscosity, the gas phase is significantly more mobile than the liquid phase in the reservoir. The gas/oil ratio is initially low, then rises to a maximum, and finally drops as most of the liberated gas is produced. Typical oil recovery due to solution gas drive could be from 10% to 25%, with an average of 16%.

Gascap drive—saturated black oil reservoirs

When the initial reservoir pressure and temperature are within the two-phase region (at A_2 in fig. 3–30), reservoirs with gas caps are encountered. Gas being lighter than oil, it rises above the oil zone due to gravity segregation. As the reservoir pressure declines with production, the gas cap expands, resulting in gascap drive.

Under gascap drive, reservoir pressure falls slowly and continuously (fig. 8–1). Initially, gas is produced due to the free gas saturation in the reservoir. At first the gas/oil ratio is low, then rises to a maximum, and finally drops. Production from the gascap reservoir is due to both solution and gas drives, resulting in higher oil recovery than the solution gas drive reservoir. Oil recovery due to gascap drive could be 15% to 35%, and averages 25%.

Volatile oil reservoirs—undersaturated and saturated

In the case of an undersaturated volatile reservoir, the initial reservoir pressure and temperature is close to the critical point but above the bubblepoint (at V in fig. 3–30). As compared to the black oil reservoir, volatile reservoirs have higher API gravity, in the range of 50° or more. The dissolved gas/oil ratio could be 1,500 scf/bbl or greater.

As in the case of black oil reservoirs, production mechanisms for volatile reservoirs are due to rock and fluid expansion above the bubblepoint. Production mechanism is solution gas below the bubblepoint.

A gas cap is encountered for the volatile reservoirs when the initial reservoir pressure and temperature are within the two-phase region. Production from the gascap reservoir is due to both solution and gas drives, resulting in a higher oil recovery than the solution gas drive reservoir.

Because of the lighter oil with lower viscosity, recoveries from the volatile oil reservoirs could be greater than those of the black oil reservoirs.

Gas condensate reservoirs with retrograde condensation

Gas condensate reservoirs are characterized by an initial reservoir temperature that lies between the critical temperature of the fluid system and the cricondentherm. Reservoir pressure is found to be above the dewpoint pressure (at R in fig. 3–30). The vertical line from point R to point R_1 on the dewpoint curve represents isothermal production.

Condensate, which appears below the dew point (R_1 in fig. 3–30), would continue to increase to a maximum value. It would then decrease again until the abandonment pressure is reached (point R_2). Isothermal retrograde condensation may result in the loss of some intermediate to heavy hydrocarbon components in the reservoir due to poor mobility.

Above and below the dew point, production is due to gas expansion or depletion drive.

Gas expansion or depletion drive—wet and dry gas reservoirs

Gas reservoirs exist in the single-phase region, with the initial temperature exceeding the cricondentherm (at G in fig. 3–30). When the gas phase undergoes isothermal depletion inside the reservoir without any condensation, it traces a path as shown from point G to point G_1 that lies in the single-phase region as well. Production is due to gas expansion or depletion drive.

When a portion of the produced gas condenses in surface separators under reduced pressure and temperature, the produced gas is referred to as wet gas. The path traced by the wet gas from reservoir conditions to the surface facilities is illustrated by the curve from point G to point G_2. Greater condensate recovery could be realized by operating the separators at lower temperatures. However, some of the condensate could be trapped in the reservoir, not producible due to lack of effective permeability.

Figure 3–30 shows the path of dry gas from the reservoir to the surface under reducing pressure and temperature conditions. The reservoir fluid would remain as a single phase in the reservoir due to isothermal depletion along the vertical line from point G to point G_1. If the gas is sufficiently lean, the produced gas remains in a single phase under reduced pressure and temperature at the surface, as shown from point G to point G_1. Dry gas reservoirs contain mostly lighter hydrocarbons, with a gas/oil ratio of more than 100,000 scf/stb of condensate. Gas recoveries could be 80% to 90% at low separator pressures. Gas pumps can be installed to raise the gas pressure to the pipeline delivery pressure.

Aquifer water drive

When an oil or gas reservoir is in communication with a surrounding (bottom or edge) active aquifer, production from the reservoir results in a pressure drop between the reservoir and the aquifer. This allows influx of water into the reservoir. A producing reservoir is referred to as bottom water drive or edge water drive reservoir, depending on the location of the adjacent aquifer providing energy for production.

Reservoir pressures in water drive reservoirs remain high. Pressure is influenced by the rate of water influx, and by the rate of oil, gas, and water productions. Gas/oil ratios remain low if pressure remains high. Downdip wells produce water earlier, and water production continues to increase. Water drive is usually the most efficient reservoir driving force in oil reservoirs. Recovery efficiencies may vary from 30% to 80%, depending upon the size and strength of the aquifer.

Recovery efficiencies for the depletion drive gas reservoirs can be 80% to 90%. However, in the case of water drive gas reservoirs, recovery efficiencies could be in the 50% to 60% range because of the bypassed gas and high reservoir pressures. Recovery from bottom water drive would be substantially affected by a water coning problem.

Recovery efficiencies of oil reservoirs worldwide are presented graphically in chapter 10. The global average of recovery factor is found to be in the range of 35% or slightly less.

Reservoir heterogeneities

Reservoir heterogeneity is largely dependent upon the depositional environment and subsequent events, along with the nature of the sediments involved. The variation in rock properties with evaluation is primarily because of differing depositional environments in time sequence. In sandstone reservoirs, the development of rock properties, e.g., porosity and permeability depends on the nature of the sediment, the depositional environment, and subsequent compaction and/or cementation. The development of rock properties in carbonates may occur similarly to sandstones, along with development as a result of processes related to solution or dolomitization, etc. In general, carbonate formations are found to exhibit various heterogeneities. These include significant variations in permeability that lead to more common occurrences of high permeability streaks. In addition, faulting and fracturing may occur in both types of rocks, leading to more complex reservoir heterogeneities.

The heterogeneities encountered in a reservoir affect the design, implementation, and performance of waterflooding and other recovery operations. Areal and vertical heterogeneities are determined by a combination of geologic, rock and fluid, and logging and coring analysis. In addition, well testing and production/injection performance analysis are also used. The heterogeneities are eventually integrated into geosciences and simulation models of the reservoir. The presence and direction of fractures critically affect the performance of secondary and tertiary recovery efforts. Therefore, their characterization is absolutely necessary early in the life of the reservoir, preferably during primary production.

The variation of vertical permeability is classically described by the Dykstra-Parsons permeability variation factor. It is based upon the lognormal permeability distribution and statistically is defined as follows:[3]

$$V = \frac{k - k_\sigma}{k} \qquad (8.1)$$

where
k = mean permeability (i.e., permeability at 50% probability), and
k_σ = permeability at 84.1% of the cumulative sample.

The permeability variation ranges from 0 (uniform) to 1 (extremely heterogeneous) in various hydrocarbon-bearing formations. The procedure to compute the Dykstra-Parsons permeability variation factor is given in chapter 16. An example set of data and a plot of permeability versus probability (percentage of total sample having higher permeability) are used in that discussion on waterflood recovery.

Reservoir Performance Analysis Techniques

Commonly used reservoir performance analysis and reserves evaluation techniques and their results include the following:

- Volumetric

- Original hydrocarbon in place (OHCIP)

- Reserves based on the assumption of ultimate recovery

- Decline curve/type curve
  - Reserves
  - Ultimate recovery
  - Reservoir properties

- Classical material balance
  - Original hydrocarbon in place
  - Natural recovery mechanism

- Well test interpretation
  - Reservoir rock properties
  - Geologic heterogeneities affecting reservoir performance
  - Characterization of reservoir boundary

- Mathematical simulation of the reservoir model
  - Original hydrocarbon in place
  - Reserves
  - Ultimate recovery
  - Well and reservoir performance prediction under various scenarios, including drilling of new and recompleted wells, and during secondary or tertiary recovery

Another categorization stems from the approaches utilized in analysis. While certain methods are deterministic, others are probabilistic. Due to inherent uncertainties associated with petroleum reservoirs, the engineer seeks to determine a range of values for a parameter with various probabilities attached to each value, rather than a unique answer. Probabilistic approaches are widely practiced in basin exploration, reserve estimation, and economic analysis, among other uses.

These techniques and approaches will be illustrated in the following chapters, often aided by computer-based tools. The accuracy of a reservoir performance analysis is dictated by the depth of understanding of the reservoir characteristics and associated flow dynamics, i.e., the reservoir model. The accuracy of the analysis is also dictated by the validity of the techniques used under various assumptions.

Table 8–1 presents the applicability and accuracy of the various techniques at the different stages of the reservoir life cycle.

Table 8–1. Applicability and accuracy of techniques. *Source: Satter, A., J. Baldwin, and R. Jespersen. 2000. Computer-Assisted Reservoir Management. Tulsa: PennWell.*

	Applicability and Accuracy of Techniques			
	Volumetric Estimate	**Decline Curve**	**Material Balance**	**Reservoir Simulation**
Exploration	Yes?[a]	No	No	No
Discovery	Yes?	No	No	No
Delineation	Yes?	No	No	Yes[a], Fair
Development	Yes, Fair	No	No	Yes, Good
Production	Yes, Good	Yes, Fair	Yes, Fair to good	Yes, Good to very good

[a] Some data based on geophysical, geological, and regional trends is required.

Table 8-2. Comparison of reservoir performance analysis techniques

Comparison of Reservoir Performance Analysis Techniques				
Data Requirements	**Volumetric Estimate**	**Decline Curve**	**Material Balance**	**Reservoir Simulation**
Geometry	A, h	No	No	A, h
Rock	ø, S	No	No	ø, S, K_r, C
Fluid	B	No	PVT (homogeneous reservoir model)	PVT (heterogeneous reservoir model)
Well	No	No	No	Locations, perforations, PI
Production and injection	No	Production only	Both	Both
Pressure	No	No	Yes	Yes
Results				
Original hydrocarbon in place	Yes	Yes[a]	Yes	Yes
Ultimate recovery	Yes[a]	Yes	Yes	Yes
Rate versus time	No	Yes	Yes, with productivity index data	Yes
Pressure versus time	No	No	Yes, with productivity index data	Yes

[a] If the recovery factor can be estimated by other methods or reasonably assumed.

The volumetric method is applicable during development and production phases with fair accuracy.

The decline curve method is applicable only during the production phase when an identifiable trend in production decline is established. Accuracy in decline curve analysis can improve when more data is available.

The material balance method, assuming a homogeneous formation, can give good results during the production phase. However, the technique requires pressure data at various stages of production.

Mathematical simulation of fluid flow incorporating known reservoir heterogeneities is the preferred method to predict reservoir performance during delineation, development, and production. The results can vary from fair to good, or even very good, accuracy. However, this method is generally resource intensive.

Table 8–2 presents data requirements and results of the various techniques. The decline curve method requires only production data, giving an estimate of ultimate recovery, which can be used to estimate original hydrocarbon in place and recovery efficiency. However, the method is applicable only when the production rate is declining with time. The volumetric method needs data on geometry, rock properties, and limited fluid properties. This technique can give an estimate of the original hydrocarbon in place, which can be used to estimate ultimate recovery with estimated recovery efficiency. Both material balance and reservoir simulator methods require data on geometry, rock and fluid properties, production/injection, and pressure. This data is needed to estimate original hydrocarbon in place and ultimate recovery. Since the reservoir simulator method considers reservoir heterogeneity, the results are expected to be more accurate.

Sensitivity of reservoir and fluid properties affecting production performance

Using the reservoir simulator of Gemini Solutions, Inc., an in-depth study was made to investigate the influence of several properties on primary oil recoveries of an example reservoir. The approach to the sensitivity study was to build a base case and then vary specific parameters to analyze the recovery performance. The properties of the base case are shown in Table 8–3. Numerical results are presented there.

Table 8-3. Properties of the example reservoir

Properties of the Example Reservoir[a]

Drainage area, acres	40
Depth, ft	5,332
Total net thickness, ft	25
Average porosity, %	16.6
Average permeability, mD	24.2
Dykstra-Parsons permeability variation factor[b]	0.5
Initial oil saturation, % PV	78
Critical gas saturation, % PV	2
Reservoir temperature, °F	123
Initial reservoir pressure, psia	2,332
Bubblepoint pressure, psia	1,855
Bottomhole production pressure, psia	150
Oil gravity, °API	33
Gas specific gravity (air = 1)	0.67
Economic oil production rate, bopd	10
Computed original oil in place, Mstbo	843.4

[a] PVT data for fluid properties and relative permeabilities for water-wet sand are derived from standard correlations in the simulator.

[b] Defined in chapter 4 and in this chapter.

Figure 8–3 shows simulated oil recovery results, including pressure versus recovery efficiency for liquid and rock expansion and solution gas, gascap, and gravity drainage drive mechanisms. The results show that performances of gas drive and gravity drainage drives are strikingly different from the ideal case shown in Figure 8–1. For example, as compared to the gas drive case, the performance of the gravity drainage case is lower than shown in Figure 8–1. The point is that performances under various drive mechanisms will be influenced by specific conditions, i.e., rock and fluid properties, and the size and strength of the gas cap and aquifer.

Figure 8–4 shows pressure versus oil recovery efficiency under depletion drive for a homogeneous reservoir. Production above the bubblepoint is due to rock and fluid expansion. Pressure drops sharply and continuously from the initial pressure to the bubblepoint, and then it drops slowly and continuously under depletion drive.

Figure 8–5 shows the gas/oil ratio versus oil recovery efficiency under depletion or solution gas drive for a homogeneous reservoir. Initially, the gas/oil ratio remains low or drops even lower until the critical gas saturation is reached. It then rises to a maximum and drops.

Figure 8–6 shows pressure versus oil recovery efficiency as influenced by oil gravity from 10° to 35°API (highly viscous oil to light oil) on primary recoveries. Recovery efficiency is relatively less with lower API gravity oils. In case of 10° API oil, reservoir pressure declines relatively quickly, and recovery is poor.

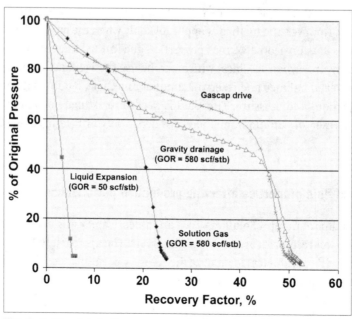

Fig. 8-3. Simulated oil recovery results for various drive mechanisms.
Courtesy of Gemini Solutions, Inc.

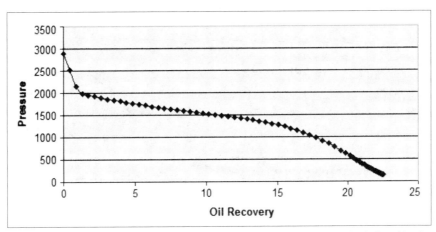

Fig. 8–4. Pressure (psia) versus oil recovery efficiency (percent) under depletion drive for a homogeneous reservoir. *Courtesy of Gemini Solutions, Inc.*

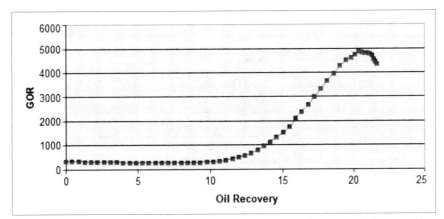

Fig. 8–5. Gas/oil ratio versus oil recovery (percent) under depletion or solution gas drive. Significant increase in the gas/oil ratio is observed at the wells once the bubblepoint is reached. When evolution of volatile components peaks, a decline in the gas/oil ratio is observed. *Courtesy of Gemini Solutions, Inc.*

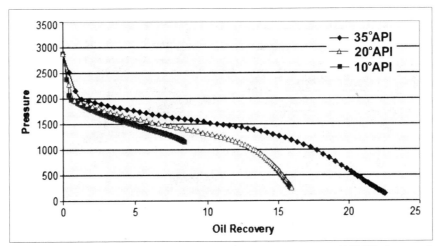

Fig. 8–6. Effect of oil gravity on primary recovery efficiency. In the case of 10° API oil, reservoir pressure declines relatively quickly, and recovery is poor. Oil recovery is plotted as percent of the total hydrocarbon volume in the pertaining figures appearing in this section. *Courtesy of Gemini Solutions, Inc.*

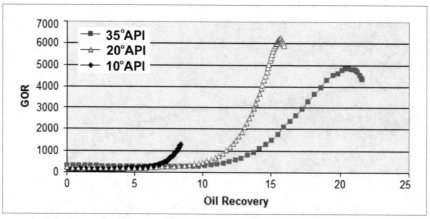

Fig. 8-7. Effects of oil gravity on the gas/oil ratio. Relatively light oil (35°API) indicates the highest recovery due to solution gas drive and low viscosity. *Courtesy of Gemini Solutions, Inc.*

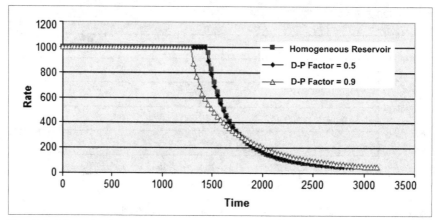

Fig. 8-8. Oil production rate (stb/d) versus time (d) as affected by the Dykstra-Parsons permeability variation factor (V). *Courtesy of Gemini Solutions, Inc.*

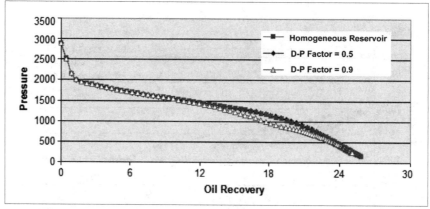

Fig. 8-9. Pressure versus oil recovery efficiency as affected by the Dykstra-Parsons permeability variation factor (V). *Courtesy of Gemini Solutions, Inc.*

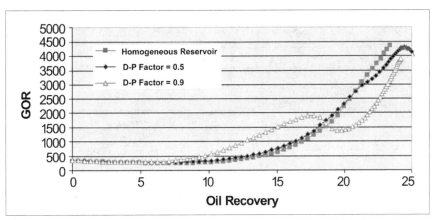

Fig. 8–10. Gas/oil ratio versus oil recovery efficiency as affected by Dykstra-Parsons permeability variation factor (V). In a highly heterogeneous formation, gas evolution occurs rather early, accompanied by a decline in oil rate as shown in Figure 8–8. *Courtesy of Gemini Solutions, Inc.*

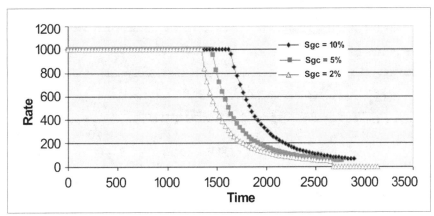

Fig. 8–11. Oil production rate versus time as affected by critical gas saturation. At low critical saturation (2%), evolved gas in the reservoir becomes mobile relatively early, leading to the decline in oil rate. The time scale represents the number of days in production. *Courtesy of Gemini Solutions, Inc.*

Figure 8–7 shows gas/oil ratio versus oil recovery efficiency as influenced by oil gravity from 10° to 35°API (highly viscous oil to light oil) on primary recoveries. The rise in the gas/oil ratio occurs earlier with lower gravity oil. Also, the peak gas/oil ratio is lower.

Figure 8–8 shows the oil production rate versus time as affected by the Dykstra-Parsons permeability variation factors of 0, 0.5, and 0.9. There is a longer time of constant production rate in the case of a homogeneous reservoir with a Dykstra-Parsons permeability variation factor of 0. A shorter time of constant production rate and longer time needed to reach a prescribed economic production rate is observed in the case of Dykstra-Parsons permeability variation factor = 0.9.

Figure 8–9 shows pressure versus oil recovery efficiency as affected by Dykstra-Parsons permeability variation factors of 0, 0.5,and 0.9. No significant variation is observed even with the heterogeneous reservoirs.

Figure 8–10 shows the gas/oil ratio versus oil recovery efficiency as affected by Dykstra-Parsons permeability variation factors of 0, 0.5, and 0.9. Earlier gas/oil ratio increase is observed in the case of the most heterogeneous formation. However, the peak of the gas/oil ratio is lower.

Figure 8–11 shows oil production rate versus time as affected by critical gas saturations of 2%, 5%, and 10%. A longer, constant production rate in the case of the 10% critical gas saturation is observed. As gas becomes mobile at lower saturations in the reservoir, oil production declines earlier.

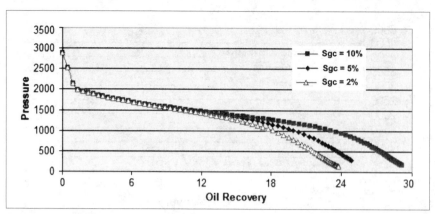

Fig. 8–12. Pressure versus oil recovery efficiency as affected by critical gas saturation
Courtesy of Gemini Solutions, Inc.

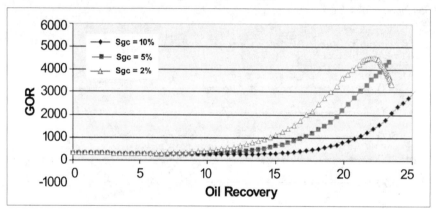

Fig. 8–13. Gas/oil ratio versus oil recovery efficiency as affected by critical gas saturation.
Courtesy of Gemini Solutions, Inc.

Figure 8–12 shows pressure versus oil recovery efficiency as affected by critical gas saturations of 2%, 5%, and 10%. Pressure and recovery efficiency are both higher in the case of the 10% critical gas saturation. When the pressure is higher, the reservoir can be allowed to go somewhat below the bubblepoint, where the oil viscosity is the lowest. This way waterflooding can be initiated without sacrificing recovery efficiency.

Figure 8–13 shows gas/oil ratio versus oil recovery efficiency as affected by critical gas saturations of 2%, 5%, and 10%. A lower gas/oil ratio and higher recovery efficiency are observed in the case of the 10% critical gas saturation.

Finally, the value of reservoir simulation and the effects of reservoir heterogeneity are highlighted in the following study. Two cases are considered. In the base case, no vertical communication exists between adjacent layers, while in the other case, reservoir fluids are allowed to flow in the vertical direction. In both cases, the reservoir is comprised of four layers, with decreasing layer permeability from top to bottom. Performance of the two reservoirs is evaluated under water injection (secondary recovery). Simulation studies are conducted routinely during primary recovery in order to efficiently design improved oil recovery (IOR) projects. The following contrast in performances was observed in the results of the simulation:

- Ultimate recovery is greater in the absence of crossflow, whereas recovery is better at the initial stages in the case where vertical communication exists (figure 8–14).
- In the other case, free gas percolates to the top of the reservoir and is blown down, which results in a much lower gas/oil ratio throughout production during waterflooding (figure 8–15).

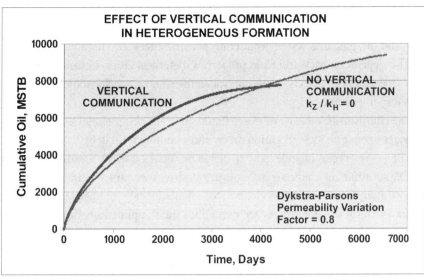

Fig. 8–14. Plot of cumulative oil recovered versus time showing sensitivity to crossflow between layers. The Dykstra-Parsons permeability variation factor = 0.8. For the reservoir considered, recovery is significantly greater when there is no crossflow between layers. However, the reservoir with crossflow initially recovers more oil. *Courtesy of Gemini Solutions, Inc.*

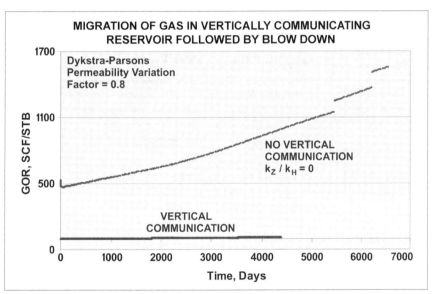

Fig. 8–15. Plot of gas/oil ratio versus time. When crossflow exists between layers, the free gas phase rises to the top of the formation and is subjected to blow down. Hence, the gas/oil ratio is significantly lower in the case where vertical communication exists. *Courtesy of Gemini Solutions, Inc.*

Based on the contrasting reservoir performance in the presence or absence of crossflow, field development and management strategies will not be the same. Moreover, economic analyses may lead to significantly different conclusions. Simulation studies are treated in detail in chapter 14, where four different scenarios are evaluated. The Dykstra-Parsons permeability variation factor and waterflooding are described in chapter 16.

Summing Up

Reservoir pressure and oil, gas, and water production performances are dictated by geological and geophysical characteristics. They also depend on rock and fluid properties, heterogeneities, natural production mechanisms, and fluid flow mechanisms. Operators can exercise some control on the reservoir performance through professional reservoir management, which is very important.

Natural producing mechanisms in oil reservoirs are liquid and rock compressibility, solution gas or depletion drive, gascap drive, aquifer water drive, gravity segregation drive, and a combination drive.

Drive mechanisms in gas reservoirs include gas expansion or depletion drive, aquifer water drive, and combination drive. The following table summarizes characteristics of various drive mechanisms in a reservoir.

Reservoir performance analysis and reserves evaluation techniques include volumetric, decline curve, material balance, and reservoir simulation. Careful considerations are needed for their applications through the different stages of the reservoir life cycle.

Reservoir production performance is sensitive to reservoir rock and fluid properties, in addition to drive mechanisms. Examples are presented to demonstrate the effects of oil gravity, critical gas saturation, and heterogeneity.

Reservoir simulation may lead to very different predictions in performance, which in turn may lead to different reservoir development and management strategies.

Table 8-4. Summary of primary drive characteristics. *Source: A. Satter and G. C. Thakur. 1994. Integrated Petroleum Reservoir Management—A Team Approach. Tulsa: PennWell.*

Primary Drive	Reservoir Pressure	Gas/Oil Ratio	Water Production	Primary Recovery[a]	Comments
Liquid expansion and pore volume reduction	Declines rapidly and continuously	Remains low and constant, with no gas liberated in the reservoir	None, except where part of the formation water is mobile	1%–5% Average 3%	Reservoir pressure is above the bubblepoint
Solution gas	Declines rapidly and continuously	At first low, then rises to a peak, and then drops	None, except where part of the formation water is mobile	10%–25% Average 16%	Reservoir pressure is below the bubblepoint
Gas cap (underlain by oil)	Falls slowly and continuously	Rises continuously in crestal or updip wells	Negligible	15%–35% Average 25% or more	
Water influx	Remains relatively high and is sensitive to oil, gas, and water production	Remains low if reservoir pressure remains high	Early water production in downdip wells. Water production rises rapidly	35%–80% Average 50%	Apparently high OHCIP calculated by material balance method when aquifer influx is ignored
Gravity drainage	Declines rapidly and continuously	Remains low in updip wells and high in downdip wells	Negligible	30%–80% Average 60%	Favorable recovery where permeability > 200 mD, formation dip > 10°, and oil viscosity < 5 cp

[a] Recovery is strongly influenced by reservoir heterogeneities, in addition to drive mechanism.

Class Assignments

Questions

1. Which major factors govern primary recovery in oil and gas reservoirs?

2. How can operators add more value to the company's assets?

3. What are the natural producing mechanisms in oil and gas reservoirs?

4. Which are the most efficient natural producing mechanisms in oil reservoirs, and under what conditions?

5. Why can natural water drive be good in oil reservoirs but not in gas reservoirs?

6. What are the techniques used for evaluating reservoir performance?

7. Are these techniques always applicable throughout the reservoir cycle? Are all the techniques necessary in evaluating a reservoir?

8. Which of the techniques appears to be more reliable, and why? Discuss the requirements of reservoir data for each application. Which technique is the most resource intensive?

9. In a reservoir under combination drive, can the relative contributions of drive mechanisms vary with production? Why or why not?

10. Are the learning objectives of this chapter fulfilled? Are there any topics are missing that should have been covered in this chapter?

Exercises

8.1. Based on a literature review, prepare case studies describing in detail the performance of the following:

   (a) Volatile oil reservoir under depletion drive

   (b) Heavy oil reservoir under gravity drainage

   (c) Gas reservoir with strong aquifer influx

   (d) Offshore field under combination drive

   Include in the description the size of the reservoir, field development and monitoring plan, number of wells, and expected recovery. Describe any plan for augmenting recovery once the reservoir is depleted of natural energy.

8.2. Distinguish between gas cap drive and solution gas drive in a black oil reservoir. Plot the likely scenarios of producing GOR and oil rate over time for the two cases and compare, given all other factors be the same. Would the analysis be any different in case of a volatile oil reservoir? Explain.

8.3. A drill-stem test conducted in an exploratory well has indicated a very good potential for oil production. Is it possible to determine the primary drive mechanism of the reservoir at this point? List the reservoir, fluid and other data that may be valuable in order to conduct a study.

References

1. Clark, N. J. 1969. *Elements of Petroleum Reservoirs.* Henry L. Doherty Series. Dallas: Society of Petroleum Engineers (of AIME).
2. Satter, A., and G. C. Thakur. 1994. *Integrated Petroleum Reservoir Management—A Team Approach.* Tulsa: PennWell.
3. Dykstra, H., and R. L. Parsons. 1950. The prediction of oil recovery by waterflooding. *In Secondary Recovery of Oil in the United States.* 2nd ed. New York: American Petroleum Institute, 160–174.

9 · Volumetric Methods in Petroleum Reservoir Analysis and Applications

Introduction

Volumetric estimation of the original oil and gas in place (OOIP, OGIP) is the first and foremost reservoir engineering function. A wide variety of data is needed to accurately determine the original oil in place and original gas in place, including geologic, geophysical, petrophysical, and reservoir rock and fluid data.

The purpose of the volumetric method is to quantify the hydrocarbon volumes in place. The most important end products of the reservoir engineers' efforts are ultimate recoveries and reserves, which can be estimated as follows:[1]

- Estimating oil or gas recoveries by multiplication of the original hydrocarbon in place (OHCIP) and estimated recovery factors.
- Determining the oil recovery efficiency factor by using empirical analogy, or better yet, API correlations (chapter 10). The performance of producing oil and gas fields located in similar geologic settings aids in recovery estimates. Again, ultimate recovery could be greater with technological innovations (chapter 19).
- Making production rate versus time forecasts based on the decline curve method or reservoir simulation (chapters 11 and 13).
- Finally, making reserves estimates by economic analysis (chapter 18).

The original oil and gas in place can be also determined by classical material balance (chapter 12) and reservoir simulation techniques (chapter 13). These techniques are more involved than the relatively simple volumetric method. However, the other methods mentioned above require dynamic reservoir data. It is a good practice to compare the answers from the different techniques, and resolve the discrepancies by using the expertise of a geosciences and engineering team. The results obtained by the three techniques should be checked against each other and validated for greater accuracy.

This chapter is devoted to the following:

- Learning the fundamentals of the deterministic and probabilistic methods used to estimate the original oil in place and original gas in place, and understanding their applications and limitations
- Knowing the variables and sources of data involved in making calculations
- Learning to calculate the original oil in place and original gas in place
- Working example problems and class exercises

Fundamentals of Volumetric Methods

The volumetric method for estimating the original hydrocarbon in place is given by the following relationship:

$$\text{OHCIP} = \frac{\text{Bulk Volume (Area} \times \text{Thickness)} \times \text{Porosity} \times \text{Saturation}}{\text{Formation Volume Factor}} \qquad (9.1)$$

It is based upon the bulk volume of the reservoir (product of the area and thickness), porosity (void space containing fluids), and the initial fluid saturation. It also depends on the formation volume factor, which is a factor to convert the reservoir volume of the fluid to standard conditions. This equation to calculate the original hydrocarbon in place, based on reservoir and fluid characteristics, was introduced previously in chapter 2.

Variables and sources of data

Satter, Baldwin, and Jespersen list the necessary data that must be integrated in order to perform a volumetric estimate of original hydrocarbon in place:[2]

- Field and lease maps
- Structure and isopach maps
- Openhole and cased-hole logs
- Core analysis
- Porosity and permeability, with cutoff values
- Elevations of oil/water and gas/water contacts
- Water saturation, with cutoff values
- PVT properties of reservoir fluids, including formation volume factor and solution gas/oil ratio

Area, thickness, oil/water contact, and gas/oil contact data is gathered from structure and isopach maps, well logs, seismic studies, and core analysis (chapter 3).

Porosity and saturation with cutoffs are obtained from well logs and core analysis (chapter 2). The cutoffs should be determined by considering the minimum effective formation permeability to allow commercial production.

Formation volume factor data is obtained from laboratory tests and correlations (chapter 3).

Associated Oil Reservoir

Oil reservoir with a gas cap

For an oil reservoir with an associated gas cap, the volumetric equation to estimate the original oil in place is given below (Equation 2.95 from chapter 2):

$$OOIP = \frac{7758 \, A \, h \, \phi \, (1 - S_{wi})}{B_{oi}}$$

where

A = reservoir area, acres,

h = oil zone thickness above the transition zone, ft,

ϕ = reservoir porosity, faction,

S_{wi} = irreducible or connate water saturation, fraction of pore volume, and

B_{oi} = initial oil formation volume factor, rb/stb.

This equation is frequently referred to in the literature as the stock-tank barrels of oil initially in place (STOIIP) equation.

The transition zone extends from above the 100% water saturation to the irreducible water saturation. This height depends upon the reservoir rock and fluid properties. If the transition zone is substantial, oil can be produced from an upper portion of the transition zone with an acceptable water cut. In that case, the original oil in place of the producible transition zone can be calculated using average thickness, oil saturation, and porosity of that producible transition zone.

Solution gas in the original oil:

$$G_{si} = NR_{si} \tag{9.2}$$

where

G_{si} = solution gas in place, scf, and

R_{si} = initial solution gas/oil ratio, scf/stb.

Gas in place in the gas cap:

$$G_{GC} = mNR_{si} \qquad\qquad\qquad (9.3)$$

where
G_{GC} = gas in gas cap, scf, and
m = volume of gas cap/volume of oil zone.

Oil reservoir without a gas cap

The original oil in place and solution gas in the oil are calculated by using Equations 9.2 and 9.3, respectively. Since there is no gas cap, the gas in place in the gas cap is zero (m = 0).

Example 9.1. Using the data given below, calculate the original oil in place and gas in solution. What would be the maximum initial reservoir pressure expected in this case?

Drainage area, acres, = 160.
Average oil zone net thickness, ft, = 50.
Average porosity, %, = 30.5.
Initial oil saturation (fraction) = 0.78.
Oil specific gravity, °API, = 33.
Gas specific gravity = 0.66. (Air = 1.)
Solution gas/oil ratio at b.p., scf/stb, = 385.
Size of gas cap = 0.15.
Reservoir temperature, °F, = 170.

Solution: The bubblepoint is first calculated based on the Standing correlation presented in chapter 3):
p_b = 1,960.5 psi

Since a free gas cap is present, the initial reservoir is either at or below the bubblepoint. The formation volume factor is obtained by Equation 3.69:
B_o = 1.241 rb/stb

Using Equation 2.96 in chapter 2, the original oil in place is estimated as follows:
OOIP = 11.9 MMstb

Gas in solution is calculated based on Equation 9.2:
G_{si} = $(11.9 \times 10^6$ stb) × (385 scf/stb)
= 4.58 bcf

Finally, the volume of gas in the gas cap is determined by Equation 9.3:
G_{GC} = 0.15 × 4.58
= 0.687 bcf

Gas Reservoir

Unassociated gas reservoir

For an unassociated gas reservoir (i.e., without an oil zone), the volumetric equation to calculate the original gas in place, as given in chapter 2, is as follows:

$$G = \frac{7758\, A\, h\, \emptyset\, S_{gi}}{B_{gi}} \tag{9.4}$$

where

h = gas zone thickness above the gas/water contact,
S_{gi} = initial gas saturation $(1 - S_{wi})$, and
B_{gi} = initial gas formation volume factor, rb/scf.

Example 9.2. Using the reservoir data from Example 9.1, calculate the original gas in place (OGIP) in a dry gas reservoir. The following data is available related to fluid saturation and PVT properties:

Initial gas saturation, % pore volume (PV), = 78.

Initial gas formation volume factor, rb/scf, = 0.00123.

Solution: An example showing estimation of the initial gas in place is shown in chapter 2. Similarly, the following can be obtained:

OGIP = 12 bcf

Calculation Methods

Methods used to calculate both stock-tank barrels of oil initially in place and original gas in place include the following:

- Deterministic. Based upon a fixed average value for each property, as illustrated previously.
- Probabilistic. Based upon a range of maximum and minimum values for each property.

Normally, a simple deterministic method is used. However, the probabilistic approach is desirable because the reservoir is not homogeneous, and there is an uncertainty in the data. In recent times, the probabilistic approach is being used with increasing frequency.

Deterministic approach

A very simple method is to use average or weighted values of thickness, porosity, saturation, and formation volume factor, and apply these to the drainage area of the reservoir. In the case of a layered reservoir, the original oil or gas in place is summed for the various layers. However, better results can be obtained by using Ah (isopach), Ahø (isovol), or even AhøS (hydrocarbon pore volume, isoHCPV) maps. Computerized mapping programs are commonly available to produce isopach maps and even to calculate original hydrocarbon in place. This can save a great deal of time and can enhance accuracy of the results. Computer-generated maps should be carefully checked to ensure geological consistency.

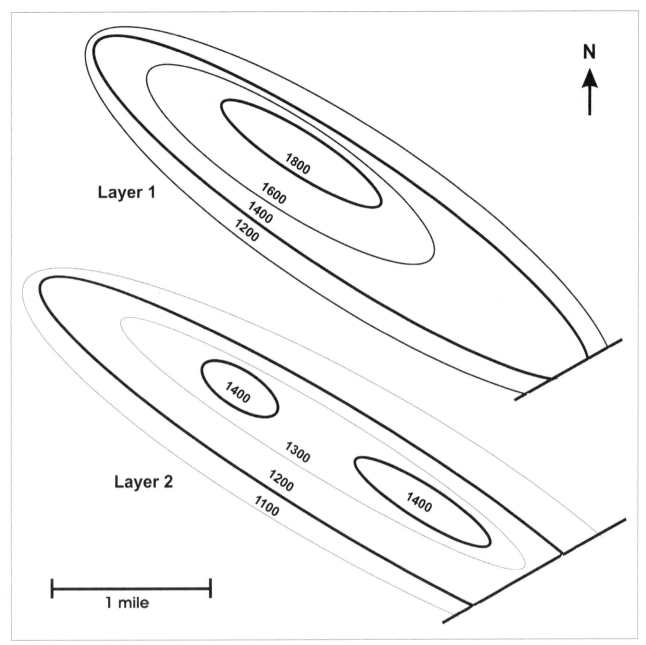

Fig. 9–1. Example isoHCPV map of two layers in an oil reservoir with a sealing fault at the southeast flank. Contours, in barrels per acre-feet, are based on porosity and saturation information obtained from dozens of wells. Contour values are calculated as: $7{,}758 \, \varnothing \, (1 - S_{wc})$.

The more frequently used method involves creating separate contour maps for each input variable with control points. Each map includes a common grid pattern. Overlays of individual variable contour maps provide accurate interpolations for each grid block. Calculated values of h, $h\varnothing$, $h\varnothing S_w$, $h\varnothing S_o$, and $h\varnothing S_g$ may then be contoured as isopachs, isovols, isoHCPV, etc. (fig. 9–1).

Input data for the net oil or net gas pay zones comes from core and well log analyses. The net sand isopach maps can be developed when there are sufficient wells available for analysis. A general rule of thumb would be that 5–10 wells (nicely dispersed) are the minimum number needed to make a reliable map. The more wells that are available, the better the isopach map, which will better reflect the reservoir characterization.

The determination of net pay is complicated by a number of factors:

- **Directional drilling.** A directionally drilled well needs to have log-measured depths converted to true vertical thickness, which accounts for the well deviation in addition to the dip and strike of the formation.

- **Inadequate porosity and permeability.** A porosity cutoff value is needed for determining net pay. There must be sufficient effective porosity with formation permeability so that the production rates of oil or gas, or both, are at commercially viable levels.

- **Fluid contacts.** Location of fluid contacts could differ from well to well as a result of varying transition zones due to the lack of hydrodynamic equilibrium or sealing faults, etc.

Examples of the volumetric estimates of oil and gas shown earlier were based on single values of reservoir properties. In reality, these estimates are based on a number of wells present in the reservoir. A large field may have hundreds of wells, and a multitude of reservoir data obtained from these wells is utilized in volumetric calculations. Data obtained from the dry wells is also incorporated to have better control on reservoir thickness, fluid saturation, and other properties.

Volumetric estimates of the hydrocarbon in place were traditionally performed with the aid of a planimeter. This was used to measure the area under each contour on a map of the volume of hydrocarbon per unit area (bbl/ft^2 or bbl/acre). With the advent of digital computing, such tasks are carried out with relative ease within a short period of time. Relevant data obtained from each well is entered, including the well location, depth, dip, formation thickness, water saturation, and the oil formation volume factor, among others. This leads to the development of a contour map of the oil volume. The plot is similar to what is shown in Figure 9–2, where the contours are in barrels per acre (bbl/acre) or similar units. Next, the total area under each contour is determined. A suitable numerical algorithm, such as Simpson's rule or the pyramidal rule, can be employed to calculate the total volume of hydrocarbon in place.

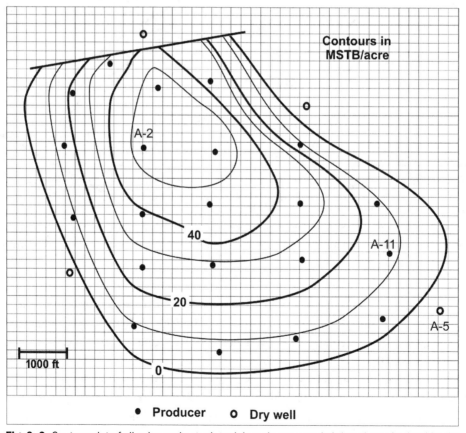

Fig. 9–2. Contour plot of oil volumes in stock-tank barrels per acre (stb/acre) as obtained by the petrophysical studies conducted for wells in the reservoir

According to Simpson's rule, a definite integral of $f(x)$ can be numerically integrated between limits a and b, provided the values of the function at intermediate points are known:

$$\int_a^b f(x) \, dx \approx \frac{1}{3} \left[f(x_0) + 4f(x_1) + 2f(x_2) + 4f(x_3) + 2f(x_4) + \dots + 4f(x_{n-1}) + f(x_n) \right] \Delta x \qquad (9.5)$$

where

$x_i = a + i \, \Delta x$,

$\Delta x = \dfrac{b-a}{n}$, and

n is an even integer.

The pyramidal rule uses the following equation for n number of contours:[3]

$$V = \left(\frac{\Delta Z}{3} \right) \left[A_1 + 2A_2 + 2A_3 + \dots + 2A_{n-1} + A_n + (A_1 A_2)^{0.5} + (A_2 A_3)^{0.5} + \dots + (A_{n-1} A_n)^{0.5} \right] \qquad (9.6)$$

where

V = volume of oil, bbl,

ΔZ = contour interval, and

A_n = area enclosed by the nth contour.

The following table (table 9–1) shows a few examples of the pertinent well information used to develop Figure 9–2.

Table 9–1. Well data used in hydrocarbon volume determination

Well number	Location (x,y)	Porosity, fraction	Thickness, ft	S_w, fraction	N, Mstb/acre	Comment
A-2	18,27	0.28	35	0.18	52.39	Crestal
A-11	43,17	0.22	13	0.35	12.12	Peripheral
....	....	....	....	....	....	.....
A-5	48,16	0.18	18	1.00	0.00	Dry

$$N = \frac{7758(0.22)(13)(1-0.35)}{1.19}$$
$$= 12{,}119 \text{ stb/acre}$$

where

$B_o = 1.19$, used in the calculation as based on the PVT studies.

Example 9.3: Determine the total volume of the oil in place based on Figure 9–2. Based on the contour plot, the area enclosed by each contour is obtained by using a planimeter, as follows (table 9–2):

Table 9-2. Area enclosed by contours

Contour in Mstb/acre	Area Enclosed by Contour, acres
0	964.19
10	771.35
20	560.15
30	360.88
40	207.53
50	92.75

The volume of the oil in place is estimated by using the pyramidal rule, as in Equation 9.5, as follows:

$$V = \left(\frac{10}{3} \frac{\text{MSTB}}{\text{acre}} \right) [964.19 + 2(771.35) + 2(560.15) + 2(360.88)$$
$$+ 2(207.53) + 92.75 + (964.19 \times 771.35)^{0.5} + (771.35 \times 560.15)^{0.5}$$
$$+ (560.15 \times 360.88)^{0.5} + (360.88 \times 207.53)^{0.5} + (207.53 \times 92.75)^{0.5} \text{ acres}]$$
$$= 24{,}128 \text{ Mstb (24.13 MMstb)}$$

Besides Simpson's rule and the pyramidal rule, any other suitable numerical integration algorithm can be employed to perform the calculation. If the area under the crestal part of the reservoir is treated as a cone or the top of a sphere, certain corrections can be applied.

Probabilistic approach

There is never enough well, log, and core data available to accurately determine the average input values, such as porosity and fluid saturation. Unlike the deterministic method, which uses average properties of the wells to calculate the original oil or gas in place, the probabilistic approach assigns a range of values for each variable. These values range from a minimum to a maximum, with some statistical distribution to compute the probability of possible answers. Reservoir parameters such as porosity, net thickness, and hydrocarbon saturation are found to fall in certain quantifiable probability distribution patterns. These patterns could be triangular, random, normal, or lognormal, among others. It is a common practice in the industry to utilize a Monte Carlo simulation based on distributions of reservoir properties. This simulation is used in order to generate a large set of values for the sought result, such as the oil in place, and then assign a range of probability values to it. It is recognized from experience that various reservoir parameters used in probabilistic analysis frequently follow lognormal distribution.

It is customary to report predicted answers for the 10%, 50%, and 90% cumulative probabilities. The technique needs a sizeable amount of data, which is not available early in the life of the reservoir.

The probabilistic approach differs from the deterministic method as follows:
- The probabilistic result is a statistically significant distribution of the possible answers.
- An S-curve plot of all the results assigns a probability to each of the possible answers.
- Predicted results are usually reported for the 10%, 50%, and 90% cumulative probabilities.
- The confidence envelope surrounding the probabilities depends on the number of possible answers (i.e., on the number of simulations).
- The analysis typically takes 500 to 10,000 simulations to generate a smooth S-curve.
- Variables are confined to limited ranges, often with a most-likely value.
- Random numbers are used to assign precise values for each input variable.
- A different set of random numbers is used for each successive calculation.

Ironically, it can take less time to develop the input property distributions and perform 10,000 simulations than to calculate the exact (deterministic) result. This is because there are never enough wells, logs, and cores to truly know the exact input values.

Example 9.4. Estimate the original oil in place for the 10%, 50%, and 90% probabilities, if the area, thickness, porosity, and oil saturation from Example 9.1 vary randomly within ±10% of the median value. Assume that the oil formation volume factor varies within ±2%.

Solution: In performing a probabilistic analysis of the original oil in place, @RISK software is used. This software is capable of generating a large number of values of relevant parameters required in original oil in place estimation. The minimum, maximum, and probability distribution pattern of individual parameters need to be known in performing the analysis. Next, values of area, porosity, thickness, saturation, and formation volume factor are generated by the thousands in a spreadsheet. This data is used to calculate a range of probable values of original oil in place, in a method widely known as Monte Carlo simulation. The Monte Carlo simulation software tool is capable of generating values of a parameter when a variety of probability distributions is assigned. Such designated distributions could be normal, lognormal, triangular, uniform, and binomial. Based on field data, the distribution of porosity could be found to be lognormal, while the distribution of formation thickness could be random within a given range. Monte Carlo simulation and typical probability distributions associated with oilfield parameters are described and illustrated in chapter 18.

In this example, 10,000 values of original oil in place were generated based on given parameter ranges and associated probability distribution patterns. The distributions used in the analysis are given in Table 9–3.

Graphical representations of these probability distribution patterns are included in chapter 18.

The results of the Monte Carlo simulation are plotted in Figure 9–3. It was found that 90% of simulated original oil in place values were 9.997 MMstb or greater. This observation leads to the prediction that there is a 90% probability that the oil in place would be 9.997 MMstb or greater. Similarly, the 50% probability line indicates an original oil in place of 11.841 MMstb, suggesting that there are equal chances of discovering oil above or below the figure. Lastly, the plot indicates that there is only a 10% probability of finding an oil volume of 13.878 MMstb or more.

Example 9.5. Do the same calculations for a gas reservoir prospect, if the area, thickness, porosity, gas saturation, and gas formation volume factor from Example 9.2 vary within ±15% of the median value.

Solution: Results of Monte Carlo simulation of probable values of gas in place, based on the above ranges of parameters, yielded results given in Table 9–4.

Table 9-3. Probability distribution of reservoir parameters

Parameter in Original Oil in Place Estimation	Assumed Probability Distribution
Reservoir area	Triangular
Formation porosity	Normal
Thickness	Uniform
Connate water saturation	Lognormal
Initial formation volume factor	Triangular

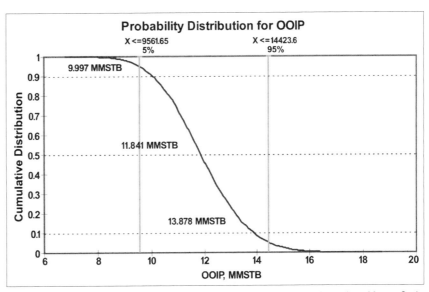

Fig. 9–3. Cumulative probability distribution of original oil in place based on Monte Carlo simulation using @RISK software. *Courtesy of Palisade Corporation*

Table 9-4. Cumulative probability distribution of original gas in place

Probability	Gas in Place, bcf
10%	15.2 or greater
50%	11.87 or greater
90%	9.09 or greater

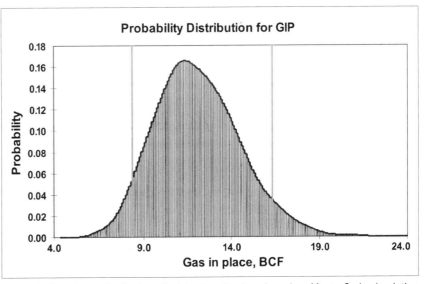

Fig. 9–4. Probability distribution of original gas in place based on Monte Carlo simulation using @RISK software. *Courtesy of Palisade Corporation*

The cumulative probability distribution plot for gas in place shows a familiar S-shaped curve, as in the Monte Carlo simulation of the original oil in place. Figure 9–4 presents the probability distribution of 10,000 simulated values of gas in place.

Summing Up

The volumetric method is used to quantify the original oil and gas in place in a reservoir. The calculations are based on area, thickness, porosity, initial fluid saturations, and formation volume factors. Data is obtained from geologic and seismic maps, well logs, cores, and reservoir rock and fluid data analyses. If laboratory data is not available, correlations can be used.

Deterministic and probabilistic approaches are used to calculate stock-tank barrels of oil initially in place and original gas in place. The deterministic method uses average properties of the wells. Determination of the total oil in place involves drawing a contour map of the oil in place, followed by numerical integration of the area enclosed by the contours. In contrast, the probabilistic approach assigns values for each variable from a minimum to a maximum, with some statistical distribution, to compute the probability of the range of values.

Class Assignments

Questions

1. Why it is so important to quantify the original hydrocarbon in place?

2. List the data needed to calculate the stock-tank barrels of oil initially in place and the original gas in place.

3. What methods are used to determine the stock-tank barrels of oil initially in place and original gas in place?

4. What are the fundamental differences in the approaches for estimating the original hydrocarbon in place?

5. Can an initial estimate of the stock-tank barrels of oil in place or original gas in place change during the life of the reservoir? Why or why not?

6. A dry gas reservoir has been discovered offshore. Based on log studies, the average formation porosity is 0.16, and the connate water saturation is 0.23. Estimate the original gas in place and reserve in standard cubic feet per acre-feet, and the water volume in barrels per acre-feet. In the case of a gas condensate reservoir, are the estimates likely to change? If the formation is tight (< 0.01 mD), describe how the reserve estimations can be affected. Make any assumptions necessary.

Exercises

9.1. In the following, use the data from Example 9.1, with initial reservoir pressure 300 psia higher than the maximum possible pressure in the saturated reservoir:

    (a) Calculate the original oil in place and gas in solution. Draw conclusions from the results obtained. Is the information provided adequate to make the estimates?

    (b) Would a gas cap be expected in this case? Why or why not?

    (c) Reservoirs in the region are known to be overpressured by about 300 psi. Can the depth of the reservoir be estimated? List any assumptions necessary.

9.2. Using the data from Example 9.1, calculate the original gas in place in a dry gas reservoir. Assume that the initial gas saturation is 78% PV.

References

1. Satter, A., and G. C. Thakur. 1994. *Integrated Petroleum Reservoir Management—A Team Approach*. Tulsa: PennWell.
2. Satter, A., J. Baldwin, and R. Jespersen. 2000. *Computer-Assisted Reservoir Management*. Tulsa: PennWell.
3. Towler, B. F. 2002. *Fundamental Principles of Reservoir Engineering*. SPE Textbook Series. Vol. 8. Richardson, TX: Society of Petroleum Engineers.

10 · Empirical Methods for Reservoir Performance Analysis and Applications

Introduction

Oil recovery efficiency can be estimated by using empirical relationships:
- Analogous reservoirs having similar characteristics
- API correlations based on a large number of case studies

These techniques can be useful for approximate estimates but are not accurate enough to make important investment decisions. They should be used with caution, depending upon how well they represent the property in question. Each reservoir is unique in a given geologic setting and may exhibit unique behavior during its producing life. Again, with the advent of new technology, such as multilateral horizontal wells and 4-D seismic studies, petroleum recovery from a specific reservoir may improve notably.

This chapter is devoted to the following:
- Explanation of the techniques related to recovery estimates, and their limitations
- Evaluation of petroleum reserves based on API correlations—deterministic and probabilistic
- Review of a global database on the recovery factor of oil fields
- Class problems

The following examples are provided in the chapter:
- Oil recovery estimate in a sandstone reservoir under solution gas drive
- Predicted performance of an oil reservoir under water drive
- Cumulative distribution of reserves based on a Monte Carlo simulation
- Sensitivity of various rock and fluid parameters to oil recovery depicted by a tornado chart

Furthermore, a plot indicating worldwide recovery trends is included in the chapter to present a global perspective of petroleum reservoir performance.

Analogous Reservoirs

Recoveries from analogous or similar reservoirs are used to estimate the recovery for a subject reservoir. Analogous reservoirs should have the following in common:
- Fluid PVT properties
- Reservoir lithology and characterization
- Nature and intensity of rock heterogeneities
- Reservoir drive mechanisms
- Primary and improved recovery methods
- Field development strategy, including well spacing and patterns
- Reservoir management and production practices

Recovery adjustments may be necessary for the differences between analogous reservoirs and the subject reservoirs.

API Correlations

American Petroleum Institute (API) correlations are based upon 312 case studies of natural depletion and water drive reservoirs.[1]

Depletion or solution drive oil reservoir at bubblepoint

API correlation for recovery efficiency for solution gas drive reservoirs (sands, sandstones, and carbonate rocks) is given by the following equation:

$$E_R = 41.815 \left[\frac{\emptyset(1-S_{wi})}{B_{ob}} \right]^{0.1611} \times \left(\frac{k}{\mu_{ob}} \right)^{0.0979} \times (S_{wi})^{0.3722} \times \left(\frac{P_b}{P_a} \right)^{0.1741} \tag{10.1}$$

where

E_R = recovery efficiency, % original oil in place at bubblepoint,

$\emptyset$ = porosity, fraction of bulk volume,

S_{wi} = interstitial water saturation, fraction of pore space,

B_{ob} = oil formation volume factor at the bubblepoint, rb/stb,

k = absolute permeability, darcies,

μ_{ob} = viscosity of the oil at the bubblepoint, cp,

p_b = bubblepoint pressure, psia, and

p_a = abandonment pressure, psia.

The correlation is based upon the following:
- Bubblepoint and abandonment pressures
- Rock properties such as porosity, permeability, and oil saturation
- Fluid properties such as viscosity and oil formation volume factor at the bubblepoint

Since the correlation is applicable to depletion drive reservoirs, recovery above the bubblepoint needs to be added to determine the total recovery.

Water drive oil reservoir

Recovery efficiency for water drive reservoirs (sands and sandstones) is given as follows:

$$E_R = 54.898 \left[\frac{\emptyset(1-S_{wi})}{B_{oi}} \right]^{0.0422} \times \left(\frac{k\mu_{wi}}{\mu_{oi}} \right)^{0.0770} \times (S_{wi})^{-0.1903} \times \left(\frac{P_i}{P_a} \right)^{-0.2159} \tag{10.2}$$

where

E_R = recovery efficiency, % original oil in place,

B_{oi} = initial oil formation volume factor, rb/stb,

μ_{wi} = initial water viscosity, cp,

μ_{oi} = initial oil viscosity, cp, and

p_i = initial reservoir pressure, psia.

The correlation is based upon the following:
- Initial and abandonment pressures
- Rock properties such as porosity, permeability, and oil saturation
- Fluid properties such as viscosity and oil fluid volume factor at the initial pressure

For both correlations, the constants and exponents in the equations can be adjusted for the reservoirs being evaluated.

Probabilistic analysis of petroleum recovery

The SPE classification of proved, probable, and possible reserves of oil and gas requires probabilistic analysis, whereby each category of reserves is based upon the cumulative probability of occurrence. This is usually accomplished by a Monte Carlo simulation, as described in chapter 18. Since the reserves are directly tied to the recovery factor, a probabilistic approach can be adopted to predict the reservoir performance, an example of which is illustrated later in the chapter.

Oil and gas reserves are described in chapters 2 and 15.

Recovery factor based on worldwide data

Figure 10–1 shows cumulative oil in place on a global basis versus recovery factor based on the ISH database.[2] Field data was ranked for recovery factor with the corresponding original oil in place of the field integrated. A flat portion of the curve at the bottom left of the plot indicates that very few fields have ultimate recovery in the low single digits. The slope of the curve becomes somewhat steep when recovery is greater than 20%, implying that a large number of fields will have a recovery greater than 20%. The curve begins to flatten again beyond 40%, which is an indication that the ultimate recovery beyond 40% will not be as common. It is obvious that there are very few fields with a recovery factor greater than 60%. It is estimated that the worldwide average of recovery factor is about one-third of the original oil in place.

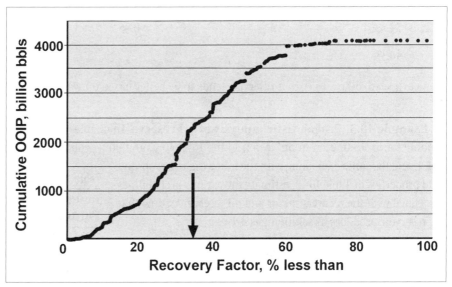

Fig. 10–1. Ranking of recovery factor against oil in place in reservoirs located worldwide. The arrow indicates the percent recovery factor, in the low to mid-thirties, which can be expected on the average. *Source: W. M. Schulte. 2005. Challenges and strategy for increased oil recovery. IPTC Paper #10146. International Petroleum Technology Conference, Doha, Qatar, November 21–23. © IPTC. Reprinted with permission.*

Examples

10.1. Sandstone reservoir with solution gas drive. Using API correlations, calculate oil recovery efficiency below the bubblepoint for a solution or depletion drive sandstone reservoir. Rock and fluid properties are provided in Table 10–1.

Table 10–1. Reservoir data for estimation of recovery

$\phi =$	porosity, fraction of bulk volume	$= 0.22$
$S_{wi} =$	interstitial water saturation, fraction of pore space	$= 0.35$
$B_{oi} =$	initial oil formation volume factor, rb/stb	$= 1.311$
$B_{ob} =$	bubblepoint oil formation vol. factor, rb/stb	$= 1.319$
$\mu_{wi} =$	initial water viscosity, cp	$= 0.5$
$\mu_{oi} =$	initial oil viscosity, cp	$= 1.032$
$\mu_{ob} =$	viscosity of oil at bubblepoint, cp	$= 1.011$
$p_i =$	initial reservoir pressure, psia	$= 3,450$
$p_b =$	bubblepoint pressure, psia	$= 2,805$
$p_a =$	abandonment pressure, psia	$= 500$
$C_o =$	oil compressibility, psi^{-1}	$= 1.24 \times 10^{-5}$
$C_w =$	water compressibility, psi^{-1}	$= 3.00 \times 10^{-6}$
$C_f =$	formation compressibility, psi^{-1}	$= 3.00 \times 10^{-6}$
$k =$	average permeability of reservoir, millidarcies	$= 350$
$R_s =$	initial gas solubility, scf/stb	$= 615$

Solution: Oil recovery efficiency below the bubblepoint is estimated based on Equation 10.1, as in the following:

$$E_R = 41.815 \left[\frac{0.22(1-0.35)}{1.319} \right]^{0.1611} \times \left[\frac{0.35}{1.011} \right]^{0.0979} \times 0.35^{0.3722} \times \left[\frac{2805}{500} \right]^{0.1741}$$
$$= 24.1\%$$

Example 10.2. Sandstone reservoir with water drive. Follow the same process for oil recovery efficiency for a water drive sandstone reservoir when the abandonment pressure is 950 psia. Use the rock and fluid properties from Example 10.1.

Solution: Oil recovery efficiency for a water drive sandstone reservoir is given in Equation 10.2. The following estimate of recovery is obtained:

$$E_R = 54.898 \left[\frac{0.22(1-0.35)}{1.311} \right]^{0.0422} \times \left[\frac{0.35 \times 0.5}{1.032} \right]^{0.0770} \times 0.35^{-0.1903} \times \left[\frac{3450}{950} \right]^{-0.2159}$$
$$= 40.3\%$$

As expected, oil recovery in a reservoir tends to be better based on water drive than on solution gas drive.

Example 10.3. Probabilistic analysis of oil reserve. In chapter 9, the original oil in place was determined by a Monte Carlo simulation in order to find the 10%, 50%, and 90% probabilities associated with the outcome (Example 9.4). Using the same data along with the following reservoir and fluid properties in Table 10–2, estimate the various probabilities associated with the recovery factor and oil reserve. Assume that the reservoir produces by solution gas drive.

Solution: In performing the probability analysis using @RISK software, Equation 10.2 is used to estimate the recovery factor. Finally, oil reserve is calculated by noting the following:

$$N = OOIP \times E_R$$

Table 10–2. Range of parameter values

Reservoir Parameter	Minimum Value	Mean Value	Maximum Value
Permeability, darcies	0.275	0.305	0.336
Oil viscosity, cp	0.618	0.800	0.998
p_b, psia	1,570	1,745	1,920
B_{ob}, rb/stb	1.235	1.252	1.273

During a Monte Carlo simulation, 10,000 values are generated for each of the following: original oil in place, recovery factor, and oil reserve. A cumulative distribution for recovery factor and oil reserve plot shows the familiar S-shaped curve, as in Figure 9–1. The results of simulation are tabulated in Table 10–3.

The table suggests that there is a cumulative probability of 90% to recover 2.6 MMstb of oil. A tornado chart is generated to analyze the sensitivity of the recovery calculation for related reservoir parameters (fig. 10–2). Sensitivity analyses point to the degree of influence of a particular parameter on the outcome when the parameter is varied within the prescribed range during simulation. Based on the correlation used along with ranges of data, recovery calculations are found to be most sensitive to variations in the initial water saturation. Variations in oil formation volume factor, on the other hand, have the least influence on the recovery calculations in this case.

Probabilistic analysis is discussed further in chapter 18.

Table 10–3. Cumulative probability distribution

Cumulative Probability	Recovery Factor, %	Oil Reserve, MMstb
10%	27.82[a]	3.7[a]
50%	26.32[a]	3.13[a]
90%	24.88[a]	2.6[a]

[a] or greater

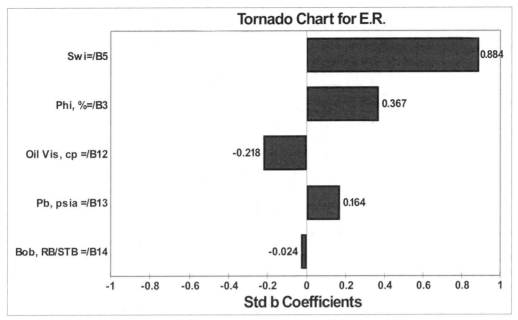

Fig. 10–2. Tornado chart for recovery factor calculation showing sensitivity to various parameters used in correlation. In probabilistic analyses of peteroleum reserves and others, the charts are generated to assess the impact of various parameters on the final outcome. *Courtesy of Palisade Corp.*

Summing Up

Oil recovery efficiency can be estimated by using the following:
- Empirical analogous reservoirs, having similar properties and production practice.
- API correlations for solution gas drive and natural water drive reservoirs. The correlations are based upon rock and fluid properties, and initial and abandonment pressures.
- Worldwide data obtained from oil fields suggests that most common recovery factors range approximately between 20% and 40%, with an average around 34%.

These techniques can be useful for approximate estimates, but they should be used with caution, depending on how well they represent the subject property.

Class Assignments

Questions

1. When should empirical analogous and API correlations be used to estimate recovery efficiency for reservoirs?

2. What are the limitations to these techniques? Which technique (empirical or API) is preferable?

3. What can be done to improve on the API correlations?

Exercises

10.1. Using data from Example 10.1, plot the following to perform a sensitivity study:

(a) Recovery factor versus interstitial water saturation between 0.2 and 0.35.

(b) Recovery factor versus oil viscosity between 1 cp and 2.5 cp.

(c) Recovery factor versus initial reservoir pressure between 2,810 psia and 5,000 psia. Keep bubblepoint pressure constant at 2,805 psia.

(d) Recovery factor versus bubblepoint pressure between 3,405 psia and 2,805 psia. Keep initial reservoir pressure constant at 3,450 psia.

(e) Recovery factor versus abandonment pressure between 100 psia and 500 psia.

(f) Recovery factor versus reservoir permeability between 15 mD and 305 mD.

Draw conclusions from the above study. In (c) and (d), certain oil properties would change as different values of reservoir pressure and bubblepoint pressure are selected for the computation of recovery. Make any assumptions necessary to calculate the fluid properties

References

1. Arps, J. J., F. Brons, A. F. van Everdingen, R. W. Buchwald, and A. E. Smith. 1967. A statistical study of recovery efficiency. *API Bulletin*. D14.
2. Schulte, W. M. 2005. Challenges and strategy for increased oil recovery. IPTC Paper #10146. International Petroleum Technology Conference, Doha, Qatar, November 21–23

11 · Decline Curve Analysis and Applications

Introduction

In earlier days, smart producers realized that they were dealing with depleting petroleum reservoirs, and well rates were bound to decline with production. When sufficient production data is available and production is declining, the past production curves of individual wells by lease or of the field as a whole can be extended to an economic rate (fig. 11–1). The objectives are the prediction of future performance and estimation of reserves. Since the graphical presentations of oil

and gas production rates over time show that the rate would eventually decline with time, the curves are known as "decline curves." Unlike other developments, graphical techniques in predicting the production decline preceded mathematical analysis techniques, which were developed later.

The applicability and accuracy of the decline curve technique through the reservoir life cycle were presented in chapter 8. The accuracy of the analysis improves when sufficient production data is available. Classical decline curve analysis was developed for oil reservoirs exhibiting a decline pattern during primary production. In many cases, the reservoirs were relatively small. However, many other reservoirs do not exhibit a definitive trend in production decline due to the existence of aquifers, geologic heterogeneities, and external fluid injection. These reservoirs require analyses that are more rigorous in nature, including numerical simulation.

Decline curve analysis techniques are used to evaluate reserves annually for more than 95% of the thousands of reservoirs in the United States. Because most of these reservoirs are small, with known drive mechanisms, high technology reservoir simulation techniques requiring significant resources may not be justified economically. Data requirements and results are given in chapter 8. The technique is relatively straightforward, since only the oil production rate and gas/oil ratio and water/oil ratio information over the production time are

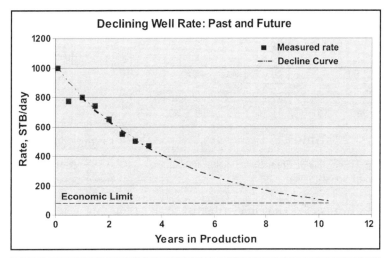

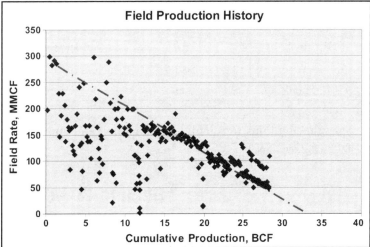

Fig. 11–1. Two views of decline. The first plot shows an early trend in production decline of an oil well, which is extrapolated to well abandonment rate based on a best-fit decline curve. The dotted line predicts future well performance. The second plot depicts the decline rate of an entire gas field (comprised of a large number of wells) throughout the major part of its life cycle, as plotted against cumulative production. A large scattering of field production rates in the early years is partly due to the development of new wells and frequent maintenance operations.

387

needed. The ultimate recovery of a petroleum reservoir can be calculated by adding past and expected future production data. Oil or gas in place can be calculated based on the estimated recovery efficiency in the specific case.

The objectives of this chapter are as follows:
- Learning the fundamentals of decline curve analysis
- Decline curve types and representative mathematical equations
- Assumptions and limitations of decline curve analysis
- Classical and advanced decline curve analysis procedures, including computer-assisted methodology
- Decline curve applications
- Factors affecting well performance during decline
- Working example problems and class exercises

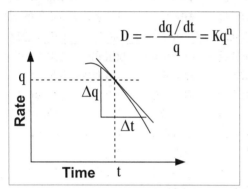

Fig. 11–2. General equation for decline in oil or gas production with time

Fundamentals of Decline Curve Analysis

Decline characteristics depend upon the rate of decline (D) and its exponent (n), as explained below (see fig. 11–2):[1,2]

$$D = -\frac{dq/dt}{q} = Kq^n \tag{11.1}$$

where
- q = production rate, barrels per day, month, or year,
- t = time, day, month, or year,
- K = constant, and
- n = exponent.

The rate of decline in Equation 11.1 can be constant or variable with time, yielding three basic types of production decline characteristics identified in classical analysis as follows:

Type of decline	n	D
Exponential	0	Constant
Hyperbolic	> 0	Variable
	< 1	
Harmonic	1	Variable

Specific equations to calculate production rate, cumulative production, and economic life are given in the following sections.

Decline Curve Equations

The type of decline is determined by the value of n as given above. Standard curve types can be of the following types:
- Exponential, where decline in production rate is a constant percentage.
- Harmonic, when decline is directly proportional to the rate.
- Hyperbolic, when the decline rate (D) varies, and the exponent (n) is more than 0 but less than 1. The hyperbolic curve models production performance in between the harmonic and exponential decline.
- Both the exponential and harmonic decline curves can be regarded as special cases of the hyperbolic decline curve. Equations for production rates (q) and cumulative productions (Q) for the various types of decline curves 1, 2, and 3 are detailed in the following discussion.

Exponential or constant decline

In case of exponential decline of well rate, n is 0 and D is a constant. Hence equation (11.1) can be written as follows:

$$D = -\frac{dq/dt}{q} = K = -\frac{\ln\left(\frac{q_t}{q_i}\right)}{t} \tag{11.2}$$

where

n = 0,

K = constant,

q_i = initial production rate, and

q_t = production rate at time t.

The rate versus time and rate cumulative relationships are given by the following:

$$q_t = q_i\, e^{-Dt} \tag{11.3}$$

$$Q_t = \frac{q_i - q_t}{D} \tag{11.4}$$

where

Q_t = cumulative production at time t.

If Δq is the rate change in the first year of production, then:

$$D' = \frac{\Delta q}{q_i} \tag{11.5}$$

In this case, the relationship between D and D' is given as follows:

$$D = -\ln\left(1 - \frac{\Delta q}{q_i}\right) = -\ln(1 - D') \tag{11.6}$$

For wells under exponential decline, the value of D is a constant and can be determined graphically by plotting the well rate versus cumulative production, or log of rate versus time as shown in Fig. 11–3.

Hyperbolic decline

The decline is proportional to the power n of the production rate, where n is a positive fraction.

$$D = -\frac{dq/dt}{q} = Kq^n \quad (0 < n < 1) \tag{11.7}$$

Note that this is the same as the general decline rate as in Equation 11.1, except for the constraint on n.

In case of wells producing under hyperbolic decline, the value of D changes with time. For the initial condition, however, the following applies:

$$K = \frac{D_i}{q_i^{\,n}}$$

The rate versus time and the rate cumulative relationships are given by the following:

$$q_t = q_i(1 + n\,D_i T)^{-\frac{1}{n}} \tag{11.8}$$

$$Q_t = \frac{q_i^{\,n}(q_i^{\,1-n} - q_t^{\,1-n})}{(1 - n)\,D_i} \tag{11.9}$$

where

D_i = initial decline rate.

Harmonic decline

The decline is proportional to production rate, as in the following:

$$D = -\frac{dq/dt}{q} = Kq \qquad (11.10)$$

where

$n = 1$.

For the initial condition, the following applies:

$$K = \frac{D_i}{q_i}$$

The rate versus time and the rate cumulative relationships are given by the following:

$$q_t = \frac{q_i}{(1 + D_i t)} \qquad (11.11)$$

$$Q_t = \frac{q_i}{D_i} \ln \frac{q_i}{q_t} \qquad (11.12)$$

Well production drops rather slowly with time under harmonic decline, in comparison to exponential or hyperbolic decline, given that all other parameters are the same. Consequently, this type of decline indicates relatively large reserves.

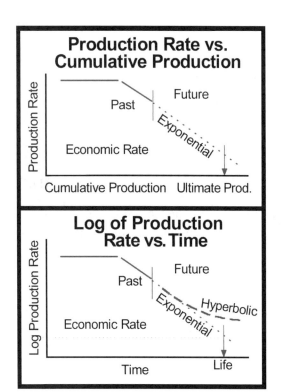

Fig. 11–3. Schematic of exponential and hyperbolic decline. *Source: G. C. Thakur and A. Satter . 1998. Integrated Waterflood Asset Management: PennWell.*

Decline Curves

The commonly used decline curves for oil reservoirs include the following:

1. Log of production rate versus time (fig. 11–3)

2. Production rate versus cumulative production (fig. 11–3)

3. Log of water cut or oil cut versus cumulative production (fig. 11–4)

Additionally, the following special curves can be used for analysis:

4. Oil/water or gas/oil contact versus cumulative production

5. Log cumulative gas production versus cumulative oil production

When type 1 and 2 plots are straight lines, they are called constant rate or exponential decline curves. Since a straight line can be easily extrapolated, exponential decline curves are most commonly utilized wherever applicable.

In the case of harmonic or hyperbolic rate decline, the plots show curvature. The hyperbolic decline pattern is illustrated in Figure 11–3. Both the exponential and harmonic decline curves are special cases of hyperbolic decline curves. Unrestricted early production from a well shows a hyperbolic decline rate. However, a constant or exponential decline rate may be reached at a later stage of production.

Type 3 curves (fig. 11–4) are employed when the economic production rate is dictated by the cost of water disposal or even a change in oil price. A straight-line extrapolation of the log of water cut versus cumulative oil production may not be reasonably applicable in the lower water-cut levels, and it may yield a conservative estimate of reserves. On the other hand, if oil cut data is used instead of water cut in the same levels, straight-line extrapolation of the log of oil-cut versus cumulative oil production may deteriorate. This could lead to overly optimistic reserve estimates.

There can be a special use for the type 4 curves. By tracking the movement of the fluid contact, and plotting the height of the contact, cumulative oil production can be determined from this curve.

If the cumulative oil production is known, a type 5 plot can be utilized to predict cumulative gas production.

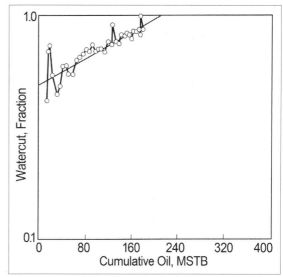

Fig. 11–4. Water cut history of a producing well. *Source: G. C. Thakur and A. Satter. 1998.* Integrated Waterflood Asset Management: *PennWell.*

Assumptions in decline curve analysis

Several important assumptions are involved in decline curve analysis, as given in the following:
- Sufficient production performance data is available, and a declining trend in well rate has been established.
- Ongoing field operations will continue in the future without interruptions.

All of the factors that influenced the curve in the past remain essentially unchanged throughout the producing life.
- Each well or a group of wells drains a constant finite reservoir area, producing essentially at full capacity.
- Flow across the reservoir boundary does not occur, including influences of an adjoining aquifer.
- Depletion is the only driving mechanism. A large gas cap or strong edge water drive or even a weak to moderate bottom water drive can affect the decline curve trends, as demonstrated in certain examples later in the chapter.

Many factors may influence production rates, and thus the decline curve performance, such as the following:
- Proration, or restricted production, which was the most prominent practice in the industry in earlier days
- Changes in bottomhole pressure affecting gas/oil ratio and water/oil ratio; water or gas breakthrough in wells
- Water influx from an adjacent aquifer
- Well treatments, such as stimulation and hydraulic fracturing
- Changes in production methods, such as gas lifting
- Initiation of pressure maintenance, waterflooding, or an enhanced recovery operation
- Workovers, such as perforating and producing from another layer, and recompletion of a vertical or deviated well as a horizontal well
- Pipeline disruptions causing shut downs of the wells, or weather or market conditions, such as hurricanes or the rise and fall of oil prices, adversely affecting production

Therefore, care must be taken in extrapolating the production curves into the future. When the shape of a decline curve changes or erratic fluctuation of the data occurs, the cause needs be determined, and its effect upon the reserves should be evaluated.

Decline curve analysis methodology

The general procedure for analysis is as follows:
- Gather available production performance data.
- Observe plots of the log of oil rate versus time and the log of water cut versus cumulative production.
- Determine reasons for production data anomalies (in both upward and downward directions).
- Evaluate various means of plotting the data.
- Select the portion of the performance data applicable for analysis.
- Perform regression analysis for history matching of past data.
- Forecast future performance with economic oil production rate and water cut.

Advances in decline curve analysis techniques

A review of the literature indicates significant advances in decline curve analysis techniques as follows:
- Ershaghi and Omoregie method using modified water-cut factor
- Fetkovich method using log-log type curves (based upon dimensional rate versus dimensional time for exponent 0 to 1)[3]
- Blasingame type curve method[4,5]
- Agarwal and Gardner type curve method[6]

The first two methods are described in the following sections.

Ershaghi and Omoregie method. Because extrapolation of the past water cut plot is often complicated, Ershaghi and Omoregie devised a method to plot recovery efficiency versus X, as defined below, which yielded a straight line:[7]

$$E_R = mX + n \tag{11.13}$$

where

E_R = overall recovery efficiency,
$X = -\left[\ln(1/f_w - 1) - 1/f_w\right]$, $\tag{11.14}$
f_w = fraction of water flowing,
m = slope, and
n = constant.

This method is designed to be more general than the classical plot of water cut versus cumulative oil production, and more applicable when water cut exceeds 0.5. Given actual water cut versus recovery efficiency data, a graph of recovery versus X would result in a straight line. This may be extrapolated to any desired water cut to obtain the corresponding recovery. The parameters m and n in Equation 11.13 can be derived from the straight-line relationship in Figure 11–5. These values then can be used in Equation 11.14 to predict water cut versus oil recovery in Figure 11–4. Figure 11–4 shows that straight-line extrapolation of water cut would result in pessimistic ultimate recovery.

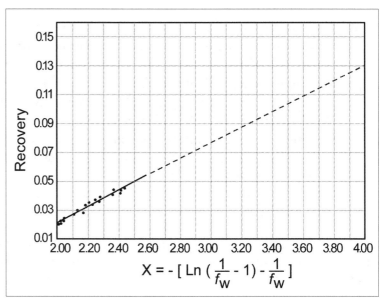

Fig. 11–5. Plot of oil recovery versus X. *Source: A. Satter, J. Baldwin, and R. Jespersen. 2000. Computer-Assisted Reservoir Management. Tulsa: PennWell.*

Fetkovich method. Fetkovich developed rate-time type curves that combined early transient flow of a well with boundary dominated late time flow behavior as represented by Arps' classical equations of decline[8]. The latter flow regime is influenced by reservoir boundary effects. Therefore, the type curves are comprised of two sections that blend both transient and boundary-dominated flow behavior. The transient flow behavior during the early time is based upon dimensionless flow rate (q_D) and dimensionless time (t_D) as encountered in well testing theory. The dimensionless quantities are expressed as in the following:

$$q_D = \frac{141.3 \, q(t) \, \mu \, B}{kh(p_i - p_{wf})} \tag{11.15}$$

$$t_D = \frac{0.00634kt}{\emptyset \, \mu \, c_t \, r_w^2} \tag{11.16}$$

where

$q(t)$ = production rate at a given time (t), stb/d,

μ = fluid viscosity, cp,

B = fluid formation volume factor, rb/stb,

k = permeability, mD,

h = thickness, ft,

p_i = initial reservoir pressure, psia,

p_{wf} = flowing well pressure, psia,

$\emptyset$ = porosity, fraction,

c_t = total rock and fluid compressibility, psi^{-1}, and

r_w = wellbore radius, ft.

Fetkovich showed that the dimensionless flow rate at late times is an exponential function of dimensionless time and the dimensionless radius of the reservoir (r_e/r_w). The assumption is that the well is producing under constant bottomhole pressure in a closed circular reservoir.

The decline curves of dimensionless time (t_{Dd}) and flow rate (q_{Dd}) in terms of reservoir variables become the following:

$$t_{Dd} = \frac{t_D}{1/2 \, [(r_e/r_w)^2 - 1][\ln \, (r_e/r_w) - 0.5]} \tag{11.17}$$

$$q_{Dd} = \frac{141.3 \, q(t) \, \mu \, B[\ln \, (r_e/r_w) - 0.5]}{kh(p_i - p_{wf})} \tag{11.18}$$

Note that flow capacity (kh) can be determined if $q(t)$, μ, B, r_e, r_w, p_i, and p_{wf} are known. The assumptions inherent in the preceding equations include a single well producing from a closed circular reservoir.

These dimensionless quantities can be expressed based on Arps' rate-time equations, as in the following:

$$q_{Dd} = q(t)/q_i \tag{11.19}$$

$$t_{Dd} = D_i \, t \tag{11.20}$$

Equations 11.16, 11.17, and 11.20 can be combined to obtain an expression for exponential decline rate ($D_i = D$) in terms of reservoir and fluid properties, as in the following:

$$D = \frac{0.00634k / (\emptyset \, \mu \, c_t \, r_w^2)}{1/2[(r_e/r_w)^2 - 1][\, \ln \, (r_e/r_w) - 0.5]} \tag{11.21}$$

Any effect of skin is also accounted for by considering the effective wellbore radius. Equation (11.21) suggests that a decline curve analysis based on Fetkovich type curves may lead to estimation of reservoir permeability, among other factors. The ultimate recovery can be obtained from the type curve analysis by assuming an abandonment rate.

The procedure for matching the decline curve with a type curve manually is as follows:

- Plot on a log-log tracing paper the actual rate versus time, using the same scale and size of the q_{Dd} versus t_{Dd} type curve to be used.
- Place the tracing paper data over the type curve to obtain the best fit of the data, keeping the coordinate axes of the two curves parallel.
- Note the value of the decline curve exponent on the type curve matching the actual data.
- Extend the actual data in the future following on the matched curve. Knowing the economic production rate, the economic life of the reservoir can then be directly determined.
- Read the values of q_{Dd} and t_{Dd} corresponding to a selected actual rate.

In computer-assisted analysis, the preceding procedure is largely automated, as illustrated in Example 11.2. Additionally, various shapes and trends observed in a log-log plot of actual rate versus time during the type curve analysis may aid in diagnosing the following:

- Degree of skin damage or enhancement in a well. The latter may result following a hydraulic fracturing operation.
- Identification of pressure support as boundary effects become dominant.
- Apparently high or low difference between initial reservoir pressure and flowing well pressure.
- Phenomenon of liquid loading.

The methodology related to type curve analysis of declining production rates is enhanced by various authors. For example, Blasingame developed a set of type curves that take into account any variations in bottomhole pressure during transient flow. These curves also consider the variations in PVT properties in the case of gas reservoirs under depletion. Agarwal and Gardner introduced a new set of rate versus time and rate cumulative type curves in order to estimate reserves. The type curves are based on dimensionless well test variables (q_D and t_{DA}). These dimensionless quantities are described in chapter 5.

Examples in decline curve analysis

In order to illustrate the methodology of decline curve analysis, several examples are presented in this chapter using decline curve analysis software. The examples are selected to highlight the following:

- Identification of decline trend and prediction of future performance by the classical method
- Application of type curves in decline curve analysis
- Analysis of wells with high water cut
- Limitations of decline curve analysis in certain situations

Example 11.1. Identification of decline trend and prediction of future performance. Table 11–1 shows the monthly production data of an oil well for the first year following completion. There is no water or gas production observed from the well. Based on this information, the ultimate reserve and life of the well can be predicted, assuming that the economic limit is 30 bbl/d.

Solution: Plots of oil rate and cumulative production versus time generated by F.A.S.T. RTA software are shown in Figure 11–6.[9] The harmonic decline pattern appears to best fit the plotted data. The following results are obtained from computer-assisted analysis:

Decline exponent = 1.0 (harmonic).

Decline rate (D) = 0.08.

Estimated total reserve, Mbbl, = 1,170.

Remaining reserve, Mbbl, = 1,106.

Production life, days, = 12,335 (or 33.8 years).

It can be noted that the production data in this example does not fall on a smooth line, which is typical of most wells.

Table 11-1. Monthly production data

Month of Production	Average Rate, bbl/d
1	199.5
2	198
3	189
4	198
5	194
6	194
7	191
8	188
9	190
10	188
11	186
12	185

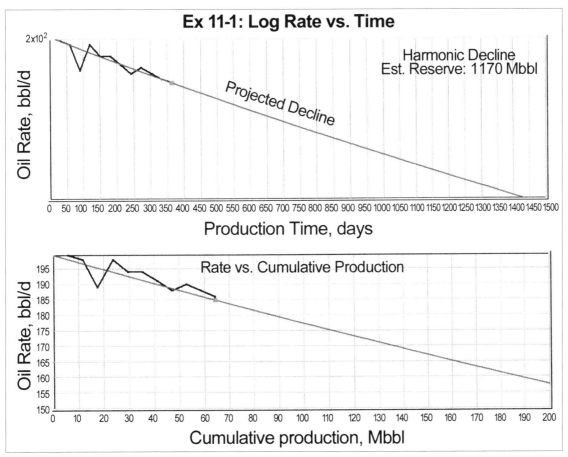

Fig. 11–6. Analysis of production data of an oil well showing harmonic decline. *Courtesy of Fekete Associates.*

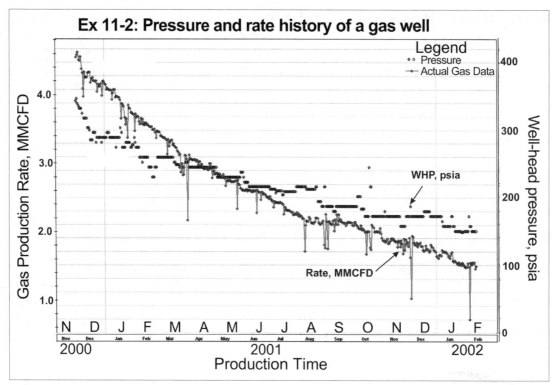

Fig. 11–7. Production history of a gas well under exponential decline. *Courtesy of Fekete Associates.*

Example 11.2. Traditional method aided by type curve analyses, including estimation of reservoir properties.
Figure 11–7 presents the production history, including the rate and wellhead pressure, of a gas well over a period of 15 months. Remaining reserve and life of the well need to be estimated. Reservoir properties, including permeability, also need to be estimated based on type curve analysis. The following reservoir, fluid, and well data is available:

Thickness of the producing formation, ft, = 65.7.

Porosity = 0.2.

Gas saturation = 0.8.

Gas specific gravity = 0.65.

Total compressibility, psi^{-1}, = 19.49×10^{-4}.

Wellbore radius, ft, = 0.35.

Drainage area, acres, = 193.8.

Solution: Traditional analysis of rate versus time and cumulative production versus time indicated an exponential decline, as shown in Figure 11–8. The following results are obtained from the analysis, assuming an economic limit of 50 Mscfd:

Decline exponent = 0 (exponential).

Decline rate (D) = 0.962.

Initial gas in place (IGIP), bcf, = 2.5.

Estimated total reserve, bcf, = 1.67.

Remaining reserve, bcf, = 0.52.

Remaining life of the well, years, = 3.4.

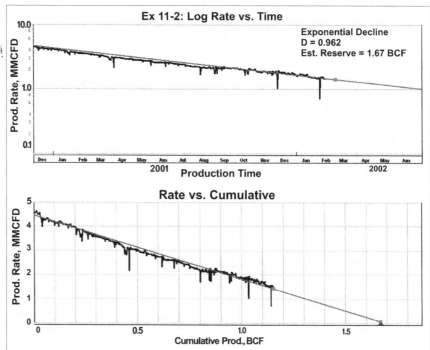

Fig. 11–8. Traditional decline curve analysis of a gas well. *Courtesy of Fekete Associates.*

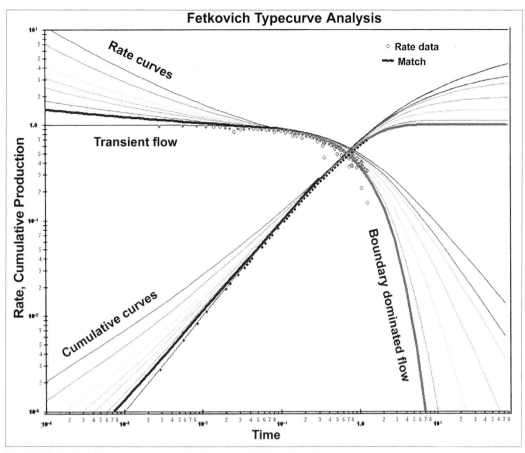

Fig. 11–9. Fetkovich type curve analysis. *Courtesy of Fekete Associates.*

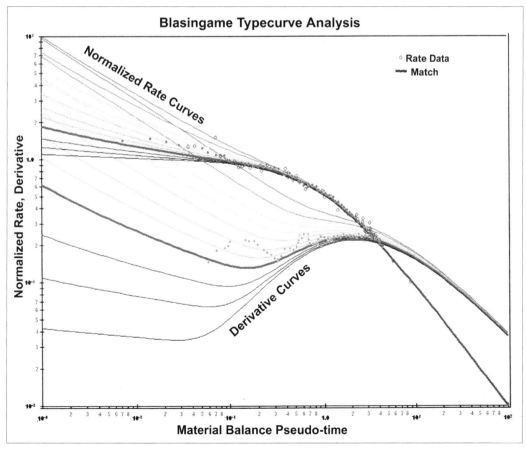

Fig. 11–10. Blasingame type curve analysis. *Courtesy of Fekete Associates.*

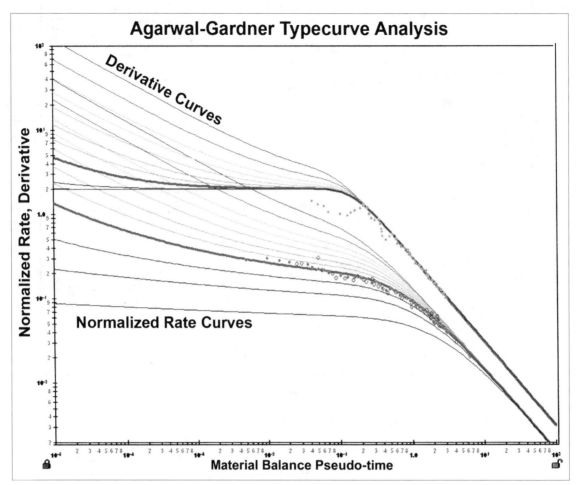

Fig. 11–11. Agarwal and Gardner type curve analysis. *Courtesy of Fekete Associates.*

Table 11–2. Summary of results

Type Curve/Reservoir Model	Initial Gas in Place, bcf	Drainage Area, acres	Total Reserve, bcf	Reservoir Permeability, mD	Skin
Fetkovich/radial	2.67	207.17	1.75	—	—
Blasingame/radial	2.25	174.61	1.76	28.3	−3.32
AG rate versus time/radial	2.51	194.35	1.96	30.2	−3.38

Once again, it is observed that the rate data does not fall on a smooth line due to various constraints associated with field operations. In performing the traditional analysis, emphasis was placed on matching the late trend observed between October 2001 and February 2002 to predict the future performance.

Furthermore, results obtained from type curve analyses performed by various methods (figs. 11–9 through 11–11) include estimates of initial gas in place, estimated reserve, drainage area, formation permeability, and skin. These are presented in Table 11–2.

Example 11.3. Decline curve analysis of a well with high water cut. This example is based on Baker-Hughes/SSI's Production Data Analysis tool. Figure 11–12 shows oil, gas, and water production performance of a well. Figure 11–13 shows oil rate performance history match and prediction.

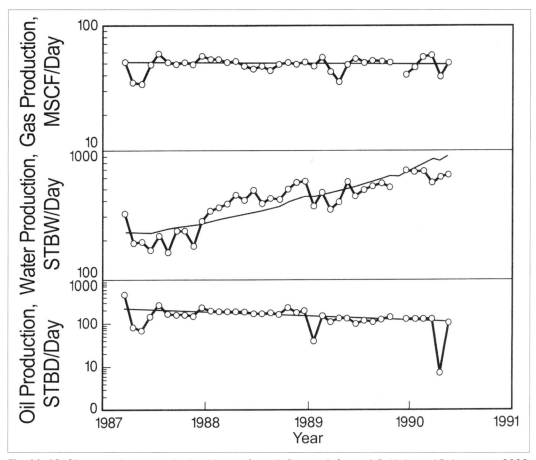

Fig. 11–12. Oil, gas, and water production history of a well. *Source: A. Satter, J. Baldwin, and R. Jespersen. 2000. Computer-Assisted Reservoir Management. Tulsa: PennWell.*

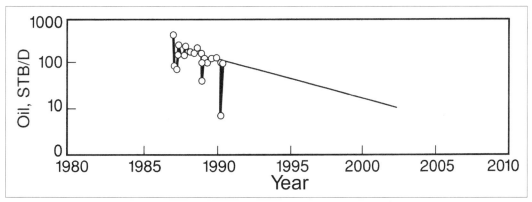

Fig. 11–13. Decline curve analysis predicting future performance. *Source: A. Satter, J. Baldwin, and R. Jespersen. 2000. Computer-Assisted Reservoir Management. Tulsa: PennWell.*

Solution: The analysis indicated an exponential decline of 21.3% per year. Based on an economic production rate limit of 100 stbo/d, the reserves were calculated to be 19.9 Mbbl of oil by the end of 1990. If the economic limit of the well is less, ultimate recovery is expected to be higher. However, it is anticipated that the water cut will exceed 90% before the end of the economic production rate limit of 100 stbo/d, as shown in Figure 11–4.

It must be emphasized that both the oil rate and water cut should be analyzed to ensure reliability in the results of decline curve analysis. In this example, a water cut limit of 90% would shut down the well earlier than anticipated.

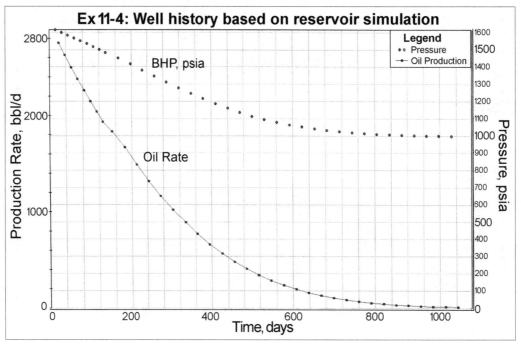

Fig. 11–14. Production decline of an oil well based on a reservoir simulation study. *Courtesy of Gemini Solutions.*

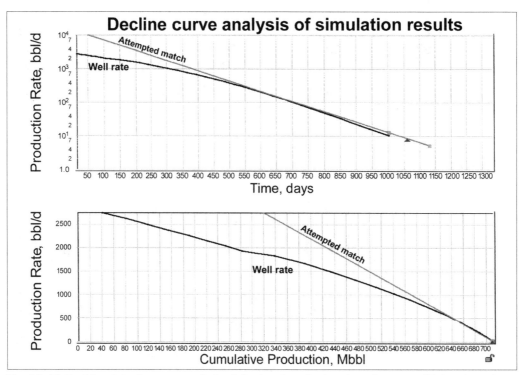

Fig. 11–15. Traditional decline curve analysis of simulated production data. *Courtesy of Fekete Associates.*

Example 11.4. Limitation of decline curve analysis technique. Figure 11–14 presents the production history of a well obtained by reservoir simulation.

The production of the well needs to be predicted by decline curve analysis.

Solution: Figure 11–15 presents the attempted analysis of the production decline as predicted by simulation. It is observed that no specific decline pattern, namely exponential, hyperbolic, and harmonic, satisfactorily matches the trend. The exponential decline curve matched only the middle portion of the decline period.

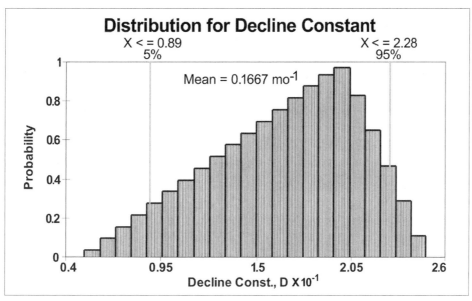

Fig. 11–16. Triangular probability distribution of decline constant. *Courtesy of Palisade Corporation.*

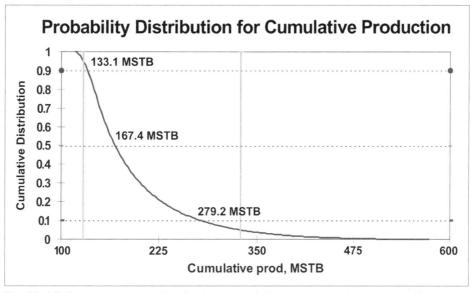

Fig. 11–17. Cumulative probability of distribution of ultimate recovery from the well. *Courtesy of Palisade Corporation.*

Example 11.5. Probabilistic analysis of well rate decline and estimation of recovery. Initial production rate of a new well located in a developed field is found to be 1,000 stb/d. Decline curve studies performed on a large number of producers in the same field suggest that the future production rates of most wells could be reasonably predicted by an exponential decline model. The decline constant is found to vary between 0.05 and 0.25 per month, the most likely value being 0.2 per month. The objective is to estimate ultimate oil recovery from the well having cumulative probabilities of 10%, 50%, and 90%.

Solution: Using @RISK software, the cumulative production from a well under exponential decline is calculated by Equation 11.4. A well abandonment rate of 40 stb/d was assumed in the analysis. A triangular probability distribution based on available data is assigned to the decline constant as shown in Figure 11–16. Results of the cumulative probability distribution obtained by a Monte Carlo simulation involving 10,000 trials are plotted in Figure 11–17.

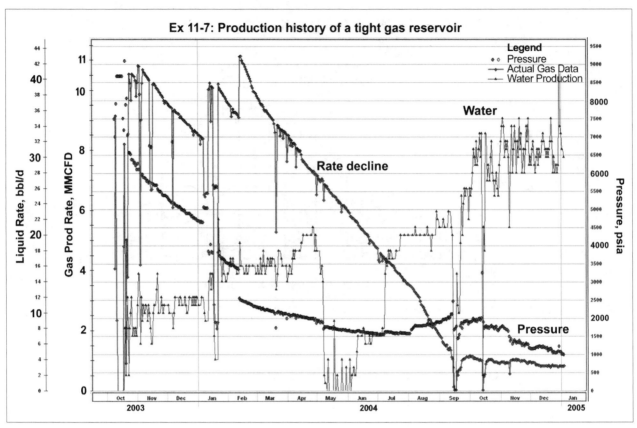

Fig. 11–18. Production history of a well in tight gas reservoir with water cut. *Courtesy of Fekete Associates.*

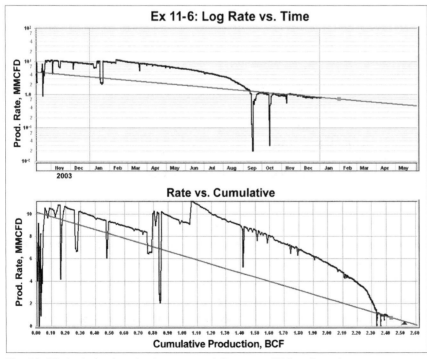

Fig. 11–19. Analysis of production rate trend. *Courtesy of Fekete Associates.*

Example 11.6. Analysis of a rapidly declining well in a low-permeability formation. Figure 11–18 presents the production history of a well, including gas and water production rates and bottomhole pressure, over a period of 15 months. The production data does not exhibit an idealized decline curve and is a typical representation of what is likely to occur in practical circumstances. The reservoir considered is found to have an average permeability between 1 and 2 mD. The first few months of production are marked by frequent drops in rate, followed by stimulation that augmented well productivity temporarily. However, the rate of decline is significant due to the low permeability of the formation. Rapidly increasing water cut is also evident.

The gas production rate of the well dropped from more than 11 MMcfd to about 1 MMcfd in just eight months. The reservoir, located at a depth of 11,875 ft, is overpressured, as the initial pressure was found to be in excess of 9,000 psia. However, the drop in bottomhole pressure during the relatively short production period is substantial, declining to about 1,000 psia. Determine the reserve based on decline curve analysis.

Solution: Figure 11–19 shows an exponential decline fit of the data during the last few months. It can be noted that the earlier decline of the well did not match the later part, as the well is observed to have produced at a low but somewhat steady rate of about 1 MMcfd during the last few months of analysis.

For the low permeability reservoir studied here, a sustainable level of production could be achieved following much higher rates of production initially. In situations where the well stabilizes to a relatively lower rate of decline at a later date, performing a decline curve analysis based on the initial rates alone may lead to the underestimation of gas reserves.

In conclusion, wells or fields located in very low permeability environments may not be ideal candidates for decline curve analysis, especially when the analysis is based on initial production data. Moreover, increasing quantities of water may be encountered in certain wells following initial production of oil or gas only, potentially rendering the earlier analysis inaccurate.

Example 11.7. Decline curve analysis of field production. The same methodology used to analyze the declining rate of an individual well can be applied for the entire field with a number of producers in certain instances. For example, in a matured gas field without any major heterogeneity present, all wells may exhibit a similar decline pattern in production and can be treated as a whole. This analysis readily points to fluid in place and ultimate recovery.

Figure 11–20 shows the number of active wells in a matured gas field over two decades. In early and late 1980s, the field saw major development when most of the wells were drilled. Since 1990, some of the wells have been temporarily or permanently abandoned, and a declining trend in field production is observable (fig. 11–21). The objective of the study is to determine the type of decline for the entire field and to estimate the gas reserve.

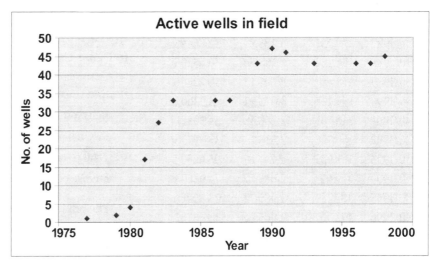

Fig. 11–20. Total number of active wells versus time

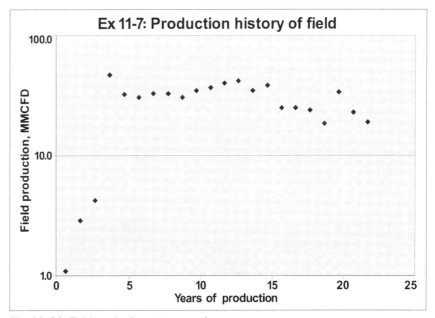

Fig. 11–21. Field production rate versus time

Table 11-3. Production and reservoir pressure versus time

Date	Cumulative Wet Gas, bcf	Pressure, psia	Average Rate, MMcfd
11/01/1977	0	—	—
12/30/1977	0.086	9,494.36	0.117808
12/30/1978	0.482	8,158.49	1.084932
12/30/1979	1.536	9,288.32	2.887671
12/30/1980	3.08	8,362.93	4.230137
12/30/1981	20.09	7,313.89	46.60274
12/30/1982	31.808	6,211.35	32.10411
12/30/1983	42.901	5,481.11	30.39178
12/30/1984	54.843	4,776.42	32.71781
12/30/1985	66.811	4,488.93	32.78904
12/30/1986	77.998	4,236.21	30.64932
12/30/1987	90.731	4,185.2	34.88493
12/30/1988	104.182	4,008.84	36.85205
12/30/1989	118.864	3,547.17	40.22466
12/30/1990	134.167	3,023.21	41.92603
12/30/1991	146.761	2,780.58	34.50411
12/30/1992	160.776	2,513.8	38.39726
12/30/1993	169.925	2,247.87	25.06575
12/30/1994	179.023	2,085.82	24.92603
12/30/1995	187.719	1,985.72	23.82466
12/30/1996	194.467	2,005.68	18.48767
12/30/1997	206.79	1,926.46	33.76164
12/30/1998	215.2	1,446.01	23.0411
12/30/1999	222.137	1,265.65	19.00548

Solution: A decline curve analysis is performed of the information available since 1990, about the time the field indicates the beginning of a declining trend, as shown in Table 11–3. The results of the analysis are summarized in the following:

Exponential decline (b=0).

D = 0.081.

Effective decline rate, %, = 7.8.

Total reserve, bcf, = 240.

Remaining reserve, bcf, = 80.

Traditional decline curve analysis, performed with the aid of RTA software, is shown in Figure 11–22. Since reservoir pressure information is available throughout the life of the field, various type curve analyses can be performed, including the one shown in Figure 11–23. Results obtained by using Agarwal-Gardner type curves indicate the following, where production data flowing since 1990 was matched:

Reservoir drainage area, acres, = 81,870.

Original gas in place, bcf, = 271.2.

Total reserve, bcf, = 230.5 (based on a recovery factor of 85%).

Average permeability, mD, = 48.

The results of this study can be compared further by utilizing other methods, such as the material balance method described in the following chapter.

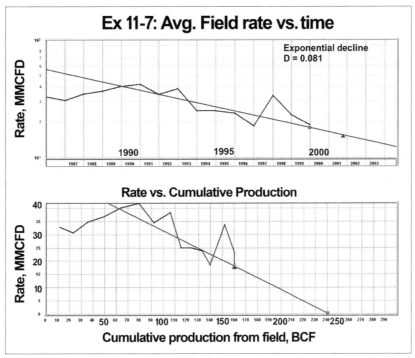

Fig. 11–22. Traditional decline curve analysis. *Courtesy of Fekete Associates.*

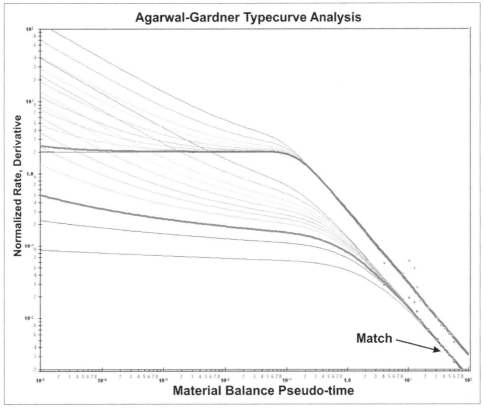

Fig. 11–23. Agarwal-Gardner type curve analysis. *Courtesy of Fekete Associates.*

Summing Up

The classical decline curve technique is quite straightforward, since only oil production rate, gas/oil ratio, and water/oil ratio information is necessary for analysis. It can be used to calculate ultimate recovery by adding past and expected future productions. The remaining life of a well or reservoir can also be predicted by decline curve method. The analysis requires the assumption of an oil or gas rate at an economic limit.

Empirical equations are used for decline characteristics involving decline rate and exponent. Standard decline curves are: exponential, with 0 exponent and constant decline rate; harmonic, with variable rate and exponent =1; and hyperbolic, with variable rate and exponent greater than zero but less than 1.

The commonly used decline curves are: oil production rate versus time in Cartesian or semilog plot, oil production rate versus cumulative oil production, and log of water cut or oil cut versus cumulative production.

The most important assumptions involved in decline curve analysis are that sufficient production performance data is available and decline in rate has been established. It is further assumed that the ongoing operations will continue without interruptions in the future. In many instances, well performance is not amenable to decline curve analysis, as the above assumptions are not valid.

There are many factors that can affect the decline curve performance. These include proration, changes in bottomhole pressure and production methods, water or gas breakthrough, pipeline disruptions, well recompletion and treatments, and weather and market conditions. Therefore, care must be taken in extrapolating the production curves into the future. When the shape of a decline curve changes or erratic fluctuation of data occurs, the cause should be determined, and its effect upon the reserves evaluated.

Ershaghi and Omoregie made enhancements in the classical decline curve analysis techniques by using a modified water-cut factor. Fetkovich has provided improvements by including transient flow regime and using log-log type curves based upon dimensional rate versus dimensional time for exponent 0 to 1. Besides the Fetkovich method, other type curves are available to analyze the decline in production, taking into account varying well pressure and fluid properties. Reservoir parameters, such as permeability and skin factor, can be estimated based on type curve analysis. Furthermore, type curves may aid in reservoir diagnostics, including boundary influence, skin damage, and effectiveness of a hydraulic fracture.

Class assignments have been provided to further understanding and application of the decline curve techniques.

Class Assignments

Questions

1. When can the traditional decline curve methods be used to calculate reserves?

2. Name the data needed for decline curve analysis methods.

3. Name the three decline curve types. How do they differ from each other?

4. Are the decline analysis techniques applicable for all wells, all of the time? Why or why not?

5. Name the procedure for normal decline curve analysis.

6. What is the basis for the Ershaghi and Omoregie method for enhancing the decline curve analysis technique?

7. What is the basis for Fetkovich type curve analysis for improving the decline curve analysis technique?

8. Describe the basic steps in type curve matching for decline curve analysis.

9. Are the learning objectives of this chapter met? If not, what information should be reviewed?

Exercises

11.1. The initial production rate of an oil reservoir was 11,300 stb/d. Over a period of a year, the rate declined to 9,500 stb/d. Using exponential, hyperbolic, and harmonic decline equations, estimate the oil reserve and remaining life of the reservoir. Assume an abandonment rate of 850 stb/d. List all necessary assumptions. Which mode of decline leads to the estimation of largest reserve?

11.2. Given the following annual data from an oil reservoir under depletion drive in Table 11–4, determine the mode of decline (exponential, harmonic, or hyperbolic) that best fits the production data. Estimate the primary reserve and projected production rates in the future. List any assumptions necessary.

Table 11-4. Annual production data

Year	Average Rate, stb/d
0.0	15,200
0.5	11,600
1.0	8,651
1.5	6,818
2.0	5,312
3.0	3,716
4.0	2,797
5.0	2,138
6.0	1,687
7.0	1,367
8.0	1,128

11.3. Using the production history of a gas field in Table 11–5, identify any trend in decline. Estimate the gas reserve. Can the drainage area be estimated? How further drilling in the field would affect the overall decline trend identified in this exercise? How would a strong water influx or frequent downtimes affect the decline in rate?

Table 11-5. Annual rate and cumulative data

Time, years	Rate, MMscf/d	Cumulative Production, bscf
0	73.45	0
1	63.22	24.8964
2	54.41	46.3250
3	46.83	64.7687
4	40.31	80.6434
5	34.69	94.3069
6	29.86	106.0671
7	25.70	116.1893
8	22.12	124.9015
9	19.041	132.4001
10	16.39	138.8543
11	14.11	144.4094
12	12.14	149.1908
13	10.45	153.3061
14	8.99	156.8483
15	7.74	159.8970
16	6.66	162.5211
17	5.74	164.7796
18	4.94	166.7236
19	4.25	168.3968
20	3.66	169.8369

References

1. Arps, J. J. 1945. Estimation of decline curves. *Transactions*. AIME. Vol. 160, 228–247.

2. Arps, J. J. 1956. Estimation of primary oil reserves. *Transactions*. AIME. Vol. 207, 182–191.

3. Fetkovich, M. J., E. J. Fetkovich, and M. D. Fetkovich. 1996. Useful concepts for decline curve forecasting, reserve estimation, and analysis. *Society of Petroleum Engineers Reservoir Evaluation*. February: p. 13.

4. Blasingame, T. A., T. L. McCray, and W. J. Lee. 1991. Decline curve analysis for variable pressure drop/variable flowrate systems. SPE Paper #21513. Paper presented at the Society of Petroleum Engineers Gas Technology Symposium, Houston, TX,. January 22–24.

5. Blasingame, T. A., and W. J. Lee. 1986. Variable-rate reservoir limits testing. SPE Paper #15028. Paper presented at the Permian Basin Oil & Gas Recovery Conference of the Society of Petroleum Engineers, Midland, TX, March 13–14.

6. Agarwal, R. G., D. C. Gardner, S. W. Kleinsteiber, and D. D. Fussell. 1998. Analyzing well production data using combined type curve and decline curve analysis concepts. SPE Paper #57916. Paper presented at the 1998 Society of Petroleum Engineers Annual Technical Conference and Exhibition, New Orleans, September 27–30.

7. Ershaghi, I., and O. Omoregie. 1978. A method for extrapolation of cut vs. recovery curves. *Journal of Petroleum Technology*. February, 203–204.

8. Fetkovich, M. J. 1980. Decline curve analysis using type curves. *Journal of Petroleum Technology*. June: pp. 1,065–1,077.

9. Fekete Associates, Inc. 2006. RTA version 3.037. Calgary, Canada: Fekete Associates, Inc.

12 · Material Balance Methods and Applications

Introduction

The material balance method finds wide application in order to estimate the original hydrocarbon in place of a reservoir by history matching the past production performance. It can also be used to predict the future performance of the reservoir and to understand the dynamics of the reservoir under various mechanisms of production, including solution gas drive and water influx. In most studies, the data requirement is limited to average reservoir pressure, cumulative production, and fluid PVT properties over various time intervals during production. A reasonable description of the aquifer may also be needed where applicable. A number of software tools are available in the industry to perform the necessary analyses in a relatively straightforward manner within a short period of time.

The classical material balance method is based upon the law of conservation of mass, which simply means that mass is conserved, i.e., it is neither created nor destroyed. The basic assumptions made in this technique consist of the following:
- The reservoir is viewed as a homogeneous tank, i.e., rock and fluid properties are the same throughout the reservoir.
- Fluid production and injection occur at single production and single injection points.
- The analysis is independent of the direction of fluid flow in the reservoir.

However, in reality, the reservoirs are not homogeneous, and production and injection wells are distributed areally. Furthermore, the wells are activated at different times. In addition, reservoir fluids flow in definite directions. Nevertheless, the material balance method became very popular with reservoir engineers due to its simple yet robust foundation. It has been found to be a valuable tool for analyzing reservoir performance with reasonably acceptable results. As discussed later, material balance studies may aid in reservoir characterization when the results are compared against others, such as those obtained from volumetric analysis and reservoir simulation.

The material balance method is more fundamental than the decline curve technique for analyzing reservoir production performance. The advantage of the material balance method is that reservoir heterogeneities need not be known in detail in order to perform a meaningful analysis. Moreover, material balance studies are not as resource intensive as multidimensional, multiphase numerical simulation. Material balance techniques are used to estimate the following:
- Original oil and original gas in place (OOIP, OGIP)
- Ultimate primary recovery
- The influence of natural production mechanisms in the reservoir, such as gascap, solution gas, or water drives

Furthermore, the results of material balance analysis can be used as verification of the hydrocarbon in place and recovery estimates obtained by other methods.

This chapter is devoted to learning the following:
- The principle of material balance as applied to petroleum reservoirs involving production, injection, and influx of various fluids
- Application of the material balance method in different types of reservoirs producing under various drive mechanisms
- Mathematical equations and graphical techniques used for oil and gas reservoir performance analysis
- Prediction of reservoir performance
- Role of material balance analysis in reservoir characterization
- Material balance analysis based on flowing bottomhole pressure
- Working example problems and class exercises

The various capabilities of the material balance method are illustrated by the following analyses in the later part of the chapter:
- Estimation of recovery factor in a newly discovered reservoir with limited data
- Assessment of original oil and gas in place of an oil reservoir with a gas cap
- Investigation of water influx characteristics affecting oil reservoir performance
- Analysis of an abnormally pressured gas reservoir under depletion drive
- Aquifer modeling for a reservoir under a strong water drive
- Gas reservoir simulation with various aquifer models
- Estimation of original gas in place of a reservoir with a small radial aquifer, and comparison of the result with a numerical reservoir simulation
- Analysis of 3-D plots as opposed to conventional 2-D plots utilized in material balance studies
- Performance review of a wet gas reservoir
- Material balance of a gas well based on flowing bottomhole pressure

Most of the analyses have been performed with the aid of various software tools available in the industry.

Material Balance of Oil Reservoirs

A producing reservoir can be considered in which the underground withdrawal of petroleum fluids and water is determined by any or all of the following:
- Changes in volume of in-situ oil, gas, and water due to a pressure decline in the reservoir
- Water influx from the surrounding aquifer
- Pore volume compressibility
- External water or gas injection for reservoir pressure maintenance, in certain instances

In this case, the general material balance equation for oil reservoir performance can be expressed as follows:

Underground withdrawal = Expansion of oil + Original dissolved gas + Expansion of gas caps
+ Reduction in hydrocarbon pore volume due to connate water expansion and decrease in the pore volume
+ Natural water influx from an adjacent aquifer.

Havlena and Odeh showed that the material balance components in the equation above can be arranged as an equation of a straight line.[1,2] The equation is solved for original hydrocarbon in place and other parameters by using relatively simple graphical techniques. The general material balance is given as follows:

$$F = N(E_o + mE_g + E_{fw}) + W_e \tag{12.1}$$

where

F = underground withdrawal, rb

$$= N_p[B_o + (R_p - R_s)B_g] + W_pB_w - W_iB_w - G_iB_g \tag{12.2}$$

$$= N_p[B_t + (R_p - R_{si})B_g] + W_pB_w - W_iB_w - G_iB_g , \tag{12.3}$$

N_p = cumulative oil production, stb,

B_o = oil formation volume factor, rb/stb,

R_s = gas in solution in oil, scf/stbo,

B_g = gas formation volume factor, rb/scf,

W_p = cumulative water production, stb,

W_i = cumulative water injection, stb,

G_i = cumulative gas injection, scf,

R_p = cumulative gas/oil ratio (cumulative gas production over cumulative oil production, scf/stb),

N = original oil in place, stb,

E_o = expansion of oil and original gas in solution, rb/stb

$$= B_t - B_{ti} , \tag{12.4}$$

$$= (B_o - B_{oi}) + (R_{si} - R_s)B_g , \tag{12.5}$$

B_t = Two-phase formation volume factor defined in chapter 3

$$= B_o + (R_{si} - R_s)B_g , \text{ rb/stb} \tag{12.6}$$

m = initial gas cap volume fraction

$$= \frac{\text{initial hydrocarbon volume of the gascap}}{\text{initial hydrocarbon volume of the oil zone}} , \text{ rb/rb} ,$$

E_g = expansion of gascap gas, rb/stb

$$= B_{oi}\left(\frac{B_g}{B_{gi}} - 1 \right), \tag{12.7}$$

E_{fw} = expansion of the connate water and reduction in the pore volume, rb/stb

$$= (1+m)B_o\left[\frac{c_w S_{wi} + c_f}{1 - S_{wi}} \right]\Delta p , \tag{12.8}$$

c_w, c_f = water and formation compressibilities, respectively, psi^{-1},

S_{wi} = initial water saturation, fraction,

Δp = pressure drop, psi,

W_e = cumulative natural water influx, bbl

$$= US(p,t), \tag{12.9}$$

U = aquifer constant, rb/psi, and

S(p,t) = aquifer function.

Wang and Teasdale list the aquifer constants and aquifer functions for various aquifers.[3]

One or more of the following aquifer models are commonly used in material balance analysis:
- **Small (pot) aquifer.** The model utilizes a simple time-independent aquifer equation. It is suitable for very small aquifers. The function S(p,t) is dependent on the pressure decline of the reservoir.
- **Steady-state aquifers.** The model assumes that aquifer pressure remains constant throughout production from the reservoir. The aquifer is rather large and replenishes any depletion in the reservoir by water influx, which is approximately constant. Schilthuis and Hurst steady-state aquifer models belong to this category.[4]
- **Unsteady-state aquifers.** The model differs from the steady-state aquifers in that the aquifer pressure is not constant during the life of the reservoir. Unsteady-state aquifer models for both linear and radial cases, as proposed by van Everdingen and Hurst, assume that S(p,t) is a function of both pressure decline and time.

It must be emphasized that aquifer characteristics are not known with certainty in most reservoirs. Keeping this limitation in mind, a typical material balance study evaluates the applicability of a specific aquifer type by analyzing the reservoir performance with various aquifer models (steady state versus unsteady state, for example). This is continued until the best match is obtained with the field data. In addition, material balance software is capable of calculating best-fit parameters for certain aquifer models.

Monitoring of reservoir performance may aid significantly in recognizing aquifer behavior. For example, if producing wells are shut in for maintenance, a rapid rise in static bottomhole pressure could indicate the presence of a strong water drive. As illustrated in this chapter, material balance methods may lead to unrealistically high values of the original hydrocarbon in place (OHCIP) if the influence of the aquifer is not considered.

For a solution gas drive reservoir, where there is no initial gas cap (m = 0), no gas and water injection ($G_i = 0$, $W_i = 0$), and no natural water influx ($W_e = 0$), Equation 12.1 can be written in a simplified form. The general material balance equation thus can be shown as the following:

$$N_p [B_t + (R_p - R_{si}) B_g] + W_p B_w = N [(B_t - B_{ti}) + B_o \{(c_w S_{wi} + c_f)/(1-S_{wi})\} \Delta p] \tag{12.10}$$

Above the bubblepoint pressure (undersaturated oil), $R_p = R_s = R_{si}$, $E_t = B_{oi}$, and $E_{ti} = B_{oi}$. Neglecting water production, the material balance equation is then reduced as follows:

$$N_p / N = (B_{oi} / B_o) \, c_e \, \Delta p \tag{12.11}$$

where

$$c_e = \frac{c_o S_o + c_w S_w + c_f}{1 - S_{wi}}, \tag{12.12}$$

S_o = oil saturation, fraction,

c_o = oil compressibility, psi^{-1},

c_w = water compressibility, psi^{-1}, and

c_f = formation compressibility, psi^{-1}.

If original oil in place (OOIP) is known, Equation 12.11 can be used to calculate future production for sequential pressure drops starting from the initial pressure.

Below the bubblepoint pressure, neglecting water production and rock compressibility, recovery efficiency is given as the following:

$$\frac{N_p}{N} = \frac{B_t - B_{ti}}{B_t + (R_p - R_{si})B_g} \tag{12.13}$$

Reservoir performance prediction

Calculation of future production requires not only the solution of Equation 12.13, but also of the subsidiary equations for liquid saturation, produced gas/oil ratio, and cumulative gas production. This requires the knowledge of fluid phase relative permeabilities and formation volume factor at each time step of calculation. The relevant equations are as follows:

$$S_o = \left(1 - \frac{N_p}{N}\right)\left(\frac{B_o}{B_{oi}}\right)(1 - S_{wi}) \tag{12.14}$$

$$R = R_s + \left(\frac{B_o}{B_g}\right)\left(\frac{\mu_o}{\mu_g}\right)\left(\frac{k_{rg}}{k_{ro}}\right) \tag{12.15}$$

$$R_p = \frac{G_p}{N_p} = \frac{\int_0^t R_d N_p}{N_p} - \frac{\Sigma R \Delta N_p}{N_p} \tag{12.16}$$

where

μ_o, μ_g = oil and gas viscosities, respectively, cp, and

k_{rg}, k_{ro} = gas and oil relative permeabilities, respectively, fraction.

Simultaneous solutions of the material balance and the subsidiary equations are required based upon time steps or corresponding pressure steps.

Figure 12–1 presents pressure and gas/oil ratio versus oil recovery for an undersaturated homogeneous oil reservoir. Hydrocarbons are in the liquid phase above the bubblepoint pressure, and pressure drops sharply due to expansion of the rock, water, and oil. Oil recovery at the bubblepoint pressure is low, and the produced gas/oil ratio remains low and constant. Below the bubblepoint pressure, the gas/oil ratio initially remains low, then rises to a maximum value, and finally drops as most of the liberated free gas is produced. Recovery below the bubblepoint is influenced by expansion of

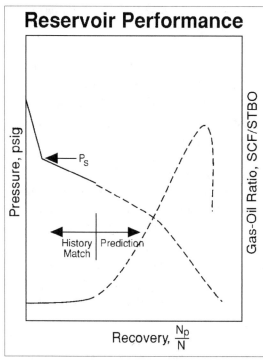

Fig. 12–1. Reservoir pressure and gas/oil ratio versus oil recovery. *Source: A. Satter and G. C. Thakur. 1994. Integrated Petroleum Reservoir Management—A Team Approach. Tulsa: PennWell.*

the free gas coming out of solution from the oil. Depending upon the rock and oil characteristics, oil recovery can be an order of magnitude or more greater than that above the bubblepoint.

The general material balance equation (12.1) for oil reservoirs contains three unknowns:

- Original oil in place
- Gascap size
- Cumulative natural water influx

The equations include production and injection data, and rock and fluid properties, which are dependent upon reservoir pressure.

Summary of material balance analysis techniques

This section summarizes the oil and gas material balance equations, along with the plots. The original oil in place can be determined by history matching the past performance using the graphical methods described in the following sections.

I. FE method.[5]

$$F = N(E_o + mE_g + E_{fw}) + W_e$$
$$= NE_t + W_e$$

where

$E_t = E_o + mE_g + E_{fw}$, and
W_e = influx of aquifer water into the reservoir.

Assume $W_e = 0$, and $F = NE_t$. Plot F versus E_t through the origin, where N = slope. Without a gas cap, $m = 0$, and the plot is shown in Figure 12–2.

With a gas cap, $m \neq 0$, and the plot is shown in Figure 12–3.

II. Gas cap method. Assume $W_e = E_{fw} = 0$. Then:

$$\frac{F}{E_o} = N + mN \left(\frac{E_g}{E_o}\right)$$

Plot F/E_o versus E_g/E_o (shown in fig. 12–4), where N = intercept on y-axis, and mN = the slope.

III. Havlena and Odeh method (water drive).

$$\frac{F}{E_t} = N + \frac{W_e}{E_t}$$
$$= N + U\frac{S}{E_t}$$

Plot F/E_t versus W_e/E_t (or S/E_t); shown in Figure 12–5, where U = slope.

IV. Campbell method.[6] Plot F/E_t versus F (fig. 12–6), where N = intercept on the y-axis.

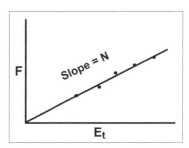

Fig. 12–2. Plot of F versus E_t

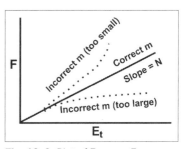

Fig. 12–3. Plot of F versus E_t

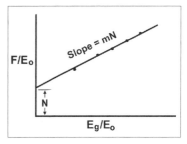

Fig. 12–4. Plot of F/E_o versus E_g/E_o

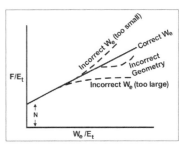

Fig. 12–5. Plot of F/E_t versus W_e/E_t

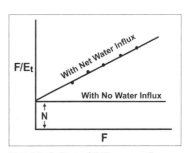

Fig. 12–6. Plot of F/E_t versus F

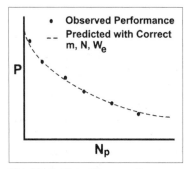

Fig. 12-7. Plot of P versus N_p

V. Pressure match method. Plot p versus N_p (fig. 12–7). Correct values of N, m, and W_e are obtained for the best match.

Each method applies to a specific type of reservoir as follows:

1. Method I is used for solution gas drive oil reservoirs (unknown is N).

2. Method II is used for gascap drive oil reservoirs (unknowns are N and m).

3. Methods III and IV are used for water drive oil reservoirs (unknowns are N and W_e).

4. Method V is used to model oil reservoirs under gascap and water drives (unknowns are m, N, and W_e).

All three unknowns can be adjusted during pressure matching.

Tehrani proposed that the unknown parameters N and U could be estimated by fitting a plane representing the material balance equation in three dimensions.[7] Values of F, E_t, and S(P,t) are plotted on a 3-D graph in z, x, and y axes, respectively. Values of N and U are then obtained from the slopes of the surfaces on the 3-D plot. This approach is viewed to be more precise than traditional analysis, where a best-fit straight line is drawn through a set of points in two dimensions. An example of Tehrani plot is presented later in the chapter.

Data required for history matching includes the following:

1. Cumulative oil, water, and gas productions from the reservoir for a series of "time points"
2. Average reservoir pressures at the corresponding time points
3. PVT data for the reservoir fluids over the expected pressure ranges

Applications of oil reservoir material balance equation

Unknown parameters as given below are determined by using material balance as an equation of a straight line for different drive mechanisms:

- **Solution gas drive oil reservoirs.** Unknown variable is original oil in place.
- **Gascap drive oil reservoirs.** Unknown variables are original oil in place and gascap size.
- **Water drive oil reservoirs.** Unknown variables are original oil in place and natural water influx.

If the original oil in place, gascap size, and aquifer strength and size are known with reasonable accuracy, material balance equations can be used to predict the future performance. Solutions for gascap drive and natural water drive reservoirs are very complex. Performance prediction above the bubblepoint is rather straightforward. However, prediction below the bubblepoint requires simultaneous solutions of the material balance equation and subsidiary liquid saturation, produced gas/oil ratio, and cumulative gas production, as indicated earlier.

Material Balance of Gas and Gas Condensate Reservoirs

The general material balance as an equation of a straight line for dry gas reservoirs is given below:

$$F = G (E_g + E_{fw}) + W_e \tag{12.17}$$

where
$\quad$ F $\;$ = underground withdrawal, rb,
$$\quad\quad = G_p B_g + W_p B_W \tag{12.18}$$
$\quad G_p$ = cumulative dry gas production, Mscf,
$\quad B_g$ = gas formation volume factor of dry gas, rb/Mscf,
$\quad W_p$ = cumulative water production, stb, and
$\quad B_W$ = water formation volume factor, rb/stb.

Original gas in place (Mscf) is given by the following (refer to chapters 2 and 9):

$$G = 7,758 Ah \, \emptyset (1 - S_{wi}) / B_{gi} \tag{12.19}$$

where

A = drainage area, acres,

h = gas zone thickness above the gas/water contact,

$\emptyset$ = porosity, fraction,

B_{gi} = initial gas formation volume factor, rb/Mscf,

S_{wi} = initial water saturation, fraction PV,

W_e = cumulative water influx, rb

    = $US(p,t)$,

U = aquifer constant,

$S(p,t)$ = aquifer function, psi,

E_g = expansion of gas, rb/scf

    = $B_g - B_{gi}$, $\tag{12.20}$

E_{fw} = expansion of connate water and reduction in the pore volume, rb/Mscf

    = $B_{gi} C_e (P_i - P)$ $\tag{12.21}$

For a depletion drive reservoir, W_e and E_{fw} can be neglected. Then considering no water and condensate production, general material balance Equation 12.17 can be reduced to a more popular form as follows:

$$p/z = p_i/z_i (1 - G_p/G) \tag{12.22}$$

where

p = current pressure, psia,

p_i = initial pressure, psia,

z = gas compressibility at the current pressure,

z_i = gas compressibility at the initial pressure, and

G_p = cumulative produced gas at the current pressure.

From Equation 12.22, the following equations can be obtained:

$$G_p = G \left[1 - (p/z) / (p_i/z_i) \right] \tag{12.23}$$
$$G_p = G \left(1 - B_{gi}/B_g \right) \tag{12.24}$$

where

B_{gi} = initial gas formation volume factor (FVF) at the initial pressure, rb/Mscf, and

B_g = gas formation volume factor at the current pressure, rb/Mscf.

A plot of p/z versus G_p should yield a straight line. The original gas in place can be obtained by extrapolating the straight line to 0. The equation can be used directly to calculate future gas production corresponding to a given pressure. If the abandonment pressure is known, this equation can be used to estimate the ultimate gas production.

For condensate reservoir:

$$F = G_w (E_g + E_{fw}) + W_e \tag{12.25}$$

where

G_w = wet gas in place, scf.

Using Equation 12.18 and replacing B_g by B_{gw} for wet gas, the following expression for F can be obtained:

$$F = G_{wp} B_{gw} + W_p B_W \qquad (12.26)$$

where

$$G_{wp} = G_{dp} + N_{pc} F_c \qquad (12.27)$$

G_{wp} = cumulative wet gas production, Mcf

G_{dp} = cumulative dry gas production, Mcf

N_{pc} = cumulative condensate production, stb,

F_c = condensate conversion factor, Mcf/stb

$\quad = 132.79 \times$ sp. gr./Mcf, $\qquad (12.28)$

sp. gr. = specific gravity of condensate (water = 1.0)

$\quad = 141.5 / (131.5 + °API), \qquad (12.29)$

M_c = molecular weight of condensate

$\quad = 6,084/ (°API - 5.9), \qquad (12.30)$

B_{gw} = rb/Mscf wet gas

$\quad = 5.04\, z_w\, T/p \qquad (12.31)$

z_w = wet gas compressibility factor,

$B_{gd} = 5.04\, z_w\, T/p * f_s, \qquad (12.32)$

f_s = dry gas mole fraction

$\quad = n_g/ (n_g + n_c), \qquad (12.33)$

$n_g = 1/379.4$ moles per scf/[oil/gas ratio (OGR), stb/scf], $\qquad (12.34)$

OGR = separator oil (condensate)/gas ratio, stbo/Mscf,

$n_c = 350 \times$ sp. gr./M_c moles bbl, $\qquad (12.35)$

$WGR = 1 - (1 - B_{gwi}/B_{gw}) \qquad (12.36)$

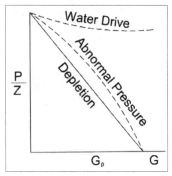

Fig. 12-8. Plot of p/Z versus G_p. *Source: A. Satter and G. C. Thakur. 1994. Integrated Petroleum Reservoir Management—A Team Approach. Tulsa: PennWell.*

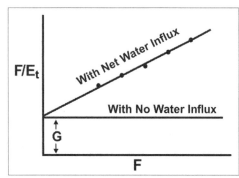

Fig. 12-9. Plot of F/E_t versus F

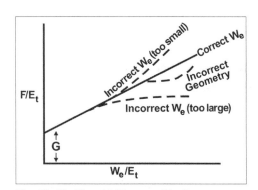

Fig. 12-10. Plot of F/E_t versus W_e/E_t

This section summarizes gas material balance as an equation of a straight line and the associated plots. The original gas in place can be determined by history matching the past performance using the graphical methods as described in the following discussion.

I. p/z method.

$$\frac{p}{z} = \left(1 - \frac{G_p}{G}\right)\frac{p_i}{z_i} \qquad (12.37)$$

Plot p/z versus G_p, where G = the intercept on the x-axis, as shown in Figure 12–8. The plot is linear if $W_e = E_{fw} = 0$.

II. Cole method. Plot F/E_t versus F, where G = the intercept on the y-axis (see fig. 12–9).

III. Havlena and Odeh method. Plot F/E_t versus W_e/E_t (or S/E_t), where G = the intercept on the y-axis, and U = the slope (see fig. 12–10).

$$\frac{F}{E_t} = G + \frac{W_e}{E_t}$$

$$= G + U\frac{S}{E_t} \qquad (12.38)$$

IV. Pressure match method (any type reservoir). Plot p versus G_p.

Adjust values of G and W_e for the best match, in a manner similar to the estimation of oil in place shown in Figure 12–7.

The p/z versus cumulative gas production relation can be used to estimate the original gas in place for depletion drive reservoirs. Other methods that account for water influx can be employed for estimating original gas in place in the case of water drive reservoirs. Types of data required for history matching include the following:

- Cumulative gas, condensate, and water productions from the reservoir for a series of time points
- Average reservoir pressures at the corresponding time points
- PVT data for the reservoir fluids over the expected reservoir pressure ranges

The future performance of gas reservoirs without water influx and water production can be calculated directly from a material balance equation, giving a straight line relationship between p/z and cumulative gas production. To calculate gas production, gas compressibility (z) versus pressure (p) and original gas in place must be known.

Applications of the gas reservoir material balance equation

Unknown parameters as given below are determined by using material balance as an equation of a straight line for different drive mechanisms:

- Depletion drive gas reservoirs. Unknown variable is the original gas in place.
- Water drive gas reservoirs. Unknown variables are the original gas in place and natural water influx.

Material balance based on flowing pressure method

The traditional p/z method to calculate the original gas in place is based on the measurement of static reservoir pressure at regular intervals throughout the life of the reservoir. However, in the case of reservoirs having low or ultra low permeability (0.1 mD or less), shut-in times are quite long to obtain a stabilized bottomhole pressure. A long shut-in period results in a significant loss in production.

The flowing material balance method is based on pseudosteady-state flow that is achieved with stabilized production from a reservoir. As explained in chapter 5, pseudosteady-state flow occurs when the rate of change of reservoir pressure is constant, as a result of continued production at a constant rate. During pseudosteady-state flow, the magnitude of the pressure drop at the wellbore is the same as the drop in average reservoir pressure in a given period (fig. 5–4). The difference between the static average reservoir pressure and the flowing bottomhole pressure remains the same, although both decrease in magnitude as production continues. The net result is that the flowing p/z line has the same slope as the static p/z, but the latter is shifted to the left (fig. 12–11). Since the initial static reservoir pressure is known at the discovery of the reservoir, original gas in place is estimated by appropriately shifting p/z line to the right. However, in most wells, the production rate of oil or gas is not constant and is found to decline with time. Hence the difference between average reservoir pressure and flowing bottomhole pressure is not the same and is a function of flow rate.

Slight modification of the classical psuedo steady-state fluid flow equation in porous media leads to the development of a flowing material balance equation for both oil and gas reservoirs, as shown by Agarwal and Gardner.[8] The approach

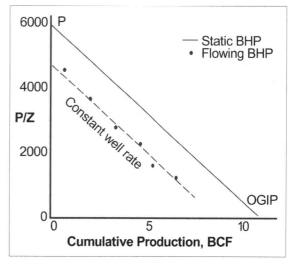

Fig. 12–11. Comparison between static and flowing bottomhole pressure in classical analysis

is applicable to both constant and varying well rates. Based on the general equation for psuedo steady-state flow in chapter 5, the following can be shown:

$$p_{av} - p_{wf} = \frac{141.2\, B_o\, \mu\, q}{kh}\left[\ln\frac{r_e}{r_w} - \frac{3}{4}\right] \tag{12.39}$$

$$p_i - p_{av} = \frac{q}{c_e N}\, t \tag{12.40}$$

where

p_{av} = average reservoir pressure as a function of flow rate and time.

Eliminating p_{av} from the preceding equations and dividing by q, the following can be obtained:

$$\frac{p_i - p_{wf}}{q} = \frac{1}{c_e N}\, t + \frac{141.2\, B_o\, \mu}{kh}\left[\ln\frac{r_e}{r_w} - \frac{3}{4}\right] \tag{12.41}$$

Equation 12.41 can be modified to account for variable rates, as given in the following:

$$\frac{\Delta P}{q} = \frac{1}{c_e N}\, t_c + \frac{141.2\, B_o\, \mu}{kh}\left[\ln\frac{r_e}{r_w} - \frac{3}{4}\right] \tag{12.42}$$

In Equation 12.42, t_c is the material balance time and is calculated as N_p/q.

Multiplying both sides by $q/\Delta p$, an equation for a straight line can finally be obtained. When $q/\Delta p$ is plotted against $Q/(c_e\Delta p)$, the line extrapolates to N, the original oil in place. A similar approach is followed to estimate original gas in place based on the flowing bottomhole pressure of the gas well and a varying well rate.

Role of material balance analysis in reservoir characterization

Material balance study of an oil or gas reservoir may aid in identifying the reservoir drive mechanisms (gascap drive, water drive, or a combination), as previously demonstrated. However, this technique is also helpful in characterizing the type of aquifer that best fits the production history. Several steady-state and unsteady-state aquifer models can be compared for a reservoir in a relatively short time by computer software. A few examples are included later in the chapter in order to illustrate the method. In an integrated reservoir study, an aquifer model that best fits the production history is confirmed by other methods, including reservoir simulation.

In certain cases, material balance analysis may raise more questions than it answers, requiring further reservoir studies. Dake cited a material balance study that indicated a significantly lower hydrocarbon volume than indicated by volumetrics.[9] Further reservoir studies revealed that oil production was primarily from a high permeability zone, while the rest of the formation was virtually untapped due to significantly low permeability. This finding is an important milestone in characterizing the reservoir in question, as future engineering efforts are necessary to produce oil from the tight zones. It is obvious that the material balance method is based on the dynamic response of a reservoir, regardless of the hydrocarbon volume calculated on the basis of drainage area, thickness, porosity, and saturation.

Limitations of material balance technique

Well-recognized limitations of the material balance technique are mentioned in the beginning of the chapter. These include the following:

- The reservoir is treated as a tank having homogeneous properties. When significant heterogeneity exists, estimation of oil in place based on material balance could be misleading.
- Fluid flow behavior in the reservoir is not considered.
- Individual well performance cannot be evaluated.

The material balance method may prove to be inadequate in studying a reservoir during fluid reinjection late the life cycle. Produced hydrocarbon fluids and encroached water from the adjacent aquifer cannot "move backward," so to speak, leading to possible errors in analysis.

Review of Material Balance Equations

Familiar material balance equations for various types of production mechanisms are reviewed in the following discussion.[10–12]

Undersaturated oil reservoir with negligible water influx

Production above bubblepoint. The controlling factors are effective compressibility of the system and changes in the oil formation volume factor:

$$N = \frac{N_p B_o + W_p B_w}{B_{oi} c_e \Delta P} \tag{12.43}$$

Note that oil production is controlled by the compressibilities of oil and rock pore above the bubblepoint in the absence of water or gas drive. Furthermore, $W_p = 0$, as the initial water saturation is not mobile in most cases.

Production below bubblepoint. The dominating factors are the high values of gas compressibility and producing gas/oil ratio.

$$N_p / N = \frac{B_t - B_{ti}}{B_t + (R_p - R_{si}) B_g} \tag{12.44}$$

Residual oil saturation in the depleted reservoir can be estimated by Equation 2.50, as presented in chapter 2:

$$\frac{N_p}{N} = \frac{(S_{oi}/B_{oi} - S_{or}/B_{or})}{S_{oi}/B_{oi}} \times 100$$

During production below the bubblepoint, the volume of free gas in the reservoir can be estimated as in the following:

Free gas volume = Initial volume dissolved in oil (NR_{si}) − Produced volume of gas ($N_p R_p$)
$$\text{− Volume remaining in solution } (N − N_p)R_s \tag{12.45}$$

Saturation of gas in a depleted reservoir pressure can be calculated as follows:

$$S_g = 1 − (1 − N_p B_o / N B_{oi})(1 − S_w) \tag{12.46}$$

S_w denotes the saturation at the initial condition, as reservoir production is not influenced by water encroachment. An example to estimate the recovery factor in an undersaturated reservoir is presented later (Example 12.1). Computation of the free gas volume and gas saturation is left as an exercise later.

Saturated reservoir with a gas cap and negligible water influx

In this case, the general material balance equation in Equation 12.1 reduces to the following:

$$\frac{F}{E_o} = N + mN \left(\frac{E_g}{E_o} \right) \tag{12.47}$$

E_o, E_g, and m are defined in Equations 12.4 through 12.7.

A plot of F/E_o versus E_g/E_o yields a straight line having a slope of mN and an intercept of N. Example 12.2 illustrates the procedure. Alternately, a trial-and-error solution can be obtained by assuming multiple values of m and plotting F versus $(E_o + mE_g)$ until a straight line is obtained that passes through the origin. Incorrect values of m lead to lines that are curved either upward or downward.

Reservoir with water influx from an adjacent aquifer

The following equation is used as a basis to estimate N and W_e by trial and error:

$$\frac{F}{E_t} = N + \frac{W_e}{E_t} \tag{12.48}$$

where

$$E_t = E_o + mE_g + E_{fw} . \tag{12.49}$$

Water influx from the aquifer (W_e) is usually estimated by assuming an aquifer model described in chapter 4.

Dry gas reservoir with no water influx

$$G_p = G\left(1 - B_{gi}/B_g\right) \tag{12.50}$$

Computer-Aided Material Balance Analysis

Computer applications that perform material balance analysis of oil and gas reservoirs are typically found to have the following capabilities:

Original oil in place
- FE method
- Campbell method
- Gascap method
- Direct approach involving curve matching techniques

Initial gas in place
- p/z method
- Ramagost method
- Coal method
- Roach method

Pseudosteady-state aquifer model
- Schilthuis
- Hurst
- Fetkovich
- Small (pot) aquifer

Unsteady-state aquifer model
- Infinite linear
- Carter-Tracy
- Finite/infinite radial
- Finite/infinite bottom water drive

In the following section, selected examples with the material balance method are analyzed with the aid of OilWat/GasWat software.★

★ OilWat and GasWat are trademarks of IHS Energy, Inc.

Examples

In order to illustrate the applications of the material balance method in petroleum reservoirs, the following cases are included:

1. Undersaturated oil reservoir with no gas cap or water influx
2. Oil reservoir with a small gas cap
3. Oil reservoir with natural water drive
4. Depletion drive gas reservoir
5. Gas condensate reservoir
6. Abnormally pressured gas reservoir

Example 12.1: Recovery estimate of an undersaturated oil reservoir based on limited data. Given just a few fluid properties, the material balance technique permits estimation of the recovery factor in the case of an undersaturated oil reservoir producing above the bubblepoint pressure. The data may be available with relative ease from fluid samples obtained from an exploratory well drilled in a new reservoir. In this case, no production data would be available, and very little would be known about any reservoir heterogeneities.

Primary recovery from an undersaturated oil reservoir, as it is depleted from initial pressure to bubblepoint pressure, depends primarily on the driving energy. This production mechanism results from the expansion characteristics of reservoir fluids and the reduction in pore volume. Hence, oil recovery is a function of the effective compressibility of the fluid-rock system as presented in Equation 12.11.

(a) Given the following reservoir and fluid properties, estimate the recovery factor of a reservoir above the bubblepoint. List the assumptions used in the analysis.

Initial reservoir pressure, psia, = 2,554.
Bubblepoint pressure, psia, = 1,745.
Initial oil formation volume factor, rb/stb, = 1.24.
Oil formation volume factor at bubblepoint, rb/stb, = 1.252.
Initial water saturation = 0.22.
Compressibility of water, psi^{-1}, = 3.1×10^{-6}.
Formation compressibility , psi^{-1}, = 3.4×10^{-6}.

(b) Estimate the recovery factor below the bubblepoint with the following data:

Initial solution gas/oil ratio, scf/stb, = 850.
Estimated abandonment pressure, psia, = 1,000.
The following are evaluated at abandonment pressure:
Oil formation volume factor, rb/stb, = 1.045.
Gas formation volume factor, rb/stb, = 0.00123.
Solution gas/oil ratio, scf/stb, = 250.
Producing gas/oil ratio, scf/stb, = 1,500.

Solution:

(a) The average oil compressibility between initial pressure and bubblepoint pressure is calculated as in the following:

$$c_o = \frac{B_{ob} - B_{oi}}{B_{oi}\, \Delta P}$$

$$= \frac{1.252 - 1.24}{1.24 \times (2,554 - 1,745)}$$

$$= 11.96 \times 10^{-6}\, psi^{-1}$$

Next, the effective compressibility of the fluid-rock system is computed based on Equation 12.12, as follows:

$$c_e = \frac{c_o(1\text{-}S_{wc}) + c_w S_{wc} + c_f}{(1\text{-}S_{wc})}$$

$$= \frac{[11.96 \times (1 - 0.22) + 3.1 \times 0.22 + 3.4] \times 10^{-6}}{1 - 0.22}$$

$$= 17.19 \times 10^{-6} \, psi^{-1}$$

Finally, the recovery factor at the bubblepoint is estimated by the following:

$$N_p/N = \frac{B_{oi}}{B_{ob}} \, c_e \, \Delta p$$

$$= \frac{1.24 \times 17.19 \times 10^{-6} \times (2{,}554\text{-}1{,}745)}{1.252}$$

$$= 0.0138 \text{ or } 1.38\%$$

The assumption inherent in the preceding analysis is that the oil recovery is influenced only by fluid and rock compressibilities. No pressure support is provided by an adjacent aquifer, external injection, or any other means. If additional driving energy is available, the recovery is expected to be greater.

Below the bubblepoint, the mechanism of recovery is dominated by the expansion of the gas phase that comes out of the liquid phase. The recovery factor can be estimated from Equation 12.13 once other information is available, including solution gas/oil ratio, gas formation volume factor, producing gas/oil ratio, and estimated abandonment pressure.

(b) Values of two-phase formation volume factor are computed at the initial pressure and at abandonment based on Equation 12.6. This is followed by estimation of recovery from a reservoir producing under solution gas drive based on Equation 12.13:

$$B_t = B_o + (R_{si} - R_s) \, B_g$$

$$= 1.045 + (850 - 250) \times 0.00123$$

$$= 1.783 \text{ rb/stb at } 1{,}000 \text{ psia}$$

$$B_{ti} = 1.24 \text{ rb/stb} \qquad (R_s = R_{si})$$

$$N_p/N = \frac{B_t - B_{ti}}{B_t + (R_p - R_{si}) \, B_g}$$

$$= \frac{1.783 - 1.24}{1.783 + (1500 - 850) \times 0.00123}$$

$$= 0.21 \text{ or } 21\%$$

Example 12.2: Analysis of an oil reservoir with gascap drive and no aquifer influence. Production and fluid properties data is available for a depleting oil reservoir as presented in Table 12–1. The primary mechanism of production is gascap drive. Water influx into the reservoir and production of water are assumed to be negligible. The objective is to estimate the original oil in place and the size of the gas cap based on the material balance method.

Table 12-1. Production history and fluid properties

Pressure, psia	N_p, MMstb	G_p, bcf	W_p, stb	R_p, scf/stb	B_o, rb/stb	R_s, scf/stb	B_g, rb/scf
2,890	0	0	0	0	1.319	526	9.289E−04
2,730	0.709	0.522	0	736	1.303	491	1.1384E−03
2,650	1.118	0.843	0	753	1.294	474	1.2555E−03
2,570	1.512	1.221	0	807	1.286	457	1.3850E−03
2,490	1.930	1.515	0	785	1.278	441	1.4472E−03
2,410	2.437	2.009	0	824	1.271	424	1.6232E−03

Solution: In this case, the general expression for material balance presented in Equation 12.1 reduces to the following:

$$F = N(E_o + mE_g)$$

This is due to the fact that the terms representing water influx, water compressibility, and pore volume reduction are negligible. The above equation can be rewritten as follows:

$$F/E_o = N + mN\, E_g/E_o$$

This indicates that a plot of F/E_o versus E_g/E_o would yield the values of m and N. Example calculations for F, E_o, and E_g at 2,730 psia are given in the following:

$$\begin{aligned}
F &= N_p[B_o + (R_p - R_s)B_g] \\
&= 0.709\,[1.303 + (736 - 491) \times 1.1384 \times 10^{-3}] \\
&= 1.122 \text{ MMstb}
\end{aligned}$$

$$\begin{aligned}
E_o &= (B_o - B_{oi}) + (R_{si} - R_s)\,B_g \\
&= (1.303 - 1.319) + (526 - 491)\,(1.1384 \times 10^{-3}) \\
&= 0.0238
\end{aligned}$$

$$\begin{aligned}
E_g &= B_{oi}\,(B_g/B_{gi} - 1) \\
&= 1.319\,(1.1384 \times 10^{-3}/9.289 \times 10^{-4} - 1) \\
&= 0.2975
\end{aligned}$$

Values of F, E_o, and E_g are tabulated in Table 12–2.
A plot of F/E_o versus E_g/E_o indicates the following values of the slope and intercept of the best-fit straight line:

Slope (mN) = 1.69 MMstb.
Intercept (N) = 25.69 MMstb.
m = 0.065.

Table 12–2. Results of calculation

Pressure, psia	F, MMstb	E_o	E_g
2,890			
2,730	1.122	0.0238	0.2975
2,650	1.839	0.0403	0.4638
2,570	2.678	0.0626	0.6477
2,490	3.427	0.0820	0.7361
2,410	4.681	0.1176	0.9859

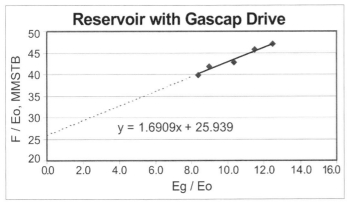

Fig. 12–12. Analysis of gascap reservoir by material balance technique using the straight-line method

Table 12–3. Reservoir performance

Production Time (days)	Reservoir Pressure (psia)	Cumulative Oil Production (MMstb)
0	2,740.0	0.00
365	2,500.0	7.88
730	2,290.0	18.42
1,095	2,109.0	29.15
1,460	1,949.0	40.69
1,825	1,818.0	50.14
2,190	1,702.0	58.20
2,555	1,608.0	65.39
2,920	1,535.0	70.74
3,285	1,480.0	74.54
3,650	1,440.0	77.43

Example 12.3: Investigation of water influx into an oil reservoir. The following production data over a period of 10 years is obtained from an oil reservoir (table 12–3). Reservoir pressure declined from 2,740 psia to 1,440 psia following the production of 77.43 MMstb. No water production was observed. The objective is to perform a material balance analysis of the reservoir production with two assumed models as follows:

- Case I: Reservoir performance without any aquifer influence
- Case II: Reservoir performance affected by water influx from an adjacent aquifer

Reservoir fluid properties as a function of pressure are given in Table 12–4.

Solution: Material balance analyses were performed for the two reservoir models mentioned above by OilWat/GasWat. Results are plotted in Figures 12–13 and 12–14. An interesting observation may be noted. The original oil in place of 210 MMstb, as estimated by assuming water influx into the reservoir, is less than one-half of that calculated by the material balance method when no aquifer influence is assumed (533.7 MMstb). An unsteady-state radial aquifer model is assumed in the latter.

Table 12–4. Fluid properties

Pressure (psia)	B_o (rb/stb)	R_s (scf/stb)	B_g (rb/Mscf)	B_w (rb/stb)
2,740.0	1.404	650	0.93	1.0026
2,500.0	1.374	592	0.98	1.0033
2,290.0	1.349	545	1.07	1.0040
2,109.0	1.329	507	1.17	1.0045
1,949.0	1.316	471	1.28	1.0050
1,818.0	1.303	442	1.39	1.0054
1,702.0	1.294	418	1.50	1.0057
1,608.0	1.287	398	1.60	1.0060
1,535.0	1.280	383	1.70	1.0062
1,480.0	1.276	371	1.76	1.0064
1,440.0	1.273	364	1.82	1.0065

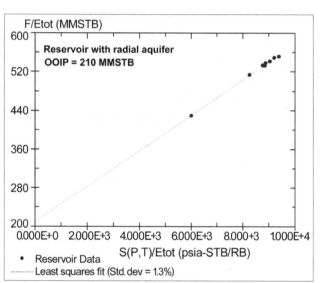

Fig. 12–14. Estimate of original oil in place by Havlena-Odeh method when a radial aquifer model is considered. *Source: D. Havlena and A. S. Odeh. 1963. The material balance as an equation of straight line. Journal of Petroleum Technology. August, 896–900. Courtesy of IHS Inc., Energy Division.*

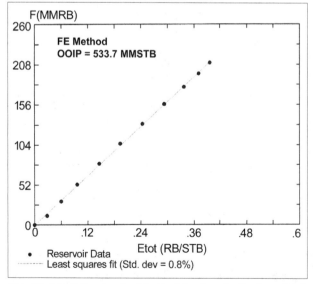

Fig. 12–13. Estimate of original oil in place by Havlena-Odeh method. No aquifer model is considered. *Source: D. Havlena and A. S. Odeh. 1963. The material balance as an equation of straight line. Journal of Petroleum Technology. August, 896–900. Courtesy of IHS Inc., Energy Division.*

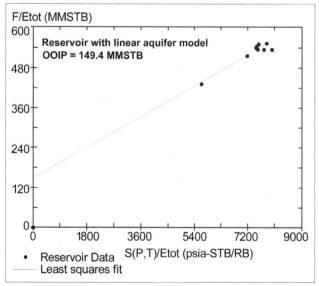

Fig. 12–15. Estimate of original oil in place by Havlena-Odeh method when a linear aquifer model is considered. *Source: D. Havlena and A. S. Odeh. 1963. The material balance as an equation of straight line. Journal of Petroleum Technology. August: pp. 896–900. Courtesy of IHS Inc., Energy Division.*

Table 12–5. Production history of abnormally pressured reservoir. *Source: B. Wang and T. S. Teasdale. 1987. GASWAT-PC: A microcomputer program for gas material balance with water influx. SPE Paper #16484. Presented at the Petroleum Industry Applications of Microcomputers meeting, Society of Petroleum Engineers, Del Lago on Lake Conroe, TX, June 23–26.*

Production Time (days)	Reservoir Pressure (psia)	Cumulative Gas Production (bcf)	Cumulative Condensate Production (MMstb)	Wet Gas Production (bcf)
0.00	9,507.00	0.0000	0.0000	0.000
69.00	9,292.00	0.3930	0.0299	0.415
182.00	8,970.00	1.6420	0.1229	1.734
280.00	8,595.00	3.2260	0.2409	3.407
340.00	8,332.00	4.2600	0.3171	4.498
372.00	8,009.00	5.5040	0.4069	5.809
455.00	7,603.00	7.5380	0.5612	7.959
507.00	7,406.00	8.7490	0.6508	9.237
583.00	7,002.00	10.5090	0.7767	11.092
628.00	6,721.00	11.7580	0.8643	12.407
663.00	6,535.00	12.7890	0.9395	13.494
804.00	5,764.00	17.2620	1.2553	18.204
987.00	4,766.00	22.8900	1.6158	24.103
1,183.00	4,295.00	28.1440	1.9134	29.580
1,373.00	3,750.00	32.5670	2.1360	34.170
1,556.00	3,247.00	36.8200	2.3078	38.552

This issue is resolved by looking into other studies involving the reservoir. These include, but are not limited to, the following:

- Geologic and geophysical studies
- Volumetric analysis
- Well test results where boundary effects are evident

Once the influence of an adjacent aquifer is established by integrated study, applicability of various aquifer models can be investigated by the material balance method. Figure 12–15 presents the results of analysis by assuming a linear aquifer model. This indicates a poor fit with the reservoir performance, hence the first model (the unsteady-state radial aquifer model) is the preferred choice.

Example 12.4: Analysis of abnormally pressured gas reservoir under depletion drive. The following data in Table 12–5 is obtained from an overpressured gas reservoir producing for more than four years.[13] The pressure gradient is found to be 0.843 psi/ft. No aquifer influence is detected in this case. The objective is to estimate ultimate gas recovery from the reservoir by the material balance method.

Relevant PVT properties at declining reservoir pressures are presented in Table 12–6.

Table 12–6. Fluid PVT properties

Pressure (psia)	z Factor	B_g (rb/Mscf)	p/z (psia)
9,507.00	1.4400	0.5537	6,602.08
9,292.00	1.4180	0.5578	6,552.89
8,970.00	1.3870	0.5652	6,467.20
8,595.00	1.3440	0.5716	6,395.09
8,332.00	1.3160	0.5773	6,331.31
8,009.00	1.2820	0.5851	6,247.27
7,603.00	1.2390	0.5957	6,136.40
7,406.00	1.2180	0.6012	6,080.46
7,002.00	1.1760	0.6139	5,954.08
6,721.00	1.1470	0.6238	5,859.63
6,535.00	1.1270	0.6304	5,798.58
5,764.00	1.0480	0.6646	5,500.00
4,766.00	0.9770	0.7493	4,878.20
4,295.00	0.9280	0.7898	4,628.23
3,750.00	0.8910	0.8685	4,208.75
3,247.00	0.8540	0.9614	3,802.11

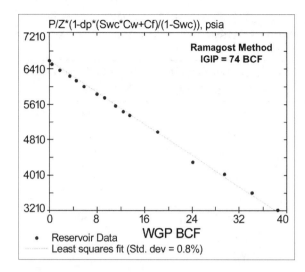

Fig. 12–16. Estimation of initial gas in place by the Ramagost method. *Courtesy of IHS Inc., Energy Division.*

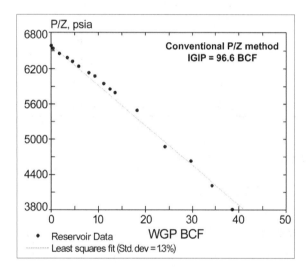

Fig. 12–17. Estimation of initial gas in place by conventional p/Z method where the effect of pore compressibility is not considered. *Courtesy of IHS Inc., Energy Division.*

Table 12-7. Production history of reservoir under water drive

Time (days)	G_p (bcf)	Wet Gas Production (bcf)	p/z (psia)	Cum. Water Production (MMstb)
0.00	0.000	0.000	4,364.50	—
283.00	23.451	23.540	4,302.83	0.0000
452.19	29.323	29.427	4,254.93	0.0010
799.50	57.343	57.491	4,111.86	0.0080
1,296.69	102.317	102.550	3,961.91	0.0270
2,172.50	165.984	166.351	3,764.00	0.0570
2,607.81	190.122	190.536	3,667.39	0.0630
2,966.00	226.208	226.688	3,644.11	0.0780
3,298.31	263.241	263.797	3,571.72	0.0930
3,663.50	297.909	298.529	3,424.86	0.1190
4,028.81	327.622	328.310	3,385.81	0.1740
4,390.00	352.843	353.578	3,348.48	0.2010
4,767.19	370.186	370.952	3,302.96	0.2070
5,156.50	387.236	388.036	3,326.75	0.2410
5,414.50	417.052	417.912	3,181.62	0.3070
5,504.69	424.861	425.735	3,165.40	0.3170
5,579.69	433.398	434.285	3,145.07	0.3430
5,810.00	449.845	450.759	3,122.64	0.4390
5,940.00	465.379	466.313	3,052.85	0.4580
6,293.31	521.950	522.951	2,727.55	0.6670
6,501.19	549.323	550.349	2,574.61	0.8610
6,807.50	575.786	576.834	2,514.40	1.1540
6,875.50	581.104	582.156	2,504.52	1.2450

Table 12-8. Composition of natural gas

Gas Composition	Mole Fraction	Mol. Wt.
N_2	0.0034	28.02
CO_2	0.0015	44.01
H_2S	0.0000	34.08
C_1	0.9711	16.04
C_2	0.0161	30.07
C_3	0.0031	44.09
i-C_4	0.0009	58.11
n-C_4	0.0007	58.12
i-C_5	0.0003	72.13
n-C_5	0.0002	72.14
C_6	0.0006	86.17
C_7+	0.0021	100.20

Solution: Since the reservoir is overpressured, the Ramagost method is utilized to estimate initial gas in place. Additionally, a comparative study is performed by estimating initial gas in place based on the conventional p/z method. The gas in place estimated by the latter method is found to be significantly higher: 96.6 bcf compared to 74 bcf calculated by the Ramagost method.

Relevant material balance plots are presented in Figures 12–16 and 12–17. Significant deviations from the expected straight-line trend are observed when the p/z method is used. This is due to the fact that the influence of pore compressibility in depleting an overpressured formation was not considered. In fact, p/z plots of abnormally pressured gas reservoirs may show dual slope. The Ramagost method is found to have a better match of reservoir data. Hence, the estimation of initial gas in place of 74 bcf is viewed as more realistic.

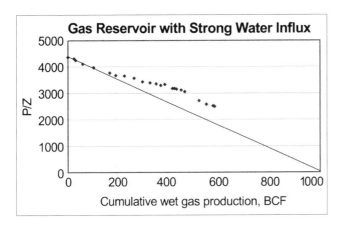

Fig. 12–18. Gas reservoir with strong water influx. The declining trend in reservoir pressure is affected by strong aquifer support.

Example 12.5: Aquifer modeling for a reservoir under strong water influx. The following production versus time data is available from an offshore gas field, including cumulative water production (table 12–7).[14] Typically, reservoir performance in the area is strongly influenced by an adjacent aquifer. The objective of the study is to model the aquifer and estimate initial gas in place by a computer-aided material balance study. The initial gas in place is estimated to be 806 bcf by the volumetric method. Gas composition and properties are also included (table 12–8).

Solution: A conventional plot of p/z over cumulative production does not fall on a straight line (fig. 12–18), as the reservoir experiences aquifer influence in the form of a strong water drive. Moreover, water encroachment is time dependent. Hence, the Cole method is used to further analyze the reservoir performance, which takes into account any water influx into the reservoir (fig. 12–19). When compared with the results of a volumetric study, the initial gas in place as calculated by the Cole method (896 bcf) appears to be more realistic.

Attempts were made in order to fit several aquifer models, steady state and unsteady state, to match reservoir performance data. It was found that the unsteady-state radial aquifer model fits the data better than the others (fig. 12–20). Models reviewed included the Hurst steady-state model (fig. 12–21), the Schilthuis steady-state model, and the infinite linear aquifer model.

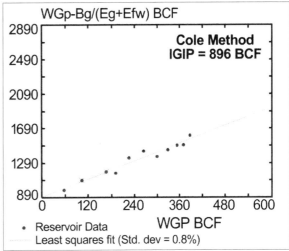

Fig. 12–19. Estimate of initial gas in place by the Cole method. *Courtesy of IHS Inc., Energy Division.*

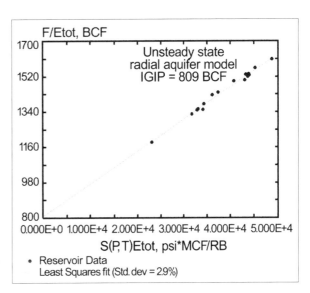

Fig. 12–20. Estimate of initial gas in place by Havlena-Odeh method that considers unsteady-state radial aquifer model. *Courtesy of IHS Inc., Energy Division.*

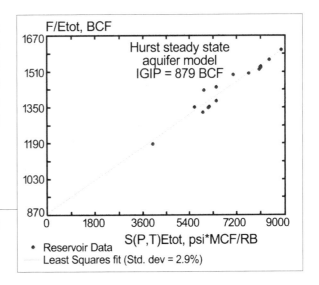

Fig. 12–21. Estimate of initial gas in place. Analysis incorporates the Hurst steady-state model. *Courtesy of IHS Inc., Energy Division*

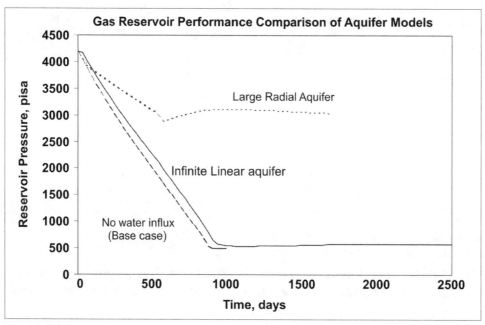

Fig. 12–22. Reservoir pressure decline under various scenarios of aquifer influence. *Courtesy of Gemini Solutions, Inc.*

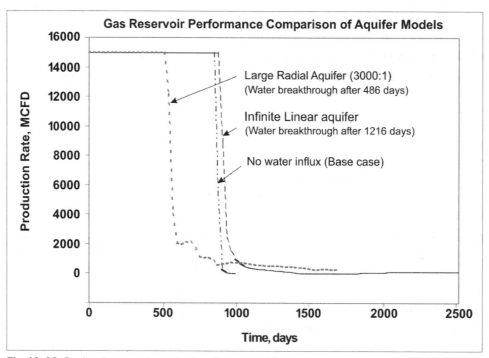

Fig. 12–23. Decline in reservoir production. *Courtesy of Gemini Solutions, Inc.*

Example 12.6: Simulation of gas reservoir performance with various aquifer models. In order to gain more insight into the influence of various aquifer models on gas reservoir performance, case studies were performed using Merlin simulation software. The following scenarios were considered in a sensitivity study:

- Reservoir under depletion drive with no water influx (base case)
- Infinite linear aquifer, in which water influx occurs from a linear direction
- Large radial aquifer (radius of aquifer/radius of reservoir = 3,000:1)

Selected reservoir model data as utilized in simulation is given in the following:

Initial gas in place, bcf, = 15.

Reservoir pressure, psia, = 4,185.

Reservoir temperature, °F, = 192.

Depth of reservoir, ft, = 9,000.

Drainage area, acres, = 160.

Number of reservoir layers = 3.

Net thickness of each layer, ft, = 10.

Porosity of each layer (fraction) = 0.30.

Average reservoir permeability, darcies, = 1.

Standard correlation is used for relative permeability data.

Gas specific gravity = 0.6096. (Air = 1.)

Initial gas formation volume factor = 0.0042.

Maximum allowable well rate, MMcfd, = 15.

Minimum well bottomhole pressure, psi, = 500.

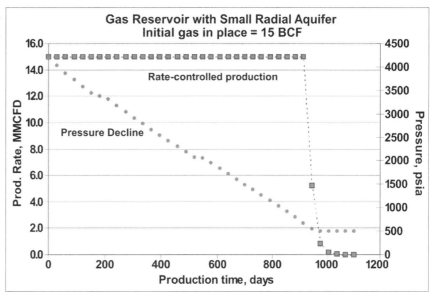

Fig. 12–24. Simulation of a gas reservoir with a small radial aquifer. Well production is under rate control at 15 MMcfd initially, followed by bottomhole pressure control. *Courtesy of Gemini Solutions, Inc.*

Solution: Figures 12–22 and 12–23 present the results of the sensitivity analyses. The following observations are made:

1. Earliest water breakthrough was observed in the case of a large radial aquifer. Strong water influx affected reservoir performance significantly. Consequently, the ultimate recovery was the lowest of the three cases considered. Due to a strong water drive, reservoir pressure was significantly greater than in the other cases. Since gas recovery is mainly driven by volumetric expansion, a less-than-satisfactory recovery is anticipated when the reservoir pressure remains relatively high at abandonment.

2. Ultimate recovery was the highest in the case of an infinite linear aquifer. Water influx was not as significant as in the large radial aquifer case which led to premature breakthrough. However, it was sufficiently strong to augment production compared to the base case. The reservoir pressure depletes more than the large radial aquifer case. Breakthrough of water occurred much later, following the recovery of most of the initial gas in place.

Example 12.7: Gas in place estimation of a reservoir with a small radial aquifer (comparison with results of reservoir simulation). Performance of a gas reservoir under the influence of a small radial aquifer is simulated. Initial gas in place was assumed to be 15 bcf. Production from the gas well was initially rate-controlled at 15 MMcfd, followed by a decline once this rate could no longer be sustained. Results of the simulation indicated that the well produced under rate control for 912 days. Following the rate control production, it quickly reached its limiting pressure of 500 psia in about six months (fig. 12–24). The ultimate recovery was more than 94%. Rather high reservoir permeability (1 darcy) was used in simulation. The radius of the aquifer was assumed to be three times the radius of the gas reservoir. The objective of this study is to calculate the initial gas in place by material balance methods with first six months of production data, and compare the result with that of simulation. Merlin simulator and OilWat/GasWat software were used in this comparative study.

Table 12–9. Production data and gas deviation factor

Time, days	Pressure, psia	z Factor	Cumulative Gas Prod., bcf
0.00	4,186.00	0.953	0.0000
30.38	4,028.00	0.944	0.4560
60.81	3,871.00	0.936	0.9120
91.31	3,729.00	0.929	1.3690
121.69	3,582.00	0.923	1.8250
152.13	3,439.00	0.917	2.2810
182.50	3,372.00	0.914	2.7370

Table 12–10. Formation volume factor of gas

Pressure, psia	z Factor	B_g, rb/Mscf
4,208.70	0.9544	0.7444
3,509.80	0.9197	0.8602
3,010.50	0.9015	0.9831
2,511.30	0.8951	1.1701
2,012.00	0.8972	1.4639
1,512.80	0.9102	1.9751
1,013.50	0.9325	3.0205
813.80	0.9439	3.8077
614.10	0.9562	5.1113
514.30	0.9629	6.1460
414.40	0.9696	7.6812
314.50	0.9765	10.1923
214.70	0.9839	15.0436
114.80	0.9962	28.4880

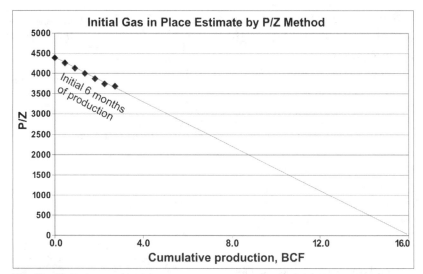

Fig. 12–25. Estimate of initial gas in place by conventional p/Z method

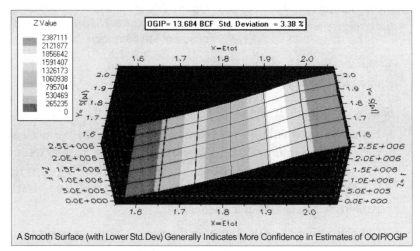

Fig. 12–26. Estimation of initial gas in place based on 3-D Tehrani plot. *Courtesy of IHS Inc., Energy Division.*

Solution: Production history obtained by numerical simulation for the initial months, along with PVT data, is tabulated in Tables 12–9 and 12–10. Results of the material balance study using various methods are also presented. Conventional p/z versus cumulative gas analysis, shown in Figure 12–25, overestimated the gas in place (16 bcf). This is a common observation in the case of gas reservoirs under water influx when analyzed by the p/z method. The aquifer model calculated the initial gas in place at 13.6 bcf, which can be viewed as a conservative estimate. Moreover, the standard deviation was smaller than that obtained by the p/z method. Figure 12–26 shows a 3-D Tehrani plot of E_t, S(p,t), and F plotted in the x, y, and z directions for the radial aquifer model, respectively. It is noted that the closest match (14 bcf) was obtained by using the Schilthuis steady-state aquifer model for the set of data used.

Table 12–11. Summary of results based on various methods

Method	Estimated IGIP, bcf	Std. deviation in regression, %
Conventional p/z	16.2	2.89
Radial aquifer	13.7	3.38
Infinite linear aquifer	12.0	3.4
Hurst steady state	16.0	4.4
Schilthuis steady state	14.0	3.4
Small (pot) aquifer	19.7	4.15

Comparative results of a material balance study, with the initial gas in place assumed in simulation to be 15 bcf, is shown in Table 12–11.

The study points to the fact that the material balance method can be used with a reasonable degree of accuracy to predict reservoir performance in relatively simple situations. Analytical methods, including material balance, are usually limited by simplifying assumptions. It is further noted that the study is based on only six months of initial production data. Accuracy in results is expected to improve once more data becomes available. The method can complement a full-blown reservoir simulation study or can be utilized solely where certain answers are needed. This allows information to be obtained without involving the huge investment of time and money typically required in detailed simulation studies.

Table 12-12. Production history and fluid properties

Reservoir Pressure, psia	z Factor	p/z	B_{gw}, rb/mcf	Condensate Gas Ratio, stb/MMcf	Condensate Sp. Gr., °API	Density of Condensate, lb/ft³
6,014.7	1.1079	5,428.92	0.6739	107.86	43.8	0.8072
4,856.7	0.9824	4,943.71	0.7423	107.86	43.8	0.8072
4,014.7	0.9078	4,422.45	0.8268	84.87	49	0.7839
3,014.7	0.8541	3,529.68	1.0337	58.67	54	0.7628
2,014.7	0.8563	2,352.8	1.5512	35.73	59	0.7428
1,014.7	0.9059	1,120.1	3.2699	25.8	61	0.7351

Last, but not least, it is noted that a traditional decline analysis is not possible in this case, as the well produced at steady rate for most of its life.

Example 12.8. Performance review of a wet gas reservoir based on p/z analysis. The production history of a wet gas reservoir, including the fraction of condensate obtained from the separator, is given in Table 12–12. PVT properties of the produced fluid are also included. The objective is to calculate the ultimate recovery of the fluid phases, gas and condensate liquid, from the reservoir.

Recovery calculations are based on determination of the mole fractions of gas and condensate in the producing stream as the reservoir is produced.

Solution: Example calculations are presented in the following.

Reservoir pressure = 6,014.7 psia.

The sp. gr. of the condensate = 141.5/(131.5 + °API)

$$= 141.5/(131.5 + 43.8)$$

$$= 0.8072.$$

The mol. wt. of the condensate is estimated as follows: M.W. = 6,084 /(°API − 5.9)

$$= 6,084 /(43.8 − 5.9)$$

$$= 160.53.$$

Number of moles of condensate = 107.86 × 5.615 × 62.4 × 0.807/160.53

$$= 189.98.$$

Moles of 1 MMcf of dry gas = 1,000,000/379.4

$$= 2,635.74.$$

Mole fraction of gas = (2,635.74)/(2,635.74 + 189.98)

$$= 0.9328.$$

Formation volume factor of dry gas: $B_{gd} = B_{gw}$ /Mole fraction of gas

$$= 0.6739/0.9328$$

$$= 0.7225.$$

Recovery of wet gas between reservoir pressure of 6,014.7 and 4,856.7 psia is calculated as:

Recovery = $1 - B_{gwi}/B_{gw}$

$$= 1 - 0.6739/0.7423$$

$$= 0.09214.$$

Table 12–13. Computation of wet gas recovery

Reservoir Pressure, psia	Mol. Wt. of Condensate	Moles of Condensate	Mole Fraction of Gas	B_{gd}, rb/Mcf	Wet Gas Recovery, fraction	Incremental Wet Gas Recovery, fraction
6,014.7	160.528	189.983	0.933	0.7225	—	—
4,856.7	160.528	189.983	0.933	0.7958	0.0921	0.0921
4,014.7	141.160	164.994	0.941	0.8786	0.1849	0.0928
3,014.7	126.486	123.860	0.955	1.0823	0.3481	0.1638
2,014.7	114.576	81.086	0.970	1.5989	0.5656	0.2175
1,014.7	110.417	60.125	0.978	3.3445	0.7939	0.2283

Table 12–14. Incremental dry gas recovery

Reservoir Pressure, psia	B_{gw}, avg.	B_{gd}, avg.	Incremental Dry Gas Recovery	Dry Gas Recovery
6,014.7	—	—	—	—
4,856.7	0.7081	0.7591	0.0921	0.0921
4,014.7	0.7846	0.8372	0.0932	0.1854
3,014.7	0.9303	0.9804	0.1659	0.3513
2,014.7	1.2925	1.3406	0.2248	0.5761
1,014.7	2.4106	2.4717	0.2387	0.8149

Table 12–15. Incremental condensate recovery

Reservoir Pressure, psia	Avg. Condensate Ratio	Condensate, avg./initial	Incremental Condensate Recovery	Total Condensate Recovery
6,014.7	—	—	—	—
4,856.7	107.93	1.0	0.0823	0.0823
4,014.7	96.4	0.8932	0.0833	0.1656
3,014.7	71.77	0.6650	0.1104	0.2759
2,014.7	47.2	0.4373	0.0983	0.3742
1,014.7	30.77	0.2850	0.0681	0.4423

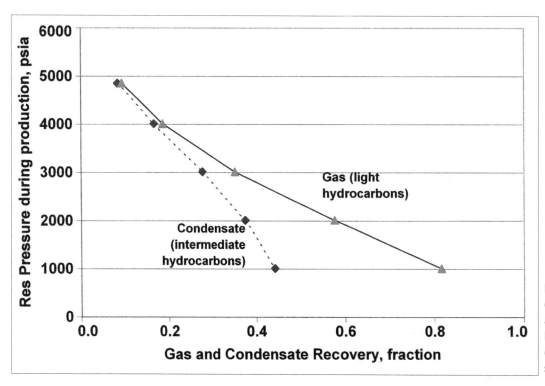

Fig. 12–27. Recovery comparison of gas and condensate liquid. The recovery of condensate is significantly lower.

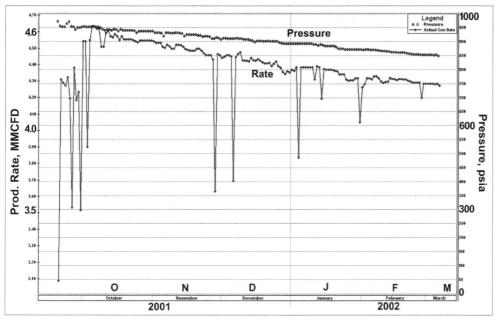

Fig. 12-28. Production history, including rate and pressure over time. *Courtesy of Fekete Associates, Inc.*

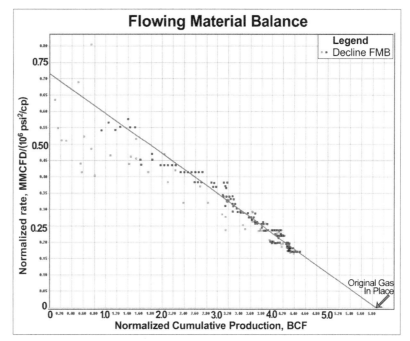

Fig. 12-29. Material balance analysis based on flowing bottomhole pressure. *Courtesy of Fekete Associates, Inc.*

Incremental wet gas recovery is calculated as the difference of the recovery between two successive pressure values. The results of the calculation for reservoir pressure between 6,014.7 and 1,014.7 psia are given in Table 12–13.

Next, the values of incremental dry gas recovery and incremental condensate recovery are computed, leading to the final results given in Table 12–14 and 12–15.

Incremental dry gas recovery = Incremental wet gas recovery × (Avg. B_{gw}/Avg. B_{gd}) × (B_{gdi}/B_{gwi})

Incremental condensate recovery = Incremental dry gas recovery × (Avg. condensate/Initial condensate)

Recovery of dry gas and condensate is plotted in Figure 12–27.

Example 12.9: Material balance analysis of a gas well based on flowing bottomhole pressure. Figure 12–28 shows the production history of a gas well. The well produced about 0.74 bcf in seven months. Both flowing bottomhole pressure and rate declined within the period. Static bottomhole pressure is not available for the well.

Solution: The test is analyzed by the flowing material balance method of Agarwal-Gardner. Figure 12–29 presents a plot of normalized rate versus normalized cumulative production based on the above. The original gas in place is

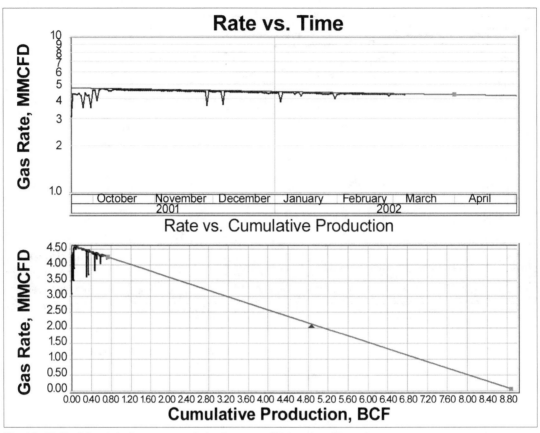

Fig. 12–30. Traditional decline curve analysis. *Courtesy of Fekete Associates, Inc.*

estimated to be about 5.86 bcf. Traditional decline curve analysis, based on rate information alone, overestimates gas recovery significantly (8.8 bcf), as seen in Figure 12–30.

Summing Up

Concepts

The material balance method is a simple yet potent tool in estimating original hydrocarbon in place and understanding the dynamics of reservoir. The classical material balance method is based upon the law of conservation of mass and a homogeneous reservoir. It does not consider direction of flow and distribution of wells. It is used to estimate original oil in place, original gas in place, and ultimate primary recovery.

Mathematical equations

According to the general material balance equation, the underground withdrawal is equal to the sum of a number of factors. These factors are the expansion of oil, original dissolved gas, expansion of gas caps, reduction in hydrocarbon pore volume due to connate water expansion and decrease in the pore volume, and natural water influx.

Graphical presentations of the material balance as an equation of a straight line provide very useful relationships to estimate the following:
- The original oil in place for solution gas drive oil reservoirs
- The original oil in place and gascap size for solution gas and gascap drive oil reservoirs
- The original oil in place and water influx for solution gas and aquifer water drive oil reservoirs

Past history matching is needed to make the estimates using PVT properties, pressure, and cumulative oil, water, and gas productions.

Material balance as an equation of a straight line is also applicable to gas reservoir performance analysis. Graphical presentations using historical gas and water productions, pressure, and PVT properties provide the following:
- The original gas in place for depletion drive gas reservoirs
- The original gas in place and water influx for depletion drive, aquifer water drive gas reservoirs

Class Assignments

Questions

1. Describe the basic principle of the material balance method in reservoir engineering studies. List the information that could be obtained about a reservoir based on the material balance technique, listing the data requirements.

2. The first well is put in production for few weeks in a newly discovered heavy oil reservoir. Can the material balance technique be applied to estimate the original oil in place with reasonable certainty? Why or why not?

3. Discuss the advantages and limitations of the material balance method. Consider the following scenarios and discuss the applicability of the method in determining initial hydrocarbon in place in each case:

    (a) A naturally fractured reservoir in which oil production occurs predominantly from the fracture system.

    (b) A relatively homogeneous formation with a long transition zone. Water production was encountered early in the life of the reservoir.

    (c) A heterogeneous reservoir in which multiple compartmentalization is suspected due to the existence of sealing faults.

    (d) A gas condensate reservoir in which a negligible volume of condensate was recovered during primary production. A gas recycling project was initiated in later years.

    (e) A matured oil field in which early production data is not available. However, required data has been collected regularly in recent years.

    (f) An overpressured gas reservoir located at a depth of 12,000 ft, in which reservoir performance was affected significantly due to the reduction in pore volume and transmissivity.

4. Based on water-cut information from producers in an oil reservoir undergoing waterflooding, it is suspected that a high permeability channel exists throughout the reservoir. Can the material balance method be used in order to confirm the heterogeneity? Explain.

    In another reservoir, vertical crossflow of injected water is suspected across two adjacent layers during waterflooding. Can the material balance method be utilized to obtain further reservoir description? (Waterflooding a reservoir to augment oil recovery is discussed in chapter 16.)

5. Studies conducted in a highly heterogeneous oil reservoir on production for a few years led to the following estimates of stock-tank oil initially in place:

    Volumetric estimate: 900 MMstbo
    Decline curve: 750 MMstbo
    Material balance: 650 MMstbo

    Propose further studies with specific details to resolve the apparent discrepancy. Note that there are plans to drill a few more wells in the reservoir in the near future.

6. Studies conducted in a gas reservoir following a few months of production yielded the following values of original gas in place:

   Volumetric estimate = 900 MMscf.

   Material balance = 1.080 bscf.

   Decline curve—Not analyzable due to rate controlled production.

   What would be the most likely reason or reasons for the apparent discrepancy?

7. Under what circumstances is flowing bottomhole pressure data used in material balance studies? Provide a case study based on a literature review

Exercises

12.1. Consider a newly discovered volatile oil reservoir under depletion drive. Based on log and core studies conducted in the exploratory well, the following reservoir and fluid properties are known:

   Depth of the reservoir, ft, = 11,500.

   Regional pressure gradient, psi/ft, = 0.449.

   Regional temperature gradient, °F/100 ft, = 1.8.

   Crude oil specific gravity, °API, = 42.9.

   Solution gas ratio (initial), scf/stb, = 920.

   Gas specific gravity (air = 1) = 0.66.

   Estimated producing gas/oil ratio, scf/stb, = 1,600.

   Estimated abandonment pressure, psia, = 1,200.

   Oil formation volume factor at abandonment, rb/stb, = 1.01.

   Initial water saturation, fraction, = 0.32.

   Dissolved solids in formation water, ppm, = 20,600.

   Formation compressibility, psi^{-1}, = 4.05×10^{-6}.

   Range of porosity, fraction, = 12.5–16.5.

   Range of permeability, mD, = 5–55.

   Lithology of the rock is dolomite.

   (a) Estimate the ultimate recovery from the reservoir. What percent of the oil is expected to be recovered before the bubblepoint is reached? Make any valid assumptions necessary and state them clearly.

   (b) Calculate residual oil and gas saturations at abandonment.

   (c) In estimating recovery from the reservoir, why was the production history not necessary in this case? Explain.

   (d) Following a year of dry production, a small water cut was periodically observed in the production stream. Water cut remained low in the following months. Discuss the most likely reasons. In the light of new information, would the analysis performed still be valid? Explain.

   (e) What should a reservoir engineer recommend to obtain a better estimate of oil recovery?

Table 12–16. Production history

Time (days)	Pressure (psig)	Cumulative Oil Production (MMSTB)	Cumulative Production GOR (SCF/STB)	Cumulative Water Production (MMSTB)
0.00	2,046.99	0.000	0.00	0.000
152.31	1,963.00	0.017	48.30	0.000
517.56	1,735.00	0.082	48.30	0.000
852.13	1,479.00	0.170	48.30	0.000
1,187.06	1,209.00	0.287	48.30	0.001

Table 12–17. Fluid PVT properties

Pressure (psia)	B_o (RB/STB)	R_s (SCF/STB)	B_g (RB/MSCF)	B_W (RB/STB)	$c_f \times 10^{-6}$ (psi)
2061.69	1.0580	48.23	1.104245	1.0067	3.37
1977.70	1.0585	48.23	1.152382	1.0070	3.37
1749.70	1.0597	48.23	1.314901	1.0080	3.37
1493.70	1.0612	48.23	1.581933	1.0090	3.37
1223.70	1.0628	48.23	1.972540	1.0101	3.37

Table 12–18. Aquifer constants

Time (days)	Pressure (psia)	$S(p,t)/E_{tot}$ (psia-stb/rb)		
		Schilthuis Steady State	Hurst Steady State	Infinite Linear Unsteady State
0	2,061.69	0	0	0
152.31	1,977.7	21,065.6	13,082.5	43,496.12
517.56	1,749.7	69,791.3	26,151.8	79,090.78
852.12	1,493.7	110,030.7	35,881.7	98,832.55
1,187.1	1,223.7	149,604.6	44,729.3	114,309.8

12.2. The production history and fluid properties data are available in Tables 12–16 and 12–17. Calculate the original oil in place by using the straight-line method. Assume various aquifer models for best fit.

(a) Based on production data, determine whether it is an undersaturated or saturated reservoir.

(b) Assume no water influx from the aquifer as a base case. Use the FE and Campbell methods to calculate the original oil in place graphically.

(c) Next, perform a material balance analysis by assuming the following aquifer models:
- Schilthuis steady-state aquifer
- Hurst steady-state aquifer
- Unsteady-state linear aquifer

Use the Havlena-Odeh method to calculate the original oil in place graphically. Values for the aquifer function S(p,t) are provided in Table 12–18.

(d) Tabulate the results obtained from all five cases. What model appears to best fit the production history?

(e) Plot cumulative water influx versus cumulative oil production for all three aquifer models. Compare the results.

(f) Would it be useful to use a gascap model in this study and compare the results? Why or why not?

References

1. Havlena, D., and A. S. Odeh. 1963. The material balance as an equation of straight line. *Journal of Petroleum Technology.* August, 896–900.

2. Havlena, D., and A. S. Odeh. 1964. The material balance as an equation of straight line—part II, field cases. *Journal of Petroleum Technology.* July, 815–822.

3. Wang, B., and T. S. Teasdale. 1987. GASWAT-PC: A microcomputer program for gas material balance with water influx. SPE Paper #16484. Presented at the Petroleum Industry Applications of Microcomputers meeting, Society of Petroleum Engineers, Del Lago on Lake Conroe, TX, June 23–26.

4. Schilthuis, R. J. 1936. Active oil and reservoir energy. *Transactions.* AIME. Vol. 118, 33–52.

5. Satter, A., and G. C. Thakur. 1994. *Integrated Petroleum Reservoir Management—A Team Approach.* Tulsa: PennWell.

6. Campbell, R. A., and J. M. Campbell, Sr. 1978. Petroleum property evaluation. Vol. 3. *Mineral Property Economics.* Norman, OK: Campbell Petroleum Series.

7. Tehrani, D. H. 1986. An analysis of a volumetric balance equation for calculation of oil in place with water influx. *Journal of Petroleum Technology.* September, 1,664–1,670.

8. Agarwal, R. G., D.C. Gardner, S. W. Kleinsteiber, and D. D. Fussell. 1998. Analyzing well production data using combined type curve and decline curve analysis concepts. SPE Paper #57916. Paper presented at the 1998 SPE Annual Technical Conference and Exhibition, New Orleans, September 27–30.

9. Dake, L. P. 1978. *Fundamentals of Reservoir Engineering.* Amsterdam: Elsevier Science.

10. Havlena, D., and A. S. Odeh. 1963.

11. Dake, L. P. 1978.

12. Mian, M. A. 1991. *Petroleum Engineering: Handbook for the Practicing Engineer.* Vol. I. Tulsa: PennWell.

13. Wang, B., and T. S. Teasdale. 1987.

14. Wang, B., and T. S. Teasdale. 1987.

15. Litvak, B. L. Texaco E&P Technology Department. Personal contact.

13 · Reservoir Simulation Fundamentals

Reservoir simulators are widely used to study reservoir performance and to determine methods for enhancing the ultimate recovery of hydrocarbons from the reservoir. They play a very important role in the modern reservoir management process and are used to develop a reservoir management plan. This plan includes the ability to monitor and evaluate reservoir performance during the life of the reservoir, from exploration and discovery to delineation, development, production, and finally abandonment.

Over the last 50 years, there have been numerous publications and conferences concerning various aspects of reservoir simulation. Pertinent and recent references are listed in this presentation.[1–8] It is worth mentioning that to our knowledge, Litvak published the first paper describing the proper representation of the matrix-to-fracture gravity flow contribution that was ignored at the time in many dual porosity simulators.[9] Litvak's gravity contribution consisted of a new gravity term corresponding to the density difference between a matrix block and the surrounding fracture acting over the average block height.

This chapter presents the fundamentals of reservoir simulation. Applications with examples are presented in the following chapter.

Learning objectives in this chapter include the following:
- Concepts and technologies
- Reservoir simulators: black oil, compositional, dual porosity, pseudocompositional
- Reservoir types: black oil, dry gas, gas condensate, volatile oil
- Simulator applications
- The value of reservoir simulation
- Mathematical basis of simulation: law of conservation of mass, Darcy's fluid flow law, and PVT behavior of fluids
- Partial differential equations
- Finite difference approximations and solutions
- Gridding techniques: rectangular and other grids
- Simulation process: input data gathering, history matching, and performance prediction
- Data requirements
- Simulator input data
- History matching
- Performance prediction
- Output results
- Abuse of reservoir simulation
- Golden rules for simulation engineers

Concepts and Technologies

Numerical simulation is still based upon material balance principles, taking into account reservoir heterogeneity and the direction of fluid flow. Unlike the classical material balance approach, a reservoir simulator takes into account the locations of the production and injection wells and their operating conditions. The wells can be turned on or off at desired times with specified downhole completions. The well rates or limiting bottomhole pressure, or both, can be set as desired.

The reservoir is divided into many small tanks, cells, or blocks to take into account heterogeneity. Computations of pressures and saturations for each cell are carried out at discrete time steps, starting with the initial time. Simulation has become a reality because of technological advances and the computational power now available.

Black oil simulators are characterized by the number of fluid phases, direction of flow, and the type of solution used for the complex fluid flow equations.

The fluid phase can be characterized as the following:
- Single phase (oil or gas)
- Two phase (oil and gas, or oil and water)
- Three phase (oil, gas, and water)

The direction of flow can be characterized as the following (fig. 13–1):
- Flow is described as 1-D linear or radial when it occurs only in one direction.
- Flow can be described as 2-D areal, cross-sectional, or radial cross-sectional when it occurs in the x-y, x-z, or r-z directions.
- Flow is considered 3-D when it occurs in the x-y-z directions.

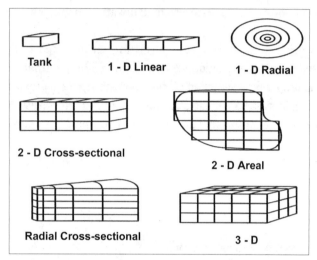

Fig. 13–1. Directions of fluid flow. *Source: A. Satter, J. Baldwin, and R. Jespersen. 2000. Computer-Assisted Reservoir Management. Tulsa: PennWell.*

Reservoir Simulators

Reservoir simulators are generally classified as black oil, compositional, thermal, or chemical, as governed by flow mechanisms:
- Black oil: fluid flow
- Compositional: fluid flow, phase composition flow
- Thermal: fluid flow, heat flow
- Chemical: fluid flow, mass transport due to dispersion, adsorption, and partitioning

In petroleum engineering, there are two principal types of simulators—a black oil simulator and a compositional simulator. More than 90% of all simulation studies can be performed with a black oil simulator.

Black oil simulators

A black oil simulator is suitable for most oil and gas reservoirs. The underlying physics are straightforward. All black oil simulators assume that the stock-tank oil and gas have multiple components, and that all of the resulting fluid PVT behavior is only a function of pressure and temperature. The equations in a black oil simulator are expressed in terms of stock-tank volumes. The various terms are normally expressed as reservoir volumes divided by the formation volume factor to yield stock-tank volumes. Stated slightly differently, the K values (defined as the mole fraction of each component in the gas phase divided by the oil phase) are only a function of pressure under isothermal conditions. A black oil simulator is equivalent to a one-component compositional simulator.

Most engineers probably are not aware of an important fundamental assumption with a black oil simulator. Regardless of pressure, the liquid/gas stock-tank compositions of the separator flash conditions are constant. In the model, different fractions of stock-tank oil and gas compositions mix to create the reservoir compositions, and reproduce compositional mixing. Although this assumption is not exactly true in the real world, it is nearly correct for all reservoir fluids within the operating pressure limits found in most reservoirs. In any case, these errors are small enough to not appreciably affect the production streams predicted by simulators.

Run times for most black oil problems represent a small fraction of the overall simulation study and do not appreciably affect the total time required to study a reservoir. Not that long ago, this was not the case, but today's everyday desktop computers are just as powerful as the world's fastest computers of 15 years ago.

Black oil simulators are very general and are capable of modeling most reservoir fluids, including dry gases, wet gases, heavy oils, and volatile oils. Many engineers are not aware of the wide range of applications that can be accurately modeled using a black oil simulator. For wet gas and volatile oil applications, the PVT data for the reservoir must include both the gas dissolved in the oil phase and the oil dissolved in the gas phase. Several authors confirm that even for very wet condensates, a modified black oil approach is very accurate.

Black oil simulators only fail in the event that lean gases, for example, carbon dioxide or methane, are cycled in the presence of a liquid phase. For these cases, a true compositional simulator is needed. The explanation for this is straightforward. When dry gases are cycled in the presence of liquids, they will preferentially pick up the light ends in the liquid first. In the case of a black oil simulator, since there is only one oil component, the variability in oil vaporization cannot be modeled. As a consequence, when black oil models are used to represent gas recycling processes, the black oil simulator will underpredict early oil vaporization in the presence of injected lean gas. In addition, it will overpredict later oil vaporization, where the reservoir oil is much heavier than the oil in the black oil simulator.

Compositional simulators

Whereas black oil simulators represent fluids at stock-tank volume conditions, compositional simulators write their equations at reservoir conditions. They are written in terms of moles of the individual oil components. Practically speaking, there are hundreds of oil components in a reservoir oil sample, and engineers normally represent the first three to eight components precisely. However, the remaining components, such as C_{9+}, are grouped into what is referred to as a pseudocomponent. The properties of the pseudocomponent represent the average of the properties of the remaining components. The fluid properties of the components are most commonly represented by cubic equations of state (EOS). When vaporization test data is available, engineers commonly modify the equation of state properties of one or more of the components to more closely represent the test data through a regression computer process.

Compositional simulators are the accepted technology for modeling lean gas cycling in the presence of oil that will vaporize in the lean gas. The multiple components in the oil allow the variability in the oil vaporization as a function of lean gas throughput to be modeled far more accurately than a one-component black oil simulator. However, it should be recalled that the oil pseudocomponent lumps all of the heavier oil components together. As a result, the compositional simulator overpredicts the vaporization of the pseudocomponent, for the same reason that the black oil one-component simulator overpredicts vaporization. A good compositional study will have enough components included to minimize the errors caused by the pseudocomponent vaporization errors. Another technique used to model the vaporization of the heaviest ends more accurately is through regression of one or more components as previously described.

Unlike black oil simulators, run times for compositional simulators can become long enough to have a significant impact on study times. To understand why, consider the fact that a 13-component compositional simulator (water plus hydrocarbon components) needs much more computational time. Such a simulator requires about two orders of magnitude more CPU time than a comparable black oil simulator with three components (oil, gas, and water).

True compositional simulators are not odiously difficult packages to use, but even for experts, compositional simulation studies can take an order of magnitude more time than a comparable black oil simulation. Because these simulators are commonly used by a small community of experts who are very familiar with key word inputs, the interfaces lag behind the black oil simulator interfaces. Compositional simulators also require the engineer to be well-versed in an equation of state package used to duplicate lab PVT behavior and produce accurate equation of state data for the simulator. Whenever possible, black oil simulators should be chosen in place of compositional simulators for these reasons. When applied properly, black oil and compositional simulators produce identical results.

Dual porosity and higher-order simulators

Black oil and compositional simulators can be written assuming one rock type per reservoir grid block or multiple rock types per grid block. If one rock type exists per grid block, the model is referred to as a single porosity simulator. If multiple rock types exist per grid block, the model is referred to as a multiple porosity simulator. In the case of two rock systems per grid block, the simulator is commonly used to model natural fracture systems. For these applications, it is called a dual porosity simulator. Within each rock type, unique porosity, permeability, and relative permeability curves are assigned.

Dual porosity simulators are commonly used to represent the interaction of fractures and matrices in a fractured reservoir. Higher ordered systems can be used to represent multiple rock types interacting among one another within the same grid block. Normally, for computational reasons, only one of the rock systems communicates between neighboring grid blocks. In the case of dual porosity simulators, the fractures are the conduits between neighboring grid blocks. The matrix-to-matrix flow between neighboring grid blocks is usually negligible and is ignored.

The Society of Petroleum Engineers published a benchmark comparison for the dual porosity simulators in the market in the early 1990s. The results were very different among the various vendors, and it was hypothesized that the physics involved was treated differently in the various simulators. At the time, many simulators did not properly represent all of the required physics for matrix-to-fracture flow.

Litvak published a paper describing the proper representation of the matrix-to-fracture gravity flow contribution that was ignored at the time in many dual porosity simulators. Litvak's gravity contribution consisted of a new gravity term corresponding to the density difference between a matrix block and the surrounding fracture acting over the average block height.

A simple experiment illustrating this force would be to place a hollow cylinder into an empty glass. The volume outside the cylinder is filled with water, and the volume inside the cylinder is filled with oil. Now if the cylinder is removed from the glass, the oil quickly moves to the top of the glass and overrides the water, because it is less dense than the water. This simple gravity component was ignored in many early dual porosity simulators.

The authors do not have extensive experience with the detailed physics in all of today's commercial dual porosity simulators. However, as a caution, they urge the reader to verify the physics in any dual porosity simulator before performing a study.

Pseudocompositional simulator

Pseudocompositional simulators are simplifications of compositional simulators that attempt to model the same fluid behavior. The most common formulation for pseudocompositional simulators is a modified black oil approach. In this approach, a mixing parameter attempts to track the varying oil volatility as dry gas passes through the different cells in the model, and the oil progressively becomes heavier. For example, a common approach is to create a 1-D model and then inject lean gas into the first cell, producing wet gas out of the last cell. The engineer tracks cumulative gas through a cell, along with changes in the oil properties, including viscosity, formation volume factor, condensate yield (volatility), and gas/oil ratio. These changes are then incorporated into a modified black oil simulator, where a new component is added to track the throughput volumes of injected gas through each cell in the model.

The pseudocompositional approach described above is a rough approximation at best. Although it offers better results than an unmodified black oil simulator, it still can fall far short of the accuracy of a true compositional simulator. The reason for the shortcoming is that the fluid behavior as lean gas cycles through a complex oil mixture correlates closely to moles and not to throughput volumes. Intuitively, this makes sense, as compositional simulators are written in terms of moles and mole fractions. It thus stands to reason that any accurate modified approach should use the same set of assumptions.

One attempt was made in the early 1990s to write a pseudocompositional model that married the black oil simulator with one compositional equation tracking the lean injected gas. The model proved very accurate and significantly faster than comparable compositional formulations. However, the compositional community is very small and very comfortable with current compositional simulator formulations. Even with the rapid increase in computer processing power over the past 15 years, a commercial development using this formulation has not proved commercially viable.

Coalbed methane (CBM) simulators are essentially black oil simulators that have included the desorption Langmuir isotherm behavior for natural gas trapped within coal seams. Some coalbed methane simulators use the dual porosity approach.

Reservoir Types

Performance of black oil, dry gas, gas condensate, and volatile oil reservoirs can be modeled by simulators.

Black oil

The color of oil varies. However, in the nomenclature, a typical oil, whether heavy or moderately light, with associated dissolved gas at initial conditions, is classified as "black" oil. The fluid properties of black oils can be represented as only a function of pressure. As the name would imply, black oils are modeled using the black oil simulator. Lighter black oils can be accurately modeled by including both gas dissolved in the oil phase and oil dissolved in the gas phase.

Dry gas

Dry gas reservoirs refer to gas reservoirs without significant dissolved condensate in the gas at initial reservoir conditions. Practically speaking, almost all reservoir gases will have some entrained liquids that exit during the flash separation to ambient conditions. In the event these liquids are a few barrels per million cubic feet of gas, the condensate is normally ignored, and the gas is represented as a dry gas with no condensate dropout. The dry gas formulation works well for gases where significant condensate dropout does not occur within the pressure range encountered during the life of the reservoir.

There are conditions where gases with a few barrels of liquid per million cubic feet of gas should be treated as a wet gas, as discussed below. One important example would be condensate dropout around the wellbore in a tight gas reservoir at very low pressures. In these conditions, the associated drop in the gas relative permeability and well productivity can become important.

Black oil simulators model dry gas reservoirs accurately.

Gas condensate

When reservoir gases have significant condensate dissolved at initial conditions, and when the effects related to condensate dropout are important, the gas is called a wet gas. These reservoirs have appreciable condensate dropout in the reservoir or around the individual wells, or both. For wet gases below the dew point, oil is dissolved in the gas, and gas is dissolved in the free oil that forms.

The black oil simulator used to develop examples for this text provides for both the gas/oil ratio and condensate yield. It models wet gases accurately except in cases where significant free oil becomes present during lean gas cycling. To model these compositional effects, a true compositional simulator with multiple oil components is necessary to properly model the varying liquid volatility.

Gas condensate has both oil dissolved in the gas phase and gas dissolved in the oil phase below the dewpoint pressure.

Volatile oil

If significant oil becomes dissolved in the gas phase upon the drop of reservoir pressure below the bubblepoint, the reservoir is usually classified as a volatile oil reservoir. For volatile oils, a black oil simulator will include both gas dissolved in the oil, and oil dissolved in the gas at various pressures. In the event that lean gas and CO_2 cycling occurs, a compositional simulator is required to model the varying fluid volatility, as explained in the wet gas description section.

Simulator Applications

The majority of simulation models built today are for complex full-field evaluations of the largest company assets. Most engineers and managers feel that these are the only applications that make economical sense for this technology. This viewpoint arose when simulation was a process that took years to learn, and months or years to complete on only one reservoir.

This viewpoint could not be further from the truth today with the advanced interfaces now available. Models can be built in a short time, and results are obtained in a timely fashion. The value of simulation and applying technology to smaller assets can equal or outweigh the benefits in larger reservoirs. In large assets, the difference between applying technology and using engineering experience represents the difference between making money and, at best, making more money. In either case, the application of technology probably does not make or break the project.

However, in marginal assets, the difference between using technologies such as simulation and not applying simulation can be the difference between making money and losing money. Marginal assets need technology to reduce the error bars and guarantee the overall success of the project.

Single-well and pattern models are underutilized in reservoir simulation. These models can be built in minutes and then used to examine well production and interference issues. In particular, flat reservoirs (reservoirs with many small individual compartments) and tight reservoirs lend themselves easily to analysis. These models are equally valuable to a geologist trying to understand the relationship between the geology and fluid flow, and a production engineer trying to complete and produce wells properly. They are also useful to a reservoir engineer in drilling wells at a spacing that maximizes economic value.

Examples of single-well and full-field simulation application results are presented in chapter 14.

The Value of Reservoir Simulation

Reservoir engineers can use a reservoir simulation model as a very powerful tool to represent fluid flow through a reservoir. It allows them to know reasonably well the reservoir pressure and production performance with time. The output from the simulator can be utilized for many purposes. For example, engineers can use the output from the simulator to propose an optimized development scenario for a newly discovered field based on the available information.

The reservoir simulation model can be considered a good representation of the real reservoir, with all the appropriate physics, reservoir properties, geology, geophysics, and wells included. The reservoir simulation model is divided into a series of interconnected blocks (or more complex objects in more complicated formulations). Within the blocks themselves, the reservoir fluids are normally treated as a stirred homogeneous mixture, with saturations and fluid properties varying from one grid block to the next.

Fundamentally, reservoir simulation is simply a marriage of the basic material balance equation (also called the equation of continuity) and Darcy's law for flow in porous media. The resulting software system attempts to verify that production versus pressure matches historical records. When this relationship falls short, professionals working with simulators adjust the parameter or parameters within the simulator that are most likely to be in error. These parameters are usually the parts of the model that cannot be directly seen—with the primary culprit being the geology of the reservoir. Fluid and rock properties, reservoir pressures, and fluid productions, including oil, gas, and water, can all be directly measured. These parameters are usually not the desired candidates for adjustment.

In the case of a new reservoir with no historical data, the simulator allows the user to generate reasonable production forecasts for different statistically possible interpretations of the reservoir. In many instances, the model chosen for identifying the optimal field development can have many inaccuracies. Even so, this model is remarkably adequate in defining the proper field development strategy, which can be modified when more data becomes available.

A reservoir simulation is the most general useful reservoir engineering analysis tool available. It is the only one that integrates data from all related engineering and geological disciplines. This data includes basic geology, geophysics, rock and fluid properties, well locations, completions and plumbing, compression/separator constraints, pressure/production data, and material balance results. As a result, for the first time, all of the basic data and other geological and engineering analyses are brought together in one place. This information can be adjusted until it is reasonable and consistent. The engineers and geologists can finally understand the reservoir and can account for errors overlooked in the piles of data gathered and stored in various files.

Those who remain unconvinced of the power and value of simulation should consider that wells are very expensive to drill. It can take many years before an accurate forecast of a well's ultimate performance and the optimal well drilling/production strategy is understood. In a simulation model, wells can be drilled with the click of a mouse, and results for the next 20 years can be forecast in minutes. The model allows a professional the luxury for the first time of discovering the optimal well drilling/production strategy to produce a reservoir. Having said this, the initial geological description is approximate at best. Thus the optimal drilling strategy will never be the best possible one until real production and pressure data is obtained from the reservoir to confirm the geology. Nevertheless, the simulator is the only tool that can calculate the optimal development strategy for the data gathered on a reservoir today, and that is a powerful fact. Finally, a simulator can be used as a crystal ball of sorts to look through many years of performance of a reservoir.

Even though reservoir simulators have been used on virtually all of the world's larger reservoirs for the past 50 years, this technology has critics. There are two common complaints regarding the use of reservoir simulation. The first is that the simulator can be made to give any answer desired, and the results are not unique. The second is that simulators have limited value because the data going into them can be wrong, and consequently, the output is also incorrect. Of this concern, it can be said, "Garbage in, garbage out."

The idea that a reservoir simulation engineer can get any answer wanted from the simulator is not true, unless the engineer puts data into the simulator that is pure fantasy. Occasionally a simulation engineer chooses to make porosities greater than 1 or to represent infinite aquifers compressed into a small area. These situations could never occur in nature. Many of these errors are simply mistakes, or an overzealous engineer is desperate to match historical data. At other times, these errors might be made deliberately to produce false results for political reasons. In either case, the advanced interfaces in many of today's simulators make these errors easy to catch. Even professionals inexperienced with simulation can in minutes learn the tools necessary to review the geology, PVT, and well completion data in any model.

A real-life example showing the value of simulation was a reservoir simulation study performed on a mid-continent reservoir. The first phase of the study took about three months. The resulting poor history match suggested that the production data for the individual wells was incorrect, as the production data was inconsistent with the reservoir geology modeled. A cursory visual review of the production data and geology looked more or less consistent. However, the simulation model, and in particular the interactive saturation fields that it generated, clearly showed the production data and geology were inconsistent. The independent operator reviewed the reservoir's production records and found numerous errors. A second history match produced similar poor results and showed the same type of problem as encountered in the first study. A second review of the production data found additional errors. The third iteration resulted in a very close history match in less than one week. The lesson learned here is that a reasonable production history match requires a reasonably accurate reservoir description.

The second argument is that "the data going into the simulator is always wrong." While some of the data may in fact be wrong, it may well be the best and only data available. Why shouldn't it be used to do the best analysis possible and create the best production strategy for the data available? Prudent engineers realize the error bars in their data and run enough sensitivity studies in the simulator to understand the best development strategy in spite of these errors. Engineers should gather data early in the drilling and production life. They should import this data into the reservoir simulation model to improve the quality of the model. In turn, this optimizes the net present value of the assets from that time forward.

Mathematical Basis for Simulation

There are a number of excellent references that describe the mathematics behind the underlying equations in a simulator. To a much lesser degree, they also describe the proper mathematical techniques for solving the resulting set of equations. The following discussion will only present a cursory overview of the mathematics and solvers. Students are referred to the references at the end of the chapter, in which the developed equations and their solutions are described.

Simultaneous flow of oil, gas, and water through a porous medium by a set of phases (oil, gas, and water) can be described by a set of nonlinear partial differential equations. The following are used:

- Law of conservation of mass
- Darcy's fluid flow law
- PVT behavior of fluids

Derivation of the diffusivity equation based on the above is treatd in chapter 4. For simplicity, the 1-D flow of oil, gas, and water can be considered in the x-direction through an arbitrary volume element. According to the law of conservation of mass:

Mass rate in − Mass rate out = Mass rate of accumulation

This can be expressed in derivative form as follows:

$$\frac{\partial w}{\partial t} = w_{in} - w_{out} \tag{13.1}$$

where

w = mass of fluid in volume element, g
w_{in} = mass rate of fluid flowing into the volume element, g/sec, and
w_{out} = mass rate of fluid flowing into the volume elemet, g/sec.

Let us consider that the fluid flowing in and out of the element is oil in this case. Note that Equation (13.1) can be expressed in terms of oil density, flowrate, formation volume factor and the dimensions of the element as follows:

$$w = A\emptyset\Delta\rho_o\, S_o/B_o \tag{13.2}$$

$$w_{in} = \rho_o\left(\frac{q_o}{B_o}\right)_X \tag{13.3}$$

$$w_{out} = \rho_o\left(\frac{q_o}{B_o}\right)_{X+\Delta x} \tag{13.4}$$

where

A = cross-sectional area of the volume element, cm^2,
$\emptyset$ = porosity, dimensionless,
x = direction of fluid flow,
Δx = length of the volume element, cm,
q_o = oil flow rate, cm^3/sec,
ρ_o = density of oil at standard pressure and temperature, g/cm^3,
S_o = saturation of oil in the volume element, and
B_o = oil FVF, standard cm^3/reservoir cm^3

Based upon Darcy's law, an expression for oil flow rate can be obtained as in the following:

$$q_o = -\frac{Akk_{ro}}{\mu_o}\left(\frac{\partial p_o}{\partial x} - \rho_o\, g\, \frac{\partial D}{\partial x}\right) \tag{13.5}$$

where

D = depth, cm,

g = acceleration due to gravity (980.7 cm/sec²),

k = permeability, darcies,

k_{ro} = relative permeability to oil, fraction,

p_o = pressure in the oil phase, atm,

ρ_o = oil density, g/cm³,

x = direction of flow, cm, and

μ_o = oil viscosity, cp.

Partial Differential Equations

Substituting Equations 13.2, 13.3, 13.4, and 13.5 in Equation 13.1, the differential equation for oil flow in one dimension is obtained as follows:

$$\frac{\partial}{\partial x}\left[\frac{Akk_{ro}}{B_o\mu_o}\left(\frac{\partial p_o}{\partial x} - \rho_o g\,\frac{\partial D}{\partial x}\right)\right] = A\,\frac{\partial}{\partial t}\left(\frac{\varnothing\, S_o}{B_o}\right) \tag{13.6}$$

Similarly, water and gas (including not only gas in the gas phase, but also the gas dissolved in the oil phase) flow equations can be obtained as given below:

$$\frac{\partial}{\partial x}\left[\frac{Akk_{rw}}{B_w\mu_w}\left(\frac{\partial p_w}{\partial x} - \rho_w g\,\frac{\partial D}{\partial x}\right)\right] = A\,\frac{\partial}{\partial t}\left(\frac{\varnothing\, S_w}{B_w}\right) \tag{13.7}$$

$$\frac{\partial}{\partial x}\left[\frac{Akk_{rg}}{B_g\mu_g}\left(\frac{\partial p_g}{\partial x} - \rho_g g\,\frac{\partial D}{\partial x}\right)\right] + \frac{AR_s kk_{ro}}{B_o\mu_o}\left(\frac{\partial p_o}{\partial x} - \rho_o g\,\frac{\partial D}{\partial x}\right) = A\,\frac{\partial}{\partial t}\left[\varnothing\left(\frac{S_g}{B_g} + \frac{S_o R_s}{B_o}\right)\right] \tag{13.8}$$

where

R_s = gas in solution, standard cm³/reservoir cm³, and

g, o, w = subscripts referring to gas, oil, and water, respectively.

Equations 13.6, 13.7, and 13.8 relate the saturations to the pressure in the phases (p_o, p_w, and p_g) and the rock ($\varnothing$, k, and k_r) and fluid properties (B, ρ, μ, and R_s).

In addition to the partial differential equations, certain auxiliary relationships must be satisfied to solve these equations. First, the sum of the volumes of the oil, gas, and water must always be equal to the pore volume at any point in the system. Therefore, the following is used:

$$S_o + S_g + S_w = 1 \tag{13.9}$$

If the rock and fluid properties are assumed to be known functions of pressure, then there are four equations and six unknowns (S_o, S_g, S_w, p_o, p_g, and p_w).

The capillary pressures at any position can be taken to be functions of saturations alone, as shown below:

$$P_{cow} = P_o - P_w = P_{cow}(S_o, S_w) \tag{13.10}$$

$$P_{cgo} = P_g - P_o = P_{cgo}(S_o, S_g) \tag{13.11}$$

where

p_c = capillary pressure, atm, and

ow, go = subscripts referring to oil-water and gas-oil, respectively.

Now, there are six equations, i.e., Equations 13.6–13.11, involving six unknowns, i.e., three saturations and three pressures. With appropriate boundary and initial conditions, the system of six equations can be solved for saturation and pressure distributions in the reservoir.

Finite Difference Approximations and Solutions

Approximate solutions of the complex equations can be obtained by using finite difference schemes. The pressures and saturations can be solved explicitly, implicitly, or by a combination method:

- Explicit—explicit pressure, explicit saturation
- Implicit—implicit pressure, implicit saturation
- Combination—implicit pressure, explicit saturation

The numerical solution of the partial differential equations by finite difference involves replacing the partial derivatives by finite difference equations. Then, instead of obtaining a continuous solution, an approximate solution is obtained at a discrete set of grid blocks or points at discrete times. A general treatment of finite difference schemes is beyond the scope of this chapter. However, certain basic concepts will be illustrated using simple examples.

Finite difference approximations of partial differential equations require spatial and time discriminations. The length of the system is divided into a discrete set of grid blocks of size intervals (Δx). Time is divided into a set of discrete time intervals (Δt). The subscript n is used to identify the time level.

The discrimination of the reservoir into blocks will depend upon the size and complexity of the reservoir, and the quality and quantity of the reservoir data. It will also depend on the objective of the simulation study and the accuracy of the solution needed. In practice, the number of blocks will be limited principally by the time available to prepare input data and to interpret results. The model should contain enough blocks and dimensions so as to represent the reservoir and simulate its performance adequately.

The life of the reservoir must also be divided into time increments. Starting at the initial time, pressure and saturations, along with other factors, are computed at each block over each of the many finite time increments. In general, the accuracy with which reservoir behavior can be calculated will be influenced by the size of the time steps and the number of grid blocks.

The partial derivatives can be evaluated using explicit or implicit procedures.

Explicit difference schemes are based upon the values of a variable known at the beginning of a time step. The end of a time step (n) is the beginning of the next time step (n + 1). The explicit difference expressions of $\partial p/\partial x$ and $\partial^2 p/\partial x^2$ are shown below:

$$\left(\frac{\partial p}{\partial x}\right)_{i-\frac{1}{2},n} = \frac{P_{i,n} - P_{i-1,n}}{\Delta x} \tag{13.12}$$

$$\left(\frac{\partial p}{\partial x}\right)_{i+\frac{1}{2},n} = \frac{P_{i+1,n} - P_{i,n}}{\Delta x} \tag{13.13}$$

$$\left(\frac{\partial^2 p}{\partial x^2}\right)_{i,n} = \frac{P_{i-1,n} - 2P_{i,n} + P_{i+1,n}}{(\Delta x)^2} \tag{13.14}$$

In implicit difference, the spatial derivatives are evaluated at the current time level (n + 1) rather than at the previously known time (n). In Equations (13.12) through (13.14), i is the index for location in the x direction.

Solutions

The finite difference approximations of the partial differential flow equations ultimately involve a set of simultaneous equations requiring matrix problem solutions.

The pressures and saturations can be solved explicitly, implicitly, or by a combination method. The method of solution will affect the following:

- Stability, the more implicit formulation generally being more stable
- Accuracy, including the truncation error, time step, and grid size, which affect the accuracy of the solution
- Cost, which involves the computer storage and run time

One of the most common formulation methods of numerical simulation is the implicit pressure-explicit saturation (IMPES) method. This involves solving first implicitly (as required for stability) for the phase pressures at each point and then solving explicitly for the saturations. Its appeal is a result of greatly reduced computing requirements, because it avoids the simultaneous implicit solution for several unknowns at each point.

The implicit pressure-explicit saturation method involves solving both the phase pressures and saturations implicitly at each point. It offers a more stable solution at the expense of more computing time. This type of solution is particularly needed for a water or gas coning problem, or both, and for a gas percolation problem.

Complex matrix solutions of the finite difference approximations of multidimensional, multiphase partial differential fluid flow equations have become a reality. This has occurred due to the computing power of mainframe high-speed computers. Even a personal computer today is powerful enough with its storage capacity and computing speed to handle a good-size reservoir simulation. Technological advances in computational techniques, data handling, report writing, and graphics have also made reservoir simulation more practical and widely used.

Gridding Techniques—Rectangular and Other Grids

Gridding is a necessary step in the simulation process. As viewed from the top down, the reservoir is divided into many interlocking small tanks, cells, or blocks to take into account layering, thickness, fluid saturations, and variations in rock and fluid properties.

The most common gridding scheme today is rectangular. In this case, looking at the reservoir from the top down, the grid system is a group of interlocking rectangular blocks. The individual rows and columns can each have different column heights and row widths.

Other gridding schemes have found their way into simulators during the past 20 years. Two of the more popular are corner point gridding and Pebi gridding. In these techniques, irregularly shaped cells are used in the model. This allows faults, reservoir boundaries, and other irregularities to be followed precisely, without the stair-step efforts required in rectangular gridding.

The use of these sophisticated gridding schemes has proven very popular. This has occurred in large part because geophysicists like to see their geological interpretations very precisely translated from geophysical and mapping packages into the simulator. Whether these techniques improve the physics of fluid flow in most applications is debatable. In the authors' opinion, in many applications considering the possible geological inaccuracies, these techniques complicate the process of reservoir modeling tremendously. In addition, they add little or no value to the end results.

An adequate grid density will represent the saturation and pressure profiles present in the reservoir in enough accuracy to calculate historical well production and pressure data with small errors. The goal of the simulation process is not to accurately represent each geologic feature or pressure/saturation change. Rather, the goal is to retain only the level of geology required to model fluid flow accurately. This is where the simulation process falls apart for many new users and experts alike. They strive to build the most accurate model possible, and in the process, forget that perhaps the complexities they are adding only make the entire study orders of magnitude longer, with no change in the ultimate results. A more complicated model is not necessarily a better model, and in fact, the reverse is often true.

Rules of thumb exist for gridding. One often heard is that grid cells with a common face should have no more than a two-fold to three-fold difference in dimension. This rule allows most models to reasonably accurately reproduce pressure and saturation changes. It should be noted that this rule is only a general guideline. In individual situations, more cells between wells may be necessary to accurately model fluid front movements.

Another good rule to follow is to start simple and then add detail as needed. It is much easier to add complexities to a model than vice versa. And when the model starts simple, it is often the case that the engineer realizes he can answer his questions adequately without the complex model. Another way of stating this might be to first look at the forest and not the individual trees—it will prevent getting lost in the forest.

In the rectangular grid case, the reservoir is represented by an interlocking set of rectangles with common side faces. From a cross-sectional view, two different types of gridding are possible—conventional and horizontal layering.

In conventional layering, the individual layers follow some geologic feature. For example, in a coning situation, an engineer might subdivide one geologic layer into six equal sublayers. The new layers will allow proper saturation and pressure initialization. Changes in these parameters in the future can also be modeled accurately. Having one cell that straddles the oil, water, and gas zones at initial conditions will certainly not produce accurate movement of these three phases during simulation.

In horizontal layering, the layer depths are picked based on preassigned depths. The top layer would be comprised of the top of a geologic unit down to some reference depth, such as the gas/oil contact. Layers 2 to 5 might comprise the oil zone, and layer 6 might comprise the geology from the oil/water contact to the bottom of the structure. The advantage of horizontal layering is that the contacts can be precisely modeled. Additionally, as the model runs into the future, any vertical coning is modeled more accurately than in conventional layering.

Simulation Process

In general, the reservoir simulation process can be divided as follows:
- Input data gathering: geological, reservoir, well completions, production, injection, etc.
- History matching: initialization, pressure match, saturation match, and productivity index match
- Performance prediction: existing operating and/or some alternative development plan

Input Data Gathering

The first step in a successful simulation study is the design of a proper data-gathering program. A simulator can be of great use in the design of the data-gathering program, as data acquisition can be one of the most expensive costs in the development of a reservoir. Gathering the right data in proper quantities can both reduce costs and drastically improve the quality of the resulting reservoir simulation model.

Most new simulation engineers think that the majority of their time will be spent running the simulator. Nothing could be further from the truth for most models run today. Computers today are orders of magnitude faster than their counterparts of 10 years ago. Even in the case of very large models that take hours to run, the CPU time is an insignificant part of the entire process. The vast majority of time an engineer spends on a simulation study will be in gathering data. It is here where the real time savings can be realized in reservoir studies.

Another misconception in data gathering and model building is that the proper time to construct a simulation model is once a wealth of data is available on the reservoir. It is true that the quality of the model and its ability to make predictions improve considerably after pressure and production data become available. However, the earlier the reservoir simulation is applied in the life of a reservoir, the greater the value of the simulation study.

The bulk of the costs are incurred early in the life of a reservoir. This is the time interval in which a better understanding of the optimal reservoir development plan can have a major impact on optimizing production and minimizing costs. Early in the life of a reservoir, there are typically large error bars in model parameters, in particular in the geology. However, most engineers do not recognize that the optimal well/field development strategy can still be reasonably understood. For example, in a tight gas reservoir with uncertain reservoir limits, the optimal well spacing can be calculated based on early well tests in delineation wells. It may be many years before the true extent of the reservoir is known, but a prudent operator using a reservoir simulator will develop the prospect in an optimal manner. Case studies have shown that the net present value of a tight gas asset with multiple wells can be doubled, or even tripled, versus the same asset developed using established well spacing.

Another important point to be made is to never wait on data before commencing or updating a reservoir model. Early models can show what data is important in the reservoir, and what data can be ignored. Early models can allow engineers the ability to have good answers on time, and not perfect answers too late.

Data Requirements

Input data consists of general data, grid data, rock and fluid data, production/injection data, and well data. Gathering the needed data can be a very time-consuming and expensive process. Ascertaining the reliability of the available data is vital for successful reservoir modeling.

Performing a proper reservoir management simulation study requires input from a cross-functional team of technical experts. The simulation data set is accumulated from a wide range of professionals. This involves the geologist, geophysicist, reservoir engineer, drilling engineer, production engineer, facility design engineer, pipeline engineer, field operations personnel, and asset management professionals. This integrated team effort leads to the effective and efficient capture of data. It also offers the experience required to construct the reservoir model and facilitate history matching to prior performance. This team effort is vital in developing the most accurate prediction forecast possible for use in asset operations optimization, field development, and management planning.

One of the most misunderstood areas in reservoir simulation is the data requirement for a proper simulation model.

Many engineers and other professionals think that most reservoirs are not suitable for simulation because the data required for simulation is not readily available. The reality is that the data that engineers work with every day is the data required for a simulation study. It is true that things like downhole pressure gauges, 3-D seismic studies, and conventional cores can improve the quality of the resulting model built. However, reasonably accurate models can be built with data at every engineer's fingertips. The following list shows the minimal data required for a typical study:

- Geological maps
- Net and gross sand thicknesses
- Oil and gas gravities
- Initial gas/oil ratio or condensate yield
- Reservoir temperature and pressure
- Initial water saturation
- Gas/oil and oil/water contacts
- Separator conditions
- Production and pressure information
- Flowing wellhead or bottomhole pressures at the economic limit

The various categories of data required for a simulator are given in the following discussion.

Geology

Geology data consists of structure maps and any other properties available, including net, gross, and net to gross thickness, and porosity and permeabilities. Sparse data such as well-calculated permeabilities can be distributed using geostatistics to yield a consistent interpretation that honors the known well points. Point-to-point continuity of the sands, commonly referred to as the correlation length, can also be varied to yield a range of potential geologic interpretations possible. These ranges of interpretations can then be used to create a range of simulation models that bracket the statistical range of expected outcomes for a new reservoir. In very fortunate cases, nearby outcrops or seismic interpreted data can yield an understanding of the variability of the data. This allows a sophisticated geostatistics approach to be applied.

Reservoir fluid data

Pressure-volume-temperature (PVT) is the common nomenclature for reservoir fluid behavior as a function of pressure and temperature. Except in the case of thermal simulation, the reservoir temperature is considered fixed, although it may be different in noncommunicating areas of the model.

PVT data can be calculated from lab tests using the original reservoir oil and gas recombined to initial reservoir bubblepoint or gas/oil ratio (chapter 3). Two common tests are performed. The first test is called a constant composition expansion test, and it requires expanding an undersaturated fluid below the bubblepoint to various pressures. The relative volume changes are recorded. The second test is called a differential liberation test, and it involves expanding a cell starting with the original reservoir fluid at the bubblepoint. At each step, the free gas is removed from the cell. With the recorded pressures for each step, and the volume of oil remaining and gas liberated, the fluid behavior can be recreated for the simulator.

Once the PVT lab data is calculated, it is then corrected using the PVT data resulting from a flash of the original reservoir fluid to the field separator conditions. For example, the resulting oil formation volume factors, oil viscosity, and gas/oil ratios from the flash test are used to correct the lab PVT data where the flash is to standard conditions. The oil formation volume factors in the lab test are corrected using a simple scaling factor, given as follows:

$$B_{o, \text{ corrected lab}} = B_{o \text{ lab}} \left[\frac{B_{o \text{ sep}} @ \text{ b.p.}}{B_{o \text{ lab}} @ \text{ b.p.}} \right]$$

where

b = bubble and

p = point.

In some cases, the original reservoir fluid is not available, and the reservoir has been depleted to a lower pressure, perhaps even significantly below the original bubblepoint pressure. Even so, good PVT data can still be calculated from the lab tests described above. Technically speaking, there are small compositional changes in the stock-tank oil and gas compositions as a reservoir is depleted under natural depletion or water drive. However, except at much lower pressures (less than 100 psi), these effects are small on the resulting PVT. So even at later stages of depletion, if the stock-tank oil and gas are recombined to the original gas/oil ratio or bubblepoint pressure, the resulting PVT data from lab tests is accurate.

Proof for the assertion that stock-tank oil and gas compositions in reservoirs below the bubblepoint can still be recombined to yield reasonable PVT information is not hard to find. The fundamental assumption in any black oil simulator, and one rarely understood, is that the stock-tank oil and gas compositions are fixed. Black oil simulators have been used successfully for the past 50 years to study reservoirs around the world.

When lab PVT data is not available, a variety of correlations can be used to estimate these factors. Each of the correlations was generated using a crude from a particular region, and it may or may not reasonably represent the crude being studied. It is recommended that when correlations are used, at a minimum, an engineer should still recombine stock-tank oil and gas to the original bubblepoint pressure or gas/oil ratio. The oil viscosity, formation volume factor, and gas/oil ratio should then be measured. This data can then be used to adjust the correlation data using the same techniques used to adjust the lab oil formation volume factors derived above.

Relative permeability

Relative permeabilities can be measured in conventional cores, derived from empirical correlations, or created using correlations and endpoints derived from cores (chapter 2).

Conventional cores are often altered from reservoir conditions and at best only represent a small fraction of the true reservoir rock. As such, inputting relative permeabilities directly from conventional core measurements is very suspect.

Relative permeabilities can be derived from correlations, and one of the most widely recognized correlations in the industry was created by Corey. The Corey correlation provides for the inputting of wettability, rock type, connate water, critical gas saturation, and residual saturations for oil in the presence of water and gas.

When gas, water, and oil are all flowing at the same time, Stone provides a relationship for interpolating the relative permeability to oil using the oil-water and oil-gas curves.[10] Two relationships were developed—Stone 1 and 2. Whenever two phases are flowing, the Stone correlations default to two-phase curves; for example, an oil-water or oil-gas curve. Practically speaking, Stone 1 and 2 provide almost identical solutions for most reservoir applications.

As a general rule, the best relative permeabilities combine a correlation such as Corey's with residual and critical saturations derived from conventional or sidewall cores.

Well production and completion data

Production from wells is often constrained by surface facility conditions. In addition, it also can be constrained by the normal pressure losses from gravity and friction at higher flow rates within the wellbore itself.

The simplest and most common technique for controlling wells in a model is to impose a reasonable sandface limiting bottomhole pressure at the economic limit for the well. For example, nodal analysis may show that for a particular gas reservoir at the economic limit, the fluid gradient in the wellbore plus the limiting wellhead surface pressure adds to 800 psi. An engineer might then control the well such that a maximum target rate controls the well flow rate. A constraint will be added so that if the sandface pressure for the well drops below 800 psi, the well switches to pressure control. From this point forward, the well is controlled such that the rate delivered corresponds to 800 psi flowing bottomhole pressure.

There are various controls available to the engineer for controlling wells. Injection or production controls can be imposed, with one of the three phases targeted for the well. For example, common well types include rate-specified water or gas injectors, oil and gas rate producers, and gas reinsertion wells that inject produced gas. The well index for the well is also specified, and all layers with well indexes of 0 are assumed to be not completed. Skin also can be imposed on individual layers.

History Matching

History matching of the past production and pressure performance consists of adjusting the reservoir parameters of a model until the simulated performance matches the observed or historical behavior. This is a necessary step before the prediction phase because the accuracy of a prediction can be no better than the accuracy of the history matching. However, it must be recognized that history matches are not unique.

The history-matching procedure consists of the following sequential steps:
1. Pressure matching
2. Saturation matching
3. Productivity matching

Average reservoir pressure matching involves the following steps (fig. 13–2):
1. Adjust rates to correct for total voidage.
2. Adjust total compressibility, porosity, permeability, thickness, and water influx from the aquifer to correct for the pressure level.
3. Adjust permeability for pressure shape.
4. Adjust total compressibility, porosity, thickness, and water influx from the aquifer to correct for individual well performances.

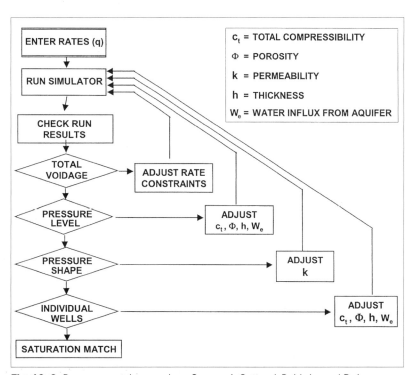

Fig. 13–2. Pressure match procedure. *Source: A. Satter, J. Baldwin, and R. Jespersen. 2000.* Computer-Assisted Reservoir Management. *Tulsa: PennWell.*

Reservoir saturation matching involves the following (fig. 13–3):

1. Adjust relative permeabilities and capillary pressures for field water/oil ratio and gas/oil ratio.
2. Adjust local relative permeabilities and capillary pressures for well water/oil ratio and gas/oil ratio.
3. Repeat pressure match.

Productivity match procedure involves the following (fig. 13–4):

1. Adjust well productivity index and injectivity index for well productivity.
2. Make final history match run followed by prediction.

Performance Prediction

The final phase of a reservoir simulation study involves predicting the future performance of a reservoir. This prediction could be for existing operating conditions or for some alternate development plan, such as infill drilling or waterflooding after primary production, and so forth. The main objective is to determine the optimum operating condition in order to maximize the economic recovery of hydrocarbons from the reservoir.

Output results

Simulators provide for output of well and field pressures, production rates, and maps of saturations and pressures at specified times. Contour maps can be viewed in a planar, cross-sectional, or 3-D perspective. Interactive maps showing production bubbles on a background reference map are most useful. Example studies in reservoir simulation are presented in chapter 14.

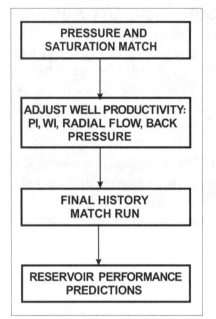

Fig. 13–3. Saturation match procedures. *Source: A. Satter, J. Baldwin, and R. Jespersen. 2000.* Computer-Assisted Reservoir Management. *Tulsa: PennWell.*

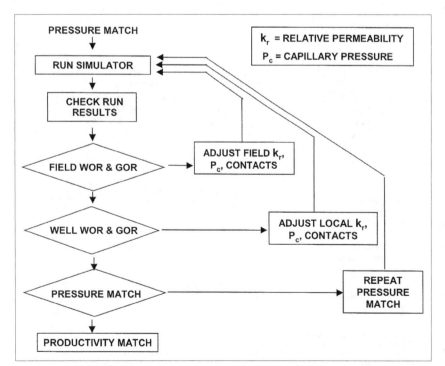

Fig. 13–4. Productivity match procedures. *Source: A. Satter, J. Baldwin, and R. Jespersen. 2000.* Computer-Assisted Reservoir Management. *Tulsa: PennWell.*

Abuse of Reservoir Simulators

The use of reservoir simulation has grown steadily over the last 25 years because of the constant improvement in simulator software and computer hardware. The rapid growth and acceptance of simulation has led to some confusion and occasional misuse of the reservoir engineering tool. This has occurred due to unrealistic expectations, insufficient justification for simulation, and unrealistic reservoir description.

Golden Rules for Simulation Engineers

There is always a danger that users will misuse sophisticated models available to them. Aziz offers the following basic rules in order to minimize this danger:[11]

1. Understand the problem and define the objectives.
2. Keep it simple—start and end with the simplest model. Understand model limitations and capabilities.
3. Understand the interaction between the different parts; reservoir, aquifer, wells, and facilities are interrelated.
4. Do not assume bigger is always better. Always question the size of a study that is limited by the computer resources or the budget, or both. The quality and the quantity of the data are important.
5. Know your limitations and trust your judgment. Remember that simulation is not an exact science. Do simple material balance calculations to check simulation results.
6. Keep expectations reasonable. Often the most that can be achieved from a study is some guidance on the relative merits of the choices available.
7. Question data adjustments for history matching. Remember that this process does not have a unique solution. Only change the data that is known with lesser certainty.
8. Do not smooth extremes—never average out extremes.
9. Pay attention to the measurements and use scales. Measured values at the core scale may not directly apply at the larger block scale, but measurements at one scale do influence values at other scales.
10. Do not skimp on necessary laboratory data. Plan laboratory work with its end use in mind.

Summing Up

Reservoir simulators are widely used to study reservoir performance and to determine methods for enhancing the ultimate recovery of hydrocarbons from the reservoir. They are used to develop a reservoir management plan and to monitor and evaluate reservoir performance during the life of the reservoir.

Concepts and technologies

Numerical simulation is still based upon material balance principles, taking into account reservoir heterogeneity and the direction of fluid flow. In addition, it considers the locations of the production and injection wells and their operating conditions.

Reservoir simulators

Reservoir simulators are generally classified as black oil, compositional, thermal, and chemical. While all types of simulators account for fluid flow mechanisms, compositional simulators additionally account for phase composition flow. Thermal simulators account for heat flow, and chemical simulators account for mass transport due to dispersion, adsorption, and partitioning.

In petroleum engineering, there are two principal types of simulators—a black oil simulator and a compositional simulator. More than 90% of all simulation studies can be performed with a black oil simulator.

Reservoir types

Performance of black oil, dry gas, gas condensate, and volatile oil reservoirs can be modeled by simulators.

Simulator applications

Most of the simulation models built today are for complex, full-field evaluations of the largest company assets. Most engineers and managers feel that these are the only applications that make economical sense for this technology.

This viewpoint arose when simulation was a process that took years to learn, and months or years to complete for only one reservoir.

Now models can be built in a short time, and results are obtained in a timely fashion. In marginal assets, the difference between using technologies such as simulation and not applying simulation can be the difference between making money and losing money.

Single-well and pattern models are underutilized in reservoir simulation. These models can be built in a short time and then used to examine well production and interference issues. These models are equally valuable to a geologist trying to understand the relationship between the geology and fluid flow and a production engineer trying to complete and produce wells properly. They are also valuable to reservoir engineers in drilling wells at a spacing that maximizes economic value.

Value of reservoir simulators

Reservoir simulation models provide a good representation of the real reservoir, with all the appropriate physics, reservoir properties, geology, geophysics, and wells included. The reservoir simulation model is divided into a series of interconnected blocks (or more complex objects in more complicated formulations). Within the blocks themselves, the reservoir fluids are normally treated as a stirred homogeneous mixture, with saturations and fluid properties varying from one grid block to the next.

The reservoir simulation is the most general good reservoir engineering analysis tool available. It is the only one that integrates data from all related engineering and geological disciplines. A simulator can be used as a crystal ball of sorts to look through many years of performance of a reservoir

Wells are very expensive to drill, and it can take many years before an accurate forecast of a well's ultimate performance and the optimal well drilling/production strategy is understood. In a simulation model, wells can be drilled with the click of a mouse, and results for the next 20 years can be forecast in minutes.

Mathematical basis and numerical solutions

Simultaneous flow of oil, gas, and water through a porous medium can be described by a set of nonlinear partial differential equations. These equations involve the law of conservation of mass, Darcy's fluid flow law, and the PVT behavior of fluids.

The numerical solutions of the partial differential equations for simultaneous flow of fluids can be solved by finite difference techniques. Instead of obtaining a continuous solution, an approximate solution is obtained at a discrete set of grid blocks or points at discrete times. Finite difference approximations of partial differential equations require spatial and time discriminations. The length of the system is divided into a discrete set of grid blocks and time intervals.

The finite difference approximations of the partial differential flow equations ultimately involve a set of simultaneous equations requiring matrix problem solutions.

The pressures and saturations can be solved explicitly, implicitly, or by a combination method. The method of solution will affect the following:
- **Stability.** The more implicit formulation is generally more stable.
- **Accuracy.** The truncation error, time step, and grid size affect the accuracy of the solution.
- **Cost.** The methods vary in their requirements for computer storage and run time.

Gridding techniques—rectangular and other grids

Gridding is a necessary step in the simulation process. Looking at the reservoir from the top down, the reservoir is divided into many interlocking small tanks, cells, or blocks. This is done in order to take into account layering, thickness, fluid saturations, and variations in rock and fluid properties.

The most common gridding scheme today is rectangular. Other gridding schemes have found their way into simulators during the past 20 years. Two of the more popular are corner point gridding and Pebi gridding. In these techniques, irregularly shaped cells are used in the model, allowing faults, reservoir boundaries, and other irregularities to be followed precisely without the stair-stepping required in rectangular gridding.

Simulation process

In general, the reservoir simulation process can be divided into three main phases:
- Input data gathering
- History matching
- Performance prediction

Input data gathering

The first step in a successful simulation study is the design of a proper data-gathering program. A simulator can be of great use in the design of the data-gathering program, as data acquisition can be one of the most expensive costs in the development of a reservoir. Gathering the right data in proper quantities can both reduce costs and drastically improve the quality of the resulting reservoir simulation model.

The vast majority of time an engineer spends on a simulation study will be in gathering data. It is here where the real time savings can be realized in reservoir studies.

Data requirements

Input data consists of general data, grid data, rock and fluid data, production/injection data, and well data. Gathering the needed data can be a very time-consuming and expensive process. Ascertaining the reliability of the available data is vital for successful reservoir modeling.

The various categories of data required for a simulator are geology, reservoir fluid PVT, relative permeability, well production and injection, tubing, flow lines, and completion.

History matching

History matching of the past production and pressure performance consists of adjusting the reservoir parameters of a model until the simulated performance matches the observed or historical behavior. This is a necessary step before the prediction phase, because the accuracy of a prediction can be no better than the accuracy of the history match. However, it must be recognized that history matches are not unique.

The history-matching procedure consists of the following sequential steps:
1. Pressure matching
2. Saturation matching
3. Productivity matching

Performance prediction

The final phase of the reservoir simulation study involves predicting the future performance of a reservoir. This could be for existing operating conditions and/or some alternate development plan, such as infill drilling or waterflooding after primary production.

Output results

Simulators provide for output of well and field pressures, production rates, and maps of saturations and pressures at specified times. Contour maps can be viewed in planar, cross-sectional, or 3-D perspective. Interactive maps showing production bubbles on a background reference map are most useful. Example studies are presented in the next chapter.

Abuse of reservoir simulation

The rapid growth and acceptance of simulation has led to some confusion and occasional misuse of the reservoir engineering tool. This has occurred due to unrealistic expectations, insufficient justification for simulation, and unrealistic reservoir description.

Golden rules for simulation engineers

There are basic rules in order to minimize the misuse of simulators. These include understanding the problem, defining objectives, keeping it simple, and knowing the limitations. It is also necessary to question the data adjustments, thus avoiding making unnecessary adjustments for history matching, and not to skimp on necessary laboratory data.

Class Assignments

Questions

1. Why is a reservoir simulator preferable to the classical material balance method for reservoir performance analysis?

2. What are the basic concepts and technologies involved in the reservoir simulation method?

3. What fundamentals comprise the mathematical basis for reservoir simulation?

4. Why are the mathematical solutions not exact?

5. What are the basic reservoir simulation types? List them with their associated flow mechanisms.

6. What are the components in a reservoir simulation process?

7. What basic data is required for the simulation process?

8. Why is a history match not unique?

9. What is the main objective for future performance prediction?

10. What are the values in applying the simulation method? Explain its limitations.

References

1. Mattax, C. C., and R. L. Dalton. 1990. *Reservoir Simulation*. SPE Monograph Series. Vol. 13. Richardson, TX: Society of Petroleum Engineers.
2. Litvak, B. L. Personal contact.
3. Turgay, E., J. H. Abou-Kassem, and G. R. King. 2001. *Basic Applied Reservoir Simulation*. SPE Textbook Series. Vol. 7. Richardson, TX: Society of Petroleum Engineers.
4. Fanchi, J. R. 2001. *Principles of Applied Reservoir Simulation*. Houston: Gulf Publishing Co.
5. Satter, A., and G. C. Thakur. 1994. *Integrated Petroleum Reservoir Management—A Team Approach*. Tulsa: PennWell.
6. Satter, A., J. Baldwin, and R. Jespersen. 2000. *Computer-Assisted Reservoir Management*. Tulsa: PennWell.
7. Merlin Simulation Software. 2005. Houston: Gemini Solutions, Inc.
8. Aziz, K. 1989. Ten golden rules for simulation engineers. Dialog. *Journal of Petroleum Technology*. November.
9. Litvak, B. L. Personal contact.
10. Stone, H. L. 1973. Estimation of three-phase relative permeability and residual oil data. *Journal of Canadian Petroleum Technology*.
11. Aziz, K. 1989

14 · Reservoir Simulation Model Applications

Historically, reservoir simulators have been used for studying full-field production performance. These studies are costly, require highly trained professionals, and are often too time-consuming for operating department environments.

Simulators on personal computer platforms can be used not only for full-field simulation, but also for single-well simulation. They allow a team to gain an understanding of the well development scheme to maximize the present net worth. Early in the life of a field, the exact reservoir extents may not be known, and for these situations, single-well models may represent a better approach than coarse full-field models. Mini-simulators can play an important role in everyday operations.[1]

Chapter 13 presents the fundamentals of reservoir simulation. These fundamentals include the mathematical basis and solutions, simulator types, applications, the simulation process (input data and history matching), and prediction.

This chapter is devoted to presenting results of computer runs using Merlin Simulator software from Gemini Solutions.[2] These computer runs include the following:

- Simulator-generated correlations
- Single-well simulation performance
- Full-field simulation

Gemini Solution's Merlin reservoir simulation model was used to provide results.

Simulator-Generated Correlations

When laboratory data is not available, simulator-generated correlation results are beneficial, even for reservoir engineering studies. Merlin PC Simulator of Gemini Solutions was used to generate correlations for relative permeability curves presented in chapter 2, and for the oil, gas, and water properties given in chapter 3. The input data used for generating correlations is given below:

Initial water saturation, % PV, = 22.
Residual oil saturation to water, % PV, = 20.
Residual oil saturation to gas, % PV, = 20.
Critical gas saturation, % PV, = 5.
Relative permeability exponents = 3.
Gas relative permeability end point = 1.
Oil relative permeability end point = 0.9.
Water relative permeability end point = 0.15.

Data used for oil, gas, and water properties is based upon the following:

Reservoir temperature, °F, = 130.
Oil gravity, °API, = 35.
Gas gravity = 0.65. (Air = 1.)
Initial gas/oil ratio, scf/stb, = 354.
Bubblepoint pressure, psia, = 2,002.

Single-Well Applications

A Merlin reservoir simulation model was used to predict reservoir performance of a producing, single-well oil reservoir, along with sensitivity to fluid flow conditions and variations in rock and fluid properties. The basic data used is given below:

Depth, ft, = 5,520.
Reservoir temperature, °F, = 169.
Initial pressure, psia, = 2,554.
Bubblepoint pressure, psia, = 1,745.
Net thickness, ft, = 50.
Porosity, % PV, = 30.5.
Permeability, mD, = 30.
Initial oil saturation, % PV, = 22.
Oil gravity, °API, = 35.
Gas gravity = 0.66. (Air = 1.)
Bubblepoint solution gas/oil ratio, scf/stb, = 385.
Initial oil formation volume factor, rb/stb, = 1.252.
Bubblepoint oil formation volume factor, rb/stbo, = 1.33.
Oil viscosity at bubblepoint, cp, = 0.80.
Oil viscosity at initial pressure, cp, = 0.846.
Water viscosity at bubblepoint = 0.393.
Oil compressibility, psi^{-1}, = 11.17×10^{-6}.
Water compressibility, psi^{-1}, = 1.31×10^{-6}.
Drainage area, acres, = 160.

Chapter 4 presents simulated results, which show changes in pressure with time at a 40-ft radial distance from the wellbore. The results also show unsteady-state pressure changes with time and radial distance.

Chapter 8 presents the simulated performance of an oil reservoir for the following:
- Pressure versus recovery efficiency under the influence of natural producing drive mechanisms
- Pressure and gas/oil ratio versus oil recovery for a homogeneous reservoir under depletion drive
- Sensitivity to 10°, 20°, and 30° API oil on pressure versus recovery efficiency
- Pressure, gas/oil ratio, and production rate versus recovery efficiency as influenced by Dykstra-Parsons permeability variation factors of 0 (homogeneous), and 0.5 and 0.9 (heterogeneous)
- Pressure, gas/oil ratio, and production rate versus recovery efficiency as affected by critical gas saturation

Chapter 16 presents primary and waterflood recovery efficiency for a 5-spot waterflood project considering various Dykstra-Parsons permeability variation factors.

Full-Field Applications

Satter, Baldwin, and Jespersen presented full-field simulation results for the following cases:[3]
1. A newly discovered offshore field development plan
2. Mature field revitalizations
3. Waterflood project development plan

Results of these studies are summarized below. (Details are available in *Computer-Assisted Reservoir Management* by Satter, Baldwin, and Jespersen.[4])

Full-field simulation examples using Merlin Simulator of Gemini Solutions, Inc., are also presented in this chapter.

Newly discovered offshore field development plan

The field considered was analogous to Gulf Coast reservoirs. The input of all disciplines, mutual understanding, and interdisciplinary communication were the keys to successfully developing an optimum plan. The team needed to address the following main unknowns in order to come up with an economically viable development and depletion strategy:
- Recovery scheme: natural depletion or natural depletion augmented by water injection?
- Well spacing: number of wells, platforms, reserves, and economics

Considering development of the field using 40-, 80-, 120-, and 160-acre well spacings, a full-field reservoir simulation model was constructed to predict depletion drive performance and also for waterflooding.

A commercial simulator was used to predict production rates and reserves forecasts. The resulting production rates and reserves forecasts showed that the larger the spacing, the longer the reservoir life, with less oil recovery.

Using estimated production, capital, operating expenses, and other financial data, economic analyses of the primary development plans with 40-, 80-, 120-, and 160-acre well spacings were made. The 160-acre well spacing case offered the lowest capital investment, development cost, and payout time, and the highest present worth index and discounted cash flow return on investment. It also showed the next highest net present value and offered the economically optimum primary development plan. Even though the 80-acre case yielded the highest net present value, the additional capital investment over the 160-acre case did not give appreciable incremental net present value.

Results of the economic analysis of the waterflood case showed the highest oil reserves, discounted cash flow return on investment, net present value, and present worth index. It also showed the lowest development costs per barrel of oil. Therefore, early waterflooding offered the most economic means to exploit this field. The platform needed to be designed such that the water injection facilities could be installed later, i.e., some deck space would be left for future water injection equipment.

Based on the economic evaluation results, the team recommended to its management that the initial 160-acre primary development be followed by 80-acre, 5-spot infill waterflooding after two years.

Mature field revitalizations

The mature North Apoi/Funiwa field, operated by Texaco Overseas (Nigeria) Petroleum Company Unlimited (TOPCON), is located offshore of Nigeria. The North Apoi field was discovered in 1973, followed by Funiwa, an extension of North Apoi, in 1978. The fields consisted of a northwest/southeast-trending anticline, with several major faults and multiple sand reservoirs. As of June 30, 1995, 64 wells were drilled, including 58 commercial wells. Primary producing mechanisms are a combination of depletion, gascap, and natural water drives. Ewinti-5, -6, and -7, and Ala-3, -5, and -7, are the major producing sands.

An integrated team of geoscientists and engineers from TOPCON, Nigerian government organizations, and Texaco's E&P Technology Department (EPTD) were charged to review the field's six largest reservoirs in three phases to evaluate and capture the upside potential of the reserves.

The first phase of study involved the Ewinti-5 and Ala-3 reservoirs, which contained more than 50% of the field's booked reserves. This phase was initiated in March and completed in June 1995 in Houston by Texaco's EPTD. The second phase of study, covering the Ewinti-7 and Ala-5 reservoirs, and the third phase of study, covering Ewinti-6 and Ala-7, were also carried out in Houston, taking three months each.

The objectives of the studies were to determine the following:
- Ultimate primary recovery
- Optimum recovery with additional vertical and horizontal wells, along with workovers, including gas lift
- Enhanced oil recovery potential

Challenges faced in the studies included the following:
- Declining production and increasing operating costs for this mature field
- An unrealistic recovery factor
- The need to enhance asset value

The approach taken involved the following:
- Review geosciences and engineering data
- Perform classical material balance analysis
- Perform reservoir simulation analysis for full-field performance history matching and prediction
- Plan strategies and forecast performance under existing conditions, with workovers and infill wells, gas lift, and water injections

The studies utilized integration/alliance of organizations, integration of data and software, and professionals working together as a team, employing their tools and technologies. Integrated geosciences and engineering models were developed using revised maps based upon reprocessing and reinterpreted 3-D seismic survey data from 1986. Well log and core analysis data, rock and fluid properties, well test data, and other engineering data, along with 20 years of field production history, were also incorporated into the reservoir description.

The classical material balance analysis was considered to be a prerequisite to reservoir simulation. The EPTD-developed OilWat material balance software was used for estimating the original oil in place and primary drive mechanisms. The material balance analysis showed that the primary production mechanism of the Ewinti 5, Ewinti-7, and Ala-5 sands was a strong water drive, with additional support from gascap drive and solution gas drive. The Ala 3 reservoir demonstrated a weak water drive plus gascap drive and solution gas drive.

Integrated Petroleum WorkBench software of Scientific Software Intercom (SSI)'s oil simulator was utilized for full-field performance history matching and forecasts. The stepwise history-matching procedure consisted of pressure matching followed by saturation matching. Pressure matching was achieved by specifying the historical total three-phase voidages for the wells, while adjusting pore volumes, aquifer strength, and fault connections. In order to validate the reservoir models, the 3-D seismic survey data was reexamined. Pressure matching ensured that the reservoirs' historical total (three-phase) voidages were duplicated both for the total reservoir and for each of the wells.

The original oil in place values estimated from classical material balance analyses and simulation techniques are comparable to each other but are substantially higher than the booked values.

Good history matches using the black oil simulator in WorkBench were achieved for most of the wells by adjusting the usual reservoir parameters within their accepted ranges of uncertainty. Difficulties matching a few wells, however, led to questions about the structure maps. Here, the interaction between the geosciences and engineering members of the team proved very beneficial. The 3-D seismic survey data was reexamined to validate the reservoir model. Ultimately, some areas of poor seismic resolution were reinterpreted, leading to successful history matches in all of the wells.

After reservoir performance history matching using the black oil simulator in WorkBench, model prediction runs were made. These were conducted under various investment scenarios for optimally draining the reservoirs, including additional take points, horizontal wells, gas lift, and water injection. Opportunities were identified for performing workovers and placing additional wells to improve drainage in the Funiwa area.

Performance forecasts for the remaining period of the contract were made under different operating scenarios in order to determine the optimum development plan as follows:
- Case 1: primary depletion with the current wells and production limitations (base case)
- Case 2: base case + infill wells + workovers
- Case 3: case 2 + gas lift
- Case 4: case 3 + water injection, applicable to the Ala-3 sand reservoir

The performance results are presented in Figures 15-5 and 15-6.

Since the Ala-3 reservoir has weak natural water drive, the studies showed that recovery could be improved with water injection. The Ewinti-5, Ewinti-7, and Ala-5 reservoirs, on the other hand, have strong water drives, and thus no water injection case was attempted.

The new estimated reserves were substantially more than the booked values. Additional drilling/completion recommendations from the studies for Ewinti-5 Ala-3 were made. Recommendations for the six reservoirs studied in phases 1, 2, and 3 included 10 horizontal wells, 4 deviated wells, 1 replacement well, and 4 workovers.

All infill and workover wells are located in the Funiwa field. Since the current drainage patterns in the North Apoi area are adequate, no additional offtake points are necessary there.

The placement of the wells was determined from the simulator-calculated fluid saturation distributions initially and throughout the producing life of the reservoirs. The oil saturation distributions initially in the Ewinti-5 layer 4 model at the time of the study (1995), plus the predicted distributions at the end of the lease expiration (2008), were utilized for the base case and for the infill/workover cases. These were used to determine the locations of the horizontal wells based upon the high remaining base case saturation predicted for 2008.

TOPCON and the Nigerian government acted quickly on the study recommendations. Within nine months from the start of the first phase of study, two successful horizontal wells were drilled and completed in the Funiwa Ewinti-5 reservoir. The first well came on production at 2,670 bopd of oil from a 700-ft horizontal section. The second well has a 1,600-ft horizontal pay section and produced at 4,020 bopd.

The role played by each partner in this alliance was essential to the successful outcome of the project. TOPCON recognized both the need for the investigation to be made and the benefits of collaboration. Their engineers and geoscientist provided all the field data, plus an in-depth knowledge of the reservoirs and of the current producing operations. They performed a majority of the technical project work themselves.

EPTD provided project coordination, computer software and hardware, software training, and specialized expertise in 3-D seismic interpretation. EPTD also provided well log analysis, reservoir simulation, 3-D visualization, and horizontal drilling.

Government engineers took an active role in the project work. Their participation ensured that all regulations would be met and that the government's interests were considered early in the planning of proposed operations. This led to rapid approval and early commencement of drilling.

SSI consultants contributed significantly to the timely completion of the projects. They arranged for the availability of extra software licenses for the project and provided technical support for this first major project at EPTD using their WorkBench product.

The joint efforts resulted in significant cycle time reduction and set an excellent example of integration and alliance.

Waterflood project development plan

A hypothetical field, akin to real-life reservoirs, was used as an example to develop a plan for a waterflood project. This field was discovered many years ago and is now depleted. It consisted of a simple domal structure, and five of the nine wells drilled were producers. Primary producing mechanisms were fluid and rock expansion (reservoir pressure above the bubblepoint), solution gas drive, and limited natural water drive. Data available was limited, and even the gas, oil, and water production data was unreliable. Reservoir pressures were not monitored.

An integrated team of geoscientists and engineers was charged by the management to review the past performance and investigate the waterflood potential of this field. The team's approach was to accomplish the following:
- Build an integrated geosciences and engineering model of the reservoir using available data and correlations
- Simulate full-field primary production performance without history matching, since no historical pressure data was available
- Forecast performance under peripheral and pattern waterflooding
- Recommend an optimum development plan based upon economic analysis

Even though the field was hypothetical, much can be learned about how to engineer a waterflood project, even with incomplete data.

The results of this study are presented in Figures 18-12 and 18-13.

Merlin reservoir simulator

A full-field model was built for a new prospect. In this case, the value of integrating real-time data early in the life of a reservoir is illustrated. This data is incorporated for reservoir optimization to maximize the value of an asset.

In the first step, the base model is built, and sensitivities are run to reservoir uncertainties, particularly the distribution of vertical layering.

The sensitivity runs will illustrate that the optimal development plan for this reservoir is very sensitive to the vertical layering distribution. In fact, an optimal development plan cannot be developed with the limited exploration well data available. Early time data is needed to improve the reservoir model and prepare an optimal development plan for this reservoir.

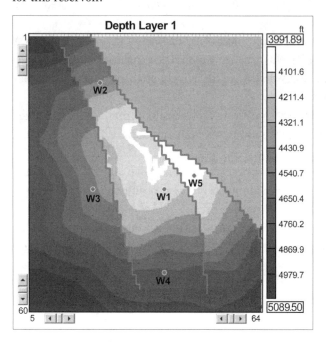

Fig. 14–1. Structure map of faulted reservoir. Progressively shaded areas indicate increasing depth. *Courtesy Gemini Solutions, Inc.*

In order to accomplish the optimal plan, continuous downhole pressure gauges are installed. The rate and flowing pressure histories are gathered for several months in the exploration wells. Using the data gathered, the reservoir model could be improved appreciably, and an accurate reservoir description and optimal development plan can be calculated. Delaying the large initial capital investments several months to understand a reservoir in detail can be shown to greatly increase the present net value of a typical asset. It will also increase the ultimate recovery and accelerate production, where economically feasible.

This exercise shows the value that can be realized through real-time integration of early reservoir data into reservoir development.

Basic data. Figures 14–1 through 14–7 show the basic data used in the construction of the reservoir model, including the basic geology, PVT data, and relative permeability curves. All figures used in this chapter are courtesy of Gemini Solutions Inc.[2]

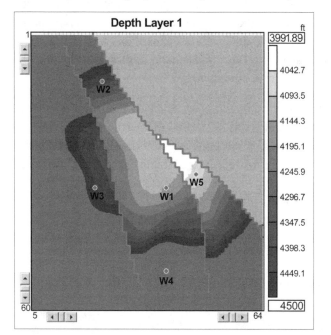

Fig. 14–2. Structure map showing area above original oil/water contact. *Courtesy Gemini Solutions, Inc.*

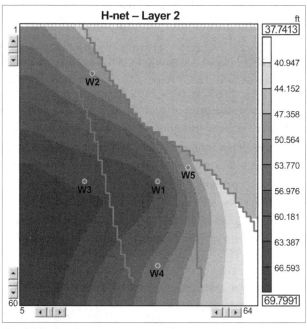

Fig. 14–3. Net thickness map. Progressively shaded areas indicate increasing net thickness of the formation *Courtesy Gemini Solutions, Inc.*

Layering assumptions. Core studies suggest a highly heterogeneous rock with a Dykstra-Parsons coefficient approaching 0.8, and possible thin vertical shale barriers that may or may not extend across the field. The high Dykstra-Parsons coefficient suggests that the reservoir may not waterflood efficiently.

Additionally, cores and well tests both suggest that the average rock permeability is in the 30–50 mD range.

As a result of the core uncertainties, the following cases were designed to look at the range of potential producing profiles and recovery factors for this reservoir. Based on the data collected during the exploration period, a set of cases was considered to encompass the range of possible outcomes.

All cases assumed a conservative Dykstra-Parsons coefficient of 0.8 (and this was consistently confirmed by core analysis).

A conservative geometric mean permeability of 30 mD was assumed.

Layering was assumed, and cases with permeabilities increasing both from top to bottom and bottom to top were analyzed. Core data suggests that in general, permeability increases from top to bottom, although the reverse trend was considered for sensitivities.

Vertical communication is uncertain between the layers. Thus for the purposes of this exercise, both sealing shales between the layers and leaking shales between the layers were assumed.

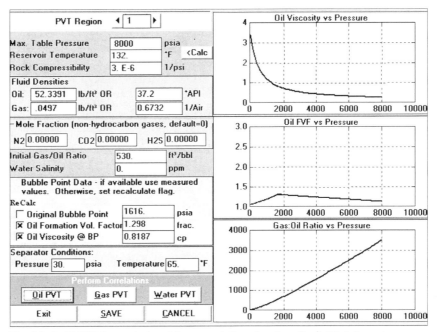

Fig. 14–4. Oil PVT data calculated from correlation. *Courtesy Gemini Solutions, Inc.*

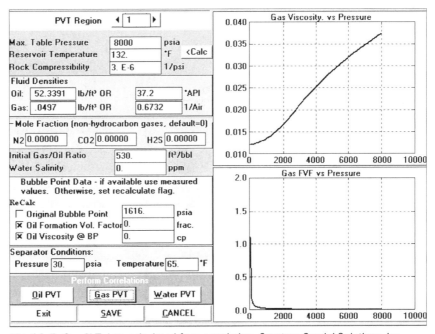

Fig. 14–5. Gas PVT data calculated from correlation. *Courtesy Gemini Solutions, Inc.*

Based on the preceding assumptions, the following sensitivity cases were developed, and corresponding simulation runs were conducted:

- Case 1. Shale between layers, permeability increasing top to bottom.
- Case 2. Shale between layers, permeability decreasing top to bottom.
- Case 3. No shale between layers, permeability increasing top to bottom.
- Case 4. No shale between layers, permeability decreasing top to bottom.

All cases assumed infinite aquifer support, as verified by offsetting reservoirs in the area of interest.

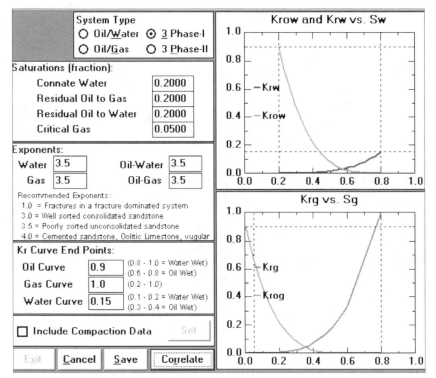

Fig. 14–6. Relative permeability curves. *Courtesy Gemini Solutions, Inc.*

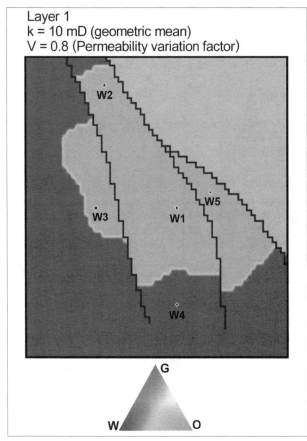

Fig. 14–7. Oil, water, and gas saturation of the top layer in plane view within the faulted reservoir. Original oil and water volumes are shown in light and in dark colors, respectively. *Courtesy Gemini Solutions, Inc.*

Dykstra-Parsons layering distributions. The following layering was used for the purposes of this problem. Four individual layers were assumed. Additional input data included the following: Dykstra-Parsons permeability variation factor, 0.8; geometric mean permeability, 30 mD; and logarithmically increasing cell sizes.

- Layer 1: 3% of total sand thickness, 759 mD
- Layer 2: 9% of total sand thickness, 339 mD
- Layer 3: 23% of total sand thickness, 117 mD
- Layer 4: 65% of total sand thickness, 17 mD

As evident from the layering scheme, a thin layer of very high permeability exists at the top of the formation. This is followed by a decrease in permeability and an increase in layer thickness toward the bottom.

Original oil in place. The original oil in place is calculated to be 52 MMbo, of which 39 MMbo are thought to be moveable in the presence of a water or gas drive. The total hydrocarbon area is 971 acres, with approximately 36,000 acre-ft of hydrocarbon-bearing volume in the reservoir.

Results of sensitivity runs. The exploration wells were produced at the maximum rate possible through the tubing and flow system for the four assumed layering cases. Nodal analysis established a practical maximum rate possible of 10,000 bopd through the facilities available.

Discussion of the sensitivity cases. Fortunately, the pressure behavior in this reservoir for the various cases is remarkably different. Additionally, it can be seen that the pressure behavior deviates very early for the various cases. In fact, the true reservoir layering for this reservoir can be inferred from producing the four available exploration and development wells for only a few months.

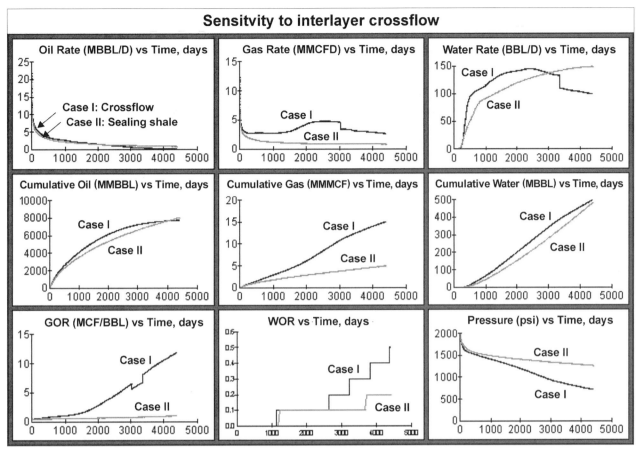

Fig. 14–8. Forecasts assuming sealing versus communicating shales between the layers. *Courtesy Gemini Solutions, Inc.*

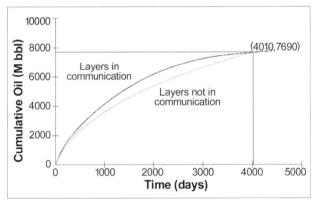

Fig. 14–9. Forecasts of cumulative oil production assuming sealing versus communicating shales between the layers. *Courtesy Gemini Solutions, Inc.*

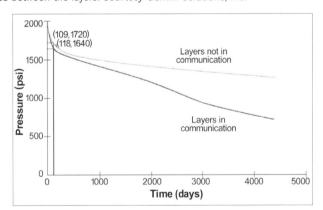

Fig. 14–10. Forecasts of pressures assuming sealing versus communicating shales between the layers. *Courtesy Gemini Solutions, Inc.*

Cases I and II in Figure 14–8 show the comparison between cases 2 and 4. In both of these cases, permeability is decreasing from top to bottom. However, a comparison is made between the case with permeable shales between the layers and the case with sealing shales between the layers. The leaking shale case allows an effective gas cap to form in the top two layers. Although the oil production rates are close for both cases, the formation of the gas cap and subsequent blowdown of free gas produces drastically different gas/oil ratios and pressure behaviors. Even more importantly, these differences are evident within the first two to three months of production.

Case II, with sealing shales between the layers, produces at lower rates early in its production life (fig. 14–9). At later times, since an effective gas cap has not formed for this case, the average reservoir pressure remains higher (fig. 14–10). For times greater than 4,000 days, the ultimate recovery is higher than the case with leaking sealing shales between layers.

Note that within 100 days, the pressure and cumulative production profiles for the two assumed cases are drastically different. This example clearly illustrates the ability to use downhole pressure data to understand the effective vertical permeability in a reservoir. This is typically a major uncertainty in reservoir simulation that is very difficult to estimate from core analysis.

Similarly, the gas production rates from the communicating versus noncommunicating cases are quite different at very early times, as illustrated by the sample comparisons at 100 days (fig. 14–11). The higher gas rates in the comparison below correspond to leaking shale units between the layers.

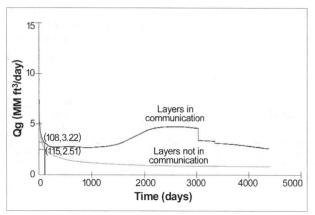

Fig. 14–11. Forecasts of cumulative gas production assuming sealing versus permeable shales between the layers. *Courtesy Gemini Solutions, Inc.*

Gas cap is formed for the leaking shale case around the end of the producing life, and majority of the top two layers is within the gas cap.

In comparison, no effective gas cap is formed even at the end of the producing life for the sealing fault run.

However, gas saturation is developed for the sealing fault case at the end of the producing life.

Figure 14–12 compares a case where permeability is decreasing from top to bottom to a case where permeability is increasing from top to bottom. Both cases assume leaking shales between the Dykstra-Parsons sublayers. The case with permeability increasing downward has significantly higher water influx from the aquifer and almost double the recovery in the first 10 years.

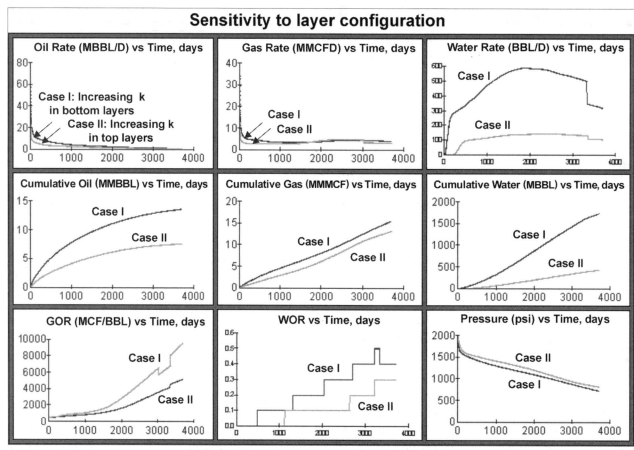

Fig. 14–12. Effect on production when layering is reversed. *Courtesy Gemini Solutions, Inc.*

A comparison of cumulative production from the two cases is shown in Figure 14–13.

The difference in the recovery for the two cases is remarkable and illustrates the need to understand the reservoir geology before the optimal well development plan can be formulated. If permeability truly increases from top to bottom for the four wells, it can be observed that the ultimate recovery is almost double what is possible from a model in which the permeability degrades as depth increases for the same number of wells. In terms of optimal well development, the two cases will require about a two-fold difference in the number of wells required for a target recovery factor. Understanding which case actually represents the reservoir could save an operator as much as 50% of the initial development costs.

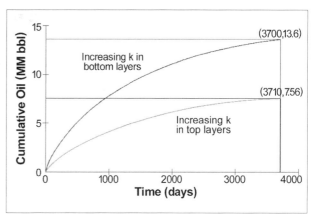

Fig. 14–13. Effect on cumulative oil production when layering is reversed. *Courtesy Gemini Solutions, Inc.*

In conclusion, the sensitivity cases illustrate that radically different optimal well development approaches are possible in this reservoir based on the range of likely sensitivities that were run.

Integrating Early Reservoir Behavior and Optimizing Field Development

Many petroleum professionals think that years of downhole data are required before an accurate reservoir description and optimal development plan can be developed for most assets. Traditionally that has been true, with pressures measured infrequently on intervals of a year or more.

Within the last 15 years, with the advent of continuous downhole pressure gauges and easy-to-use reservoir simulation systems, even large assets can be accurately characterized in weeks or months. These advancements make real-time reservoir optimization, including optimization of the initial field development, a reality.

Summing Up

Simulator-generated correlations, single-well simulation performance, and full-field simulation results are presented.

When laboratory data is not available, simulator-generated correlation results are beneficial even for reservoir engineering studies. The Merlin PC Simulator of Gemini Solutions can be used to generate correlations for relative permeability curves and oil, gas, and water properties.

The Merlin reservoir simulation model can be used to predict reservoir performance of a producing single well and sensitivity to fluid flow conditions, along with variations in rock and fluid properties.

Example results for full-field simulations using a commercial simulator include a newly discovered offshore field development plan, mature field revitalizations, and a waterflood project development plan.

For the newly discovered offshore field plan case, recommendations to the management were made based on the economic evaluation results. These results included the initial 160-acre primary development, followed by an 80-acre, 5-spot infill waterflooding process after two years.

For the mature field revitalization case, the placement of the infill wells was determined from the simulator-calculated fluid saturation distributions initially and throughout the producing life of the reservoirs.

The roles played by the operator, the technology department, and the software company professionals were essential to the successful outcome of the project.

Government engineers took an active role in the project work. Their participation ensured that all regulations were met and that the government's interests were considered early in the planning of proposed operations. This led to rapid approval and early commencement of drilling.

From the waterflood project development plan case, much can be learned about how to engineer a waterflood project, even with incomplete data.

A full-field model using Gemini Solution's Merlin simulator for a new real-life prospect was built to illustrate the value of integrating early life real-time data into reservoir optimization to maximize the value of an asset.

In the first step, the base model was built, and sensitivity runs were made to account for reservoir uncertainties, particularly the vertical layering distribution.

The sensitivity runs illustrated that the optimal development plan for this reservoir was very sensitive to the vertical layering distribution. In fact, an optimal development plan could not be developed with the limited exploration well data available. Early time data was needed to improve the reservoir model and to develop an optimal development plan for this reservoir.

In order to accomplish the optimal plan, continuous downhole pressure gauges were installed. The rate and flowing pressure histories were monitored for several months in the exploration wells. Using the data gathered, the reservoir model could be improved appreciably, and an accurate reservoir description and optimal development plan made.

In this real-life example, the two cases with different layering assumptions showed radically different field pressure behavior within 100 days. A continuous downhole pressure gauge for this reservoir would establish both the hydrocarbons in place and the likely distribution of layering within this short time. Once this is accomplished, the reservoir engineer can, by trial and error, establish the optimal well development strategy. At the same time, it must be noted that this is a reasonably large asset of 50+ MMbo.

References

1. Satter, A., J. Baldwin, and R. Jespersen. 2000. *Computer-Assisted Reservoir Management*. Tulsa: PennWell.
2. Merlin Simulation Software. 2005. Houston: Gemini Solutions, Inc.
3. Satter, A., J. Baldwin, and R. Jespersen. 2000.
4. Satter, A., J. Baldwin, and R. Jespersen. 2000.

15 · Fundamentals of Oil and Gas Reserves and Applications

Introduction

Petroleum, consisting of naturally occurring oil and gas, is a valuable underground resource. Petroleum is generated following very long periods of geological and other activities in nature. Not all of this vital resource can be recovered with present-day technology. Hence, the million-dollar question is, How much oil and gas can be recovered, and how much is left underground? The amounts of oil and gas that are producible economically, known as petroleum reserves, involve a high degree of uncertainties. Government regulations and unknown reservoir heterogeneity, among other factors, contribute to the uncertainties. However, estimates of petroleum in subsurface reservoirs are necessary for regulatory, operational, and financial purposes.

Ultimate oil recovery is controlled by reservoir rock properties, fluid properties, heterogeneities, and more importantly, by natural reservoir energies. The latter includes liquid and rock expansion drive, solution gas drive, gascap drive, natural water influx, and combination drive processes. Additional oil recoveries can be made by secondary and enhanced oil recovery methods. Secondary methods include water, natural gas, and gas/water combination floods, which are discussed in chapters 16 and 17. Enhanced oil recovery methods include thermal and nonthermal techniques. Thermal techniques include steam flooding, hot water flooding, and in-situ combustion. Nonthermal techniques include chemical floods, miscible floods, and gas drives. Enhanced oil recovery processes are described in chapter 17.

Recovery of natural gas is also controlled by reservoir rock and fluid properties, heterogeneities, and by natural drive mechanisms. Such mechanisms include gas expansion, natural water influx, and combination drive processes. Recovery factors are generally far greater than those obtained from oil reservoirs, implying that a significant portion of the gas in place is considered as reserves.

The United States does not produce oil and gas in sufficient quantities to meet its daily consumption. As a result, the country is heavily dependent upon importing oil and gas from OPEC (Organization of Petroleum Exporting Countries) and non-OPEC member nations.

This chapter is devoted to the fundamentals of petroleum reserves to provide a general perspective. The objective is to learn about the following:

- Reserves
- U.S. and world reserves
- World resources of petroleum
- History of reserves definitions
- Reserves classifications
- Reserves determination techniques
- Pitfalls in reserves estimations
- Case study
- Reserves growth
- Class assignments

The information presented in the first five items is based on Society of Petroleum Engineers (SPE) and Energy Information Administration (EIA) publications.[1–3]

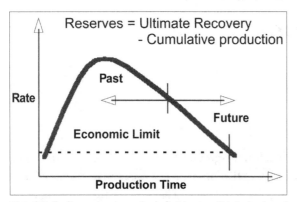

Fig. 15–1. Reserves based on field rates (historical and predicted) versus time.

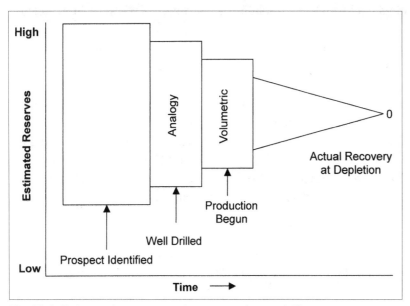

Fig. 15-2. Estimate of reserves at various stages of reservoir life

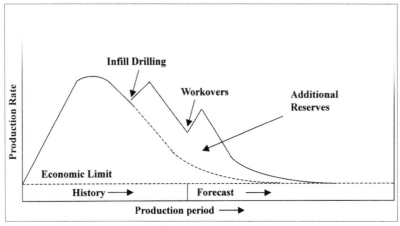

Fig. 15-3. Reserves improvement

Reserves

Simply stated, the following relationship is given (fig. 15–1):

$$\text{Reserves} = \text{Ultimate economical recovery} - \text{Cumulative production}$$

Estimates of reserves vary during the various stages in the life of the field. It is usually the maximum at the identification of a prospect. In reality, the actual recovery is known with certainty only when the field is depleted, as depicted in Figure 15–2.

The economic production rate, which is influenced by the price of oil or gas, is a very important factor in determining the ultimate recovery. The ultimate recovery from a reservoir could be enhanced by operational changes, such as infill drilling, well recompletion, and workovers, as shown in Figure 15–3.

Reserves estimates depend upon the integrity, skill, and judgment of the evaluator, and involve a lot of uncertainties. The results are dependent upon the reliability of the available data at the time the estimates are being made, the interpretation of the data, and the techniques used. The following approaches are used:

- **Deterministic.** Based upon known geological, engineering, and economic data.
- **Probabilistic.** Based upon the ranges of geological, engineering, and economic data.

The probability approach is the preferred method, considering the inherent uncertainties.

Reserves are classified as proved, probable, and possible, accounting for uncertainties in the estimates.

U.S. reserves

The United States is the world's largest energy consumer. In 2004, 40% of U.S. energy consumption was provided by crude oil and natural gas liquids combined. Natural gas provided 23%, amounting to 63% of the total. U.S. energy consumption was about 21 million bbl of oil and natural gas liquids and 61 bcf of gas per day.

In the past, the U.S. government and the public relied heavily upon industry estimates of proved reserves. However, the industry ceased publication of its reserve estimates after its 1979 report. In response to a recognized need for creditable annual proved reserves estimates, the U.S. Congress in 1977 required the Department of Energy (DOE) to prepare such estimates. The Energy Information Administration of the DOE established a unified, verifiable, comprehensive, and continuing annual statistical program series for proved reserves of crude oil and natural gas. Natural gas liquids were added to the reporting in 1979.

Table 15-1. U.S. proved reserves of crude oil, 1976–2004. *Source:* U.S. Crude Oil, Natural Gas, and Natural Gas Liquids Reserves. *1977 through 2004 annual reports. Washington, D.C.: Energy Information Administration. DOE/EIA-0216 (1–27).*

(Million Barrels of 42 U.S. Gallons)

Year	Adjustments[a] (1)	Net Revisions (2)	Revisions[b] and Adjustments (3)	Net of Sales and Acquisitions (4)	Extensions (5)	New Field Discoveries (6)	New Reservoir Discoveries in Old Fields (7)	Total[c] Discoveries (8)	Estimated Production (9)	Proved[d] Reserves 12/31 (10)	Change from Prior Year (11)
1976	–	–	–	–	–	–	–	–	–	[e]33,502	–
1977	[f]-40	386	346	NA	496	168	130	794	2,862	31,780	-1,722
1978	366	1,390	1,756	NA	444	267	116	827	3,008	31,355	-425
1979	337	437	774	NA	424	108	104	636	2,955	29,810	-1,545
1980	219	1,889	2,108	NA	572	143	147	862	2,975	29,805	-5
1981	138	1,271	1,409	NA	750	254	157	1,161	2,949	29,426	-379
1982	-83	434	351	NA	634	204	193	1,031	2,950	27,858	-1,568
1983	462	1,511	1,973	NA	629	105	190	924	3,020	27,735	-123
1984	159	2,445	2,604	NA	744	242	158	1,144	3,037	28,446	711
1985	429	1,598	2,027	NA	742	84	169	995	3,052	28,416	-30
1986	57	855	912	NA	405	48	81	534	2,973	26,889	-1,527
1987	233	2,316	2,549	NA	484	96	111	691	2,873	27,256	367
1988	364	1,463	1,827	NA	355	71	127	553	2,811	26,825	-431
1989	213	1,333	1,546	NA	514	112	90	716	2,586	26,501	-324
1990	86	1,483	1,569	NA	456	98	135	689	2,505	26,254	-247
1991	163	223	386	NA	365	97	92	554	2,512	24,682	-1,572
1992	290	735	1,025	NA	391	8	85	484	2,446	23,745	-937
1993	271	495	766	NA	356	319	110	785	2,339	22,957	-788
1994	189	1,007	1,196	NA	397	64	111	572	2,268	22,457	-500
1995	122	1,028	1,150	NA	500	114	343	957	2,213	22,351	-106
1996	175	737	912	NA	543	243	141	927	2,173	22,017	-334
1997	520	914	1,434	NA	477	637	119	1,233	2,138	22,546	529
1998	-638	518	-120	NA	327	152	120	599	1,991	21,034	-1,512
1999	139	1,819	1,958	NA	259	321	145	725	1,952	21,765	731
2000	143	746	889	-20	766	276	249	1,291	1,880	22,045	280
2001	-4	-158	-162	-87	866	1,407	292	2,565	1,915	22,446	401
2002	416	720	1,136	24	492	300	154	946	1,875	22,677	231
2003	163	94	257	-398	426	705	101	1,232	1,877	21,891	-786
2004	74	420	494	23	617	33	132	782	1,819	21,371	-520

[a]Includes operator reported corrections for the years 1978 through 1981. After 1981 operators included corrections with revisions.

[b]Revisions and adjustments = Col. 1 + Col. 2.

[c]Total discoveries = Col. 5 + Col. 6 + Col. 7.

[d]Proved reserves = Col. 10 from prior year + Col. 3 + Col. 4 + Col. 8 - Col. 9.

[e]Based on following year data only.

[f]Consists only of operator reported corrections and no other adjustments.

– = Not applicable.

Notes: Old means discovered in a prior year. New means discovered during the report year. The production estimates in this table are based on data reported on Form EIA-23, "Annual Survey of Domestic Oil and Gas Reserves". They may differ from the official Energy Information Administration production data for crude oil contained in the *Petroleum Supply Annual*, DOE/EIA-0340.

The 2004 EIA report published proved reserves that were based upon large, intermediate, and a select group of small operators on oil and gas wells. Tables 15–1, 15–2, and 15–3 present U.S. proved reserves of crude oil, 1976–2004, U.S. proved reserves of dry natural gas, 1976–2004, and U.S. proved reserves of natural gas liquids, 1978–2004, respectively. The tables include adjustments, net revisions, revisions and adjustments, net of sales and acquisitions, and extensions.

Table 15–2. U.S. proved reserves of dry natural gas, 1976–2004. *Source: U.S.* Crude Oil, Natural Gas, and Natural Gas Liquids Reserves. *1977 through 2004 annual reports. Washington, D.C.: Energy Information Administration. DOE/EIA-0216 (1–27).*

Year	Adjustments[a] (1)	Net Revisions (2)	Revisions[b] and Adjustments (3)	Net of Sales and Acquisitions (4)	Extensions (5)	New Field Discoveries (6)	New Reservoir Discoveries in Old Fields (7)	Total[c] Discoveries (8)	Estimated Production (9)	Proved[d] Reserves 12/31 (10)	Change from Prior Year (11)
1976	–	–	–	–	–	–	–	–	–	[e]213,278	–
1977	[f]–20	-1,605	-1,625	NA	8,129	3,173	3,301	14,603	18,843	207,413	-5,865
1978	2,429	-1,025	1,404	NA	9,582	3,860	4,579	18,021	18,805	208,033	620
1979	-2,264	-219	-2,483	NA	8,950	3,188	2,566	14,704	19,257	200,997	-7,036
1980	1,201	1,049	2,250	NA	9,357	2,539	2,577	14,473	18,699	199,021	-1,976
1981	1,627	2,599	4,226	NA	10,491	3,731	2,998	17,220	18,737	201,730	2,709
1982	2,378	455	2,833	NA	8,349	2,687	3,419	14,455	17,506	201,512	-218
1983	3,090	-15	3,075	NA	6,909	1,574	2,965	11,448	15,788	200,247	-1,265
1984	-2,241	3,129	888	NA	8,299	2,536	2,686	13,521	17,193	197,463	-2,784
1985	-1,708	2,471	763	NA	7,169	999	2,960	11,128	15,985	193,369	-4,094
1986	1,320	3,572	4,892	NA	6,065	1,099	1,771	8,935	15,610	191,586	-1,783
1987	1,268	3,296	4,564	NA	4,587	1,089	1,499	7,175	16,114	187,211	-4,375
1988	2,193	-15,060	-12,867	NA	6,803	1,638	1,909	10,350	16,670	168,024	-19,187
1989	3,013	3,030	6,043	NA	6,339	1,450	2,243	10,032	16,983	167,116	-908
1990	1,557	5,538	7,095	NA	7,952	2,004	2,412	12,368	17,233	169,346	2,230
1991	2,960	4,416	7,376	NA	5,090	848	1,604	7,542	17,202	167,062	-2,284
1992	2,235	6,093	8,328	NA	4,675	649	1,724	7,048	17,423	165,015	-2,047
1993	972	5,349	6,321	NA	6,103	899	1,866	8,868	17,789	162,415	-2,600
1994	1,945	5,484	7,429	NA	6,941	1,894	3,480	12,315	18,322	163,837	1,422
1995	580	7,734	8,314	NA	6,843	1,666	2,452	10,961	17,966	165,146	1,309
1996	3,785	4,086	7,871	NA	7,757	1,451	3,110	12,318	18,861	166,474	1,328
1997	-590	4,902	4,312	NA	10,585	2,681	2,382	15,648	19,211	167,223	749
1998	-1,635	5,740	4,105	NA	8,197	1,074	2,162	11,433	18,720	164,041	-3,182
1999	982	10,504	11,486	NA	7,043	1,568	2,196	10,807	18,928	167,406	3,365
2000	-891	6,962	6,071	4,031	14,787	1,983	2,368	19,138	19,219	177,427	10,021
2001	2,742	-2,318	424	2,630	16,380	3,578	2,800	22,758	19,779	183,460	6,033
2002	3,727	937	4,664	380	14,769	1,332	1,694	17,795	19,353	186,946	3,486
2003	2,841	-1,638	1,203	-10,092	16,454	1,222	1,610	19,286	19,425	189,044	2,098
2004	-114	744	630	1,844	18,198	759	1,206	20,163	19,168	192,513	3,469

[a]Includes operator reported corrections for the years 1978 through 1981. After 1981 operators included corrections with revisions.

[b]Revisions and adjustments = Col. 1 + Col. 2.

[c]Total discoveries = Col. 5 + Col. 6 + Col. 7.

[d]Proved reserves = Col. 10 from prior year + Col. 3 + Col. 4 + Col. 8 - Col. 9.

[e]Based on following year data only.

[f]Consists only of operator reported corrections and no other adjustments.

[g]An unusually large revision decrease to North Slope dry natural gas reserves was made in 1988. It recognizes some 24.6 trillion cubic feet of downward revisions reported during the last few years by operators because of economic and market conditions. EIA in previous years carried these reserves in the proved category.

– = Not applicable.

Notes: Old means discovered in a prior year. New means discovered during the report year. The production estimates in this table are based on data reported on Form EIA-23, "Annual Survey of Domestic Oil and Gas Reserves," and Form EIA-64A, "Annual Report of the Origin of Natural Gas Liquids Production". They may differ from the official Energy Information Administration production data for natural gas contained in the *Natural Gas Annual,* DOE/EIA-0131.

They also include new field discoveries, new reservoir discoveries in old fields, total discoveries, estimated production, proved reserves, and change from the prior year.

Available data shows that U.S. crude oil reserves continued to decrease from 33.5 billion bbl in 1976 to 21.4 billion bbl in 2004. Dry gas reserves also decreased from 213 tscf in 1976 to 193 tscf in 2004. However, natural gas liquids actually increased from 6.7 billion bbl in 1978 to 7.9 billion bbl in 2004.

Table 15–3. U.S. proved reserves of natural gas liquids, 1978–2004. *Source:* U.S. Crude Oil, Natural Gas, and Natural Gas Liquids Reserves. *1977 through 2004 annual reports. Washington, D.C.: Energy Information Administration. DOE/EIA-0216 (1–27).*

(Million Barrels of 42 U.S. Gallons)

Year	Adjustments[a] (1)	Net Revisions (2)	Revisions[b] and Adjustments (3)	Net of Sales and Acquisitions (4)	Extensions (5)	New Field Discoveries (6)	New Reservoir Discoveries in Old Fields (7)	Total[c] Discoveries (8)	Estimated Production (9)	Proved[d] Reserves 12/31 (10)	Change from Prior Year (11)
1978	–	–	–	–	–	–	–	–	–	[e]6,772	–
1979	[f]64	-49	15	NA	364	94	97	555	727	6,615	-157
1980	153	104	257	NA	418	90	79	587	731	6,728	113
1981	231	86	317	NA	542	131	91	764	741	7,068	340
1982	299	-21	278	NA	375	112	109	596	721	7,221	153
1983	849	66	915	NA	321	70	99	490	725	7,901	680
1984	-123	142	19	NA	348	55	96	499	776	7,643	-258
1985	426	162	588	NA	337	44	85	466	753	7,944	301
1986	367	223	590	NA	263	34	72	369	738	8,165	221
1987	231	191	422	NA	213	39	55	307	747	8,147	-18
1988	11	453	464	NA	268	41	72	381	754	8,238	91
1989	-277	123	-154	NA	259	83	74	416	731	7,769	-469
1990	-83	221	138	NA	299	39	73	411	732	7,586	-183
1991	233	130	363	NA	189	25	55	269	754	7,464	-122
1992	225	261	486	NA	190	20	64	274	773	7,451	-13
1993	102	124	226	NA	245	24	64	333	788	7,222	-229
1994	43	197	240	NA	314	54	131	499	791	7,170	-52
1995	192	277	469	NA	432	52	67	551	791	7,399	229
1996	474	175	649	NA	451	65	109	625	850	7,823	424
1997	-14	289	275	NA	535	114	90	739	864	7,973	150
1998	-361	208	-153	NA	383	66	88	537	833	7,524	-449
1999	99	727	826	NA	313	51	88	452	896	7,906	382
2000	-83	459	376	145	645	92	102	839	921	8,345	439
2001	-429	-132	-561	102	717	138	142	997	890	7,993	-352
2002	62	31	93	54	612	48	78	738	884	7,994	1
2003	-338	-161	-499	30	629	35	72	736	802	7,459	-535
2004	273	97	370	112	734	26	54	814	827	7,928	469

[a]Includes operator reported corrections for the years 1978 through 1981. After 1981 operators included corrections with revisions.

[b]Revisions and adjustments = Col. 1 + Col. 2.

[c]Total discoveries = Col. 5 + Col. 6 + Col. 7.

[d]Proved reserves = Col. 10 from prior year + Col. 3 + Col. 4 + Col. 8 - Col. 9.

[e]Based on following year data only.

[f]Consists only of operator reported corrections and no other adjustments.

– = Not applicable.

Notes: Old means discovered in a prior year. New means discovered during the report year. The production estimates in this table are based on data reported on Form EIA-23, "Annual Survey of Domestic Oil and Gas Reserves," and Form EIA-64A, "Annual Report of the Origin of Natural Gas Liquids Production". They may differ from the official Energy Information Administration production data for natural gas liquids contained in the *Natural Gas Annual*, DOE/EIA-0131.

World reserves

International oil and natural gas reserves, published in the *Oil & Gas Journal* in 2006 and reprinted by EIA, are given in Table 15–4. Saudi Arabia with 264 billion bbl has the highest reserves, followed by Canada, Iran, Iraq, Kuwait, the United Arab Emirates, Venezuela, Russia, Libya, and Nigeria. The United States ranks 11th, with 21.4 billion bbl. The table also reflects the vast hydrocarbon reserves in oil sands in Alberta, Canada, estimated at 178.8 billion bbl. The additional reserves put Canada from fifth to second place, next only to Saudi Arabia in recent years. Innovative techniques to extract highly viscous bitumen from oil sands are described briefly in chapters 17 and 19.

Table 15–4 also shows international gas reserves, in addition to oil reserves. Russia with 1,680 tcf has the largest gas reserves, followed by Iran, Qatar, Saudi Arabia, the United Arab Emirates, the United States, Nigeria, Algeria, Venezuela, and Iraq.

World crude oil production figures are presented in Table 15–5. Total production to date is between 73 MMbopd and 74 MMbopd. Top producing countries are Saudi Arabia, Russia, the United States, and Iran. North Sea production is also significant.

Table 15–6 shows the top 18 giant fields of the world, the year of discovery, and reserves.

Several interesting points can be noted from this table. The majority of giant fields appear to be concentrated in the Persian Gulf region. No giant field of comparable reserves has been discovered since the discovery of Rumalia N&S and the Cantarell Complex in 1976. Estimates of ultimate recovery vary by a wide margin in certain fields, indicating a high degree of uncertainty.

The following is an estimate of ultimate oil reserves, in billions of barrels, which can be produced by conventional means:[4]

Middle-East/OPEC		Other countries	
Proved:	743	Proved:	550
Discovered/unproved:	150	Discovered/unproved:	250
Undiscovered:	250	Undiscovered:	550

OPEC members include Algeria, Indonesia, Iran, Iraq, Kuwait, Libya, Nigeria, Qatar, Saudi Arabia, the United Arab Emirates (UAE), and Venezuela. On a regional basis, the percentage of total proved reserves in the Persian Gulf (Middle East) region is estimated to be more than 60%. About 40% of world oil production comes from OPEC member countries, the average gravity of the crude oil being 32.7°API. It must be noted that more than 1 trillion bbl of oil have already been produced from the existing reservoirs of the world.

Worldwide proved reserves of natural gas are estimated to be about 6,100 tcf. Natural gas reserves are found to be concentrated in the Persian Gulf region and the former Soviet Union countries, including Russia.

World Resources of Petroleum

Resources consist of hydrocarbon accumulations, discovered and undiscovered, that have economic value. Besides proved, probable, and possible reserves of oil and gas, resources include the following:

- Discovered accumulations of hydrocarbons that are not commercially viable with current technology
- Prospects of oil and gas that are not yet discovered, and are usually subject to analogy, hypothesis, and speculation

Total world resources of petroleum are estimated to be 9 to 13 trillion bbl according to the following categories:[5]

Conventional—30%

Heavy oil—15%

Extra heavy oil—25%

Oil sands and bitumen—30%

Table 15-4. International petroleum reserves. *Source:* Oil & Gas Journal, *2006.*

Major (Top 20) Oil- and Gas-Producing Countries

Oil		Gas	
Country	**Reserves, billion bbl**	**Country**	**Reserves, tcf**
Saudi Arabia	264.3	Russia	1,680.000
Canada	178.8	Iran	971.150
Iran	132.5	Qatar	910.520
Iraq	115.0	Saudi Arabia	241.840
Kuwait	101.5	UAE	214.400
UAE	97.8	United States	192.513
Venezuela	79.7	Nigeria	184.660
Russia	60.0	Algeria	160.505
Libya	39.1	Venezuela	151.395
Nigeria	35.9	Iraq	111.950
United States	21.4	Indonesia	97.786
China	18.3	Norway	84.260
Qatar	15.2	Malaysia	75.000
Mexico	12.9	Turkmenistan	71.000
Algeria	11.4	Uzbekistan	66.200
Brazil	11.2	Kazakhstan	65.000
Kazakhstan	9.0	Netherlands	62.000
Norway	7.7	Egypt	58.500
Azerbaijan	7.0	Canada	56.577
India	5.8	Kuwait	56.015
Other countries	68.1	Other countries	600.87
World total	**1,292.6**	**World total**	**6,112.14**

Table 15-5. World production. *Source: DOE/EIA;* Journal of Petroleum Technology, *December 2006.*

Country	Production[a], Mbopd
Russia	9,330
Saudi Arabia	9,300
United States	5,155
Iran	4,035
China	3,670
Mexico	3,252
UAE	2,702
Kuwait	2,550
Venezuela	2,490
Canada	2,438
Nigeria	2,430
Norway	2,430
Iraq	2,203
Algeria	1,805
Brazil	1,703
Libya	1,700
Angola	1,468
United Kingdom	1,198
Indonesia	1,015
Qatar	885
Oman	727
Argentina	697
Malaysia	685
India	650
Egypt	630
Ecuador	542
Colombia	534
Australia	470
Syria	340
Gabon	237
Other countries	6,401
Total	**73,672**

[a] As of August 2006

Table 15-6. Ultimate recoverable reserves in giant fields. *Source: AAPG/*Oil & Gas Journal*/EIA; and F. Robelius. 2005. Giant oil fields of the world. Presented at the AIM Industrial Contact Day, May 23.*

Field	Country	Year Discovered	Ultimate Recoverable Reserve, MMMbo
Ghawar	Saudi Arabia	1948	66–100
Burgan Greater	Kuwait	1938	32–60
Safaniya	Saudi Arabia	1951	21–36
Bolivar Coastal	Venezuela	1917	14–36
Berri	Saudi Arabia	1964	10–25
Rumalia N&S	Iraq	1976	22
Zakum	Abu Dhabi	1964	17–21
Cantarell Complex	Mexico	1976	11–20
Manifa	Saudi Arabia	1957	17
Kirkuk	Iraq	1927	16
Gashsaran	Iran	1928	12–15
Abqaiq	Saudi Arabia	1941	10–15
Ahwaz	Iran	1958	13–15
Marun	Iran	1963	12–14
Samotlor	Russia	1961	6–14
Agha Jari	Iran	1937	6–14
Zuluf	Saudi Arabia	1965	12–14
Prudhoe Bay	Alaska	1969	13

History of Reserves Definitions

There has been a growing awareness worldwide of the need for a consistent set of reserves definitions for use by the government and industry. Over the past 60 years, numerous technical organizations, regulatory bodies, and financial institutions have introduced nomenclatures for the classification of petroleum reserves. The history of reserves definitions is outlined as follows:

- 1936 to 1964: American Petroleum Institute (API) standard
- 1946: API and American Gas Association (AGA) annual publication of crude oil, natural gas liquids, and natural gas proved reserves
- 1964: Society of Petroleum Engineers (SPE) reserves definitions
- 1979: U. S. Security and Exchange Commission (SEC) reserves definitions
- 1981: SPE updated reserves definitions
- 1983: World Petroleum Congress reserves definitions
- 1985: Society of Petroleum Evaluation Engineers (SPEE) interim reserves definitions
- 1987: SPE and SPEE probable and possible reserves category
- 1987: World Petroleum Council (WPC) independent reserves definitions
- 1997: SPE- and WPC-approved reserves definitions for industry use worldwide
- 2001: SPE/WPC/AAPG guidelines for the evaluation of petroleum reserves and resources.
- 2007: SPE/WPC/AAPG/SPEE Petroleum Resources Management System (PMRS) defining petroluem reserves and resources

The proved reserves as defined by the SEC are not necessarily the same as those of the SPE, WPC, or AAPG. The SPE allows the potential use of a broader range of technologies to verify proved reserves and permits use of average price rather than price on the last day of the year. The current goal of the SPE is to increase educational efforts among members, industry professionals, and the public.

In 2001, the SPE made changes in its voluntary standard pertaining to estimating and auditing of reserves information. The changes were made to be compatible with the 1997 joint SPE/WPC definition of petroleum reserves.

In June 2006, the SPE signed a memorandum of understanding (MOU) with the United Nations Economic Commission for Europe (UNECE). This agreement was to develop one globally applicable harmonized standard for reporting fossil energy reserves and resources.

In September 2006, the SPE, AAPG, WPC, and SPEE published a draft proposal on the classification, definitions, and guidelines of petroleum reserves and resources for industry review.[6] The proposed definitions of reserves include broad guidelines that are applicable to conventional as well as unconventional resources. Furthermore, reserves categories are based on defined projects. They take into account the conditions of critical parameters in the future, including oil and gas prices, technological innovations, and environmental and regulatory issues, among other factors.

In March 2007, SPE approved the Petroleum Resources Management System, which consolidates, builds on and replaces the previous definitions pertaining to petroleum reserves. The broad-based system recognizes that the hydrocarbon occurrences in earth's crust fall into major categories of (a) commercial reserves, (b) sub-commercial contingent resources and (b) prospective resources, the latter being yet to be discovered. Hydrocarbon volume in each category is further classified according to the range of uncertainty associated with recovery (low, best and high). It is also classified according to project maturity. SPE also published standards for estimating and auditing reserves information. The co-sponsors of the system are WPC, AAPG, and SPEE.

Reserves Definitions and Classifications

According to the 2007 SPE/WPC/AAPG.SPEE Petroleum Resources Management System, petroleum reserves "are those quantities of petroleum anticipated to be commercially recoverable from known accumulations from a given date forward under defined conditions."[7] This set of conditions must be defined to support the estimate of reserves. Petroleum reserves must satisfy the following criteria:

- Reserves of hydrocarbon accumulations must be discovered.
- They are recoverable by certain technological means.
- They are subject to commercial production.
- Remaining reserves are based on applicable development projects.

In addition to reserves, the universe of hydrocarbon accumulations includes contingent and prospective resources. Contingent resources "are those quantities of petroleum estimated, as of a given date, to be potentially recoverable from known accumulations, but which are not currently considered commercially recoverable."[8] Examples would include hydrocarbon accumulations where commercial production is subject to technology under development or where no viable market currently exists. Another example would be "if the evaluation of the accumulation is still at an early stage."[9] Contingent resources may be considered commercially producible, i.e., petroleum reserves, if the organization claiming commerciality is committed to developing and producing them with reasonable certainty based upon a timetable of development and sound economic criteria, among other factors. Prospective resources "are those quantities of petroleum estimated, as of a given date, to be potentially recoverable from undiscovered accumulations by future development projects."[10] But these are associated with chance of discovery as anticipated in a prospect or play in a basin. Once discovered, prospective resources are potentially recoverable.

It should be noted that the 1997 SPE/WPC reserves definitions established that proved reserves can be determined by either deterministic or probabilistic methods. With the former, "reasonable certainty" is the criterion for the proved classification. With probabilistic methods, the proved quantities are identified as having "at least a 90% probability that the quantities actually recovered will equal or exceed the estimate."[11]

Given this duality of proved reserves definitions, it seems quite likely that proved reserves estimates will differ for the same property according to the reserves estimation method. Furthermore, the current definitions do not specify the aggregation level (well, reservoir, field, company, etc.) to which the 90% probability will be applied when using probabilistic methods.

The SPEE voted at its annual meeting in June 1997 not to adopt any specific set of reserves definitions. Instead, the recommendation was that members include in their reports a full disclosure of the reserves definitions they have used.

According to the 2007 SPE/WPC/ AAPG/SPEE Petroleum Resources Management System, probable reserves "are those additional reserves that are less certain to be recovered than proved reserves."[12] A 50% probability is attached to these quantities. Possible reserves "are those additional reserves that are less certain to be recovered than probable reserves."[13] A 10% probability value is assigned to possible reserves.

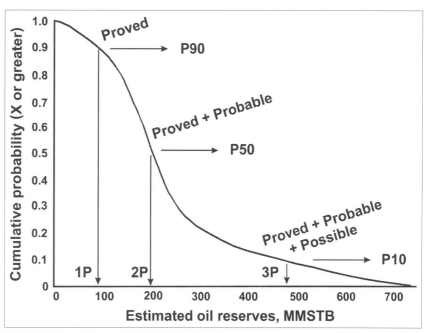

Fig. 15–4. Cumulative probability distribution of petroleum reserves

Reserves are also denoted as 1P (proved), 2P (proved + probable), and 3P (proved + probable + possible) quantities, as depicted in Figure 15–4. Again, referring to Figure 9–3 in chapter 9, the following can be concluded:

Reserves in MMstb

Proved:	9.997
Proved + Probable:	11.841
Proved + Probable + Possible:	13.878

When using a deterministic method, the analysis should consider low, best, and high scenarios, and approximately reflect the same probabilities mentioned above (10%, 50%, and 90%). The best estimate corresponds to the sum of proved and probable reserves (2P).

Reserves overview

Reserves are those quantities of petroleum that are recoverable economically in the future using commercial methods and government regulations. Reserves represent estimates that are subject to revisions during the life of a field, and they can be made for varying recovery processes:

- Primary reserves
- Secondary reserves
- Tertiary reserves

According to the Monograph I, second edition, published by the Society of Petroleum Evaluation Engineers, reserves are classified as follows:[14]

1. Proved reserves
   - Developed
   - Developed nonproducing
   - Undeveloped
   - Improved recovery

2. Probable reserves

3. Possible reserves

It should be noted that the earlier definitions of petroleum reserves differ from the 2006 proposed definitions. In evaluating proved, probable, and possible reserves, technical uncertainties associated with each category must be defined. The new definitions of reserves are to be adopted in 2007 following industry review.

Proved reserves

Proved or proven reserves are those quantities of petroleum that by analysis of geologic and engineering data can be estimated with reasonable certainty to be commercially recoverable in the future. This recovery is from known reservoirs and is expected under current economic conditions, operating methods, and government regulations. Probability of recovery should be at least 90%.

Proved reserves can be determined by either of the following:

- Deterministic methods, with "reasonable certainty" implying a high degree of confidence as the criterion for the proved classification
- Probabilistic methods, with the proved quantities being identified as having at least a 90% probability that the quantities actually recovered will equal or exceed the estimate

The preceding suggests that proved reserves estimates based on deterministic and probabilistic methodologies will be different for the same field. Furthermore, the current definitions do not specify the well, reservoir, field, company, etc., to which the 90% probability will be applied when using probabilistic methods.

Proved reserves pertain to the area of the reservoir delineated by drilling and defined by fluid contacts, such as the oil/water contact. Adjacent undrilled portions of the reservoir can be included on the basis of available geological, geophysical, and engineering studies that indicate continuity and similar reservoir characteristics. When fluid contact is not known, lowest known hydrocarbon indicated by well data is used unless indicated otherwise by definitive geophysical, geological, and engineering analyses.

In general, reserves are considered proved if commercial producibility of the reservoir is supported by actual production of formation tests. The term proved refers to the estimated volume of reserves and not just to the productivity of the well or reservoir.

Proved developed reserves. Proved developed reserves are those estimates of proved reserves that will be recovered from existing wells using existing facilities or requiring only minor additional expenditures. Developed producing reserves are expected to be recovered from completion intervals that are open and producing at the time of estimate.

Developed reserves may require further capital expenditures or additional equipment to be produced. Oil wells may require artificial lift, and gas wells may require compressor facilities to deliver to the pipeline. The developed reserves category implies that the equipment and operating practice technology are known and are being generally applied. It also implies that the equipment is commercially available and the additional costs are relatively insignificant.

Proved developed nonproducing reserves. In some cases, there are serious limiting factors to production. These could include lack of a market, inadequate gas reserves to support building a pipeline, mechanical problems, or waiting on stimulation treatments. In these cases, assignment of proved reserves category can be questionable. In fact, when large reserves will be required to justify a pipeline and secure a market, proved reserves should not yet be assigned. This should not occur until wells have been drilled or there is adequate supporting data of an acceptable nature to reasonably confirm the economics of the project. An exception would be in certain areas where large reserves have been discovered and will not be fully developed until the operator needs the deliverability to meet market or contract requirements. In this case, the reserves would be classified as both proved developed nonproducing and proved undeveloped.

Proved undeveloped reserves. Proved undeveloped reserves are assigned only to locations considered proved if available geological, geophysical, and engineering data support a geologic demonstrable hydrocarbon occurrence. These are quantities expected to be recovered through future investments. These include, but are not limited to, drilling of new wells, recompletion of existing wells, and installation of production or transportation facilities.

Proved undeveloped reserves can be assigned to other locations when the geological/geophysical interpretation is relatively certain. A continuous reservoir up dip from the lowest known hydrocarbon level should be indicated, and the estimator should expect the area to be productive when drilled. The area of the reservoir considered proved includes that portion delineated by drilling down to the lowest known occurrence of hydrocarbons. This guideline is applied unless definitive engineering or geological evidence is available to demonstrate that a lower structural level is appropriate.

Proved improved recovery reserves. Improved recovery reserves can only be classified as proved after the technique has been demonstrated to be commercially viable in the geologic formation in the immediate area. This must be accomplished by either a pilot project or the operation of an installed program that has confirmed through production response that increased recovery will be achieved. Improved recovery reserves are not considered developed until the installation of the project has been completed.

Unproved reserves—probable and possible

Unproved reserves are those categorized to include and distinguish probable and possible reserves from proved reserves. Unproved reserves are not to be added to proved reserves because of the different levels of uncertainty between proved, probable, and possible reserves. If a client or company requires that they be added, an explanation of the varying degrees of risk should be included as a footnote to the table in which the proved and unproved (probable and possible) reserves are totaled.

Probable reserves

Probable reserves are those unproved reserves that geologic and engineering data suggest are more likely than not to be recoverable. Probability of recovery should be at least 50% or more of the sum of the estimated proved plus probable reserves. It is equally likely that the actual remaining quantities of petroleum recovered are either greater or less than the sum of the proved and probable reserves (2P).

Probable reserves should be in a formation that is a known producer in the general area or geologic province. In the absence of commercial production, tests or other data should indicate the high likelihood of hydrocarbons being present in commercial quantities in the reservoir in question. Such supporting information could include production, drillstem, or formation tests, along with well log data or other geologic information.

Possible reserves

Possible reserves are those unproved reserves that geologic and engineering data suggests are less likely to be recovered than probable reserves. Probability of recovery should be at least 10% or more of the sum of the estimated proved plus probable plus possible reserves. Possible reserves are to be determined on the basis of engineering, geological, and geophysical analyses that indicate the possible existence of recoverable hydrocarbons. However, this is not to the level of proof required for probable reserves.

Speculative, potential, prospective, or "exploratory" reserves are not acceptable in this reserves category. The following requirements must be met for the reserves to merit the classification of possible:

1. Possible reserves are in reservoirs associated with known accumulations.

2. Reserves are located in a formation that has produced commercial quantities of oil or gas in the general area or geologic province.

3. Reserves are in formations that appear to be hydrocarbon-bearing based on logs or cores but that may not be productive at commercial rates at prevailing prices and costs.

4. There is a favorable indication of petroleum in the reservoir, i.e., an oil or gas show, by the least one of the following techniques: openhole or cased-hole logs, mud logs, cores, formation tests, drillstem tests, or production tests.

5. Geological or geophysical interpretations of the reservoir indicate the following:
    - Favorable structural position
    - Absence of faults, pinchouts, or other flow barriers between the area of the possible reserves and the petroleum shows or production
    The use of high-quality, quantitative (calibrated with well control) 2-D or 3-D seismic data should significantly improve the reservoir description over that possible with subsurface data alone. This information is often quite helpful in distinguishing between probable and possible reserves.

Possible reserves are less certain to be recovered than probable reserves and can be estimated with only a low degree of certainty. The supporting information is insufficient to indicate whether they are more likely to be recovered than not.

In general, possible reserves may include the following:
- Reserves suggested by structural or stratigraphic extrapolation beyond areas classified as probable, based on geologic or geophysical interpretations
- Reserves in formations that appear to be hydrocarbon-bearing based on logs or cores but that may not be productive at commercial levels

Reserves Determination Techniques

Reserves determination techniques are listed below:
- Volumetric: original hydrocarbon in place (OHCIP).
- Ultimate recovery = OHCIP X Recovery efficiency.
- Reserves = Ultimate recovery – Cumulative production.
- Decline curve: ultimate recovery and reserves.
- Classical material balance: ultimate recovery and reserves.
- Reservoir simulation: original hydrocarbon in place, ultimate recovery, and reserves.

These techniques are discussed in chapters 2, 9, 10, 11, 12, and 13, with examples. Probability distribution of reserves is also illustrated by example. Applicability and accuracy of these techniques through the reservoir life cycle are given in chapter 8.

The accuracy of the performance analysis is dictated by the depth of understanding of the reservoir characteristics, i.e., the reservoir model, and the quality of tools and techniques used.

The decline curve method is applicable only during the production phase where production decline is established. Accuracy can improve when more data is available. The volumetric method is applicable during the development and production phases with fair accuracy. The material balance method can give good results during the production phase but requires pressure data. Mathematical simulation, which considers reservoir heterogeneity, is the preferred method to use during delineation, development, and production with fair to very good accuracy.

Case Study of Asset Enhancement in a Matured Field[15]

This example presents the results of a mature field performance analysis and opportunities for adding more value to the asset. It demonstrates the following:

- Application of the reservoir management process and methodology
- Integration of professionals, tools, technologies, and data
- Multidisciplinary professionals working as a well-coordinated team
- Application of geosciences and engineering computer software to history match and predict reservoir performance under various scenarios
- Identification of opportunities for performing workovers and placing infill horizontal wells to improve recovery.

The mature offshore North Apoi/Funiwa field was operated by Texaco Overseas (Nigeria) Petroleum Company Unlimited (TOPCON). North Apoi was discovered in 1973, followed by Funiwa, an extension of the North Apoi field. As of June 30, 1995, 64 wells, including 58 commercial wells, were drilled. Primary producing mechanisms at the time were a combination of depletion, gascap, and natural water drives. The major producing sands were Ewinti-5, -6, and -7, and Ala-3, -5, and -7.

Table 15–7 presents basic reservoir data pertaining to Ewinti-5 and Ala-3.

In 1995, an integrated team of geoscientists and engineers was charged to review the past performance of the field. The team consisted of TOPCON, Nigerian government organizations, and Texaco's E&P Technology Department (EPTD). The goal was to capture the upside potential of the reserves.

Table 15–7. North Apoi/Funiwa field data. *Source: Y. Akinlawon, T. Nwosu, A. Satter, and R. Jespersen. 1996. Integrated reservoir management doubles Nigerian field reserves.* Hart's Petroleum Engineer International. *October.*

Reservoir/Fluid Characteristics	Ewinti-5	Ala-3
Depth, ft SS	5,000	7,000
Trap	Structural/faulted	Structural/faulted
Rock type	Unconsolidated sand	Unconsolidated sand
Gross thickness, ft	70–130	60–170
Porosity, %	30	20–25
Permeability, mD	1,500	500–1,500
Initial reservoir pressure, psig	2,200	3,000
Reservoir temperature, °F	165	222
Initial solution gas/oil ratio, scf/stb	364	940
Initial formation volume factor, rb/stb	1.2	1.6
Oil viscosity, cp	1.5	0.5
Oil gravity, °API	28	40
Gas sp. gr. (Air = 1)	0.6	0.7
Primary drive mechanism	Gascap/strong water drive	Gas cap/weak water drive
Original oil in place, MMstb	293	214
Cumulative production (12/94), MMstb	85	47

The studies were conducted in three phases, as follows:

Phase 1: Ala-3 and Ewinti-5 formations, containing 52% of the field's booked reserves

Phase 2: Ala-5 and Ewinti-7 formations, with 29%

Phase 3: Ala-7 and Ewinti-6 formations, with 10%

The objectives of the studies were to determine ultimate primary recovery, additional recovery with more vertical and horizontal wells, workovers, gas lift, and also enhanced oil recovery potential.

The approaches adopted included the following:

- Review of geosciences and engineering data
- Analysis of reservoir performance using the following methods:
  - Classical material balance
  - Decline curve analysis
  - Reservoir simulation analysis for full-field performance history match and full-field performance forecasts for improved recoveries

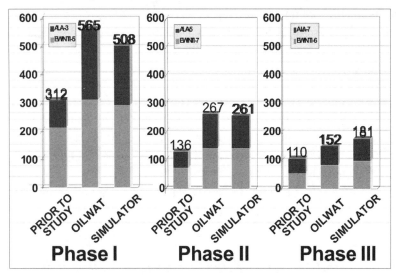

Fig. 15–5. Original oil in place in North Apoi/Funiwa field. *Source: A. Satter, J. Baldwin, and R. Jespersen. 2000. Computer-Assisted Reservoir Management. Tulsa: PennWell.*

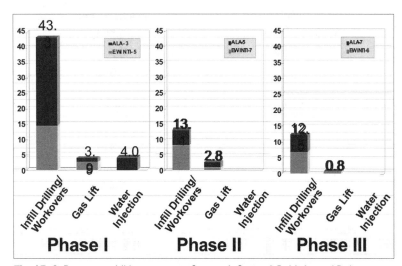

Fig. 15–6. Reserves addition summary. *Source: A. Satter, J. Baldwin, and R. Jespersen. 2000. Computer-Assisted Reservoir Management. Tulsa: PennWell.*

These analyses were conducted to provide an improved reservoir description, updated original oil in place, and potential reserves additions utilizing workovers, infill wells, gas lift, and water injection.

The results of the reservoir studies were highly significant. Newly available 3-D seismic survey data showed that the field was larger than originally estimated. The original oil in place values estimated from classical material balance analyses and simulation techniques were comparable to each other. However, they were substantially higher than the previously booked values, as shown in Figure 15–5.

Good performance history matches using a commercial black oil simulator were achieved. The reserves increases (over and above the booked values) due to infill drilling, workovers, gas lift, and water injection were determined by the simulator. Figure 15–6 presents the results of the three-phase studies.

As stated before, the new estimated reserves were substantially more than the previously booked values. Recommendations were made for the six reservoirs studied in phases 1, 2, and 3, and included 10 horizontal wells, 4 deviated wells, 1 replacement well, and 4 workovers.

TOPCON and the Nigerian government acted quickly on the study recommendations. Within nine months from the start of the phase 1 study, two successful horizontal wells were drilled and completed in the Funiwa Ewinti-5 reservoir. The first well came on production at 2,670 bopd from a 700-ft horizontal section. The second well had a 1,600-ft horizontal pay section and produced at 4,020 bopd.

The conclusions made from the integrated study included the following:
- The 3-D seismic survey data improved the reservoir description.
- The original oil in place was revised by more than 50%.
- The upside potential of reserves additions was significant.
- Recovery factors were estimated to be 30% to 55%.
- Teamwork and integration were critical to the success of the project.

Mistakes and Errors

Harrell, Hodgin, and Wagenhofer discussed the most common mistakes, errors, and guidelines associated with reserves estimates concerning pay and future production.[16]

Geosciences-based mistakes and errors occur in estimating net and gross pays by incorrectly utilizing the following:
- Structure maps for top and bottom effective pay zones accounting for stratification and lateral variation
- Isopach maps by not contouring the net to gross ratio from actual well data to account for lateral variation

Reservoir engineering–based mistakes and errors for estimating future production performance are associated with the following:
- Oil production decline curves
- Gas production decline curves
- Analogs
- Reservoir simulation
- Operating costs estimates

Guidelines to reduce mistakes in oil production decline curves include the following:
- Avoid assuming exponential decline in reservoirs that may indicate a hyperbolic decline trend.
- Conversely, avoid assuming hyperbolic decline leading to optimistic reserves in reservoirs where exponential decline would better fit the performance.
- Always attempt to estimate performance decline on a well or completion level for best results.
- Check well work and consistent trend in well counts.
- Always check gas/oil ratios and water cuts.
- Use analogous fields or more mature wells to establish typical decline behavior.
- Understand reservoir properties and the depositional environment in order to exercise better judgment to expect exponential or hyperbolic decline.
- Attempt to combine various types of evaluation techniques with decline curve analysis for consistency in results.

Guidelines to reduce the risk of overestimating gas in place and ultimate recovery using pressure/compressibility (P/Z) versus cumulative gas production (G_p) include the following:
- Review other or more mature fields in the area to investigate p/z versus production decline behavior and observed abandonment pressure.
- Be aware of possible water influx that may not show up during very early production.
- Be aware of overpressured reservoirs that would need modified p/z versus G_p analysis.

Guidelines to reduce mistakes using analogies include the following:
- Give preference to analogies in areal proximity to the target field.
- Ensure that key parameters are similar to the analogous reservoirs.
- Review and design for operational similarity, particularly well density.

Guidelines to reduce mistakes in the reservoir simulation method include the following:
- Ensure that the simulation model is based upon the integration of the geosciences and engineering components.
- Ensure that the simulation-derived original hydrocarbon in place compares favorably with the value from the volumetric method.
- Ensure a reasonable past performance history match, avoiding arbitrary and unrealistic adjustments of parameters. Remember that history match is not unique.

Guidelines to reduce operating costs mistakes include the following:
- Ensure that future operating costs are in close agreement with the historic costs.
- Attempt to separate costs into fixed and variable components.
- Account for changes in costs due to new recovery processes.

Guidelines to reduce the frequency of mistakes include the following:
- Always review the potential fieldwide implications of new data.
- Do not assume that only poor locations are being drilled and the good ones are yet to come.
- Exercise caution in estimating reserves of undeveloped locations where drive mechanisms or efficiencies are not known with certainty.

Reserves Growth

As worldwide demand for oil and gas increases at a significant rate, the principal objective of the petroleum industry is to attain sustained growth. The industry's goal is to discover additional reserves and add value to existing assets based on technological innovation. Processes related to exploration, discovery, development, and appropriate reservoir management convert possible or probable reserves to proved reserves. Some of the key elements for the future growth of petroleum reserves are listed below:

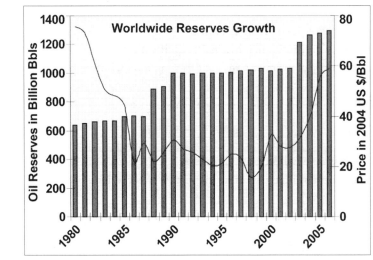

- Detailed reservoir description, characterization, and visualization based on a team approach, leading to better recovery
- Review of existing field performance to optimize oil recovery
- Locating bypassed oil in matured fields and developing a recovery strategy
- Implementation of an intelligent field that incorporates an integrated reservoir team, smart wells, robust database systems, and decision making in real-time or near real-time
- Evolution of the best reservoir management practices in order to optimize the recovery of oil and gas
- Innovative and cost-effective methods in improved oil recovery and enhanced oil recovery
- Quest for oil and gas in hitherto unexplored territories, including deepwater and arctic drilling
- Development of nonconventional recovery processes for oil sands, bitumen, heavy oils, coalbed methane, etc.

Fig. 15–7. Oil reserves growth between 1985 and 2006 due to continued exploration for oil, advances in drilling and completion, reservoir characterization, innovations in primary and enhanced recovery methods, and integrated reservoir management. Annual data on oil price per barrel is also included as a dotted line. *Source: Energy Information Administration. 2006. International Energy Outlook 2006. Washington, D.C.: Energy Information Administration, Department of Energy. June.*

Figure 15–7 presents the worldwide growth of oil reserves for 21 years due to the implementation of many of the measures mentioned. Oil price history, expressed in 2004 U.S. dollars, is also included in the plot.

Summing Up

Reserves are those quantities of petroleum that are recoverable economically in the future using commercial methods and government regulations. Reserves determination involves a great deal of uncertainty. However, estimates are necessary for regulatory, operational, and financial reasons.

Estimates of reserves vary during the various stages of the field, being at their maximum at prospect identification, but the actual recovery is known only when the field is depleted.

In 2007, SPE adopted the Petroleum Resources Management System that broadly classified the hydrocarbon accumulations in earth's crust (including unconventional resources) into three major categories: petroleum reserves, contingent resources and undiscovered resources, each of which was further classified according to the range of uncertainty involved: low, best and high. Resources and reserves were also classified according to project maturity, as shown in Figure 15–8.

Reserves (Discovered & commercial)	On production
	Approved for development
	Justified for development
Contingent Resources (Discovered & sub-commercial	Development pending
	Development unclarified or on hold
	Development not viable
Prospective Resources (Undiscovered)	Prospect
	Lead
	Play

Fig. 15–8. Classification of hydrocarbon reserves and resources according to project maturity and commercial viability as documented in Petroleum Resources Management System. Resources contain unrecoverable hydrocarbons, which are not shown.

Approaches taken for reserves determination can be deterministic or probabilistic. The deterministic approach is based upon known geological, engineering, and economic data. The probabilistic approach is based upon the ranges of geological, engineering, and economic data. It is the preferred method, considering the uncertainties.

The United States is the world's largest energy consumer. In 2004, 40% of U.S. energy consumption was provided by crude oil and natural gas liquids combined. Natural gas provided another 23%, for a total of 63%. U.S. energy consumption was about 21 million bbl of oil and natural gas liquids and 61 bcf of natural gas per day. The United States does not produce enough oil and gas to meet its consumption and is heavily dependent upon importing oil and gas.

U.S. crude oil reserves continued to decrease from 33.5 billion bbl in 1976 to 21.4 billion bbl in 2004. Dry gas reserves also decreased from 213 tscf in 1976 to 193 tscf in 2004. However, natural gas liquids actually increased from 6.7 billion bbl in 1978 to 7.9 billion bbl in 2004.

Most of the oil and natural gas reserves are located in the Middle East. Saudi Arabia, with 264 billion bbl, has the highest reserves. The United States ranks 11th, with 21.4 billion bbl.

A growing awareness worldwide recognized the need for a consistent set of reserves definitions for government and industry use. In response to the recognized need for creditable annual proved reserves estimates, the U.S. Congress in 1977 required the DOE to prepare such estimates. The EIA then established a unified, verifiable, comprehensive, and continuing annual statistical program series for proved reserves of crude oil and natural gas. Natural gas liquids were included in 1979. The SPE, WPC, and AAPG are actively involved in improvements and in setting standards for reserves definitions and evaluations.

Reserves, which are subject to revisions during the life of a field, can be made for varying recovery processes such as primary, secondary, and tertiary reserves.

Reserves are generally classified as proved, probable, and possible reserves. Proved reserves can be subclassified as produced developed, produced developed nonproducing, proved undeveloped, and produced improved recovery processes. Probability plays a very important role in reserves determination.

Reserves determination techniques include volumetric, decline curve, classical material balance, and reservoir simulation methods.

The decline curve method is applicable only during the production phase, where production decline is established. Accuracy can improve when more data is available. The volumetric method is applicable during the development and production phases, with fair accuracy. The material balance method can give good results during the production phase but requires pressure data. Mathematical simulation, which considers reservoir heterogeneity, is the preferred method to use during delineation, development, and production, with fair to very good accuracy.

The case study on the Nigerian North Apoi/Funiwa field presents the results of this mature field performance analysis. Analysis was conducted to determine opportunities to add value to the asset. This study demonstrates the following:
- Application of the reservoir management process and methodology
- Integration of professionals, tools, technologies, and data
- Multidisciplinary professionals working as a well-coordinated team
- Application of geosciences and engineering computer software to history match and predict reservoir performance under various scenarios
- Identification of opportunities for performing workovers and placing infill horizontal wells to improve recovery

Pitfalls in estimating reserves are generally rooted in the misinterpretation of geologic trends or reservoir performance. Some of the common mistakes in the calculation of reserves are associated with the following issues:
- Identification of the top and bottom of the oil-bearing structure
- Estimation of net to gross pay thickness
- Interpretation of production decline trend (exponential, hyperbolic, harmonic, etc.)
- Detailed reservoir description and production history, including water/oil ratio, gas/oil ratio, well counts, etc.
- Analysis of p/z versus cumulative production for gas reservoirs, with particular attention to reservoirs with water influx and abnormally pressured reservoirs
- Analysis based on analogous fields requiring correct identification of similarities of key parameters
- Validation of the reservoir simulation model based on history matching
- Correct estimates of operating costs based on historical data and new recovery techniques

On a global level, sustained growth in petroleum reserves is attained by exploration and discovery of oil and gas accumulations in hitherto unexplored territories. On a reservoir level, the key to adding value to reserves includes best reservoir management practices, teamwork, and the implementation of intelligent fields to make decisions in real-time or near real-time. In both cases, technological innovation and market economics play vital roles.

Class Assignments

Questions

1. Why are oil and gas reserves so important?

2. What is required in order to determine oil and gas reserves?

3. Why is the probability method for estimating reserves preferred to the deterministic approach?

4. How are reserves classified, and on what basis?

5. Why does the United States need to import petroleum products?

6. What are the organizations involved in revising and setting standards for reserves determination?

7. What are the items needed for reserves accounting?

8. What are the methods available for reserves evaluations?

9. What is the simplest practical method for reserves evaluations? What type of data is needed, and when?

10. How can the volumetric method be used to estimate reserves? Describe the impact of the following on petroleum reserves:
    (a) Porosity
    (b) Absolute permeability
    (c) Two-phase and three-phase relative permeabilities
    (d) Connate water saturation and oil-water contact
    (e) Aquifer influx
    (f) Oil gravity
    (g) Depth and location of the field
    (h) Naturally occurring fractures in a tight matrix
    (i) Harmonic versus hyperbolic decline in production
    (j) Reservoirs with limited transient pressure test information
    (k) Wells with severe water-cut or sanding issues
    (l) A reservoir where enhanced oil recovery feasibility studies are not performed
    (m) Integration of seismic, log, core, and numerical simulation studies

11. Describe the role of reservoir engineers in estimating reserves, giving two examples. In addition, consider a case where geologic studies and dynamic test data are in apparent conflict in order to calculate reserves.

12. What are the most notable characteristics of the majority of the world's oil reserves? Mention at least two new technologies that may succeed in recovering the major portion of these resources.

13. A new reservoir has been discovered in a basin where other reservoirs produce from a highly faulted unit with zonal discontinuities. Describe a strategy to estimate the reserves. List the major pitfalls the reservoir engineer needs to be aware of in estimating reserves at discovery.

14. A volatile oil reservoir has produced for four years with an average of five new wells drilled per year. Describe the data that must be acquired in order to estimate the reserves. What are the mistakes that can affect the estimates if all the data is not acquired?

Exercises

15.1. List the major reasons for revising reserves of an oil field and illustrate the answer with a field example.

15.2. Describe the approach a reservoir engineer can adopt in improving the reserves of a heterogeneous reservoir where early production is less than satisfactory. Geologic heterogeneities, such as extensive faulting, are detected by a seismic survey conducted in one section of the reservoir.

15.3 Based on the review of the Petroleum Resources Management System (2007) adopted by SPE, describe three possible scenarios in a petroleum basin where certain hydrocarbon accumulations cannot be reported as proved reserves. In each case, discuss the technical measures that can be undertaken in order to determine proved reserves with better accuracy.

15.4 How unconventional resources differ from conventional resources according to the Petroleum Resources Management System (2007)? What criteria are applied to report an unconventional resource as proved reserve? Cite an example.

15.5. What is the outlook for world petroleum reserves in the coming decades? The basis of the response should include a study of past trends and advancements in technology.

References

1. Society of Petroleum Evaluation Engineers. Reserves Definitions Committee. 1998. *Guidelines for Application of Petroleum Reserves Definitions*. Monograph I. 2nd ed. Society of Petroleum Evaluation Engineers. October.
2. Society of Petroleum Engineers. Web pages on petroleum reserves and resources definitions, 2003–2006, accessed at www.spe.org.
3. Energy Information Administration. 2005. *U.S. Crude Oil, Natural Gas, and Natural Gas Liquids Reserves: 2004 Annual Report*. Washington, D.C.: Energy Information Administration (DOE). November.
4. International Energy Agency. 2005. *Resources to Reserves: Oil & Gas Technologies for the Energy Markets of the Future*. Paris: International Energy Agency.
5. Belani, A. 2006. It's time for an industry initiative on heavy oil. *Journal of Petroleum Technology*. June. Vol. 58, no. 6, 40–42.
6. Society of Petroleum Engineers/American Association of Petroleum Geologists/World Petroleum Council/Society of Petroleum Evaluation Engineers. 2006. *Petroleum Reserves and Resources: Classification, Definitions, and Guidelines*. September. Accessed January 16, 2006 at www.spe.org
7. Society of Petroleum Engineers/American Association of Petroleum Geologists/World Petroleum Council/Society of Petroleum Evaluation Engineers. 2007. Petroleum Resources Management System. Accessed September 12, 2007 at http://www.spe.org.
8. SPE/AAPG/WPC/SPEE. 2007.
9. SPE/AAPG/WPC/SPEE. 2007.
10. SPE/AAPG/WPC/SPEE. 2007.
11. Society of Petroleum Evaluation Engineers. 1998.
12. SPE/AAPG/WPC/SPEE. 2007.
13. SPE/AAPG/WPC/SPEE. 2007.
14. Society of Petroleum Evaluation Engineers. 1998.
15. Akinlawon, Y., T. Nwosu, A. Satter, and R. Jespersen. 1996. Integrated reservoir management doubles Nigerian field reserves. *Hart's Petroleum Engineer International*. October.
16. Harrell, D. R., J. E. Hodgin, and T. Wagenhofer. 2004. Oil and gas reserves estimates: recurring mistakes and errors. SPE Paper #91069. 2004 SPE Annual Technical Conference and Exhibition, Houston, TX, September 26–29.
17. Broome, J. 2005. Personal communications regarding reserves.

16 · Improved Recovery Processes: Fundamentals of Waterflooding and Applications

Introduction

Improved oil recovery processes broadly encompass all of the measures aimed towards increasing ultimate recovery from a petroleum reservoir, including both conventional and emerging technologies. Measures related to more effective reservoir surveillance and management, leading to an increase in recovery efficiency, are also viewed as part of the process. Most reservoirs are subjected to improved oil recovery (IOR) processes following primary recovery. Natural reservoir energies control the ultimate recovery of petroleum during primary production. Such drive mechanisms include liquid and rock compressibility drive, solution gas drive, gascap drive, natural water influx, and combination drive processes. Primary recovery from oil reservoirs is influenced by reservoir rock properties, fluid properties, and geologic heterogeneities, as discussed in chapters 2, 3, and 8. Secondary production methods based on fluid injection provide further energy in order to augment or sustain the production level once well rates decline during primary recovery. Fluid injection could involve water or gas-water combination floods. Slugs of water and gas are injected sequentially, which is referred to as water alternating gas injection (WAG). Simultaneous injection of water and gas (SWAG) in a reservoir is also practiced in the industry. In many oil fields, improved recovery operations begin long before all the natural energies are depleted in order to maximize recovery.

Ranges of primary recovery from oil and gas reservoirs are provided in chapter 8. Experience has shown that about one-third of the original oil in place, or less, is recovered from reservoirs all over the world, while the remainder of hydrocarbon is left behind. In unfavorable geologic settings, or in the case of heavy oil, recovery could be substantially below average. Hence successful design, implementation, and continued optimization of improved oil recovery processes constitute a major challenge for reservoir engineers in order to maximize the asset (the petroleum reserves). Based on current technology and economics, about 10%–30% of the original oil in place is likely to be recovered by waterflooding in most fields with favorable fluid and rock properties. This leaves room for technological innovations in the future. The need for successful engineering of improved recovery processes is gaining further significance. Most of the hitherto discovered giant oil fields are maturing, while the worldwide demand for oil is increasing. For example, a 10% increase in ultimate recovery from these oil fields would lead to a significant increase in the world's petroleum reserves. This incremental amount would surpass the current reserves of some of the highest producing countries.

It must be noted that only marginal recovery is accomplished by primary and secondary methods in many circumstances, such as with highly viscous heavy oil reservoirs, tar sands, and oil shales. It is thus apparent that enhanced oil recovery techniques need to be employed to recover additional oil. It is a common experience in the industry that most oil fields are subjected to tertiary recovery processes once secondary recovery by waterflooding becomes marginal. Enhanced oil recovery processes include all methods that use external sources of energy or materials to recover oil that cannot be produced economically by conventional means. These processes are discussed in the following chapter. It should be noted that the cost of production by enhanced oil recovery processes is markedly higher than waterflooding in most cases.

This chapter is devoted to learning about the following:
- History of waterflooding
- Waterflood process
- Waterflood design
- Screening criteria
- Waterflood patterns and well spacings
- Water injection rates
- Waterflood strategies and life of the waterflood project
- Waterflood recovery efficiency
- Performance prediction methods
- Waterflood surveillance and management
- Case studies
- Example problems

This chapter presents four field case studies highlighting the tools and techniques available to reservoir engineers in order to successfully manage a field under waterflooding in order to maximize assets, i.e., oil reserves. These include the following:

- A classical approach to waterflood operations, including decades of water injection in a reservoir in West Texas, followed by infill drilling and an enhanced oil recovery [carbon dioxide (CO_2) injection] process to optimize recovery.
- Integrated methodology in managing a waterflood operation in a low permeability reservoir, including petrophysical studies, an injection profile survey, reservoir simulation, and economic evaluation.
- Waterflooding a highly heterogeneous reservoir in Kuwait, having a large number of faults, compartments, and formation stratifications.
- Implementation of an efficient waterflood surveillance scheme and analysis of waterflood data to augment ultimate recovery in a faulted reservoir with high permeability streaks located in Alaska.

The following examples are illustrated in the chapter based on analytical, graphical, and empirical methods in waterflood performance analysis:

- Calculation of the recovery factor based on the water/oil ratio and mobility ratio
- Estimation of the vertical sweep and water/oil ratio in a stratified formation
- Analysis of waterflood performance based on the frontal advance theory
- Prediction of the breakthrough of the injected water bank at the producer

History of Waterflooding[1]

As early as 1865, waterflooding occurred as a result of accidental injection of water in the Pithole City area in Pennsylvania. Leaks from shallow water sands and surface water entered drilled holes, resulting in much of the early waterflooding. In the late 1880s, water was primarily injected to maintain reservoir pressure and prolong reservoir life, thereby improving oil recovery.

In 1924, the first 5-spot pattern flood was attempted in the Bradford field in Pennsylvania. The technique became popular in the following decades. (Waterflood well patterns are described later in the chapter.) Waterflooding grew from Pennsylvania to Oklahoma in 1931 in the shallow Bartlesville sand, and then in 1936 to the Fry Pool of Brown County, Texas. Waterflooding found widespread applications in oil industry in the early 1950s.

Overview of the Waterflood Process

Waterflooding consists of injecting water into certain wells while producing from the surrounding wells. It maintains reservoir pressure and physically displaces oil with water moving through the reservoir from the injector to the producer. Throughout the decades, waterflooding has been the most widely used postprimary recovery method in the United States and other petroleum regions of the world. In most reservoirs, regardless of their type, a carefully engineered waterflood process is expected to contribute substantially to field production and reserves.

Reasons for the success of waterflooding include the following:

- Water is an efficient agent for displacing oil of light to medium gravity.
- Water is relatively easy to inject into oil-bearing formations.
- Water is generally available and inexpensive.
- Waterflooding involves relatively lower capital investment and operating costs, leading to favorable economics.

Saturated versus undersaturated oil reservoir

In many circumstances, water injection is initiated early in the life of an unsaturated oil reservoir. Early water injection ensures that the reservoir pressure is always maintained above the bubblepoint, and no free gas phase develops in the reservoir. Waterflood experience has shown that ultimate recovery is much better as long as the dissolved gas remains in solution. This concept also is supported by comparison of scenarios based on reservoir simulation. If the reservoir pressure falls below the bubblepoint pressure, free gas evolves. In addition, the liquid phase is expected to become more viscous and less compressible due to significant shrinkage. As a result, ultimate oil recovery is lower. In many cases, highly depleted reservoirs are not feasible candidates for waterflooding. Literature review indicates that even in large offshore fields where the costs associated with waterflood projects are high, an early start of injection leads to optimal production. Comparison between waterflooding a saturated reservoir and an unsaturated reservoir is illustrated later in the chapter based on a simulation study.

When waterflooding is initiated in a depleted or nearly depleted reservoir, in which the reservoir pressure is below the bubblepoint during primary production, a gas cap is likely to be present. Pressure is restored as the gas-filled pore volume is refilled with the injected water, dissolving the free gas back into the liquid phase. Production response at the wells occurs after the gas space is filled up, resulting in a delay in waterflood response at the producers. This delay adversely affects the net present value and payout period of a project.

Economic limit

With continued injection, a peak production rate is reached, after which the injected water eventually breaks through at the producing wells. Once breakthrough occurs, the general trend is an increase in water/oil ratio and a decline in oil rate until the well reaches its economic limit due to a very high water/oil ratio. The reservoir engineer's goal is to recover as much oil as possible before the limit is reached. The typical cost of oil production by waterflooding ranges from several cents to several dollars per barrel. In contrast, the cost of enhanced oil recovery processes ranges widely, depending on the specific process. It is significantly higher in most cases.

Factors affecting waterflood performance

The waterflood response can be seen in the form of enhanced oil rates followed by eventual water breakthrough at the producing wells. The timing of this response and the magnitude of the peak production rates are governed by the injection rate, well spacing, fluid properties, and reservoir heterogeneities. In some cases, dry oil is not produced following water injection. A fraction of the produced liquid is found to be water. In certain other cases, measurable quantities of water begin to show up in the production stream, following a few months or years of dry production.

Fluid saturation profile during waterflooding

Figure 16–1 shows the saturation profiles of water, oil, and gas in the reservoir during waterflooding. Three possible scenarios associated with waterflooding are depicted:

(a) The oil reservoir is undersaturated, with no gas cap.

(b) Free gas is present in the reservoir, and the oil bank leaves behind trapped gas in the reservoir.

(c) Free gas dissolves into the liquid phase due to increased reservoir pressure.

Depending on reservoir conditions, both viscous and gravity forces can influence oil recovery during waterflooding. Reservoir simulation is used extensively to study the movement of fluids and their saturations, which change dynamically during waterflood operations.

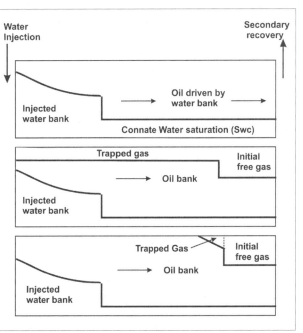

Fig. 16–1. Idealized saturation profile of oil, gas, and water during waterflooding undersaturated and saturated reservoirs

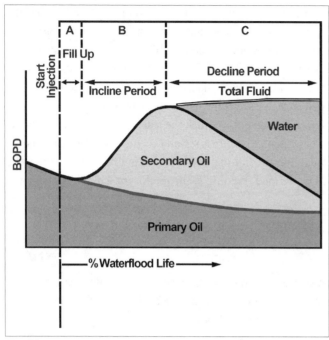

Fig. 16-2. Typical example of a successful waterflood performance. *Source: G. C. Thakur. 1991. Waterflood surveillance techniques—a reservoir management approach. Journal of Petroleum Technology. October: pp. 1,180-1,188. © Society of Petroleum Engineers. Reprinted with permission.*

During waterflooding, oil and water saturations in the reservoir change with time and space. This change is controlled by the fractional flow characteristics of water displacing in-situ oil, among other fluid and rock properties. Again, the fractional flow of individual phases depends on the relative permeabilities of the oil and water, as discussed in chapters 2 and 4. At the leading edge of the advancing water bank, a sharp interface between the two immiscible fluids is envisaged. Movable oil, both at the front and the back of the water bank, is "pushed" (displaced) towards the producer. Well production is free of water cut as long as the injected water bank does not move all the way to the producer. In dipping reservoirs, gravitational effects may predominate as water injected at downdip wells rises with time to sweep in-situ oil. A time-lapse study of water saturation increasing from the bottom up is shown later in the chapter.

Figure 16–2 is a typical plot of the oil production rates versus waterflood life for a successful waterflood performance in a reservoir with a gas cap. It presents the filling up of pore spaces initially occupied by free gas, and the incline and decline of secondary oil saturation periods.[2, 3] Fill-up volume is defined as the pore volume that is occupied by injected water where a gas cap existed prior to waterflooding.

Waterflood Design

Effective design for waterflooding a field involves an integrated team approach and major effort. There are many requirements, including, but not limited to, the following:

- Detailed reservoir description, including the identification of flow units
- Monitoring and interpretation of past reservoir performance
- Drilling and completion data from existing wells
- Laboratory analysis
- Development and validation of reservoir models
- Evaluation of "what-if" scenarios
- Pilot projects
- Economic analysis

Various tests at the well and reservoir level may be conducted to reduce the uncertainties in the design. As water injection is initiated, a methodical reservoir surveillance program is put in place. (Reservoir and well surveillance are discussed later.) The original design is updated on a regular basis as information is processed from the actual waterflood operation. Strategies adopted to enhance the performance of an ongoing waterflood operation are outlined in this chapter and in chapter 19.

Objectives

The major objectives of waterflood design include the following:
- Maximize secondary oil recovery within the economic, technological and regulatory framework.
- Maximize the contact with oil, specifically where zones of high residual oil saturation exist in a heterogeneous formation. A waterflood project is judged by its effectiveness in covering all the areas of the reservoir, followed by efficient displacement of in-situ oil.
- Minimize injection water cycling through the formation and handling of the water at the surface facilities.
- Optimize water injection.
- Efficiently schedule conversion of producers to injectors.
- Drill infill wells to increase areal coverage and augment production.
- Minimize capital expenditures related to drilling new injectors or producers.
- Maximize the net present value of the asset based on returns within a relatively short time horizon.

In attaining these objectives, reservoir engineers principally rely on the studies, analyses, and experiences mentioned in the preceding section. Many objectives or elements in design may not be achievable in every field. Various constraints may affect the outcome, including the geologic setting of the reservoir, issues related to water handling, and economic considerations. There are uncertainties associated with unknown reservoir heterogeneities, expected oil recovery, added capital expenditure due to new drilling, water handling, and construction of surface facilities. These are some of the critical factors in waterflood design, and they interact in a complex manner.

Elements in design

Essential elements in the design of waterflooding include, but are not limited to, the following:

1. A detailed reservoir description is needed, with an emphasis on reservoir heterogeneities that may affect the efficiency of waterflooding in a significant manner. A case in point is the existence of high permeability channels in the formation, which may be observed in carbonate reservoirs in greater frequency. Care must be taken to avoid injecting water into rapid pathways that may directly lead to the producer.

2. Flow units of similar rock characteristics, geologic continuity, and depositional history are identified in rock characterization studies, which aid significantly in formulating an effective strategy to produce from the zones of good reservoir quality. Certain measures can also be undertaken to recover from other zones where the reservoir quality is not as good, such as the low permeability formations. Reservoir characterization is described in chapter 2.

3. The design of waterflooding, specifically the alignment of injectors and producers, should be based on reservoir heterogeneities that control the direction of fluid flow. Water breakthrough is likely to occur much sooner when the injector and producer are aligned with the principal direction of the naturally occurring fractures or the permeability trend. Hence, the wells are aligned in a manner such that fluid flow occurs transverse to highly conductive pathways for better coverage by injected water.

4. Many reservoirs have distinct zones (strata) or flow units of highly contrasting permeabilities, with varying degrees of crossflow between adjacent zones. These require detailed studies based on reservoir characterization and simulation in order to maximize recovery. In certain formations, marked gradations in areal permeability exist, typically toward the flanks.

5. Identification must be made of areas and zones in the reservoir where in-situ oil remains unrecovered during primary production. The integrated study involves knowledge of rock properties, reservoir fluid dynamics, evaluation of individual well performance, and water-oil displacement studies in the laboratory. It also involves time-lapse seismic studies and field-wide production history matching by reservoir simulation, among other inputs.

6. The number, location, pattern, and scheduling of the wells must be designed to optimize recovery. Optimized waterflooding represents maximum recovery at minimum cost, among other factors. Based on reservoir simulation leading to the best-case scenario, producers are converted to injectors according to a time schedule during waterflooding. In other circumstances, new wells are drilled to reduce the spacing between injectors and producers.

7. Injection and production rates must be optimized to ensure maximum coverage, efficient displacement of oil by water, and ultimate recovery. However, limitations in injection rate arise due to the fracture gradient and the amount of water that can be recycled economically.

8. Reservoir pressure must be optimized during waterflooding based on reservoir simulation. Waterflooding can be initiated at various stages of primary production as reservoir pressure depletes. Low pressure regions in a reservoir may not lead to the maximum oil recovery in a given time frame. Primary production below the bubblepoint pressure in undersaturated oil reservoirs may alter the waterflood scenario significantly.

9. The response of a reservoir to water injection is not known beforehand, as many uncertainties are involved. Waterflood surveillance is regarded as the key to successful reservoir management during secondary recovery. A detailed waterflood monitoring and evaluation plan is needed. This usually leads to continuous evaluation of the initial design and implementation of certain changes in the design throughout the life of the waterflood operation. This frequently involves selective waterflooding in targeted zones, drilling of new wells, recompletion, and pattern realignment, among other tactics.

In addition to the above, Craig mentions two other rock characteristics that need to be considered in waterflood design.[4] These are crossbedding, which may affect the degree of communication between injectors and producers, and planes of weakness or fractures. The latter are not detected during primary production but may open at high injection pressures.

Tools and techniques in waterflood design

Laboratory studies involving oil-water displacement in core samples are conducted routinely, and they aid significantly in designing a waterflood project. However, reservoir-scale rock heterogeneities cannot be replicated in much-smaller core samples, among other constraints. A sharp contrast between oil recoveries attained in various laboratory studies and in the field, as commonly observed, attests to that fact.

Earlier waterflood designs included pilot floods involving a small segment of the field as a means of studying the recovery potential and engineering aspects of waterflooding. The valuable experience gained from the pilot recovery could then be utilized in full-field water injection operations. With the advent of the digital age, reservoir simulation is extensively used to predict and evaluate various waterflood scenarios, in addition to any pilot projects. Again, integrated models are developed that include reservoir performance under waterflooding, surface facilities, and economics, including the projected price of oil and capital expenditures (CAPEX).

The following information is typically sought in reservoir simulation studies:
- Expected ultimate recovery based on various waterflood scenarios
- Scheduling of wells for conversion to injectors
- Need for drilling infill wells
- Optimum well spacing
- Optimum injection rate and pressure
- Expected production rate of individual fluid phases over time
- Anticipated time of water breakthrough
- Water/oil ratio per well versus time following breakthrough
- Sensitivity studies examining the effects of the injection rate, well conversion schedule, and spacing

- Effects of uncertainties in rock properties, specifically in the areas of permeability, relative permeability, fractures, faults, layering, and the degree of crossflow between layers
- Effects of any uncertainties related to original oil in place estimation
- Life of the waterflood project

Reservoir simulation is part of a continuous learning process during waterflooding. An example of reservoir development based on a scenario-building exercise that included waterflooding is illustrated in chapter 19. Besides reservoir simulation, which could be resource intensive, relatively simple approaches can be adopted in aiding recovery predictions in less complex circumstances. These include analytical models, graphical techniques, and empirical correlations.

As mentioned earlier, the design of an efficient waterflood project requires an integrated team approach. This includes data collection from various sources, such as geosciences, petrophysical, production, and well testing. Regional and global experience of waterflooding similar types of reservoirs may provide significant insight in the design phase. A literature review indicates that many waterflood operations had limited success due to a lack of information about various reservoir heterogeneities and their effects on fluid flow behavior. Besides reservoir engineering issues, there are other factors that must be considered. These include fluid lifting costs, the sourcing and compatibility of injected water, and the design of surface facilities (fluid handling and treatment, equipment sizing, etc.).

Waterflood technology broadly encompasses both reservoir and production engineering.[5,6] Reservoir engineers are responsible for waterflood design, performance prediction, and reserves determination. They share responsibilities with the production engineers for the implementation, operation, and evaluation of the waterflood project. Geosciences professionals participate in ongoing reservoir characterization studies based on the response from waterflood operations. These may include time-lapse seismic studies, review of production data obtained by continuous downhole monitoring, and interference well test analysis, among others.

Screening criteria

The screening criteria for waterflooding a reservoir are described in the following discussion.

Residual oil saturation. There are a number of methods to estimate the average residual oil saturation of the reservoir following primary recovery. Some of these are listed in the following:

- Laboratory investigation of core samples
- Material balance and volumetric estimates based on original oil in place and the amount of hydrocarbon produced
- Wireline logs
- Tracer studies
- Reservoir simulation

Based on production data obtained from the field during primary recovery, estimation of average oil saturation at the start of waterflooding is straightforward, given as follows:

$$S_o = (1 - S_{wi})\left(1 - \frac{N_p}{N}\right)\left(\frac{B_o}{B_{oi}}\right) \tag{16.1}$$

where

S_o = oil saturation following the production of N_p bbl, fraction of pore volume,

S_{wi} = initial water saturation, fraction,

N_p/N = cumulative oil production at the start of waterflooding as a fraction of original oil in place, N, both expressed in stb,

B_o = oil formation volume factor at start of waterflooding, rb/stb, and

B_{oi} = initial oil formation volume factor, rb/stb.

Accurate determination of residual oil saturation is very important in evaluating the feasibility of waterflooding a reservoir. A reservoir with less than 40% oil saturation following primary depletion may not be the best prospect for waterflooding. Furthermore, the relative permeability to oil at low saturations is comparatively less. Relatively low saturations do not lead to the formation of an oil bank, as observed in fields. In a water-oil system, water would dominate flow in such circumstances.

A prudent waterflood design focuses on the areas of formation where relatively large volumes of oil are not produced during primary recovery. Typical examples are tight hydrocarbon-bearing zones, compartments formed due to faulting, and untapped areas in the reservoir. In order to gain knowledge about the distribution of the remaining oil, detailed reservoir description and characterization are necessary. Development of geologic models, validated by reservoir performance during primary production, is the first step. In many cases, seismic studies have identified further opportunities for oil recovery.

Oil gravity and viscosity. Reservoirs with oil gravity more than 25°API, and oil viscosity less than 30 cp, are good waterflooding prospects. A highly viscous fluid, such as heavy oil, is displaced less efficiently by injected water, which is relatively less viscous. Other recovery methods, chiefly thermal, are utilized to recover heavy oil.

Reservoir heterogeneity. Various factors may adversely impact waterflood performance. These include the presence of fractures, high permeability channels, unidentified crossflow between layers, and low transmissibility in certain zones. Compartments created by sealing faults and other heterogeneities also may adversely impact performance. Bypassing of in-situ oil in significant quantities poses a major challenge in a waterflood operation.

Waterflooding is more predictable, and more likely to succeed, in a relatively homogeneous formation where fluid displacement as well as flood coverage may occur as expected. Close well spacing aids in recovering more oil in both homogenous and heterogeneous formations. The modification of injection and production profiles can be attempted to selectively produce from more desirable zones in a formation. Various methods utilized in augmenting waterflood performance are discussed later in this chapter and in chapter 19.

Lithology. Both sandstone and carbonate reservoirs are likely candidates for improved oil recovery by waterflooding. However, certain rock heterogeneities, including secondary porosity, fractures, and conductive channels, are frequently observed in the latter, leading to poor recovery.

Compatibility of injected water. Injected water needs to be compatible with the reservoir water to minimize formation damage. Incompatible water may lead to issues related to injectivity.

Effect of aquifer. Reservoirs experiencing strong water influx may not be good candidates for waterflooding, as the ongoing natural process of water displacing oil may lead to marginal added benefits. However, reservoirs with weak water influx have been waterflooded successfully.

Bottom water zone. In reservoirs with a bottom water zone, injected water is found to "slump down" from the upper to the lower zone where good vertical communication exists. This can lead to poor waterflood performance in some instances.

Gas cap. In reservoirs where a gas cap exists, displaced oil may enter pores previously occupied by gas. This is due to the increased reservoir pressure created by water injection. Consequently, a portion of oil migrating to the gas zone cannot be produced as dictated by the residual oil saturation characteristics of the reservoir rock.

Injection pressure. Reservoirs located at a shallow depth or tight reservoirs may have limitations of injectivity. (Well injectivity is defined in chapter 4.) Injection pressure is kept below the fracture pressure of the formation to ensure that rapid pathways are not created for water channeling. In many cases, limited injection pressure and injection rate translate into less-than-optimum recovery. Generally speaking, a low injection rate leads to a delayed response at the producer, affecting the net present value of the asset.

Waterflood pattern and well spacing

Commonly observed flood patterns, i.e., injection/production well arrangements, are given as follows:

- Direct-line drive, involving injectors and producers on a direct line
- Staggered-line drive, involving staggered injectors and producers
- Regular 5-spot drive, including four injectors at the corners and the producer at the center
- Inverted 5-spot drive, including four producers at the corners and the injector at the center
- Regular 7-spot drive, including six injectors at the corners and the producer at the center
- Inverted 7-spot drive, including six producers at the corners and the injector at the center
- Regular 9-spot drive, including eight injectors at the corners and the producer at the center
- Inverted 9-spot drive, including eight producers at the corners and the injector at the center

In regular patterns, producers are located in the central location, surrounded by injectors. In inverted patterns, injectors are drilled in the middle of the pattern, and producers are at the corners.

Injection wells can be positioned around the periphery of a reservoir, which is referred to as peripheral injection. In contrast, crestal injection involves positioning of the wells along the crests of small reservoirs with sharp structural features.

In a dipping reservoir, water injection wells are located down dip to take advantage of gravity segregation. If a gas cap exists in the reservoir, produced gas may be reinjected through updip wells to maintain reservoir pressure.

Various well configurations implemented in reservoir waterflooding are shown in Figure 16–3.

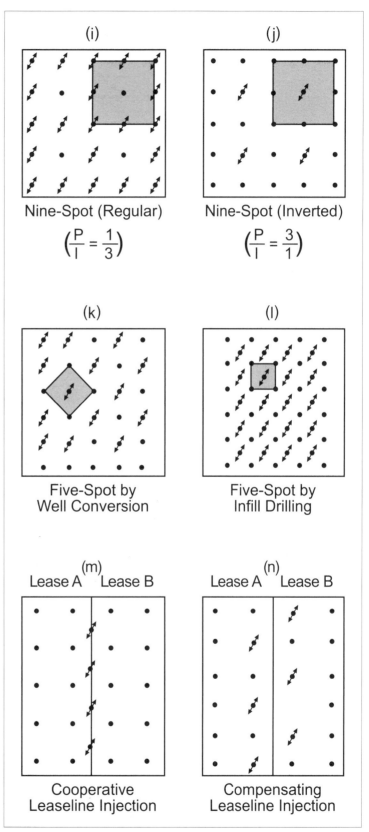

Fig. 16–3. Familiar injection-production well patterns (configurations) utilized in waterflood design. *Source: A. Satter and G. C. Thakur. 1994.* Integrated Petroleum Reservoir Management—A Team Approach. *Tulsa: PennWell.*

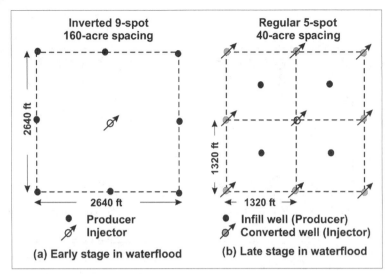

Fig. 16–4. Changes in well pattern and spacing at (a) earlier and (b) later stages of the life of a waterflood project. The objective is to contact and displace more oil in order to increase ultimate recovery.

Table 16-1. Comparison of waterflood patterns

Pattern	P/I Regular	P/I Inverted	d/a	E_A%	Geometry
Direct-line drive	1	—	1	56	Rectangle
Staggered-line drive	1	—	1	78	
5-spot drive	1	1	0.5	72	Square
7-spot drive	1/2	2	0.866		Equilateral triangle
9-spot drive	1/3	3	0.5	80	Square

Note: P = number of production wells

I = number of injection wells

d = distance from an injector to the line connecting the two producing wells

a = distance between wells in line in regular patterns

E_A = areal sweep efficiency at water breakthrough at a producing well for unit water/oil mobility ratio (M = 1). Formation is assumed to be homogeneous.

During the life of a waterflood project, the injector/producer pattern and well spacing changes in many instances, with the objective of maximizing oil recovery. Based on reservoir simulation studies and economic analyses, producers are converted to injectors at certain stages of recovery. Infill wells are drilled, and relatively dense well spacing is implemented. The development plan hinges on the expected increase in recovery and whether the incremental oil justifies the capital expenditure and operating costs. For example, a waterflood operation may start with an inverted 9-spot pattern and gradually transform to a 5-spot pattern following conversion of the wells and infill drilling (fig. 16–4).

Producing a well first and then converting it to an injector is usually preferred. The benefits include the oil recovery from a new well for a certain period of time and the development of a depleted (low pressure) zone around the well. The latter may lead to an increased injection rate following conversion. Other beneficial effects may include the natural cleaning of the wellbore after completion and transport of any fine particles from the adjacent formation into the producing stream, thus limiting skin damage. However, it is noted that producers with less-than-satisfactory performance may turn out to be poor injectors.

The characteristics of various waterflood patterns are given in Table 16–1.

A wide variation in well spacing is observed from field to field during waterflooding, which is dictated by the following:
- Formation transmissibility
- Oil viscosity
- Reservoir heterogeneities
- Optimum injection pressure
- Targeted time frame for recovery

Literature review indicates that 20-acre to 40-acre well spacing is implemented frequently in a large number of waterflood operations. In enhanced oil recovery (tertiary recovery) processes, closer well spacing is commonplace. Poor rock permeability, a high degree of geologic heterogeneity, heavy oil, and relatively low permissible injection pressure usually require closer well spacing for good waterflooding results. Consequently, capital and operating expenses, including the cost of production per barrel of oil, tend to be high.

In heterogeneous and tight reservoirs, peripheral waterflooding has not always been very effective. This is due to poor injectivity at the reservoir flanks, the relatively long distance between injectors and producers, and the existence of geologic heterogeneities. Close well spacing (20 acres or less) based on 5-spot patterns has been highly successful in such circumstances, as indicated by the historical waterflood performance.

Injection rate

The rate of oil recovery, and therefore the life of a waterflood project, depend upon the water injection rate into a reservoir. Knowledge of well injectivity is required in waterflood design. Transient pressure falloff (PFO) tests are usually employed to determine the injectivity of a well. The water injection rate, which can vary throughout the life of the project, is influenced by many factors. The variables affecting the injection rates include the following:

- Rock and fluid properties. Low injectivities are associated with tight rocks, skin, and viscous fluids.
- Mobility of fluids.
- Areas related to swept and unswept regions.
- Oil geometry, i.e., well pattern, spacing, and wellbore radii.

For an injection well, the optimum injection rate ensures maximum contact with residual oil and recovers oil within the desired time frame. In most waterflood operations, the objective is to attain maximum injectivity. However, shallow reservoirs have limitations in the maximum achievable injection pressure. Another important point to note is that the injectors are operated below the fracture gradient in order to avoid creating highly conductive microchannels that may bypass oil. Craft and Hawkins provide a discussion of any influence of high production rates on recovery.[7]

Injectivity of water is defined as the rate of water injection over the pressure differential between the injector and the producer. It has the unit of barrels per day per pounds per square inch (bbl/d/psi). Decline in water injectivity is observed during the early stages of injection into a reservoir depleted by solution gas drive. This occurs as pore spaces initially occupied by free gas are gradually filled up. Following fill-up, the injectivity of water depends upon the mobility ratio. As shown in Figure 16–5, it remains constant in the case of unit mobility ratio and increases when the mobility ratio is greater than unity (unfavorable for displacing oil). It decreases when the ratio is less than unity (favorable for displacing oil). Again, well injectivity may deteriorate noticeably during the life of a waterflood project as a consequence of formation damage around the wellbore.

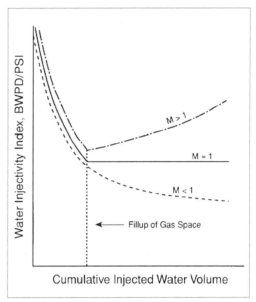

Fig. 16–5. Water injectivity variations in a radial system. In formations where the mobility of water is relatively high, well injectivity is greater. *Source: F. F. Craig, Jr. 1971.* The Reservoir Engineering Aspects of Waterflooding. *SPE Monograph. Vol. 3. Richardson, TX: Society of Petroleum Engineers. © Society of Petroleum Engineers. Reprinted with permission.*

The mobility ratio is defined as the mobility of the displacing phase (water) over that of displaced phase (oil), as discussed in chapter 4. This is given as follows:

$$M = \frac{[k_{rw}/\mu_w]}{[k_{ro}/\mu_o]} \tag{16.2}$$

The relative permeabilities are based on two different and separate regions in the reservoir during waterflood. Craig suggested calculating the mobility ratio prior to water breakthrough, i.e., k_{rw} at the average water saturation in the swept region, and k_{ro} in the unswept zone.[8]

Well injectivity is a function of the distance between the injector and producer, along with the pressure drop between the wells. Well injectivity is also a function of the formation thickness, oil viscosity, and effective permeability to the displaced fluid, among other factors. For various waterflood patterns, it can be estimated as given in the following:[9,10]

Direct-line drive:

$$i = \frac{1.538 \times 10^{-3}\, k\, k_{ro}\, h\, \Delta p}{\mu_o\, [\log (a/r_w) + 0.682\, d/a - 0.798]}, \quad \text{for } d/a \geq 1 \tag{16.3}$$

5-spot pattern:

$$i = \frac{1.538 \times 10^{-3} \, k \, k_{ro} \, h \, \Delta p}{\mu_o \, [\log (d/r_w) - 0.2688]}$$

(16.4)

7-spot pattern:

$$i = \frac{2.051 \times 10^{-3} \, k \, k_{ro} \, h \, \Delta p}{\mu_o \, [\log (d/r_w) - 0.2472]}$$

(16.5)

Inverted 9-spot pattern:

$$i = \frac{1.538 \times 10^{-3} \, k \, k_{ro} \, h \, \Delta p'}{\mu_o \, [(1+R)/(2+R)][\log (d/r_w) - 0.1183]}$$

(16.6)

where

i = water injection rate, bbl/d,

d = distance between adjacent rows of wells, ft,

a = distance between like wells in a row, ft,

k = absolute permeability, mD,

k_{ro} = relative permeability to oil,

Δp = pressure drop between the injector and producer, psi,

$\Delta p'$ = pressure drop between the injector and producer located at the corner of the inverted 9-spot pattern, psi, and

R = ratio of the production rate of the corner well over the side well.

The injection rate and pressure differential are evaluated at base (initial) conditions.

Time frame for waterflooding

Reservoir simulation, based on a detailed reservoir description, is utilized to predict the life of a waterflood operation in a field. The duration of a waterflood project in a reservoir depends on the length of time the operation can be conducted with economic benefits. Maximization of recovery within a relatively short time frame is desirable. Some factors significantly influence the duration of a waterflood project. These include reservoir performance under water injection, the injectivity and productivity of the wells, and the water cut in the producing streams. Capital expenditures in drilling new wells, operating costs, and incremental oil recovery over time also affect the duration of the project. In general, efforts are made to produce oil above the economic limit and delay excessive water cut as long as possible. Problematic zones responsible for early water breakthrough are isolated at injectors and producers. Existing wells are recompleted or sidetracked to avoid water production and to enhance productivity. Infill wells are drilled in areas where relatively high oil saturation exists. These are some of the strategies reservoir engineers adopt in optimizing a waterflood operation. The strategies are described with field cases in chapter 19.

Many large fields have been waterflooded for decades, an example of which is included later in the chapter.

Waterflood Recovery Efficiency

The efficiency of a waterflood operation primarily depends on the following:
- Injected water is expected to provide wide coverage in contacting in-situ oil within the injector/producer pattern area and vertically across all flow units that may exist in the targeted formation.
- Once in-situ oil is contacted, the injected water should efficiently displace oil as much as possible toward the producers, leaving minimum oil saturation in the reservoir.

In light of the above requirements, ultimate oil recovery from a reservoir undergoing water injection is determined by the following:

- The displacement efficiency of water displacing oil, a function of rock and fluid characteristics, including relative permeability and viscosity of the fluid phases
- The areal sweep efficiency, i.e., the fraction of the reservoir area contacted by injected water, dependent on reservoir heterogeneity in the horizontal direction, relative location of wells, and distance between the wells, among other factors
- The vertical sweep efficiency, primarily controlled by flow units having different characteristics, including vertical permeability across the flow units

The recovery efficiency is a measure of the fraction of the in-situ oil at the start of waterflooding that would be recovered from the reservoir. In this context, the amount of oil recovered during primary production is not considered. The equation for overall waterflood recovery efficiency (E_R) is given by the following:

$$E_R = E_D \times E_V \tag{16.7}$$

where

E_R = overall recovery efficiency based on waterflooding, fraction or percent,

E_D = displacement efficiency within the volume "swept" by water, fraction or percent, and

E_V = volumetric sweep efficiency, the fraction of the reservoir volume actually swept by water, fraction or percent.

It is important to note that the vertical efficiency is customarily cited as E_I, and it should not be confused with the volumetric sweep efficiency, E_V.

Displacement efficiency

Pore-to-pore displacement efficiency under reservoir conditions (E_D) is given by the following:

$$E_D = \frac{\text{oil saturation before water flood - oil saturation after water flood}}{\text{oil saturation before water flood}} \tag{16.8}$$

In estimating E_D, the average values of saturation are used. Note that in calculating the recovery efficiency from a reservoir, any changes in oil formation volume factor (B_o) need to be incorporated in Equation 16.8, as follows:

$$E_D = 1 - \frac{(S_o / B_o)}{(S_{oi}/B_{oi})} \tag{16.9}$$

In this equation, subscript i denotes initial conditions at the start of the waterflood process. When the difference between the initial and final values of the formation volume factor is large, the displacement efficiency is less for the same initial and residual oil saturation.

When free gas is present at the start of waterflooding, the initial oil saturation at the start of the waterflooding is expressed as the following:

$$S_{oi} = 1 - S_{wi} - S_{gi} \tag{16.10}$$

At the end of waterflooding, the gas saturation is zero, as free gas is redissolved due to elevated pressure, and $S_o = 1 - S_w$. Hence, displacement efficiency can be expressed in terms of water and gas saturations, as follows:

$$E_D = 1 - \frac{(1 - S_w)B_o}{(1 - S_{wi} - S_{gi})B_{ti}} \tag{16.11}$$

B_{ti} is the two-phase FVF in rb/stb at the initiation of waterflood.

The effects of various factors on displacement efficiency during waterflooding are clearly evident from the fractional flow of water versus saturation as obtained from relative permeability data of the oil and water. The fractional flow equation in a two-phase flow system as described earlier in chapter 4 is usually expressed as in the following:

$$f_w = \frac{q_w}{q_w + q_o} \qquad \qquad (16.12)$$

where

f_w = fractional volume of water or water cut, bbl/bbl,

q_w = flow rate of water, bbl/d, and

q_o = flow rate of oil, bbl/d.

In this equation, all parameters are evaluated at reservoir conditions. Assuming linear horizontal flow geometry and applying Darcy's law, it can be shown that the fractional flow of a fluid phase is a function of effective permeability (or relative permeability) and viscosity. Hence, Equation 16.12 can be expressed in the final form as follows:

$$f_w(S_w) = \frac{1}{1 + (k_{ro}/\mu_o)/(k_{rw}/\mu_w)} \qquad \qquad (16.13)$$

The effective and relative permeabilities of reservoir fluids are discussed in chapter 2. Basically, the effective permeability of a fluid phase is a measure of its ability to flow in porous media in the presence of other fluids. Relative permeability is the ratio of effective permeability to absolute permeability. Fluid viscosity, including its effect on production, is discussed in chapter 3. In Equation 16.13, oil and water relative permeabilities are evaluated at the saturation of water for which fractional flow is sought. It must be kept in mind that the effective (and relative) permeabilities of the fluid phases are strongly dependent on fluid saturation in a given two-phase system. In cases where the effective permeability to water increases sharply as a consequence of a relatively small increase in water saturation, the displacement efficiency would be less.

A typical fractional flow versus saturation plot is shown Figure 16–6. Based on the Buckley and Leverett frontal advance theory discussed later, reservoir performance can be studied by fractional flow plots. These plots are generated from

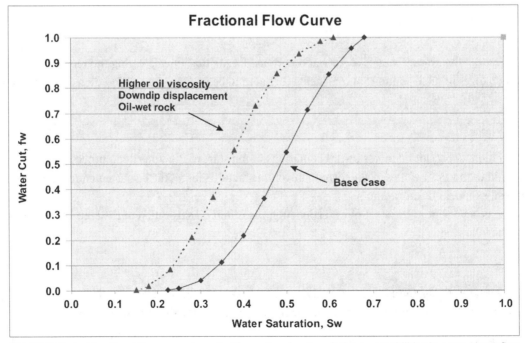

Fig. 16–6. Typical fractional flow curve generated from oil and water relative permeabilities in a two-phase flow system. Two-phase relative permeability curves are presented in Figure 4–6 in chapter 4. Fractional flow curves tend to shift to the left when the displacement efficiency is poor. The curve approaches a vertical configuration in the case of near-pistonlike displacement.

relative permeability curves by applying Equation 16.11. Relative permeability data is obtained from laboratory studies or from available correlations discussed in chapter 2. Alternately, fractional flow curves can be generated by noting the following relationship between the relative permeability ratio and fluid saturation:[11]

$$(k_{ro}/k_{rw})_{Sw} = ae^{bSw} \tag{16.14}$$

where

a = constant in $\log(k_{ro}/k_{rw})$ versus S_w plot, and
b = slope of $\log(k_{ro}/k_{rw})$ versus S_w plot.

Hence, the equation of fractional flow of water and its derivative can be written as in the following:

$$(f_w)_{Sw} = \frac{1}{1 + \mu\, a\, e^{bS_w}} \tag{16.15}$$

$$(df_w/dS_w)_{Sw} = \frac{-\mu\, a\, b\, e^{bS_w}}{[1 + \mu\, a\, e^{bS_w}]^2} \tag{16.16}$$

where

$\mu = \mu_w/\mu_o$.

Displacement efficiency is influenced by rock and fluid properties, as detailed in the following discussion.

Oil viscosity. Relatively viscous oil is displaced less efficiently by injected water. The recovery factor is generally low in heavy oil reservoirs, as evident from Table 3–10 in chapter 3. The mobility ratio is significantly higher than unity and is viewed as unfavorable in the case of water displacing heavy oil. As the viscosity ratio of oil to water increases, the fractional flow curve becomes steep and shifts to the left. This indicates that the flow of the water phase tends to be more pronounced, given that other factors remain the same.

Inclined flow in porous media. Since water has a higher density than oil, a reduction in the fractional flow of water over oil occurs when flowing up dip, resulting in better displacement efficiency. In inclined reservoirs, injectors are generally located down dip, and oil producers are located up dip, in order to take the advantage of gravity forces.

Wettability of the rock. The effects of rock wettability on waterflooding are discussed in chapter 2. In general, water-wet rocks perform better during the displacement of oil by injected water. In oil-wet rocks, however, there is a tendency of in-situ oil to adhere to the pore surface, and the resulting displacement efficiency is poor.

Interfacial tension. When interfacial tension is relatively low, displacement efficiency is generally found to be higher. This principle is applied to a number of enhanced recovery processes, including surfactant flooding.

Besides the above, displacement efficiency is also influenced by throughput, i.e., pore volumes injected. However, the economic limit may be reached before large volumes of water can be injected to produce incremental oil.

Displacement efficiency can be either determined or estimated by the following:
- Laboratory core floods
- Frontal advance theory
- Empirical correlations

Laboratory core floods, using representative formation cores and actual reservoir fluids, are usually conducted to obtain residual oil saturation following waterflooding. Laboratory studies aid in understanding the mechanism of the waterflood process. However, field-scale heterogeneities and actual water injection conditions can seldom be simulated or replicated in the laboratory. Thus results of waterflood studies tend to be optimistic.

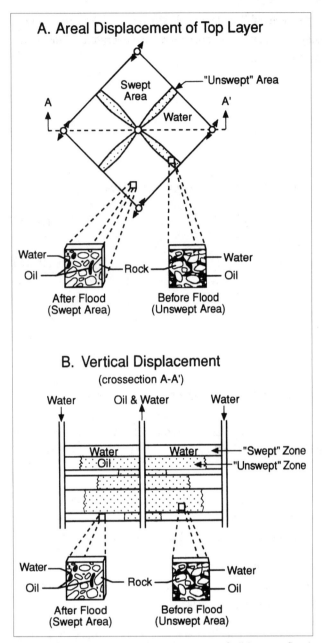

Fig. 16-7. Areal and vertical displacement of oil by waterflood. *Source: A. Satter and G. C. Thakur.* Integrated Petroleum Reservoir Management—A Team Approach. *Tulsa: PennWell.*

Various methods are employed to improve the displacement efficiency of the waterflood process. These include, but are not limited to, the following:

- Addition of a water-soluble polymer to attain a favorable mobility ratio
- Addition of a surfactant to reduce interfacial tension
- Addition of wettability alteration agents
- Water-alternating-gas processes that attain miscibility between injected gas and in-situ oil

These methods are discussed in chapter 17.

Areal and vertical sweep efficiency

Volumetric sweep efficiency of waterflooding has two components:

- Areal sweep efficiency
- Vertical sweep efficiency

Volumetric sweep represents the fraction (or percent) of pore volume in porous media that is swept by injected water. The actual amount of oil that is displaced and produced from that fraction of pore volume is estimated by the displacement efficiency of the waterflooding, which is discussed in the preceding section. Volumetric sweep efficiency is defined by the following:

$$E_V = E_A \times E_I \qquad \textbf{(16.17)}$$

where

E_A = areal sweep efficiency, or the fraction of the pattern area that is swept by the displacing fluid, i.e., water, and

E_I = vertical (or invasion) sweep efficiency, or the fraction of the pattern thickness that is swept by the displacing displacing fluid, i.e., water.

Factors that determine areal or pattern sweep efficiency are the flooding pattern type, mobility ratio, and throughput and reservoir heterogeneity in a lateral direction. The areal sweep efficiency is affected significantly when the injector and producer are aligned with the principal orientation of naturally occurring fractures. Areal sweep can be limited by the existence of physical boundaries in a highly faulted formation, the reservoir being referred to as compartmental. A favorable mobility ratio (M < 1) and relatively homogeneous rock are expected to lead to better sweep during waterflooding. Relative locations of wells, and the distance between them, also influence areal sweep.

Vertical sweep efficiency, as shown in Figure 16-7, is influenced by variations in layer permeability due to varied depositional environments that occurred millions of years ago. Rock permeability varies from layer to layer, and injected water is found to move preferentially through zones of higher permeability. In a preferentially water-wet rock, water is imbibed into the adjacent lower permeable zones from the higher permeable zones because of capillary forces. Also, injected water tends to flow to the bottom of the reservoir due to gravity segregation. The net effect of these factors is to influence the vertical sweep efficiency of a waterflood project.

Figure 16–8 depicts common scenarios in which the areal and vertical sweep efficiencies are adversely affected due to the presence of directional permeability and high permeability streaks, respectively.

Laboratory Investigations of Waterflooding

Craig presented a detailed compilation of laboratory studies on waterflooding that shed light on ultimate recovery, sweep efficiencies, and waterflood mechanisms.[12] Results of selected studies are described briefly in the following sections.

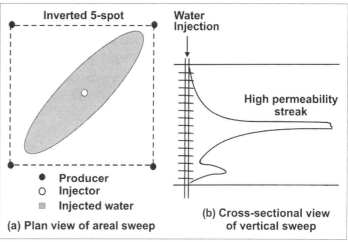

Fig. 16–8. Limited sweep in heterogeneous reservoirs during waterflooding: (a) areal and (b) vertical.

Effects of gravity, capillary, and viscous forces[13–15]

During waterflooding, the effects of gravity forces are evident in certain cases. This is due to the marked difference between the specific gravity of displaced and displacing fluid phases, i.e., oil and water. Injected water tends to "underrun" in-situ oil when the rate of injection is lower than optimum. This phenomenon is also referred to as "tonguing." Consequently, water breakthrough occurs earlier than expected, and oil recovery is likely to be less than satisfactory. Again, the degree of gravity segregation is influenced by the vertical permeability of the rock.

Viscous forces arise due to the pressure differential between two points in the fluid flow path following injection of water at high pressure. Viscous forces may act in both directions, laterally and vertically, depending on the difference in fluid potential from one point to the other in porous media. At sufficiently high injection rates, viscous forces predominate over gravity forces. This results in better volumetric sweep efficiency and oil recovery.

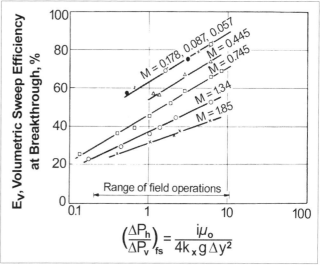

Fig. 16–9. Volumetric sweep efficiency as influenced by viscous and gravity forces for 5-spot uniform systems. *Source: F. F. Craig, Jr., J. L. Sanderlin, D. W. Moore, and T. M. Geffen. 1957. A laboratory study of gravity segregation in frontal drives. Transactions. AIME. Vol. 210: pp. 275–282. © Society of Petroleum Engineers. Reprinted with permission.*

Figure 16–9 presents the effects of gravity and viscous forces on volumetric sweep efficiency at different mobility ratios. The detrimental effects of an unfavorable mobility ratio (as with water displacing heavy oil) and relatively low values of the ratio of viscous to gravity forces (due to low injection rates) are highlighted.

Capillary effects develop in porous media as water tends to imbibe through a porous network in a water-wet rock. Capillary imbibition is discussed in chapter 2. When the water injection rate is relatively low, water may imbibe into relatively tight zones, augmenting recovery in preferentially water-wet formation. However, imbibition of water is less effective in zones that are relatively thick. Any increase in recovery by capillary imbibition is not expected in oil-wet rock. Furthermore, gravity forces depending on rock, fluid, and waterflood characteristics may counteract the capillary effects to varying degrees.

Results of two-layer, 5-spot waterflood model studies have indicated that the sweep efficiency at breakthrough is significantly greater when the more permeable layer is at the top. In the studies, the two layers were in flow communication throughout the flooded area.

Phenomenon of viscous fingering[16,17]

At high viscosity ratios between oil and water, certain laboratory investigations have encountered a phenomenon called viscous fingering. A large bank of water moving uniformly as expected in an ideal waterflood operation is not observable. Instead, water moves in thin, irregular streams, or "fingers," toward the producing end. Displacement of viscous oil by injected water in the porous medium is no longer stable. The net effect of viscous fingering is poor sweep and less oil recovery during waterflooding.

Effect of crossflow between layers[18–20]

Crossflow may occur due to the various forces listed above, as fully or partially communicating vertical pathways can exist between two adjacent layers. The phenomenon of crossflow in a stratified reservoir is observed in a large number of reservoirs, many of which are carbonates. This is because the intervening shale is either permeable or discontinuous. Laboratory investigations have revealed that the effect of crossflow is a strong function of the mobility ratio and permeability contrast between the layers. Sweep efficiency during waterflooding is improved when the mobility ratio is less than unity. It is detrimental to recovery when the mobility ratio is greater than unity.

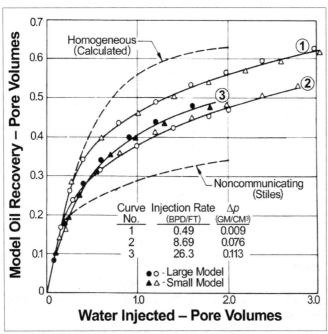

Fig. 16–10. Plot of waterflood response in stratified systems illustrating the effects of crossflow between adjacent layers. *Source: C. W. Carpenter, Jr., P. T. Bail, and J. E. Bobek. 1962. A verification of waterflood scaling in heterogeneous communication flow models. Society of Petroleum Engineers Journal. March, 9–12. © Society of Petroleum Engineers. Reprinted with permission.*

Figure 16–10 presents the results of a waterflood study based on a two-layer, 5-spot model where the effects of gravity, capillary, and viscous forces were simulated. When the mobility ratio is unity and the gravities of the fluids are the same, oil recovery from a stratified system with crossflow is found to be greater than that of a comparable system without crossflow. Homogeneous systems with uniform rock properties, however, lead to the highest recovery.

The crossflow index is a measure of the performance of a stratified system with crossflow, as given in the following:

$$\text{Crossflow Index} = \frac{N_{p(cf)} - N_{p(ncf)}}{N_{p(u)} - N_{p(ncf)}} \quad (16.18)$$

where

$N_{p(cf)}$ = cumulative production from a stratified system with crossflow between layers, bbl,

$N_{p(ncf)}$ = cumulative production from a stratified system with no crossflow, bbl, and

$N_{p(u)}$ = cumulative production from a homogeneous system of uniform permeability, bbl.

As the index approaches unity, the response from the stratified system approaches that of a homogeneous formation. On the other hand, a zero value for the crossflow index represents a stratified formation in which the layers or flow units are not in communication with each other.

Effect of trapped gas saturation[21]

Waterflood commences in many reservoirs where a gas cap is present. A gas cap is encountered in a saturated reservoir at discovery. When an unsaturated reservoir is allowed to deplete below the bubblepoint pressure, free gas evolves, leading to the formation of a gas cap. Following water injection, the reservoir is repressurized, and free gas is compressed and redissolved into solution. A portion of the free gas is displaced by the oil bank, and the rest is trapped.

In strongly water-wet rock, residual oil saturation is found to be lower when the trapped gas saturation is higher. In an oil-wet formation, no simple relationship between residual oil saturation and trapped gas saturation is observed. This is because residual saturation depends on the structure of the pores, the viscosity of the oil, and the injected water volume.[22]

Methods of Predicting Waterflood Performance

Prior to the wide utilization of multiphase, multidimensional reservoir simulators in studying reservoir fluid behavior under water injection, reservoir engineers frequently resorted to analytical and graphical methods. These were aided by laboratory investigations and pilot projects in order to predict waterflood performance. Classical methods for predicting waterflood performance include the following:

- Dyes, Caudle, and Erickson[23]
- Dykstra-Parsons[24]
- Stiles[25]
- Buckley-Leverett[26]
- Craig-Geffen-Morse[27]
- Prats-Matthews-Jewett-Baker[28]

Most of the prediction methods are based on either laboratory investigations or simplified analytic solutions of waterflooding in an ideal or near-ideal situation. Such assumptions of an ideal situation include a porous medium of uniform rock properties. Certain methods do not consider the relative permeabilities of the oil and water as a function of saturation. Waterflooding is seen as a process of pistonlike displacement of oil. Certain other methods are based on linear flow geometry and do not incorporate the adverse effects of lateral variations in rock properties. Stratified reservoir models may assume independently acting flow units without any crossflow between layers, which in reality may not be the case. However, the methods are highly valuable in conceptualizing the mechanism of oil displacement by waterflooding, as influenced by various factors.

The following sections describe some of the familiar graphical and analytical methods to predict waterflood performance.

Areal sweep efficiency

Areal sweep efficiency for various patterns has been studied using both physical and mathematical models. For the 5-spot pattern, one of the most familiar correlations is proposed by Dyes, Caudle, and Erickson.[29] Figure 16–11 presents areal sweep efficiency correlated with the reciprocal of the mobility ratio (M) and water cut (f_w) as a fraction of the total flow coming from the swept portion of the pattern. It should be

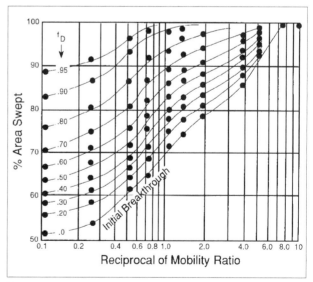

Fig. 16–11. Areal sweep efficiency versus the reciprocal of the mobility ratio. At favorable values of the mobility ratio toward the left of the plot, the percent area swept is relatively large. With continued injection, both water cut and areal sweep are found to increase. *Source: A. B. Dyes, B. H. Caudle, and R. A. Erickson. 1954. Oil production after breakthrough as influenced by mobility ratio. Transactions. AIME. Vol. 201, 81–86. © Society of Petroleum Engineers. Reprinted with permission.*

noted in the correlations that the areal sweep efficiency increases after water breakthrough for all mobility ratios. For favorable mobility ratios (i.e., M ≤ 1), areal sweep efficiency approaching 100% can be obtained with prolonged injection of water.

Based on Figure 16–11, estimates of areal sweep can be obtained from the time of breakthrough of water to a water-cut value of 95%. It is usually observed that the economic limit of a waterflood operation is reached when the water cut reaches 90% or higher. However, before any breakthrough is encountered, the areal sweep efficiency is directly proportional to the volume of water injected in the formation as long as the vertical coverage is the same.

Recovery factor based on variations in rock permeability

Dykstra and Parsons presented waterflood recovery correlations, which are based on laboratory investigation.[30] The authors introduced the permeability variation factor as a measure of vertical permeability stratification. Reservoirs are not uniform in their properties such as permeability, porosity, pore size distribution, wettability, connate water saturation, and PVT properties of in-situ fluids. The variations can be areal as well as vertical. Reservoir heterogeneity is attributed to the depositional environments and subsequent events, as well as to the nature of the particles constituting the sediments. The Dykstra-Parsons permeability variation factor is the widely used method to characterize vertical permeability stratification and to predict waterflood recovery influenced by the degree of heterogeneity. The permeability variation factor, abbreviated as V, is based upon the lognormal permeability distribution (permeability versus percent of total sample having higher permeability). Statistically, it is defined as the following:

Table 16–2. Dykstra-Parsons permeability variation factor based on 22 core samples

Index	Core Permeability, mD	Number of Samples	Cumulative Number of Samples	Percent (greater than)
1	29	1	1	—
2	22	2	3	4.5
3	15	3	6	13.6
4	12	2	8	27.3
5	10	3	11	36.4
6	8	3	14	50.0
7	7	2	16	63.6
8	5	2	18	72.7
9	4	2	20	81.8
10	3	2	22	90.9

$$V = \frac{k_{50} - k_{84.1}}{k_{50}} \qquad (16.19)$$

where

k_{50} = log mean permeability, i.e., permeability at 50% probability, and

$k_{84.1}$ = permeability at 84.1% of the cumulative sample.

Values of V range from 0 (uniform) to 1 (extremely heterogeneous). Table 16–2 presents an example set of data.

Figure 16–12 shows the determination of the permeability variation factor (V).

Based on the Dykstra-Parsons method, Johnson presented plots of recovery factor as a function of mobility ratio, initial water saturation, and water/oil ratios of 1, 5, 25, and 100 bbl/bbl (fig. 16–13).[31] The correlated recovery values represent total primary and secondary production. The following assumptions are made in predicting oil recovery by the Dykstra-Parsons method:

- There is linear flow geometry in a stratified or layered reservoir.

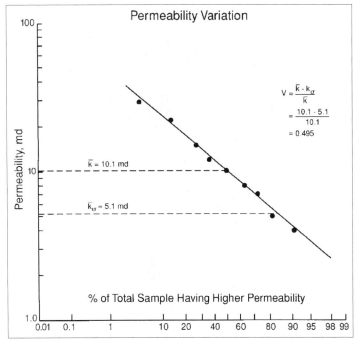

Fig. 16–12. Dykstra-Parsons permeability variation factor analysis. *Source: A. Satter and G. C. Thakur. 1994. Integrated Petroleum Reservoir Management—A Team Approach. Tulsa: PennWell.*

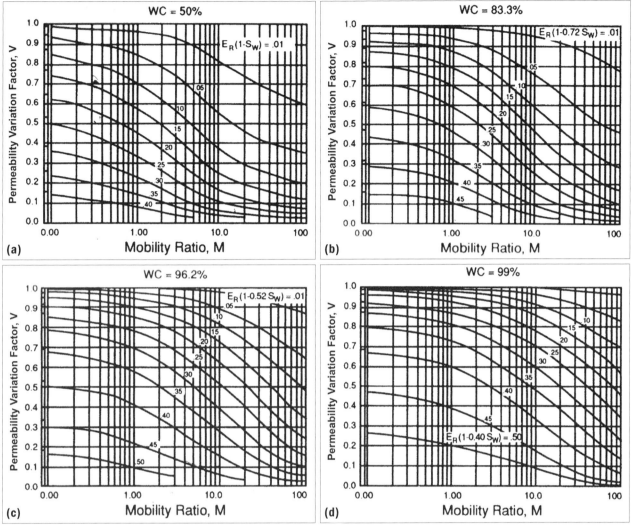

Fig. 16–13. Plots of recovery factor at water/oil ratios of (a) 1, (b) 5, (c) 25, and (d) 100. *Source: C. E. Johnson, Jr. 1956. Prediction of oil recovery by water flood—a simplified graphical treatment of the Dykstra-Parsons method. Transactions. AIME. Vol. 207, 345–346. © Society of Petroleum Engineers. Reprinted with permission.*

- There is lognormal distribution of permeability, which is widely observed in reservoir rocks.
- Waterflooding is based on pistonlike displacement of oil by the injected water.
- Recovery is a function of the mobility ratio.
- No crossflow exists between adjacent layers.
- The pressure drop is constant across the layers.
- Individual layers in the waterflooded formation have equal thicknesses.

The following observations can be readily made from the four plots:

- Oil recovery from waterflooding is expected to decrease with increasing reservoir heterogeneity in a significant manner. Waterflood performance is inversely correlated to the Dykstra-Parsons permeability variation factor.
- Unfavorable mobility ratios (M > 1) lead to poor recovery. Relatively high values of mobility are associated with viscous oil as well as low effective permeability to oil in an oil-water system.
- With continued water injection, both water/oil ratio and cumulative oil recovery increase. In practice, a limiting water cut is reached, usually in the range of 80% to 90% or even higher, when large volumes of water recycling and handling are no longer feasible.

Example 16.1. Estimation of the recovery factor based on the water/oil ratio and the mobility ratio.
The following data is given for a waterflood operation:

Permeability variation factor $= 0.5$.

Water/oil ratio, bbl/bbl, $= 5$.

Mobility ratio $= 1$.

Initial water saturation, fraction, $= 0.2$.

(a) Estimate the recovery factor.

(b) Recalculate the recovery factor in waterflooding a reservoir where the oil has a viscosity six times greater than that in (a).

(c) Cores obtained from recently drilled wells indicate that the reservoir is more heterogeneous than previously thought. Estimate the changes in recovery factor predictions in (a) and (b) by assuming a permeability variation factor of 0.7.

Solution: (a) Based on the plot representing water/oil ratio $= 5$ in Figure 16–13, the following is obtained:

$$E_R(1 - 0.72\, S_w) = 0.285$$
$$E_R = 0.333$$

where

E_R = oil recovery factor, dimensionless.

(b) The mobility ratio is 6 in this case. The recovery factor is calculated as follows:

$$E_R(1 - 0.72\, S_w) = 0.175$$
$$E_R = 0.204$$

(c) The changes in recovery factor in the two cases from (a) and (b) can be calculated as follows:

Case I: $M = 1$

$$E_R(1 - 0.72\, S_w) = 0.198$$
$$E_R = 0.231$$

Case II: $M = 6$

$$E_R(1 - 0.72\, S_w) = 0.083$$
$$E_R = 0.097$$

The preceding estimates highlight the detrimental effects of reservoir heterogeneity and high oil viscosity on waterflood performance. Methods to improve oil recovery may include closer well spacing and selective zonal injection, among others. Waterflooding issues and related remedial strategies are discussed in this chapter and in chapter 19.

Vertical sweep efficiency and water/oil ratio in stratified flow

Reservoirs often exhibit stratified flow characteristics under waterflooding due to variations in rock properties as discussed earlier. Oil-bearing formations are generally found to be comprised of multiple flow units of varying transmissibility and storavity. One such system comprised of two flow units in a reservoir is depicted in Figure 2–20 in chapter 2. Stiles modeled waterflood performance as a function of individual layer transmissibility, where water injection occurs through independent flow units without any communication between adjacent layers.[32] Individual values of core permeability are viewed to represent discrete and noncommunicating layers. Hence, the conceptualization of the waterflood process is straightforward. The most permeable layer will experience the earliest breakthrough, followed by other layers of decreasing permeability. The layer with the least permeability would be the last to break through. Consequently, the vertical sweep efficiency is a function of individual layer permeability and thickness, as follows:[33]

$$E_I = \frac{k_i(h_1 + h_2 + ... + h_i) + (k_{i+1}h_{i+1} + k_{i+2}h_{i+2} + ... + k_n h_n)}{k_i h_t} \qquad (16.20)$$

where

subscript i = the layer experiencing the latest breakthrough,

k_i = the permeability of ith layer,

h_i = the thickness of ith layer,

h_t = the total thickness of the formation, and

n = the total number of layers in the flow system.

For convenience in calculation, layers are ordered from highest to lowest permeability. In other words, $k_1 > k_2 > k_3$, and so forth.

The water/oil ratio (WOR) at the surface is calculated as follows:

$$WOR = \left[\frac{k_{rw}/\mu_w}{k_{ro}/\mu_o} \frac{B_o}{B_w} \right] \left[\frac{k_1h_1 + ... + k_ih_i}{k_{i+1}h_{i+1} + ... + k_nh_n} \right] \tag{16.21}$$

where

k_{rw} = relative permeability to water at residual oil saturation, fraction,

k_{ro} = relative permeability to oil at initial water saturation, fraction,

B_o = oil formation volume factor, rb/stb, and

B_w = water formation volume factor, rb/stb.

The assumptions in deriving the above equations include the following:
- Stratified fprmation having uniform porosity and linear flow geometry
- Steady-state flow
- Unit mobility ratio
- No communication between adjacent layers
- Piston-like displacement of oil by water
- Equivalent oil relative permeabilities ahead of and behind the flood front

Example 16.2. Estimation of the vertical sweep and water/oil ratio in a stratified formation. Estimate vertical sweep efficiency and water/oil ratio during waterflooding of the following five-layer system:

Layer	Permeability, mD	Thickness, ft
1	200	2
2	50	20
3	30	20
4	20	10
5	10	4

The following rock and fluid data is given:

$k_{rw} = 0.35$.

$k_{ro} = 0.88$.

μ_o, cp, = 2.5.

μ_w, cp, = 0.5.

B_o, rb/stb, = 1.18.

B_w, rb/stb, = 1.012.

Solution: Based on Equations 16.20 and 16.21, Table 16–3 is prepared.

Example calculations for layer 3 (k = 30 mD, h = 20 ft) are given as follows:

$$E_I = \frac{30 \times (2 + 20 + 20) + (20 \times 10 + 10 \times 4)}{30 \times 56}$$

$$= 0.893$$

$$WOR = \left[\frac{0.35/0.5}{0.88/2.5} \frac{1.18}{1.012} \right] \left[\frac{400 + 1000 + 600}{200 + 40} \right]$$

$$= 2.318 \times 8.333$$

$$= 19.32 \text{ bbl/bbl}$$

Table 16–3. Calculation of vertical sweep and water/oil ratio

k_i	h_i	k_ih_i	Cumulative k_ih_i	Cumulative h	$K_i \times$ Cumulative h	k_ih_t	E_I	WOR	Water cut, %
(1)	(2)	(1) × (2)	(3)	(4)	(1) × (4)				
200	2	400	400	2	400	11,200	0.200	0.50	33.3
50	20	1,000	1,400	22	1,100	2,800	0.693	3.86	79.4
30	20	600	2,000	42	1,260	1,680	0.893	19.32	95.1
20	10	200	2,200	52	1,040	1,120	0.964	127.49	99.2
10	4	40	2,240	56	560	560	1.000	—	—

The preceding calculations illustrate the effect of a thin, highly conductive layer on waterflood performance. When the first water breakthrough occurs from the high permeability (200 mD) layer, which is only 2 ft thick, water cut at the producer is estimated to be 33.3%. This is one-third of the total production rate. Following breakthrough from the third layer, water cut increases substantially and is in excess of 95%. In most cases, high water cut is likely to render the water injection and cycling inefficient. However, oil from the remaining two layers of relatively low permeability remains virtually unrecovered.

Fractional flow theory

The fractional flow theory provides an expression for the fraction of individual fluid phases, i.e., oil and water, during displacement as shown earlier in Equation 16.12:

$$f_w = q_w / q_t$$

where

q_t = total flow rate of oil and water phases ($q_w + q_o$), bbl/d.

Note that f_w, q_w, and q_o are evaluated at reservoir conditions. For flow rates measured under stock-tank conditions:

$$f_w = q_{w,s} B_w / (q_{o,s} B_o + q_{w,s} B_w) \qquad (16.22)$$

where

subscript s denotes stock-tank volume, given in stb.

By manipulating the above equations, the following can be shown:

$$f_{w,s} = \frac{B_o}{B_w(1/f_w-1)+B_o} \qquad (16.23)$$

where

$f_{w,s}$ = water cut at the surface, fraction.

Since the water/oil ratio is defined as q_w/q_o, one can obtain a relationship between water cut and the water/oil ratio as follows:

$$WOR = \frac{f_w}{1 - f_w} \qquad (16.24)$$

It can be further shown that the water/oil ratio at the surface is related to the water/oil ratio under reservoir conditions as follows:

$$WOR_s = WOR_{res} \left(\frac{B_o}{B_w} \right) \qquad (16.25)$$

The fractional flow equation is derived in the following manner:

1. Flow rates of individual fluid phases, namely q_w and q_o, are estimated by Darcy's law.
2. Pressure gradients in the oil and water phases are correlated by the capillary pressure that exists between the two phases: $P_c = P_o - P_w$.
3. The fractional rate of oil, q_o/q_t, is set to $1 - f_w$.

The fractional flow equation, presented earlier in chapter 4, has the following final form, considering viscous, capillary, and gravity forces:

$$f_w = \frac{1 + 1.127 \times 10^{-3}[A/q_t][k_o/\mu_o][P_c' - 0.433\Delta\rho \, \text{Sin}(\alpha)]}{1 + [k_o/\mu_o]/[k_w/\mu_w]} \tag{16.26}$$

where

A = fluid flow cross section, ft^2,

k = effective permeability to fluid, mD,

μ = fluid viscosity, cp,

p_c' = capillary pressure gradient over flow length, psi/ft,

$\Delta\rho$ = difference in density between water and oil, g/cm^3, and

α = angle of inclination at which fluid flow occurs, degrees.

In Equation 16.26, subscripts o and w denote oil and water phases, respectively. It should be noted that the equation is generic and is valid for any other fluid system, such as in-situ oil displaced by immiscible gas or surfactant liquid.

Darcy's law and capillary pressure are described in chapter 2. The capillary effect is usually small compared to viscous forces. Neglecting this effect, and noting that the angle of inclination is zero in horizontal systems, the above expression for fractional flow reduces to Equation 16.13, presented earlier. Another point to note is that the ratio of the effective permeabilities of oil and water is equal to that of the relative permeabilities, and k_o/k_w is replaced by k_{ro}/k_{rw} in Equation 16.13.

Derivation of this equation is left as an exercise at the end of the chapter.

Buckley and Leverett frontal advance theory[34]

Buckley and Leverett proposed the frontal advance theory of waterflooding, which enables the determination of displacement efficiency at and after water breakthrough in a homogeneous formation.[35] The procedure, later modified by Welge, utilizes oil-water relative permeability, fluid viscosity, and formation volume factor.[36]

Important equations related to fractional flow and frontal advance of fluids during waterflooding are presented in chapter 4 and in the following. The method requires the determination of the fractional flow of water (f_w) and its derivative (df_w/dS_w) from relative permeability data. Either the graphical method, shown in Figure 16–14, or Equations 16.15 and 16.16 can be utilized to determine f_w and df_w/dS_w, respectively. An incompressible flow of fluids is assumed in the analysis.

Based on material balance, it can be shown that the distance of the frontal edge of the water bank during waterflooding is a function of d_{fw}/d_{sw}, as follows:

$$x, S_{wf} = \frac{5.615 \, i_w \, t}{\emptyset \, A}\left(\frac{df_w}{dS_w}\right)_{S_{wf}} \tag{16.27}$$

where

S_{wf} = frontal saturation of the advancing water bank determined by drawing a tangent from the initial water saturation, as shown in Figure 16–14,

x = distance traveled by the leading edge of the injected front, ft,

i_w = water injection rate, bbl/d,

t = time injection started, days,

$\emptyset$ = formation porosity, fraction, and

A = cross-sectional flow area, ft^2.

The preceding suggests that the time for breakthrough to occur following the injection of water can be estimated when the reservoir pore volume is known. Rearranging Equation 16.27, the following can be obtained:

$$t,_{breakthrough} = \frac{\emptyset\,A\,L}{5.615\,i_w}\left[1/\left(\frac{df_w}{dS_w}\right)_{Swf}\right] \qquad \textbf{(16.28)}$$

where

L = distance traveled by injected front between the injector and producer, ft, and

$$\frac{\emptyset\,A\,L}{5.615} = PV = \text{reservoir pore volume given in bbls.} \qquad \textbf{(16.29)}$$

It can be further deduced that the pore volume of water injected at the time of breakthrough can be estimated from the above equation as follows:

$$PVWI = \frac{\text{Injected water volume at breakthrough}}{\text{Reservoir pore volume}} = \frac{i_w t_{breakthrough}}{\emptyset\,A\,L/5.615} \qquad \textbf{(16.30)}$$

$$= \left[1/\left(\frac{df_w}{dS_w}\right)_{Swf}\right]$$

where

PVWI = pore volume of water injected.

Figure 16–14 illustrates the methodology in analyzing waterflood performance, as given in the following discussion.

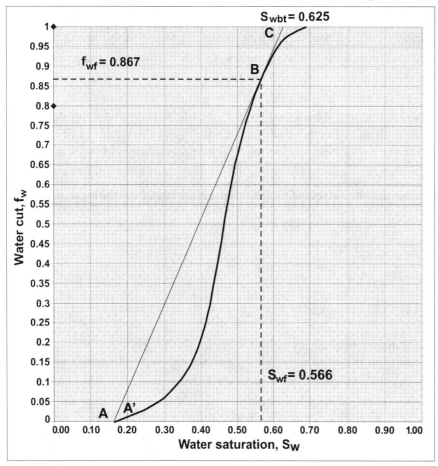

Fig. 16–14. Interpretation of fractional flow curve

At breakthrough of injected water.

- Frontal saturation of the advancing water bank (S_{wf}) is determined by drawing a tangent on the fractional flow curve (point B) from the connate water saturation located on the fractional flow curve (point A). If movable water is present in the reservoir, $S_{wi} > S_{wc}$, and the tangent is drawn from the initial water saturation located on the fractional flow curve (point A').

- The corresponding water cut in the producing end is found from the value of f_{wf} read from the y-axis.

- The average water saturation in the reservoir at breakthrough is found by extending the tangent to point C on the $f_w = 1$ line at the top.

- The value of the derivative (df_w/dS_w) is obtained from the slope of the tangent drawn to determine frontal saturation. The following relationships apply at breakthrough:

$$S_{w,BT} - S_{wc} = \left(\frac{1}{df_w/dS_w}\right)_{S_{wf}} \tag{16.31}$$

$$= \frac{S_{wf} - S_{wc}}{f_{wf}} \tag{16.32}$$

where

f_{wf} is the fractional flow of water at the flood front.

Analysis of waterflood performance following breakthrough.
Following breakthrough, saturation of water and water cut are observed to increase with continued water injection. At water saturation values greater than the frontal saturation at breakthrough (S_{wf}), tangents are drawn successively, as shown in Figure 16–15. For example, when water saturation reaches a value of 0.58 at the producing end, average saturation behind the front is 0.635, and the water cut is estimated to be 0.91. The value of df_w/dS_w, as obtained from the slope of the tangent (marked A in figure), is 1.5. Tangents B and C, drawn subsequently at higher saturations, indicate increasing values of water cut and average saturation in the reservoir, and decreasing values of the derivative df_w/dS_w. Once the value of df_w/dS_w is known for a specific saturation, the position of the latter can be determined by using Equation 16.27. The procedure is illustrated in Example 16.4.

After breakthrough of the injected water, the following relationships apply:

$$S_{w,avg} - S_{w2} = \left(\frac{1 - f_{w2}}{df_w/dS_w}\right)_{S_{w2}} \tag{16.33}$$

The pore volume of water injected following breakthrough ($PVWI_2$) can be estimated by Equation 16.30. The term df_w/ds_w is evaluated at values of water saturation that are greater than the saturation at breakthrough (S_{wf}). Hence, Equation 16.30 takes the following form:

$$PVWI_2 = \left(1 / \left(\frac{df_w}{dS_w}\right)_{S_{w2}}\right) \tag{16.34}$$

where

$PVWI_2 > PVWI$,

$S_{w2} > S_{wf}$, and

$\left(\frac{df_w}{dS_w}\right)_{S_{w2}} < \left(\frac{df_w}{dS_w}\right)_{S_{wf}}$.

With increasing S_w, the pore volume of the water injected is greater, and values of the derivative df_w/dS_w become smaller.

At and following breakthrough, the water cut can be determined from Figure 16–14, which leads to the determination of the water/oil ratio based on Equations 16.24 and 16.25.

The displacement efficiency during waterflooding can be estimated by Equation 16.11. Assuming negligible gas saturation, the following can be obtained:

$$E_D = \frac{S_w - S_{wi}}{1 - S_{wi}} \tag{16.35}$$

where

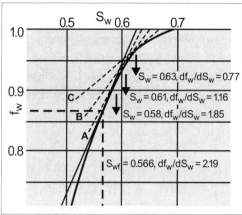

Fig. 16–15. Fractional flow analysis following breakthrough of injected water. Figure 16-14 is enlarged to highlight the section of the plot following breakthrough, where tangents A, B, and C are drawn at increasing water saturations, 0.58, 0.61, and 0.63, respectively, observed at the producing end. The tangents lead to additional information, including average water saturation behind the flood front and the pore volume of the injected water.

S_w = average water saturation in the reservoir at or following breakthrough, fraction.

Assuming that the volumetric sweep efficiency is 100%, Equation 16.32 provides an estimate of recovery efficiency of the waterflooding. In practice, based on the degree of rock heterogeneity, well pattern, operating conditions, and field experience, a reasonable value of volumetric sweep efficiency may be assumed in waterflood performance calculations. The likely range of volumetric sweep efficiency is from 0.4 to 0.8.

Finally, the amount of oil produced during secondary recovery can be estimated as the following:

$$N_p = N_{i,wf} \times E_D \times E_V \tag{16.36}$$

where

N_p = volume of oil produced during waterflooding, bbl,

$N_{i,wf}$ = initial volume of in-situ oil at the start of waterflooding, bbl, and

$E_V = E_A \times E_I$, dimensionless.

Besides the graphical technique illustrated here, Equations 16.15 and 16.16 can be used to calculate the values f_w and df_w/dS_w as continuous functions of water saturation (S_w).

A limitation of the method is that only one set of oil-water relative permeability data is utilized in predicting waterflood performance, which may not be representative of the entire reservoir. It must be borne in mind that relative permeability characteristics can vary widely in a reservoir, depending on the nature of heterogeneities in rock. Numerical reservoir simulation attempts to model the reservoir by incorporating significant variations in rock properties, including the utilization of pseudorelative permeability curves.

The following example demonstrates the utilization of the Buckley-Leverett method in evaluating waterflood performance in a homogeneous formation of uniform rock properties.

Example 16.3. Analysis of waterflood performance based on the frontal advance theory. Based on the fractional flow curve presented in Figure 16–14, estimate the following at breakthrough of injected water:
- Frontal saturation of the water bank
- Average water saturation in the reservoir
- Water cut and water/oil ratio under reservoir conditions and at the surface, assuming B_o = 1.28 rb/stb and B_w = 1.02 rb/stb
- Pore volume of water injected (PVWI)
- Displacement efficiency
- Secondary recovery of in-situ oil, percent, assuming horizontal and vertical efficiencies of 0.75 and 0.85, respectively

Solution: As the first step, a tangent is drawn from the initial water saturation as shown, in order to obtain the following:

Water saturation at the flood front at breakthrough = 0.566.

Average water saturation behind the front = 0.63.

Fractional flow or water cut = 0.867.

Value of the derivative df_w/dS_w = (0.867-0)/(0.566-0.17) = 2.19.

Equation 16.23 is used to calculate the water cut at the surface as follows:

$$f_{w,s} = \frac{1.28}{1.02(1/0.867-1)+1.28}$$

$$= 0.891.$$

The water/oil ratios at reservoir conditions and at the surface are computed by Equations 16.24 and 16.25, respectively:

$\text{WOR} = 0.867/(1 - 0.867) = 6.52.$

$\text{WOR}_s = 6.52(1.28/1.02) = 8.18.$

The pore volume of water injected at breakthrough is determined from the derivative df_w/ds_w, evaluated at S_{wf}, as follows (fig. 16–14):

$\text{PVWI} = 1/2.19 = 0.457.$

The cumulative water injection volume can be obtained as follows:

Injected volume = Pore volume in bbl × PVWI, fraction × Volumetric efficiency, fraction

It can be noted that the average saturation behind the flood front is 0.63 at breakthrough, and connate water saturation is 0.17 (read from the plot). The displacement efficiency thus can be determined by Equation 16.35:

$$E_D = \frac{0.63 - 0.17}{1 - 0.17}$$

$$= 0.55.$$

Finally, the recovery efficiency at breakthrough is estimated as follows:

$$E_R = 0.55 \times 0.75 \times 0.85 = 0.35, \text{ or } 35\%.$$

where

E_R is the recovery efficiency.

The predicted performance is found to be good, judging from the current performance of waterflooded reservoirs in general. In reality, areal and vertical sweep could be lower than the estimates used in this example. Again, the water cut is calculated to be 0.867 at breakthrough based on the fractional flow curve utilized here. In many reservoirs, however, the water cut at breakthrough is generally much lower (less than 0.1) and then increases with time. This is due to the fact that only highly permeable channels break through first, which usually comprise a small fraction of the formation thickness.

Before breakthrough of the injected water, the oil production volume is simply equal to the volume of water injected based on material balance. Following breakthrough, displacement efficiency needs to be calculated at various saturation values, which are greater than S_{wf}, in order to estimate recovery efficiency based on Equation 16.36. However, the analysis requires the assumption that any change in volumetric efficiency is negligible throughout the waterflooding process.

Example 16.4. Prediction of the breakthrough of the injected water bank at the producer. Calculate the frontal advance of the injected water bank with time in the previous example. Figure 16–15 is an enlarged version of Figure 16–14, which highlights the procedure of drawing tangents through saturations behind the leading edge of the water bank. Assume the following:

    Injection rate, bbl/d, = 1,250.
    Porosity, fraction, = 0.28.
    Net thickness of waterflooded formation, ft, = 25.
    Distance between injector and producer, ft, = 933.
    Cross-sectional flow area, ft², = 42,000.

Solution: In analyzing the injected fluid front behavior, Equations 16.27 and 16.28 are used. Noting that $df_w/dS_w = 2.19$ at breakthrough, the corresponding time is estimated as follows:

$$t,\text{ breakthrough} = \frac{(0.28)(42,000)\ 933}{(5.615)(1,250)}\left(\frac{1}{2.19}\right)$$

$$= 714 \text{ days}$$

Table 16–4. Calculation of frontal advance

		Frontal Advance, ft			
S_w	df_w/dS_w	@ days = 90	180	365	714
0.566	2.19	118	236	477	933
0.58	1.85	99	199	403	788
0.61	1.16	62	125	253	494
0.63	0.77	41	83	168	328

Following 90 days of injection, the location of a specific saturation, $S_w = 0.58$, for example, can be obtained by noting that the corresponding value of df_w/dS_w is 1.85:

$$x,\ _{Sw=0.58} = \frac{(5.615)(1,250)(90)}{(0.28)\ (42,000)}\ (1.85)$$

$$= 99\ \text{ft}$$

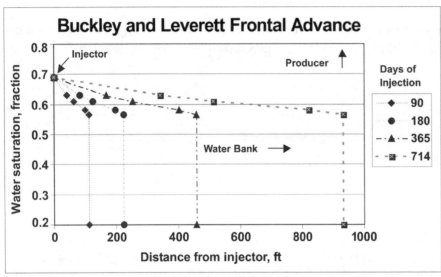

Fig. 16–16. Analysis of waterflooding based on Buckley and Leverett frontal advance theory. The saturation profile is determined by the fractional flow curve.

Based on Equation 16.27, locations of various saturations at 90, 180, 365, and 714 days are tabulated in Table 16–4. The results are plotted in Figure 16–16.

Estimates of empirical performance

- The ratio of recovery by waterflooding over primary oil recovery usually varies between 0.5 and 1, depending upon rock and fluid properties, geologic heterogeneity, the Dykstra-Parsons permeability variation factor, and well spacing. However, there are instances where oil recovered by waterflooding exceeds primary recovery.
- Before peak production rate, oil production is about 50% of the remaining primary and waterflood recovery.
- After peak production rate, oil production is likely to decline at a rate of 10% to 25% per year.
- The total production rate is about 80% of the water injection rate.
- Ultimate water production is likely to vary between 50% and 75% of the ultimate injection volume.

The API correlation is used to predict reservoir performance under water drive, as presented in chapter 10. Guthrie and Greenberger proposed the following correlation for estimating the recovery efficiency due to waterflooding based upon the study of 73 sandstone reservoirs.[37] Some of the fields produced under solution gas drive, in addition to water drive.

$$E_R = 0.2719 \log k + 0.25569\ S_w - 0.1355 \log \mu_o - 1.538\ \phi - 0.0003488\ h + 0.11403$$

In 75% of the cases, the deviation from the correlation was found to be within 9%.

It must be emphasized that the figures presented here are estimates only, based on the experience gained from waterflooding reservoirs over the years. Field-specific studies are required to design, implement, and evaluate waterflood projects, aided by reservoir simulation studies discussed in the following section. With the technological innovations in many areas, including horizontal wells, reservoir characterization, and waterflood surveillance, matured fields nearing abandonment are being rejuvenated. Consequently, recovery efficiency from waterflooding a reservoir is likely to increase.

Reservoir Simulation

In recent decades, reservoir simulators have been utilized extensively to aid in planning, execution, evaluation, and optimization of waterflood operations in the most comprehensive manner. Black oil simulators are generally used in predicting waterflood performance. In comparison to the previously described analytic methods, numerical models permit inclusion of a detailed reservoir description. They also allow the inclusion of laboratory-measured rock and fluid properties for achieving better accuracy in results. One of the advantages is the ease of studying the effects of alternate waterflood strategies. These include, but are not limited to, the following:

- Pattern type and size (5-spot, 9-spot, etc.)
- Infill drilling (including cost-benefit ratio)
- Effect of irregular patterns (depending on well locations and reservoir heterogeneities)
- Well scheduling (conversion of producers to injectors)
- Injection rate optimization
- Lifting capacities (incremental production)
- Zonal completions (selective waterflooding in target zones)

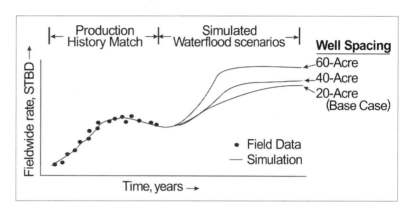

Fig. 16–17. Three scenarios generated by reservoir simulation: plot of waterflood performance prediction for wells at 20-, 40-, and 60-acre spacings based on the history matching of primary production data

Many years of project life can be simulated under different field management and operating strategies in a few seconds or minutes with the aid of high-speed computers. The results of the simulation study are integrated with reservoir economics to maximize the value of the asset. Economic analyses of various waterflood strategies as a part of reservoir development are included in chapters 18 and 19.

The role of reservoir simulation in waterflooding is illustrated in the following case studies.

Case I: Sensitivity to well spacing. Figure 16–17 presents a typical scenario in waterflood design. The reservoir simulation model is history matched with fieldwide primary production data, followed by evaluation of future waterflood (and asset) performance based on various development strategies:

1. 60-acre spacing based on very few infill wells (base case)

2. 40-acre spacing based on a moderate number of infill wells

3. 20-acre spacing based on a large number of infill wells

The incremental oil recovery, as predicted by the close spacing of the wells, is evaluated against capital expenditures (CAPEX) and other costs associated with an intensive drilling program. This is carried out in order to optimize the waterflood design.

Not shown in Figure 16–17 is the eventual decline in oil production rates from the producers. When close well spacing is implemented in waterflooding, cumulative recovery within a given time frame is greater, but a decline in oil rate from the producers is anticipated earlier.

Case II: Sensitivity to timing of waterflooding. The following study highlights the importance of commencing a waterflood operation at the most appropriate time during the life of a reservoir. In the case of an unsaturated reservoir, the time to start waterflooding is above the bubblepoint pressure due to favorable oil viscosity and compressibility, as mentioned earlier.

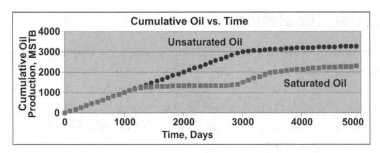

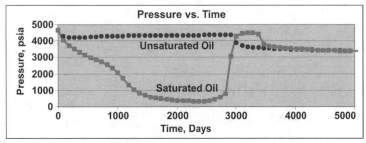

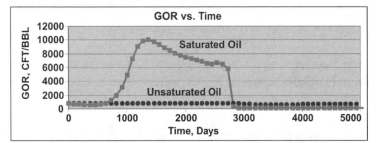

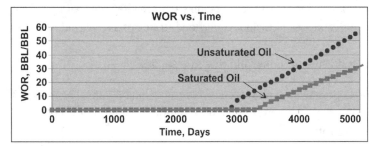

Fig. 16–18. Plots highlight waterflood performance above and below bubblepoint pressure in an unsaturated oil reservoir. Recovery is significantly higher when the waterflood operation is started earlier in the life of a reservoir, above the bubblepoint pressure. *Courtesy of Gemini Solutions, Inc.*

Figure 16–18 presents the comparison of reservoir performance under waterflooding in two cases in which the waterflooding starts above and below the bubblepoint. These cases are referred to as the unsaturated and saturated cases, respectively. The following observations are made from the various plots:

- Once the reservoir pressure is allowed to decline below the bubblepoint, the recovery of oil declines. The recovery efficiency cannot match the case where water injection is started earlier, above the bubblepoint pressure.
- Gas production and consequently the gas/oil ratio can be significant in waterflooding a saturated reservoir.
- The water cut from the producers shows up later in the case of a saturated reservoir, as the initial volume of injected water fills up the rock pores containing free gas. However, overall recovery performance is better in the case of an unsaturated reservoir.

Case III: Visualization of oil displacement mechanism. Figure 16–19 depicts the time-lapse study of an advancing water front in a dipping reservoir over decades of injection. Effects of gravity play a significant role in oil recovery. Water is injected through peripheral injectors located down dip. Oil is recovered through producers located up dip due to bottom-up sweep. Historical production data and information obtained from reservoir surveillance are utilized to simulate oil displacement.

Case IV: Simulation of a matured reservoir for optimum recovery. Todd and Arzola studied the improved recovery potential of a highly depleted giant field in Venezuela.[38] The field is compartmental, where the geologic barriers are the results of the depositional process and subsequent diagenesis. Barriers and baffles to both vertical and horizontal flow, due to the presence of shale laminations, facies changes, and faults, were detected. Reservoir simulation was preceded by extensive geosciences and reservoir studies, including reservoir characterization, leading to the identification of 16 flow units.

The reservoir produces well below the bubblepoint pressure. In relation to waterflooding, the detrimental aspects are cited as follows:

- Significant shrinkage of the oil following dissolution of the gas
- Insufficient volume of gas in the reservoir to swell the remaining oil following any water injection
- Development of secondary gas caps separating the flow units

The simulation results indicated that without initial fillup, the displacement efficiency of the injected fluids, such as water alternated by gas, would be rather poor. The scenario is further affected by the presence of significant reservoir heterogeneities. However, once the injected water fills up the pore space followed by gas injection, recovery estimates improve significantly to about 11% OOIP, in addition to primary recovery. In another scenario where a 2:1 water-alternating-gas (WAG) process was simulated, the estimate of enhanced recovery further improved to 16% OOIP. The most optimistic scenario included the drilling of infill wells, where the results of simulation indicated that as much as 23% OOIP can be recovered from the matured and heterogeneous field.

The case study demonstrates the power of reservoir simulation in optimizing improved oil recovery and enhanced oil recovery processes in a reservoir. The scenarios generated by simulation may not only involve waterflooding, but also other recovery processes, such as water alternating gas and infill drilling. The water-alternating-gas injection process is treated briefly in chapter 17.

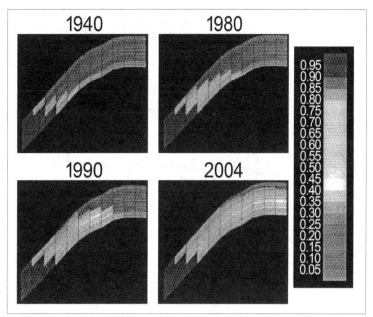

Fig. 16–19. Simulation of oil recovery by peripheral injection in a dipping reservoir over several decades. Light-colored grids indicate the gradual rise of injected water, resulting in bottom-up sweep of in-situ oil. *Source: A. H. al-Huthali, A. A. al-Awami, D. Krinis, Y. Soremi, and A.Y. al-Towailib. 2005. Water management in North Ain Dar, Saudi Arabia. SPE Paper # 93439. Presented at the 14th Annual SPE Middle East Oil and Gas Show and Conference, Bahrain, March 12–15. © Society of Petroleum Engineers. Reprinted with permission.*

Case V: Waterflood sensitivity studies. Based on the review of waterflood literature, some of the results of the waterflood sensitivity studies are summarized in the following, including studies made by Thakur and Satter.[39] The factors given in the following discussion were investigated.

Timing of waterflooding. Oil recovery is significantly greater if waterflooding is started at or above the bubblepoint. This is preferable to waiting for waterflood operations at some point below the bubblepoint or at the very end of primary depletion. Based on reservoir simulation, Kolbikov and coauthors studied the optimum reservoir pressure for waterflooding the Kogalym field in Siberia.[40] The reservoir is located at a depth of 7,874 ft (2,400 m), with average permeability and porosity of 50 mD and 0.19, respectively. The bubblepoint of the in-situ oil is 1,232 psi (8.5 MPa). Sensitivity analyses performed to determine optimum waterflooding pressure indicated the following scenarios:

1. Waterflooding the reservoir at 2,656 psia (19 MPa) leads to the best economic case. The optimum pressure is 75% of the original reservoir pressure and is about 1,424 psi higher than the bubblepoint pressure.

2. However, waterflood pressure of 2,176 psi (15 MPa) maximizes oil recovery due to better sweep.

This study brings attention to a few points to consider. First, waterflooding needs to be initiated in the early stages of reservoir development, well above the bubblepoint pressure. Second, the most desirable economic scenario may not coincide with maximum sweep by injected water in every circumstance. In this case, intensive waterflooding at the early stage resulted in a quicker payout period. The authors further noted that field development experience supported these conclusions, underscoring the importance of reservoir simulation and subsequent waterflood surveillance.

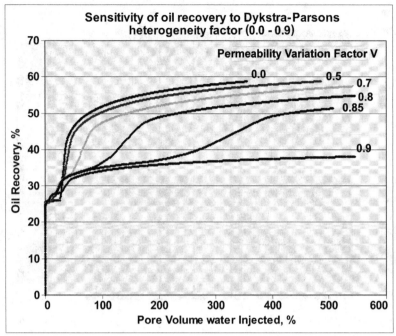

Fig. 16–20. Improved oil recovery performance as influenced by permeability variation factor. In severely heterogeneous cases (V > 0.8), recovery is significantly less. *Courtesy of Gemini Solutions, Inc.*

Variation in layer permeability. Figure 16–20 presents the sensitivity of oil recovery to the Dykstra-Parsons permeability variation factor during waterflooding of a 5-spot pattern based on numerical reservoir simulation. Merlin reservoir simulation software was utilized to predict oil recovery efficiency with the permeability variation factor ranging between 0 and 0.9.[41] Data used in the simulation is similar to the rock and fluid properties provided in Table 8–3 in chapter 8. In a highly heterogeneous formation, the recovery is markedly poor, as zones of oil are bypassed. The injected water preferentially travels the path of least resistance, which in this case channels through the high permeability zones.

Effects of interlayer communication and permeability ordering of layers. Ultimate primary and waterflood recoveries are likely to increase with higher vertical to horizontal permeability ratios due to good vertical sweep within a layer. Increased recovery as influenced by crossflow between the formation layers is significant when the layer permeability increases from top to bottom. Furthermore, the water/oil ratios are greater in reservoirs where crossflow between layers exists. Quantifying the effects of crossflow is one of the most difficult aspects in waterflooding. This difficulty arises because the intervening shale layers may exhibit lateral discontinuities, resulting in varying degrees of communication between the layers.

Oil gravity. Primary as well as waterflood oil recoveries are significantly lower for relatively viscous oil with lower API gravity due to an adverse mobility ratio. In the case of heavy oil with high viscosity, water breakthrough occurs relatively early. Consequently, the water/oil ratio in a producer reaches an economic limit sooner. Thermal methods are usually employed in producing heavy oil (< 22°API gravity). Thermal enhanced oil recovery processes are described in chapter 17.

Effect of critical gas saturation. Greater critical gas saturation yields higher oil recovery, as observed from simulation studies.

Factors affecting recovery

A literature review indicates the following major factors that can impede the success of waterflood operation:[42]
- The presence of high permeability channels leads to early breakthrough of injected water.
- The existence of directional permeability can lead to rapid propagation of the flood front in a preferred direction, while substantial areas remain unswept.
- Viscous fingering of injected water can occur in the case of relatively viscous oil.
- Slumping of injected water in a thick formation can occur due to high vertical permeability. Water underruns the in-situ oil.
- A fracture network in the formation may provide conduits for injected water.
- Inadequate oil saturation could allow the formation of an oil bank, with oil resaturation of the gas cap.
- Unknown geologic barriers could hinder contact between the injected water and in-situ oil.

Reservoir management practices to address some of the issues mentioned above are discussed in this chapter and in chapter 19.

Waterflood projects may face certain operational challenges. These include improper design of wells and surface facilities, equipment failures leading to downtime and expensive remedial work, and water disposal issues, among other factors. Economic aspects also play a vital role, as certain projects may not be attractive due to prevailing market conditions.

Waterflood Surveillance and Reservoir Management

Intensive monitoring of the waterflood operation, followed by implementation of mid-course corrections whenever required, is vital to a successful waterflood project. In an integrated surveillance program, all of the components of a waterflood project, including the performance of the reservoir and the individual wells, are monitored and evaluated on a regular basis. In fact, current trends in the industry strive for continuous data collection from injection, production, and observation wells based on permanent downhole gauges (PDG). This allows decision making in real-time or near real-time and optimization in rates in selective zones by deploying an intelligent well system (IWS) or "smartwell" technology. Robust computer models optimize the waterflood operation by taking into account related economic aspects, including oil prices. The ultimate aim of reservoir management at the well level is automated reservoir surveillance and subsequent analysis. This leads to dynamic control of injection and production without human intervention. Currently, time-lapse seismic studies are conducted at certain intervals in order to monitor the advance of an injected fluid front and changes in the water/oil contact. Another emerging technology is 4-D seismic survey, which is discussed in chapter 19.

A typical waterflood surveillance program followed by reservoir management includes, but is not limited to, the following:

1. Monitoring of the wells and reservoir by a comprehensive data-gathering program.
2. Evaluation and analysis of all available data. By this process, the waterflood surveillance data is turned into information.
3. Diagnosis of existing or potential issues.
4. Corrective actions at the well and reservoir level, some of which are given in the following reservoir management practices discussion.

Modeling and simulation

Reservoir models, both static (based on geosciences data) and dynamic (based on numerical simulation of fluid flow), are continuously updated. Data is obtained by waterflood surveillance, including 4-D seismic studies in certain instances, which enables reservoir engineers to predict reservoir performance fairly accurately. In fact, all three models, namely static, dynamic, and seismic, are iterated until convergence is achieved. When a match between actual and predicted waterflood performance cannot be obtained, it may point to the need for a better reservoir description, review of data quality, and a change in waterflood strategy.

Waterflood management strategy

The following are a few examples of reservoir management practices based on monitoring and surveillance:
- Various reservoir evaluation techniques, including 4-D seismic surveys, are capable of identifying locations of bypassed oil, which may necessitate manipulation of injection or production rates, well conversion, or infill drilling.
- Identification of thoroughly drained areas may require shutting off injectors or reducing the injection rates.

- Failure to achieve target pressures during waterflooding may indicate the loss of injected water into an unintended zone.
- When excessive water cut is detected in a producer, choking down the well, a water shut-off job, or horizontal sidetracking of the well may be necessary.
- No changes in reservoir pressure in an area may indicate reservoir compartmentalization and the need for infill drilling in isolated sections.
- Prior to breakthrough, the focus of waterflood surveillance could be on voidage control to ensure adequate injection of water in response to enhanced oil production in various wells. Following breakthrough, however, the focus usually shifts to the control of water cut in wells and better sweep.

Satter and Thakur provided a comprehensive list of waterflood surveillance tasks as part of integrated reservoir management.[43] Mechanical issues related to well performance during waterflooding, such as poor cement bonding, flow behind casing, and wellbore cleanup, are not discussed here. The following presents important aspects of waterflood monitoring and surveillance from a reservoir point of view.

Reservoir pressure during waterflooding

Fieldwide pressure surveillance is a requirement to balance injection and production rates for maximization of pattern coverage and recovery. Average reservoir pressure in the drainage region can be obtained by transient well tests at selected time intervals, ranging from several months to a few years. Static bottomhole pressures (SBHPs) are recorded at observation wells. As mentioned earlier, current reservoir management strategies include deployment of permanent recording devices that continuously monitor bottomhole well pressure. Estimates of average drainage region pressure, as obtained by transient well tests, also aid in the studies. Time-lapse study of fieldwide pressure, with the aid of reservoir simulation, is performed regularly as part of the waterflood surveillance.

Injection and production data

Injection and production data includes oil, water, gas, water cut, and gas/oil ratio data, followed by analysis of identifiable trends. Plots frequently utilized by reservoir engineers in evaluating performance of waterflood include, but are not limited to, the following:
- Cumulative water injection, and oil and water production versus time (individual patterns and total field).
- Water/oil ratio versus time (individual wells).
- Decline curve analysis and x-plot technique described in chapter 11. The use of x-plots is applicable at relatively high water saturations.
- Log of water/oil ratio versus recovery.
- Ratio of injection over withdrawal volumes (individual patterns and total field).
- Cumulative injection pressure versus cumulative water injection (Hall plot).
- Idealized radial water front surrounding injectors (bubble map).
- Water and oil saturation contours based on analytical procedure and numerical simulation.

Allocation of rates

Ideally, a waterflood operation is considered to be the most effective when all the wells recover the remaining amount of oil simultaneously and reach their economic limits. This suggests that producers located in drainage areas having large pore volumes should be produced at relatively high rates. Hence, wells are allocated injection and production rates according to the pore volumes, with an objective of minimizing project duration and operating costs. Well allocation factors (WAFs) play a crucial role in voidage replacement. Each injector is assigned a well allocation factor, which is related to the ratio of the injected water volume to the volume of oil produced at offset wells.

Pattern balancing

One of the principal objectives of waterflood surveillance is to attain a balance between injection and production of wells within a pattern and minimize oil migration to adjacent patterns. An effective pattern balancing leads to better areal sweep, and along with well realignment, can result in higher recovery. Reservoir simulation, based on reservoir pressures and well rates, is usually employed in pattern-balancing studies.

Figure 16–21 compares the effects of an unbalanced pattern in an inverted 5-spot pattern in a dipping reservoir against a base case. Darker shades represent higher water saturation radiating from the injector located in the center of the pattern. In the base case of a horizontal and homogeneous reservoir, the injection is balanced, as shown in (a). However, in a dipping reservoir shown in (b), the injected front advances unevenly due to gravity effects. A similar asymmetrical pattern frequently results from an injection/production imbalance or from heterogeneity of the rock. The net result is poor sweep, premature breakthrough, and high water cycling.

Based on reservoir surveillance and simulation, injection and production rates are fine-tuned in order to achieve pattern balance as much as possible.

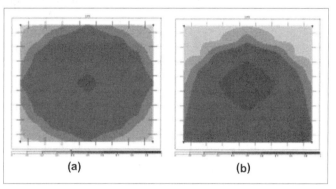

(a)   **(b)**

Fig. 16–21. Comparison between (a) balanced and (b) unbalanced patterns during waterflooding. Based on reservoir surveillance and simulation, injection and production rates are fine-tuned in order to achieve pattern balance as much as possible. *Courtesy of Gemini Solutions, Inc.*

Streamline simulation

The numerical simulation technique tracks the advance of injected water towards the producer or producers, leading to visualization of the fluid flow as a series of streamlines at various stages of waterflooding. The path of the streamlines between a pair of wells is influenced by injection and production rates, well locations, and reservoir heterogeneity. Streamline simulation allows optimized allocation between the injector/producer pair by manipulating rates. The ultimate objective is to achieve a perfectly balanced pattern that leads to maximum recovery with minimum water recycling. This approach offers the advantage of conceptualizing the advancing water front as a series of streamlines and tracking their paths to neighboring wells. In addition, streamline simulation usually requires less resources in time and effort in simulating waterflooding than conventional models. Theory and field applications of streamline simulation are widely available in the literature.

Analysis of pressure and injection data

Figure 16–22 shows a Hall plot of cumulative injection pressure versus cumulative water injection, which provides a wealth of information during waterflood surveillance.[44] This is referred to as Hall plot, following the name of the author. Any deviation from the regular trend in the plot serves as a diagnostic of reservoir conditions during waterflooding. The following features are identifiable:

- If free gas is present in the reservoir, a change in slope is evident. The line is concave upwards during the initial stages of injection as the rock pores containing gas are filled up with water, and the gas is dissolved.
- A marked increase in slope at later stages may point to severe skin damage around the well and compatibility issues

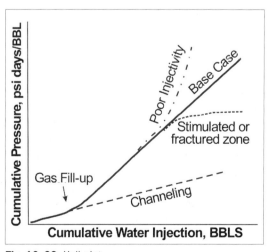

Fig. 16–22. Hall plot

related to the injected water. Similarly, a decrease in slope may indicate negative skin and water injection above fracturing pressure.

• A very low slope may indicate water channeling through a high transmissibility zone or injection into an unintended zone.

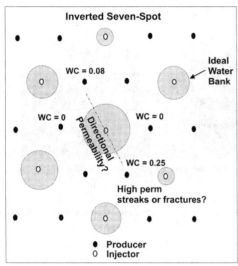

Fig. 16–23. Bubble map showing injected water volue and observed water-cut in producers

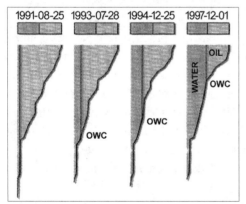

Fig. 16–24. Time-lapse study of the vertical profile of oil and water production based on flowmeter surveys conducted over several years. The upper zones have contributions historically. Due to the effects of gravity forces in the dipping reservoir, oil is produced by bottom-up sweep, indicated by the gradual rise of the oil/water contact and an increase in the water cut. *Source: A. H. al-Huthali, A. A. al-Awami, D. Krinis, Y. Soremi, and A.Y. al-Towailib. 2005. Water management in North Ain Dar, Saudi Arabia. SPE Paper # 93439. Presented at the 14th Annual SPE Middle East Oil and Gas Show and Conference, Bahrain, March 12–15. © Society of Petroleum Engineers. Reprinted with permission.*

Bubble map with water-cut information

Figure 16–23 is a map prepared from an ongoing waterflood project where idealized water banks are depicted as "bubbles." The radius of the water bank around a well is based on cumulative injection volume, which expands with the progress of waterflooding. The radius is not the same for each injector, as injection starts at different times at different locations, based on optimization studies. Moreover, well injectivity varies from one well to another. Assuming the uniform radial spread of the injected water around an injector, the following equation can be used to estimate the radius of an idealized water bubble:

$$Q_{inj} = \frac{\pi r^2 h \ \emptyset \ (1 - S_{or} - S_{wc}) \ E_I}{5.615 \ B_w} \qquad (16.37)$$

where

Q_{inj} = volume of water injected at surface conditions, bbl,
r = radius of the water bubble, ft, and
E_I = vertical sweep efficiency.

In reality, a perfectly radial shape of the bubbles is never expected around the injector due to the inherent reservoir heterogeneities.

Water cut, if detected in a well, is also included in the bubble map as a vital source of information. When premature breakthrough occurs in the producers, it can point to critical information about the reservoir, including the existence of directional permeability and high permeability streaks in the general area. Furthermore, bubble maps can point to the unswept regions in a reservoir.

Production and injection logging

These logging techniques include spinner, temperature, and tracer surveys to identify zones of fluid movement in disproportionate quantities. Ideally, a uniform injection and production profile across the formation thickness (vertical conformance) is desirable in order to ensure maximum recovery. The ultimate objective is to inject water in proportion to the hydrocarbon pore volume per zone. Application of polymers, gels, and cement squeezing to shut off high permeability zones and channels may aid in attaining conformance. In chapter 19, a case study is presented concerning conformance control in wells during waterflooding in Eunice Monument, New Mexico.

Smartwell technology allows selective zonal injection or production, while a problematic zone in the same formation is isolated. Low productivity zones can be selectively stimulated to improve waterflood performance. Figure 16–24 presents the results of a time-lapse study of flowmeter surveys indicating the gradual increase of water cut in a producer over a period of several years.

Transient well testing

Transient well tests are discussed in detail in chapter 5. There are many types of pressure tests employed in waterflood surveillance, including, but not limited to, the following:

- Pressure buildup (PBU) test
- Pressure falloff (PFO) test
- Step rate test
- Pulse and interference test
- Dynamic formation testing

Pressure falloff testing of injectors is carried out at regular intervals in order to assess skin damage, loss of injectivity, and the radial extent of the injected water bank. Transient tests are discussed in detail in chapter 5. Similarly, pressure falloff tests of the producers yield a wealth of valuable information, including skin damage, average drainage region pressure, and productivity index. Pulse and interference tests are utilized to determine connectivity between producers and injectors, which relates to waterflood sweep efficiency. Formation testing tools, such as the Modular Dynamics Tool (MDT), are utilized in new or recompleted wells to characterize the reservoir, including detection of high permeability streaks and vertical communication. This may provide valuable information in successfully engineering a waterflood project.

Tracer injection study

In heterogeneous reservoirs characterized by faults, fractures, and directional permeability, reservoir surveillance involves injection of a tracer at the injector, followed by tracking of tracer particles at nearby producers. The objectives are to determine connectivity between the wells, characterize the flood pattern, analyze the effects on volumetric sweep efficiency, and identify problematic injectors.[45] A positive response at a well completed in a different zone indicates connectivity between the two zones, which may lead to significant changes in the injection strategy. A small transit time between a set of injector and producer wells may indicate the existence of fractures or a permeability trend. The injected tracer may be detected in other wells much later, or not at all. In such cases, better sweep can be achieved by realigning the injection and production wells, or by manipulating well rates. Again, formation transmissibility is usually higher in the transverse direction to a fault, causing a change in flow direction, as identified by tracer surveys conducted in the vicinity of a fault (fig. 16–25). Tracer injection studies may aid in pattern balancing as the relative amount of injected water can be ascertained as illustrated in Figure 16–26.

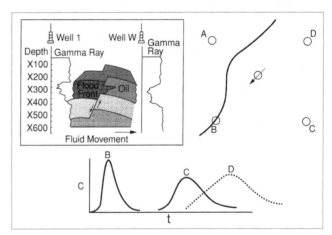

Fig. 16–25. Results of tracer survey in a pattern intersected by a fault. *Source: G. C. Thakur and A. Satter. 1998. Integrated Waterflood Asset Management. Tulsa: PennWell.*

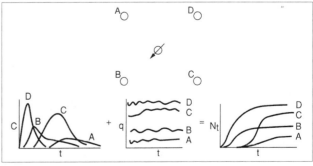

Fig. 16–26. Tracer survey to aid in pattern balancing. *Source: G. C. Thakur and A. Satter. 1998. Integrated Waterflood Asset Management. Tulsa: PennWell.*

Waterflooding guidelines summary

Based on the review of a large number of waterflood operations in West Texas, Gulick and McCain summarized the best management practices in waterflood design and operation, most of which were given earlier:[46]

1. Start waterflooding early in the reservoir life cycle.
2. Understand the reservoir's geology.
3. Plan infill drilling to reduce lateral pay discontinuities.
4. Develop the field with pattern waterflooding, with an injector to producer ratio of 1:1.
5. Have all the pay open in both producers and injectors in order to contact and displace the maximum amount of oil in place.
6. Keep all the producing wells pumped off.
7. Inject below the fracture gradient so as not to create water channels.
8. Inject clean water.
9. Operate waterflooding based on injection well tests.
10. Conduct a waterflood surveillance program.

Case Studies in Waterflooding

I. Decades of waterflooding in Means San Andres Unit[47,48]

The Means Field is an excellent example of successful oil recovery by primary, waterflood, and carbon dioxide processes, aided by infill drilling. The field, discovered in 1934, is located in the Midland Basin in West Texas. It has continued to meet ever-changing economic and technical challenges for more than 70 years.

The field is a north-south trending anticline separated into a north dome and a south dome (fig. 16–27).

Table 16–5 lists reservoir and fluid property data. The field produced from the Grayburg and San Andres carbonate formations at depths ranging from 4,200 ft to 4,800 ft.

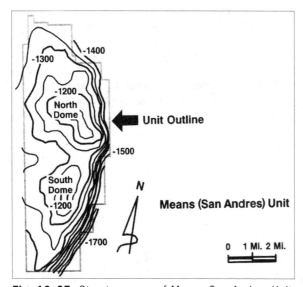

Fig. 16–27. Structure map of Means San Andres Unit. *Source: L. H. Stiles and J. B. Magruder. 1992. Reservoir management in the Means San Andres Unit. Journal of Petroleum Technology. April, 469–475. © Society of Petroleum Engineers. Reprinted with permission.*

Table 16–5. Reservoir and fluid properties of Means San Andres Unit. *Source: L. H. Stiles and J. B. Magruder. 1992. Reservoir management in the Means San Andres Unit. Journal of Petroleum Technology. April, 469–475.*

Formation	San Andres
Lithology	Dolomite
Reservoir area, acres	14,328
Reservoir depth, ft	4,400
Gross thickness, ft	300
Average net pay, ft	54
Average porosity, %PV	9 (upper limit: 25)
Average permeability, mD	20 (upper limit: 1,000)
Average connate water saturation, %	29
Primary drive	Weak water drive
Average original pressure, psig	1,850
Saturation pressure, psig	310
Stock-tank oil gravity, °API	29
Oil viscosity, cp	6.0
Oil formation volume factor, rb/stb	1.04

Figure 16–28 is a type log showing the various zonations.

After 30 years of primary production under weak natural water drive, a study was conducted to evaluate the application of a secondary recovery process. Highlights of this study included Humble's (Exxon) full-field computer simulation. A structural cross section (fig. 16–29) aided in the design of an initial waterflood pattern.

In 1963, the field was unitized, and a peripheral waterflood operation was initiated involving the north and south domes. It was later realized that the peripheral patterns did not provide adequate pressure support. In 1969, reservoir engineering and geological studies were conducted to determine an improved depletion plan to offset the pressure decline. Interior injection with a 3:1 line drive was recommended. The recommendation was implemented, resulting in a substantial oil production

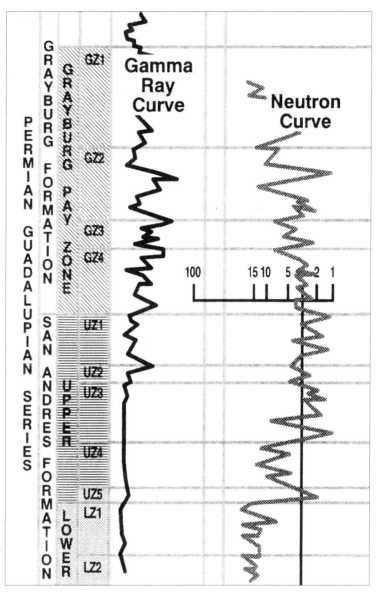

Fig. 16–28. Means San Andres Unit type log. *Source: L. H. Stiles and J. B. Magruder. 1992. Reservoir management in the Means San Andres Unit.* Journal of Petroleum Technology. *April, 469–475. © Society of Petroleum Engineers. Reprinted with permission.*

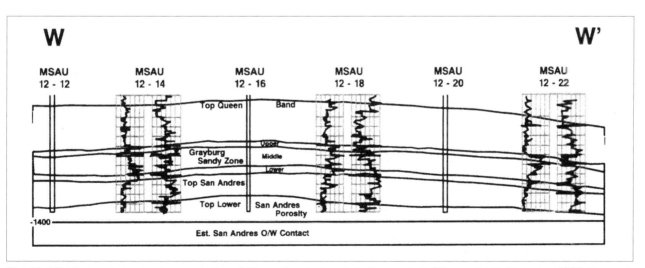

Fig. 16–29. West-east structural cross section of Means San Andres Unit. *Source: L. H. Stiles and J. B. Magruder. 1992. Reservoir management in the Means San Andres Unit.* Journal of Petroleum Technology. *April,. 469–475. © Society of Petroleum Engineers. Reprinted with permission.*

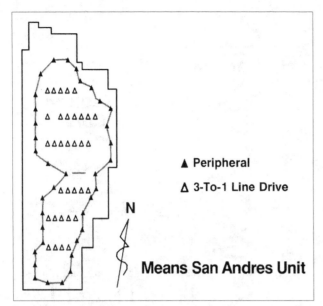

Fig. 16–30. Injection well patterns in Means San Andres Unit. *Source: L. H. Stiles and J. B. Magruder. 1992. Reservoir management in the Means San Andres Unit.* Journal of Petroleum Technology. *April, 469–475. © Society of Petroleum Engineers. Reprinted with permission.*

rate. Figure 16–30 shows the peripheral injectors and 3:1 line drive in the field.

After reaching a peak rate in 1972, production began to decline. An in-depth geological study in 1975 showed a lack of lateral and vertical distributions of pay. This study led to major infill drilling with pattern densification, improving recovery by increasing areal and vertical sweep efficiencies. (Infill drilling, i.e., drilling of additional wells after primary or secondary development of a field, enhances connectivity between the injectors and producers. It essentially attempts to eliminate or minimize the adverse effects of areal heterogeneity.) Maintenance of adequate reservoir pressure necessary for efficient waterflooding was also attained by closely spaced injection wells.

Figure 16–31 shows oil recovery with and without infill drilling in the Means San Andres Unit. The dotted line represents the production decline curve without the infill wells. The area above the decline curve represents additional oil recovery due to drilling of infill wells. More than 500 infill wells were drilled. Well spacing was reduced from 40-acre to 20-acre spacing during

waterflooding, and finally to 10-acre spacing during tertiary recovery by CO_2 injection. Figure 16–31 shows Means San Andres oil recovery in the years 1970–1990, accounting for 20-acre and 10-acre infill wells. Volumetric sweep efficiency increased from 59% to 85%.

Figure 16–32 shows the effects of infill drilling on the water/oil ratio. Closer well spacing improved oil production, while the upward trend in water production was checked.

A CO_2 tertiary recovery study was conducted during the years 1981 and 1982, and a pilot flood was initiated with extensive laboratory and simulation works. Figure 16–33 shows the performance of the tertiary project area, with oil and gas productions, water/oil ratio, and gas injection.

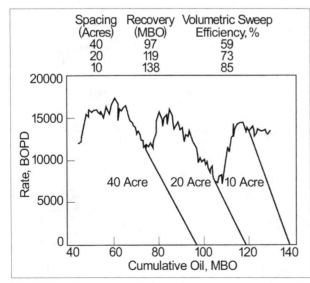

Fig. 16–31. Means San Andres oil recovery with infill wells. *Source: J. M. Goodwin. Infill drilling and pattern modification in the Means San Andres Unit.*

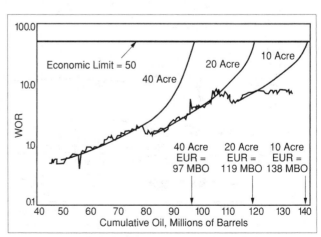

Fig. 16–32. Effect of infill drilling on water cut in Means San Andres. *Source: J. M. Goodwin. Infill drilling and pattern modification in the Means San Andres Unit.*

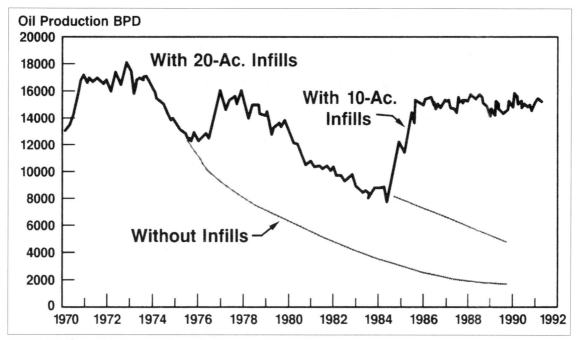

Fig. 16–33. Means San Andres enhanced oil recovery (CO_2) performance. *Source: L. H. Stiles and J. B. Magruder. 1992. Reservoir management in the Means San Andres Unit. Journal of Petroleum Technology. April, 469–475. © Society of Petroleum Engineers. Reprinted with permission.*

II. Integrated approach in waterflood management[49]

The following case study highlights an integrated approach to improve reservoir performance for a field located in the North American continent. This approach was formulated when the results of waterflooding a massive carbonate formation did not meet expectations in certain waterflooded areas. In a waterflood operation where performance issues are encountered, this study exemplifies the key ingredients in providing a necessary solution.

Poor recovery by waterflooding was attributed to the following:
- Low to very low formation permeability (usually less than 1 mD).
- Injection water channeling due to fracturing during primary production.
- Lack of perforations in certain pay intervals.
- Inadequate conformance control at the injectors and producers. (This topic is treated in detail in chapter 19.)

The gross thickness of the formation varies from 100 ft to 650 ft or more. Porosity values range between 5% and 20%. Core permeabilities were found to be as low as 0.05 mD, with an upper limit near 5 mD. Oil production from the field began in 1957. The primary production mechanisms were expansion of a small gas cap and solution gas drive. The field was initially developed with 1,320-ft triangular spacing, completed in 1965. Water injection was started in 1968 based on an inverted 7-spot pattern. About 100 wells were converted to injectors. Infill wells were also drilled in the 1980s in order to enhance the sweep efficiency.

Five original 7-spot patterns were studied in detail. All relevant information, including geologic cross-section and production/injection data, both from pattern and adjacent wells, were included in the study. The integrated reservoir study included data from several sources, as detailed in the following.

Log and core studies. Twelve zones were identified from logs, the upper zones having greater porosity and permeability. The units located at the upper sections of the formation indicated better reservoir quality and were continuous throughout the study area. The lower zones, however, disappeared progressively due to pinchouts. Air permeability profiles based on laboratory studies were generated. Net pay and floodable pay were mapped, based on porosity cutoff and microlog separation, the latter being indicative of adequate permeability for fluid to flow.

Injection profile survey. Flowmeter surveys conducted in the past were reviewed to analyze water entry into various zones. It was found that some zones were not taking water at all in certain injectors. The study recognized that certain remedial measures, including selective stimulation, were needed to address the issue.

Reservoir simulation. Various reservoir development scenarios were evaluated based on a 3-D, three-phase reservoir simulation model. This was matched with field data based on the following:

- Reservoir pressure and gas/oil ratio during primary production
- Water breakthrough time and water cut at individual wells during subsequent waterflooding

The results of the simulation indicated that an infill drilling program would be the most effective means of attaining incremental recovery. The conclusions are as expected in a tight reservoir.

Economic evaluation. An economic analysis of the scenarios generated above indicated that opening up the net unperforated pay zone would be the most desirable option from the standpoint of cash flow and net present value. Drilling of new wells became an attractive option when a modest increase in oil price was assumed. Conducted in the 1990s, the economic study assumed a much lower oil price than current levels.

Enhanced oil recovery screening. Enhanced oil recovery methods were found to be less likely to succeed according to the screening criteria established in the industry and presented in chapter 17. The oil is relatively heavy for gas injection, and the formation permeability is rather low for chemical flooding. Again, economics play an important role in planning enhanced oil recovery operations in a field.

III. Waterflood challenges in a highly heterogeneous reservoir[50]

Reservoir heterogeneities for a field located in North Kuwait include the following:

- **Existence of severe faulting.** The geosciences study pointed to the existence of 42 major faults in the reservoir, resulting in poor connectivity between different regions (compartmentalization). The compartments were found to have dissimilar water/oil contacts and varying aquifer support.
- **Stratification.** The geologic layering scheme indicated 18 subzones, subdivided into 64 layers.
- **Drive mechanism.** The primary drive mechanism was depletion. However, weak aquifer support was detected at the flanks, which was inadequate to sustain reservoir pressure at relatively high rates of production.
- **Production performance.** During primary production, wide variations in reservoir pressure were observed due to compartmentalization. A substantial decline in pressure, from initial pressure of 3,800 psia to 2,500 psia, occurred in the crestal part of the reservoir, where most of the production took place. Reservoir pressure in the crestal area was found to be near the bubblepoint pressure (2,450 psia), while the pressure was found to be several hundreds of pounds per square inch higher toward the flanks.
- **Fluid properties.** Variations in fluid properties were encountered both in areal and vertical directions due to poor connectivity. In-situ oil gravity was found to be 25°–28°API in the crestal region. However, it became considerably heavier toward the flanks (20°API).
- **PVT properties.** The variation in PVT properties, along with extreme reservoir heterogeneities, made the adoption of a waterflood strategy across the field complicated. Pattern waterflood was not planned. Placement of wells was determined by the location of oil sands as obtained from the geosciences model. Four pattern injectors, drilled in small compartments, rapidly pressured up the formation in locations where aquifer support existed, while the rest of the reservoir continued to experience a decline in pressure.

In the light of these uncertainties, the following strategy was adopted:

1. Development of a detailed reservoir description, focused on faulting and compartmentalization
2. Update of the full-field simulation model based on the static model, followed by history matching
3. Evaluation of various injection scenarios based on the full-field model, leading to an effective plan to repressurize each compartment

IV. Prediction of recovery based on waterflood surveillance[51]

This case study highlights the utilization of empirical performance techniques to predict ultimate recovery from the Kuparuk River field in the North Slope of Alaska. Data from early stages of waterflooding was used. Decline curve analysis, x-plot technique, and plots of water/oil ratio versus cumulative production were utilized to study waterflood performance in two different sands where flooding rates were significantly different. Inadequate pattern performance and opportunities to recover more oil by shutting in the high water-cut wells were also identified in the study.

The field consists of two physically separated sands of contrasting transmissibility (1,000 mD-ft and 5,000 mD-ft). The two sands are connected at the wellbore by a single string, resulting in commingled production. The reservoir is characterized by faults aligned in a north-south direction, with a density of about three faults per square mile. It was anticipated that the faults would affect the formation continuity. Most of the waterflood patterns were east-west line drive. Wells were drilled at a spacing of 160 acres. Production contributions from individual sands were estimated by tests conducted in wells with single-zone completions, and by production logging. The material balance method and reservoir simulation were employed to check the consistency in estimating the individual zonal contributions per well.

Based on the Buckley and Leverett frontal advance theory, it can be shown that a semilog plot of water/oil ratio versus cumulative production yields a straight-line relationship, given as follows:[52]

$$\log(\text{WOR}) = \left(\frac{b}{PV}\right)Q_o + n \qquad (16.38)$$

where

Q_o = cumulative oil production, stb,

$$n = \log_{10}\left(\frac{a\mu_o}{\mu_w}\right) + bS_{wc} - \left(\frac{1}{\ln 10}\right), \text{ and} \qquad (16.39)$$

b, a = slope and constant in a semilog plot of k_{ro}/k_{rw} versus S_w, respectively, as in Equation 16.14.

It is to be noted that the slope in Equation 16.39 is inversely proportional to the pore volume contacted by the injected water.

Figure 16–34 shows the water/oil ratio versus cumulative oil production of two wells. Assuming the same water/oil ratio of 30 at abandonment, the well with the higher slope indicates poor recovery. This led to the investigation of injection imbalance and the presence of a thief zone or discontinuity in the pattern.

Figure 16–35 presents a plot showing a significant drop in water/oil ratio in a producer when an adjacent producer was shut in. The production from the well also increased significantly. This was attributed to a change in flow pattern that led to areal sweep in previously unswept areas. This strategy can be used to enhance oil recovery in similar situations.

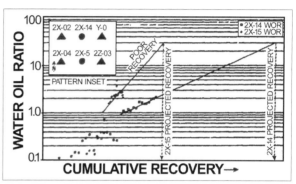

Fig. 16–34. Plot of water/oil ratio versus cumulative recovery, showing comparison of well performance. *Source: J. H. Currier and S. T. Sindelar. 1990. Performance analysis in an immature waterflood: the Kuparuk River field. SPE Paper #20775. Presented at the 65th Annual Technical Conference and Exhibition, New Orleans, LA, September 23-25. © Society of Petroleum Engineers. Reprinted with permission.*

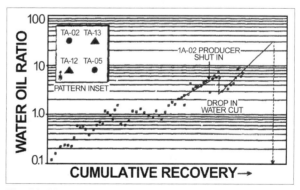

Fig. 16–35. Plot of water/oil ratio versus cumulative recovery, showing the effect of shutting down an adjacent producer. *Source: J. H. Currier and S. T. Sindelar. 1990. Performance analysis in an immature waterflood: the Kuparuk River field. SPE Paper #20775. Presented at the 65th Annual Technical Conference and Exhibition, New Orleans, LA, September 23-25. © Society of Petroleum Engineers. Reprinted with permission.*

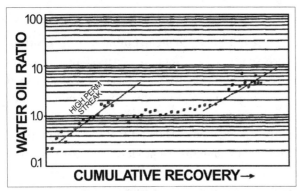

Fig. 16–36. Effect of breakthrough of a high permeability channel on water/oil ratio. *Source: J. H. Currier and S. T. Sindelar. 1990. Performance analysis in an immature waterflood: the Kuparuk River field. SPE Paper #20775. Presented at the 65th Annual Technical Conference and Exhibition, New Orleans, LA, September 23–25. © Society of Petroleum Engineers. Reprinted with permission.*

Figure 16–36 demonstrates the dual layer behavior of the upper sand, based on water/oil ratio data from several wells. An early increase in water/oil ratio is encountered, followed by a prolonged period of plateau production. Finally, a second increasing trend in water/oil ratio is observed. This led to modeling of the reservoir in two subzones, the first one consisting of a thin, high permeability channel responsible for early breakthrough. The rest of the zone is of lower permeability, leading to water breakthrough much later.

Summing Up

Overview

Most reservoirs are subjected to waterflooding and other improved recovery processes once natural reservoir energies can no longer sustain production. The objective is to increase oil recovery from the reservoirs and add value to the asset (petroleum reserves). Based on current technology and economics, about 10% to 30% of the original oil in place is expected to be recovered by waterflooding in most cases.

History of waterflooding

Waterflooding, which was discovered accidentally in 1865, found widespread application in the early 1950s.

Waterflood process

Waterflooding consists of injecting water into a set of wells while producing from the surrounding wells. It maintains reservoir pressure and displaces oil from the injector to the producer.

Water is an inexpensive and efficient agent for displacing light or medium gravity oil. Its success lies in low capital investments and operating costs, with favorable economics.

Traditionally, waterflooding has been initiated in depleted or nearly depleted reservoirs with a free gas phase present. Injected water fills up the pores previously occupied by gas, which is redissolved in solution, and the reservoir pressure is restored. Modern waterflooding practices, however, often involve start of injection above the bubblepoint of oil in the early stages of reservoir life cycle in order to optimize recovery.

Typical waterflood response is characterized by an increase in oil rate, followed by a decline, and an eventual breakthrough of injected water at the producers. The water/oil ratio continues to rise with time, and the economic limit is reached when the water production becomes excessive. Various tertiary methods are available to further recover oil once waterflooding has run its course. Injection rate, well spacing, fluid properties, and reservoir heterogeneities influence the response to waterflooding.

During waterflooding, oil and water saturations change with time and location in the reservoir, as the injected water bank "pushes" or displaces the oil bank toward the producer. Changes in fluid saturations are controlled by relative permeability characteristics and other fluid and rock properties.

Waterflood design

The design of a waterflood project involves an integrated team approach, with a detailed reservoir description and monitoring and interpretation of past reservoir performance. It also involves drilling and completion data from existing wells, laboratory analysis, development and validation of reservoir models, evaluation of what-if scenarios, pilot projects, and economic analysis. Waterflood design and implementation are critically dependent on a methodical reservoir surveillance program that must be put in place at the initiation of the waterflood project.

Waterflood objectives

Waterflooding is primarily designed to accomplish the following:
- Maximize oil recovery
- Maximize contact with oil, specifically where zones of high residual oil saturation exist
- Minimize injection water cycling and handling
- Optimize water injection
- Ensure efficient scheduling of well conversion and infill drilling
- Minimize capital expenditure related to drilling of new injectors or producers
- Maximize the net present value of the asset based on maximum returns within a relatively short time horizon

There are several critical factors in waterflood design. These include uncertainties associated with unknown reservoir heterogeneities, expected oil recovery, and added capital expenditures due to drilling of infill wells and construction of surface facilities.

Waterflood design considerations

Design considerations include, but are not limited to, the following:
- Detailed knowledge of reservoir heterogeneities, including faults, fractures, and high permeability streaks that dictate the location and alignment of wells
- Identification of zones having high residual oil saturation or untapped reservoir areas during primary recovery
- Optimization of injection and production rates
- Optimization of reservoir pressure

Tools and techniques in design

Besides laboratory studies and pilot flood projects, reservoir simulation is used extensively in designing a waterflood project. Based on scenario building, reservoir simulation can point to the best strategy related to the following:
- Ultimate oil recovery
- Scheduling of wells for conversion to injectors
- Infill wells
- Optimum well spacing
- Injection rate
- Time to breakthrough
- Life of the waterflood project

Screening criteria

The following provides a general guideline for screening reservoirs before waterflooding is implemented:

- Residual oil saturation should be high, in excess of 40%. Relatively low water saturations do not lead to formation of an oil bank, as observed in fields.
- Reservoirs with oil gravities greater than 25°API and oil viscosities less than 30 cp are good waterflooding prospects.
- Waterflooding is more likely to succeed in relatively homogenous sandstones as well as carbonate formations. Natural fractures, faults, directional permeability, and compartments create some of the challenges to efficient waterflood design and management.
- Reservoirs located at a shallow depth or tight reservoirs may have limitations of injectivity.
- Reservoirs with strong water influx from an adjacent aquifer may not be good candidates for waterflooding.
- In reservoirs where a gas cap exists, displaced oil may enter pores that are previously occupied by gas, leading to lower-than-expected recovery.

Water injection rates

The rates of oil recovery, and therefore the life of a waterflood project, depend upon the water injection rate into a reservoir. A drastic decline in water injectivity occurs during the early period of injection into a reservoir depleted by solution gas drive. After fill up, the injectivity variation depends upon the mobility ratio. It remains constant in the case of unit mobility ratio (M = 1), increases if M > 1 (unfavorable), and decreases if M < 1.

The variables affecting the injection rates are as follows:

- Rock and fluid properties. Poor well injectivity is associated with a tight reservoir, skin damage, and relative viscosity of fluids.
- Mobility of fluids.
- Areas related to swept and unswept regions.
- Oil geometry. Well pattern, spacing, and wellbore radii.

Waterflood patterns and well spacing

Commonly observed injection-production well arrangements are: line drives (direct and staggered), and 5-spot, 7-spot, and 9-spot patterns. A regular 5-spot pattern consists of a producer located in the center of a square, with the four injectors located at the corners. In regular patterns, producers are located in the middle, while in inverted patterns, injectors are drilled in the middle of the pattern, and the producers are at the corners.

Injection wells can be positioned around the periphery of a reservoir, which is referred to as peripheral injection. On the other hand, crestal injection involves positioning of the wells along the crest of small reservoirs with sharp structural features. In a dipping reservoir, water injection wells are located down dip to take advantage of gravity segregation.

Well spacing is dictated by rock and fluid characteristics, reservoir heterogeneities, optimum injection pressure, time frame for recovery, and economics, among other factors.

Mobility ratio

Injectivity and displacement efficiency are dependent on the mobility ratio. It is defined as the mobility (effective permeability/viscosity) of the displacing fluid (i.e., water) in the water-contacted portion of the reservoir, divided by the mobility of the displaced fluid, i.e., oil in the oil bank. In essence, it is determined by the oil-water relative permeability and the viscosity ratios.

Recovery efficiency

The overall waterflood recovery efficiency is given by the product of pore-to-pore displacement efficiency and volumetric sweep efficiency.

Displacement efficiency, which is influenced by rock and fluid properties, and throughput (pore volumes injected) can be determined by laboratory core floods, the frontal advance theory, and empirical correlations. It depends on fluid viscosity, reservoir dip, wettability of the rock, and interfacial tension.

Volumetric sweep efficiency is given by the product of areal and vertical sweep efficiencies. The areal sweep efficiency is influenced by the flooding pattern type, mobility ratio, and throughput and reservoir heterogeneity. Vertical (or invasion) sweep efficiency is influenced by layer permeability variations and the mobility ratio.

Laboratory studies

Results of laboratory investigations that shed light on the mechanisms of waterflooding and factors that influence recovery are summarized in the following:
- Injected water tends to underrun oil when gravity forces predominate, leading to poor recovery.
- At sufficiently high rates, viscous forces predominate, resulting in better sweep efficiency and oil recovery.
- With an unfavorable mobility ratio (M > 1), volumetric sweep is poor.
- Water may imbibe into tight rock, augmenting recovery due to capillary forces. This phenomenon may be observed in water-wet rock at low water injection rates.
- At high viscosity ratios between oil and water, the injected water is found to move in thin streams, leading to poor sweep and displacement. This phenomenon is referred to as viscous fingering.
- In stratified reservoirs, adjacent layers can be partially or fully communicating, leading to crossflow. Sweep efficiency is improved due to crossflow when the mobility ratio is favorable. In contrast, crossflow leads to poor waterflood performance in the case of an unfavorable mobility ratio.
- In strongly water-wet rock, higher trapped gas saturation leads to lower residual oil saturation. In oil-wet rock, no straightforward relationship is observed.

Waterflood performance prediction methods

Classical methods for predicting waterflood performance include Dykstra-Parsons; Stiles; Prats-Matthews-Jewett-Baker; Buckley-Leverett, and Craig-Geffen-Morse. However, these methods have many restrictive assumptions noted later in the section.

Reservoir properties vary areally as well as vertically due to changes in the depositional environment and subsequent events in geologic times. The Dykstra-Parsons permeability variation factor is a widely used method to characterize vertical permeability stratification. It is based upon lognormal permeability distribution (permeability versus the percent of the total sample having higher permeability).

Stiles modeled waterflood performance as a function of individual layer transmissibility. The most permeable layer will experience the earliest breakthrough, followed by the layer having the second highest permeability, and so forth. No crossflow is assumed between the layers. Consequently, the vertical sweep efficiency is a function of individual layer permeability and thickness.

Buckley and Leverett proposed the frontal advance theory of waterflooding. This enables the determination of displacement efficiency, oil recovery, water cut, and the volume of water injected at and after water breakthrough in homogeneous formations. The procedure, based on fractional flow theory, utilizes oil-water relative permeability, fluid viscosity, and formation volume factor.

Most of the prediction methods are based on either laboratory investigations or simplified analytic solutions of waterflooding in an ideal or near-ideal situation. These assumptions include one or more of the following:

- Porous medium of uniform rock properties
- Pistonlike displacement of oil by water
- Linear flow geometry in which areal heterogeneities are not considered
- Insulated layers in a stratified reservoir model
- Incompressible flow

However, the methods are highly valuable in conceptualizing the mechanism of oil displacement by waterflooding as influenced by various factors. The most comprehensive waterflood performance prediction tool is the reservoir simulator. Today, black oil simulators are used extensively for predicting performance for pattern as well as full-field waterflooding.

Reservoir surveillance

A reservoir surveillance program incorporates monitoring of reservoir and individual well performance on a regular basis. Intensive monitoring of the waterflood operation, followed by implementation of mid-course corrections whenever required, is the key to a successful waterflood project. Permanent downhole gauges are employed to collect rate, pressure, and fluid composition data. Fieldwide data is evaluated and analyzed to diagnose existing or potential issues. Finally, corrective actions are taken at the well and reservoir level based on best reservoir management practices. Reservoir models, both static and dynamic, are continuously updated based on the information obtained by waterflood surveillance, which enables reservoir engineers to adopt a better management strategy.

The essential elements of reservoir surveillance include the following:

- Measurement of reservoir pressure throughout the waterflooded area
- Continuous monitoring of well rates, bottomhole pressure, and water cut
- Vertical profiling of injection and production at wells, followed by conformance control
- Water injection rate control and waterflood pattern balancing to recover maximum oil
- Design and implementation of various pressure transient tests, tracer injection, and time-lapse seismic surveys in certain cases
- Analysis of waterflood performance based on various plotting techniques, such as Hall plots, x-plots, and bubble maps
- Update of reservoir models, static and dynamic, based on waterflood surveillance
- Field development plan, including further well conversion and future infill wells

Case studies

This chapter presents four field case studies highlighting various aspects of waterflood management. These include, but are not limited to, the following:

- A case study of the Means field in the Midland Basin of West Texas offers an excellent example of successful oil recovery by primary, waterflood, and carbon dioxide processes. A classical approach to a waterflood operation is presented. Waterflooding was initiated with a peripheral flood, followed by successful interior pattern floods with 20-acre and 10-acre infill drillings. Decades of water injection in a reservoir in the Midland basin of West Texas were followed by infill drilling and an enhanced oil recovery process to maximize recovery. The field has continued to meet ever-changing economic and technical challenges for more than 70 years.
- Integrated methodology in managing a waterflood operation in a low permeability reservoir, including petrophysical studies, an injection profile survey, reservoir simulation, and economic evaluation
- Waterflooding a highly heterogeneous reservoir in Kuwait, having a large number of faults, compartments, and formation stratifications
- Implementation of an efficient waterflood surveillance scheme and analysis of waterflood data to augment ultimate recovery in a faulted reservoir with high permeability streaks located in Alaska

Class Assignments

Questions

1. What is waterflooding? Why is waterflooding implemented in most oil reservoirs?

2. Describe the basic principle of waterflooding. What force or forces (viscous, gravity, and capillary) may dominate in a successful waterflood operation?

3. Name at least five reasons why waterflooding is the most desirable option once the natural driving energies are depleted in a reservoir.

4. Consider the start of water injection into a reservoir with a large gas cap. Would the producing gas/oil ratio be expected to increase, decrease, or stay the same? Explain.

5. Distinguish between waterflooding a volatile oil reservoir above and below the bubblepoint. What timing of waterflood operations holds the potential for better recovery? Explain.

6. As water injection is started, describe the response at the producers at various stages of waterflooding. Under what circumstances is the operation usually terminated?

7. What are the objectives to attain when designing a waterflood operation in terms of production, injection, and efficiency? List the most important factors in designing a waterflood operation.

8. What rock and fluid properties primarily control the success of waterflooding? Explain.

9. Consider three reservoirs at various depths and locations having the following rock and fluid properties. With limited resources, which of the three reservoirs should be waterflooded first, all other factors being the same? Explain why.

    (a) Depth: 3,500 ft. Permeability: 20–40 mD. Oil gravity: 23°API. Onshore location.

    (b) Depth: 7,500 ft. Permeability: 14–18 mD. Oil gravity: 28°API. Onshore location.

    (c) Depth: 11,050 ft. Permeability: 10–500 mD. Oil gravity: 31°API. Offshore location.

    List all other information that should be gathered before making a final decision.

10. How does the mechanism of waterflooding differ between oil-wet and water-wet rocks? In what case is waterflooding an imbibition process? How does high vertical permeability affect waterflood performance?

11. Under what circumstances can gravity forces play a major role in waterflooding? Explain with an example.

12. How can various reservoir heterogeneities affect recovery efficiency in a waterflood operation? Give three examples.

13. What are the factors that influence the injector/producer pattern and spacing? During the life of a typical improved oil recovery or enhanced oil recovery operation, why is the well spacing reduced?

14. What is infill drilling? Describe its role in waterflood and other recovery processes, with examples. Name at least two types of reservoir characteristics that indicate that infill drilling would significantly increase recovery.

15. Describe the role of reservoir simulation in designing a waterflood project. Name at least five parameters related to waterflooding that can be optimized by reservoir simulation. Why is a history match of primary production performance necessary in the study?

16. In a waterflood operation, reservoir pressure observed in the field was found to be significantly higher than computer model predictions near the southern flank. Discuss possible reasons for the anomaly.

17. Why is optimization of the injection rate necessary? Describe the consequences of too high or too low an injection rate in a waterflood project. Describe in detail how the upper limit of the injection rate in a formation is determined. Does reservoir depth influence the maximum rate?

18. What is a pilot flood? Based on a literature review, prepare a case history of a pilot flood project. Include how the experience was utilized in implementing a field-scale flood project.

19. Name at least six important screening criteria in designing a waterflood project. Why is the accurate determination of residual oil saturation critical prior to waterflooding?

20. Based on reservoir and fluid properties of Ewinti-5 and Ala-3 in Table 15–7 in chapter 15, are the reservoirs good candidates for waterflooding? Why or why not?

21. Describe in brief the effectiveness of waterflooding in the following reservoirs:

    (a) Depletion drive reservoir operating above bubblepoint pressure

    (b) Reservoir with a large gas cap

    (c) Combination drive reservoir (depletion and water influx)

    (d) Highly depleted saturated reservoir with significant oil shrinkage

    (e) Stratified reservoir with moderate permeability contrast and no crossflow between the layers

    (f) Stratified reservoir with significant permeability contrast and partial communication between the layers

    (g) Reservoir with thin, high permeability channels

    (h) Shallow viscous oil reservoir

    (i) Very thick formation having high vertical permeability

    (j) Reservoir with a strong bottom water drive

    (k) Reservoir with very low permeability

    (l) Highly faulted reservoir (sealing faults)

    (m) Fractured formation (production from the fracture network alone)

22. Consider the following stratified reservoirs as waterflood candidates. Would reservoir A be expected to perform better? Why or why not? The thicknesses of the layers are the same in both cases. Assume partial communication between the layers.

    Reservoir A: Permeability of upper layer: 1 darcy. Permeability of lower layer: 50 millidarcies.

    Reservoir B: Permeability of upper layer: 50 millidarcies. Permeability of lower layer: 1 darcy.

23. True or false?

    (a) In the two-layer system described above, the Dykstra-Parsons permeability variation factor (V) is greater in the first case.

    (b) The Stiles model of stratified flow will not be appropriate to predict waterflood performance in the above case.

24. How does a seismic study and review of old log data aid in designing a waterflood project? Explain.

25. True or false?

   (a) A 5-spot pattern consists of five producers and five injectors.

   (b) A 7-spot pattern consists of more producers than a 5-spot pattern.

   (c) In an inverted 9-spot pattern, the producer to injector ratio is 1/9.

   (d) Peripheral wells are more likely to be converted to injectors.

   (e) Infill wells are always drilled as producers.

   (f) In efficient waterflood design, problem wells are always converted to injectors.

   (g) The injector/producer ratio is less in a 9-spot pattern than in an inverted 7-spot pattern.

   (h) Certain injectors or producers may be shut down during the course of waterflooding to improve recovery.

   (i) Well spacing is primarily influenced by formation transmissibility and economics.

   (j) During waterflooding, realignment of wells leads to changes in flow direction and reductions in water cut.

   (k) More infill wells are needed in waterflooding a compartmental reservoir.

   (l) Fewer infill wells are needed when a number of high permeability channels exist.

   (m) An inverted 7-spot pattern cannot be converted to a regular 5-spot pattern regardless of changes in well spacing.

   (n) For the same spacing, the recovery from a direct-line drive and a staggered-line drive are expected to be the same in a reservoir.

   (o) In a perfectly homogenous formation of high transmissibility, infill wells are not necessary to augment recovery. Wells with larger spacing will ultimately recover the same amount of oil.

   (p) In an inclined reservoir, water is injected through downdip wells and oil is produced up dip, as good reservoir quality is generally found in the updip portion.

   (q) As more oil is recovered from an inclined reservoir, injectors are progressively moved up dip by converting producers in certain cases.

   (r) Optimized waterflooding is defined as the maximization of oil recovery in the shortest possible time frame.

26. Define mobility ratio and describe its significance in waterflooding. Which of the following displacements is characterized by the most favorable mobility ratio?

   (a) Water displacing volatile oil

   (b) Water displacing heavy oil

   (c) Oil displacing gas

27. In determining the mobility ratio, why are the relative permeabilities of the fluids obtained from two different points in the reservoir? Explain.

28. Define well injectivity in oilfield units. How does it affect waterflood performance? Can well injectivity change during waterflooding? What well test can be conducted to measure injectivity? True or false: Well injectivity is expected to decrease after gas volume fill-up when the mobility ratio is less than unity.

29. Define displacement, areal, vertical, and volumetric efficiency of waterflooding. List what efficiency can be impacted, and how, under the following circumstances:

    (a) Closer well spacing based on infill drilling

    (b) Variations in oil viscosity from the crestal part of the reservoir to the flanks

    (c) Facies change

    (d) High permeability streak

    (e) Crossflow from upper to lower layer

    (f) Reservoir of mixed wettability

    (g) Naturally occurring fractures

    (h) Presence of free gas in the reservoir

    (i) Water/oil ratio near the economic limit

30. Is the displacement efficiency calculated under reservoir conditions the same as that estimated under surface conditions? Explain.

31. Discuss the effects of capillary, viscous, and gravity forces on the displacement of oil by water. How do these affect volumetric sweep? Describe the phenomena of viscous fingering and crossflow between layers.

32. Compare the methods of predicting waterflood performance described in this chapter in light of the inherent assumptions. What are the advantages and limitations of these methods in relation to reservoir simulation?

33. List the reasons for an unsuccessful waterflood operation. Why is it important to correctly estimate residual oil saturation at the end of primary recovery? Would residual oil saturation be expected to be uniform? Explain. Describe briefly the reservoir management strategies that should be adopted as a reservoir engineer addresses the relevant issues in waterflooding.

34. Why is a carefully planned waterflood surveillance program necessary? Describe the most important tasks in waterflood surveillance. Give three examples of how waterflood surveillance can aid in enhancing recovery from waterflooding. How can areal sweep be determined from waterflood surveillance?

35. Briefly describe the following and their significance in waterflood surveillance:

    (a) Hall plot

    (b) Pattern balancing

    (c) Injection profile

    (d) An x-plot based on water cut

    (e) Vertical and areal conformance

    (f) Well allocation factor

    (g) Voidage ratio

Exercises

16.1. In a 40-acre 5-spot pattern, estimate the minimum injection rate that would be necessary to maintain an average pressure differential of 1,500 psia. The following data is available:

Rock permeability, mD, = 25,
Relative permeability to oil = 0.6.
Formation thickness, ft, = 28.
Oil viscosity, cp, = 0.68.
Radius of wellbore, ft, = 0.27.

Redo the estimate of well injection rate for each of the following six cases while other factors remain the same. Compare the results with the base case and explain the difference in predicted injection rate, if any:

(a) 20-acre well spacing, 5-spot pattern

(b) 40-acre well spacing, 7-spot pattern

(c) Formation thickness, ft, = 45.

(d) Oil viscosity, cp, = 1.1.

(e) Radius of wellbore, ft, = 0.32.

(f) Mobility ratio is about 50% of the base case.

Calculate the minimum depth of the reservoir in the base case where the calculated injection rate should not cause any accidental fracturing of the formation. State any assumptions made in the course of the study.

16.2. Figure 16–37 presents the relative permeability curves of oil and water, assumed to be representative of a porous medium where waterflooding is planned. Prepare the following plots:

(a) Fractional flow curve

(b) Log (k_{ro}/k_{rw}) versus S_w

(c) df_w/dS_w versus S_w based on Equation 16.16

Assume $\mu_o = 3.8$ cp and $\mu_w = 0.55$ cp.

16.3. Plot waterflood recovery as a function of water/oil ratio for the following:

(a) $M = 0.5, V = 0.7$

(b) $M = 1.0, V = 0.7$

(c) $M = 5, V = 0.7$

(d) $M = 10, V = 0.1$

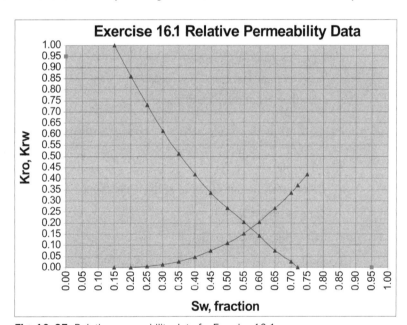

Fig. 16–37. Relative permeability data for Exercise 16.1

Draw conclusions from the above. Include all necessary assumptions made in the study and discuss their limitations in an actual field.

16.4. Determine the following based on relative permeability and fractional flow data in Exercise 16.2:

    (a) Connate water saturation

    (b) Residual oil saturation

    (c) Displaceable hydrocarbon pore volume (DHPV)

    (d) Water saturation at the front of the water bank

    (e) Time to breakthrough

    (f) Water cut at breakthrough

    (g) Average water saturation at breakthrough

    (h) Pore volume of water injected at breakthrough

16.5. Determine waterflood performance after breakthrough by preparing the following plots. Use Equation 16.16 to compute the derivative df_w/dS_w at various S_w.

    (a) Log of water/oil ratio versus cumulative recovery

    (b) Cumulative recovery versus pore volume of water injected

    (c) An x-plot of E_R versus $- [\ln(1/f_w - 1) - 1/f_w]$ for $f_w > 0.5$ (Equation 11.14)

What is the ultimate recovery when the water cut reaches 90%?

16.6. Draw a frontal advance profile at various times, starting from 30 days following injection until the breakthrough period. The following data is available:

    Injection rate, bbl/d, = 1,000.
    Porosity, fraction, = 0.24.
    Net thickness of waterflooded formation, ft, = 25.
    Distance between injector and producer, ft, = 933.
    Cross-sectional flow area, ft^2, = 25,000.

Make any other assumptions necessary.

16.7. Derive Equation 16.26. State all necessary assumptions. Considering the flow of injected water at an inclination of 20° up dip, and oil viscosity of 1.8 cp, recalculate the following and compare with the base case analyzed previously:

    (a) Water saturation at the front of the water bank

    (b) Time to breakthrough

    (c) Water cut at breakthrough

Plot cumulative recovery versus pore volume of the water injected and compare with the base case

References

1. American Petroleum Institute. 1961. *History of Petroleum Engineering*. Washington, DC: American Petroleum Institute.

2. Satter, A., and G. C. Thakur. 1994. *Integrated Petroleum Reservoir Management*. Tulsa: PennWell.

3. Thakur, G. C., and A. Satter. 1998. *Integrated Waterflood Asset Management*. Tulsa: PennWell.

4. Craig, F. F., Jr. 1971. *The Reservoir Engineering Aspects of Waterflooding*. SPE Monograph Vol. 3. Richardson, TX: Society of Petroleum Engineers.

5. Craig, F. F., Jr. 1971.

6. Rose, S. C., J. F. Buckwalter, and R. J. Woodhall. 1989. *The Design Engineering Aspects of Waterflooding*. SPE Monograph Series. Richardson, TX: Society of Petroleum Engineers. p. 11.

7. Craft, B. C., and M. Hawkins. 1990. *Applied Petroleum Reservoir Engineering*. 2nd ed. Revised by R. E. Terry. Englewood Cliffs, NJ: PTR Prentice-Hall.

8. Craig, F. F., Jr. 1971.

9. Muskat, M. 1950. *Physical Principles of Oil Production*. New York: McGraw-Hill Book Co., Inc.

10. Deppe, J. C. 1961. Injection rates—the effect of mobility ratio, areal sweep, and pattern. *Society of Petroleum Engineers Journal*. June, 81–91.

11. Ahmed, T. H. 2001. *Reservoir Engineering Handbook*. 2nd ed. Houston: Gulf Professional Publishing Co.

12. Craig, F. F., Jr. 1971.

13. Craig, F. F., Jr., J. L. Sanderlin, D. W. Moore, and T. M. Geffen. 1957. A laboratory study of gravity segregation in frontal drives. *Transactions*. AIME. Vol. 210, 275–282.

14. Gaucher, D. H., and D. C. Lindley. 1960. Waterflood performance in a stratified 5-spot reservoir—a scaled model study. *Transactions*. AIME. Vol. 219, 208–215.

15. Carpenter, Jr., C. W., P. T. Bail, and J. E. Bobek. 1962. A verification of waterflood scaling in heterogeneous communicating flow models. *Society of Petroleum Engineers Journal*. March, 9–12.

16. Engelberts, W. L., and L. J. Klinkenberg. 1951. Laboratory experiments on the displacement of oil by water from packs of granular materials. Proceedings of the 3rd World Petroleum Congress. Vol. II: p. 544.

17. van Meurs, P. 1957. The use of transparent three-dimensional models for studying the mechanism of flow processes in oil reservoirs. *Transactions*. AIME. Vol. 210, 295–301.

18. Craig, F. F., Jr. 1971.

19. Carpenter, Jr., C.W., P. T. Bail, and J. E. Bobek. 1962.

20. Hutchinson, Jr., C. A. 1959. Reservoir inhomogeneity assessment and control. *Petroleum Engineer*. September. Vol. 31, no. 10, B19–26.

21. Craig, F. F., Jr. 1971.

22. Kyte, J. R., R. J. Stanclift, S. C. Stephan, Jr., and L. A. Rapoport. 1956. Mechanism of waterflooding in the presence of free gas. *Transactions*. AIME. Vol. 207, 215–221.

23. Dyes, A. B., B. H. Caudle, and R. A. Erickson. 1954. Oil production after breakthrough as influenced by mobility ratio. *Transactions*. AIME. Vol. 201, 81–86.

24. Dykstra, H. and R. L. Parsons.1950. The prediction of oil recovery by waterflooding. In *Secondary Recovery of Oil in the United States*. 2nd ed. Washington, DC: American Petroleum Institute. pp. 160–174.

25. Stiles, W. E. 1949. Use of permeability distribution in water flood calculations. *Transactions of AIME*. Vol. 186, 9–13.

26. Buckley, S. E., and M. C. Leverett. 1942. Mechanisms of fluid displacement in sands. *Transactions*. AIME. Vol. 146, 107–116.

27. Craig, Jr., F. F., T. M. Geffen, and R. A. Morse. 1955. Oil recovery performance of pattern gas or water injection operations from model tests. *Transactions*. AIME. Vol. 204, 7–15.

28. Prats, M. 1959. Prediction of injection rate and production history for multifluid five-spot floods. *Transactions*. AIME. Vol. 216, 98–105.

29. Dyes, A. B., B. H. Caudle, and R. A. Erickson. 1954.

30. Dykstra, H., and R. L. Parsons. 1950.

31. Johnson, C. E., Jr. 1956. Prediction of oil recovery by waterflood—a simplified graphical treatment of the Dykstra-Parsons method. *Transactions*. AIME. Vol. 207, 345–346.

32. Stiles, W. F. 1949.

33. Ahmed, T. H. 2001.

34. Buckley, S. E., and M. C. Leverett. 1942.

35. Buckley, S. E., and M. C. Leverett. 1942.

36. Welge, H. J. 1952. A simplified method for computing oil recovery by gas or water drive. *Transactions*. AIME. Vol. 195, 91–98.

37. Guthrie, R. K., and M. H. Greenberger. 1955. The use of multiple-correlation analyses for interpreting petroleum engineering data. *Drilling and Production Practice*. Washington, D.C.: American Petroleum Institute. pp. 130–137.

38. Todd, M., and E. Arzola. 2007. Rejuvenating a mature supergiant field, VlC-363, Block III Field, Lake Maracaibo, Venezuela: numerical simulation. March. Web literature, accessed at www.interp3.com.

39. Thakur, G. C., and A. Satter. 1998.

40. Kolbikov, S. V., H. F. Vaughn, A. A. Usmanov, and S. E. Chalov. 2000. Improved oil recovery based on optimal waterflood pressure. SPE Paper #65172. Presented at the SPE European Petroleum Conference, Paris, France, October 24–25.

41. Merlin Reservoir Simulator. 2006. Houston: Gemini Solutions Inc.

42. Jackson, R. 1968. Why waterfloods fail. *World Oil*. March: p. 65.

43. Satter, A., and G. C. Thakur. 1994.

44. Hall, H. N. 1963. How to analyze waterflood injection well performance. *World Oil*. October, 128–130.

45. Thakur, G. C., and A. Satter. 1998.

46. Gulick, K. E., and W. D. McCain, Jr. 1998. Waterflooding heterogeneous reservoirs: an overview of industry experiences and practices. SPE Paper #40044. SPE International Petroleum Conference and Exhibition of Mexico, Villahermosa, Mexico, March 3–5.

47. Stiles, L. H. and J. B. Magruder. 1992. Reservoir management in the Means San Andres Unit. *Journal of Petroleum Technology*. April, 469–475.

48. Goodwin, J. M. Infill Drilling and Pattern Modification in the Means San Andres Unit.

49. Lopez, J. D. Y., H. Ferreira, and A. C. Carnes, Jr. 1998. An integrated reservoir study to improve reservoir performance. SPE Paper #39843. Presented at the SPE International Petroleum Conference and Exhibition of Mexico, Villahermosa, Mexico, March 3–5.

50. Chetri, H. B., D. Sturrock, A. Al-Roweyh, H. Al-Dashti, M. Al-Mufarez, and R. Clark. 2005. Water flooding a heterogeneous clastic upper Burgan reservoir in North Kuwait: tackling the reservoir management and injectivity challenges. SPE Paper #93548. Presented at the 14th SPE Middle East Oil & Gas Show and Conference, Bahrain, March 12–15.

51. Currier, J. H., and S. T. Sindelar. 1990. Performance analysis in an immature waterflood: the Kuparuk River field. SPE Paper #20775. Presented at the 65th Annual Technical Conference and Exhibition, New Orleans, LA, September 23–25.

52. Lo, K. K., H. R. Warner, Jr., and J. B. Johnson. 1990. A study of post-breakthrough characteristics of waterfloods. SPE Paper #20064. Presented at the 1990 SPE California Regional Meeting, Ventura, CA, April 4–6.

17 · Improved Recovery Processes: Enhanced Oil Recovery and Applications

Introduction

Enhanced oil recovery relates to advanced processes to further augment oil recovery beyond secondary recovery (by waterflooding or natural gas injection) in a reservoir. Enhanced oil recovery processes include all methods that use external sources of energy and/or materials to recover oil that cannot be produced economically by conventional means. These recovery processes can be broadly classified as given in the following:[1]

- Thermal: steam flooding, hot water flooding, and in-situ combustion
- Nonthermal: chemical flood, miscible flood, and gas drive

Certain water-alternating-gas (WAG) processes are assisted by foam injection. Thermal and nonthermal methods are frequently referred to as tertiary oil recovery. Nonthermal processes also include microbial enhanced oil recovery (MEOR) processes, among others, which have been found to be of limited success to date.

Worldwide experience has shown that only a fraction of original oil in place could be recovered economically by primary and secondary recovery methods. In optimistic scenarios where rock and fluid properties are favorable, the recovery factor is likely to range between one-third and one-half of the original oil in place. On the other hand, recovery by primary or secondary methods from viscous heavy oil reservoirs, oil sands, and oil shales is far less satisfactory. Some of these reservoirs will not produce at all unless an efficient enhanced oil recovery scheme is engineered and implemented. A comprehensive review based on petroleum reservoir performance worldwide indicates average recovery from petroleum reservoirs of only about 35%. It is thus apparent that the various enhanced oil recovery techniques and future innovations hold the promise for recovering significant quantities of conventional and unconventional hydrocarbon resources. Economic considerations, including the prevailing price of petroleum and cost of new technology, play a critical role in implementing enhanced oil recovery operations in a reservoir.

Enhanced oil recovery processes are reservoir-specific in relation to the fluid properties and geologic setting of the reservoir. A thorough understanding of reservoir geology is vital. The effects of rock heterogeneities such as existing fractures, high permeability streaks, crossflow between layers, and isolated compartments must be analyzed before any enhanced oil recovery project is implemented. Some of these may not be identifiable during primary production. In the case of chemical injection, rock mineralogy plays an important role in quantifying the degree of absorption of injected material in a porous medium.

This chapter is devoted to learning about the following:
- Enhanced oil recovery process concepts
- Basic knowledge of enhanced oil recovery processes
- Oil recovery mechanisms
- Active U.S. enhanced oil recovery projects
- U.S. and global enhanced oil recovery production
- Screening criteria of enhanced oil recovery methods
- Case studies, including design, implementation, and monitoring
- Emerging recovery technologies for unconventional resources

The following field case studies are presented to provide an overview of the various enhanced oil recovery processes:
- Thermal recovery processes and reservoir management in Duri field, Indonesia—world's largest steam flooding operation
- CO_2 flooding in a West Texas field
- Low-tension waterflooding in Salem field, Illinois

- Recovery of oil sands in Canada based on emerging technologies, including steam-assisted gravity drive (SAGD) and cyclic steam stimulation (CSS)

Various enhanced oil recovery (miscible, thermal, and others) processes are described in detail in the literature, including monographs and reprint series published by the Society of Petroleum Engineers.

Enhanced Oil Recovery Process Concepts[2,3]

Since a considerable amount of oil is left after primary and secondary production methods, the ideal goal of enhanced oil recovery processes is to mobilize the "residual" oil throughout the entire reservoir. This can be achieved by enhancing microscopic oil displacement and volumetric sweep efficiencies. Oil displacement efficiency can be increased by decreasing oil viscosity using thermal floods or by reducing capillary forces or interfacial tension with chemical floods. Volumetric sweep efficiency can be improved by increasing the drive water viscosity using polymer compounds.

Oil displacement efficiency can be increased by improvement of the mobility ratio or by increasing the capillary number, or both.

The mobility ratio (M) was defined in earlier chapters. It is the ratio of the mobility of displacing fluid over that of the displaced fluid. The mobility ratio can be made relatively favorable (lower) by lowering the viscosity of the oil or increasing the viscosity of the displacing fluid (water). Thermal methods, including steam stimulation, steam flooding, hot water drive, and in-situ combustion are primarily used to heat the crude oil. This reduces its viscosity and thereby reduces the mobility ratio. The addition of a polymer in the water will make the solution more viscous, affecting the mobility ratio favorably.

If the mobility of the displacing phase (water) is greater than the phase being displaced (oil), the mobility ratio is unfavorable for improving oil recovery efficiency.

The capillary number, as discussed in chapter 2, plays a very important role in enhanced oil recovery efficiency. It is a dimensional group expressing the ratio of viscous to interfacial forces, as shown in Equation 2.58, given previously:[4]

$$N_{ca} = C \frac{k_w \Delta p}{\emptyset \, \sigma_{ow} L}$$

As the capillary number in an enhanced oil recovery process is increased by lowering the interfacial tension and oil viscosity, the residual oil saturation decreases. For miscible displacement, the interfacial tension approaches zero, and the oil displacement efficiency on the microscopic scale is very good.

There is no single process that can be considered a "cure-all" for recovering additional oil from every reservoir. Each process has its specific application. Before initiating an enhanced oil recovery process, reservoir rock and fluid properties and past production history should be analyzed. It is also important to review the preceding secondary recovery process in order to determine the principal reasons why the residual oil was left in that reservoir. Factors that strongly affect the success of a waterflood project will usually also affect the success of a subsequent tertiary project.

Enhanced Oil Recovery Processes

Over the past several decades, the petroleum industry has been engaged in research and development of various enhanced oil recovery processes needed to produce oil left behind by conventional methods. Exploitation of this enormous untapped energy source is the greatest challenge ever faced by the oil industry.

The following enhanced oil recovery methods are discussed, including process description, mechanisms, limitations, and problems: [5–15]

1. Thermal methods: steam stimulation, steam flooding, and in-situ combustion
2. Chemical methods: surfactants, polymer, micellar-polymer, and caustic alkaline
3. Miscible methods: hydrocarbon gas, carbon dioxide, and nitrogen. In addition, flue gas and partial miscible/immiscible gas floods may be also considered.

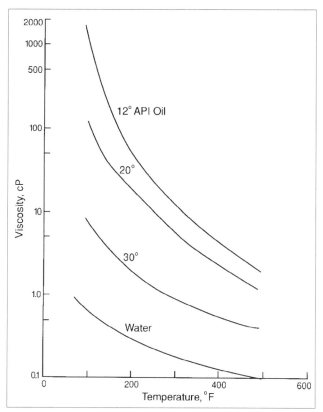

Fig. 17–1. Variation of oil viscosity with gravity and reservoir temperature. *Source: A. Satter and G. C. Thakur. 1994. Integrated Petroleum Reservoir Management—A Team Approach. Tulsa: PennWell.*

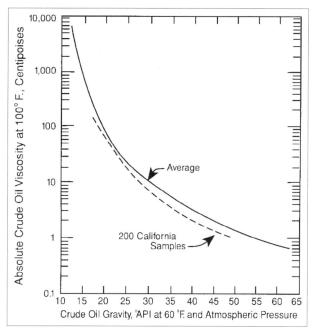

Fig. 17–2. Reduction of oil viscosity with increasing API gravity. *Source: A. Satter and G. C. Thakur. 1994. Integrated Petroleum Reservoir Management—A Team Approach. Tulsa: PennWell.*

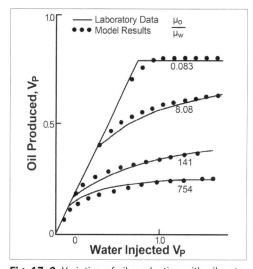

Fig. 17–3. Variation of oil production with oil-water viscosity ratio for 5-spot waterflood. Source: A. Satter and G. C. Thakur. 1994. *Integrated Petroleum Reservoir Management—A Team Approach.* Tulsa: PennWell.

Thermal methods

Many reservoirs contain viscous crude oil. Attempts to produce such oils with waterflooding will yield very poor recoveries, as the water-oil mobility ratio is quite large.[16] Application of heat is often the only feasible solution to recovery from such reservoirs. The basic idea is to make the viscous oil relatively mobile in order to facilitate its production. Thermal methods are primarily used for heavy viscous oils (10°–20° API) and tar sands. It is estimated that about 60% of all enhanced oil recovery oil production is due to the application of thermal recovery processes.

Figure 17–1 illustrates the sensitivity of crude oil viscosity to gravity and reservoir temperature. Figure 17–2 shows viscosity reduction of oil with increasing oil gravity. Oil recovery is affected significantly by an increase in the oil-water viscosity ratio, as seen in Figure 17–3.

Basically, oil recovery by steam injection includes steam soak (huff-and-puff steam flood) and the direct steam drive process.

Steam soak or huff-and-puff steam flood. Steam soak, cyclic steam injection, or "huff-and-puff" is the most successful thermal process. Steam is injected into a single well at a high rate for a short period of time (a few weeks). Next the steam is allowed to soak in for a few days, and then the well is allowed to flow back and is pumped. The oil rate increases initially, then drops off. When the rate becomes low, the entire process is repeated. This process is repeated many times until the well becomes uneconomic to produce, or in some cases, it is converted from steam stimulation to direct steam flooding.

Cyclic steam injection or huff-and-puff is considered to be a well stimulation process. In the stimulation process, the steam fingers through the oil around the wellbore and heats the oil. The soak period permits the oil to be heated even further. During the production cycle, the mobilized oil flows into the wellbore, as a result of pressure drop, gravity, and other mechanisms.

This process is most effective with highly viscous oils in reservoirs with good permeability. The performance of this method drops as more and more cycles are carried out. Oil recovery is generally very small in this process, because only a fraction of the formation is affected.

Steam flooding. In the steam flooding process, steam is continuously introduced into injection wells to reduce the oil viscosity and mobilize oil towards the producing wells. The injected steam forms a steam zone that advances slowly. The injected steam at the surface may contain about 80% steam and 20% water, i.e., an 80% steam quality. When steam is injected into the reservoir, heat is transferred to the oil-bearing formation, the reservoir fluids, and some of the adjacent cap and base rock. Due to this heat loss, some of the steam condenses to yield a mixture of steam and hot water.

Ahead of the steam zone, an oil bank forms and moves towards the producing well. In many cases, the injected steam overrides the oil due to gravity. This behavior can create some problems. When steam breakthrough occurs, the steam injection rate is reduced by recompletion of wells or shutting off steam-producing intervals. Steam reduces the oil saturation in the steam zone to very low values (about 10%, more or less). Some oil is transported by steam distillation.

The normal steam flooding practice is to precede and accompany the steam drive by a cyclic steam stimulation of the producing wells.

Steam flooding recovers crude by accomplishing the following:
- Heating the crude oil and reducing its viscosity
- Steam distillation
- Solvent/extraction effects
- Supplying pressure to drive oil to the producing well

The limitations of the steam flooding process including the following:
- Oil saturations must be quite high, and the pay zone should be more than 20 ft thick to minimize heat losses to adjacent formations.
- Lighter, less viscous crude oils can be steam flooded, but normally it will not be applicable if the reservoir will respond to an ordinary waterflood operation.
- Steam flooding is primarily applicable to viscous oils in massive, high permeability sandstones or unconsolidated sands.
- Because of excessive heat losses in the wellbore, steam-flooded reservoirs should be as shallow as possible, as long as pressure for sufficient injection rates can be maintained.
- Steam flooding is not normally used in carbonate reservoirs. Both bottom water and gas caps are undesirable.
- Since about one-third of the additional oil recovered is consumed in the generation of the required steam, the cost per incremental barrel of oil is high.

The issues in this process include adverse mobility ratios and channeling of steam. Case studies are presented later in this chapter and in chapter 19.

In-situ combustion. In-situ combustion or fire flooding involves starting a fire in the reservoir and injecting air to sustain the burning of some of the crude oil. The most common technique is forward combustion. In this process, the reservoir is ignited at the bottom of the injection well by a special heater, and air is injected to propagate the combustion front away from the well. A significant amount of fluid is burned (as much as 10% of the original oil in place) to generate heat. The lighter ends of the oil are carried forward ahead of the burned zone, upgrading the crude oil. The heavy ends of the crude oil are burned. Heat is generated within a combustion zone at a very high temperature, about 600°C. As a result of burning the crude oil, large volumes of flue gas are produced. Steam, hot water, combustion gas, and distilled solvent produced in the process further aid in driving oil toward the wellbore.

One of the variations of this technique is a combination of forward combustion and waterflooding (COFCAW).

A second technique of the in-situ combustion process is a reverse combustion. In this process, a fire is started in a well that will eventually become a producing well, and air injection is then switched to adjacent wells. No successful field trials have been known to be completed with reverse combustion.

In-situ combustion recovers crude oil by the following:
- Application of heat, which is transferred downstream by conduction and convection, thus lowering the viscosity of the crude
- Products of steam distillation and thermal cracking, which are carried forward to mix with and upgrade the crude
- Burning coke that is produced from the heavy ends of the crude oil
- Pressure supplied to the reservoir by the injected air

The limitations of the process include the following:
- If sufficient coke is not deposited from the oil being burned, the combustion process will not be sustained.
- If excessive coke is deposited, the rate of advance of the combustion zone will be slow, and the quantity of air required to sustain combustion will be high.
- Oil saturation and porosity must be high to minimize heat loss to the rock.
- The process tends to sweep through upper part of reservoir, and thus sweep efficiency is poor in thick formations.

The problems in this process include the following:
- The mobility ratio can be adverse.
- The process is complex, difficult to control, and requires large capital investments.
- Produced flue gases can present environmental problems.
- Operational problems can occur. These include severe corrosion caused by low pH hot water, serious oil-water emulsions, increased sand production, deposition of carbon or wax, and pipe failures in the producing wells as a result of the very high temperatures.

Chemical methods

Chemical flooding processes include polymer, surfactants, micellar-polymer, and caustic alkaline. These processes require conditions favorable to water injection, as they are modifications of waterflooding.

Polymer flooding. Polymer-augmented waterflooding consists of adding water-soluble polymers to the water before it is injected into the reservoir. Low concentrations (often 250–2,000 mg/L) of certain synthetic or biopolymers are used. The objective of polymer flooding is to enhance volumetric sweep efficiency.

Polymers improve recovery by accomplishing the following:
- Increasing the viscosity of water
- Decreasing the mobility of water
- Contacting a larger volume of the reservoir
- Reducing the injected fluid mobility to improve areal and vertical sweep efficiencies

It should be noted that the addition of a polymer in the injected water does not lower the residual oil saturation.

The oil displacement is more efficient in the early stages as compared to a conventional waterflood process. As a result, more oil will be produced in the early life of the flood. This is the primary economic advantage, since it is generally accepted that ultimate recovery will be the same for polymer flooding as for waterflooding.

Polymer flood limitations include the following:
- If oil viscosities are high, a higher polymer concentration is needed to achieve the desired mobility control.
- Results are normally better if the polymer flood is started before the water/oil ratio becomes excessively high.
- Clays increase polymer adsorption.
- Some heterogeneities are acceptable, but for conventional polymer flooding, reservoirs with extensive fractures should be avoided. If fractures are present, the cross-linked or gelled polymer techniques may be applicable.

Problems associated with polymer flooding include the following:
- The injectivity is lower than with water alone, and this can adversely affect the oil production rate in the early stages of the polymer flood.
- Acrylamide-type polymers lose viscosity due to shear degradation or increases in salinity and divalent ions.
- Xanthan gum polymers cost more, are subject to microbial degradation, and have a greater potential for wellbore plugging.

Surfactant/polymer flooding. Surfactant/polymer flooding is also called micellar/polymer or microemulsion flooding. It consists of injecting a slug that contains water, a surfactant, an electrolyte (salt), usually a cosolvent (alcohol), and possibly a hydrocarbon (oil). The size of the slug is often 5%–15% PV for a high surfactant concentration system, and 15%–50% PV for low concentrations. The surfactant slug is followed by polymer-thickened water. Concentration of the polymer often ranges from 500mg/L to 2,000 mg/L. The volume of the polymer solution injected may be 50% PV, more or less, depending on the process design.

Surfactant/polymer flooding recovers oil by accomplishing the following:
- Lowering the interfacial tension between the oil and water
- Solubilization of oil
- Emulsification of oil and water
- Mobility enhancement

This process, generally applicable to light oils, suffers from the following limitations:
- An areal sweep of more than 50% on a waterflooding is desired.
- A relatively homogeneous formation is preferred.
- High amounts of anhydride, gypsum, or clays are undesirable.
- Available systems provide optimum behavior over a very narrow set of conditions.
- With commercially available surfactants, formation water chlorides should be less than 20,000 ppm, and divalent ions (Ca^{++} and Mg^{++}) less than 500 ppm.

Problems in this process include the following:
- Complexity and expense of the system
- Possibility of chromatographic separation of chemicals
- High adsorption of surfactant
- Interactions between surfactant and polymer
- Degradation of chemicals at high temperatures

Caustic flooding. Caustic or alkaline flooding involves the injection of chemicals such as sodium hydroxide, sodium silicate, or sodium carbonate. These chemicals react with organic petroleum acids in certain crude oils to create surfactants in situ. They also react with reservoir rocks to change wettability.

Oils in the API gravity range of 13°–35° are normally targets for alkaline flooding. One desirable property for the oils is to have enough organic acids so that they can react with the alkaline solution. Another such property is moderate oil gravity, so mobility control is not a problem.

Sandstone reservoirs are generally preferred for this process, since carbonate formations often contain anhydride or gypsum, which can consume a large amount of alkaline chemicals. The alkali is also consumed by clays, minerals, or silica. In addition, the consumption is high at elevated temperatures. Another problem with caustic flooding is scale formation in the producing wells.

Presently there are no active caustic projects in the United States.

Miscible methods

Miscible methods include hydrocarbon gas, carbon dioxide, and nitrogen. In addition, flue gas and partial miscible/immiscible gas floods may be also considered.

Miscible flooding involves injecting a gas or solvent that is miscible with the oil. As a result, the interfacial tension between the two fluids (oil and solvent) is very low. Very efficient microscopic displacement efficiency takes place.

Hydrocarbon miscible flooding. Hydrocarbon miscible flooding consists of injecting light hydrocarbons through the reservoir to form a miscible flood. Three different methods are as follows:

- One method uses a slug of liquefied petroleum gas (LPG) of about 5% PV, such as propane, followed by lean gas. Sometimes water is injected with the chase gas in a water-alternating-gas mode to improve the mobility ratio between the solvent slug and the chase gas. Variations of the process involve simultaneous water and gas (SWAG) injection and foam combined with water-alternating-gas injection, referred to as FAWAG. A detailed review of reservoir performance based on water-alternating-gas processes in large number of fields is provided by Christensen, Stenby, and Skauge.[17]
- A second method is called enriched (condensing) gas drive. It consists of injecting a 10%–20% PV slug of natural gas that is enriched with ethane through hexane (C_2 to C_6), followed by lean gas (dry, mostly methane) and possibly water. The enriching components are transferred from the gas to the oil. A miscible zone is formed between the injected gas and the reservoir oil, and this zone displaces the oil ahead.
- The third method, called high pressure (vaporizing) gas drive, consists of injecting lean gas at high pressures. This allows the C_2 through C_6 components to vaporize from the crude oil being displaced, resulting in multiple contact miscibility.

Hydrocarbon miscible flooding recovers crude oil by accomplishing the following:
- Generating miscibility (in the condensing and vaporizing gas drive)
- Increasing the oil volume (swelling)
- Decreasing the viscosity of the oil

The limitations of the process include the following:
- The minimum depth is set by the pressure needed to maintain the generated miscibility. The required pressure ranges from about 1,200 psi for the LPG process to 3,000–5,000 psi for the high-pressure gas drive, depending on the oil composition.
- A steeply dipping formation is very desirable to permit some gravity stabilization of the displacement that normally has an unfavorable mobility ratio.

The problems of hydrocarbon miscible floods include the following:
- Viscous fingering results in poor vertical and horizontal sweep efficiency.
- Large quantities of expensive products are required.
- The solvent may be trapped and not recovered.

Carbon dioxide flooding. Carbon dioxide (CO_2) flooding is carried out by injecting large quantities of CO_2 (15% or more of the hydrocarbon PV) into the reservoir. Although CO_2 is not truly miscible with the crude oil, it extracts the light to intermediate components from the oil. If the pressure is sufficiently high, miscibility develops to displace the crude oil from the reservoir.

Miscible displacement by CO_2 is similar to that in a vaporizing gas drive. The only difference is that a wider range of components, C_2 to C_{30}, is extracted. As a result, the CO_2 flood process is applicable to a wider range of reservoirs at lower miscibility pressures than those for the vaporizing gas drive.

CO_2 is generally soluble in crude oils at reservoir pressures and temperatures. It swells the net volume of oil and reduces its viscosity even before miscibility is achieved by the vaporizing gas drive mechanism. As miscibility is approached as a result of multiple contacts, both the oil phase and the CO_2 phase (containing intermediate oil components) can flow together because of the low interfacial tension. One of the requirements of the development of miscibility between the oil and CO_2 is the reservoir pressure.

In this process, about 20%–50% of the CO_2 slug is followed by chase water. Water is generally injected with CO_2 in a water-alternating-gas mode to improve the mobility ratio between the displacing phase and the oil.

CO_2 recovers crude oil by accomplishing the following:
- Generation of miscibility between in-situ oil and injected gas
- Swelling the crude oil
- Lowering the viscosity of the oil
- Lowering the interfacial tension between the oil and the CO_2-oil phase in the near-miscible regions

The limitations of CO_2 floods include the following:
- The very low viscosity of the CO_2 results in poor mobility control.
- The process can be limited by availability of the CO_2.

The problems associated with the process include the following:
- Resultant problems from early breakthrough of CO_2
- Corrosion in the producing wells
- The necessity of separating CO_2 from saleable hydrocarbons
- Repressuring of CO_2 for recycling
- A high requirement of CO_2 per incremental barrel produced

A case study of CO_2 flooding as a tertiary recovery process is presented later.

Nitrogen and flue gas flooding. Nitrogen and flue gas flooding are oil recovery methods that use these inexpensive nonhydrocarbon gases to displace oil. The resulting displacement may be either miscible or immiscible, depending on the pressure and oil composition. Because of their low cost, large volumes of these gases may be injected. Nitrogen or flue gas can also be considered for use as a chase gas in hydrocarbon-miscible and CO_2 floods.

Both nitrogen and flue gas are inferior to hydrocarbon gases (and much inferior to CO_2) from an oil recovery point of view. Nitrogen has a lower viscosity, poor solubility in oil, and requires a much higher pressure to generate or develop miscibility.

Nitrogen and flue gas flooding recover oil by accomplishing the following:
- Vaporizing the lighter components of the crude oil and generating miscibility, given sufficient pressure
- Providing a gas drive whereby a significant portion of the reservoir volume is filled with low-cost gases

The process limitations include the following:
- Developed miscibility can only be achieved with light oils and at high pressure; therefore, deep reservoirs are needed.
- A steeply dipping reservoir is desired to permit gravity stabilization of the displacement, which has a very unfavorable mobility ratio.

Problems associated with this process include the following:
- Viscous fingering results in poor vertical and horizontal sweep efficiency.
- Corrosion can cause problems in the flue gas method.
- The nonhydrocarbon gases must be separated from the saleable produced gas.

U.S. Enhanced Oil Recovery Projects and Production

Active U.S. enhanced oil recovery projects and production in the years 1980–2004 are presented in Tables 17–1 and 17–2, respectively.

While the number of steam flood projects reached a maximum of 181 in 1986, the number was 46 in 2004. Oil production rate reached a maximum of 468,692 bopd in 1986. The production rate was 340,253 bopd in 2004.

Combustion processes have not been successful. There were only seven projects in the United States in 2004, producing about 1,901 bopd. The maximum numbers were reported to be 21 in 1982, producing 10,228 bopd.

Table 17-1. Active U.S. enhanced oil recovery projects

	1980	1982	1984	1986	1988	1990	1992	1994	1996	1998	2000	2002	2004
Thermal													
Steam	133	118	133	181	133	137	119	109	105	92	86	55	46
Combustion in situ	17	21	18	17	9	8	8	5	8	7	5	6	7
Hot water				3	10	9	6	2	2	1	1	4	3
Total Thermal	**150**	**139**	**151**	**198**	**190**	**141**	**145**	**124**	**117**	**112**	**97**	**92**	**62**
Chemical													
Micellar-polymer	14	20	21	20	9	5	3	2					
Polymer	22	55	106	178	111	42	44	27	11	10	10	4	4
Caustic/alkaline	6	10	11	8	4	2	2	1	1	1			
Surfactant							1						
Total Chemical	**42**	**85**	**138**	**206**	**124**	**50**	**49**	**30**	**12**	**11**	**10**	**4**	**4**
Gas													
Hydrocarbon (miscible and immiscible)	9	12	16	26	22	23	25	15	14	11	6	7	8
CO_2 miscible	17	28	40	38	49	52	52	54	60	66	63	66	70
CO_2 immiscible		1	18	28	8	4	2	1	1		1	1	1
Nitrogen	1	4	7	9	9	9	7	8	9	10	4	4	4
Flue gas (miscible and immiscible)	3	3	3	3	2	3	2						
Other			2										
Total Gas	**30**	**48**	**86**	**104**	**90**	**91**	**88**	**78**	**84**	**87**	**74**	**78**	**83**
Other													
Microbial	0	0	0	1	0	0	2	1	1	1	0	0	0
Total Other	**0**	**0**	**0**	**1**	**0**	**0**	**2**	**1**	**1**	**1**	**0**	**0**	**0**
Total All	**222**	**272**	**375**	**404**	**282**	**282**	**232**	**213**	**210**	**181**	**174**	**149**	**0**

Table 17-2. U.S. enhanced oil recovery production (bopd)

	1982	1984	1986	1988	1990	1992	1994	1996	1998	2000	2002	2004
Thermal												
Steam	288,396	358,115	468,692	455,484	444,137	454,009	415,801	419,349	439,010	417,675	365,717	340,253
Combustion in Situ	10,228	6,445	10,272	6,525	6,090	4,702	2,520	4,485	4,760	2,781	2,384	1,901
Total Thermal	**298,624**	**364,560**	**478,964**	**462,009**	**450,227**	**458,711**	**418,321**	**423,834**	**443,770**	**420,456**	**368101**	**342154**
Chemical												
Micellar-polymer	902	2,832	1,403	1,509	617	254	64	0	0	0	0	0
Polymer	2,927	10,232	15,313	20,992	11,219	1,940	1,828	139	139	1,598	0	0
Caustic/alkaline	580	334	185									
Surfactant					20					60	60	60
Total Chemical	**4,409**	**13,398**	**16,901**	**22,501**	**11,856**	**2194**	**1892**	**139**	**139**	**1,658**	**60**	**60**
Gas												
Gas	0	14,439	33,767	25,935	113,072	99,693	96,263	102,053	124,500	95,300	96,300	
Hydrocarbon (miscible and immiscible)	0	31,300	28,440	64,192	144,973	161,486	170,715	179,024	189,493	187,410	205,775	
CO_2 Miscible	0	702	1,349	420	95				66	66	102	
CO_2 Immiscible	0	7,170	18,510	19,050	22,580	23,050	28,017	28,117	14,700	14,700	14,700	
Flue gas (miscible and immiscible)	0	29,400	26,150	17,300	11,000							
Immiscible												
Other	0				6,300	4,400	4,350	4,350	0	0	0	
Total Gas	**0**	**83,011**	**108,216**	**126,897**	**298,020**	**288,629**	**299,345**	**313,544**	**328,759**	**297,476**	**317,877**	**0**

Where applicable, steam flooding is now routinely used on commercial basis. In the United States, a majority of the field applications have occurred in California. Many of the shallow, high oil-saturated reservoirs, with the high viscosity crude oils that are found there, are good candidates for thermal recovery. The world's largest steam flood project was in the Duri field in Indonesia, producing about 500,000 bopd at one time.

There were only 10 active polymer projects in the United States in 2000, producing about 1,600 bopd.

After thermal recovery, miscible flooding contributes the most among various enhanced oil recovery methods. There were 83 active gas flood projects in 2004 in the United States, out of which 70 were CO_2 miscible floods.

More than 40% of the total enhanced oil recovery production has been by gas miscible/immiscible flooding.

CO_2 flooding is the fastest growing enhanced oil recovery method in the United States, and field projects continue to show good incremental oil recovery in response to CO_2 injection. The CO_2 flooding method works well as either a secondary or tertiary operation. However, most large CO_2 floods are tertiary projects in mature reservoirs that have been waterflooded for many years.

Offshore enhanced oil recovery operations

Producing petroleum fields located offshore, especially deepwater fields, need to have relatively large reserves in order to operate from economic point of view. The volume of oil left behind following primary or secondary recovery is also rather large. Bondor, Hite, and Avasthi list the following issues in planning enhanced oil recovery operations offshore:[18]

- Very large well spacing. A typical scenario is based on a handful of long multilateral horizontal wells drilled in the reservoir.
- Lack of detailed information about reservoir geology in the large areas that exist between the wells.
- Unknown degree of continuity between various portions of the reservoirs.
- Necessity for detailed reservoir description and development of robust reservoir model based on available information.
- Availability and cost of injection materials; weight and space constraints.
- Technical issues specific to offshore fields.
- Significant capital investment.

Worldwide Production Statistics and Cost of Recovery

Global statistics of various enhanced oil recovery processes indicate the following:[19]

	Percent of Total EOR Production	**Concentration of Project Locations**
Thermal methods:	41	California, Indonesia Canada, China, and Venezuela
Injection of hydrocarbons:	25	Alaska and Algeria
Injection of nitrogen:	19	
Polymer/chemical methods:	8	China
CO_2 flood:	7	Texas

As of 2005, worldwide production based on all enhanced oil recovery processes was estimated at 2.93 million bopd. An increase of about 7% to 15% in overall recovery is expected from a reservoir undergoing an enhanced oil recovery process based on current technology and economics. In recent years, enhanced oil recovery focus has shifted to extracting huge oil sand deposits by novel technologies described later.

The cost of an enhanced oil recovery process can be wide ranging, depending on the specific process and field. A typical range would be $10–$30 per barrel of oil produced by thermal, polymer, and CO_2 injection processes. The cost of surfactant flooding and novel enhanced oil recovery processes could run substantially higher.

Enhanced Oil Recovery Screening Criteria

All of the processes described in this chapter have limitations in application. These limitations have been derived partly from theory, partly from laboratory experiments, and partly from field experiences. Prospect screening consists of the following:

1. Evaluating available information about the reservoir, oil, rock, water, geology, and previous performance
2. Supplementing available information with certain brief laboratory screening tests
3. Selecting those processes that are potentially applicable and eliminating those that definitely are not

A candidate reservoir for one or more enhanced oil recovery processes should not be discarded because it does not satisfy one or two criteria. Each prospect should be evaluated on its own merits by analyzing the many reservoir operational and economic variables.

Screening is the first step in the enhanced oil recovery implementation sequence. The next step would be a further evaluation of candidate processes if more than one satisfies the screening criteria. Subsequent steps could include a pilot test design, pilot test implementation, pilot test evaluation/scale-up forecast, and a commercial venture.

Table 17–3 presents screening criteria based upon oil properties for application of various enhanced oil recovery processes. The criteria include the gravity, viscosity, and saturation of the oil.

Table 17–3. Screening criteria for enhanced oil recovery methods based on oil properties

Process	Gravity °API	Viscosity (cp)	Composition	Oil Saturation
Waterflooding	> 25	< 30	N.C.	>10% mobile oil
Hydrocarbon	> 35	< 10	High % of C_2–C_7	> 30% PV
Nitrogen & flue gas	> 24 Nitrogen > 35 Flue gas	< 10	High % of C_1–C_7	> 30% PV
Carbon dioxide	> 26	< 15	High % of C_5–C_{12}	> 20% PV
Surfactant/polymer	> 25	< 30	Light to intermediate desired	> 30% PV
Polymer	> 25	< 150	N.C.	> 10% PV mobile oil
Alkaline	13–35	< 200	Some organic acids	Above waterflood residual
Combustion	< 40 (10–25 normally)	< 1,000	Some asphaltic components	> 40%–50% PV
Steam flooding	< 25	> 20	N.C.	> 40%–50% PV

Note: PV = pore volume; N.C. = not critical.

Steam flooding is primarily applicable to viscous oils in massive, high permeability sandstones or unconsolidated sands. It is limited to shallow formations due to heat losses from the wellbore. Heat is also lost to the adjacent formations once steam contacts the oil-bearing formation. Hence, sufficiently high steam injection rates are needed to compensate for heat losses.

The minimum miscibility pressure for effective CO_2 flooding ranges widely. The required pressure can be 1,200 psi for high gravity oil (more than 30°API) at lower temperatures to more than 4,500 psi for heavy crudes at higher temperatures. To satisfy this requirement, the reservoir has to be deep enough to achieve the minimum miscibility pressure. For an example, the minimum miscibility pressure for West Texas CO_2 floods is around 1,500 psi at depths of more than 2,000 ft. On the other hand, more than 4,500-ft deep reservoirs are needed for effective NO_2 and high-pressure hydrocarbon miscible floods.

Table 17-4. Screening criteria for enhanced oil recovery methods based on reservoir characteristics

Process	Formation Type	Net Thickness (ft)	Average Permeability (mD)	Depth (ft)	Temp (°F)
Waterflood	Sandstone or carbonate	N.C.	N.C.	N.C.	N.C.
Hydrocarbon	Sandstone or carbonate	Thin unless dipping	N.C.	>2,000 (LPG) >5,000 (H.P. gas)	N.C.
Nitrogen & flue gas	Sandstone or carbonate	Thin unless dipping	N.C.	>4,500	N.C.
Carbon dioxide	Sandstone or carbonate	Thin unless dipping	N.C.	>2,000	N.C.
Surfactant/polymer	Sandstone preferred	>10	>20	<8,000	<175
Polymer	Sandstone preferred; carbonate possible	N.C.	>10 (normally)	<8,999	<200
Alkaline	Sandstone preferred	N.C.	>20	<9,000	<200
Combustion	Sand or sandstone with high porosity	>10	>100	>500	>150 preferred
Steam flooding	Sand or sandstone with high porosity	>20	>200	300–5,000	N.C.

Note: N.C. = not critical.

Table 17–4 presents screening criteria based upon reservoir characteristics for application of the various enhanced oil recovery processes. The criteria include formation type, net thickness, average permeability, depth, and temperature.

Thermal floods are primarily applicable to heavy viscous oils. Steam floods are used for oil with gravity less than 25°API, viscosity more than 20 cp, and oil saturation more than 40% PV. Higher viscosity with less than 100 cp may be applicable for combustion floods.

Hydrocarbon, nitrogen, carbon dioxide, and surfactant floods are applicable to higher oil gravities and lower oil saturations than those needed for steam floods.

Screening of those processes that are potentially applicable for enhanced oil recovery processes is a necessary step, thus eliminating those that definitely are not. A candidate reservoir for one or more enhanced oil recovery processes should not be discarded because it does not satisfy one or two criteria. Each prospect should be evaluated on its own merits by analyzing the many reservoir operational and economic variables.

Case Studies

Thermal recovery from the Duri field[20-23]

Introduction. To date, the Duri steam flood (DSF) project is the largest thermal recovery operation in the world. The Duri field, the second largest in Indonesia, covers approximately 140-km^2 area at a relatively shallow depth of 400 ft to 700 ft true vertical depth (TVD). Structurally, the field is a faulted anticline. Following about a decade of steam injection, field production reached 300,000 bopd in 1995. Production was reported to be about 210,000 bopd in 2005. The Duri field, divided into 13 sections, has more than 4,000 producers, 1,600 injectors, and 450 temperature observation wells. The average production rate is 60 bopd per well. The original oil in place is estimated to be in the billions of barrels.

Rock and fluid properties. The rock and fluid characteristics of the Duri field are summarized in the following:
- Average oil gravity is about 21°API, with heavier oil (about 17°API) found toward the northern flank.
- Reservoir permeability ranges from 100 millidarcies to 4 darcies. Permeability degradation is observed toward the periphery of the field. Porosity of the formation varies between 15% and 45%.
- Reservoir pressure and temperature prior to steam flood are 100 psia and 100°F, respectively.
- Initial oil saturation is about 55%. Irreducible water saturation is 40%.
- Oil viscosity is 330 cp at 100°F and decreases to about 8 cp at 300°F.

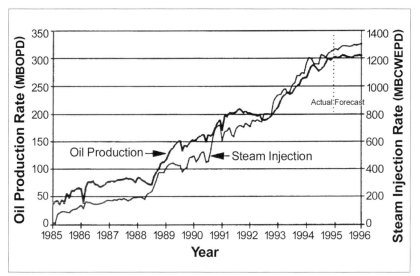

Fig. 17–4. Increase in production rate due to steam flooding in Duri field. *Source: B. T. Gael, S. J. Gross, and G. J. McNaboe. 1995. Development planning and reservoir management in the Duri steam flood. SPE Paper #29668. Presented at the SPE Western Regional Meeting, Bakersfield, CA, March 8–10. © Society of Petroleum Engineers. Reprinted with permission.*

- Solution gas ratio of the oil is 15 scf/stb, and the formation volume factor is 1.02. Both the properties are characteristics of viscous oil having low API gravity. Heavy oil is overlain by a small gas cap in one of the zones.
- Reservoir thermal conductivity is 27.4 Btu/ft-d-°F. Heat capacity of the rock is 33.2 Btu/ft^3-°F.
- Oil properties do not vary significantly in the vertical direction.
- The target of the steam drive is two major intervals (hydraulic units), namely the Pertama and Kedua sands, which contain two-thirds of the original oil in place. The average net pay thickness is 140 ft.

Primary recovery mechanisms, enhanced oil recovery pilot, and field-scale implementation. Primary production from the Duri field began in late 1950s and reached its peak (about 65 million bopd) in the mid-1960s. Production declined rather rapidly due to the relatively poor mobility of the heavy oil and low solution gas/oil ratio. A weak aquifer provided marginal support, and production due to gravity drainage was not significant. Cyclic steam stimulation began in 1967, followed by initiation of a steam flood pilot project in 1975. The pilot project involved 16 inverted 5-spot patterns and recovered about 30% of the original oil in place. Another enhanced oil recovery pilot involved injection of sodium hydroxide into the heavy oil reservoir (caustic flood), which did not succeed. Based on encouraging results of steam injection, a field-scale operation started in 1985. After 14 years of steam flooding, a certain area in the field achieved a recovery factor as high as 64%, with an ultimate recovery target of 69%. Reservoir simulation studies indicated that about 1 PV of injected fluid would be required for optimum recovery. The production and steam injection history of the Duri field is presented in Figure 17–4.

Key factors in steam flood management. In order to optimize recovery of heavy oil, with a regard to economic analysis, the following strategy was adopted:
- A detailed characterization of the reservoir was developed to realistically predict enhanced oil recovery performance in all areas of the reservoir.
- Selection of various injection patterns was based on the reservoir simulation and field experience. An inverted 7-spot pattern on 11⅝-acre spacing dominated the selection.
- Based on further studies, inverted 9-spot patterns were selected where the pay thickness was greater than 100 ft in newer areas. Again, an inverted 5-spot pattern was selected where pay thickness was less than 100 ft. In both cases, pattern spacing was 15½ acres. The economic limit was estimated at a minimum of 50 ft of pay.
- Steam was injected according to the $h\phi S_{oi}$ product of a layer. Injection rate was set at up to 1.2 bscwepd/naf, or the maximum allowable below the fracture gradient.
- Planning of steam flooding in unrecovered areas of the field was based on the availability of steam and constraints in the construction schedule of the facilities. Steam supply was "shifted" from the previously flooded sections to the new areas based on overall optimization (including breakthrough at certain producers). This was due to the fact that total steam generation at the field was fixed.

Steam flood process. During steam flooding in the Duri field, the following field observations were made, aided by reservoir simulation studies:

- Heavy oil was driven towards the producers by injected steam due to the reduction in oil viscosity and the high pressure gradient (viscous forces). The oil production rate showed the potential to increase by as much as 500% compared to the rate before steam flooding.

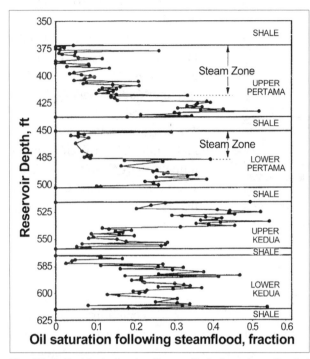

Fig. 17–5. Residual oil saturation profile in four steam-flooded zones based on core study. *Source: B. T. Gael, S. J. Gross, and G. J. McNaboe.1995. Development planning and reservoir management in the Duri steam flood. SPE Paper #29668. Presented at the SPE Western Regional Meeting, Bakersfield, CA, March 8–10. © Society of Petroleum Engineers. Reprinted with permission.*

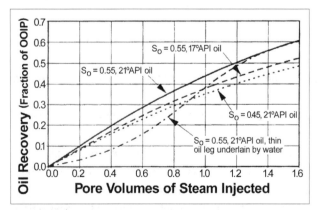

Fig. 17–6. Performance prediction of zones having different fluid properties and geologic setting in the Duri field. *Source: B. T. Gael, S. J. Gross, and G. J. McNaboe. 1995. Development planning and reservoir management in the Duri steam flood. SPE Paper #29668. Presented at the SPE Western Regional Meeting, Bakersfield, CA, March 8–10. © Society of Petroleum Engineers. Reprinted with permission.*

- The preceding was accompanied by reservoir pressurization to a significant degree, and a pressure gradient in the horizontal direction ranged between 0.7 psi/ft to 1.0 psi/ft.
- Once steam broke through, the horizontal pressure gradient decreased to about 0.2 psi/ft, which was a fraction of the gravity gradient. Hence, forces due to gravity become predominant in displacing oil.
- Injected steam broke through in the most permeable layer first. Core studies indicated that the residual oil saturations following steam flooding tended to be higher toward the bottom of a zone. This indicated gravity override by steam and limited vertical sweep efficiency (fig. 17–5).
- Reservoir simulation studies further indicated that enhanced oil recovery was adversely affected by relatively low oil saturation. Enhanced oil recovery was also affected by low API gravity (heavier hydrocarbons) and the presence of an underlying water zone where thermal energy was lost (fig. 17–6).

Monitoring of steam flooding. Reservoir and well monitoring during the Duri steam flood were found to play a crucial role in maximizing the asset. Conventional methods of monitoring included the following:

- Tracer studies
- Temperature logs
- Collection of pressure, temperature, and fluid composition data at injection, production, and observation wells

In addition to conventional techniques, time-lapse seismic monitoring was introduced in 1995. This was initiated in order to locate the steam front in vertical and horizontal directions during highly dynamic displacement. The underlying principle is that the velocity of seismic waves is influenced by fluid temperature, pressure, and phase saturations. Moreover, fiber optic distributed temperature sensing (DTS) technology was utilized to identify the zones of steam breakthrough by pulsing laser light through the fiber placed in the wellbore.

North Ward Estes field CO_2 flood project[24, 25]

The North Ward Estes field, located in Ward and Winkler counties in West Texas, was discovered in 1929. The dominant producing formation includes up to seven major reservoirs and is composed of very fine-grained sandstones to siltstones, separated by dense dolomite beds. Within the 3,840-acre project area, the following is known:

Average properties:	Initial conditions:
Depth, ft, = 2,600.	Water saturation, PV%, = 50.
Reservoir temperature, °F, = 83.	Reservoir pressure at gas/oil contact, psia, = 1,400.
Dykstra-Parsons permeability variation factor = 0.85.	Oil viscosity, cp, = 1.4.
Porosity, PV%, = 16.	Saturation pressure, psia, = 1,400.
Permeability, millidarcies, = 37.	Oil formation volume factor, rb/stb, = 1.2.
Oil gravity, °API, = 37.	Solution gas/oil ratio, scf/stb, = 500.

Waterflooding in the field began in 1955 and has continued. Cumulative oil production is more than 320 million bbl (25% of stock-tank oil initially in place). CO_2 flooding was implemented in early 1989 in a project area comprised of six sections, located in the better part of the field. Flood patterns in sections 3 and 6 to 8 were 20-acre 5-spots, and sections 9 and 10 were 20-acre line drives. Estimated pure CO_2 minimum miscibility pressure was 937 psia.

The comprehensive approach taken for designing the CO_2 flood project included the following:

1. **Laboratory work.** Extensive laboratory work was conducted to support the evaluation of the CO_2 flooding for the North Ward Estes field project, including black oil PVT and oil/CO_2 phase behavior, and slim-tube experiments for determining minimum miscibility pressure.

2. **History matching.** History matching was accomplished by reservoir simulation. Historical production rates for the years 1929 to 1986 were utilized in the study. The reservoir simulator computed gas and water production rates and reservoir pressures. The matches were obtained largely by layer permeabilities. During history match, oil and water relative permeability curves were adjusted. To improve the prediction of the estimated time for CO_2 breakthrough at the producers, particular attention was paid to water breakthrough time after the initiation of waterflooding.

3. **Simulation of CO_2 flood.** A finite-difference, four-component, black oil simulator was selected for history matching both primary and waterflood performance. This was followed by prediction of CO_2 flood performance for a multilayer 5-spot pattern.

4. **CO_2 injectivity test.** A CO_2 injectivity test showed no reduction in the injection rates during or after injection, and no significant changes in injection profile during or after injection. CO_2 falloff data was in agreement with laboratory measurements from CO_2 core floods.

5. **Optimum economic slug size.** Optimum economic slug size was found to range between 38% and 60% hydrocarbon pore volume (HCPV) of CO_2 injected.

Enhanced oil recovery simulation results. Performance predictions for the entire project area were based upon the scale-up of the average pattern simulation results. The results are given below:

> Recovery as of 1990 = 29% of original oil in place.
> Primary and secondary waterflood recovery = 31% of original oil in place.
> CO_2 flood recovery = 8% of original oil in place.
> CO_2 slug size = 38% hydrocarbon pore volume.
> Water-alternating-gas injection ratio = 1:1.
> CO_2 injection/water-alternating-gas cycle = 2.5% hydrocarbon pore volume.
> CO_2 utilization:
> Gross 12 Mscf/stbo
> Net   4 Mscf/stbo

Low-tension waterflood.[26, 27] A 5-acre, 5-spot pilot test of a low-tension waterflood process was undertaken in a previously waterflooded Benoist sand in the Salem Field, Marion County, Illinois. This was joint Texaco Inc./Mobil Oil Company test utilizing a Mobil-licensed process.

The Benoist sand in the pilot area is separated into upper and lower segments by a thin shale stringer. The pilot was conducted in the upper layer only. The average properties of the pilot area are:

> Porosity, PV%, = 14.8.
> Permeability, millidarcies, = 87.
> Upper layer pay thickness, ft, = 26.
> Calculated oil saturation after waterflooding, PV%, = 30.
> Oil viscosity at reservoir conditions, cp, = 3.6.
> Oil in place in the 5.8-acre pilot area, stbo, = 50,000.

The process was comprised of the following:

1. 0.519 PV preflush of softened fresh water to displace the formation brine
2. 0.285 PV petroleum sulfonate surfactant slug
3. 0.305 PV polymer drive slug
4. 1.0 PV Salem field injection brine to provide the final drive

The pilot pattern is a 5.8-acre regular 5-spot inside a 20-acre regular 5-spot. The 20-acre pattern was used as four backup injection wells for the pilot pattern producer surrounded by four chemical injectors (fig. 17–7). Two observation wells were located 67 ft and 164 ft from the injector in the east quadrant. These were used to periodically collect samples for analysis of the chemical tracers injected at the four injectors.

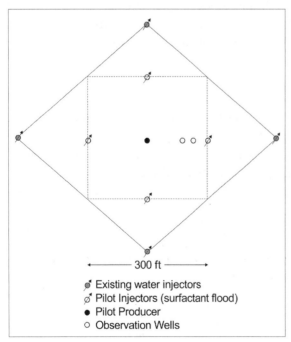

Fig. 17–7. Enhanced oil recovery pilot pattern

- 300 ft
- ⌀ Existing water injectors
- ⌀ Pilot Injectors (surfactant flood)
- ● Pilot Producer
- ○ Observation Wells

A computer model simulating tracer and surfactant flooding was used to evaluate pilot performance. The 5-spot multilayer model accounted for the following:

- Chemical transport involving dispersion, adsorption, and partitioning
- Flow of oil and water considering high tension (immiscible) or low tension (miscible-like), depending on the chemical environment
- Non-Newtonian flow of polymer solution and permeability reduction because of polymer adsorption on the rock

The comprehensive analysis of the pilot performance consists of the following:

- Analysis of the tracer concentration data to verify quadrant flow patterns, fraction of production contributed by each quadrant, and vertical reservoir heterogeneity
- Estimation of chemical consumption based upon tracer and chemical breakthrough volumes and concentration profiles at the observation and production wells
- Comparison of simulated recovery performance with the measured results

Results of the pilot performance evaluation are given as follows:

- Observation concentration data established that the low-tension waterflood process is capable of displacing essentially all of the oil in place under the proper formation environment.
- Only 25% of the original expected recovery volume of the tertiary oil will be recoverable.

- Inadequate preflushing and greater petroleum sulfonate retention than indicated in the laboratory tests contributed to lower-than-expected oil recovery (fig. 17–8).

Even though pressure wave tests showed similar communication in all quadrants, only three of four chemical tracers were detected at the producer.

Unfavorable sodium bromide retention might have been the reason for its not showing up in the producer.

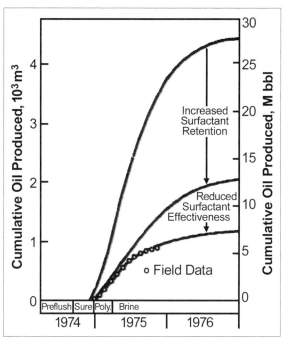

Fig. 17–8. Cumulative oil production versus time. *Source: R. H. Widmyer, A. Satter, G. D. Frazier, and R. H. Graves. 1977. Low-tension waterflood pilot at the Salem Unit, Marion County, Illinois—part 2: performance prediction. Journal of Petroleum Technology. August, 933–938. © Society of Petroleum Engineers. Reprinted with permission*

Recovery of Oil Sands —Emerging Technologies

As noted in chapter 15, much of the world's petroleum resources are based on oil sands. These are composed of highly viscous (> 50,000 cp under standard conditions), tarlike compounds known as bitumen. Hence, considerable focus has been directed toward extracting bitumen and extra heavy oils in recent years. The total initial hydrocarbon in place in the Alberta oil sands is estimated to be 1,699 billion bbl. Oil sands are located at relatively shallow depths (900 ft–1,600 ft). Certain quantities of bitumen occur at or near the surface, which can be extracted by open-pit mining techniques. The remaining reserves from oil sands are estimated at 174 billion bbl, resulting in Canadian reserves being the second largest in the world. Besides Canada, extra heavy oil and bitumen are also in abundance in Venezuela.

Bitumen does not flow naturally in porous media due to its very high viscosity; it has a molasses-like consistency under standard conditions. The typical range of bitumen API gravity is 8°–12°API. The percentage of bitumen in oil sands is about 10%–12%. Bitumen requires special treatment (upgrade to synthetic crude oil) prior to refining and transportation through pipelines. In 2004, bitumen production from oil sands exceeded 1 million bopd. A small percentage of highly viscous crude can be recovered by utilizing conventional techniques. However, it is believed that up to 80% of the initial hydrocarbon in place can be ultimately recovered by various innovative methods. These include, but are not limited to, the following elements or their combinations:[28]

- Injection and production through horizontally drilled wells
- Injection of steam or air, each method being tailored to specific geology
- Solvent extraction
- Oil recovery by gravity drainage

The emerging technologies related to enhanced oil recovery processes as currently applied to oil sands are as follows:

- Cyclic steam stimulation through vertical or horizontal wells
- Steam-assisted gravity drainage
- Expanding solvent–steam-assisted gravity drive process
- Vapor extraction process
- In-situ combustion utilizing toe-to-heel air injection (THAI) through a horizontal wellbore

In a cyclic steam stimulation process, high pressure steam is injected into the formation. This is followed by viscosity reduction of the bitumen and finally production of more mobile hydrocarbons through the wellbore for weeks or months.

Once production starts to decline, the three-stage process is repeated for further recovery. Multilateral horizontal wells may be employed to inject steam and then produce the less-viscous oil by pumping. Besides mobility enhancement, production is augmented as high pressure steam injection creates certain channels in the porous media. About 3,200 wells operate under the process in Cold Lake, Alberta. Two above-ground pipelines are installed; one of them delivers the required steam, and the other collects the heavy petroleum for processing.

A steam-assisted gravity drive process that employs a pair of horizontal wells placed one above the other is illustrated in chapter 19. In the Athabasca sands, Alberta, the horizontal wells are 2,300 ft long and vertically separated by 16.4 ft. Steam is injected in the upper horizontal well. This is followed by production of hydrocarbons of reduced viscosity through the lower horizontal well due to gravity drainage.

In the vapor extraction process, solvent hydrocarbons such as ethane or propane are injected into the formation through vertical or horizontal wells. A vapor chamber is created in the porous media, which facilitates the flow of heavy hydrocarbons by the mechanism of gravity drainage. In a variation of the process, the solvent is co-injected along with steam.

As of this writing, the typical range of recovery by using in-situ methods varies between 25% and 60%. Cyclic steam stimulation and steam-assisted gravity drive processes appear to be more successful. There are challenges with the thermal processes. These include large requirements of injection fluids (water and gas) and the adverse effects of greenhouse gases (CO_2, for example) that are produced.

Relatively light bitumen can be extracted "cold," i.e., without employing a thermal process. Progressive cavity pumps are utilized to produce bitumen that is capable of flowing through the wellbore, along with coproduction of sand. The primary production process, known as cold heavy oil production with sand (CHOPS), is used to recover both heavy oil and bitumen. The method works well where the sand grains are not cemented, a large amount of gas is dissolved in hydrocarbons, and no bottom water drive is present.[29]

Summing Up

Enhanced oil recovery processes include all methods that use external sources of energy or materials to recover oil that cannot be produced economically by conventional means. Enhanced oil recovery processes include the following:

- Thermal methods: steam stimulation, steam flooding, and in-situ combustion.
- Chemical methods: surfactants, polymer, micellar-polymer, and caustic alkaline.
- Miscible methods: hydrocarbon gas, CO_2, and nitrogen. In addition, flue gas and partial miscible/immiscible gas flood may be also considered.

Enhanced oil recovery processes are utilized to mobilize the residual oil throughout the entire reservoir after primary and secondary recovery processes. This can be achieved by enhancing microscopic oil displacement and volumetric sweep efficiencies. Oil displacement efficiency can be increased by decreasing oil viscosity using thermal floods or by reducing capillary forces or interfacial tension with chemical floods. Volumetric sweep efficiency can be improved by increasing the drive water viscosity using a polymer flood.

Many reservoirs contain viscous crude oil. Attempts to produce such oils with water flooding will yield very poor recoveries. Application of heat is often the only feasible solution to recovery from such reservoirs. Thermal methods, particularly steam floods, are effectively used for heavy viscous oils (10°–20° API). Steam floods are used commercially in California's heavy oil reservoirs.

Chemical flood processes, which are applicable to lighter oils, require conditions favorable to water injection, as they are modifications of waterflooding. Even though they showed promise earlier as a viable enhanced oil recovery process, chemical floods were not really successful. They are no longer utilized.

Among the miscible floods, CO_2 miscible floods applicable to lighter oils have been commercially successful. They are utilized widely in West Texas.

Enhanced oil recovery processes require heavy financial investments initially and have high operating costs. Response and returns of capital investments come several years down the road. Statistics show that active U.S. enhanced oil recovery projects and productions are declining.

Offshore enhanced oil recovery operations require consideration of certain issues. These include detailed reservoir description, cost and space requirements for injected material, unique technical risks, and high capital expenditure.

Screening of enhanced oil recovery processes for potential application in the field is a necessary step. The screening criteria are based upon rock and fluid properties of the reservoirs. There is no cure-all process for recovering residual oil after primary and secondary recovery processes. After screening, the subsequent steps would be further theoretical and experimental evaluation of the candidate processes, and possibly a pilot test in the field. Also, pilot test evaluation/scale-up of forecast and commercial venture are necessary.

The Duri steam flood in Indonesia has been very successful. After 14 years of steam flooding, a certain area in the field achieved a recovery factor as high as 64%, with an ultimate recovery target of 69%. Reservoir simulation studies indicated that about 1% PV of injected fluid is required for optimum recovery.

Ultimate primary and waterflood recovery in the North Ward Estes field was estimated to be 31% original oil in place, as compared to 29% to date. CO_2 flood recovery was expected to be 8% original oil in place, with a CO_2 slug size of 38% HCPV and net CO_2 utilization of 4 MMcf/stbo.

Evaluation of the Salem low-tension waterflood pilot performance results shows that the process is capable of displacing essentially all of the in-place oil under the proper formation environment. However, the tertiary oil recovery was significantly lower than expected due to inadequate preflushing and greater-than-expected petroleum sulfonate retention.

Emerging technologies in enhanced oil recovery processes are currently geared toward extracting bitumen from oil sands. Bitumen is a highly viscous and heavy tar-like compound of hydrocarbons. The largest deposits of oil sands containing bitumen are found in Canada and Venezuela. In those locations, emerging technologies have the potential of recovering hundreds of billions of barrels of bitumen, which is subsequently upgraded and refined. In-situ thermal recovery processes include, but are not limited to, cyclic steam stimulation (CSS), steam-assisted gravity drive (SAGD), and vapor extraction processes (VAPEX). These technologies are based on air, steam, or solvent injection, followed by the production of hydrocarbons through the mechanisms of mobility enhancement or gravity drainage, among others. Both horizontal and vertical wells are employed in recovery. For relatively light bitumen, a nonthermal method known as cold heavy oil production with sand (CHOPS) is used. In this method, the bitumen is extracted along with coproduction of unconsolidated sand.

Class Assignments

Questions

1. Why are enhanced oil recovery processes needed? Describe in general the reservoir conditions that are most suitable for enhanced oil recovery processes.

2. What are the fundamental concepts involved in enhanced oil recovery processes?

3. List the various types of applicable enhanced oil recovery processes and their specific applications.

4. What do the mobility ratio and capillary number have to do with enhanced oil recovery processes?

5. What criteria are used to screen for applications of enhanced oil recovery processes? On what are these criteria based?

6. What are the follow-up steps for applications of the selected process in the field?

7. Is there a "cure-all" process for recovering residual oil after primary and secondary recovery processes?

8. What are the commercial, i.e., economically viable, enhanced oil recovery processes that the industry has today?

9. Where and why are steam floods utilized in the United States?

10. What is the largest steam flood project known in the world? Describe a few management strategies adopted in the field.

11. Where and why are CO_2 miscible floods utilized in the United States?

12. Describe briefly the mechanisms of oil recovery in the various enhanced oil recovery processes discussed in this chapter, including the extraction of oil sands

Exercises

17.1. Reviewing the data given in Table 8–3 in chapter 8, select a potentially applicable enhanced oil recovery process for the field. Describe the key factors that must be considered in the selection process. Provide a range for ultimate recovery of oil. Assume the field is located onshore, and secondary recovery has been reasonably successful. Make any other necessary assumptions.

17.2. Prepare case studies related to the following enhanced oil recovery projects based on a literature review:

(a) Huff-and-puff process

(b) CO_2 injection

(c) Polymer or surfactant flooding

(d) Enhanced oil recovery in fractured formations

(e) Application of innovative technology in enhanced oil recovery

(f) Steam-assisted gravity drive in oil sands

Include the following in the case studies prepared:
- Location and size of the field
- Any known heterogeneities
- Integrated reservoir studies
- Design aspects, including rate, pressure, and well pattern
- Enhanced oil recovery monitoring, including logs and well tests, if any
- Duration of the project
- Incremental oil recovery
- Cost aspects, if available
- Field problems and lessons learned

17.3. Compare two case studies in large-scale enhanced oil recovery projects from the literature, one performed in the current decade and the other in the 1970s or 1980s. Highlight any differences observed in the design and management of the projects.

17.4. Discuss emerging and experimental technologies in developing oil sands that are not described in the chapter. Describe how oil shales are extracted and discuss the potential for this method of production.

References

1. Satter, A., and G. C. Thakur. 1994. *Integrated Petroleum Reservoir Management—A Team Approach*. Tulsa: PennWell.

2. Satter, A., and G. C. Thakur. 1994.

3. Craig, F. F., Jr. 1971. *The Reservoir Engineering Aspects of Waterflooding*. SPE Monograph. Vol. 3. Richardson, TX: Society of Petroleum Engineers.

4. Moore, T. F., and R. L. Slobod. 1956. The effect of viscosity and capillarity of the displacement of oil by water. *Producers Monthly*. (August).

5. Satter, A., and G. C. Thakur. 1994.

6. Society of Petroleum Engineers. 1985. *Thermal Recovery Processes*. SPE Reprint Series No. 7. Richardson, TX: Society of Petroleum Engineers.

7. Society of Petroleum Engineers. 1972. *Thermal Recovery Techniques*. SPE Reprint Series No. 10. Richardson, TX: Society of Petroleum Engineers .

8. Prats, M. 1982. *Thermal Recovery*. SPE Monograph 7. Richardson, TX: Society of Petroleum Engineers.

9. Boberg, T. C. 1988. *Thermal methods of oil recovery*. Exxon Monograph. New York: John Wiley & Sons.

10. Society of Petroleum Engineers. 1971. *Miscible Processes*. SPE Reprint Series No. 8. Richardson, TX: Society of Petroleum Engineers.

11. Society of Petroleum Engineers. 1985. *Miscible Processes II*. SPE Reprint Series No. 18. Richardson, TX: Society of Petroleum Engineers.

12. Stalkup, Jr., F. I. 1983. *Miscible Displacement*. SPE Monograph 8. Richardson, TX: Society of Petroleum Engineers.

13. Klins, M. A. 1984. *Carbon Dioxide Flooding*. Boston: IHRDC.

14. Taber, J. J., and F. D. Martin. 1983. Technical screening guides for the enhanced recovery of oil. SPE Paper #12069. Presented at the 1983 Annual Technical Conference and Exhibition, San Francisco, CA, October 5–8.

15. Martin, F. D., and J. J. Taber. 1992. Carbon dioxide flooding. *Journal of Petroleum Technology*. April, 396–400.

16. Abrams, A. 1988. The influence of viscosity, interfacial tension, and flow velocity on residual oil saturation left by waterflood. *Surfactant/Polymer Chemical Flooding*. Vol. II, SPE Reprint Series No. 24.

17. Christensen, J. R., E. H. Stenby, and A. Skauge. 2001. Review of WAG field experience. *SPE Reservoir Evaluation & Engineering*. April. Vol. 4, no. 2, 97–106

18. Bondor, P. L., J. R. Hite, and S. M. Avasthi. 2005. Planning EOR projects in offshore oil fields. SPE Paper #94637. Presented at the SPE Latin American and Caribbean Petroleum Engineering Conference, Rio de Janeiro, June 20–23.

19. Schulte, W. M. 2005. Challenges and strategy for increased oil recovery. IPTC Paper #10146. International Petroleum Technology Conference, Doha, Qatar, November 21–23.

20. Gael, B. T., S. J. Gross, and G. J. McNaboe. 1995. Development planning and reservoir management in the Duri steam flood. SPE Paper #29668. Presented at the SPE Western Regional Meeting, Bakersfield, CA, March 8–10.

21. Fuaadi, I. M., J. C. Pearce, and B. T. Gael. 1991. Evaluation of steam-injection design for the Duri steamflood project. SPE Paper #22995. Asia-Pacific Conference, Perth, Australia, November 4–7.

22. Sigit, R., D. Satriana, J. P. Peifer, and A. Linawati. 1999. Seismically guided bypassed oil identification in a mature steamflood area, Duri field, Sumatra, Indonesia. Asia Pacific Improved Oil Recovery Conference, Kuala Lumpur, Malaysia, October 25–26.

23. Nath, D. K., 2005. Fiber optic used to support reservoir temperature surveillance in Duri Steam Flood. SPE Paper #93240. Asia Pacific Oil and Gas Conference and Exhibition, Jakarta, Indonesia, April 5–7.

24. Winzinger, R., J. L. Brink, K. S. Patel, C. B. Davenport, Y. R. Patel, and G. C. Thakur. 1991. Design of a major CO_2 flood—North Ward Estes field, Ward County, Texas. *SPE Reservoir Evaluation*. February, 11–16.

25. Thakur, G. C. 1990. Implementation of a reservoir management program. SPE Paper #20748. Presented at the 1990 SPE Annual Technical Conference and Exhibition, New Orleans, LA, September 23–26.

26. Whiteley, R. C., and J. W. Ware. 1977. Low-tension waterflood pilot at the Salem Unit, Marion County, Illinois—part 1: field implementation and results. *Journal of Petroleum Technology*. August, 925–932.

27. Widmyer, R. H., A. Satter, G. D. Frazier, and R. H. Graves. 1977. Low-tension waterflood pilot at the Salem Unit, Marion County, Illinois—part 2: performance prediction. Journal of Petroleum Technology. August, 933–938.

28. Cunha, L. B. 2005. Recent in-situ oil recovery—technologies for heavy- and extraheavy-oil reserves. SPE Paper #94986. SPE Latin American and Caribbean Petroleum Engineering Conference, Rio de Janeiro, Brazil, June 20–23.

29. Centre for Energy. 2006. Web literature. Canada

18 · Fundamentals of Petroleum Economics, Integrated Modeling, and Risk and Uncertainty Analysis

Introduction

All petroleum ventures, from basin exploration to matured field revitalization, require capital investment, with an objective of generating profits. Investments are usually substantial, requiring careful and detailed economic study. For example, the overall cost of developing a large offshore complex (exploration, production, surface handling facilities, transportation, and others) may run into billions of dollars. The goal of reservoir management is to maximize the economic profitability of a project. Making sound business decisions requires that the project will be economically viable, generating profits that meet or exceed the economic goal of the enterprise. In recent years, a reservoir team is viewed by the management as more of an "asset team," and is expected to add value to the asset (petroleum reserves). The view is that all technical initiatives of a reservoir team must be integrated with the overall asset management goals. This chapter provides a review of commonly used economic criteria, and a working knowledge of analyzing project economics.[1–8]

The learning objectives of this chapter include the following:
- Objectives of economic analysis
- Integrated economic model
- Risk and uncertainty in the petroleum industry
- Economic analysis procedure
- Data requirement for economic analysis
- Economic decision criteria
- Discounted cash flow analysis
- Example cash flow analysis and sensitivity
- Probabilistic approach to economic evaluations
- Decision tree analysis with example
- Monte Carlo simulation with examples

Objectives of Economic Analysis

This chapter outlines the typical approach adopted in the industry for analyzing and evaluating capital investments in the exploration, production, and development of oil and gas fields. Economic optimization, including competitive production costs, is the ultimate goal of sound reservoir management. It involves building multiple scenarios or alternative approaches in order to arrive at the optimum solution. Issues in oilfield development include, but are not limited to, the following:
- Exploration strategy: the optimum number of exploratory wells to drill in a new basin
- Recovery scheme: natural depletion augmented by fluid (water or gas) injection
- Well spacing: the number of wells and platforms
- Drill high-density wells or initiate an enhanced oil recovery project?
- Return on investment/rate of return under best-case and worst-case scenarios
- Correlation of capital investment with field size as new reserves are discovered

The resulting economic analyses and comparative evaluation of scenarios can provide the required answers to make the best business decisions. This may lead to the maximum value added to the petroleum asset given the available technology, expected reservoir performance, and market conditions.

Integrated Economic Model

An integrated approach to develop and manage oil and gas fields requires the evaluation of all relevant technological and economic aspects. These include, but are not limited to, the following:

- Optimum scheduling of wells and fields to be developed, based on economic analysis and contractual obligations, in addition to reservoir simulation.
- Optimum scheduling of construction of necessary infrastructure, including surface facilities and pipelines.
- Technical, operational, and financial constraints imposed on field development. Examples include the following:
  · Number of wells that can be drilled from an offshore platform
  · Uncertainties in future reservoir performance
  · Penalty imposed by contract for nondeliverability of required volumes
  · Logistics (involving personnel, material, etc.)

In optimizing an integrated model, the following three aspects should be considered:[9,10]

- **Design decisions.** The results are either true or false, or integer valued. Examples are: fields closest to the platforms to be developed first, number of wells to be drilled, number of compressors to be installed, etc.
- **Operational decisions.** The results are continuous in nature and may assume any value. Examples are: optimum oil or gas production rate, bottomhole pressure–controlled production, water injection rates, etc.
- **Nonlinearity in physical system.** The relationships between well rates, reservoir pressure, operating conditions, and other related parameters are highly nonlinear.

Given this complexity, the need to develop an integrated workflow is apparent. It is vital in order to optimize all activities during the entire life cycle of the reservoir. It is used to coordinate capital expenditures, construction of surface facilities and pipelines, scheduling of wells, design of an enhanced recovery project, and abandonment. This effort leads to maximization of asset performance under existing contractual obligations or market conditions. Optimization of an integrated economic model requires a multidisciplinary approach, evaluation of a large number of what-if scenarios, and modeling of uncertainties involved in various elements. The latter is discussed in the following.

Risk and Uncertainty in the Petroleum Industry

The activities related to exploration and production of petroleum are inherently associated with a myriad of risks and uncertainties.[11] Consider the following scenarios in predicting the success or failure in oil and gas property investments:

- Exploratory wells drilled in a new basin or field may or may not turn out to be productive or economically feasible.
- An oil or gas field may not generate revenue as expected following initial production. The issue could be rooted in unidentified geologic complexities, among other factors.
- Future oil and gas prices could move unpredictably in a rapidly changing world of supply and demand.
- Unforeseen events such as political unrest, regional conflict, or natural calamity may adversely affect the demand, production, and transport of petroleum.
- New governmental policies, regulations, and taxes may significantly influence the way a petroleum company conducts business.
- The inflation factor or other economic indicators in the future cannot be known with certainty.
- As oil and gas prices increase due to ever-increasing world demand, alternate sources of energy may become economically attractive.
- Environmental considerations may play a role in weighing other options of energy in a specific industry or region.

Obviously, economic analysis of a petroleum venture requires the recognition and quantification of risk and uncertainties in wide-ranging areas. In conclusion, the feasibility of a petroleum field may be critically affected by a myriad of factors. Most of them cannot be controlled by reservoir professionals. Thus not all of the influencing factors are within the scope of this book.

Table 18–1. Data requirements summary

Data	Source/Comment
Oil and gas rates versus time	Reservoir engineers/Unique to the project
Oil and gas prices	Finance and economic professionals/Strategic planning interpretation
Capital investments (tangible and intangible), and operating costs	Facilities, operations, and engineering professionals
Royalty/production sharing	Unique to each project
Discount and inflation rates	Unique to each project
State and local taxes (production, severance, ad valorem, etc.)	Finance and economic professionals/Strategic planning interpretation
Federal income taxes, depletion, and amortization schedules	Accountants

Economic Analysis Procedure

The tasks in project economic analysis require team efforts consisting of the following:

1. **Setting economic objectives based on the company's economic criteria.** Members of the asset team, including reservoir engineers, are responsible for developing the economic justification, with input from the management.

2. **Formulating scenarios for project development.** These include, but are not limited to, risk and uncertainties inherent in most reservoir engineering projects. Engineers and geologists are primary contributors, with guidance from the management.

3. **Collecting production, operation, and economic data.** The basic data required for economic analysis includes production, injection, investment, price, operating costs, discount and inflation, production sharing, and taxes. Table 18–1 presents a list of pertinent data, its source, and comments.

4. **Performing economic calculations.** These may be either deterministic or probabilistic, or both. Engineers and geologists are primarily responsible.

5. **Performing sensitivity analyses and choosing an optimum project.**[12] Both engineers and geologists are primarily responsible for analysis. Engineers, geologists, operations staff, and management work together to decide on the optimum project.

In essence, sound estimates of hydrocarbon in place, reservoir performance forecasts, capital investment, and operating expenses are indispensable in any economic analysis.

Figure 18–1 presents the steps involved in economic optimization and analysis.

Economic Decision Criteria

Making a sound business decision requires yardsticks for measuring the economic value of proposed investments and financial opportunities. Each company has its own

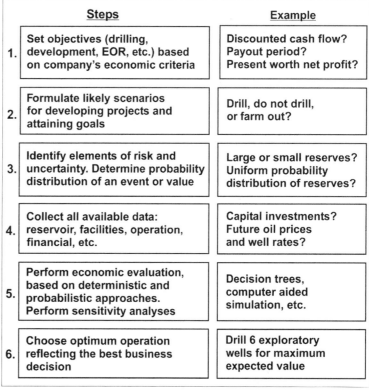

Steps	Example
1. Set objectives (drilling, development, EOR, etc.) based on company's economic criteria	Discounted cash flow? Payout period? Present worth net profit?
2. Formulate likely scenarios for developing projects and attaining goals	Drill, do not drill, or farm out?
3. Identify elements of risk and uncertainty. Determine probability distribution of an event or value	Large or small reserves? Uniform probability distribution of reserves?
4. Collect all available data: reservoir, facilities, operation, financial, etc.	Capital investments? Future oil prices and well rates?
5. Perform economic evaluation, based on deterministic and probabilistic approaches. Perform sensitivity analyses	Decision trees, computer aided simulation, etc.
6. Choose optimum operation reflecting the best business decision	Drill 6 exploratory wells for maximum expected value

Fig. 18–1. Economic optimization in petroleum ventures

economic strategy for conducting business profitably. Standards are set based on certain minimum values in outcome in order to either accept and pursue a venture or reject it altogether. In the petroleum industry, the business ventures include, but are not limited to, basin exploration, field development, and productivity enhancement. Commonly used economic criteria in the industry are outlined in the following discussion.

Payout time. The time needed to recover the investment is defined as the payout time. It is the time when the undiscounted or discounted cash flow is equal to zero. Cash flow may be defined as the following:

Cash flow (CF; in $) = Revenue − Capital investment − Operating expenses

The shorter the payout time, the more attractive the project. In the petroleum industry, the payout time is generally considered to be two to five years. In other words, the cost of drilling and operating wells should be realized back within the above time frame. Although it is an easy and simple criterion, it does not measure the ultimate lifetime profitability of a project and should not be used solely for assessing the economic viability of the project.

The time value of money is not recognized in the case of undiscounted cash flow.

Discounted cash flow, as opposed to undiscounted cash flow, means that a deferment or discount factor is used to account for the time value of money. The future value or worth of money is converted to its present worth (PW) in accordance with the specified discount rate.

Considering that revenues are received once a year at the midpoint of the year, a familiar practice in the petroleum industry, the discount factor (DF) is given by the following:

$$DF = 1/(1 + i)^{(t - 0.5)} \tag{18.1}$$

where

t = the time in years, and
i = the annual discount rate, fraction.

Hence, the discounted cash flow (DCF) in any given year can be computed based on the following:

$$DCF = CF \times (1 + i)^{-(t - 0.5)} \tag{18.2}$$

Profit to investment ratio. The profit to investment ratio is the total undiscounted cash flow, without capital investment, over the total investment. Unlike the payout time, it reflects total profitability. However, it does not recognize the time value of money.

Present worth net profit (PWNP). Present worth net profit is the present value of the entire cash flow discounted at a specified rate.

Investment efficiency. Also known as the present worth index or profitability index, investment efficiency is the total discounted cash flow divided by the total discounted investment.

Discounted cash flow return on investment (DCFROI). The discounted cash flow return on investment is also called the internal rate of return (IRR). It is the maximum discount rate that must be charged for the investment capital to produce a break-even venture. In other words, it is the discount rate at which the present worth net profit is equal to zero. This can be expressed in the following:

$$0 = -C + \frac{CF(1^{st} \text{ yr})}{(1+i)^{0.5}} + \frac{CF(2^{nd} \text{ yr})}{(1+i)^{1.5}} + ... + \frac{CF(n^{th} \text{ yr})}{(1+i)^{n-0.5}} \tag{18.3}$$

where

C = initial capital investment, $,
CF = net cash flow, $, and
i = discounted cash flow or internal rate of return.

In Equation 18.3, it is assumed that the entire amount (cash flow) is received at the midpoint in any given year.

Discounted Cash Flow Analysis

The results of economic analysis are subjected to many restrictive assumptions in forecasting recoveries, oil and gas prices, investment and operating costs, and the inflation rate.

In the petroleum industry, the procedure used in calculation of discounted cash flow generated by production and sales, before federal income tax (BFIT), is outlined as follows:

1. Calculate annual revenues using oil and gas sales from production and unit sales prices.
2. Calculate year-by-year total costs, including capital investments (drilling, completion, facilities, and abandonment, etc.), operating expenses, and production taxes.
3. Calculate annual undiscounted cash flow by subtracting total costs from the total revenues.
4. Calculate annual discounted cash flow by multiplying the undiscounted cash flow by the discount factor at a specified discount rate.

Examples are provided in later sections to illustrate this procedure, including a computer-aided probability analysis of economic outcome.

Probabilistic Approach to Economic Evaluations

Economic evaluation of petroleum ventures requires recognition of risk and uncertainties rooted in relevant technical, economic, and political conditions, as mentioned previously. Risk relates to the likelihood of a given venture or investment encountering possible losses or failures in the future. A "high-risk" project bears the connotation that the probability of failure of the project is substantial, and the invested capital may not bring the expected return.

Uncertainty can also be expressed in terms of a range of probability of occurrences of a particular event. For instance, it could address the degree of certainty or uncertainty that the next well drilled in a large field will have an initial production of 1,000 bopd or more. The reservoir engineer may perform a simple probability analysis based on the production records of previously drilled wells. The engineer could determine that about 45% of the new producers had an initial rate of 1,000 bopd or more. However, there are other relevant factors to consider before assigning a probability value. These include the quality of the reservoir at the drill location (crest versus periphery) and the type of well (vertical versus multilateral horizontal). Other factors include declining reservoir pressure (as compared to reservoir pressures at the time of drilling earlier wells) and an advancing waterflood front (possible breakthrough in nearby wells).

There are situations, however, in which only a few wells have been drilled, and most of the reservoir is unexplored. How can a probability value be assigned to a specific outcome if the field is new? In this case, attempts must be made to glean information from nearby fields located in a similar geologic environment. In any case, the degree of uncertainty will be greater, and the confidence attached to the probability analysis will be low.

Distribution of Probability

In reservoir-related studies, the degree of uncertainty is introduced in calculations by assigning a range of probabilities attached to a parameter or event. A probability distribution function can be developed for a relevant parameter, such as reservoir porosity or field size. This function is based on the frequency of occurrence of various values of the parameter by observation, experience, rational belief, or intuition. The same is true in the case of assigning a probability distribution to an event, such as encountering a dry hole. Examples of probability distributions commonly observed in typical reservoir studies are presented later in this chapter. In addition, cumulative distribution plots, as shown previously in Figures 9–2 (chapter 9) and 11–17 (chapter 11), are frequently utilized to depict the ranges of probability attached to an outcome. The y-axis for cumulative probability in the plots ranges from 0 to 1. A cumulative distribution function is often found to have a lazy S-shape due to the relatively low probabilities encountered towards the far ends of a parameter range.

Selected case studies are presented in the following sections to describe the most familiar types of probability distribution encountered in economic evaluations. Probability distributions are broadly classified into discrete and continuous. Certain distributions are inherently discrete and involve the counting of a particular event, such as rolling dice or drilling exploratory wells in a new field or basin. Other distributions are continuous, such as petroleum reserve, porosity, permeability, connate water saturation, recovery factor, and cash flow, to name a few.

Certain built-in spreadsheet functions are available to simulate the most familiar probability distributions. However, Monte Carlo simulation tools, such as @RISK of Palisade Corp., feature a large array of probability distribution functions that can be used in most situations.

Table 18–2. Binomial probability distribution in oil and gas exploration

Scenario	Probable Outcome of Three Wildcats							
Well No.	1	2	3	4	5	6	7	8
X	Prod	Prod	Prod	Prod	Dry	Dry	Dry	Dry
Y	Prod	Prod	Dry	Dry	Prod	Prod	Dry	Dry
Z	Prod	Dry	Dry	Prod	Prod	Dry	Prod	Dry

Table 18–3. Probability of successful drilling in a petroleum basin

No. of Successful Strikes	Probability
0	0.316
1	0.422
2	0.211
3	0.047
4	0.004
Total:	1.0

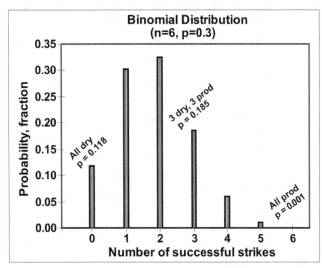

Fig. 18–2. Binomial distribution of outcome following the drilling of 6 wells when the success rate is 3 in 10.

Table 18–4. Cumulative probability distribution

Cumulative Distribution	No. of Successful Strikes
1	0–4
0.684	1–4
0.262	2–4
0.051	3–4
0.004	4

Binomial distribution

Binomial distribution is based on counting of the discrete number of successes or failures. It can be utilized to describe the discrete probability distributions associated with exploratory drilling. Consider an example in which an oil company plans to drill three exploratory wells. Assuming an equal chance between a producer (Prod) and dry hole (Dry), eight outcomes are possible, as shown in Table 18–2.

Based on the above, the probability of drilling all three producers is one in eight (scenario 1), and two producers and a dry well is three in eight (scenarios 2, 4, and 5). The probability of drilling one producer and two dry wells is also three in eight (scenarios 3, 6, and 7). Finally, the probability of all three being dry holes is one in eight.

A generalized equation for binomial probability distribution is given as follows:

$$B\,(x,n,p) = \frac{n!}{[x!(n-x)!]}\,p^x\,(1-p)^{(n-x)} \qquad (18.4)$$

where

    x = number of successful outcomes,

    n = total number of trials, and

    p = probability of success known from experience, fraction.

Example 18.1: Estimate of probability of success in drilling exploratory wells. Consider a petroleum basin in which the probability of success is found to be one in four. Four exploratory wells are planned for the future. What is the probability that none of them will be producers? What is the probability that one, two, three, or all of them will be successful? Tabulate the cumulative distribution. Assume that the probability of success or failure follows a binomial distribution.

Solution: Using Equation 11.4, and noting that p = 0.25, n = 4, and x = 0, 1, 2, 3, and 4, the probability distribution shown in Table 18–3 is obtained.

Figure 18–2 shows the probability distribution of drilling 0 to 6 producers out of 6 planned wells when the historic success rate is 3 in 10.

The cumulative distribution function is calculated as given in Table 18–4.

Normal distribution of field sizes in a petroleum basin

Figure 18–3 presents a histogram of the percent of the field discovered in a basin and the respective hydrocarbon in place. A normal distribution approximately fits the pattern and can be utilized in conducting probability analyses in future exploration.

The normal distribution of a variable is expressed by the following equation:

$$f(x,\mu,\sigma) = \frac{1}{(2\pi\sigma^2)^{\frac{1}{2}}} \exp\left[-\frac{(x-\mu)^2}{2\sigma^2}\right] \qquad \textbf{(18.5)}$$

where

    x = variable in a population or data set,
    μ = mean value of x,
    σ = standard deviation of x, and
    f = normal probability distribution of x.

An inspection of Equation 18.5 indicates that the probability of x is the highest when x = μ, then declines exponentially in either direction in a symmetrical fashion.

As more information is obtained from a field or petroleum basin, the probability distribution of a specific parameter such as the one considered here needs to be updated. For example, once most of the large fields are discovered in a matured basin, the probability distribution of future

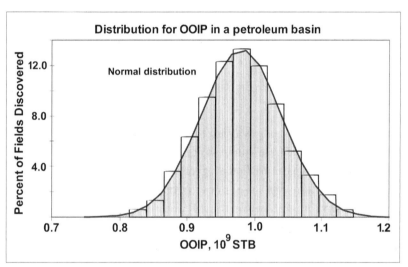

Fig. 18-3. Normal distribution of original oil in place in fields discovered in a petroleum basin

field sizes may shift to the left, indicating the possibility of smaller reserves. Even the probability distribution pattern itself may change with time from normal to another distribution pattern.

Triangular distribution of production from a future well

In another study, the following observations are made based on dozens of wells drilled in a basin:

- Of all the wells drilled, 25% were dry.
- Cumulative production from each successful well during the life of the reservoir is estimated to be between 100,000 and 600,000 bbl as indicated by various reservoir studies.
- For the wells reviewed in the study, the most likely production is found to be around 250,000 bbl.
- The number of wells having cumulative production less or greater than 250,000 bbl decreases steadily until the limiting productions, namely 100,000 bbl and 600,000 bbl, are reached.

Figure 18–4 depicts the probability distribution that can be assigned for a new well to be drilled in the same basin. The "spike" at left represents the probability of drilling a dry well, while the triangular distribution represents the probable outcome of a producing well in terms of ultimate recovery.

The value of a variable with triangular probability distribution can be generated by the following equations:

$$x = x_L + [(x_M - x_L)(x_H - x_L) R_N]^{1/2} \quad \text{when } R_N \leq (x_M - x_L)/(x_H - x_L) \tag{18.6}$$

$$x = x_H - [(x_H - x_M)(x_H - x_L)(1 - R_N)]^{1/2} \quad \text{when } R_N > (x_M - x_L)/(x_H - x_L) \tag{18.7}$$

where

x = randomly generated value of an input variable, such as drainage area or recovery factor used in estimating oil reserve,

x_L = lower limit of x,

x_M = most likely value of x,

x_H = upper limit of x, and

R_N = a random number between 0 and 1, generated by the built-in spreadsheet function.

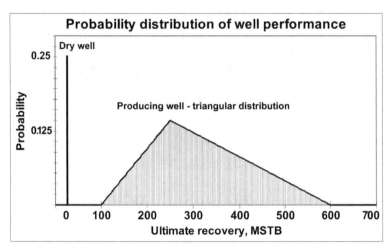

Fig. 18–4. Probability distribution of ultimate recovery from a future well (dry or producer), including triangular distribution of ultimate recovery from the latter

Lognormal distribution of rock permeability

In yet another study, values of core permeability obtained from hundreds of cores in a tight gas field appear to follow a lognormal distribution pattern (fig. 18–5). Most of the permeability values fall between 1 mD and 2.4 mD. However, a small number of cores exhibited permeability values in the range of 5 mD to 6 mD and above. In fact, a lognormal distribution pattern is reported to be common based on field observations relating porosity, permeability, water saturation, formation thickness, and ultimate recovery, among other factors.

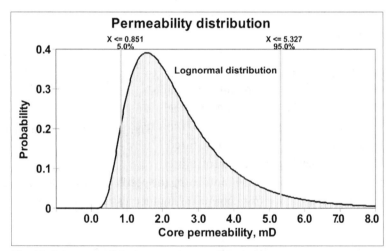

Fig. 18–5. Lognormal distribution of core permeability

Uniform distribution

In certain cases, a reservoir parameter can exhibit a uniform probability of occurrence, in addition to the other distribution patterns. Each value of the parameter in the given range appears to have an equal chance of occurrence, without exhibiting any preference for a particular region within that range. A uniform distribution appears as a rectangle or square on the probability distribution plot. With the aid of a random number generator, the value of a variable (x) having uniform distribution within the upper and lower limits can be generated as follows:

$$x = x_L + R_N (x_H - x_L) \tag{18.8}$$

Conditional probability

In certain other cases, the probability of success or failure in a trial may be dependent on the outcome of the previous trial to some degree. For example, once a new well is found to be dry, does it enhance or reduce the chances of encountering another dry hole in the next trial? Of course, it depends on the specific geology of the target formation at the location, which may or may not be similar to the regional trend. An important observation in basin exploration lies in the fact that most large fields are discovered early due to their large structures, identified by geophysical studies with relative ease. Hence, the probability of discovering fields with very large reserves may diminish with future discoveries. In conducting studies related to risk and uncertainty, conditional probabilities must be considered if it appears that all of the trials (such as drilling a well) are not entirely independent of each other.

Correlation between reservoir parameters

In assigning probability distribution, relevant studies must identify the relationship between two or more parameters encountered in a field. For example, a distinct relationship may exist between formation porosity and permeability, or between porosity and water saturation in a reservoir. In making probabilistic estimates of hydrocarbon in place, or any other factor, any relationship identified between two reservoir parameters must be taken into account. For example, a probability study must recognize that a low porosity zone is likely to be associated with relatively high water saturation in porous media. Assigning an independent probability distribution for connate water saturation may lead to incorrect association of low porosity values with relatively less water saturations. Instead, a mathematical relationship is used to calculate the value of water saturation for a specific porosity assumed during the course of probability estimates.

Figure 18–6 presents the relationship between porosity and permeability in a reservoir. However, not all reservoirs are expected to show identifiable trends.

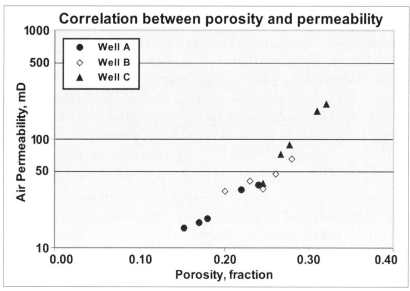

Fig. 18-6. Correlation between porosity and permeability

Assignment of upper and lower bounds of parameter values

Certain distributions, such as normal or lognormal, are continuous in nature and theoretically range from negative infinity to positive infinity. It is true that probabilities calculated by these distributions are miniscule beyond certain values in either direction. However, a more accurate approach would be to truncate the distribution, forcing the probability to be zero. For example, the value of porosity cannot be negative. Another application of truncation of probability may involve the determination oil and gas reserve. The normal probability distribution curve for porosity could be truncated below 5% (the cutoff value) and 35% (the maximum value obtained from hundreds of cores). The probability of encountering a porosity value beyond these ranges is virtually none in this specific case. Typical Monte Carlo simulation tools, including @RISK software, allow assignment of probability bounds in an analysis.

Table 18–5. Porosity range and corresponding sampling frequency based on laboratory studies

Porosity Range (Class)	Number of Samples in Class	Sampling Frequency
>26–27	3	0.01705
>27–28	13	0.07386
>28–29	21	0.11932
>29–30	30	0.17045
>30–31	35	0.19886
>31–32	33	0.1875
>32–33	21	0.11932
>33–34	14	0.07955
>34–35	6	0.03409
Total:	**176**	**1.0**

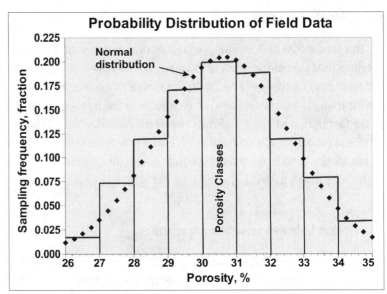

Fig. 18–7. Probability distribution fit of porosity values obtained from 176 core samples

Example 18.2: Probability distribution modeling of field data. In any economic study involving risk and uncertainty, the probability distribution patterns of all parameters need to be quantified or modeled wherever certain risks or uncertainties may exist. For example, in determining the original oil in place in a reservoir with fairly good accuracy, the variation of porosity throughout the reservoir needs to be known, among other factors. In a deterministic model, certain averaging techniques are utilized to determine original oil in place, as discussed in chapter 2. However, when extensive field data is not available, as often is the case, a probability distribution model for porosity may be obtained to perform probability analyses of original oil in place. This can be done for other factors, as well. In chapter 8, one such study is performed based on normal distribution of porosity.

In a thick formation, dozens of core samples (or even more) can be obtained from a single well. The objective of this study is to establish a probability distribution model for porosity values obtained from 176 samples in a field.

Solution: The values of porosity are tabulated in accordance with their ranges and the number of core samples from which they are obtained. In statistical studies, ranges are expressed as classes, where a value of porosity can be greater than x but less than or equal to y. The frequency of a class is calculated by the number of samples associated with porosity values falling in that class over the total number of samples in the study. In Table 18–5, the entire range of porosity is divided into nine classes. The highest frequency is observed in a class having porosity greater than 30%, but less than or equal to 31%.

In effect, the classes of porosity and their frequency as obtained from the study are depicted by binomial distribution (fig. 18–7). However, it is well recognized that normal distribution can be used to approximate the above. In this study, a reasonable fit was obtained by the following:

Distribution is normal.
Mean = 30.65%.
Standard deviation = 1.95.

Various software-based tools, including @RISK, allow regression analysis of available data to obtain the best fit with a host of probability distributions. It must be mentioned that the field data may or may not follow an established probability distribution pattern. "Outliers" may be identified as soon as the data is plotted in an appropriate manner. This would require review of quality control procedures in collecting field data and examination of possible reservoir heterogeneities, among other possible considerations.

Decision Tree Analysis

The economic viability of a petroleum venture, based on probable successes or failures at each phase, is often analyzed by constructing visually appealing decision tree diagrams.[13–15] Two examples of decision tree diagrams are presented in Figure 18–8, in which each square-shaped node represents a decision-making milestone. The decision could be to drill or not to drill, or to run further surveys. The round nodes in the diagram, referred to as chance nodes, indicate a point where two or more probable outcomes of an event can occur. For example, the probabilities associated with drilling four wells in Example 18.1 may constitute a chance node.

Example 18.3: Decision tree analysis in oil and gas exploration. Consider a relatively simple scenario of a decision tree analysis based on Example 18.1, in which an oil company plans to drill four exploratory wells. This decision tree is shown in the upper diagram in Figure 18–8. The cost of drilling each well is about $1.5 million. The probability of drilling zero to four successful producers, based on binomial distribution, is tabulated in Example 18.1. As indicated by the production potential of previously drilled producers, the net present worth of each well is expected to be about $10 million. It is the objective of the study to analyze the venture by using the decision tree method.

Solution: The decision to drill hinges upon the expected value (EV) of the outcome, as given in the following, with "m" representing millions:

$$EV (\$) = (-\$1.5m)(4) + 0.316(0) + 0.422(\$10m \times 1) + 0.211(\$10m \times 2) + 0.047(\$10m \times 3) + 0.004(\$10m \times 4)$$
$$= \$4.01m$$

The proposed venture appears to be attractive, as the expected value of the venture is positive. However, the venture may not be attractive if the expected value is either zero or negative. This can occur due to many reasons, a few of which are noted in the following:

- Ultimate recovery from the wells turns out to be much lower than expected, as oil zones are of significantly lower transmissibilities or have higher water saturations than anticipated.
- The probability of drilling dry holes increases dramatically as unexpected geologic discontinuities are encountered during drilling.
- Future uncertainty in oil prices exists, owing to supply and demand.

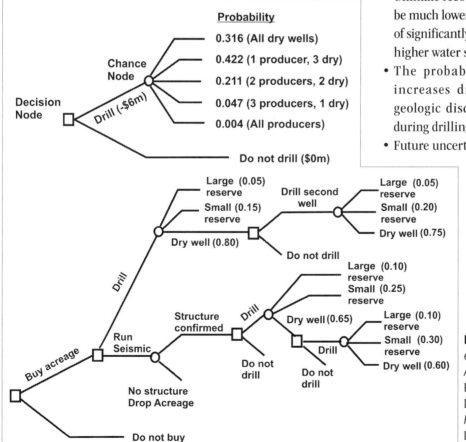

Fig. 18-8. Decision tree diagrams to evaluate various options in exploration. *Adapted from A. W. McCray. 1975.* Petroleum Evaluations and Economic Decisions. *Englewood, NJ: Prentice-Hall; and P. Newendorp. 1975.* Decision Analysis for Petroleum Exploration. *Tulsa: PennWell.*

Based on a conservative assumption that the present net worth of each well is rather $6 million, the following is obtained:

EV ($) = $0

In real-life ventures, the decision tree may take into account the benefits of further geosciences study. It may consider the possibility of farming out the petroleum prospect to an independent operator. It also might include the probability of success associated with the drilling of a second exploratory well, or future petroleum prices, among other considerations. The lower diagram in Figure 18–8 depicts a complex decision tree with several branches (paths) consisting of various choices (decisions) and corresponding chances (probabilities). The branches show outcomes (large reserve, small reserve, and dry hole) and their probabilities in parentheses. What is not shown is the dollar amount associated with each outcome. Decision tree analysis rather starts from the far right, and the expected values at each chance node (circle) are calculated. Working towards the left, choice nodes are encountered. Evaluation of all the branches from the right to the left, based on the expected values associated with each, would result in the identification of the most desirable branch to follow. Decision tree analyses depend critically on the probabilities or "chances" assigned to an outcome, as in any other economic evaluation.

Monte Carlo Simulation

In a deterministic approach, the reservoir engineer estimates one value of original oil in place based on a unique set of values of relevant parameters. This set could consist of data related to reservoir geometry, formation porosity, and saturation of hydrocarbons. The input parameters can vary in vertical or horizontal directions in a reservoir, as estimated from various reservoir studies. However, a range of probable values at a particular point in the reservoir, based on the distribution pattern of a parameter throughout, is not used. A simple example of a volumetric calculation illustrating the deterministic approach is included in chapter 2.

In a probabilistic approach, however, the engineer is interested in a range of values of original oil in place and their probabilities of occurrences. This requires good knowledge of the probability distribution of basic field data that enters into the calculations. Such data includes reservoir area, thickness, porosity, fluid saturation, and fluid formation volume factor. A Monte Carlo simulation technique is utilized to accomplish the above based on thousands of trials. Steps in a Monte Carlo simulation are summarized as follows:

1. Assign probability distribution pattern, range, and cutoff, if any, for the input parameters. Examples of probability distributions are described earlier. For estimation of original oil in place, the set of values consists of A, h, ø, S_w, and B_{oi}. When two parameters are correlated, such as porosity and water saturation, the values of one of the parameters (porosity) are generated independently. The values of the second parameter are calculated according to the correlation (saturation) during simulation.

2. Generate a set of values of individual input variables according to the probability distribution and ranges assigned. (*Note:* Steps 1 and 2 constitute the first trial in a Monte Carlo simulation.)

3. Compute the original oil in place for the first set by the volumetric method:

    $$OOIP = 7{,}758 \, Ah\varnothing S_w / B_{oi}$$

    The original oil in place in trial n = f (random values generated in the nth trial for reservoir area, thickness, porosity, saturation, and recovery factor within the given ranges).

4. Repeat steps 1 and 2 hundreds of times (or more) to obtain a large set of original oil in place values. When a sufficient number of trials has been completed, the probability associated with a particular value of original oil in place does not change significantly with further trials. Typically, a few thousand trials are more than sufficient for reservoir studies. Most of the examples shown in this book are based on 10,000 trials.

5. Based on thousands of values of original oil in place obtained in step 4, it is possible to perform a frequency distribution study similar to what has been illustrated in Example 18.2. The only difference between the two is that the previous example was based on 176 values of porosity instead of 10,000 probable values of original oil in place.

Table 18-6. Input data and results based on Monte Carlo simulation. *Courtesy of Palisade Corp.*

Input Parameter	Porosity, %	Drainage Area, acres	Formation Thickness, ft	Oil Saturation	Oil FVF, rb/stb	
Range	27.5–33.6	144–176	45–55	0.76–0.80	1.235–1.273	
Most likely	30.5	160	50	0.78	1.254	
Distribution	Normal	Triangular	Triangular	Normal	Triangular	
Trial Number	**Values of Parameters Generated by Simulation**					**Simulation Results OOIP, Mstb**
1	28.10	157.2	50.5	0.76	1.270	10,378.0
2	29.11	169.5	48.6	0.78	1.268	11,426.8
3	30.79	164.0	45.9	0.77	1.247	11,169.5
4	32.72	146.3	48.0	0.77	1.270	10,838.4
5	28.49	167.1	49.0	0.78	1.269	11,186.7
6	32.15	175.7	46.2	0.77	1.246	12,519.9
7	28.23	144.5	52.0	0.78	1.270	10,056.4
8	32.89	167.7	54.2	0.78	1.272	14,289.0
—	—	—	—	—	—	—
9,999	33.05	148.1	53.6	0.76	1.247	12,441.6
10,000	28.68	159.8	47.4	0.78	1.271	10,301.8

Results of a few trials (out of 10,000) performed during a Monte Carlo simulation for estimating original oil in place are shown in Table 18–6.

The application of Monte Carlo simulation, first introduced in chapter 9, has been demonstrated in earlier chapters with the following objectives:

- Determination of cumulative probability distribution of oil and gas in place based on volumetrics (chapter 9)
- Probability estimate of recovery factor based on empirical correlation (chapter 10)
- Sensitivity analysis of factors influencing recovery as used in simulation (chapter 10)
- Estimate of cumulative production of a new well under exponential decline having a range of probable decline constants (chapter 11)

A spreadsheet-based Monte Carlo simulation software, @RISK, was used to generate the probability distribution of sought parameters.[16]

Example 18.4: Cash flow analysis of a future well—comparison of deterministic and probabilistic models. In this example, the newly drilled well analyzed in Example 11.5 of chapter 11 is considered. Monte Carlo simulation was performed to determine the probability distribution of ultimate recovery based on lognormal distribution of the exponential decline constant. The objective of the study here is to estimate the discounted cash flow generated by the well in future years based on deterministic as well as probabilistic methods. Again, @RISK software was used to perform the latter.

At first, a single value is assumed for each parameter that enters into the necessary computation of the discounted cash flow calculations. The parameters include the following:

- Capital investment involved in drilling the well
- Annual production from the well, based on an assumed initial rate and subsequent decline with time
- Estimated gas/oil ratio, based on reservoir-specific PVT data
- Forecasted price of oil and gas in future years, including annual escalation
- Annual expenses to produce and maintain the well, including escalation of cost
- Annual discount factor

Table 18–7 summarizes the data used. It is followed by computation of annual production of oil and gas, revenues generated by sales, calculation of annual cash flow, and finally the discounted cash factor for a time frame of 10 years (table 18–8). Given the significant uncertainties involved in predicting oil and gas prices, any economic analysis further into the future is of limited significance.

In performing the cash flow analysis based on a deterministic method, it is easy to recognize the significant uncertainties involved in the calculations. Uncertainties can affect calculations concerning the future performance of the well (maintenance downtimes, unexpected breakthrough of water or gas, well recompletion, etc.). Uncertainties also surround the forecasting of petroleum prices in a dynamic climate. In this case, a probabilistic analysis based on a Monte Carlo simulation provides the "big picture" for the analyst and the company management. The probabilistic approach considers the following:

- A triangular distribution of exponential decline constant was assumed, as in Figure 11–16.
- Effects of future oil and gas prices are analyzed by performing two scenarios, described as follows:
  - (a) Oil and gas prices would increase monotonically by a predetermined value, as in the deterministic model.
  - (b) Oil and gas prices would fluctuate within certain ranges according to triangular probability distributions to oil and gas escalation factors. Data concerning these factors is given in Table 18–9.

An escalation factor less than 0 signifies that the oil price actually decreases compared to that of the previous year. This has been observed in oil industry for many years, but traditional and deterministic approaches do not often account for the above fluctuations in price.

Discount factor and escalation factor for annual expenses are held constant as in the case of a deterministic model. Appropriate probability distributions can be assigned to the above in a study.

Taxes are not included in the calculations. Moreover, it is assumed that the gas/oil ratio remains steady as the reservoir is produced above the bubblepoint pressure.

Table 18-7. Input parameters used in deterministic model

Parameter	Unique Value
Decline constant (D), mo^{-1}	0.2
Gas/oil ratio	875
Oil price, $/bbl	60
Escalation factor (EF), %	8
Gas price, $/Mcf	12
Escalation factor (EF), %	10
Capital investment, million $	2.0
Annual expenses, $	50,000
Escalation factor (EF), %	6
Discount factor (i), %	12

Table 18-8. Discounted cash flow analysis over a period of 10 years

Mos., t	Rate, stb/d	Cum Oil, stb	Annual Oil Prod., stb	Annual Gas Prod., Mcf	Gross Revenue[a]	Investment and Expenses	Net Revenue	Discounted Cash Flow @ i=0.12/yr
	$q = q_i \exp(-Dt/30.4)$	$Q = (q_i - q)/(D/30.4)$	ΔQ	$GOR \times (4)$	$(4) \times \$/bbl \times (1+EF_{oil})^{(t/12-1)}$ $+ (5) \times \$/Mcf \times (1+EF_{gas})^{(t/12-1)}$	Expenses $\times (1.06)^{(t/12-1)}$	$(6) - (7)$	$(8) \times (1+i)^{(-t/12)}$
(1)	(2)	(3)	(4)	(5)	(6)	(7)	(8)	(9)
0	1,000	0	—	—	—	—	—	—
12	924	11,539	11,538.54	10,096.22	$813,467.00	$2,050,000.00	($1,236,533.00)	($1,104,047.32)
24	854	22,201	10,662.63	9,329.80	$814,091.94	$53,000.00	$761,091.94	$606,737.84
36	789	32,054	9,853.22	8,621.56	$814,752.59	$56,180.00	$758,572.59	$539,936.98
48	729	41,160	9,105.24	7,967.09	$815,449.47	$59,550.80	$755,898.67	$480,387.27
60	674	49,574	8,414.05	7,362.30	$816,183.14	$63,123.85	$753,059.29	$427,306.07
72	623	57,349	7,775.33	6,803.41	$816,954.15	$66,911.28	$750,042.87	$379,995.06
84	575	64,534	7,185.09	6,286.96	$817,763.07	$70,925.96	$746,837.12	$337,831.18
96	532	71,174	6,639.66	5,809.70	$818,610.48	$75,181.51	$743,428.97	$300,258.49
108	491	77,309	6,135.64	5,368.68	$819,496.96	$79,692.40	$739,804.55	$266,780.94
120	454	82,979	5,669.87	4,961.14	$820,423.10	$84,473.95	$735,949.15	$236,955.93

[a] EF is the escalation factor of oil and gas prices.

Table 18-9. Range of escalation factors used in probabilistic analysis

Escalation Factor	Minimum	Most Likely	Maximum
Oil, %	−5	8	21
Gas, %	−5	10	25

Figures 18–9 and 18–10 present the comparison between the deterministic and probabilistic approaches in terms of cumulative production over time and cumulative cash flow (discounted) over time. The prices of oil and gas were allowed to increase monotonically by the same factor in both methods. The probabilistic approach reflects the uncertainties associated with well performance by developing a band around the results obtained by the deterministic method. A probability distribution pattern can be skewed, such as the triangular distribution of exponential decline rate utilized in this case. Thus the 50% probability line may not coincide with the results obtained by the deterministic method. In all of the cases (90%, 50%, and 10% cumulative probability, and deterministic approach), the payout period is between two and three years. The probabilistic envelope encompasses large areas above the discounted cash flow predictions by the deterministic model, leading to a more optimistic view of the future.

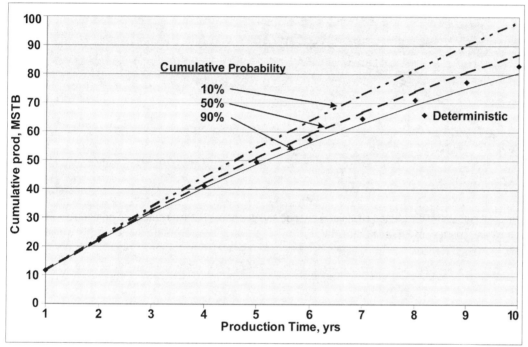

Fig. 18–9. Cumulative production predicted by deterministic and probabilistic models @RISK software, courtesy of Palisade Corp.

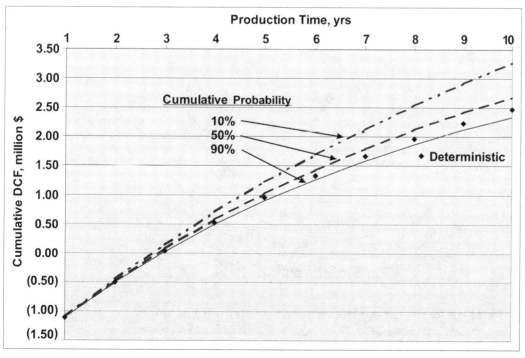

Fig. 18–10. Comparison of cumulative discounted cash flow obtained by deterministic and probabilistic models. Oil and gas prices increase monotonically. @RISK software, courtesy of Palisade Corp.

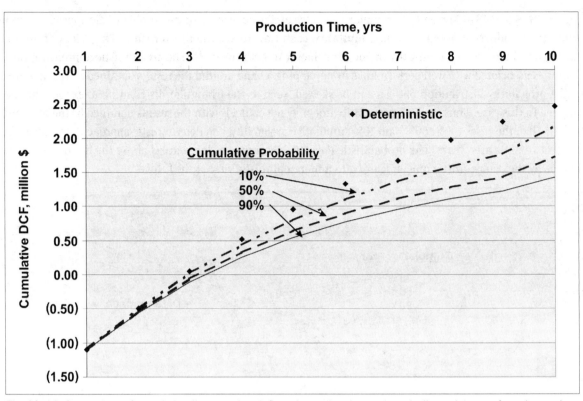

Fig. 18–11. Comparison of cumulative discounted cash flow where oil and gas prices decline or increase from the previous year's levels. @RISK software, courtesy of Palisade Corp.

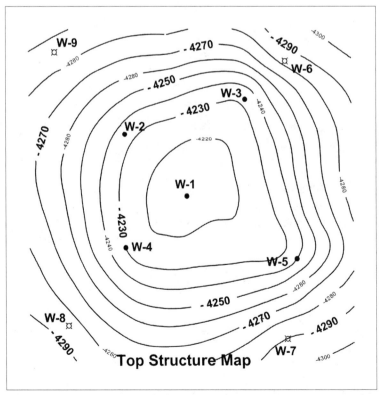

Fig. 18–12. Top structure map of waterflood prospect reservoir showing wells.
Source: G. C. Thakur and A. Satter. 1998. Integrated Waterflood Asset Management. Tulsa: PennWell.

Figure 18–11 compares the discounted cash flow expected from the well when oil and gas prices vary within a wide range, including a modest decrease, rather than a monotonic increase. The results of the probabilistic study indicate that the cash flow could be significantly less than what the deterministic model predicts. The payout period as indicated by the probabilistic study is longer than three years. In this case, the probabilistic model appears to be less optimistic.

Example 18.5: Economic optimization of a waterflood project. A field that was discovered many years ago is now depleted. It consisted of a simple domal structure, and five of the nine wells drilled were producers (fig. 18–12). Primary producing mechanisms were fluid and rock expansion (reservoir pressure above the bubblepoint), solution gas drive, and limited natural water drive. Data available is limited.[12] Even the gas, oil, and water production data is unreliable. Reservoir pressures were not monitored.

An integrated team of geoscientists and engineers was charged by the management to review the past performance and investigate the waterflood potential of this field.

Solution: The team's approach was to accomplish the following:
- Build an integrated geosciences and engineering model of the reservoir using available data and correlations
- Simulate full-field primary production performance without history matching, since no historical pressure data is available
- Forecast performance under peripheral and pattern waterflooding (fig. 18–13).

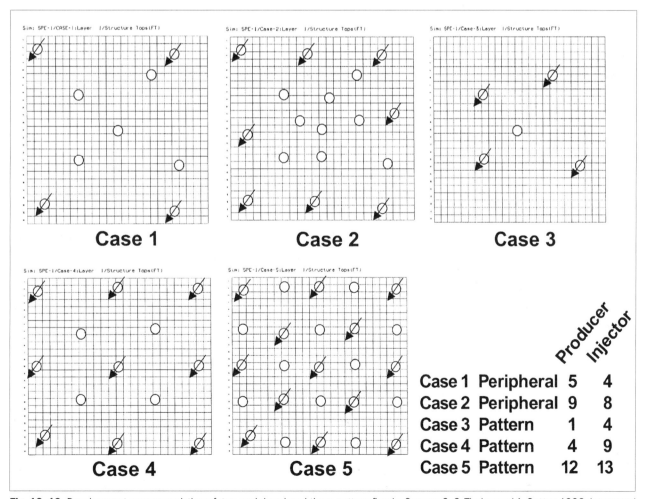

Fig. 18–13. Development cases consisting of two peripheral and three pattern floods. *Source: G. C. Thakur and A. Satter. 1998. Integrated Waterflood Asset Management. Tulsa: PennWell.*

Table 18–10. Economic evaluation of waterflood project case 2

Year	Oil Prod. (MSTB)	Oil Price ($/STB)	Oil Revenue ($MM)(1)×(2)/1000	Gas Prod. (MMSCF)	Gas Price ($/MSCF)	Gas Revenue ($MM)(4)×(5)/1000	Total Revenue ($MM)(3)+(6)	Producing Tax ($MM)(7)×Tax Rate
1997		19.50	0.00		2.10	0.000	0.000	0.000
1998	275.21	19.50	5.37	1610.5	2.10	3.382	8.749	1.750
1999	189.7	19.50	3.70	503.02	2.10	1.056	4.755	0.951
2000	451.54	19.50	8.81	89	2.10	0.187	8.992	1.798
2001	656.27	19.50	12.80	74.12	2.10	0.156	12.953	2.591
2002	605.71	19.50	11.81	69.86	2.10	0.147	11.958	2.392
2003	515.63	19.50	10.50	60.27	2.10	0.127	10.181	2.036
2004	460.57	19.50	8.98	54.43	2.10	0.114	9.095	1.819
2005	407.65	19.50	7.95	48.8	2.10	0.102	8.052	1.610
2006	351.97	19.50	6.86	42.8	2.10	0.090	6.953	1.391
2007	305.06	19.50	5.95	37.63	2.10	0.079	6.028	1.206
2008	235.47	19.50	4.59	29.77	2.10	0.063	4.654	0.931
2009	227.9	19.50	4.44	28.89	2.10	0.061	4.505	0.901
2010	227.89	19.50	4.44	28.89	2.10	0.061	4.505	0.901
2011	227.9	19.50	4.44	28.89	2.10	0.061	4.505	0.901
Total	5138.47		100.200	2706.87		5.684	105.885	21.177

	Interest Rate (%)	Discounted Cash Flow ($MM)
Value for Example Above =	12.00	34.702
Starting Interest Rate =	10.00	38.954
	20.00	22.781
	30.00	14.455
	40.00	9.675
	50.00	6.699
	60.00	4.728
	70.00	3.360
	80.00	2.373
	90.00	1.640
	100.00	1.081
	110.00	0.646
	120.00	0.302
	130.00	0.026
	140.00	-0.199
	150.00	-0.384
	160.00	-0.537
	170	-0.665
	180	-0.773
	190	-0.864
Ending Interest Rate =	200	-0.942

Five scenarios were evaluated, depending on waterflood pattern. The computational procedure is illustrated in a spreadsheet calculation for case 2 in Table 18–10. Note that federal income taxes are not taken into account. The values of several economic parameters for the case 2 example case at a 12% discount rate are given below:

Payout time, years, = 1.78.
Profit-to-investment ratio = 16.44.
Present worth net profit, million $, = 34.702.
Present worth index = 7.781.
Discounted cash flow return on investment, %, = 131.5.

The results of the economic analysis for the five waterflood cases are presented in Table 18–11, summarized here as follows:

- All the patterns considered for the waterflood project appear to be very favorable in terms of return.
- Case 3 gives the lowest amount of investment, reserves, and development costs, and yet a very favorable discounted cash flow return on investment. It also gives the highest profit to investment ratio.
- Case 2 for the peripheral flood and case 5 for pattern flood show the most promising scenarios.
- Case 5 indicates the highest present worth net profit. However, this requires 80% more capital than case 2, which gives about the same present worth net profit and better discounted cash flow return on investment.

It should be realized that for the case 2 peripheral flood, recovery estimates could be optimistic because the reservoir layers were considered to be homogeneous and continuous. This may not represent the real situation.

Table 18–10. (cont.)

	(9)	(10)	(11)	(12)	(13)	(14)	(15)	(16)	
Year	Capital Investment ($MM)	Discount Factor @ 12%	Discounted Capital Investment ($MM)(9)*(10)	Operating Cost ($MM)	Total Cost ($MM) (8)+(9)+(12)	Undiscounted Cash Flow ($MM)(7)-(13)	Discounted Cash Flow @ 12% ($MM)(10)*(14)	Cumulative Discounted Cash Flow @ 12% ($MM)	Time (years)
1997	4.678	0.9449	4.420		4.678	-4.678	-4.420	-4.420	1
1998		0.8437	0.000	0.318	2.068	6.681	5.636	1.216	2
1999		0.7533	0.000	0.318	1.269	3.486	2.626	3.842	3
2000		0.6726	0.000	0.318	2.116	6.876	4.624	8.467	4
2001		0.6005	0.000	0.318	2.909	10.044	6.032	14.498	5
2002		0.5362	0.000	0.318	2.710	9.248	4.959	19.457	6
2003		0.4787	0.000	0.318	2.354	7.827	3.747	23.204	7
2004		0.4274	0.000	0.318	2.137	6.958	2.974	26.178	8
2005		0.3816	0.000	0.318	1.928	6.123	2.337	28.515	9
2006		0.3407	0.000	0.318	1.709	5.245	1.787	30.302	10
2007		0.3042	0.000	0.318	1.524	4.504	1.370	31.673	11
2008		0.2716	0.000	0.318	1.249	3.405	0.925	32.598	12
2009		0.2425	0.000	0.318	1.219	3.286	0.797	33.395	13
2010		0.2165	0.000	0.318	1.219	3.286	0.711	34.106	14
2011	0.204	0.1933	0.039	0.318	1.423	3.082	0.596	34.702	15
Total	4.882		4.460	4.452	30.511	75.374	34.702		

Producing Tax Rate (%) = 20.00

On the other hand, case 5 for the pattern flood case is better suited to treat reservoir heterogeneity and reservoir discontinuity.

The selection of the optimum case will depend on the availability of capital, technical considerations, and the risk involved. The sensitivity analysis discussed below shows that case 2, even if recovery is 20% lower, can still be the best choice.

Sensitivity Analysis

Figure 18–14 shows the sensitivity of discounted cash flow return on investment to oil price, oil production, investment, and operating costs. The analysis shows that discounted cash flow return on investment is affected more drastically by oil price, oil production, and investment than by the operating costs. Present worth net profit is most sensitive to oil price and oil production.

It is again emphasized that the results of the economic analysis are subject to many restrictive

Table 18–11. Waterflood project: economic evaluation results summary

	Case 1	Case 2	Case 3	Case 4	Case 5
Capital investment, $MM	1.853	4.882	0.973	3.484	8.799
Reserves, MMstbo	1.965	5.138	1.378	3.176	5.105
Project life, yrs	15	15	15	15	15
Payout, yrs	2.58	1.78	2.44	2.74	2.28
DCFROI, %	69.64	131.15	80.12	87.83	104.84
PWNP, $MM	9.454	34.702	7.013	18.721	35.184
Profit to investment ratio	16.88	16.44	23.32	13.91	8.74
Present worth index	5.66	7.78	8.02	5.90	4.35
Development costs, $/stbo	1.10	1.72	0.71	1.1	1.72

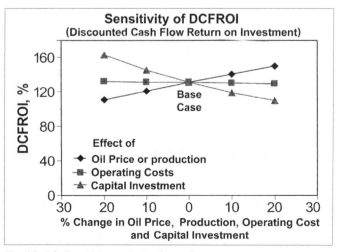

Fig. 18–14. Sensitivity analysis of case 2

assumptions. These relate to forecasting recoveries, oil and gas prices, investment and operating costs, and inflation rate.

Example 18.6: Optimization of offshore gas field development.[18] The objective is to develop a large offshore gas complex in an optimized manner that maximizes the value of the petroleum asset. The resources and constraints, typical of many offshore gas field development projects worldwide, are described as follows:

- There are three main gas reservoirs (original gas in place = 1.25 tcf) and a satellite well.
- Reservoirs are developed from offshore platforms, each costing between $100 and $150 million. Each platform has several slots for producing wells. Drilling cost is between $10 and $20 million per well.
- Gas is transported to the terminal via pipelines 8–10 mi long. Capacity of the surface handling facility is 200 MMscfd. The minimum delivery pressure is 900 psia.
- There is a penalty for nondelivery of contract gas volume. Compressors will be deployed to boost reservoir performance when necessary.

Several development scenarios, both traditional and economic, are analyzed as detailed in the following discussion.

Traditional approaches:

1. Reservoirs are developed sequentially; either the larger reservoirs first or the lower-cost reservoirs first.

2. A satellite well is brought to production following the development of the fields.

3. Compressors are deployed once production declines below the plateau production rate.

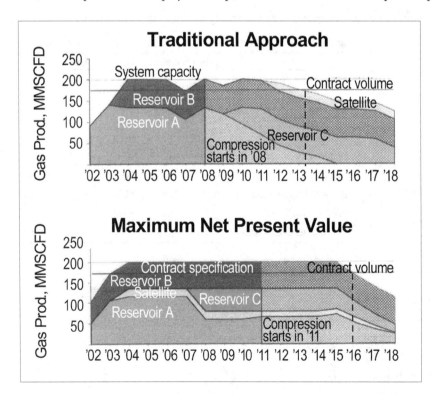

Net present value models:

1. Maximizes net present value of the project by considering all possible development scenarios.

2. Scheduling of the satellite well and the compressor are integrated in the overall optimization process.

The analysis showed interesting differences between the two approaches (fig. 18–15). In the traditional case, the reservoirs are developed sequentially. In contrast, the net present value model indicates that two of the reservoirs need to be developed concurrently, and the satellite well should be produced earlier. This arrangement allows the delay of the compressor deployment. The optimized development plan sustains the plateau rate for a longer period, resulting in maximizing the net present value.

Fig. 18–15. Development strategy of a large offshore natural gas complex. The traditional approach is compared to an integrated economic model where net present value is maximized. The gas delivery period as per contract is about 2½ years longer based on the latter, as indicated by the dotted lines. *Source: S. Vasantharajan, R. al-Hussainy, and R. F. Heinemann. 2006. Applying optimization technology in reservoir management. SPE Paper #87836. Journal of Petroleum Technology. May, 82–88.*

Summing Up

All technical activities related to the reservoir are closely tied to the overall economic goal of the enterprise. Present-day management views the reservoir team more as an "asset team," which is expected to add value to the asset (reserves) in any project undertaking.

Capital investment in petroleum ventures is substantial. Furthermore, there are wide-ranging elements of significant risk and uncertainty in the petroleum industry, making economic evaluations quite challenging.

Objectives of the economic analysis include, but are not limited to, the following:
- Exploration and development strategy: the number of wells to be drilled and platforms to be built
- Recovery scheme: evaluation of waterflood and various enhanced oil recovery processes
- Well spacing: costs involved in a dense well pattern versus increased production
- Scheduling: spread of capital outlays over time for a given discovery

The elements of risk and uncertainty, as usually experienced in the industry, are outlined in the following:
- Drilling of dry or noncommercial wells in new frontiers
- Less-than-expected well performance due to the presence of unknown geologic heterogeneities
- Unpredictability in future oil and gas prices due to changing scenarios in world demand and supply
- Unforeseen events, such as geopolitical unrest or natural calamities affecting petroleum production and distribution
- New governmental regulations, inflation, taxes, alternate energy, and environmental issues

Major steps involved in the economic analysis of a typical petroleum venture are as follows:
1. Set objectives of a new project based on established criteria, such as rate of return.
2. Formulate likely scenarios in project development.
3. Identify elements of risk and uncertainty. Determine the probability distribution of uncertain elements used in the economic evaluation.
4. Collect all pertinent information used in the analysis (reservoir, operation, financial, etc.). (Most will require forecasting.)
5. Perform evaluations based on deterministic and probabilistic approaches where uncertainties are anticipated.
6. Choose the optimum operation leading to the most value added to the asset.

Economic evaluations are usually based on the following criteria:
- Payout period
- Profit to investment ratio
- Present worth net profit (PWNP)
- Present worth index or profitability index (PWI)
- Discounted cash flow return on investment (DCFROI)

Optimization of the overall reservoir management process requires detailed consideration of the following elements:
- Design decisions
- Operational decisions
- Nonlinearity in physical systems, such as well rate as a function of bottomhole pressure

In economic analyses, uncertainty is often expressed in the form of a probability distribution. For example, it is not known with certainty how much a newly drilled well would produce throughout its life. However, based on past experience from similar wells in the field or region, a range of probabilities may be assigned to the expected production potential, such as 90%, 70%, or 50%.

Common probability distributions utilized in economic analyses are the following:
- Binomial
- Normal
- Lognormal
- Triangular
- Uniform

Binomial distribution is referred to as a discrete probability distribution, as it is based on the counting of successes or failures in a given trial. Drilling of exploratory wells in a new basin is often represented by binomial distribution.

Normal, lognormal, triangular, and uniform distributions are continuous distributions. These relate to the variability of a reservoir property or reserve, as these values are usually continuous in a given range. For example, porosity in a reservoir can be of any value within the observed range as obtained from core samples.

Out of the five distributions mentioned in this chapter, lognormal distribution is known to occur frequently for a host of variables. Such variables include porosity, permeability, fluid saturation, and recovery.

Plots of probability distribution or cumulative probability distribution usually depict various probabilities attached to a variable, such as original oil in place. Examples of these plots are presented in several instances in connection with the results of probabilistic studies.

Certain correlations may exist between two or more reservoir parameters, such as porosity and permeability, or porosity and connate water saturation. In a probabilistic study, any correlations identified between two parameters in the reservoir must be taken into account in order to perform a meaningful analysis.

Continuous distributions theoretically extend to positive and negative infinity. However, in reality, upper and lower bounds exist for the reservoir parameters, such as drainage area or fluid saturation. In assigning probability distributions to reservoir parameters, truncations at both ends of probability distribution are necessary.

Chances of failure or success of certain outcomes in a trial are dependent on the outcome of the previous trial. This is referred to as conditional probability. In exploratory drilling, when the first well turns out to be dry, it may augment or reduce the chance of drilling a successful well next.

A decision tree diagram is a logical approach to evaluating the economic feasibility of petroleum-related activities punctuated by known probabilities of success or failure at various stages. A decision tree consists of two or more branches (paths) consisting of choice (decision) and chances (probable outcomes). The most desired path is the one that results in the maximum expected value in dollars. Decision tree diagrams find application in evaluating various exploration strategies, including drilling of wildcats, farming out the prospects, and running seismic surveys before drilling.

The most familiar method in performing probabilistic studies in petroleum evaluations is based on a Monte Carlo simulation. A typical Monte Carlo simulation consists of thousands of trials in which the values of the input parameters are randomly generated to calculate the output variable. Depending on the probability distributions and ranges of input parameters, the output variable indicates its own probability distribution and range. An example would be an original oil in place calculation where 10%, 50%, and 90% probabilities are assigned to certain volumes of oil expected in a given reservoir.

Case studies have shown that the approach to development of a large oil or gas complex may differ substantially when economic models are employed. One objective might be to maximize the net present value, for example. Overall optimization of the reservoir management workflow may point to rescheduling of field development and construction of surface facilities, among other possibilities.

Class Assignments

Questions

1. Why do reservoir engineers need to get involved in economic evaluation of petroleum ventures? Name five major reservoir-related activities that require economic analysis. Who might be involved in the team that conducts the economic analyses?

2. A capital-intensive enhanced oil recovery project spanning more than a decade is being planned for a large field. Briefly describe at least six elements of risk and uncertainty that can affect the viability of the project.

3. Describe the steps involved in economic optimization. What economic criteria are used to select one option out of many? What advantages or disadvantages are associated with using these criteria? Briefly explain at least two criteria that are used in illustrated examples in this chapter.

4. How are the risk and uncertainty associated with any element in a project quantified and introduced in the economic analysis?

5. What is a probability distribution? Describe at least five probability distribution functions that are frequently observed in petroleum evaluations. Distinguish between discrete and continuous functions by giving field examples.

6. List data requirements for a detailed economic study in the following cases:
   (a) Drill one or more exploratory wells in a new prospect where most fields are marginal.
   (b) Perform a hydraulic fracturing operation in a well located in a very tight gas reservoir.
   (c) Evaluate an enhanced oil recovery project (chemical flood) as opposed to infill drilling.
   (d) Develop the peripheral region of a reservoir where the reservoir quality is poor. High connate water saturation and low permeability are suspected in the oil zone.

7. Economic analyses are being performed of planned waterflood projects in two fields. One is large, with extensive production history, and the other is small, with limited available information. Are there any fundamental differences in approach? Describe the strategy that should be adopted in performing economic analyses in both cases.

Exercises

18.1. A company plans to drill exploratory wells in a basin where the historical success rate is one in five. What are the chances of having at least two successful strikes when the company drills (a) three wells, (b) five wells, and (c) seven wells? If the first wildcat turns out to be a dry hole, should a seismic survey be proposed? What factors should be considered in comparing the following two options:
   • Drill subsequent wells based on regional trend alone
   • Run a seismic survey in the specific location before drilling the second well

18.2. The reserve of an oil field is being estimated by a probabilistic method. As a part of the study, the reservoir properties are being reviewed for any possible existing correlations between properties. The following porosity and saturation data is available from 15 core samples:

Porosity, %	Sw, fraction	Porosity, %	Sw, fraction	Porosity, %	Sw, fraction
14.5	0.240	17.3	0.220	17.6	0.227
20.6	0.203	19.7	0.214	17.3	0.209
15.1	0.230	20.4	0.228	17	0.211
15.6	0.235	14.9	0.225	17.9	0.206
16.6	0.220	17.5	0.214	18.5	0.200

(a) Plot the data and perform a regression analysis by using a built-in spreadsheet function. Calculate the standard deviation between the regressed line and the actual data.

(b) Can any data be identified that does not appear to follow the general trend? What would be the best approach to handle any data with a large deviation?

(c) Are the two parameters correlated? If so, is there any physical explanation for the correlation?

18.3. The following values of rock permeability are obtained from petrophysical studies performed on core samples of the same field:

Sample Number	Air Permeability, mD
1	173.39
2	64.68
3	59.79
4	40.79
5	24.52
6	21.12
7	17.90
8	17.00
9	16.10
10	12.99
11	12.67
12	12.51
13	12.21
14	7.44
15	6.83

Prepare a frequency distribution diagram having several classes and identify any probability distribution pattern. Plot cumulative frequency versus permeability.

18.4 The objective is to estimate the reserve of a new discovery. Based on data obtained from nearby fields, the following table is developed:

Rock/Fluid Property	Porosity, fraction	Reservoir Area, acres	Formation Thickness, ft	Sw, fraction	Oil FVF, rb/stb
Range	14.5–18.5	3,000–8,500	12–20	0.20–0.24	1.35–1.46
Most Likely	17.0	—	15	—	1.39
Distribution	Triangular	Uniform	Triangular	Uniform	Triangular

The oil recovery factor is known to vary between 21% and 29% in the region and appears to follow a uniform distribution pattern. The analysis should include the following:

(a) Best and worst case scenarios of oil reserves. Is it possible to identify a most likely case?

(b) Reserves associated cumulative probabilities of 90%, 70%, and 50%.

(c) Recalculate (b) using the correlation obtained between porosity and saturation in Example 18.2.

(d) Compare results by drawing cumulative distributions of oil reserve for (b) and (c) in one plot. What are the deviations between the two cases at 90%, 70%, and 50% probabilities?

Hint: The analyses required in steps (b) and (c) can be performed by using built-in spreadsheet functions without the aid of any specific Monte Carlo simulation tool. For example, the RAND function in an MS Excel spreadsheet generates random numbers between 0 and 1. Based on Equation 18.8, a random value of water saturation can be generated, as shown in the following:

$$S_w, \text{trial } 1 = 0.20 + \text{RAND} \times (0.24 - 0.20)$$

Hundreds or thousands of trials are usually required to generate a complete probability profile of S_w. This is accomplished by copying the preceding formula cell over the desired number of rows in the spreadsheet. Similarly, the triangular distributions of porosity and other parameters can be generated by using Equations 18.6 and 18.7. The format of the spreadsheet should be similar to that illustrated earlier in the section describing the steps involved in a Monte Carlo simulation of original oil in place. Once the set of values for estimated reserve is obtained, a frequency distribution study can be performed by using the available spreadsheet functions. This will lead to identification of the three specific values of reserves associated with 90%, 70%, and 50% probabilities.

About 500 to 1,000 trials (or rows in MS Excel) should be sufficient for the purpose of this exercise.

References

1. Satter, A., and G. C. Thakur. 1994. *Integrated Reservoir Management—A Team Approach*. Tulsa: PennWell.

2. Thakur, G. C., and A. Satter. 1998. *Integrated Waterflood Asset Management*. Tulsa: PennWell.

3. Stermole, F. J., and J. M. Stermole. 1987. *Economic Evaluation and Investment Decision Methods*. 6th ed. Golden, CO: Investment Evaluation Corporation.

4. Seba, R. D. 1987. Determining project profitability. *Journal of Petroleum Technology*. March: pp. 263–71.

5. Garb, F. A. 1988. Assessing risk in estimating hydrocarbon reserves and in evaluating hydrocarbon-producing properties. *Journal of Petroleum Technology*. June, 765–68.

6. Rose, S. C., J. F. Buckwalter, and R. J. Woodhall. 1989. *The Design Engineering Aspects of Waterflooding*. SPE Monograph Series. Richardson, TX: Society of Petroleum Engineers. p. 11.

7. Hickman, T. S. 1991. The evaluation of economic forecasts and risk adjustments in property evaluation in the U.S. *Journal of Petroleum Technology*. February, 220–25.

8. Mian, M. A. 2002. *Project Economics and Decision Analysis*. Vols. I and II. Tulsa: PennWell.

9. Vasantharajan, S., R. al-Hussainy, and R. F. Heinemann. 2006. Applying optimization technology in reservoir management. SPE Paper #87836. *Journal of Petroleum Technology*. May, 82–88.

10. van den Heever, S., I. E. Grossman, S. Vasantharajan, and K. Edwards. 2001. A Lagrangian decomposition heuristic for the design and planning of offshore hydrocarbon field infrastructure with complex economic objectives. *Industrial & Engineering Chemistry Research*. Vol. 40, no. 13, 2,857.

11. Iqbal, G. 2002. Course notes on reservoir economics. Petroleum engineering seminar, Petrobangla, Dhaka, July 8–13.

12. Satter, A., J. Baldwin, and R. Jespersen. 2000. *Computer-Assisted Reservoir Management*. Tulsa: PennWell.

13. A. W. McCray. 1975. *Petroleum Evaluations and Economic Decisions*. Englewood, NJ: Prentice-Hall.

14. Newendorp, P. 2000. *Decision Analysis for Petroleum Exploration*. 2nd ed. Tulsa: PennWell.

15. Harbaugh, J. W., J. C. Davis, and J. Wendebourgh. 1995. *Computing Risk for Oil Prospects: Principles and Programs*. Oxford: Pergamon Press.

16. Murtha, J. 2000. *Decisions Involving Uncertainty—An @RISK Tutorial for the Petroleum Industry*. Newfield, NY: Palisade Corp.

17. Jones, C. Personal communication. Texaco, Inc.

18. Vasantharajan, S., R. al-Hussainy, and R. F. Heinemann. 2006.

19 · Operational Issues in Reservoir Development and Management

This chapter discusses briefly the operational issues reservoir engineers typically deal with in the course of their professional work. The issues are wide ranging, from a routine assignment focusing on the productivity of a specific well to an integrated study related to the development of the entire reservoir. In-depth understanding of fluid flow dynamics in porous media, phase behavior, reservoir characterization, visualization, and prior experience of specific well and reservoir performance are vital. Best management practices, teamwork, staying informed about innovative technologies, and "thinking outside of the box" also are essential ingredients in order to approach any operational challenges.

Reservoir engineers are in a unique position, as most reservoirs are buried at depths of thousands of feet and can neither be seen nor scouted. These reservoirs can only be probed and sampled to a very limited extent, and consequently must be conceptualized and simulated. The situation is further complicated by the fact that very little is known about a reservoir at its discovery in order to develop it in the most efficient manner. A large amount of production and other data related to the reservoir is only available at later stages, when the reserves have declined following suboptimal development. As pointed out in earlier chapters, ultimate recovery from most oil reservoirs is rather limited. Extensive data gathering at the onset of field development is often emphasized by experienced professionals, once the "lessons were learned" at later stages. Another point to keep in mind is that no single solution works for all reservoirs, although the issues may apparently be similar in nature. Each reservoir is unique in its character and performance, and needs to be evaluated on an individual basis.

This chapter is devoted to learning about the following:
- Issues and challenges faced by reservoir engineers in the field
- Tools and techniques available to deal with the issues
- Metrics to evaluate the success of reservoir engineering solutions
- Case studies in problem solving and better field management

Operational Issues

Typical operational issues related to managing oil and gas reservoirs stem from the following:
- **Reservoir and well data.** Collection, validation, integration, and incorporation into the decision-making loop, some in real-time or near–real-time.
- **Production and well problems.** High water cut, gas/oil ratio, gas and water coning, workovers, gas lifting, sand production, etc.
- **Field development.** Infill drilling, horizontal wells, facilities, and economic optimization.
- **Augmented oil recovery (improved oil recovery and enhanced oil recovery).** Poor displacement efficiency, limited horizontal and vertical sweep, pattern imbalance, thief zone, conformance issues, early water breakthrough, viscous fingering, gravity override, etc.
- **Reservoir development in challenging geologic settings.** Low permeability and highly heterogeneous reservoirs; well stimulation and other issues.
- **Uncertainties in original hydrocarbon in place estimation.** Can occur with complex geology.
- **Marginal and matured fields.** Development techniques and economics.

The preceding are part of the larger issues of project development and management, which include the following:

- **Project development.** Primary, secondary, and tertiary recovery.
- **Project management.** Design, implementation, monitoring, and evaluation.

Tools and Techniques

The tools and techniques that are at a reservoir engineer's disposal include, but are not limited to, the following:

- **Intensive monitoring of reservoir performance.** Pressure and temperature profiles, fluid flow rates, and fluid phase and density in wells, followed by corrective actions. Deployment of permanent downhole gauges to continuously monitor the above.
- **Obtaining better reservoir description.** Well logs (open and cased hole), core sampling, transient well tests, 3-D and 4-D seismic studies, etc. Tools and techniques utilized in reservoir characterization are listed in chapter 2.
- **Drilling and recompletion.** Field development based on multilateral horizontal well technology and recompletion of existing wells, including sidetracking.
- **Initiation of an enhanced oil recovery program at optimum reservoir conditions.** An EOR project usually deploys fluid injection or thermal methods following detailed reservoir simulation studies and scenario building.
- **Identification and remediation of breakthrough zones.** Zones where injected fluids (water, gas, steam, or a chemical substance) break through.
- **Hydraulic fracturing and acidization.** Processes to boost well productivity.
- **Implementing intelligent well systems (IWS).** Selectively shut off, open, or choke down a target interval during injection or production from a remote location.
- **Establishing best management practices.** Based on continuous optimization of asset enhancement processes.

Performance Metrics

Saleri compiled a list of metrics in evaluating the overall outcome in managing a reservoir, given in the following discussion.[1]

Target versus actual:

- Reservoir production and surveillance
  - Production by field, section, or well
  - Annual changes in reservoir pressure
  - Annual decline rate in production
  - Water cut and gas/oil ratio (on reservoir and well basis)
  - Mobile oil/gas recovery efficiency and ultimate recovery
- Cost analysis
  - Unit costs related to production, drilling, and workover per well
- Economic evaluation
  - Rate of return and net present value

Topics related to major operational issues, innovative techniques, and management practices are discussed in this chapter, illustrated with selected case studies of petroleum reservoirs worldwide.

Case Studies

Case studies that illustrate various issues with fields and specific solutions include the following:

- Horizontal drilling of a thin, viscous-oil rim with water and gas coning issues in the Niger delta
- Thermal recovery from a dipping heavy oil reservoir in Kern County, California
- Waterflood performance enhancements in a large oilfield in New Mexico
- Optimal development of a tight gas reservoir based on early data acquisition, reservoir characterization, and model simulation
- Infill drilling to augment production in a West Texas field
- Development strategy of naturally fractured reservoirs, including a case study of a heavy oil reservoir in the Persian Gulf region
- Redevelopment of a matured field in Bahrain based on well recompletion, infill drilling, horizontal wells, and improved reservoir characterization, among others
- Development of a marginal oilfield in offshore Qatar based on horizontal wells with very long laterals
- Zonal control of fluid injection using smart well technology in a Texas field
- Deployment of intelligent well systems in matured fields in Oman
- Application of 4-D seismic studies and well testing to efficiently recover hydrocarbons from a compartmental reservoir located offshore from the United Kingdom

Besides the above, several field case studies on improved or enhanced oil recovery operations coupled with infill drilling are described in chapters 16 and 17. It must be mentioned that some of the reservoir engineering solutions are closely associated with efficient well completion and production methodologies practiced in the industry. However, a detailed discussion of the topic is beyond the scope of this book.

Horizontal Wells

In the 1970s or earlier, virtually all producers and injectors were drilled as vertical or deviated wells in oil and gas reservoirs. In recent times, however, a new era in the petroleum industry emerged with the development of horizontal well technology. These wells contact thousands of feet of net pay in the horizontal direction, and contribute substantially to well productivity. Horizontal wells with multilateral branches have become very popular in augmenting recovery. Various configurations of the horizontal wells are capable of reaching multiple sections of pay in different directions from a single location at the surface (fig. 19–1). Early horizontal wells were single-lateral, followed by drilling of multilateral branches as the technology matured. Maximum reservoir contact (MRC) wells may be drilled for several miles into a pay zone with multilaterals in stacked, fork, wishbone, or other configurations resulting in a PI exceeding 100 bbl/d/psi. Horizontal well productivity is usually found to be 500% to 2,000% more than that of vertical wells, while drilling costs increase only two- or three-fold. In favorable circumstances, the drilling cost per barrel of produced oil is reduced by 80% or more.

Reservoir engineers readily recognize the outstanding capabilities of horizontal wells in producing from a much larger drainage area. These wells can contact zones of high residual oil saturation left behind by vertical producers and can be used to develop relatively tight reservoirs, while avoiding watered-out zones. The total number of wells needed to develop a reservoir is relatively fewer. The advantage is easily seen with heavily faulted or compartmentalized formations. In these formations, each vertical

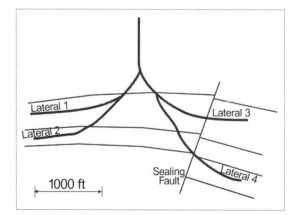

Fig. 19–1. A multilateral horizontal well producing from multiple reservoirs across a sealing fault. The trajectory of the laterals is determined by maximum contact with in-situ oil and away from watered-out zones. Laterals are typically a few thousand feet to tens of thousands of feet long in some instances, resulting in a 5-fold to 20-fold increase in productivity.

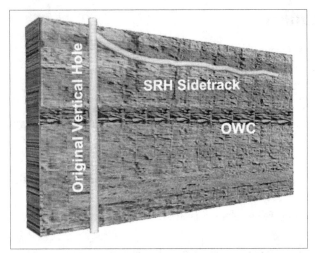

Fig. 19–2. Example of horizontal sidetracking in a reservoir where oil is produced by bottom-up sweep. Due to high water cut of the vertical well, a single-lateral horizontal well is drilled in the upper part of the formation in order to maximize oil production and minimize water production. The same strategy is adopted where gas coning is an issue. Horizontal well trajectories are of short-, medium-, and long-radius curvature, depending on formation geometry and the anticipated water saturation profile under drawdown. *Source: A. H. al-Huthali, A. A. al-Awami, D. Krinis, Y. Soremi, and A. Y. al-Towailib. 2005. Water management in North Ain Dar, Saudi Arabia. SPE Paper #93439. Presented at the 14th SPE Middle East Oil and Gas Show and Conference, Bahrain, March 12–15. © Society of Petroleum Engineers. Reprinted with permission.*

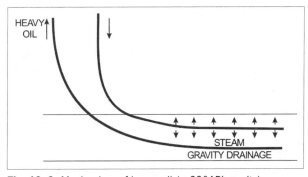

Fig. 19–3. Mechanism of heavy oil (< 22°API gravity) recovery by steam-assisted gravity drive

well can only produce from one compartment, while single laterals or multilateral horizontal wells can reach across physical boundaries. Other advantages include, but are not limited to, the following:[2]

- A significant increase in exposure in relatively thin stratigraphic sections and in thin oil rims with a large gas cap
- Connectivity to a much larger portion of the naturally occurring fracture network and multiple conduits of flow
- Production from more than one reservoir with single-lateral or multilateral horizontal completion
- Relatively fewer effects of water or gas coning, as horizontal wells can operate at relatively low drawdown compared to vertical wells
- Efficient field development where directional permeability predominates or multiple compartments exist
- Waterflooding a reservoir with zones of contrasting permeability
- Alternative to infill drilling or multiple vertical wells that can reduce capital expenditures substantially, including the cost of drilling per barrel of oil produced
- Exploitation of additional zones in the reservoir
- Reduced water cut due to the placement of the horizontal wellbore away from problematic areas
- Development of tight reservoirs
- Optimization of the pressure distribution near the wellbore and in the reservoir

A direct correlation between the length of horizontal wellbore and well productivity enhancements has been observed in numerous field studies. A large database with data from 1,306 horizontal wells in 230 fields indicated that the highest productivity enhancements were achieved in fractured reservoirs.[3] In these reservoirs, up to a 12-fold increase in rate was observed. The database was developed in 1990s, and the industry has seen significant strides in the design, deployment, and operation of horizontal well technology since then. Horizontal wells are successfully drilled in many reservoirs in which the existing vertical well is sidetracked to be completed away from the water/oil contact. This has resulted in a substantial reduction in water cut (fig. 19–2). Similarly, the gas/oil ratio is minimized in many wells by completing the horizontal section at the farthest possible distance from the gas cap and maintaining the drawdown at an optimum level.

Industry experience indicates that horizontal well technology leads to innovative techniques in oil and gas production. A case in point is thermal recovery by steam-assisted gravity drainage (SAGD). A pair of horizontal wells is drilled in a formation containing heavy oil that cannot be recovered economically by classical methods. One well is "stacked" on top of the other, separated vertically by several feet of formation (fig. 19–3). Steam is injected into the reservoir through the horizontal well located at the top, which results in a significant reduction in oil viscosity. The oil becomes mobile, "drains" downward due to gravity, and is eventually produced by the lower well.

Horizontal drilling and completion technology offers an array of potent tools in the hands of reservoir engineers, providing effective solutions to wide-ranging problems. Case studies are presented in the following to highlight the efficient utilization of horizontal wells.

Case studies

I. Redevelopment of a thin, viscous-oil rim with water and gas coning issues.[4] The following is an example of redevelopment of a reservoir having a thin oil rim overlain by a massive gas cap. A 10-fold increase in both production rate and potential reserves was attained following implementation of horizontal technology. The ultimate recovery is expected to be around 35%.

Ebughu field, located in the Niger delta, is a heavy oil (20°API) reservoir produced from an oil rim below a large gas cap. Oil viscosity is 3.6 cp, and the liquid phase is fully saturated. The vertical extent of the oil rim is 63 ft, overlain by a 365-ft thick gas cap. Reservoir quality is found to be excellent, with formation porosity exceeding 30% and permeability in darcies. The producing zone is comprised of unconsolidated sand, which leads to sand production and requires appropriate remedial measures.

The primary production mechanism is a combination drive, including water influx from an aquifer and energy from the gas cap. Initial development of the reservoir was limited to two vertical wells due to poor performance at the initial stage. The issues included the following:

- Rapid gas and water coning as viscous oil was produced from a thin rim overlain by the gas cap
- Sand production and poor cement isolation
- Production limited to 300–500 stb/d despite excellent reservoir quality and combination drive

The field redevelopment strategy included the following:
- Formation of an integrated project team.
- Detailed review of reservoir data.
- A 3-D seismic survey and material balance study, which indicated a much-larger reservoir area.
- Collection of new fluid PVT data and resolution of discrepancies.
- Reservoir simulation based on planned horizontal wells.
- Drilling of seven appraisal wells, which resulted in five new horizontal wells with 2,500-ft laterals. One existing well was sidetracked.

Wells were drilled in a 10-ft window in the middle of the oil rim in order to minimize oil and gas coning. Gas production was counteracted by choking back the horizontal wells and maintaining the gas/oil ratio within a predetermined range (0.2 Mscf/stb to 0.5 Mscf/stb).

The study is an excellent example of how reservoir engineering issues were approached, including the following:
- Integrated reservoir studies, including seismic, PVT, material balance, and simulation, which resulted in drilling of horizontal wells
- Implementation of a new technology (horizontal wells), as older methods of production did not succeed
- Design of the horizontal well trajectory in the midsection of the rim in order to keep a maximum distance from the water/oil contact and gas/oil contact
- Optimization of the production rate to minimize the gas/oil ratio and gas coning

In another case, the reservoir had a thin oil rim overlain by a small gas cap. The reservoir also experienced strong water influx. Based on reservoir simulation studies, it was decided to blow down the gas cap. This was followed by the drilling of horizontal wells in the thin oil zone to effectively produce the reservoir without the detrimental effects of gas coning.

II. Thermal recovery from a dipping reservoir.[5] The Midway Sunset field, located in Kern County, CA, was discovered in the early 1890s at a depth between 500 ft and 1,700 ft. It consists of several units that dip northeast at 40°–70°. The producing formation is a massive, poorly consolidated sandstone. The trap is formed by a stratigraphic unconformity that truncates the oil-bearing zones. Formation porosity and permeability are excellent (30% and greater than 1 darcy, respectively). The ratio of vertical to horizontal permeability is extremely low, about 0.0001.

The heavy oil (13°–14°API) reservoir was first produced in 1917. Thermal recovery by steam cycling commenced in the 1960s. Traditionally, oil recovery has been accomplished by vertical wells with ¼-acre spacing. The total number of active producers is about 1,200 in the field.

The wells were completed as close as possible to the water/oil contact due to poor vertical permeability. Although the cyclic steam operation was found to be generally very effective, certain areas located at the bottom of the steeply dipping reservoir could not be produced. Moreover, a large "steam chest" developed in the upper portion of the reservoir with continued injection, resulting in gravity override by steam. The steam zone, hundreds of feet in thickness, was responsible for low steam cycle efficiency.

In order to provide solutions for these issues, a reservoir simulation study was performed to evaluate horizontal well technology. This was undertaken with the objective of producing from previously unrecoverable portions of the reservoir. The results of the study were encouraging, and 50 horizontal wells were drilled. The wells were placed as low as possible in the oil column to take advantage of steam-assisted gravity drainage (SAGD) of the oil. The wellbore was oriented towards the dip of the reservoir. Moreover, the horizontal wells reduced the number of replacement wells to be drilled.

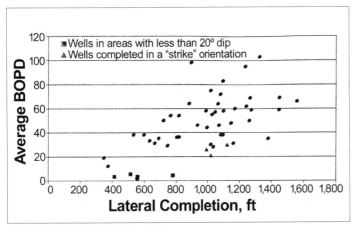

Fig. 19–4. Production rate versus horizontal wellbore length in Midway Sunset field. *Source: C. McKay, J. Jones, and J. Promerene. 2003. Successful horizontal producers in Midway-Sunset thermal operations.* Journal of Petroleum Technology. *November, 43. © Society of Petroleum Engineers. Reprinted with permission.*

Results of horizontal drilling are highlighted in the following:

- The oil production rate from a horizontal producer was found to increase by 400%–500% compared to the average production rate from a vertical well.
- The cost of drilling a horizontal wellbore was about twice that of a vertical well.
- A direct correlation between the horizontal well length and the production rate was evident (fig. 19–4).
- Horizontal wells completed in zones of relatively shallow dip (20°) did not produce as much as the horizontal wells having relatively steep dip, as the primary production mechanism is gravity drainage.
- Again, horizontal wells completed in the strike orientation were of limited success, as the wells were exposed to less sand volume.

Waterflood Conformance

Sedimentary deposition occurring over millions of years is strongly influenced by varying prevailing conditions. Hence reservoir properties are seldom uniform either vertically or horizontally. Consequently, water entry into the formation through an injector is not found be uniform over the entire vertical cross section of the formation. This leads to poor sweep, high water cycling, and less recovery. Conformance projects attempt to shut down or minimize excessive water entry points and thief zones by selectively applying various water blocking chemicals. Similarly, problematic zones responsible for high water/oil ratio or gas/oil ratio are isolated. A literature review indicates that certain producers are selectively fractured in tight zones in order to enhance oil recovery from waterflooding. Conformance control relates to efforts designed to result in improvements in a water injection project or a well's production profile. Well testing and reservoir engineering, including simulation of well performance, play a vital role in characterizing and combating various conformance-related issues. These are described in the following:[6]

- Effects of reservoir stratification
- Control of effective permeability to water
- Water and gas coning
- Channeling of injected water
- Viscous fingering

Case study: Eunice Monument field[7]

Introduction. The Eunice Monument field, located in New Mexico, is produced under a large waterflood scheme based on hundreds of injectors and producers in 5-spot patterns. The lithology of the formation is dolomite, and the reservoir has a combination (structural and stratigraphic) trap. The source for water supply is an underlying water drive reservoir. Once the waterflood project was implemented, oil production decreased to 70% of the original rate. The decline in reservoir performance was attributed to the poor waterflooding conformance encountered all over the field. The carbonate formation was characterized as heterogeneous to a large degree due to the existence of natural fractures and high permeability zones.

Reservoir performance studies. The following approach was adopted in studying the waterflooding, which was beset with significant performance issues:

- Based on reservoir and well data, cross-sectional plots of conformance were developed for each waterflood pattern. A cross-sectional plot is developed by illustrating the injection or production profile of each well in series, correlating problematic zones. The objectives were to verify the strong correlation between certain injectors and producers and to identify thief zones. Studies further revealed that a substantial portion of the injected water entered the gascap zone.
- Maps showing the high permeability streaks were developed based on core analysis, core descriptions, and log studies. Geostatistical models were utilized where necessary data was not available.
- Cumulative water injection volume was calculated for each zone in a pattern based on injection profiles obtained from the wells. This led to the identification of zones where the injected volume exceeded the pore volume of movable hydrocarbon.
- A marked increase in the water/oil ratio (WOR), accompanied by a decrease in the oil rate, served as the criteria for selecting candidate wells for remedial measures. The decline in oil production was attributed to the presence of highly conductive channels leading to substantial water breakthrough. It adversely affected the lifting capacity of the producing well, which led to an inability to produce the low pressure, low permeability zones.

Remedial measures. The remedial measures were adopted based on reservoir characteristics and well conditions, with an objective of improving conformance. These measures include the following:

- Cement squeezes were used to block injected water penetration into high permeability streaks and the gas cap.
- Natural fractures were treated by flowing gel.
- A polymer gelant was used for deep penetration into the rock matrix when crossflow was indicated between the target zone and the high water cycling zone.

Results. Following the implementation of these measures, production increased between 30% and 60% in a number of wells. The full impact on field performance was yet to be seen at the time of publication of the technical paper.

Low Permeability Reservoirs

Introduction. Study of petroleum basins worldwide indicates that vast amounts of oil and gas occur in formations of low to very low permeability. Generally speaking, oil reservoirs having permeability values less than 5 mD may fall in this category. In the case of gas reservoirs, formations having permeability of less than 0.1 mD are viewed as "tight." Holditch defines a tight gas reservoir as "a reservoir that cannot be produced at economic flow rates nor recover economic volumes of natural gas unless the well is stimulated by a large hydraulic fracture treatment or produced by use of a horizontal wellbore or multilateral wellbores."[8]

This definition is self-explanatory. First, "tight" reservoir is not defined on the basis of rock characteristics alone, such as permeability, but is related closely to economic evaluation. The author points out that certain deep, thick, and high-pressured formations, with permeability in microdarcies (10^{-6} darcies), have been produced successfully. Second, low permeability reservoirs require special measures, including massive hydraulic fracturing or drilling of horizontal wells to be economically feasible. A third option is the drilling of infill wells. Low permeability reservoirs with highly conductive, naturally occurring fractures have the potential of performing better. This occurs as oil or gas flows with relative ease through the high permeability network.

A comprehensive strategy for development of low permeability reservoirs is based on multiple sources of input. These include a detailed reservoir description, utilization of fracture propagation models, reservoir simulation, and economic analysis. One of the challenges in reservoir description is the estimation of the drainage area due to the very low permeability of the formation. The shape and size of the drainage region are influenced by the depositional environment and the length and orientation of the hydraulic fracture or the horizontal well trajectory. As indicated earlier, the options and strategies usually available to reservoir engineers include the following:

- Relatively close well spacing (infill drilling).
- Horizontal drilling, single-lateral or multilateral.
- Well stimulation by hydraulic fracturing and acidization.
- Reservoir management based on intensive monitoring of well performance, including deployment of permanent downhole gauges. This leads to evaluation of well productivity and formation damage on a regular basis.

The role of horizontal well technology in augmenting recovery in various circumstances was outlined previously. The remaining options are described in the following sections.

Optimal Field Development

In developing tight gas reservoirs, relatively dense well spacing is usually required. Optimum well spacing is achieved by maximizing gas recovery while minimizing capital expenditure due to drilling. The optimization must be achieved rather early in the life of the field by a coordinated plan for data acquisition, reservoir characterization and model simulation, and surveillance.[9] The methodology is applicable to the development of any reservoir, tight or conventional, and is outlined next.

- **Early acquisition of static reservoir data.** This involves data obtained from various studies related to cores, fluids, and logs. Certain key wells are designated for core analysis at the onset of field development.
- **Reservoir characterization.** Identification of geologic and rock features, both small and large scale, that control recovery mechanisms. This is of vital significance in optimal field development.
- **Reservoir surveillance.** Monitoring of rates and flowing and static bottomhole pressures as production is commenced. Individual layer pressure should also be included in the plan. Production logging is recognized as an important tool in surveillance.
- **Reservoir model development.** Based on statistical methods and probability, rock properties and geologic structures are modeled, as only a small part of the reservoir is sampled. Referred to as stochastic models, these are developed on small grids and strive to describe the reservoir in fine detail. On the other hand, deterministic models are developed based on coarser grids in which various rock and fluid properties are assigned. The above

leads to fluid flow simulation models, which are used to match production history and predict future performance. A stochastic model with a finer grid is integrated with a deterministic model with a coarser grid by upscaling. In the course of integration, multiple realizations of geological and petrophysical systems are obtained. The realizations considered to be the most accurate are finally selected for predicting future performance.

- **Simulation.** Reservoir models are used to validate reservoir characterization based on the history of reservoir performance. Simple models related to wells, sectors, and full fields are developed first, followed by complex models as more data becomes available. Reservoir models must be updated continuously based on field surveillance.

- **Economic analysis.** The elements in economic analysis include costs in acquiring a detailed reservoir description and in reservoir monitoring. These are evaluated against minimization of risks involved in the development of tight gas reservoirs.

- **Conclusions.** Close monitoring of reservoir performance must occur from the start. Coupled with a detailed reservoir description, it is the key to early development of a tight gas reservoir in which economic considerations are crucial. In older "serial development" programs, drilling with initial spacing was completed in the first phase. This was followed by several years of production evaluation and slowdown in drilling activities before the infill wells were put in place. In contrast, a tight gas reservoir can be developed completely and optimally within a significantly accelerated time frame when proven strategies are employed. These strategies include reservoir characterization, integrated reservoir model simulation, and reservoir surveillance.

Infill Drilling

The decision for infill drilling, which leads to closer well spacing in a field, primarily depends on the economic analysis, as in any other field development project. The incremental recovery of oil and gas attained by infill drilling of new wells is evaluated against capital expenditure (CAPEX) and other costs related to drilling additional wells. In oil fields, infill drilling may be considered as part of a larger scenario that includes waterflooding and enhanced oil recovery operations. Based on reservoir simulation, increased oil recovery due to drilling of wells at various spacings, coupled with waterflooding and enhanced oil recovery processes, is predicted.

Infill drilling finds wide application in a variety of circumstances including, but not limited to, the following:
- Reservoirs having ultra-low, low, and moderate permeability
- Reservoirs where large pockets of hydrocarbons are left behind due to the existing well spacing and alignment of wells
- Reservoirs where closer well spacing aids improved oil recovery and enhanced oil recovery processes significantly
- Highly heterogeneous formations in which the areal sweep is rather limited during waterflooding or tertiary recovery processes
- Highly faulted or compartmentalized formations in which the well drainage area is limited by physical boundaries
- Accelerated field development and production

Case study: Salt Creek field[10]

Introduction. The Salt Creek field was discovered in 1950 in the Permian basin in West Texas at a depth of 6,300 ft. The producing formation is heterogeneous limestone. The areal extent of the field is more than 12,000 acres, with two reservoirs that are not in communication. The initial pressure of the reservoir was 2,900 psi, and the discovery well flowed at a rate of 2,184 stb/d. The crude oil gravity is 39°API, and the bubblepoint pressure is approximately 1,250 psi.

Field history. The field was initially developed on 80-acre spacing, producing about 10,000 stb/d, which was the maximum allowed by the state regulatory agency. Reservoir pressure was maintained by water and gas injection. In 1967, the allowable production was increased to 37,800 stb/d. Once production began to decline in the 1970s, infill wells were drilled to reduce well spacing to 40 acres in portions of the field. In the 1980s, well spacing was further reduced to 20 acres by drilling of additional wells to augment recovery. Furthermore, the flood pattern was changed from 9-spot to 5-spot.

In the 1990s, a CO_2 injection project was initiated in phases based on compositional reservoir simulation studies. The enhanced oil recovery project involves about 130 injectors and 170 producers. CO_2 flooding was expanded to the residual oil saturation zone below the oil/water contact. Total hydrocarbon production is about 18,000 stb/d. All produced water is recycled.

Recovery. The ultimate recovery from the Salt Creek field was expected to be about 60% of original oil in place, with 50% recovery being attained by 2003. Reservoir studies indicated that the primary recovery with 80-acre well spacing was around 24%, while the drilling of infill wells and secondary recovery by water and gas injection led to a recovery of 50%. Finally, tertiary recovery based on CO_2 flooding was implemented in order to further augment hydrocarbon recovery.

Key elements in reservoir management. The Salt Creek field project succeeded in achieving high recovery in a heterogeneous formation with several high permeability zones. This outcome is attributed to successive implementation of infill drilling and improved recovery processes, namely waterflooding, gas injection, and CO_2 injection. The key ingredients of the effective reservoir management process included the following:
- A 3-D seismic study to locate infill and step-out wells
- Enhanced sweep efficiency and increased injection well density by infill drilling
- Reservoir surveillance, including close monitoring of injected fluid breakthrough in high permeability zones
- Flood balancing within patterns
- Improvement in individual injection and production profiles (conformance control)
- Reservoir pressure maintenance
- Minimization of surface facilities cost
- Decision making in real time

Well Productivity Enhancement

Most reservoir management programs include measures to enhance the productivity of vertical and horizontal wells. These techniques include acid stimulation, hydraulic fracturing, intelligent completion, and sand control, among others. The issues related to productivity decline are addressed by a team having expertise on drilling, completion, production, and the reservoir and its geology. Well management issues are treated in detail in various sources in the literature.

During the life of a well, productivity is generally monitored by the following:[11]
- Production decline trend analysis
- Pressure transient tests, including buildup test analysis in order to evaluate the effects of changing skin, among others
- Analysis of produced water in water-cut wells
- Fluid compatibility analysis
- Core studies related to formation damage
- Analysis of oil samples

Based on a performance review of several horizontal wells in fields located in Sumatra, Indonesia, the factors affecting well productivity were found to be wide ranging. Factors included the following:
- Wellbore damage due to mud filter cake and scale deposition
- Incompatible completion fluid
- Poor geosteering of horizontal laterals
- Reservoir heterogeneities along the well trajectory
- Inadequate pressure support

Remedial actions included use of appropriate drilling fluids to reduce formation damage during drilling of new wells and acid stimulation of existing wells, among others. It is emphasized that identification of the actual cause of the well productivity decline holds the key in addressing the issue.

Hydraulic Fracturing[12,13]

Hydraulic fracturing is a potent tool and the most widely used method in augmenting well productivity. A common application of hydraulic fracturing is with wells producing from tight formations. It is also used where the surrounding formation is damaged, as indicated by positive skin and unsatisfactory well performance. Design of a hydraulic fracture involves the following:

- Evaluation of well and production history.
- Study of openhole, cased hole, and production logging results.
- Determination of the reservoir properties of pressure, net pay thickness, and skin, among others.
- Optimization of fracture half-length, as long hydraulic fractures with relatively high conductivity are likely to enhance well productivity and increase revenue. However, they are associated with higher costs.
- Selection of appropriate fracture fluid and fracture propping agent.
- Simulation of a fracture propagation model to determine injection rate, optimum pad volume, and location of perforations, etc.

Agiddi compiled a list of uncertainties involved in designing a fracturing operation in the Steven reservoir of Elk Hills field in California, given as follows:[14]

- Possible damage in conductivity of fractures
- Attainment of full vertical coverage in target zone
- Flowback of proppants and sand problems
- Optimal hydraulic fracturing treatment

Post fracture evaluation. As described in chapter 5, the success of a hydraulic fracturing operation is evaluated by conducting a transient well test. Properly designed well tests can lead to the estimation of fracture half-length and conductivity, negative skin, and flow efficiency, among other results. Typical flow regimes that are observed in a hydraulically fractured well include linear, bilinear, and pseudoradial flow. However, depending on the formation and fracture characteristics, pseudoradial flow may not be observable for months or even years. The linear or bilinear flow period is an inverse function of permeability and is a direct function of fracture length. Example 5.9 presents a typical drawdown test performed on a hydraulically fractured well with finite fracture conductivity.

Naturally Fractured Reservoirs

Reservoirs with naturally occurring fractures pose unique challenges in development and management. As mentioned in chapter 2, fractured reservoirs are composed of two distinct systems, namely, rock matrix and an interbedded fracture network. Reservoirs are generally viewed as either dual porosity or dual porosity dual permeability systems based on dynamic fluid flow behavior during depletion. In a dual porosity system, hydrocarbons are produced only through the fracture network; however, reservoir fluid can flow from the matrix to the fracture. Matrix permeability is usually very low, sometimes in microdarcies. In dual porosity dual permeability systems, flow into the wellbore can occur from both the rock matrix and the fracture. The porosity of the matrix can be either high or low. Fractured reservoirs usually exhibit augmented fluid flow in an otherwise tight formation or unexpected water breakthrough within a relatively short time of water injection.

Allan and Sun list the important differences between naturally fractured reservoirs and nonfractured reservoirs:[15]

- Fracture networks are generally of very high transmissibility, resulting in a relatively low pressure drop around the wellbore. Hence, pressure gradients may not play a major role.
- Performance of fractured reservoirs is dictated by matrix-fracture interaction during fluid flow. Fluid expansion, gravity drainage, and capillary imbibition are responsible for continuous replenishment of hydrocarbon fluids from the rock matrix to the fracture network.
- In well-managed reservoirs, the gas/oil ratio tends to be low. The free gas phase tends to rise to the crest through the fractures and form a gas cap, rather than flow towards the wellbore.

- Due to the high transmissibility of the fractures, the oil/water contact is rather sharp, and transition zones are not encountered as in conventional reservoirs. Moreover, fluid PVT properties are uniform throughout due to convective circulation.
- Performance of a fractured reservoir is rate sensitive, as producing the reservoir at a higher-than-optimum rate leads to a high gas/oil ratio. This is accompanied by a significant pressure decline.

The authors cite two contrasting examples in developing naturally fractured reservoirs in two different parts of the world. One case has a higher-than-optimum production rate, and in the other, the rate was carefully controlled. In both cases, the fracture network was connected to the aquifer. In the first case, the excessive production rate led to a rapid pressure decline. The flow of oil from the tight matrix into the fractures could not "keep up" with the pace at which the reservoir was drained. Water injection was started to reverse the decline, as would have been done in conventional reservoirs, only to compound the problem with early water breakthrough. Ultimate recovery from the reservoir was less than 20%. In the other case, any well reaching a water cut of 2% was choked back to reduce the production rate. A primary recovery factor of greater than 45% was achieved.

In essence, the recovery efficiency of a naturally fractured reservoir can be disappointing if all of the factors influencing the mechanism of recovery are not recognized correctly at the onset. It is important to correctly characterize the fractures, including their density, transmissibility, and orientation. Coupled with identification of factors that influence the matrix-fracture interaction, it is essential in effectively managing naturally fractured reservoirs. Thompson emphasizes the collection and integration of information related to fractures and faults, some of which are listed below:[16]

- Data collection during drilling of exploratory and delineation wells
- Investigation of small fluid losses during drilling
- Wellbore image logs showing fractures and faults
- Transient pressure testing
- Interference test
- Imbalances in injection and production
- Production logging
- Seismic and geochemical studies showing faults and compartmentalization

These lead to development of a reservoir model based upon the characterization of fractures, and realistic prediction of reservoir performance under nonideal conditions.

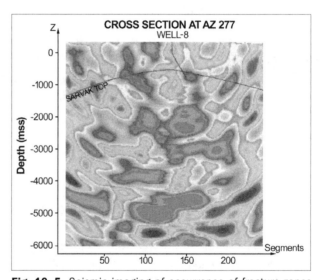

Fig. 19–5. Seismic imaging of occurrence of fracture zones in the carbonate formation. Shaded areas represent intense fracturing. *Source: M. Tabibi and S. Mousavi Mirkalei. 2005. A successful case study on development of a giant, highly fractured carbonate heavy-oil reservoir in Iran. SPE Paper #97890. SPE/ PS-CIM/CHOA, International Thermal Operations and Heavy Oil Symposium, Calgary, Canada, November 1–3. © Society of Petroleum Engineers. Reprinted with permission.*

Case study: development of a naturally fractured heavy oil reservoir[17]

The intensely fractured carbonate reservoir is located in the Persian Gulf region at a measured depth of 1,190 ft. Initial reservoir pressure is 1,408 psi. The reservoir is member of a giant field having an anticlinal structure. The oil, having a gravity of 10°–16°API, did not flow naturally to the surface. Formation porosity is 15%, and rock matrix permeability ranges between 0.2 mD and 1.4 mD. Characterization of the fractures was accomplished by core, mud loss, and seismic studies, and the latter included 2-D, 3-D, and side-view seismic location (SVSL) studies. Analysis and interpretation of the fracture system revealed that lighter hydrocarbons escaped by vertical migration through open fractures in geologic times. This was followed by closures of the fractures at later stages. Consequently, heavier hydrocarbon components were left behind in the reservoir. Due to the presence of intense fracturing, the southwest part is targeted for development, which should facilitate the flow of heavy oil in a reservoir of tight matrix permeability. Areas with low fracture intensity were considered less promising. Figure 19–5 shows fracture intensity in the carbonate formation based on seismic studies.

The oil production strategy consisted of a swab test as part of the pilot study, in which light hydrocarbons were periodically injected in order to lift heavy oil in swab cups. The productivity index of the well was estimated be 0.656 stb/d psi^{-1}. The following strategies were considered for the development of the field:

- Workover of the existing well
- Deployment of a pump to lift the oil
- Drilling of a new well for a target zone of intense fracturing as indicated by seismic survey
- Well stimulation
- Development of a reservoir simulation model based on well production data
- Evaluation of various options, including gas lift and treatment by solvent
- Evaluation of thermal recovery methods, such as steam injection, huff and puff, etc.

Redevelopment of Matured Fields

Matured fields are generally considered to include the petroleum reservoirs that are past their peak production levels and well on their way to abandonment. In most cases, the maturity of a field is reached when improved oil recovery and enhanced oil recovery processes, including waterflooding and thermal or nonthermal recovery methods, have run their course. Neither the conventional technologies nor the price of petroleum apparently supports further investment in the field. However, worldwide recovery statistics point to the fact that the majority of the oil is likely to be left behind following abandonment of a field. The average ultimate percent recovery is around 35% (fig. 10–1 in chapter 10).

Matured fields exhibit less-than-satisfactory performance, indicated by one or more of the following:

- Reservoir pressure is declining and is rapidly approaching abandonment pressure.
- There is a declining production rate, approaching the economic limit of the wells.
- Increasing water cut occurs in most or all of the wells, rendering waterflooding virtually ineffective.
- There is a substantial breakthrough of steam, gas, or chemicals injected, resulting in ineffective cycling.
- The oil recovery is marginal despite continued fluid injection.
- The cost of production, including the cost of operating the surface facilities, is no longer economically justifiable due to the current market trends.

Hence, recovering at least part of the remaining oil from a matured reservoir poses a formidable challenge to reservoir engineers. Despite the above, a literature review indicates that numerous oil fields have been rejuvenated to achieve sustainable performance. This has occurred due to a multidisciplinary approach, use of state-of-the-art tools and techniques, and integrated reservoir management. Again, it is a common observation in the oil industry that many abandoned or nearly abandoned fields draw renewed interest when the economic environment becomes more favorable.

In general, the key to approaching the issue of matured fields is based on the detailed description of geologic heterogeneities and recovery mechanisms. It also depends on reevaluation of the past performance of the reservoir and locating target areas where commercial quantities of hydrocarbon are left behind. As explained previously, a wide spectrum of tools and techniques exists in order to enhance reservoir performance and add value to the asset.

Numerous field studies are documented in the literature, suggesting implementation of one or more of the following:

- A fresh look at "old data" pertaining to the geology and production in order to formulate a new strategy by an integrated reservoir team. Detailed evaluation of various redevelopment and/or remediation scenarios is based on an updated reservoir model and experience gained during prior production.
- Infill drilling can be highly effective in heterogeneous or low permeability formations.
- Horizontal wells are a proven technology with wide-ranging applications, as noted earlier, including production enhancement and abatement of high water cut and coning.
- Infill drilling combined with waterflooding or an enhanced oil recovery operation offer a common approach to prolonging the life of a maturing field and attaining maximum recovery.

- Recompletion of existing wells is widely practiced in the industry in order to enhance productivity by avoiding problematic areas and addressing well-related issues.
- Realignment of injection and production wells during waterflooding could improve areal sweep in heterogeneous and tight reservoirs.
- Optimization of injection and production rates may lead to voidage control and better recovery.
- Studies using 3-D seismic technology could help to identify untapped regions or extensions in a mature field.
- Screening of enhanced oil recovery processes at the end of secondary recovery could be helpful.
- Hydraulic fracturing can boost well productivity. It is highly effective in low permeability formations and where formation damage is encountered.
- Well stimulation by acidization is a technique that performs well in carbonate formations.
- Gas lifts and other methods can be used in order to produce oil from low energy reservoirs.

Babadagli has compiled a list of various mature field revitalization projects, which can be categorized as detailed in the following descriptions.[18]

Optimization of waterflood operation. This is based on reservoir surveillance and model simulation. The tasks include study of bubble maps, injected water streamlines, and overall pattern performance. Simulation studies may consider various scenarios, such as infill wells, pattern flood realignment, conversion of producers to injectors, and phasing out certain injectors. The ultimate objective is to increase sweep efficiency and recovery at a minimum cost.

Infill drilling campaign. This can be used to rejuvenate field performance towards the end of primary, secondary, or tertiary recovery in a significant manner. Recent industry practices focus on single-lateral or multilateral horizontal wellbore technology wherever applicable.

Implementation of enhanced oil recovery processes. Waterflooding can lead to high water cut and low oil rates in producing wells in some cases. In certain fields, more than one type of tertiary recovery method can be implemented with an objective of recovering incremental amounts of oil. However, enhanced oil recovery processes are usually associated with longer time frames to attain the same recovery of hydrocarbons compared to the drilling of infill wells.

Case study

Bahrain field.[19] The following case study is presented to focus on redevelopment issues associated with a large, matured field having 20 reservoirs, among which 17 are oil, and the rest are gas. The reservoirs produce by various drive mechanisms. Significant variations in lithology and rock and fluid properties are encountered across the field. These required the design and implementation of a wide range of strategies in rejuvenating the reservoir. Furthermore, the field is located in a complex geologic setting, as the limestone formation is highly faulted (fig. 19–6) and various zones are in communication through the faults.

The Bahrain field was discovered in the 1930s. It reached its peak production (about 80,000 bopd) in the late 1970s, after which the decline in production averaged 6.9% annually. However, with a number of reservoir management strategies, more oil was produced, resulting in a decline rate of only 1.3%. The authors attribute the following to successful management of a large, complex field on production decline:

- Integrated approach in data acquisition and analysis, including studies based on geological, seismic, reservoir, and production data.
- Detailed reservoir characterization, including geostatistical modeling.
- Improved accuracy in reservoir simulation.
- Drilling of new wells and initiation of improved oil recovery techniques in certain cases, including the most productive formations.
- Implementation of horizontal well technology in two reservoirs.
- Dual string completion for shallow reservoirs previously undeveloped for economic reasons.
- Recompletion of existing wells for changing target zones and gas/oil ratio control.

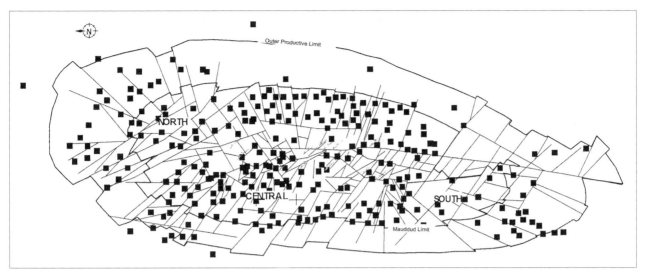

Fig. 19-6. Bahrain field showing numerous faulted structures and development of wells. Implementation of a wide array of reservoir management strategies has been successful in arresting steep production decline in this matured field. *Source: C. R. K. Murty and A. al-Haddad. 2002. Integrated development approach for a mature oil field. SPE Paper #78531. 10th Abu Dhabi International Exhibition and Conference, Abu Dhabi, UAE, October 13–16. © Society of Petroleum Engineers. Reprinted with permission.*

- Annual update of reserves based on information obtained from newly drilled wells and production data. Due to gas injection, oil was found to transfer across the fault to a juxtaposed zone. Hence, reserves are upgraded regularly for certain zones.
- Implementation of well management techniques to enhance production, including gas lift, pumping, and gas/oil ratio control devices.

Some of the specific strategies to develop the remaining reserves of the Bahrain field are described in the following:

- **Development of tight zones.** The program involved the drilling of wells in three phases. First, step-out wells were drilled on the periphery with a view to delineating the productive limit of the pay zones. Next, infill wells were drilled to produce the zones. In the final phase, more infill wells were drilled with close spacing suitable for enhanced oil recovery processes.
- **Shallow zones of low productivity.** Dual zone completions were implemented in a large number of wells, as stand-alone completions in these zones were not economically feasible. The short string produced the low productivity upper zone, while the long string was completed in the lower zone of higher productivity.

Marginal Reservoirs

Marginal reservoirs are generally referred to as reservoirs where development and production are not economically feasible, or have marginal potential, based on current technology. All matured reservoirs become marginal at some point in time in their lives. However, a newly discovered reservoir can be marginal due to anticipated poor reserves or poor production potential. Marginal reserves in a field can generally be attributed to one or more causes, as detailed in the following.

Newly discovered reservoirs:
- Poor reservoir quality and significant geologic complexity
- Relatively low volume of original hydrocarbon in place, including thin oil or gas zone and lenticular sand
- Unfavorable fluid properties, such as highly viscous oil of low API gravity
- Anticipated low recovery during primary production, coupled with poor prospects for improved oil recovery processes

- High cost of field development, as in the case of smaller fields located offshore
- High cost of production, such as deepwater heavy oil reservoirs
- High cost of transportation, such as smaller fields located in remote areas onshore or offshore

Matured reservoirs:
- High water cut and water recycling during waterflooding, leading to uneconomic operation
- High costs related to waterflood facilities for offshore fields
- Incremental recovery by enhanced oil recovery processes not found to be economically justifiable
- Drilling of infill wells not cost-effective due to anticipated low recovery

Two case studies involving marginal reservoirs are discussed briefly in this section: the Bahrain field and the al Shaheen field.

Bahrain field[20]

Reservoir management issues associated with the matured Bahrain field were presented earlier. Two zones in the field were regarded as marginal due to low matrix permeability and lower-than-normal reservoir pressure. Acid stimulation was performed in order to enhance well productivity. Wells drilled near the existing faults were found to be better producers. Reservoir quality maps were prepared along with fault locations to target drilling locations.

al Shaheen field[21]

Development of the al Shaheen field, discovered offshore of Qatar in the 1970s, was not considered to be economically feasible for the following reasons:
- Production would be from a series of thin, stacked carbonate reservoirs of low permeability. Good formation permeability was encountered only in limited areas.
- Well tests conducted in appraisal wells did not lead to satisfactory conclusions concerning their production potential.
- The high-viscosity oil would lead to low rates in production by artificial lift. Large lateral variations in oil viscosity were encountered.
- Due to lateral pressure gradients, distribution of fluids was characterized by a nonhorizontal fluid contact. This led to uncertainties in the original oil in place estimates.
- Individual reservoirs were spread over large areas. The development would necessitate a large number of platforms.

Development of the marginal reservoir began in early 1990s, principally based on horizontal drilling technology. Very long laterals (15,000 ft to 20,000 ft) were drilled and completed in order to cover the thin reservoirs interspersed throughout the area. About 131 horizontal wells were drilled, producing about 200,000 bopd. Waterflooding was implemented to augment recovery. Horizontal drilling transformed the field from its marginal status to an economically attractive operation due to a development strategy based on long horizontal wells. The advantages cited include the following:
- The number of offshore platforms and wells required for field development was significantly reduced as compared to a vertical well case.
- Long laterals led to higher productivity in tight formations. Oil was produced without the aid of artificial methods for relatively long periods.
- Waterflooding based on long horizontal wells led to better sweep in near line drive patterns.
- Reservoir appraisal information could be gathered at long distances from the offshore platforms.

Smart Well Technology in Well and Reservoir Management

In recent years, the oil and gas industry witnessed yet another innovation: smart well or intelligent well completions. These offer the ability to control well configuration and performance based on continuous production data received from the wells. The technology is viewed as part of the total system whereby certain well and reservoir management decisions can be made in real time or preprogrammed. Intelligent wells hold promise in many areas, including the following:[22,23]

- Injection or production can be selectively controlled in targeted zones within a formation based on interval control valves (ICVs). These devices are used to improve the performance of waterflood or other enhanced oil recovery operations, including shutting off zones of high water or gas production.
- Fluid injection can be balanced throughout vertical or horizontal sections of the well, including forced injection into low permeability zones. This results in better sweep and ultimate recovery.
- In multireservoir environments where production takes place through a single well, a low permeability reservoir can be produced effectively by choking back the more permeable reservoir. Crossflow between reservoirs through the wellbore can also be minimized by selectively shutting off an interval.
- With the rise of water level due to continued production, the lower section of the formation can be choked down or shut off to minimize water cut at the producers.
- In conventional systems, lower and upper zones of contrasting permeability need to be produced sequentially to avoid crossflow. Smart well technology allows alternate production of both zones, increasing the net present value of the asset.
- Real-time optimization of gaslift operation is based prevailing conditions.
- A reduction of costs related to surface facilities, fluid volumes, and human intervention is realized.

Case studies

Zonal control in CO_2 injection.[24] Smart well technology was implemented in a pilot CO_2 injection project in the Kelly-Snyder field in Texas. The objectives were to reduce the cycling of injected fluid, improve the sweep, and augment oil production based on the enhanced oil recovery process. Downhole flow control devices were deployed to optimize the injection profile in the injectors as well as to restrict production from the zones where the breakthrough of CO_2 was encountered. Zonal control of injection and production during improved or enhanced oil recovery processes has significant potential in geologically complex cases, including stratified and compartmentalized reservoirs. Initial results of the pilot study indicate that the smart wells are effective in reducing CO_2 production from the wells and producing oil at economic rates.

Wide-ranging application of intelligent wells in matured petroleum reservoirs.[25] Smart well technology is in the process of evaluation and deployment in a large number of fields (about 100) in Oman. In most cases, the reservoirs have reached secondary recovery phase and are produced by artificial lifts. Intelligent well technology gains more significance in potential applications as multiple zones are produced from a single well. In many instances, the technology is applied to multilateral horizontal wells. Some of the intended applications are listed in the following:

- Monitoring, measurement, and control of fluid flow, pressure, and other parameters
- Two-phase metering in order to measure water cut encountered in dual-lateral horizontal wells
- Digital hydraulic control valves to control production of multilateral wells
- Distributed temperature sensing to monitor wells during waterflooding and steam flooding
- Continuous monitoring and optimization of artificial lift systems to achieve greater efficiency

Time-Lapse Seismic Studies

Time-lapse 3-D seismic studies, which are often referred to as 4-D seismic studies, hold the promise of characterizing fluid flow in the reservoir. These studies are being utilized in the industry with increasing frequency. In reservoir engineering, 4-D seismic is known to have the following applications:[26–29]

- Identification of bypassed oil zones following primary, secondary, and tertiary recovery
- Monitoring of injection fluid fronts
- Assessment of heat flow in thermal processes
- Monitoring of water encroachment from an aquifer in lateral and vertical directions, leading to changes in new well trajectory
- Identification of compartments in the reservoir with no change in pressure, which may pinpoint further infill drilling opportunities
- Identification of extensively drained areas in the reservoir that may point to anticipated water breakthrough locations and subsequent manipulation of injection rates
- Fluid flow characterization across faults
- Reservoir surveillance as 3-D surveys are carried out at regular intervals in certain fields
- Correction of reservoir volumes in light of reservoir behavior
- Update of reservoir models in light of new information following each time-lapse study
- Better match of reservoir production history in simulations by integrating time-lapse seismic information

Seismic response, in the form of acoustic impedance, is evaluated between successive 3-D seismic studies. This is done to detect any changes caused by changes in fluid saturation, pressure, temperature, or rock properties. In order to reduce the uncertainties associated with the interpretation, results of the study are usually compared with the results of reservoir simulation. Fine-tuning of static (geologic) and dynamic (simulation) models follows in order to match the results obtained by seismic study. The principle of 4-D seismic study is based upon comparing the response of a monitor (current) survey with that of a base survey conducted previously.

Case study: data integration based on dynamic tests and 4-D seismic[30]

The following illustrates the application of 4-D seismic studies and data integration from multiple sources. A reservoir having multiple barriers is ideally suited for time-lapse studies in order to track the movement of injected and reservoir fluids. Schiehallion field, located in offshore UK, is a compartmental reservoir (with faults, channel margins, and other barriers) with an estimated original oil in place of more than 2 billion bbl. Effective water injection is the key to recovery, as the reservoir has relatively low natural energy (low initial gas/oil ratio, weak aquifer support, and relatively shallow depth). However, reservoir connectivity, a critical requirement in providing good pressure support by water injection, was recognized as an issue from the onset. Due to poor connectivity between the injection and production wells in various parts of the reservoir, water injection was not able to provide adequate support. This resulted in a rapid increase in the gas/oil ratio. Infill wells were drilled later, and the reservoir was monitored based on various tools and techniques, including 4-D seismic studies, which led to effective reservoir management.

In order to characterize the reservoir and understand the connectivity between various sections, a number of tests were conducted, as described in the following discussion.

Extended well test. An extended well test, conducted during the appraisal period, apparently indicated the reservoir was well connected. This was due to the fact that the test results were influenced by the presence of a small gas cap not identified at the time.

4-D seismic tests. Between 1999 and 2004, four repeat seismic surveys were performed. Results of the time-lapse seismic tests are presented in Figure 19–7, which shows paths of water movement (darker shades) from the injectors. Due to the compartmental nature of the reservoir, water movement cannot be predicted by conventional reservoir models. In addition, locations of free gas in a reservoir are clearly identified by 4-D seismic studies.

Tracer tests. By conducting tracer tests between injectors and producers, it was possible to model the water-cut trend in the producing wells. Furthermore, a supposedly sealing fault located between two wells was found not to be the case. The finding was supported by the results of a time-lapse seismic study, which indicated water influx across the fault.

Repeat formation tester data. Vertical pressure profiles were obtained by use of Schlumberger's Repeat Formation Tester (RFT) or Modular Dynamics Tool (MDT) studies conducted during the drilling of new wells. These were an excellent source for history matching and understanding connectivity.

Transient well tests. Transient tests, conducted at regular intervals, are used to detect any changes in skin, permeability, effective horizontal length, and extrapolated reservoir pressure (p*). (These parameters are described in chapter 5.) The transient test results are used to identify sanding problems and other potential issues. These tests also complement analytical and numerical simulation models.

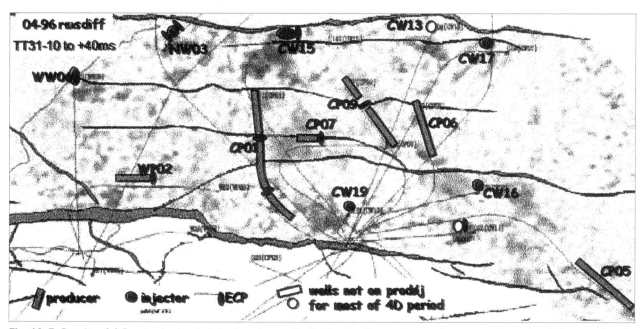

Fig. 19–7. Results of 4-D seismic showing shades of chang in fluid saturation and reservoir pressure. In a color plot, gas and water saturations can be shown in different colors, distinguishing between various fluids in the compartmental reservoir. Horizontal wells are shown as thick lines. *Source: A. Govan, T. Primmer, C. Douglas, N. Moodie, M. Davies, and F. Nieuwland. 2005. Reservoir management in a deepwater subsea field—the Schiehallion experience. SPE Paper #96610. Presented at SPE's Offshore Europe Conference, Aberdeen, Scotland, UK, September 6–9. © Society of Petroleum Engineers. Reprinted with permission.*

Interference testing. Continuous recording of bottomhole pressures in producers as nearby injectors are online and offline leads to the evaluation of connectivity between various well pairs. This is shown in Figure 19–8. Dedicated interference tests can be conducted by initially closing a production well and surrounding injectors, followed by sequential startup of the injectors. The injector known to have weakest connectivity with the producer is brought online first.

This is an excellent example of a reservoir team utilizing every bit of information obtained from wide-ranging sources in order to effectively manage the reservoir. As it was recognized in the course of field development, some data raised more questions than answers. This underscored the need for data integration, aided by material balance and simulation studies.

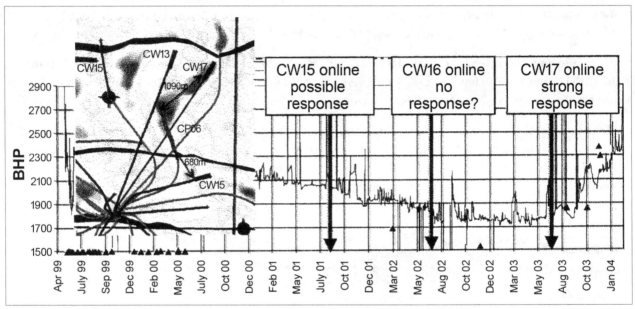

Fig. 19–8. Continuous monitoring of bottomhole pressure at the producer indicates varying degrees of influence of nearby injectors. *Source: A. Govan, T. Primmer, C. Douglas, N. Moodie, M. Davies, and F. Nieuwland. 2005. Reservoir management in a deepwater subsea field—the Schiehallion experience. SPE Paper #96610. Presented at SPE's Offshore Europe Conference, Aberdeen, Scotland, UK, September 6–9. © Society of Petroleum Engineers. Reprinted with permission.*

Summing Up

Reservoir engineers typically deal with a variety of issues in efficiently developing and managing a petroleum reservoir. In most circumstances, the issues stem from the following:

- Absence of a robust reservoir surveillance and monitoring program at the onset of the development of an asset
- Lack of detailed analysis and a real-time or near–real-time decision-making process
- Presence of unknown reservoir heterogeneities and inadequate reservoir characterization
- Uncertainties in estimating oil and gas reserves due to a large number of unknowns
- Suboptimal well productivity issues related to formation damage, high water cut and gas/oil ratio, coning, workovers, gas lifting, sand production, and other factors
- High residual oil saturations in certain layers or pockets following primary or secondary recovery
- Inefficient water and gas injection, leading to large volume of fluid recycling
- Limited recovery from compartmental, stratified, and fractured reservoirs
- Development of marginal fields
- Revitalization of matured reservoirs
- Lack of teamwork and integrated reservoir studies

The tools and techniques that are at reservoir engineers' disposal include, but are not limited to, the following:
- Intensive monitoring of reservoir performance: pressure and temperature profiles, fluid flow rates, fluid phase and density in wells. Deployment of permanent downhole gauges (PDG) for real-time monitoring and analysis.
- Detailed reservoir characterization and modeling based on all available data. Tools and techniques utilized in reservoir characterization are listed in chapter 2.
- Multilateral horizontal well drilling to significantly augment productivity.
- Infill drilling and recompletion of wells.
- Identification and remediation of zones where injected fluids (water, gas, steam, or chemical substance) break through.
- Hydraulic fracturing and acidization to enhance well productivity.
- Implementing intelligent well systems (IWS) to selectively shut in, open, or choke down a target interval during injection or production from a remote location.
- Establishing best management practices based on field-specific experience.

The metrics that can be utilized to evaluate reservoir performance and management practices can be listed as in the following:

Target versus actual
- Reservoir surveillance:
  - Production by field, section, or well
  - Annual changes in reservoir pressure
  - Annual decline rate in production
  - Water cut and gas/oil ratio (on a reservoir and well basis)
  - Mobile oil/gas recovery efficiency and ultimate recovery
- Cost analysis:
  - Unit costs related to production, drilling, and workover per well
- Economic evaluation:
  - Rate of return and net present value (NPV)

Horizontal wells have ushered in a new era of oil and gas productivity in recent decades. With innovative drilling technology, most horizontal wells are drilled as multilaterals, having varying configurations in order to maximize production. Some of the significant advantages offered by horizontal drilling are noted here:
- Exposure of the well to a significantly larger reservoir area, usually in thousands of feet
- Horizontal recompletion in order to avoid watered-out zones
- Development of thin oil rims and marginal reservoirs
- Utilization of the same well to produce from multiple hydrocarbon zones or reservoirs located across faults or other physical boundaries
- Connectivity to a larger portion of the fracture network in a naturally fractured formation
- Minimization of water and gas coning due to relatively low drawdown pressure
- Enhanced production from tight formations
- Reaching for hitherto unexploited areas with large remaining oil saturations
- Relatively fewer wells drilled in onshore and offshore fields
- Significantly lower production cost per barrel of oil

Horizontal wells have found wide applications in developing heavy oil reservoirs and unconventional resources, such as oil sands, coupled with appropriate thermal processes.

Waterflood conformance is comprised of techniques to improve injection or the production profile. Excessive water entry into a thief zone is minimized by selectively applying various water-blocking chemicals, such as gels and polymers. Similarly, problematic zones responsible for water or gas production are isolated. Conformance issues typically arise due to reservoir stratification, presence of high permeability streaks, channeling of injected water, water and gas coning, and viscous fingering. The net effect of the above is a marked decline in well performance and poor ultimate recovery.

This chapter describes the development of a low permeability reservoir. The development process involved early acquisition of static reservoir data, reservoir monitoring and surveillance, better reservoir description, and integrated model development. It also included economic analysis and drilling of wells at relatively close spacing. This methodology is applicable to any reservoir, tight or conventional.

Infill drilling is a proven method to increase ultimate recovery from a petroleum reservoir. Such drilling is often part of a larger scenario whereby a waterflooding or enhanced oil recovery operation is initiated. Examples of infill drilling coupled with improved oil recovery or enhanced oil recovery processes have been described in this chapter and in earlier chapters. The decision for infill drilling depends on an economic analysis of capital expenditure and incremental return.

Most reservoir management programs include measures to enhance the productivity of vertical and horizontal wells. This could include acid stimulation, hydraulic fracturing, intelligent well completion, and sand control, among others. Well productivity is monitored by the following:

- Analyzing an observable decline in production
- Conducting well tests to determine skin and productivity index
- Analyzing produced water and oil
- Analyzing fluid compatibility
- Conducting core studies related to formation damage

Hydraulic fracturing is the most widely used method in augmenting well productivity. Wells having positive skin or wells producing from tight formations are ideal candidates for hydraulic fracturing. Certain wells are acidized to improve productivity.

Fractured reservoirs are comprised of two distinct systems: the rock matrix and an interbedded fracture network. In a dual porosity fractured reservoir system, fluid can flow from the matrix to the fracture, but hydrocarbons are only produced through the fracture network. In a dual porosity dual permeability system, fluid flow into the wellbore can take place from both the matrix and the fracture network. Fluid flow dynamics associated with fractured reservoirs are markedly different than those of conventional reservoirs in which no fractures are present. Performance of fractured reservoirs is dictated by the matrix-fracture interaction during fluid flow. Fluid expansion, gravity drainage, and capillary imbibition are responsible for continuous replenishment of hydrocarbon fluids from the rock matrix to the fracture network.

Development and efficient management of a fractured reservoir may require a unique strategy, including detailed reservoir characterization and production at an optimum rate. For example, a fractured reservoir was produced at a relatively high rate, resulting in a rapid decline in pressure and poor recovery. This resulted as the flow of oil from the low permeability matrix to the fracture network could not "keep up" with the rate of production. A waterflooding operation was initiated, which only compounded the problem with rapid water breakthrough, as fractures provided good conduits for the injected water. In another field, the production rate from the wells was choked back when a certain water cut was reached. This led to significantly higher ultimate recovery.

Matured fields are generally past their peak production level and approaching abandonment. The performance of a matured field is associated with one or more of the following:

- Declining production
- Rapidly rising water cut or gas/oil ratio
- High cost of production and surface facilities operation

Redevelopment of a matured field may require implementation of multiple strategies. This chapter describes the revitalization of a matured field with aid of the following:

- Review of old reservoir data and formulation of a new strategy based on integrated reservoir model.
- Conducting seismic surveys to locate untapped regions or extensions of the matured field.
- Drilling of step-out wells to accurately delineate the producing fields.
- Horizontal drilling and recompletion to increase well productivity and minimize water production.
- Optimization of waterflood operation with the aid of a bubble map, streamline simulation, and study of pattern performance.
- Realignment of injectors and producers in waterflood operations in order to attain better sweep efficiency.
- Screening of enhanced oil recovery processes to find a strategy best suited for the waterflooded reservoir. Infill drilling combined with improved oil recovery or enhanced oil recovery processes to increase recovery.
- Well stimulation by acidization in a carbonate formation.

Marginal reservoirs are generally referred to as reservoirs where development and production are not economically feasible or where the potential is marginal based on current technology. Marginal reservoirs are often associated with one or more of the following:

- Poor reservoir quality
- Low volume of original hydrocarbon in place
- Unfavorable fluid properties such as extra heavy oil
- Significant geologic complexity
- High costs associated field development, production, and transportation

A typical example of a marginal field would be a relatively small reservoir located offshore at a considerable depth. Again, all matured fields become marginal at some point in their life cycle.

The techniques that are employed to develop marginal reservoirs often include the implementation of horizontal well technology. For example, just one or few long horizontal wells may be drilled in a marginal reservoir to bring down the cost of production. Hydraulic fracturing and horizontal recompletion are also commonplace to augment recovery from marginal reservoirs. Implementation of a carefully designed waterflooding or enhanced recovery operation may also aid in successfully producing a marginal reservoir.

One of the major innovations in the oil and gas industry is smart well or intelligent well completions. It offers the ability to control a well operation based on the analysis of production data on a real-time or near–real-time basis. The technology is viewed as part of the total system in which certain well and reservoir management decisions can be made without human intervention. Intelligent wells have been deployed in diverse applications, some of which are listed in the following:

- Selective control of injection or production in targeted zones within a formation. In waterflood operations, smart wells are capable of shutting off zones of high water or gas production.
- With the rise of water level due to continued production, the lower section of the formation can be choked down or shut off to minimize water cut at the producers.
- Fluid injection can be balanced throughout vertical or horizontal sections of the well, including forced injection into low permeability zones.
- A selected zone can be choked back in order to augment recovery from another zone of differing permeability.
- Crossflow between reservoirs through the wellbore can also be minimized by selectively shutting off an interval.
- Real-time optimization of the gaslift operation can be achieved based on prevailing conditions.
- Reduction of costs is realized related to surface facilities, fluid volumes, and human intervention.

Several case studies have been presented in this chapter to provide a general guide related to tools and techniques available to the reservoir engineer in order to successfully manage a field. It is emphasized that each reservoir is unique, requiring a unique solution based on detailed reservoir surveillance, data integration, teamwork, innovative technology, and economic analysis.

References

1. Saleri, N. G. 2002. Learning reservoirs: adapting to disruptive technologies. *Journal of Petroleum Technology*. March, 57.

2. Vij, S. K., S. L. Narasaiah, A. Walia, and G. Singh. 1988. Adopting multilateral technology. *Journal of Petroleum Technology*. July, 50.

3. Levitan, M. M., P. L. Clay, and J. M. Gilchrist. 2004. Do your horizontal wells deliver their expected rates? SPE Paper #87402. *SPE Drilling and Completion*. Vol. 19, no. 1. March, 40–45.

4. Clark, R. A., A. A. Afemari, T. E. Ezeukwu, and H. Awi. 2001. Case study: redevelopment of Ebughu field. *Journal of Petroleum Technology*. February, 44.

5. McKay, C., J. Jones, and J. Promerene. 2003. Successful horizontal producers in Midway-Sunset thermal operations. *Journal of Petroleum Technology*. November, 43.

6. Azari, M., M. Soliman, and N. Gazi. 1997. Reservoir engineering applications to control excess water and gas production. SPE Paper #37810. Presented at SPE's 10th Middle East Oil Show and Conference, Bahrain, March 15–18.

7. Love, T., A. McCarty, M. J. Miller, and M. Semmelbeck. 1999. Diagnosis, treatment design, and implementation process improves waterflood conformance. *Journal of Petroleum Technology*. January, 54.

8. Holditch, S. A. 2006. Tight gas sands. *Journal of Petroleum Technology*. June, 86.

9. McKinney, P. D., J. A. Rushing, and L. A. Sanders. 2003. Maximizing reserves growth and profitability in tight gas sands. *Journal of Petroleum Technology*. April, 63.

10. Genett, D. B., C. A. Whitaker, D. B. Smith, and L. M. Price. 2003. Improved recovery processes and effective reservoir management maximize oil recovery at Salt Creek. *Journal of Petroleum Technology*. December, 42.

11. Gunarto, R., T. B. Tran, B. Widyantoro, and T. Kritzler. 2004. Production improvement for horizontal wells in Sumatra. SPE Paper #86545. Presented at the 2004 SPE International Symposium and Exhibition on Formation Damage Control, Lafayette, LA, April 18–20, pp. 295–304.

12. Holditch, S. A., 2006.

13. Norris, M. R., S. N. Gulrajani, A. K. Mathur, J. Price, and D. May. 2002. Hydraulic fracturing for reservoir management. *Journal of Petroleum Technology*. March, 47.

14. Agiddi, D. O. 2005. A decision analysis approach to hydraulic fracture optimization in the W31S Stevens oil zone, Elk Hills field, California. SPE Paper #93989. Presented at the SPE Western Regional Meeting, Irvine, CA, March 31–April 2.

15. Allan, J., and S. Q. Sun. 2003. Fractured reservoir recovery factors: lessons from 100 fields. *Journal of Petroleum Technology*. October, 51–52.

16. Thompson, L. B. 2000. Fractured reservoirs: integration is the key to optimization. SPE Paper #56010. *Journal of Petroleum Technology*. February.

17. Tabibi, M., and S. Mousavi Mirkalei. 2005. A successful case study on development of a giant, highly fractured carbonate heavy-oil reservoir in Iran. SPE Paper #97890. SPE/PS-CIM/CHOA, International Thermal Operations and Heavy Oil Symposium, Calgary, Canada, November 1–3.

18. Babadagli, T. 2005. Mature field development—a review. SPE Paper #93884. SPE Eurocpec/EAGE Annual Conference, Madrid, Spain, June 13–16.

19. Murty, C. R. K., and A. al-Haddad. 2002. Integrated development approach for a mature oil field. SPE Paper #78531. 10th Abu Dhabi International Exhibition and Conference, Abu Dhabi, UAE, October 13–16.

20. Murty, C. R. K., and A. al-Haddad. 2002.

21. Thomasen, J., I. A. al-Emadi, R. Noman, N. P. Ogelund, and A. Damgaard. 2005. Realizing the potential of marginal reservoirs: the al Shaheen field offshore Qatar. IPTC Paper #10854. Presented at the International Petroleum Technology Conference, Doha, Qatar, November 21–23.

22. Welldynamics, Inc. 2006. Web literature. Spring, TX: Welldynamics Inc.

23. Robinson, M. 2003. Intelligent well completions. *Journal of Petroleum Technology*. August, 57.

24. Brnak, J., B. Petrich, and M. R. Konopczynsk. 2006. Application of smartwell technology to the SACROC CO2 EOR project: a case study. SPE Paper #100117. SPE/DOE Symposium on Improved Oil Recovery, Tulsa, OK, April 22–26.

25. al-Khodhori, S. M. 2003. Smart well technologies in PDO for production and reservoir management & control. SPE Paper #81486. Middle East Oil Show, Bahrain, June 9–12.

26. Lumey, D. E., and R. A. Behrens. 1998. Practical issues of 4-D seismic reservoir monitoring: what an engineer needs to know. SPE Paper #53004. *SPE Reservoir Evaluation & Engineering*. December: pp. 528–538.

27. Kawar, R., P. Hatchell, R. Calvert, and M. Khan. 2003. The workflow of 4-D seismic. Middle East Oil Show, Bahrain, June 9–12.

28. Mikkelsen, P. L., K. Guderian, and G. du Plessis. 2005. Improved reservoir management through integration of 4-D seismic interpretation, Drugen field, Norway. SPE Paper # 96400. Presented at SPE's Offshore Europe Conference, Aberdeen, Scotland, UK, September 6–9.

29. Waggoner, J. R. 2002. Quantifying the economic impact of 4-D seismic projects. JPT Paper #77969. *SPE Reservoir Evaluation and Engineering*. Vol. 5, no. 2. April.

30. Govan, A., T. Primmer, C. Douglas, N. Moodie, M. Davies, and F. Nieuwland. 2005. Reservoir management in a deepwater subsea field—the Schiehallion experience. SPE Paper #96610. Presented at SPE's Offshore Europe Conference, Aberdeen, Scotland, UK, September 6–9.

20 · Class Projects

This final chapter will give students the opportunity to utilize what they have learned to offer potential solutions and strategies in reservoir engineering and management. In their profession, reservoir engineers are expected to make vital contributions to the development and execution of the reservoir engineering and management projects as members of a multidisciplinary team. Students are encouraged to form small collaborative teams in doing the class projects. In arriving at a solution, all options must be investigated and debated among the team members. Again, students are expected to do literature searches extensively in the course of their class projects.

The broad objectives of the class projects include, but are not limited to, the following:
- Gather geosciences and engineering data
- Define the problem
- Investigate various solutions, including innovative technologies
- Choose the most economically viable solution
- Prepare effective presentations and obtain management approval
- Eventually work on the problem to deliver timely solutions
- Observe best reservoir management practices

Suggested problems could be the following:
- Development plan for (a) a newly discovered offshore oil field, (b) a marginal field, and (c) highly heterogeneous reservoirs
- How to enhance recovery from a primary depleted oil reservoir by using appropriate technologies, such as by horizontal well planning, infill drilling, improved oil recovery (IOR) operations, reservoir monitoring, information management, and real-time decision making, etc.
- Mitigate water and gas coning and various well productivity issues
- Investigate the potential of enhanced oil recovery processes from an oil reservoir following primary and secondary production

As members of an integrated team, students would have the objective of investigating the optimum development plan, considering the following:
- Geosciences and engineering data
- Integration of geosciences and engineering data to develop an integrated reservoir model
- Recovery schemes, such as natural depletion or water drive
- Well spacing, such as the number of wells and platforms
- Performance prediction
- Economic evaluation, including analysis of risk and uncertainty

The focus of the reservoir engineer would be on contributing to the development and execution of the projects. The projects in this chapter, or other exercises in the previous chapters, hardly have complete data sets with which to work. This book is written keeping in mind that reservoir engineers in their profession often deal with major uncertainties. They must make valid assumptions and build likely scenarios throughout their careers.

Presented on the following pages are several class projects that may require several weeks of detail-oriented study followed by a solution.

Specific Class Projects

1. Development of a faulted reservoir with dip

A sandstone reservoir is discovered with two faults (shown as dotted lines in Figure 20–1) as indicated by 3-D seismic survey. The faults are not characterized by conducting dynamic tests (sealing, nonsealing, etc.). The reservoir has a dip of 60° towards the northwest. Aquifer support is in the form of bottom water drive. No gas cap was encountered in the updip well based on logs. Initial reservoir pressure at datum was found to be 3,330 psia.

Two vertical wells are drilled within a few months. Production data indicates the following:

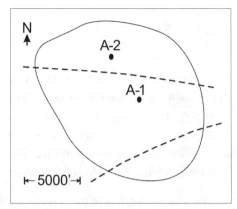

Fig. 20–1. Outline of reservoir with faults

Well A-1	**Well A-2**
PI: 0.68 stb/d/psi.	PI: 0.52 stb/d/psi.

The wells did not show any significant decline in rate during the first several months of production.

Formulate a strategy to develop the reservoir. Information obtained from core and log studies is given in Table 20–1. Include the following, with schematic diagrams, equations, tables, and any relevant information obtained from the literature:

(a) Estimate of hydrocarbon volume in stock-tank barrels per acre-feet (stb/acre-ft) and in units of thousand stock-tank barrels (Mstb)

(b) Expected recovery efficiency based on API correlation. Make any assumptions necessary.

(c) Proposed location of the next three wells and their depths

(d) Flexibility in the development program in case a dry hole is encountered

(e) Role of dynamic tests and production history in characterizing a fault and compartmental behavior of the reservoir

(f) Evaluation of horizontal drilling

(g) Estimated number of producers and expected production rates

(h) Applicability of analytical methods in predicting reservoir performance

(i) Optimum timing for secondary recovery

(j) Location of injectors during waterflooding

(k) Reservoir monitoring and surveillance plan throughout the reservoir life cycle, including before and after water breakthrough

(l) Realignment of injectors and producers during waterflooding

(m) Possible utilization of intelligent well technology

(n) Data requirements and integration for reservoir simulation in light of the reservoir heterogeneity and bottom water drive

(o) Role of reservoir simulation throughout the reservoir life cycle in this specific case

List any valid assumptions made in building the development strategy.

Table 20–1. Core, log, and PVT data

	Well A-1	**Well A-2**
Depth, ft TVD	6,510	?
Net thickness, ft	55	42
Formation permeability, mD	50–65	35–55
Permeability variation factor (V)	0.55	0.75
Porosity of formation, fraction	24%–26%	19%–23%
Connate water saturation, fraction	0.19	0.23
Oil gravity, °API	28.3	28.0
Bubblepoint, psia	1,820	1,805
Solution gas gravity	0.668	0.663
Solution GOR at b.p., scf/stb	?	518
Flowing bottomhole pressure, psia	2,482	2,456
Number of PBU tests performed	1	0

2. Revitalization of a channel sand

Figure 20–2 represents a channel-shaped reservoir located offshore. Following a 2-D seismic survey conducted in the mid-1990s, two vertical wells were drilled in certain parts of the reservoir for appraisal purposes. In well B-1, drilled toward the western part, two oil zones were encountered. The top zone was very thin (6 ft) but with good permeability (> 150 mD). The thickness of the lower zone was 20 ft, but much lower permeability was encountered (< 10 mD). In well B-2, drilled approximately 10,000 ft to the northeast, only the lower oil zone was encountered. The thickness of the zone was reduced to 15 ft in this location. Due to the poor reservoir quality of the lower sand, relatively low API gravity of the oil (24°), and shallow depth, well B-2 ceased to flow naturally following a few weeks of production. Well B-1 stopped flowing after a year. The field was subsequently abandoned. Artificial lift methods and infill wells were not planned because of marginal reserve estimates, the depressed price of oil in that period, and the high cost of offshore field development.

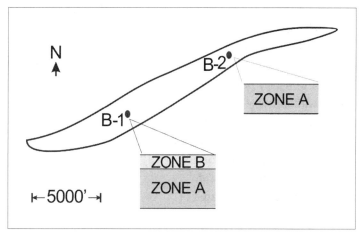

Fig. 20–2. Outline of channel sand and layer(s).

In recent years, a new discovery with much better reservoir quality (both in transmissibility and storativity) is being developed in the same basin. This has renewed interest in further exploring the abandoned field. List the data that should be collected from the previous development and production of the field. In order to ascertain whether the low permeability zone produced at all in well B-1, what kind of test results or analysis would be helpful? What data is needed to make a rough estimate of remaining oil saturation (ROS) in the more permeable layer? Why could this information be important in the new development plan?

Devise a detailed strategy to revitalize the marginal field with a minimum number of wells due to constraints related to the availability of offshore platforms. Include the following:
(a) A method to characterize the reservoir, specifically the continuity of the upper sand
(b) Number and type of new or sidetracked wells
(c) Effects of crossflow between the two layers, if any, on the fraction of oil produced from each zone
(d) Impact of facies change and other types of discontinuities on drilling and production
(e) Evaluation of the artificial lift method
(f) Selection of an enhanced oil recovery method at the end of primary production
(g) Estimated ultimate recovery and cost of offshore drilling and recompletion

Make necessary assumptions to formulate the development strategy. Include example calculations wherever appropriate to support these conclusions. The following data is available:

Zone A
Porosity: 19%. Initial water saturation: 30%.

Zone B
Porosity: 27%. Initial water saturation: 24%.

3. Waterflood design in a heterogeneous carbonate reservoir

Figure 20–3 shows a carbonate reservoir that has been under primary production for a few years. Oil production from the field has declined considerably within a short period of time due to weak aquifer support and low initial reservoir pressure. The current reservoir pressure is 2,000 psia. The depth of the reservoir is 6,600 ft.

The permeability in the crestal part of the reservoir varies between 40 mD and 70 mD. However, significant permeability degradation occurs toward the flanks. The eastern part is not developed due to apparently poor reservoir quality (< 5 mD). The reservoir was developed with 160-acre well spacing in most areas. Primary recovery from the reservoir is 16% to date. Oil gravity is 27.6°API, and viscosity is 2.9 cp. The bubblepoint pressure is 1,560 psi.

A pilot waterflood study was initiated in the shaded area shown in the figure. An existing producer with a low productivity index was converted to an injector for the study. Four new wells were drilled to complete an inverted 5-spot pattern of closer spacing of 80 acres. Reservoir simulation studies indicated that infill wells would be needed eventually to sustain production from the low permeability, high viscosity oil reservoir.

Results of a pilot waterflood study, as observed after several months of injection, are also included in the figure. Initial performance of the wells in terms of oil rate and water cut appears to differ significantly, suggesting a high degree of heterogeneity in the dolomite formation. A number of observations were made, as detailed in the following.

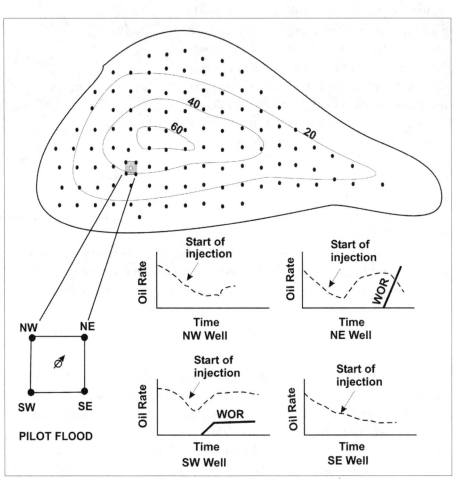

Fig. 20–3. Reservoir showing well locations and results of pilot flood.

NW well. A delayed response was observed following water injection. The increase in oil production was marginal.

NE well. The well responded quickly to injection. However, water breakthrough followed in short order, resulting in an ever-increasing water/oil ratio and an eventual decrease in oil production. Cumulative production during primary recovery from this well was better than many wells in the field.

SW well. This was the first well to register a response following water injection. As with the previous well, water breakthrough was encountered soon afterwards. However, the water/oil ratio remained low and steady, without significantly affecting recovery.

SE well. The well continued to decline in production rate without showing any effect of waterflooding. The well had been a poor producer from the start.

Based on the information given, address the following:

(a) Based on the production history alone, discuss the possible heterogeneities that may exist in the reservoir. Be as specific as possible. Describe methodologies to improve volumetric sweep efficiency in this case.

(b) Describe a possible scenario in case a high productivity index well is selected as an injector for the pilot flood.

(c) Provide a rough estimate of initial oil in place and remaining oil saturation.

(d) In Figure 20–3, contours of formation thickness are shown in feet. Assume a net to gross thickness ratio of 0.8 and an average porosity of 17.5%.

(e) Two-phase relative permeability data representative of several core studies is presented in Figure 20–4. Based on the saturation values shown in the plot, describe the possible displacement efficiency and rock wettability.

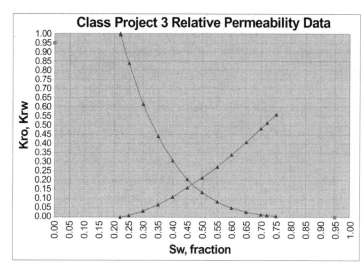

Fig. 20–4. Plot of oil-water relative permeability data

(f) Prepare a feasibility study for waterflooding the reservoir, including the drilling of infill wells, as follows:

    (i) Further tests and studies needed for reservoir characterization in order to attain maximum recovery.

    (ii) Probable range of ultimate recovery that may be achievable in the heterogeneous formation with two out four wells showing early breakthrough.

    (iii) Estimates of recovery factor at various water/oil ratios based on the core data provided in Table 20–2.

    (iv) Example calculations showing well injectivity, time to breakthrough, and water/oil ratio versus time following breakthrough. Assume a pressure difference of 1,200 psi between the injector and producer.

    (v) Waterflood surveillance.

    (vi) Any development strategy for the eastern flank of the reservoir.

State all the assumptions made in the study.

Table 20–2. Summary of core permeability data

Number of Samples	Range of Core Permeability, mD
3	1–5
9	6–15
15	16–25
20	26–35
14	36–45
12	46–55
4	56–60

4. Fractured reservoir

Prepare a detailed plan to characterize the fracture network, evaluating all the tools and techniques that can be employed, including static data and dynamic tests.

How would the reservoir performance be affected in case of the following scenarios?
(a) Single porosity system
(b) Dual porosity system
(c) Dual porosity/dual permeability system

Create a field development strategy.

Distinguish between the reservoir development and management strategies in the reservoir for the following cases:

Case I
Oil gravity: 39°API. Reservoir depth: 9,600 ft.

Case II
Oil gravity: 19°API. Reservoir depth: 1,600 ft.

5. Management of a stratified reservoir through its life cycle

An oil discovery is made in a basin where three hydrocarbon-bearing sands are known to exist. The initial 3-D seismic survey indicates an asymmetrical anticlinal trap. Sequence stratigraphy studies performed in nearby fields indicate that all three sands, separated from each other by a distance of 50 to 150 ft, are stratified with several subzones of contrasting permeability. However, well log information from other parts of the basin indicates that the reservoirs are not continuous throughout due to the existence of several faults and pinchouts.

A vertical appraisal well was drilled only to the top sand, where the initial reservoir pressure was found to be 2,550 psia. Oil gravity is 38.4°API. Bubblepoint pressure is 1,950 psia. Initial water saturation is 0.22. The initial gas/oil ratio was 900 scf/stb. Formulate a strategy to develop the field by evaluating various scenarios and options. Include the following:
- How could the other two sands in the field be explored?
- What type of wells would be best suited to explore the various sands and possible discontinuities of the reservoirs?

Devise a contingency plan, including a potential need for more surveys if the exploratory well turns out to be dry in the bottom sand.

Provide a list of data that should be gathered in drilling wells at the initial stage. Describe how the data could be integrated across many disciplines, with at least three examples. Describe in detail the logging studies (including logging-while-drilling tools) and other tests that should be used to characterize the middle and bottom sands for the exploratory well. Include the following:
- Hydrocarbon volume in barrels per acre-feet
- Formation transmissibility
- Dykstra-Parsons permeability variation factor
- Number of layers in each pay zone
- Vertical communication between the layers
- Possible existence of high permeability streaks
- Oil and gas reserves

What drilling strategy (step-out, peripheral, or both) should be adopted to delineate the field in light of the possible existence of faults and pinchouts? Provide a list of tests and surveys that could be used to identify and characterize any reservoir boundaries and their transmissibilities.

What tools and techniques should be employed in characterizing the influence of the aquifers? Describe any differences in reservoir development strategy in each of the following scenarios:
(a) Edge water drive, weak support from only one corner of the reservoir.
(b) Bottom water drive, strong support.
(c) The reservoir has multiple compartments; some do not exhibit any aquifer support.

In addition, estimate a range for the primary recovery factor.

Develop qualitatively various scenarios for secondary recovery in light of the reservoir characteristics. Include permeability, degree of heterogeneity between layers, effect of crossflow and gravity, depletion characteristics, and aquifer support. Include the following, with explanations:
- Time to start secondary recovery
- Optimum injection rate
- Maximum sweep in various layers

At the end of secondary recovery, evaluate the feasibility of various tertiary recovery processes, including the following:
- Water-alternating-gas (WAG) drive
- Surfactant flooding
- CO_2 injection
- Thermal processes

6. Integrated model development: Offshore field

Based on a Monte Carlo simulation, develop a spreadsheet to generate the probability distribution functions (PDFs) of an oil field at a depth of 11,000 ft in the Gulf of Mexico (GOM). Include the following:
- Original hydrocarbon in place
- Number of wells to be drilled
- Field production rate over years
- Ultimate recoverable reserves
- Discounted cash flow
- Net present value

The field is currently under appraisal. The following should be included in the examination of the reservoir:
- A literature survey of the rock and fluid properties and areal extents of reservoirs discovered in the area at similar depths. This could aid in the volumetric estimation of original oil in place and calculation of reserves by empirical correlation.
- High and low ranges in cumulative production from individual wells
- The decline trend of the wells
- Improved oil recovery methods practiced in the area
- Typical capital and operating costs for offshore field development in the Gulf of Mexico
- Escalation factors for oil price and expenses

Input variables for the spreadsheet program will include the following:
- Range of rock and fluid properties and their probability distributions
- Range of reservoir area typical of the Gulf of Mexico
- Decline curve parameters
- Oil and gas price escalation in the future

Most other parameters would be generated by the spreadsheet as intermediate calculations. All data, correlations, and assumptions introduced in the application should be representative of Gulf of Mexico reservoirs as much as possible. Make necessary assumptions where data is unavailable.

Based on the spreadsheet application, what would be the minimum reserves of a field in the Gulf of Mexico in order to make the project feasible? Describe the most critical components of field development from a reservoir engineering point of view. Perform sensitivity analyses on the following:
- Depth of the reservoir
- Number of wells drilled
- Horizontal versus vertical wells
- Improved oil recovery processes

List the challenges associated with deepwater exploration and development of oil and gas fields in the Gulf of Mexico. Example 18.6 provides a study of integrated model development for an offshore gas field, which can be used as a guide.

Glossary of Selected Key Terms

abandoned well. A well that would no longer produce oil or gas economically at the end of the reservoir life cycle. Certain wells, including dry holes, are plugged and abandoned early if production in commercial quantities is not possible. *See also dry hole, economic limit, and reservoir life cycle.*

abandonment pressure. *See pressure—abandonment pressure.*

abnormal pressure. *See pressure—abnormal pressure.*

absolute open flow potential (AOFP). The maximum flow rate that is theoretically achievable in a well. This condition occurs when the well is produced at zero bottomhole flowing pressure. Deliverability tests are routinely conducted in gas wells to determine the absolute open flow potential. It is also a requirement by regulatory agencies. *See also flow-after-flow test.*

absolute permeability. Pertains to a rock where only a fluid phase is present. *See also permeability.*

absolute porosity. Measure of porosity based on the total pore space in the rock. *See also porosity.*

acidization. Well productivity is enhanced by treating the well with certain acids (generally hydrochloric or hydrofluoric acid) that enlarge pores in the rock by reacting with the soluble substances in the rock matrix in a sandstone formation. In carbonates, the treatment acid dissolves the rock matrix. The net result is an improvement in the fluid conductivity of the rocks. The procedure is often referred to as acid stimulation or well stimulation. The other method to stimulate a well involves hydraulic fracturing. See also carbonate reservoir, hydraulic fracturing, and stimulation.

acoustic log. Recording of the travel time of sound waves in a borehole, which is influenced by the rock matrix and pore fluids. Used to measure formation porosity. Also referred to as a sonic log. *See also logging.*

afterflow. When a producing well is shut in, fluid flow from the reservoir continues into the wellbore due to the compressibility of the fluid for a certain period before equilibrium is reached. This phenomenon is a consequence of wellbore storage effects, as is evident during pressure transient tests. *See also wellbore storage.*

altered zone. The characteristics of the formation immediately surrounding the well are altered by a variety of factors. Such factors include the invasion of drilling fluids, changes in stress, migration of fines, and injection of an incompatible fluid. The net effect of an altered zone is usually a loss in well productivity. The altered zone, also referred to as the damaged zone, could be several inches thick. Well stimulation can lead to the improvement of rock permeability in the vicinity of the wellbore, resulting in enhanced well productivity. *See also fines migration, permeability, skin factor, and stimulation.*

anisotropic formation. Geologic formations are seldom uniform in physical characteristics, and many formations exhibit a directional trend or preference in rock permeability and fracture orientation. An example would be a reservoir where injected fluid does not flow uniformly in all directions but tends to propagate in a NW-SE direction. Formation anisotropy may have a profound impact on enhanced oil recovery (EOR). *See also enhanced oil recovery and directional permeability.*

anticline. A geologic structure formed by folding, which is convex upward and appears in the shape of an arch. It can be either symmetric or asymmetric. Many petroleum reservoirs are structurally based on anticlinal traps. Geosciences exploration studies routinely attempt to identify anticlinal structures in the quest for oil and gas. *See also fold and groundwater.*

API gravity. Crude oil is commonly classified by its API gravity. The measure was developed by the American Petroleum Institute and is defined in chapter 3. A specially graduated hydrometer can be employed to measure the API gravity of crude. Heavy oils have relatively low API gravity ($< 22.3°$), while light crude may have an API gravity of $33°$ or greater. An intermediate crude is classified as having an API gravity between light and heavy crude. Crude oil gravity depends on the relative abundance of the light or heavy hydrocarbon constituents. *See also crude oil and heavy oil.*

aquifer. Subsurface geologic formation that stores interstitial water in the rock pores, the water being usually movable and producible under a pressure differential. Aquifers may be encountered at all depths below the water table. Reservoir engineers are interested in aquifers that are part of oil and gas reservoirs. These are located at the edge or bottom of an oil or gas accumulation. Water drive reservoirs produce petroleum as water influx provides natural energy. *See also bottom water drive, constant pressure boundary, edge water drive, and water influx.*

areal sweep efficiency. Expressed as a fraction or percent, the area of the reservoir swept by injected water over total reservoir area during waterflooding. In a heterogeneous formation, the areal sweep efficiency is anticipated to be less.

artificial lift. In many instances, natural reservoir energy is not adequate to efficiently lift crude oil through the wellbore to the surface. Hence, energy is added to artificially lift reservoir fluids. Common artificial systems include rod pump, continuous or intermittent gas lift, and electrical submersible pump (ESP). *See also gas lift.*

asphaltenes. Insoluble organic compounds in crude oil that may precipitate as deposits due to a drop in pressure, such as during matrix acidization. The net result of asphaltene precipitation is a loss in productivity. *See also crude oil and paraffin.*

asset team. A multidisciplinary team comprised of engineers and geoscientists, among others, assigned to a petroleum reservoir. Based on an integrated workflow, the team works together to maximize oil and gas reserves and optimize production. The team strives to implement the best reservoir management practices based on cutting-edge technology. *See also reserves.*

associated gas. Light hydrocarbons present in the gas phase in a gas cap overlying the oil zone in a reservoir, and the volatiles that evolve out of the liquid phase in the reservoir as the reservoir pressure declines below the bubblepoint.

average reservoir pressure. *See pressure—average reservoir pressure.*

backflow. In a layered or stratified reservoir, the flow of oil or gas that may occur through the borehole from a high-pressured zone to a relatively low-pressured zone.

back pressure. Pressure created in a flowing fluid system by friction. Back pressure can be artificially induced by deploying a back pressure valve that constricts the flow of fluids.

backpressure test. *See flow-after-flow test.*

barrels of oil equivalent (BOE). A unit of energy. It is the amount of energy released by burning 1 bbl of crude oil. One barrel of oil equivalent is equal to 5,487 ft^3 of gas. This ratio is based on the energy content of natural gas. Furthermore, it can be noted that 1 ton of oil equivalent (TOE) contains approximately 7.3 BOE for an oil having API gravity of 33°.

basin. A geologic feature indicating the massive accumulation of sediments resulting in a synclinal structure (a depression in the earth's crust) due to various geologic activities. When the conditions are right, such as the existence of source rock and structural or stratigraphic traps, and given adequate pressure and temperature through geologic times, an accumulation of petroleum fluids may result. The study of geologic basins is extremely important in the exploration of oil and gas. *See also offset well, sequence stratigraphy, source rock, stratigraphic trap, and structural trap.*

bed. The smallest stratigraphic unit (layer of rock), distinguished from upper and lower beds by recognizable physical characteristics and boundaries. Bedding planes mark the separation between adjacent beds or strata.

bilinear flow. Develops through hydraulically created fractures of finite conductivity that augment fluid flow to the well. Bilinear flow can be identified by a quarter-slope line appearing in a log-log plot of pressure response versus time in well test interpretation, as illustrated in chapter 5. *See also linear flow and pseudoradial flow.*

bitumen. Highly viscous and heavy, tarlike hydrocarbon compounds that can be extracted from oil sands. Bitumen has a molasseslike consistency at room temperature. Huge deposits of oil sands are found in Canada and Venezuela. It is estimated that the reserves based on oil sands in Alberta alone are 174 billion bbl, second only to conventional reserves in Saudi Arabia. Emerging technologies to extract oil sands are described briefly in chapter 17. *See also hydrocarbons, oil sands, synthetic crude oil, and upgrading.*

blowdown. Venting of free gas through the well, which may accumulate in the subsurface due to reduced reservoir pressure. In certain situations, efficient recovery of oil requires blowdown of free gas through wells. In certain other situations, however, gas is injected into the reservoir as part of the recovery scheme.

blowout. Unanticipated eruption of subsurface fluids (oil, gas, and drilling fluid), accompanied by fire, due to the penetration of a high-pressured geologic formation during drilling. Blowouts generally occur when the downhole pressure at certain depths is much higher than that in the borehole, and this is unknown previously.

bottomhole pressure (BHP). Pressure measured downhole by placing a temporary or permanent gauge at a known depth. Usually the point of measurement is located at or close to the top of the perforation of the well string. Bottomhole pressure can be either static or flowing, depending on whether the well is active or inactive. Static bottomhole pressure may approach the average reservoir pressure in the drainage area once the well is shut in for a relatively long period of time to reach equilibrium. Certain cases require that the flow through the well be stabilized or the well be shut in for a considerable period of time before pressure measurements are made. Permanent downhole gauges record bottomhole pressure on a continuous basis during production or injection. *See also average reservoir pressure, drainage area, permanent downhole gauge, stabilization, and static bottomhole pressure.*

bottomhole shut-in pressure. *See pressure—bottomhole shut-in pressure.*

bottom water drive. Natural energy that a petroleum reservoir may experience due to water influx from an underlying aquifer. The water/oil contact is found to rise with production. *See also aquifer, edge water drive, and oil/water contact.*

bounded reservoir. A petroleum reservoir in which all the boundaries are impermeable. A bounded reservoir is not in pressure communication with an aquifer and does not receive external energy for production due to water influx. Pressure transient tests are capable of identifying the specific characteristics of the reservoir boundaries, such as no flow or constant pressure. *See also constant pressure boundary, water influx, and well test.*

breakthrough. Usually denotes the first appearance of injected water or gas in a production well. The net result of water or gas breakthrough in wells is reduced oil production and eventual abandonment of the well, unless remedial measures are taken. Reservoir engineers design petroleum recovery systems in a manner whereby water or gas breakthrough is delayed, or at least the volume of produced water is controlled. *See also parts per million and water cut.*

British thermal unit (Btu). The quantity of heat required to raise the temperature of 1 lb of water by 1°F.

Btu. *See British thermal unit.*

bubblepoint. As an undersaturated oil reservoir is produced, it is accompanied by a decline in reservoir pressure. As a result, volatile hydrocarbons dissolved in crude oil begin to come out of solution as bubbles of free gas at some point within the porous medium. The pressure at which the first bubble appears is known as the bubblepoint of the oil. However, a saturated oil reservoir is produced at or below the bubblepoint pressure. Moreover, a free gas cap may overlie the oil zone in a saturated oil reservoir. The bubblepoint of a crude oil influences major decisions that reservoir teams need to make in effectively managing the reservoir. Bubblepoint data is generally obtained by laboratory studies of crude samples or is based on available correlations when samples are not available. Fluid properties, including the bubblepoint, are described in chapter 3. *See also critical point, critical saturation, crude oil, flash vaporization, gas cap, hydrocarbons, pressure-volume-temperature properties, and saturated oil reservoir.*

buildup test. A widely practiced type of well test in which the well is usually produced at a constant rate to achieve stabilization, followed by a buildup period. Well pressure response is collected throughout the test and is analyzed to obtain important information related to well productivity and reservoir characteristics. Well tests are described in chapter 5. *See also drillstem test, fracture half-length, pressure buildup analysis, reservoir pressure, well productivity index, and well test.*

bulk density. The density of rock, which is composed of both grain and pores in petroleum reservoirs. The grain density in a porous rock is greater than its bulk density.

capillary number. The ratio of viscous forces to forces arising out of interfacial tension. For enhanced oil recovery processes where oil recovery is augmented by reducing interfacial tension of fluids in rock pores, as in a surfactant flood, a relatively high capillary number indicates that the viscous forces dominate over interfacial tension. This leads to better recovery. *See also enhanced oil recovery, interfacial tension, recovery, and viscous forces.*

capillary pressure. The difference in pressure between the nonwetting and wetting fluid phases is known as capillary pressure. The wettability of a rock is discussed in chapter 2. Pressure exerted by the nonwetting phase is higher than that exerted by the wetting phase. This condition is necessary as the nonwetting phase with higher pressure, such as migratory oil, enters the pores of rock initially filled with the wetting phase at a lower pressure, such as formation water. Capillary forces, among other factors, dictate the distribution of oil, water, and gas saturations in a petroleum reservoir, including transition zone characteristics. *See also petrophysical model, pressure, and transition zone.*

cap rock. Provides a seal for hydrocarbon accumulations. Cap rock is virtually impermeable and overlies the oil- and gas-bearing formation, preventing any upward migration and loss of hydrocarbons. Cap rocks are made of shale, evaporites, and salt, which are virtually impermeable. *See also hydrocarbons.*

carbonate reservoir. Carbonate reservoirs are composed of rocks with predominant constituents of calcite (calcium carbonate) and dolomite. The chemical formulas of calcite and dolomite are $CaCO_3$ and $CaMg(CO_3)^2$, respectively. Limestone, dolomite, and chalk formations containing hydrocarbons are generally referred to as carbonate reservoirs. In the period following deposition, various geochemical processes may occur in carbonate rocks, including dolomitization, leading to the alteration of rock characteristics that influence the reservoir performance. Many giant petroleum reservoirs in the world are of carbonate origin, and these are generally more difficult to characterize due to the abundant presence of fractures and vugs, among other heterogeneities. *See also acidization and formation.*

casing pressure. *See pressure—casing pressure.*

cementation. Process of binding sedimentary grains into a rock by chemical precipitation of cementing materials. The degree of cementation influences the porosity and permeability of a porous rock in a significant manner. *See also permeability and porosity.*

chalk. Some reservoirs are composed of chalk, which is a variety of limestone originating from the shells of single-cell life forms. *See also limestone.*

chemical flooding. In order to augment the recovery of oil following primary production and waterflooding, certain chemicals may be injected when all of the reservoir screening criteria are met. Injected chemicals chiefly consists of alkaline or polymer compounds. Alkaline floods reduce surface tension between the oil and water in the rock pores, while polymer floods increase the viscosity of the water and improve sweep efficiency.

choke. A device containing an orifice that controls the flow of fluid in wells. A choke can be either fixed or adjustable. The latter is employed in certain situations to control pressure, as the fluid enters the wellbore at high pressure and is transported to *the surface.*

clay swelling. When external fluids, as opposed to in-situ fluids, come into contact with the formation during drilling, completion, or stimulation, swelling of the clay present in the reservoir rock can occur. The net effect is a reduction in pore size and permeability in the vicinity of the wellbore, which adversely affects well productivity. *See also in-situ fluid and well productivity index.*

CO_2 injection. Carbon dioxide (CO_2) is injected into certain reservoirs to augment oil recovery by creating complete or partial miscibility between the injected gas and the in-situ oil. The viscosity of the oil is also reduced in the process due to swelling. It is one of the widely known enhanced oil recovery methods in the industry.

CO_2 sequestration. An emerging technology for storage of carbon dioxide into subsurface formations in order to abate global warming trends. Sources of carbon dioxide and other greenhouse gas emissions into the atmosphere include fossil fuel–burning power plants and autos. Subsequent applications of stored carbon dioxide (CO_2) include enhanced oil recovery and enhanced coalbed methane recovery based on CO_2 injection. *See also enhanced oil recovery.*

coalbed methane (CBM). Natural gases, chiefly methane, trapped in coal deposits underground, as found in most instances. Unregulated venting of methane from coal mines results in greenhouse effects. Coal beds are found to have relatively low permeability. Coalbed methane can be recovered by drilling wells and by employing hydraulic fracturing. *See also natural gas and methane.*

combination of forward combustion and waterflooding (COFCAW). A thermal recovery process whereby water is injected simultaneously or alternately with air into a formation. This aids in the recovery of thermal energy that would be lost to the formation if the air was injected alone. The combination of forward combustion and waterflooding is also known as wet combustion. *See also thermal recovery.*

commingled production. Production of oil and gas from multiple zones located at various depths in a reservoir. The producing zones are not in communication within the reservoir due to the existence of impermeable barriers between them. Fluid communication between the zones occurs only through the wellbore. *See also crossflow.*

compaction drive. Relates to the primary production of oil and gas based on compaction or reduction of the pore spaces as fluids are driven from the pores to the wells. The compaction process is due to the decline in reservoir pressure.

compartmentalization. Some reservoirs are not laterally continuous throughout, as certain geologic barriers exist, leading to isolated compartments. These reservoirs require special approaches for efficient recovery, including the proper location of producers and injectors, and monitoring of the fluid fronts following water or gas injection. Reservoir characterization is the key to successful management of compartmental reservoirs. An example of compartmental reservoir management is presented in chapter 19. *See also infill drilling, reservoir, and reservoir management.*

completion. *See well completion.*

compressibility. Change in volume per unit change in pressure. The unit of compressibility is given per pounds per square inch (psi^{-1}). In most reservoir engineering studies, knowledge of the fluid and rock compressibility is required. In certain cases, the reservoir rock is found to be highly compressible, leading to pore collapse and a significant reduction in productivity once the reservoir pressure declines. Total compressibility of a rock fluid system is obtained by summing the individual fluid compressibilities of the oil, gas, and water present in the pores multiplied by the respective fluid saturations plus the formation compressibility. Effective compressibility of a particular fluid phase is obtained by dividing the total compressibility by the saturation of that phase in the rock pores. *See also core, flash vaporization, pressure-volume-temperature properties, and pseudopressure.*

condensate. *See gas condensate.*

conformance. *See profile modification.*

coning. Water or gas coning may occur in an oil well when the well produces at a higher-than-optimum rate in the presence of a bottom water zone or an overlying gas cap, respectively. As a consequence, the water beneath the oil zone cones up, or a gas cap cones down, near the perforations due to high viscous forces that develop in the vicinity of the perforations. The net effect of water and gas coning is to hinder oil production. Remedial measures include the recompletion of the well further from the watered-out zone or horizontal sidetracking, among others techniques. *See also gas cap and horizontal sidetracking.*

connate water. Water that fills the pores of the rock during the formation of the rock in geologic times. Connate water is found to have a higher mineral content than sea water. During migration of the oil, much of the water is expelled from the pores. The small percentage of connate water remaining in the pores in a petroleum reservoir is usually immobile and is referred to as the irreducible water saturation. In contrast, interstitial water is simply the water present in the rock pores, including the water that may have entered the pores at a later time. *See also interstitial water.*

constant composition expansion. *See flash vaporization.*

constant pressure boundary. Boundary of a reservoir in which the pressure does not decline with production. The situation typically arises when a reservoir experiences a strong water influx from an adjacent aquifer. *See also aquifer, bounded reservoir, and water influx.*

contingent resources. *See resources—contingent resources.*

contour map. Contours are lines of equal value of a reservoir characteristic (porosity, permeability, reservoir thickness, depth, etc.) on a map of the reservoir. Contour maps are generated from both static (geosciences) and dynamic (production, well testing, etc.) data. The reservoir team relies on the dynamic contour maps of fluid saturation and reservoir pressure, among other factors, in order to effectively manage a reservoir. Contour maps are usually generated by a computer and are typically 2-D or 3-D. Well locations are typically included in the maps. A contour map depicting the barrels of oil in place per acre-feet is shown in Figure 9–1 in chapter 9.

core. Samples of a reservoir rock obtained during drilling that are subjected to detailed laboratory investigation in order to determine various rock characteristics. These characteristics include the lithology, porosity, permeability, wettability, compressibility, presence of fractures, and formation damage, among others. Certain cores can be full diameter, i.e., the diameter of the drill bit, while other samples have a much smaller diameter. The latter are obtained by sidewall coring following the drilling of a well. *See also compressibility, lithology, permeability, petrophysical model, porosity, and wettability.*

cricondenbar. Upper limit of pressure where liquid and vapor can coexist in equilibrium. At pressures above the cricondenbar, the fluid is essentially in liquid phase.

cricondentherm. The highest temperature above which hydrocarbon fluid can exist solely as gas.

critical flow rate. Rate of flow above which sand production increases significantly in a well. This relates to the phenomenon of sand production along with oil and gas. *See also sand control.*

critical point. Pressure and temperature at which the liquid and vapor phases are indistinguishable and are in equilibrium. In a phase diagram, as illustrated in chapter 3, the bubblepoint and dewpoint lines converge at the critical point. *See also bubblepoint and dew point.*

critical saturation. Threshold saturation of a fluid in porous media to be mobile. For example, as reservoir pressure declines below the bubblepoint, volatile hydrocarbons begin to evolve out of solution. However, the liberated gas does not begin to flow instantaneously until a critical value is reached. The effect of critical gas saturation on oil recovery is shown in chapter 8. *See also bubblepoint.*

crossflow. Flow between two adjacent layers that may occur due to the pressure differential existing between the layers. The intervening shale barrier between the layers is either discontinuous or not entirely impervious. Many reservoirs perform in an unexpected manner, especially under fluid injection, when crossflow between layers is either not identified or not fully understood. *See also commingled production, shale, and stratified reservoir.*

crude oil. Oil as produced from the reservoir in liquid phase prior to necessary treatments in refineries. A major classification of crude oils is based on API gravity, namely light, intermediate, and heavy. API gravity of crude oil mostly varies between 9° and 55°. "Sweet" crude has relatively low sulphur content (<0.5% by weight). Crude oil is refined to various grades before end use, including gasoline for autos and aviation fuel for airplanes. *See also API gravity, asphaltenes, bubblepoint, dry oil, and upgrading.*

cumulative gas/oil ratio (GOR). Cumulative gas production over cumulative oil production from a well or the reservoir up to a certain period in time during the life of a reservoir. *See also gas/oil ratio.*

cutoff porosity. Limiting value of porosity used in assessing the reserves of a field. Portions of the rock having porosities below the cutoff value do not contribute to production in any significant manner. *See also porosity and net to gross thickness.*

cyclic steam stimulation (CSS). A method utilized extensively to recover heavy oil and bitumen from oil sands. The thermal recovery process is carried out in three phases. Steam is first injected into the formation through horizontal or vertical wells, followed by in-situ viscosity reduction of the hydrocarbons for several weeks, and finally production through the wellbore. Due to injection at high pressure, certain channels may be created in the porous media to augment productivity. *See also well productivity index.*

damaged zone. *See altered zone.*

darcy. Unit of rock permeability, named after the famous engineer Henry Darcy. When fluid having a viscosity of 1 cp flows at a rate of 1 cm³/sec under a pressure differential of 1atm through a cross-sectional flow area of 1 cm² and a length of 1 cm, the porous medium is said to have an absolute permeability of 1 darcy. It is assumed that the fluid fills 100% of the pore space. A large number of reservoirs have permeability much lower than 1 darcy; hence rock permeability is conveniently expressed in millidarcies (mD). *See also permeability.*

Darcy's law. Darcy's law predicts the fluid flow rate that can be achieved in a porous medium of known permeability when the pressure differential between inlet and outlet points is also known. Alternately, the permeability of a porous rock can be determined (both in the laboratory and in the field) when the flow rate and the associated pressure drop are known. Darcy's law constitutes the basic building block in understanding and analyzing the fluid flow behavior in a reservoir. The law is based on several important assumptions, as discussed in chapter 2. *See also laminar flow and non-Darcy flow.*

datum. A reference in depth on which various measurements are based, such as the depth of a petroleum reservoir. A typical datum would be the oil/water contact in an oil reservoir. Regardless of various well elevations, all recorded pressure values are corrected to the same datum level based on the known fluid pressure gradient in the reservoir. The datum depth is usually reported in TVD-SS.

dead oil. Oil devoid of any volatiles. Reservoir oil becomes "dead oil" due to the dissolution of volatile components under reduced pressure as it is transported to the surface.

depletion drive. *See solution gas drive.*

depositional environment. Environment under which sediments are transported and deposited, including deltaic, alluvial, and marine, etc. The basic controlling factors are wind and water currents. Rock characteristics are usually influenced by the depositional and postdepositional environments, including permeability anisotropy and vugular porosity. *See also permeability anisotropy and vugular porosity.*

derivative plot. In well test interpretation, the derivative of the transient pressure response against time is plotted on a log-log scale to diagnose various flow regimes and reservoir characteristics. A derivative plot is also referred to as a diagnostic plot. The technique, introduced in recent decades following the advent of digital computation, has enhanced well test interpretation capabilities significantly. *See also linear flow, pinchout, and transient pressure response.*

deviated well. The drilled hole is not vertical but is deviated to some degree. The angle of deviation or inclination is measured between the wellbore axis and the vertical axis pointing downward. In deviated holes, the measured depth (MD) is greater than the true vertical depth (TVD).

dew point. As a gas condensate reservoir is produced, accompanied by a decline in reservoir pressure, liquid droplets begin to condense out of the gaseous phase and form a thin film of liquid in the pores of the rock at a certain pressure. This pressure is referred to as the dewpoint pressure of the gas condensate system. The process whereby heavier hydrocarbons condense due to a decline in the reservoir pressure is referred to as retrograde condensation. It is best understood with the aid of a phase diagram, as illustrated in chapter 3. *See also critical point, hydrocarbons, and phase diagram.*

diagenesis. The process that results from sedimentary rocks being subjected to physical, chemical, or biological changes at relatively low pressure and temperature at subsequent periods following original deposition. An example would be the dissolution of certain minerals in carbonate rocks and the development of secondary porosity. In sandstones, diagenesis usually results in the reduction of porosity. *See also secondary porosity.*

diagnostic plot. *See derivative plot.*

differential vaporization. Involves immediate removal of the vapor phase following evolution from crude oil under declining pressure. This is conducted to emulate the processes occurring in porous media as evolved gas is driven rapidly towards the wellbore. The solution gas/oil ratio, relative oil volume, density of the oil, gas deviation factor, and gas formation volume factor are some of the fluid properties determined by this method. *See also gas deviation factor and gas formation volume factor.*

diffusivity coefficient. Also referred to as hydraulic diffusivity, the diffusivity coefficient is a function of fluid viscosity, rock porosity and permeability, and total compressibility of the system. The diffusivity coefficient is expressed as follows:

$$\eta = \frac{2.637 \times 10^{-4}\, k}{\phi\, \mu\, c_t}$$

where η is in ft^2/hr, k is in mD, μ is in c_p, c_t is in psi^{-1}, and ϕ is a fraction. The coefficient is a part of the hydraulic diffusivity equation used to describe the transient fluid flow in porous media in oilfield units. The rate of propagation of fluid pressure in a porous medium is proportional to the diffusivity coefficient. *See also diffusivity equation.*

diffusivity equation. Partial differential equation for fluid flow in porous media, which is based on the equation of continuity, Darcy's law, and an equation of state. The diffusivity equation is the founding block of pressure transient test analysis and reservoir simulation. *See also diffusivity coefficient.*

directional permeability. A characteristic of an anisotropic porous medium, which tends to transmit fluid in a preferred direction. The heterogeneous property of the rock is responsible for premature breakthrough of the injected water in certain wells during waterflooding, while certain areas in the reservoir remain largely unswept. *See also anisotropic formation, permeability, and permeability anisotropy.*

displacement efficiency. Ratio of the volume of oil in the rock pores displaced by an injected fluid, such as water, to the original volume of oil found at the start of the improved recovery process, including waterflooding. Displacement efficiency and sweep efficiency determine the success of waterflooding and other improved oil recovery processes in a reservoir. Displacement efficiency is a function of several rock and fluid properties, including fluid viscosity, wettability, interfacial tension, and formation dip. *See also interfacial tension, improved oil recovery, sweep efficiency, viscosity, and wettability.*

distributed temperature sensing (DTS) tool. Identifies zones of steam breakthrough during thermal recovery based on fiber optics technology. *See also thermal recovery.*

dolomitization. A geochemical process in which the calcium ions in calcite are replaced by magnesium ions that originate from the evaporation of seawater. Carbonate rocks may be subjected to this process during deep burial diagenesis, which leads to alteration of the rock characteristics affecting accumulation and production of petroleum. *See also carbonate reservoir.*

down dip. Indicates the location of a well in an inclined reservoir at a relatively lower elevation. Downdip wells are closer to the oil/water contact. In the case of bottom water drive, downdip wells are watered out first. The antonym of down dip is up dip. In a reservoir undergoing gas injection, updip wells can be utilized to maintain reservoir pressure and augment oil recovery.

downhole sensor. Monitoring device placed downhole to record pressure, temperature, flow rate, fluid composition, and liquid holdup, among others. Certain devices can be deployed permanently and are usually referred to as permanent downhole gauges (PDGs). Based on the continuous feed of downhole data, efforts are made to manage the wells in real time or near real time.

drainage. The phenomenon whereby the saturation of a wetting phase fluid is reduced. An example would be the expulsion of water from a water-wet rock by a nonwetting fluid, such as oil. *See also imbibition, water-wet reservoir, and wettability.*

drainage area. In a developed reservoir with multiple active producers, each well drains or produces from its own area. Well tests conducted in a well lead to the determination of average permeability and average reservoir pressure in the drainage area, among other information. *See also bottomhole pressure.*

drawdown. In a producing well, the flowing bottomhole pressure is less than the average reservoir pressure, leading to oil and gas production. The difference in pressure is referred to as drawdown. Generally speaking, greater drawdown results in a higher flow rate until a limiting value is reached.

drawdown test. A common well test conducted by first shutting in a well to reach stabilization, followed by flowing the well at a constant rate. The objective is to obtain important reservoir properties and assess skin damage. The test is typically conducted in a newly discovered reservoir to ascertain initial reservoir pressure and formation transmissibility, among other factors. *See also drillstem test, fracture half-length, skin factor, transmissibility, and well test.*

drilling program. A drilling program includes well trajectories, detailed geology of the formations to be drilled, casing design, drilling fluids, pore pressure estimations, and blowout control plans, among other topics. Reservoir engineers provide target locations where the vertical or horizontal well would contact the pay zone(s).

drillstem test (DST). Openhole test usually conducted to determine the feasibility of a newly discovered petroleum reservoir. The test is performed with the drill string still in the hole. The test sequence includes relatively short periods of drawdown and buildup to determine initial reservoir pressure and formation transmissibility, among others. Fluid samples are also collected during the tests. *See also buildup test, drawdown test, flow test, initial reservoir pressure, transmissibility, and well test.*

dry gas. Natural gas that has virtually no liquid condensate. Dry gas is primarily composed of light hydrocarbons, and the gas/oil ratio (GOR) typically exceeds 100,000 scf/stb. *See also natural gas.*

dry hole. A wellbore that is unable to produce oil or gas in commercial quantities.

dry oil. Crude oil treated to minimize the sediment and water content. *See also crude oil.*

dual completion. Well completed in two zones or reservoirs that are vertically separated. The reservoirs may have distinctly different properties, including pressure. Production takes place through long and short strings.

dual permeability. *See dual porosity.*

dual porosity. Certain naturally fractured reservoirs are classified as dual porosity, where the production of reservoir fluids takes place through the fracture system alone. Fluids present in the rock matrix flow to the fracture network prior to production. In dual porosity/dual permeability reservoirs, however, the fluid can be produced to the wellbore from both the rock matrix and the fracture network. Evidently, the matrix has sufficient permeability to produce directly. In a single porosity system, however, hydrocarbons in the fracture network are produced alone without any contribution from the rock matrix, which has negligible porosity.

economic limit. Limiting rate below which a well cannot be operated commercially. The term also applies to oil or gas fields. *See also abandoned well and field.*

edge water drive. Reservoir production is driven by water influx from the aquifer located at the edge of the reservoir, providing natural energy in the form of pressure support. *See also aquifer and bottom water drive.*

effective permeability. Effective permeability to a specific fluid depends on the individual fluid saturations in a multiphase fluid system, such as the presence of oil and water in the rock pores. *See also permeability.*

effective porosity. Indication of the pore spaces that are interconnected to form continuous flow channels. *See also porosity.*

empirical. Certain useful equations are developed on the basis of observation and available data, such as prediction of the recovery factor in an oil reservoir under waterflooding as described in chapter 10. These equations, referred to as empirical, are not derived from theory, largely due to the complexity of the physical processes and related uncertainties.

enhanced oil recovery (EOR). As the name implies, enhanced oil recovery processes attempt to augment the recovery of petroleum from a reservoir above that which can be accomplished by natural reservoir energy alone. Enhanced oil recovery processes can be broadly classified into the following categories: thermal recovery, chemical flooding, and miscible displacement. These are discussed in chapter 17, with field examples. Following primary or secondary recovery, oil reservoirs are routinely screened to evaluate the suitability of a specific enhanced oil recovery process. Additional recovery of oil is accomplished by various mechanisms, including, but not limited to, the reduction of oil viscosity, improvement of mobility, attainment of miscibility between fluids, reduction in interfacial tension, and reservoir pressure enhancement. An enhanced oil recovery process is typically initiated at the end of waterflooding and hence is frequently referred to as tertiary recovery. With increasing demand for oil, enhanced oil recovery processes are of great significance, as primary recovery from most oil reservoirs is far less satisfactory. *See also anisotropic formation, capillary number, CO_2 sequestration, heterogeneous formation, interfacial tension, line drive, miscible displacement, recovery, secondary recovery, stratified reservoir, tertiary recovery, and ultimate recovery.*

equation of continuity. A material balance equation upon which the theory of fluid flow in porous media is based. It considers an elemental volume in the reservoir through which the mass of fluid entering per unit time must be equal to the mass leaving the element and the rate of accumulation within the element.

equation of state (EOS). An equation that correlates the density of the fluids, including multicomponent hydrocarbon mixtures, as a function of pressure and temperature. Equations of state are extensively used in reservoir simulation, where certain hydrocarbon components undergo phase changes (liquid to vapor and vice versa) in porous media during production. *See also hydrocarbons and reservoir simulation.*

estimated ultimate recovery (EUR). According to the Society of Petroleum Engineers' definition, estimated ultimate recovery is based on the quantities of petroleum that are estimated to be potentially recoverable from an accumulation on a given date. It also includes the quantities that are already produced. *See also recovery.*

evaluation. Estimates of quantities of hydrocarbons that can be recovered and sold based on multidisciplinary reservoir studies. A cash flow analysis is also performed as part of the evaluation.

exploration. Quest for oil and gas reserves based on geosciences study, basin analysis, and exploratory drilling, among other inputs. Petroleum exploration usually requires huge resources, including high-risk investment and technical expertise. Successful ventures are known to pay large dividends. *See also offset well, sequence stratigraphy, and wildcat.*

facies. Rock characteristics in a bed that reflect the conditions during the depositional environment. A change in the depositional environment leads to certain changes in the rock properties, such as a degradation in porosity and permeability. *See also heterogeneous formation.*

falloff test. A common well testing method for injection wells whereby fluid is injected at a constant rate for a period of time in order to attain stabilization, following which the well is shut in. Consequently, the elevated reservoir pressure around the injector "falls off" or declines with time. The trends identified in the transient pressure response are interpreted to determine the characteristics of the well and reservoir.

farmout. A contract with an oilfield operator to drill a well in exchange for working interest from the oil or gas property.

fault. A geologic feature in which a section of rock is displaced against another in an upward, downward, or lateral direction due to intense geologic activities in ancient times. A fault can be normal, reverse, thrust, or strike-slip, among others. Faults are detected by abrupt changes in lithology, among other indicators, and can be sealing or nonsealing. Fault characteristics are of significance to reservoir engineers in relation to the potential movement of in-situ and injected fluids across the fault boundary. *See also in-situ fluid and lithology.*

field. Petroleum reservoir with an accumulation of hydrocarbons in subsurface geologic formations. A field may consist of multiple zones or reservoirs of hydrocarbon accumulation in vertical sequence. Reservoirs in a field share common structural and stratigraphic features. At the surface, an oil or gas field usually points to the wellhead locations and related installations. An oil or gas field usually produces on a commercial basis. *See also formation, hydrocarbons, and zones.*

fines migration. During production of oil and gas, fine particles of clay and quartz are transported towards the well due to drag forces, resulting in clogging of pore throats and a reduction in permeability in a zone surrounding the wellbore. The phenomenon is usually more evident in an unconsolidated formation producing at a high rate. Damaged zones are treated with acid to either dissolve or disperse the fine particles. *See also altered zone, permeability, sand control, and stimulation.*

fingering. A phenomenon usually observed during waterflooding and other injection operations, where the injected fluid moves in narrow streams, or fingers, largely bypassing the viscous oil. The net result of viscous fingering is poor sweep efficiency, premature breakthrough in the producing wells, and less-than-satisfactory oil recovery. This phenomenon is also referred to as viscous fingering.

fire flooding. A thermal recovery (enhanced oil recovery) process to recover heavy oil in which a fire is initiated at the wellbore and is subsequently maintained by injecting air. The fire front propagates towards the producer wells and reduces the viscosity of the oil, which becomes more mobile. Steam is also produced as the formation water is heated. Steam, hot water, combustion gas, and distilled solvent produced in the process further aid in driving the oil towards the wellbore. This process is also referred to as in-situ combustion. *See also combination of forward combustion and waterflooding.*

5-spot pattern. Configuration of four injectors and one producer used in waterflooding or enhanced oil recovery operations. Various well patterns, including 5-spot, are illustrated in chapter 16. *See also enhanced oil recovery and waterflooding.*

flash vaporization. Also referred to as constant composition expansion (CCE). Pressure of a sample fluid is gradually lowered in a closed chamber, causing expansion in fluid volume but not permitting change in overall fluid composition. Any gas evolved during the process remains in contact with the liquid phase until equilibrium between the two phases is reached. The bubblepoint pressure, specific volume at saturation pressure, coefficient of thermal expansion, and isothermal compressibility of the liquid above the bubblepoint are measured. *See also bubblepoint and compressibility.*

flow-after-flow test. A pressure transient test usually conducted to ascertain deliverability of a gas well. The well is flowed at a predetermined rate until stabilized conditions are attained. In the next sequence, the flow rate is changed, and the well is flowed until stabilized conditions are attained again. The sequence of flow and stabilization is successively conducted a few times (usually four) at different well flow rates. The test is also referred to as a backpressure or four-point test. *See also absolute open flow potential, multiple rate test, and stabilization.*

flow efficiency. A measure of well performance, obtained by dividing the actual productivity index of the well by the productivity index that would be obtained when the well is neither damaged nor stimulated. In damaged wells where well productivity is affected adversely, the flow efficiency is less than 1. Following successful stimulation of a well, the flow efficiency is expected to be greater than 1. *See also skin factor, stimulation, and well productivity index.*

flowing bottomhole pressure. *See pressure—flowing bottomhole pressure.*

flow regime. During a pressure transient test, multiple flow regimes may develop that are identifiable by various plots of pressure versus time (e.g., log-log plot, semilog plot, or derivative plot). For example, a radial flow regime is identified by a horizontal line on a derivative plot, while a half-slope line is observed on both log-log and derivative plots in the case of flow through an infinitely conductive fracture. Correct identification of flow regimes is vital in interpreting a well test, leading to reservoir characterization and assessment of well performance. *See also well test and reservoir characterization.*

flow test. Usually refers to well tests conducted to determine the viability and productivity of the oil- and gas-bearing formation. Flow tests include drillstem tests and flow-after-flow tests, among others. *See also drillstem test and well productivity index.*

flow unit. *See reservoir characterization.*

fluid potential. The fluid potential, also referred to as the potential function, combines the gravity term with fluid pressure and is defined as follows:

$$\Phi = p - 0.433 \, \gamma \, \Delta z$$

where

Φ = fluid potential, psia,

γ = specific gravity of the flowing fluid, and

Δz = difference in elevation between an arbitrary datum level and the point in the reservoir where fluid potential is determined.

By convention, z is taken to be positive downward. In generalized form, the introduction of potential function in Darcy's law accounts for the flow of fluid between two points located at different elevations, leading to inclined flow or vertical flow. The effects of gravity on fluid flow are evident in inclined reservoirs. *See also Darcy's law.*

fluid saturation. Ratio of the volume of fluid present in the rock pores to the volume of the pores. In undersaturated oil reservoirs, a typical range of oil saturation could be 0.7–0.8, while the rest of the pore space is filled by formation or interstitial water. If the water is not movable, the reservoir is said to have an irreducible water saturation between 0.2–0.3. The fluid saturation changes dynamically as a reservoir is produced. For example, when the reservoir produces below the bubblepoint, free gas evolves, and a finite gas saturation in the pores is encountered. In water drive reservoirs, the water saturation is found to increase with time. At the end of the life of a reservoir, appreciable oil saturation is still found in rock pores, usually due to low recovery of oil. Reservoir engineers are interested to know the remaining oil saturation in a reservoir in order to recover more oil. *See also formation factor, grains, hysteresis effect, logging, petrophysical model, and remaining oil saturation.*

flushed zone. As mud filtrates enter the zone in the immediate vicinity of the borehole during drilling, in-situ fluids such as oil and water are displaced or flushed. Logging tools are designed to obtain values of the in-situ fluid saturation beyond the flushed zone. *See also in-situ fluid and mud.*

fold. A geologic feature whereby rocks deform to a wavelike structure due to compressional stress. Folds can be typified as anticlines and synclines, among other types. *See also anticline.*

formation. A body of rock that is continuous and has sufficiently distinctive properties. Formations are mapped as distinct sections of rock or depositional sequences. Commercial oil and gas reservoirs usually consist of either sandstone or carbonate formations having adequate porosity and permeability. *See also carbonate reservoir, field, limited entry, net to gross thickness, porosity, permeability, and sandstone.*

formation factor. The ratio of the resistivity of a rock that is completely filled with water to the resistivity of the water. The formation factor (F) is correlated with the porosity of the rock and the water saturation by equations discussed in chapter 2. *See also porosity and fluid saturation.*

formation pressure. *See pore pressure.*

formation tester. Wireline tool that employs transient tests of short duration in order to determine the horizontal and vertical permeabilities. It also records the formation pressure at various depth intervals within the formation when a new well is drilled. One of the objectives of the test is the investigation of the vertical communication between adjacent layers. The tool is also known by various brand names as marketed by service companies, such as MDT (Modular Dynamics Tool), which replaced the older generation RFT (Repeat Formation Tester). The above-mentioned tools are generically referred to as wireline formation testers.

formation water. *See interstitial water.*

4-D seismic study. Time-lapse seismic study in 3-D conducted in a petroleum reservoir that points to changes in fluid saturation, reservoir pressure, and thermal characteristics with time. The study aids in obtaining a snapshot of the fluid fronts in a reservoir through time and identifying any geologic heterogeneities, among other uses. *See also seismic study.*

four-point test. *See flow-after-flow test.*

frac gradient. Expressed in pounds per square inch per foot (psi/ft), the gradient relates to the pressure at which a formation would be fractured and can transmit fluid in significantly large quantities. For example, a formation located at a depth of 5,000 ft and having a frac gradient of 0.7 psi/ft is expected to part at a pressure of 3,500 psi. The frac gradient can be obtained by conducting a step rate test. This information is vital in designing a hydraulic fracturing operation and determining the limiting injection pressure during waterflooding.

fractional flow. Relates to the flow of one fluid in the presence of another as a function of individual fluid phase saturation. The fractional flow relationship is used extensively in predicting waterflood performance. During water injection, the fraction of water injected into the reservoir increases with an increase in its saturation until it reaches 100%, when the oil is no longer movable. The fractional flow curve, derived from relative permeability data, is illustrated in chapters 4 and 16. *See also relative permeability and waterflooding.*

fracture half-length. Oil and gas wells are hydraulically fractured to enhance productivity, where the created fracture is visualized to extend on two sides of the wellbore in a symmetric manner. The distance between the wellbore and one tip of the induced fracture is known as the fracture half-length, usually determined by conducting a buildup or drawdown test. *See also buildup test, drawdown test, hydraulic fracturing, and wellbore.*

fracture porosity. A type of secondary porosity that can develop due to the fracturing of a rock as a result of tectonic activities. In most cases, fractures are highly conductive but have much less porosity in comparison to the matrix porosity of the rock.

fracture pressure. *See frac gradient.*

frontal advance theory. Predicts waterflood performance, including time to breakthrough and change in water cut with time, among others. The theory is based on the immiscible displacement of oil by water and the relative permeabilities of the fluids.

gas cap. In a saturated oil reservoir where free gas is present, a gas cap overlies the oil zone due to the much smaller gravity of the gaseous hydrocarbons. *See also bubblepoint and coning.*

gascap drive. A primary production mechanism with which the reservoir pressure is partly maintained by the overlying gas cap during oil and gas production.

gas condensate. Certain hydrocarbon fractions that remain in the gas phase initially in the reservoir due to elevated pressure. However, as reservoir pressure declines below the dew point, the hydrocarbons condense into liquid in the pores of the rock. Significant condensation may occur near the wellbore as a result of a large drop in pressure. Condensates are also encountered in gas treatment facilities due to the reduced pressure at the surface. The API gravity of condensates ranges between 50° and 120°. *See also retrograde condensation.*

gas coning. *See coning.*

gas cycling. A widely recognized procedure to efficiently produce gas condensate reservoirs. Reservoir pressure is maintained by gas reinjection in order to minimize the condensation of heavier hydrocarbons ("rich" gas) within the reservoir. Gas cycling operations result in better recovery from gas condensate reservoirs. *See also gas condensate, hydrocarbons, and recovery.*

gas deviation factor. Frequently abbreviated as z factor, it accounts for the nonideal aspects of the PVT relation of various gases and hydrocarbon mixtures in gas phases. Values of the gas deviation factor at various reservoir pressures and temperatures can be obtained for natural gases if the composition or specific gravity of the gas is known. The gas deviation factor is described in chapter 3. *See also differential vaporization.*

gas formation volume factor (FVF). Ratio of the volume of in-situ natural gas under reservoir pressure and temperature over the volume of the same gas at standard conditions, usually at 14.65 psia and 60°F. In oilfield units, it is expressed in reservoir barrels per standard cubic feet in order to conform to the unit of oil formation volume factor in reservoir barrels per standard barrels. Since gas is highly compressible, values of the gas formation volume factor are usually very small and are reported in reservoir barrels per thousand standard cubic feet (rb/Mscf). In the gas industry, the expansion factor is sometimes used, with a unit of standard cubic feet per cubic feet (scf/cft). *See also differential vaporization.*

gas gravity. Ratio of the molecular weight of natural gas over the molecular weight of air. A typical value of dry gas gravity would be about 0.66–0.67. The gravity of natural gas is greater when condensates are present in significant amounts.

gas lift. An artificial lift method whereby gas is injected through the tubing or annulus to provide additional energy. The gas is injected either continuously or intermittently. *See also artificial lift.*

gas/liquid ratio (GLR). The ratio of the volume of the gas produced to the volume of the liquids produced, i.e., oil and water. Measurement of the gas/liquid ratio is associated with a well producing all three phases, namely oil, gas, and water.

gas/oil contact (GOC). In a typical oil reservoir where free gas is present, vertical segregation between the oil and gas occurs due to differences in the fluid gravity. A plane within the reservoir exists above which the fluid is predominantly gas, underlain by an oil zone. A transition zone may exist between the oil and gas phases, however, as a gradual increase in oil saturation may be observed. *See also gas/water contact.*

gas/oil ratio (GOR). The ratio of the volume of gas produced to the volume of oil produced, expressed as standard cubic feet per stock-tank barrels (scf/stb). Crude, having an abundance of volatile hydrocarbon components, exhibits a relatively high gas/oil ratio. Reservoir management strategy, including the design of surface facilities, depends on knowledge of the gas/oil ratio anticipated in a field. *See also cumulative gas/oil ratio.*

gas solubility. The ratio of the volume of the gas that is liberated from a known volume of oil, both measured at standard conditions. The unit of the gas solubility factor is standard cubic feet per stock-tank barrels. As the reservoir is produced below the bubblepoint, the gas comes out of solution in the reservoir in increasing amounts. Hence the gas solubility of the crude oil decreases accordingly.

gas solubility factor. *See gas solubility.*

gas/water contact (GWC). In rock pores, gas would segregate over the interstitial water due to differences in the fluid gravity. The location of the interface between the gas and water phases, referred to as the gas/water contact, is readily detected by a wireline log. However, the interface is not always sharp, and a transition zone may exist between zones of 100% water and maximum gas saturation. This topic is discussed in detail in chapter 2. Information related to the gas/water contact and the oil/water contact (OWC) is vital in completing a well for optimum performance. A gas/oil contact (GOC) also will be present in an oil reservoir overlain by a gas cap. *See also gas/oil contact, interstitial water, logging, oil/water contact, and transition zone.*

gel. Jellylike substance that may flow like a fluid when agitated. Widely used in profile modification of wells.

geostatistical method. Statistical approach in characterizing a reservoir. The method aids in obtaining reservoir properties, such as porosity and permeability, at locations in the reservoir where no direct measurement is possible. *See also permeability and porosity.*

geothermal gradient. Temperature gradient encountered in the subsurface, which is necessary in estimating reservoir temperature when direct measurement is not available. The gradient varies from region to region. Typical values of the gradient would be about 1.2°F/100 ft. In thermally active formations, however, abnormal temperatures are encountered.

grains. The solid portion of a porous rock. Reservoir rocks are usually made up of grains and pores. *See also fluid saturation, porosity, and reservoir.*

gravel pack. Fine gravel placed around a perforated or slotted liner in order to restrict the flow of sand particles into the wellbore. The borehole is enlarged where the gravel pack is placed. *See also wellbore.*

gravity drainage. Certain steeply dipping reservoirs located at shallow depths produce by the mechanism of gravity drainage. The formation could be highly transmissible. *See also viscous forces.*

groundwater. Subsurface water present in aquifers, occupying a rock's pore spaces. *See also aquifer.*

heavy oil. Crude oil with API gravity less than 22.3° is classified as heavy oil. The viscosity of heavy oil is usually found to be greater than 10 cp. Moreover, heavy oil composition exhibits low hydrogen/carbon ratios and high asphaltene content. Some authors refer to tar and other highly viscous hydrocarbons as ultra or extra heavy oil. Efficient recovery of heavy oil, present in abundance in many petroleum regions of the world, is of considerable interest to reservoir engineers. *See also API gravity.*

hemiradial flow. An early flow regime that may develop during horizontal well testing when the vertical distances from the horizontal section of the well to the upper and lower limits of the formation are not equal. The development of hemiradial flow is depicted in chapter 5. *See also horizontal drilling and pseudoradial flow.*

heterogeneous formation. Virtually all geologic formations, including those bearing hydrocarbons, are heterogeneous in nature. It implies that the rock properties are not uniform from one location to another, even within a short interval. Typical examples of reservoir heterogeneities are stratification or layering, natural fractures, vugular porosity, high permeability streaks, facies changes, and geologic barriers to flow. Adequate characterization of reservoir heterogeneities holds the key to successful reservoir management, leading to optimum recovery. *See also facies, high permeability streaks, infill drilling, numerical simulation, stratified reservoir, and vugular porosity.*

high permeability streaks. A type of rock heterogeneity encountered frequently in petroleum reservoirs. Due to changes in the depositional environment in ancient times, thin intervals or streaks may develop in a geologic formation having permeabilities that are one to several orders of magnitude greater than the rest of the formation. The streaks may or may not be continuous throughout the reservoir, and can be detected by utilizing formation testers as described in chapter 2. The presence of high permeability streaks leads to unexpected and premature breakthrough of the injected fluid, such as water or gas, during enhance oil recovery operations. High permeability streaks adversely affect reservoir performance during improved oil recovery or enhanced oil recovery processes. See also enhanced oil recovery, heterogeneous formation, improved oil recovery, permeability, and water cut.

high-pressure/high-temperature (HP/HT). Pertains to wells drilled in extreme environments, with temperatures of at least 300°F and pore pressures greater than 0.8 psi/ft, or where a blowout preventer (BOP) with a rating higher than 10,000 psi is required.

history matching. In reservoir simulation, the past production history of a field is matched against what is predicted by a reservoir model. This production history could include changes in the pressure, the flow rate of the various fluid phases, and the gas/oil ratio over time. History matching is vital in validating the conceptual model of the reservoir. As the reservoir is produced further, history matching efforts continue in order to update and fine-tune the reservoir model. *See also model.*

homogeneous formation. A body of rock having uniform properties such as porosity and permeability. Although homogeneous rocks are hardly encountered in petroleum reservoirs, many fluid flow models in reservoir engineering are based on the assumption of a homogeneous and isotropic formation. These models are valuable in conceptualizing the flow process in porous media. *See also isotropic formation, model, and reservoir quality.*

horizontal drilling. When the angle between the wellbore trajectory and the vertical direction exceeds 80°, the well drilling is considered horizontal. Most recent horizontal wells are multilaterals, meaning that multiple horizontal branches are drilled from one vertical or deviated hole. The mother hole and the branches are usually a few thousand feet long. The length of certain horizontal holes exceeds 20,000 ft. The era of horizontal drilling has seen significantly increased recovery in a large number of fields. The significant advantages of horizontal wells are described in chapter 19.

horizontal sidetracking. With the advent of horizontal drilling technology, many vertical or deviated wells are recompleted as horizontal wells by sidetracking the original hole at some depth. This results in significant increases in well productivity, abatement of water production, and reduction in coning effects, among other benefits. *See also coning and well productivity index.*

hydrates of gas. Solid crystals formed by natural gas and water, resulting in clogging of gas pipelines and valves at high pressures and low temperatures above the freezing point.

hydraulic diffusivity. *See diffusivity coefficient.*

hydraulic fracturing. Hydraulic fracturing is a stimulation process in wells located in low permeability reservoirs in order to commercially produce oil and gas. Furthermore, the productivity of damaged wells is also improved by hydraulic fracturing. Special fluids are pumped at high pressures (exceeding the fracture gradient) in order to part the formation and create a fracture that extends in two opposite directions. Most fractures are vertically oriented due to the nature of stress in the formation. Certain proppants are mixed with the fracturing fluid in order to maintain the fracture conductivity for production. *See also acidization, fracture half-length, productivity, and stimulation.*

hydrocarbons. Chemical compounds comprised of hydrogen and carbon, formed under appropriate conditions of pressure and temperature in ancient times from fossil and other sources. Hydrocarbons range from simple to highly complex compounds, and appear as solid (coal), liquid (oil), and gas (natural gas). Hydrocarbons present in natural gas are of relatively low molecular weight, such as methane and ethane. Viscous oils contain significantly heavier hydrocarbons consisting of long and complex hydrocarbon-carbon chains. *See also bitumen, bubblepoint, cap rock, dew point, equation of state, field, gas cycling, lithology, methane, oil-wet reservoir, and resources.*

hydrodynamic trap. Occurs due to the movement of water in the formation in a manner that prevents upward escape of the associated hydrocarbons. Hydrodynamic traps are associated with other types of traps.

hydrostatic head. The height of a column of fresh water (specific gravity = 1) at a given depth that exerts a pressure. The pressure is represented by the hydrostatic head as inches of H_2O or feet of H_2O.

hysteresis effect. A phenomenon whereby certain dynamic rock properties based on rock-fluid interaction, such as capillary pressure, not only depend on fluid saturation in the pores but also on the history of the saturation change (drainage or imbibition). Hysteresis effects are described in chapter 2. *See also fluid saturation.*

image well. Image wells are conceptualized in well test theory in order to correctly interpret test results when a barrier or constant pressure boundary exists near the well affecting the pressure response. This analysis is based on the principle of superposition.

imbibition. The phenomenon of absorption or the increase in saturation of the wetting fluid phase into a porous medium. Conversely, a decrease in the saturation of the wetting phase is known as drainage. *See also drainage and wettability.*

impermeable rock. Rocks that do not transmit any fluid due to a lack of permeability. Examples are cap rock and intervening shale between two hydrocarbon-bearing layers. *See also salt dome.*

improved oil recovery (IOR). Any method to augment recovery from a petroleum reservoir following primary production, which is driven by natural reservoir energy. At the end of the primary recovery process, the most common improved oil recovery process is waterflooding, and hence it is referred to as secondary recovery. At later stages in the life of a field, tertiary recovery methods, also part of improved oil recovery, are implemented. *See also displacement efficiency, enhanced oil recovery, heterogeneous formation, infill drilling, injection well, matured reservoir, mobility, reservoir management, water-alternating-gas injection, and waterflooding.*

infill drilling. Relates to the drilling of wells in between the existing wells to augment recovery from the field. The decision for infill drilling, which leads to closer well spacing, primarily depends on the economic evaluation of increased recovery against the capital expenditure (CAPEX) involved in drilling. During improved oil recovery, infill wells are usually drilled to increase areal sweep and reduce the detrimental effects of rock heterogeneities that may lead to bypassing of the oil. Infill drilling also finds wide application in low permeability, heterogeneous, and compartmental reservoirs. *See also compartmentalization, heterogeneous formation, improved oil recovery, and remaining oil saturation.*

inflow performance relationship (IPR). Correlation between flowing bottomhole pressure of a well and its rate of production. An inflow performance relationship curve is also referred to as a well deliverability plot, which aids in analyzing well performance and design. Inflow performance relationship curves are described in chapter 4.

initial reservoir pressure. *See pressure—initial reservoir pressure.*

injection well. Most oil reservoirs are subjected to water or gas injection, or both, in order to provide additional energy and augment recovery. Producing wells are usually converted to injection wells during improved oil recovery processes. *See also improved oil recovery, injectivity, and reservoir.*

injectivity. Relates to the volume of fluid that can be injected for a pressure differential created downhole between the well and the reservoir during injection. Wells must have good injectivity in order to conduct an efficient waterflood operation. The unit of injectivity is barrels per day–pounds per square inch [bbl/(day-psi)]. *See also injection well.*

injectivity test. A procedure to ascertain a safe injection pressure and rate in injection wells prior to waterflood operation. Care is taken not to exceed the frac gradient during injection, which would result in water flow through the fracture, bypassing oil and resulting in poor sweep efficiency.

in-situ combustion. *See fire flooding.*

in-situ fluid. Refers to oil, gas, or water present in the porous rocks of a reservoir. *In-situ* is Latin for in place. In-situ fluid analysis, including measurement of the physical properties of the fluid and compositional analysis, is conducted by downhole measurement tools, as mentioned in chapter 3. This procedure has the advantage of measuring the fluid properties, including composition, under actual reservoir conditions. *See also clay swelling, fault, flushed zone, miscible displacement, and sweep efficiency.*

intelligent oil field. With the advent of the digital age, traditional oil fields are undergoing significant transformation in the areas of continuous data collection. Technology includes permanent downhole gauges (PDG) and deployment of robust information systems. It allows real-time or near–real-time decision making by a multidisciplinary asset team and automated control of injection and production wells without human intervention, among other innovations. Intelligent fields are also known as i-fields, digital oilfields, and smart fields. *See also pressure gauge and smart well technology.*

interfacial tension (IFT). At the interface of two immiscible fluids, the force exerted by the fluid phases is dissimilar, leading to the phenomenon of interfacial tension. Many enhanced oil recovery processes attempt to recover more oil by reducing the interfacial tension between fluid phases. *See also capillary number, displacement efficiency, enhanced oil recovery, J function, and wettability.*

interference test. A transient well test whereby a perturbation (in fluid flow) is created in a source well, followed by monitoring of the pressure response in a nearby observation well or wells. Interference tests are conducted to characterize the reservoir, including the degree of connectivity between injection and production wells. An example of interference test design is shown in chapter 5. *See also well test.*

interstitial water. Water found in the pore spaces of the rock, which may or may not have been present in ancient times during the formation of the rock. It is also referred to as the formation water. *See also connate water and gas/water contact.*

interwell tomography. A technique to acquire seismic data between two wells based on an acoustic signal transmitted from the source well to a receiver well. The data is interpreted to track changes in fluid saturation and pressure over time. It is also referred to as cross-well tomography.

inverted 5-spot, 7-spot, or 9-spot pattern. Relates to the injection and production well configuration during waterflooding. In an inverted pattern, there is only one injection well located at the center of the pattern, while the rest of the wells are producers located at the corners. For example, in an inverted 7-spot pattern, there are six producers along with a single injector at the center. In regular 7-spot pattern, however, the opposite is true. There is one producer located at the center, surrounded by six injectors at the corners. Well injection patterns are illustrated in chapter 16.

isochronal test. A well test to assess the deliverability of a well. Usually conducted in gas wells, an isochronal test is comprised of a series of drawdown and buildup sequences at different rates. Drawdown periods are of equal duration, followed by a period of buildup until pressure stabilization is achieved. In tight reservoirs, the time to stabilize may be quite long. Hence, a modified isochronal test is conducted in which the buildup periods are of equal duration. The well is flowed at a new rate for the following sequence before pressure stabilization is attained. *See also multiple rate test and stabilization.*

isopach. Contour of equal stratigraphic thickness. Isopach maps are usually generated from well log information obtained at various locations in the reservoir. Determination of the hydrocarbon in place is based on isopach maps, among other factors.

isotropic formation. A formation in which rock properties are the same in all directions. Many analytic solutions of fluid flow in porous media are based on the assumption of a homogeneous and isotropic formation. *See also homogeneous formation.*

J function. A function that is based upon porosity, permeability, interfacial tension, and wettability angle. For a specific porous medium under study, the J function is plotted against water saturation. *See also interfacial tension, permeability, porosity, and wettability.*

kerogen. A chemical mixture of certain organic materials that originates from remnants of plant life. Kerogen, found in source rocks, is the principal source of oil when subjected to high temperatures over long periods of time. *See also oil shale and source rock.*

kilopascal (kPa). A unit of measurement of pressure in SI (International System of Units). To convert kilopascals to pounds per square inch in oilfield units, multiply by a factor of 0.14504.

kriging. A statistical technique to interpolate a value at a point based on values known at nearby points. Kriging is used to generate contour maps of porosity and permeability of a reservoir. *See also permeability, porosity, and reservoir.*

laminar flow. Flow of fluid in a straight line without any turbulence. It is a prerequisite for applying Darcy's law. *See also Darcy's law.*

limestone. A large number of petroleum reservoirs are in limestone, a sedimentary rock chiefly composed of calcite (calcium carbonate). Limestone is mostly of biologic origin. Dolomite and clay are usually present in small quantities. *See also chalk.*

limited entry. The phenomenon of limited fluid entry arises due to partial completion of a well where the entire thickness of the hydrocarbon-bearing formation is not perforated for production. One of the objectives of a partially completed well is to minimize the unwanted entry of gas or water into the wellbore. In certain cases, the entire oil or gas zone is not drilled. *See also formation, partial completion, wellbore, and well completion.*

linear flow. Due to flow geometry, fluids in a reservoir move in approximately straight lines, resulting in linear flow, as opposed to radial flow. Linear flow is encountered in flow through hydraulic fractures, in horizontal wells, and in channel-shaped reservoirs. In pressure transient tests, linear flow is identified by a half-slope line on a diagnostic plot of pressure versus time in a log-log scale. *See also bilinear flow, derivative plot, and pseudoradial flow.*

line drive. In waterflooding or enhanced oil recovery processes, injection wells are located in a straight line parallel to an adjacent line where the producers are located. Line drive can be direct or staggered. In the former configuration, injectors face the producers directly, while in the latter, the injector and producer face each other at an angle. Various injector-producer configurations are illustrated in chapter 16. *See also enhanced oil recovery.*

line-source solution. Well test theory implements the line-source solution of fluid flow equations, in which the well is conceptualized as a vertical line. This approach leads to satisfactory approximations of the pressure response for engineering purposes and is widely accepted in the industry.

liquefied natural gas (LNG). Natural gas, chiefly methane and ethane, converted to a liquid phase under extremely low temperatures for transportation purposes. Liquefied natural gas tankers transport large quantities of natural gas across continents. *See also methane and natural gas.*

liquefied petroleum gas (LPG). Propane and butane in liquid form, which are usually bottled to use domestically or industrially. *See also miscible displacement.*

lithology. Rock characteristics in terms of mineral content, texture, physical appearance, and grain size, etc. Hydrocarbon-bearing formations are usually described in terms of their lithology. *See also core, fault, and hydrocarbons.*

logging. Wireline logs are used to determine the geological characteristics and physical properties of the formation following the drilling of a borehole for oil and gas. The data is collected continuously against the depth of the drilled formation and is sent to the surface for processing by telemetry systems. Logging incorporates the identification of the rocks and the fluids in the rock pores. It also allows the measurement of rock properties and the determination of fluid saturation around the wellbore by various downhole tools and techniques. The principle of logging involves the measurement of formation resistivity, gamma ray, and acoustic waves, among others, as the responses of these are influenced by the rock and fluid properties. Besides the determination of lithology, porosity, and fluid saturation, the major objectives of well logging include correlation of the stratigraphy between wells. Wireline log studies are routinely conducted following the drilling of a well. Furthermore, measurements-while-drilling (MWD) and logging-while-drilling (LWD) techniques introduced in recent decades are designed to collect necessary information such as pressure, wellbore trajectory, porosity, and resistivity from subsurface geologic formations during drilling. Certain logs are run in cased holes. A summary of wireline logging systems is tabulated in chapter 2. *See also acoustic log, fluid saturation, gas/water contact, logging while drilling, marker bed, measurements while drilling, oil/water contact, petrophysical model, resistivity log, and wellbore.*

logging while drilling (LWD). Logging-while-drilling tools collect similar information about the subsurface formation as obtained by wireline logging tools. *See also logging and measurements while drilling.*

lognormal distribution. Many important reservoir-related parameters, such as the size of the fields in a petroleum basin, rock permeability, and the recovery factor, are found to have a lognormal probability distribution. The various types of probability distributions used in reservoir studies are described in chapter 18. *See also permeability and recovery.*

Lorenz coefficient. A measure of rock heterogeneity determined by plotting the permeability-thickness (kh) data against the porosity-thickness (øh) data as obtained from petrophysical studies of cores. Points are plotted in decreasing order of k/ø. In the case of a perfectly homogeneous formation, the points fall on a diagonal line drawn from the upper right to the lower left corner of the plot. However, in the more likely scenario where the formation is heterogeneous, plotted points are located away from the diagonal to the left. The Lorenz coefficient is defined as the ratio of the area enclosed by the plotted points and the diagonal over the area enclosed by the diagonal and the lower right corner of the plot. In the ideal case of uniform rock properties, the value of the Lorenz coefficient is 0. However, the value increases with increasing heterogeneity and approaches unity in extreme cases.

marginal reservoir. Petroleum reservoir in which development and production are not economically attractive in general, or the reservoir is of very limited potential based on current technology. Besides new reservoirs with poor reserves, all matured reservoirs become marginal at some point in time as the oil and gas reserves are depleted. Management of marginal and matured reservoirs is described with field examples in chapter 19. *See also matured reservoir.*

marker bed. A distinct geologic unit that can be correlated over the entire area of the reservoir. A marker bed leaves unique and identifiable signatures on logs obtained from various wells. *See also logging.*

matured reservoir. A producing reservoir is considered to reach maturity when it is past its peak production level. Matured reservoirs have limited remaining reserves and are usually produced by pressure maintenance or improved oil recovery process. Reservoir engineers strive to add value to the asset of such reservoirs by obtaining a detailed reservoir description and implementing innovative recovery techniques. *See also improved oil recovery and marginal reservoir.*

measured depth (MD). The depth along the drilled hole path. Well depths are reported in either true vertical depth or measured depth, or both. In a vertical well, the true vertical depth and the measured depth are the same, while in an inclined well, the measured depth is greater than the true vertical depth. *See also true vertical depth.*

measurements while drilling (MWD). Tool designed to monitor formation pressure, temperature, and wellbore trajectory. A logging-while-drilling tool may also be incorporated measuring various rock properties downhole, such as resistivity, porosity, density, and gamma ray emission. The tool is placed above the drill collar during drilling. Collected data is stored in the tool and then transmitted to the surface through pressure pulses in the mud system. *See also logging and logging while drilling.*

mechanical skin. A measure of the formation damage due to mechanical issues, such as debris obstructing the flow of fluid into a well. *See also skin factor.*

methane. A light hydrocarbon that occurs predominantly in natural gas. Properties of methane are provided in chapter 3. *See also coalbed methane, hydrocarbons, liquefied natural gas, and natural gas.*

minifrac. Precursor to the main hydraulic fracturing operation in order to obtain vital design data.

miscible displacement. An enhanced oil recovery process that strives to attain miscibility between the injected fluid and the in-situ oil for reducing interfacial tension, resulting in efficient recovery of oil. When two fluids are completely miscible, they are indistinguishable from each other. Injected fluids include liquefied petroleum gas (LPG), nitrogen, and carbon dioxide (CO_2). CO_2 injection is widely regarded as a major tertiary recovery process following waterflooding in certain oil regions, chiefly due to its availability. See enhanced oil recovery, in-situ fluid, liquefied petroleum gas, and tertiary recovery.

mobility. The ratio of the permeability over the viscosity of a reservoir fluid. Mobility is directly related to the ease of production of the reservoir fluids. Tight formations and the high viscosity of heavy oils lead to unfavorable mobility. In waterflooding and related improved oil recovery processes, a favorable or unfavorable mobility ratio, obtained as the ratio of displacing fluid to displaced fluid, largely dictates the success of oil recovery. *See also improved oil recovery, permeability, tight reservoir, and viscosity.*

model. Refers to reservoir simulation models, which are based on appropriate mathematical equations to reflect fluid flow characteristics in porous media. Simulation of models leads to the prediction of reservoir pressure and saturation over time, among other results. Reservoir models are validated by history matching of past production data. The models are as good as the quality of reservoir characterization study performed prior to model development. *See also history matching, homogeneous formation, numerical simulation, reservoir characterization, and reservoir simulation.*

model simulation. *See model.*

Monte Carlo simulation. A probabilistic study in which a model is developed based on the probability of occurrences of influencing parameters. A large number of trials are then performed to determine the various probabilities of the outcome. For example, the probability values associated with newly discovered petroleum reserves can be obtained by Monte Carlo simulation when the probabilities associated with related parameters, such as the porosity, thickness, and areal extent of the reservoir, are known. An example of a Monte Carlo simulation is provided in chapter 9.

movable oil saturation. Indicates the volume of oil that can displaced in porous media. Knowledge of the movable oil saturation, usually obtainable by laboratory studies, is important information in assessing the ultimate oil recovery from a reservoir following fluid injection. It can be estimated by subtracting the residual oil saturation from the initial oil saturation.

mud. Widely used to refer to water-based, oil-based, or synthetic fluids used during the drilling of a well. *See also flushed zone.*

multilateral horizontal well. *See horizontal drilling.*

multiphase fluid flow. Simultaneous flow of more than one fluid phase in porous media or in the wellbore. For example, oil, gas, and water can be produced simultaneously from one well. Multiphase fluid flow introduces additional complexity in effectively producing oil from a reservoir. *See also numerical simulation.*

multiple rate test. Gas wells are usually subjected to multiple rate tests in order to ascertain deliverability of the well. *See also flow-after-flow test and isochronal test.*

natural gas. Light hydrocarbons, chiefly methane and ethane, produced from petroleum reservoirs in a gaseous form. Propane, butane, pentane, hexane, and heavier hydrocarbons may be found in smaller quantities. Natural gas can be dry or wet, as certain heavier components may condense out in the surface facilities in the latter case. Furthermore, the natural gas can be sweet or sour, depending on the presence of certain impurities. Natural gas liquids (NGLs) are heavier hydrocarbons that exist in liquid form in the surface facilities, such as propane and butane. The typical composition of natural gas is included in chapter 3. *See also coalbed methane, dry gas, liquefied natural gas, methane, and sour gas.*

net to gross thickness. In a typical geologic formation, not all of the sections at various depths are conducive to production due to poor reservoir quality, such as low porosity and permeability. The net thickness of the hydrocarbon-bearing formation is usually less than its gross thickness. Hence knowledge of the net to gross thickness ratio is required to estimate petroleum reserves. *See also cutoff porosity and formation.*

non-Darcy flow. In certain gas wells producing at high rates, turbulence develops in the vicinity of the wellbore that creates an additional pressure drop not accounted for by Darcy's law. Non-Darcy flow can be identified by testing a well at multiple rates to evaluate the dependence of skin on the flow rate. *See also Darcy's law.*

nonconformity. A geologic discontinuity in which the older rock is overlain by much younger and dissimilar rock, indicating erosion over the geologic ages. The phenomenon points to the lack of a continuous geologic record over the ages. A nonconformity is a type of unconformity that occurs between igneous or metamorphic rock and the overlying sedimentary rock. *See also unconformity.*

numerical simulation. Mathematical simulation involving high-speed computation of a numerical reservoir model to predict reservoir performance in the future under various scenarios (number and location of wells, fluid injection, etc.). Numerical models are usually employed when the physical system is too complex to be handled by analytic models, which are usually based on simplified assumptions. Such complex systems are typically encountered in oil reservoirs with wide-ranging heterogeneities and multiphase fluid flow. *See also heterogeneous formation, model, and multiphase fluid flow.*

observation well. Certain wells, usually of low productivity, are converted to observation wells to monitor the reservoir pressure and injected fluid front in a reservoir. Information obtained from observation wells is utilized in reservoir simulation, among other applications. *See also well productivity index.*

offset well. Nearby well or wells in a location where a new well is planned to be drilled. Geologic and production information obtained from the offset wells aid enormously in designing, drilling, and completion of the new well. In exploratory wells, however, there is no offset well, and well design is based on geosciences studies and prior experiences in the same basin. *See also basin, exploration, and well completion.*

oil and gas lease. A contract between the owner of the mineral rights of a property and an oil company operator to develop and produce oil and gas by paying a royalty to the owner.

oilfield units. Units traditionally used to report reservoir pressure, well rate, fluid gravity, rock permeability, and various other physical and PVT properties. A list of oilfield units along with conversion factors is provided in Table 1–2 (chapter 1). *See permeability and pressure-volume-temperature properties.*

oil formation volume factor (FVF). An important property of crude oil. It is determined as the ratio of the volume of oil plus dissolved gas at reservoir conditions over the volume of oil at surface conditions once the volatiles have evolved out of solution. The latter occurs due to the reduction in pressure. Volatile oils and light crude oils tend to exhibit a relatively high formation volume factor due to the abundance of lighter hydrocarbons. However, the factor is usually closer to 1 in the case of heavy to extra heavy oils, where volatile components are less. The oil formation volume factor of a crude oil decreases monotonically once the reservoir pressure declines below the bubblepoint following dissolution of the volatile components. Typical values of the oil formation volume factor may range between 1.02 and 2.7. Correlations for the formation volume factor are provided in chapter 3. *See also pressure-volume-temperature properties.*

oil migration. Movement of oil from the source rock to a geologic trap for accumulation in geologic time.

oil sands. Deposits of semisolid bitumen, water, sand, and clay that constitute huge resources of energy. Bitumen typically comprises 10%–12% of oil sands and has high viscosity and density. Major occurrences of oil sands are found in Canada and Venezuela. In Canada, oil sands are encountered at relatively shallow depths (900 ft to 1,600 ft) in three regions of Alberta: Athabasca, Cold Lake, and Peace River. Oil sands are typically extracted by mining, in-situ thermal processes, and nonthermal processes. Also called tar sands. *See also bitumen, synthetic crude oil, and upgrading.*

oil shale. Geologic formation encountered at relatively shallow depths with accumulations of kerogen. Due to the shallow depth of burial, the appropriate temperature did not exist during geologic times in order to transform the kerogen to oil. However, kerogen can be subjected to thermal energy to produce shale oil. *See also kerogen.*

oil/water contact (OWC). Defined by a plane in an oil reservoir below which the reservoir fluid is predominantly water. Segregation between the oil and water occurs due to differences in fluid gravity. The depth at which the oil/water contact exists in a reservoir is readily detected by log studies. However, depending on the fluid and rock properties, a transition zone of measurable height may occur between the oil and water zones. The transition zone is usually longer in low permeability formations and where the density differences between the oil and water are relative less. In a transition zone, both oil and water phases are mobile. Knowledge of the oil/water contact, and its possible shift in the upward direction during the life of the reservoir, is vital in effectively producing a well. In some cases, wells are recompleted above the oil/water contact to produce oil without water. However, certain reservoirs are known to produce from long transition zones, as a substantial amount of oil is found. *See also gas/water contact, logging, and transition zone.*

oil-wet reservoir. Some reservoirs are found to be oil wet, meaning that the oil has a greater tendency to adhere to the surface of the rock pores than does the water. Oil-wet reservoirs generally lead to poor recovery in comparison to water-wet reservoirs during waterflooding, given that all other factors are the same. Some water-wet rocks may turn oil-wet due to the deposition of certain hydrocarbon compounds on the pore surface. Reservoirs exhibiting mixed wettability characteristics are also encountered. *See also hydrocarbons, water-wet reservoir, and wettability.*

open flow potential. See absolute open flow potential.

openhole completion. A type of well completion where no casing or liner is set in the producing zone.

operator. The company responsible for financing and decision making in drilling operations. The operator has the largest stake in an oil and gas venture in most cases.

original oil in place (OOIP). The volume of oil present in the reservoir at the time of discovery. The original oil in place is estimated based on knowledge of the reservoir volume, porosity, oil saturation, and oil formation volume factor. The estimation of the original oil in place is generally subjected to probabilistic analysis, as all of these parameters are not known with certainty when only one or a few wells are drilled.

overburden pressure. *See pressure—overburden pressure.*

paraffin. A waxlike hydrocarbon compound that tends to deposit from crude oil due to changes in pressure and temperature as oil is transported to the surface. Paraffin deposition in tubing may adversely affect production. *See also asphaltenes and crude oil.*

partial completion. Completion of a well in a certain section of the pay zone. *See also limited entry, spherical flow, and well completion.*

parts per million (ppm). Measure of concentration. In a typical analysis of produced water, minerals are reported in parts per million. During water injection, a noticeable decrease in mineral concentration in the produced water indicates the advent of injected seawater and eventual breakthrough. See also breakthrough.

pattern balancing. During a waterflooding operation, an areal pattern consisting of a number of wells, typically five to nine, balanced by manipulating injection and production rates. In a perfectly balanced pattern, injected water does not migrate outside the pattern, ensuring maximum areal sweep efficiency and recovery from the pattern area. Pattern balancing studies are usually conducted by reservoir simulation models. See also areal sweep efficiency, recovery, reservoir simulation, and waterflooding.

pay zone. Indicates the zone in the reservoir that would produce oil or gas on a commercial basis.

permeability. Measure of the ability of porous medium, such as an oil-bearing rock, to transmit fluid. Petroleum reservoirs require two most basic properties, porosity and permeability, in order to store and produce oil and gas. Rock permeability, usually expressed in darcies or millidarcies, can vary widely from one reservoir to another and even within a reservoir. Permeability in certain tight gas reservoirs is found to be in microdarcies, while certain other formations may have permeabilities in several darcies. Efficient reservoir management requires a good knowledge of rock permeability and its variations throughout the reservoir in both vertical and horizontal directions. *See also absolute permeability, altered zone, cementation, core, darcy, directional permeability, effective permeability, fines migration, formation, geostatistical method, high permeability streaks, J function, kriging, lognormal distribution, mobility, petrophysical model, porosity, tight reservoir, tortuosity, and transmissivity ratio.*

permeability anisotropy. In general, horizontal permeability in a hydrocarbon-bearing formation is found to be greater than the vertical permeability by one order of magnitude or more due to conditions prevailing during deposition of the sediments. This phenomenon is referred to as permeability anisotropy. Directional permeability is a form of permeability anisotropy. *See also depositional environment and directional permeability.*

permeability damage. *See altered zone and skin.*

permeability variation factor (V). Measure of the degree of rock heterogeneity based on variations in permeability as found in core samples. The factor, ranging from 0 (uniform rock) to 1 (extremely heterogeneous rock), is based upon the lognormal permeability distribution, i.e., permeability versus percent of total sample having higher permeability. A further definition of permeability variation factor is included in chapter 16.

Petroleum Resources Management System. In March 2007, SPE adopted a broad-based system of defining hydrocarbon reserves and resources that consolidates, builds on and replaces the previous definitions. Hydrocarbon accumulations in earth's crust (including unconventional resources) are divided into three major categories in descending order of commercial and technical feasibility, namely, petroleum reserves, contingent resources, and undiscovered resources. Hydrocarbons in each category can further be classified according to the range of uncertainty associated with recovery. The system of definitions is jointly sponsored by WPC, AAPG, and SPEE. Petroleum reserves and resources are treated in chapter 15.

petrophysical model. Reservoir model based on core and wireline log data, including porosity, permeability, fluid saturation, wettability, capillary pressure, and porosity-permeability relationship, among others. The model contributes significantly to reservoir characterization. *See also capillary pressure, core, fluid saturation, logging, permeability, porosity, reservoir characterization, and wettability.*

phase behavior. Relates to the behavior of fluids, including a phase change from liquid to gas and vice versa, as influenced by dynamic changes in pressure, temperature, and fluid composition. Since such changes are inevitably associated with production, the phase behavior of a petroleum fluid plays a critical role in the overall performance of a reservoir.

phase diagram. A phase diagram pertaining to reservoir engineering depicts the typical phase behavior of fluids or a fluid system in a petroleum reservoir whereby temperature is plotted on the x-axis and pressure on the y-axis. A typical phase diagram consists of two major regions: a single-phase region that lies outside an envelope, and a two-phase region that is enclosed by the envelope. The critical point, bubblepoint, and dew point of the fluid system are also indicated on the diagram. The plot is used to predict reservoir fluid behavior, including vaporization and condensation, based on changes in pressure and temperature during oil and gas production.

pinchout. Reduction of reservoir thickness, or pinching out against a sealing rock that forms a stratigraphic trap. Pinchout boundaries can be detected by geosciences study and transient pressure test analysis. Pinchout boundaries leave a distinct signature on a diagnostic plot obtained from a well test. An example of well test analysis in a pinchout formation is presented in chapter 5. *See also derivative plot and stratigraphic trap.*

planimeter. A mechanical device used to compute the area of any closed shape, regular or irregular. It basically integrates the path traced by its wheel into area. The object is usually drawn on a piece of paper for various purposes, including the design of buildings or machinery. In reservoir engineering, planimeters were traditionally used to obtain the hydrocarbon in place and reserves prior to the advent of the digital age.

plateau rate. Used in describing a constant or near-constant production level of oil and gas. In a gas field, production at a certain plateau needs to be maintained for meeting contractual obligations. In oil reservoirs, a plateau may be maintained as long as possible by infill drilling, water injection, and other methods as stipulated by management policy.

play. Potential prospects of hydrocarbon accumulation in a basin that require further data acquisition and evaluation. Petroleum exploration plays could be identified by a similar geologic structure located in the region producing oil or gas. Future steps for a play include drilling of exploratory wells.

pore pressure. Pressure of in-situ fluid within the pores of the rock, usually estimated from the pressure gradient of the formation water and formation depth. It represents the hydrostatic head of the formation found in the pores of the rock, which are in communication with the water at the surface. However, pore pressure can be higher than expected in geopressured formations, where fluids are trapped in the pores during compaction of the sediments in ancient times. Pore pressure can be also be lower than what is normally expected. Pore pressure is used interchangeably with formation pressure. *See also pressure, pressure—reservoir pressure, and pressure—formation pressure.*

pore volume (PV). Volume of pores in the rock. The pore volume of a reservoir is obtained by multiplying the bulk volume of the rock with porosity.

porosity. Rock porosity is a measure of the pore volume of the rock over its bulk volume. In most producing oil reservoirs, typical values of porosity range between 12% and 35%. Certain pores of a reservoir rock are interconnected, forming continuous channels, while others are isolated due to excessive cementation or bonding between surrounding grains. Formation porosity can be measured by the use of acoustic logs. *See also absolute porosity, acoustic logs, cementation, core, cutoff porosity, effective porosity, formation, formation factor, geostatistical method, grains, J function, kriging, petrophysical model, primary porosity, secondary porosity, styolite, and vugular porosity.*

possible reserves. Unproved reserves that geologic and engineering data suggest are less likely to be recovered than probable reserves. Probability of recovery should be at least 10% or more of the sum of the estimated proved plus probable plus possible reserves. *See also proved reserves, probable reserves, and reserves.*

potential function. *See fluid potential.*

pressure. Reservoir engineers deal with various types of pressure data, as summarized in the following:

abandonment pressure. Reservoir pressure at which commercial recovery of oil and gas is no longer feasible from a reservoir. The well is plugged and abandoned (P&A).

abnormal pressure. Pressure in a subsurface formation that is greater or less than the pressure corresponding to the gradient of the formation water.

average reservoir pressure. The pressure that can be obtained by ceasing all flow of fluids in the reservoir and attaining equilibrium conditions. Average reservoir pressure in the drainage area of a producing well can be obtained by conducting a buildup test, among other well test methods, at any time during the life of the reservoir.

bottomhole shut-in pressure. The bottomhole pressure obtained by shutting in a producing or injecting well for a certain period of time.

casing pressure. Pressure measured in the annulus (between tubing and casing).

flowing bottomhole pressure. Bottomhole pressure at the time when the well is flowing.

formation pressure. *See pore pressure.*

fracture pressure. *See frac gradient.*

initial reservoir pressure. Obtained by downhole gauges at the time of discovery.

overburden pressure. The combined pressure exerted by the formation rock and the fluid that exists in the pore spaces of the rock. A typical value of overburden pressure is 1 psi/ft.

pressure at datum. *See datum.*

reservoir pressure. In normally pressured formations, the reservoir pressure can be obtained from the depth of the formation and the hydrostatic gradient of the formation water. It is actually the reservoir fluid pressure. The initial reservoir pressure is obtained by downhole gauges at the time of discovery. *See also buildup test and pore pressure.*

sandface pressure. Pressure at the interface of the formation and the wellbore.

wellhead pressure. Pressure measured at the wellhead. In a producing well, the wellhead pressure is less than the bottomhole pressure due to the hydrostatic column and frictional losses during flow.

pressure at datum. *See datum.*

pressure buildup analysis. Interpretation of buildup test data based on well test theory. *See also buildup test.*

pressure falloff test. Pertains to well tests conducted in injectors. Following injection at a steady rate and well stabilization, the pressure is allowed to fall off by shutting in the well. The resulting response is monitored and subsequently analyzed.

pressure gauge. Device to monitor fluid pressure at the wellhead and at bottomhole conditions. A device used to measure pressure. The quality of well test interpretation depends on the resolution, accuracy, and frequency of the bottomhole pressure data obtained during the test. Pressure gauges can be analog or digital. Certain electronic gauges are capable of monitoring downhole pressure continuously and transmitting data to the surface for analysis and corrective actions, if necessary. *See also intelligent oil field.*

pressure gradient. Rate of change in reservoir or formation pressure per foot of depth in a reservoir. Knowledge of the pressure gradient of formation water, usually about 0.435 psi/ft–0.5 psi/ft, is required in estimating reservoir pressure. The pressure gradient can be calculated by multiplying the specific gravity of the fluid with the pressure gradient of fresh water (0.433 psi/ft). Due to the relatively low specific gravity of petroleum fluids, pressure gradients are less than water.

pressure maintenance. Involves fluid (water or gas, or both) injection into a reservoir to provide energy for maintaining reservoir pressure to sustain oil production. A pressure maintenance operation is usually started early in the life of an undersaturated oil reservoir, before the bubblepoint pressure is reached. Reservoir studies indicate that pressure maintenance above the bubblepoint leads to optimum recovery. *See also waterflooding.*

pressure transient test. *See well test.*

pressure-volume-temperature (PVT) properties. Properties of oil and gas that are dependent on pressure, volume, and temperature as hydrocarbons are normally subjected to changes during production from the reservoir. Examples are the bubblepoint, compressibility, oil formation volume factor, solution gas/oil ratio, and viscosity. Empirical equations predicting these properties are known as PVT correlations. *See also bubblepoint, compressibility, oil formation volume factor, and viscosity*

primary porosity. Porosity that initially developed in the reservoir rock during its deposition in geologic times. *See also porosity.*

primary production. Oil and gas reservoirs produce initially by certain natural energy such as rock and fluid expansion, solution gas drive, water drive, gascap drive, or a combination drive. Some reservoirs are produced by gravity forces. Primary production mechanisms are discussed in chapter 8.

primary recovery. Relates to the fraction of hydrocarbons that can be recovered due to natural energy present in the reservoir. In most oil reservoirs, the percentage of recovery is rather dismal, with an average value around 35%. Recovery from gas reservoirs, however, usually exceeds 80% due to the very high mobility of natural gas. The extent of oil and gas recovery from a reservoir is alternately expressed as percent recovery, recovery efficiency (ER), or recovery factor (RF). *See also primary production, recovery, reservoir monitoring, and water drive.*

principle of superposition. The principle states that mathematical solutions of two or more relatively simple systems are added or superposed to obtain solution of a complex system. For example, the pressure response of a well located at a certain distance from a boundary can be obtained by superposing the response of an image well located twice the distance away. No special treatment of the effects of boundary is required.

probability. The chance of occurrence of a particular event or attribute. Estimates of oil and gas reserves, among others, are subjected to probability studies, since the related parameters, such as storavity, fluid saturation, and recovery efficiency are not known with certainty.

probable reserves. Unproved reserves that geologic and engineering data suggest are more likely than not to be recoverable. Probability of recovery should be at least 50% or more of the sum of the estimated proved plus probable reserves. *See also possible reserves, proved reserves, and reserves.*

production log. Suite of logs run in a completed well in order to identify a zonal inflow or outflow profile against the formation depth based on flow meters, radioactive tracers, and temperature sensors, among other tools. The information is vital in fluid injection and thermal recovery operations. A premature breakthrough may occur through a thin section of the formation, while oil from the rest of the formation is left unswept. Production logging tools have a myriad of applications to improve well and reservoir performance, including analysis of liquid holdup in the wellbore or detection of channeling behind the pipe. *See also production logging tool survey.*

production logging tool (PLT) survey. Testing conducted by running a production logging tool in a well to profile the selective flow of fluids from various layers or zones of the formation into wellbore. Similar surveys are run in injectors. Common survey tools include temperature log, flowrate, spinner and radioactive tracers. PLT surveys are frequently used to identify the high permeability streaks, zones of water breakthrough, and non-producing intervals, among others. *See also production log.*

productivity. *See well productivity index.*

productivity index (PI). *See well productivity index.*

profile modification. Methods that attempt to reduce water production from a well by blocking water-producing zones. Profile modification involves application of a polymer, gel, or other chemical to reduce the permeability of a targeted section. It is routinely performed in wells with adverse effects of water coning, or wells in which breakthrough has occurred in thin sections of the formation. Injection profile modification is also conducted to control the loss of injected fluid through undesirable intervals of a formation.

prospect. Potential accumulation of hydrocarbons that is identified by geophysical and other studies. Exploration wells are drilled in a prospect to discover oil and gas.

prospective resources. *See resources—prospective resources.*

proved reserves. Quantities of petroleum, by analysis of geologic and engineering data, that can be estimated with reasonable certainty to be commercially recoverable in the future from known reservoirs and under current economic conditions, operating methods, and government regulations. The probability of recovery should be at least 90% or more. In certain literature sources, these are referred to as proven reserves. Unproved reserves are based on geologic and/or engineering data similar to that used in estimates of proved reserves. However, technical, contractual, economic, or regulatory uncertainties preclude such reserves from being classified as proved. Unproved reserves may be further classified as probable reserves and possible reserves. Undiscovered reserves are subject to hypothesis, analogy, and speculation. *See also possible reserves, probable reserves, and reserves.*

pseudopressure. A function that replaces the actual pressure in a gas well test interpretation. Pseudopressure is a function of pressure, viscosity, and a gas deviation factor, as defined in chapter 5. Since the compressibility and viscosity of a gas are strong functions of pressure, the classical pressure transient test theory developed for oil wells is modified for gas wells. This is done in order to account for the significant variations in the physical properties of the natural gas encountered during a pressure transient test. *See also compressibility and viscosity.*

pseudoradial flow. During production, pseudoradial flow develops around a hydraulically fractured or horizontal well, as opposed to an essentially radial flow pattern as encountered around a vertical well. *See also bilinear flow, hemiradial flow, and linear flow.*

pseudosteady-state flow. A condition in which the rate of change in the pressure is the same at every point within the reservoir or well drainage area. Pseudosteady-state flow is typically observed and interpreted in well test analysis to determine reservoir properties. *See also steady-state flow and transient flow.*

PVT properties. *See pressure-volume-temperature properties.*

radial flow. Pattern of flow usually envisaged around a well during production or injection. A minimum duration of a well test usually requires that the infinite-acting radial flow regime be reached following the initial period dominated by wellbore storage and skin. The pressure response indicates an infinite-acting reservoir, i.e., the pressure disturbance created at the well has not reached the reservoir boundary. Formation transmissibility is determined from a conventional well test based on this flow regime. Once the value of permeability is known, the skin factor is computed. In horizontal well testing, however, it is usually not practical to conduct the test for a long period of time in order to achieve radial flow. *See also transmissibility.*

radioactive tracer survey. Usually injected in formations to monitor the flow path of fluids from the injector to the producers in interwell tracer surveys. Connectivity between an injector with surrounding producers is studied by monitoring tracer concentration at the producers, as detected by gamma ray emissions.

radius of investigation. Radial distance from the wellbore that can be investigated in order to evaluate reservoir pressure and rock characteristics during transient well tests. The radius of investigation is a direct function of the formation transmissibility and the duration of the well test. *See also transmissibility.*

realization. In a reservoir, important rock properties such as porosity and permeability are not known except at the wells. Based on geostatistical modeling, also referred to as stochastic modeling, many realizations of a reservoir description can be generated for further study. For example, hundreds or thousands of realizations of rock properties may be used to produce a range of values of original hydrocarbon in place or ultimate recovery. From the large set of results, values having various probabilities (e.g., 50%, 70%, and 90%) can be obtained.

recovery. The fraction of hydrocarbons produced from a reservoir on the basis of initial oil or gas in place from a reservoir. The primary responsibility of reservoir engineers involves adding value to oil and gas assets, i.e., augmenting the recovery of petroleum by implementing various available tools and technologies. The recovery from a reservoir depends on a host of rock and fluid properties and on the degree of heterogeneity that may exist in the rock formation. The term is often expressed as recovery factor or recovery efficiency. *See also capillary number, gas cycling, lognormal distribution, pattern balancing, primary recovery, remaining oil saturation, reservoir management, secondary recovery, tertiary recovery, ultimate recovery, and well allocation factor.*

recovery efficiency. *See recovery.*

recovery factor. *See recovery.*

relative permeability. In a multiphase fluid flow system, it is the ratio of effective permeability to absolute permeability. *See also fractional flow.*

remaining oil saturation (ROS). Saturation of oil left behind following primary, secondary, or tertiary recovery. Reservoir studies, including seismic surveys and simulation, aim to identify areas of high remaining oil saturation in order to recover more oil utilizing fluid injection, infill drilling, and horizontal completion or recompletion of wells. Also referred to as residual oil saturation. *See also infill drilling, recovery, reservoir simulation, residual oil saturation, and seismic study.*

reserves. Volume of hydrocarbons that can be economically recovered from a reservoir based on current technology. Reserves are calculated from estimates of the original oil or gas in place and the expected recovery factor. A definition of petroleum reserves by the Society of Petroleum Engineers is available at the following Web link (among others): www.spe.org. Petroleum reserves are classified as proved reserves, probable reserves, and possible reserves. Reserves are also a function of market conditions, as production by an enhanced oil recovery process from certain fields may be commercially feasible only above a particular price point of oil. Petroleum reserves in a field may increase when new accumulations are discovered. Reserves are customarily reported in thousand, million, or billion barrels. However, other units are also used, including metric tons, also known as tonnes, and abbreviated as t. The density of the hydrocarbons needs to be known to report hydrocarbon reserves in this unit. *See also Petroleum Resources Management System, possible reserves, probable reserves, and proved reserves.*

reservoir. Generally speaking, a subsurface geologic unit with an accumulation of petroleum fluid made possible by a certain trapping mechanism. Oil and gas reservoirs are usually composed of sedimentary rock (sandstone or carbonate). Rocks must have fluid storage capacity as well as transmissibility to produce the accumulated petroleum. Petroleum fluid can be produced naturally or by adding external energy. Reservoirs are chiefly classified as oil, gas, and gas condensate. Oil reservoirs may further be classified according to the API gravity of the crude (heavy, light, intermediate, etc.). *See also compartmentalization, grains, injection well, and kriging.*

reservoir characterization. As the name implies, oil-bearing rocks in a reservoir are identified as consisting of one or more flow units characterized by their storavity, transmissibility, and capillary pressure, among other traits. These flow units have the same position in the depositional sequence and are continuous areally and vertically. The flow units may or may not coincide with the geologic units of the reservoir. The objectives of reservoir characterization include the development of a robust simulation model, realistic prediction of reservoir performance, and formulation of appropriate strategies to optimize production. For example, reservoir characterization may point to areas of bypassed oil or potential crossflow between adjacent layers. *See also flow regime, model, petrophysical model, reservoir monitoring, reservoir simulation, storavity, transmissibility, and visualization.*

reservoir life cycle. Consists of various phases in the life of a reservoir, from exploration to abandonment. The reservoir life cycle is depicted in chapter 1. *See also abandonment and exploration.*

reservoir management. Comprises the entire spectrum of activities to add value to the asset, i.e., the oil and gas reserves. Objectives and elements of reservoir management include, but are not limited to, the optimization of oil and gas recovery. It involves a reservoir team approach, data gathering and integration based on intelligent hardware and software systems, and reservoir development in the most efficient manner. It also involves reservoir monitoring, surveillance of improved oil recovery processes, and implementation of best management practices, among other factors.

reservoir model. *See model.*

reservoir monitoring. An essential component of reservoir management. Involves collection and analysis of dynamic well data such as production and injection rates, bottomhole pressure and fluid composition, evaluation of primary recovery or improved oil recovery process, and well test analysis. It also involves petrophysical studies based on newly drilled wells and reservoir characterization in order to attain better performance. *See also primary recovery and reservoir characterization.*

reservoir pressure. *See pressure.*

reservoir quality. A measure related to a flow unit in the rock indicating how efficiently and in what quantity oil can be recovered from the unit. Relatively homogeneous rocks having good storavity and transmissibility are usually regarded as of excellent reservoir quality. In certain studies, each reservoir unit is assigned a reservoir quality index (RQI). *See homogeneous formation, storavity, and transmissibility.*

reservoir simulation. Involves the analysis, prediction, and history matching of reservoir performance in terms of pressure, fluid saturation, and well rates, among other factors, by simulating a suitable mathematical model for fluid flow in porous media. The simulation model represents relevant rock and fluid characteristics at different locations in the reservoir. Due to the complexity of the physical processes involved, a numerical model based on 1-D, 2-D, or 3-D grids is usually simulated in a computer. *See also equation of state, model, pattern balancing, remaining oil saturation, reservoir characterization, visualization, and waterflood surveillance.*

residual oil saturation. Oil saturation in the reservoir following primary or enhanced oil recovery. Residual oil saturation, as determined by core flooding in the laboratory, may indicate the movable oil saturation and ultimate recovery. A successful waterflooding operation requires identification of zones of relatively high oil saturation following primary recovery. Some authors refer to the oil left behind following primary or secondary production in terms of remaining oil saturation (ROS). *See also remaining oil saturation and waterflooding.*

resistivity log. Wireline logging tool that measures water saturation at various depths within the hydrocarbon-bearing formation, among other measurements. It is based on the principle that hydrocarbons do not conduct electricity, while formation water does. A suite of resistivity logs is run in the open hole, with a small to large radius of investigation. *See also logging.*

resources. In the context of natural sources of energy, resources are hydrocarbon accumulations that have economic value. Hydrocarbon resources become producible reserves based on market economics and technological innovations. World resources, discovered and undiscovered, are estimated to be 9 to 13 trillion barrels of oil equivalent (BOE) at the time of writing. A sizeable portion of these include heavy oil and oil sands. The following categorization of resources by the Society of Petroleum Engineers is also noted:

contingent resources. Quantities of hydrocarbon in place that are discovered but currently not considered to be commercially recoverable.

prospective resources. Quantities of hydrocarbon in place that are yet to be discovered but are potentially commercial. *See also barrels of oil equivalent and hydrocarbons.*

retrograde condensation. In a producing gas condensate reservoir, certain heavier hydrocarbons condense into droplets of liquid when the reservoir pressure reaches the dewpoint pressure of the fluid system in the reservoir. This phenomenon is termed as retrograde, as it is in contrast to the behavior of pure substances, which vaporize upon reduction in prevailing pressure. *See also gas condensate.*

salt dome. A geologic feature in which the intrusion (upward movement) of relatively buoyant salt into the overlying rock leads to a dome-shaped structure. Salt domes are overlain by an impermeable cap rock. Petroleum accumulations are commonly encountered in the geologic formations around salt domes due to the abundance of traps. *See also impermeable rock.*

sand control. Methods that attempt to restrict the flow of sand into the wellbore or the adjacent formation. Sand problems are rather common and are found to affect well performance adversely, in addition to causing formation damage. Sand issues also have detrimental effects on surface equipment. Gravel pack is frequently employed to control sand migration. *See also critical flow rate, fines migration, and wellbore.*

sandface pressure. *See pressure—sandface pressure.*

sandstone. Petroleum accumulations are discovered in sandstone formations in vast numbers. Sandstones are sedimentary rocks with grains having a diameter of about 0.06 mm to 2 mm. A major constituent of sandstone is quartz, a mineral composed of silicon and oxygen. Sandstone usually possesses two of the most important physical properties required in commercially producing a reservoir: storavity and transmissibility. *See also formation, storavity, and transmissibility.*

saturated oil reservoir. A reservoir in which the prevailing pressure is at or below the bubblepoint of the oil. The liquid phase is saturated with dissolved gas. The reservoir development strategy could be quite different depending on whether the initial reservoir pressure is above or below the bubblepoint. *See also bubblepoint and undersaturated oil reservoir.*

saturation. *See fluid saturation.*

secondary porosity. A type of porosity in reservoir rock that may develop after the depositional period due to the circulation of certain solutions, dolomitization of carbonate rocks, and development of fractures in the rock matrix. The process leads to significant alteration of the rock characteristics affecting reservoir performance. Common examples of secondary porosity are vugs or cavities observed in limestone formations. *See also diagenesis.*

secondary recovery. Process or processes involving external fluid injection that are implemented in reservoirs to attain further oil recovery following primary production. In most cases the injected fluid is water. However, gas injection is not uncommon. *See also recovery and waterflooding.*

sedimentary rock. Petroleum reservoirs are mostly based on sedimentary rocks. Sedimentary rocks are primarily formed from erosion products of sand, silt, and clay from rock surfaces that were subsequently transported by water or wind and deposited as sedimentary layers. As these layers became relatively dry, various chemical compounds present in the water provided cementation between the grains, resulting in sedimentary rock. Rocks formed due to a sedimentation process are broadly classified into sandstone, limestone, and shale. Sandstone and shale are formed from the fragments of other rocks that are transported by wind or water currents. Limestone can be formed by secretions of organisms or inorganic precipitation of calcium carbonate in some circumstances.

seismic study. Study of rocks based on seismic energy response in the form of P-waves and S-waves. Seismic studies are routinely conducted in the exploration of oil and gas. The objective is to identify geologic structures, such as anticlinal traps, that can have potential accumulations of hydrocarbons. Borehole seismic studies point to heterogeneities present in the hydrocarbon-bearing formation. Seismic studies find valuable application in the characterization of fluid flow in a reservoir during improved recovery. *See also 4-D seismic study and remaining oil saturation.*

sequence stratigraphy. Integrated study based on log, core, and seismic data to interpret the depositional sequence of a sedimentary basin. The information is highly valuable in characterizing a petroleum reservoir, as well as in exploration activities. Sequence stratigraphy plays an important part in developing an earth model for a reservoir. *See also basin and exploration.*

shale. Sedimentary rock with negligible permeability formed by consolidation of clay and silt. Petroleum reservoirs are replete with shale layers located between producing intervals. *See also crossflow and stratified reservoir.*

simulation. *See reservoir simulation.*

skin factor. Measure of damage or improvement in a formation located in the immediate vicinity of a production or injection well. A positive skin factor, as obtained by a well test, indicates formation damage in the form of a reduction in permeability. Skin results in an additional pressure drop in the damaged zone. Under such circumstances, well productivity is less than satisfactory. Conversely, stimulated wells usually exhibit negative skin factor, as the permeability is enhanced by acid treatment or hydraulic fracturing, where highly conductive pathways are created to augment oil and gas production. In the vicinity of gas wells where turbulence may develop, an additional pressure drop is encountered, leading to an apparent skin factor, which is dependent on the well rate. *See also altered zone, drawdown test, flow efficiency, mechanical skin, and well productivity index.*

smart well technology. Oil and gas wells that perform a variety of intelligent applications that were not practiced in classical wells. These include, but are not limited to, the following: (a) continuous measurement of downhole data, (b) selective control of injection or production in targeted zones within a formation based on interval control valves (ICVs), (c) forced injection into low permeability zones resulting in better sweep and ultimate recovery, and (d) auto shutoff or choke down of watered-out zones. Significant reductions in downtime and human intervention are attained in many instances. Deployment of smart wells is part of the intelligent field technology that currently continues to evolve. *See also intelligent oil field.*

solution gas drive. A primary production mechanism that occurs when the reservoir is produced below the bubblepoint pressure. Gas is liberated from in-situ oil and provides the necessary energy to drive oil toward the wells. Solution gas drive is also referred to as depletion drive. *See in-situ fluid.*

sonic log. *See acoustic log.*

source rock. Source of petroleum, which is usually found in rocks rich in organic matter. At sufficient pressure and temperature in geologic times, oil formed from the organic matter and was expelled from the source rock due to intense pressure. The oil subsequently migrated to a reservoir where it accumulated under certain trapping mechanisms, such as entrapment by an overlying cap rock. *See also basin and kerogen.*

sour gas. Natural gas that contains sulfur compounds and carbon dioxide as impurities. Following production, sour gas is treated in surface facilities to remove or reduce various impurities. *See also natural gas.*

spherical flow. A fluid flow pattern that can develop near a partially completed well in a formation. Spherical flow can be identified on a diagnostic plot during well test analysis. *See also partial completion.*

spontaneous potential (SP). Naturally occurring electrical currents that are generated when fluids of different salinities are in contact. A spontaneous potential log is used to measure the resistivity of the formation water.

stabilization. A shut-in well is said to be stabilized when the bottomhole pressure approaches a constant value. In an active well, stabilization is achieved when the flow rate and pressure appear to be steady. Well stabilization, which may take a few hours to a few days in most circumstances, is necessary prior to conducting well tests. Wells are also stabilized to record the bottomhole shut-in pressure. *See also bottomhole pressure, flow-after-flow test, and isochronal test.*

static bottomhole pressure. A well approaches static bottomhole pressure once it is shut in for a period of time, which generally varies between several hours and several days or even weeks. Static pressure is indicative of the average reservoir pressure at a particular time and location, which is vital information in reservoir management and model simulation. Static bottomhole pressure is also monitored by observation wells. *See also bottomhole pressure, model simulation, and reservoir management.*

steady state flow. A condition in which the pressure and flow rate do not change with time. *See also pseudosteady-state flow and transient flow.*

steam-assisted gravity drainage (SAGD). A thermal recovery process involving two horizontal wells drilled several feet apart in the same vertical plane. The upper well injects steam. Consequently, heated oil of reduced viscosity becomes mobile and is drained through the lower well. The process is applied successfully in producing oil sands and is illustrated in chapter 19. *See also thermal recovery.*

step rate test. A type of well test conducted in injection wells to determine the fracture gradient of the geologic formation. Water is injected at an increasing rate until a fracture is created in the formation. This is detected by a significant change in the trend of the monitored bottomhole injection pressure. *See also well test.*

stimulation. Method to enhance the productivity of a well. Usually, stimulation operations are performed in damaged wells or in wells producing from a low permeability reservoir. *See also acidization, altered zone, fines migration, flow efficiency, hydraulic fracturing, skin factor, and well productivity index.*

storage capacity. One of the most important properties of a hydrocarbon-bearing rock, based upon formation thickness and porosity. *See also porosity and storativity.*

storativity ratio. The ratio of the storage of a fracture system over the storage of the total system (fracture and matrix) in a naturally fractured reservoir, as obtained by well test interpretation. *See also storativity, transmissivity ratio, and well test.*

storativity. Product of porosity, formation thickness, and total compressibility. The unit of storativity is given in units over feet–pounds per square inch (ft-psi^{-1}). Storativity and formation transmissibility are the two most important parameters in assessing reservoir quality. Storativity is also known as storage capacity. *See also reservoir characterization, reservoir quality, sandstone, storativity ratio, and transmissibility.*

stratified reservoir. A reservoir consisting of multiple geologic layers that are usually separated by relatively thin shale beds. Shale is either impermeable or semipermeable, and crossflow between the layers takes place in the latter case. Stratification is rather common due to changes in the depositional environment during geologic times. Two adjacent layers having highly contrasting transmissibility and crossflow would require special techniques in effectively producing them during primary or enhanced recovery. *See also crossflow, enhanced oil recovery, heterogeneous formation, shale, and transmissibility.*

stratigraphic trap. A type of hydrocarbon accumulation trap that results from the variations in lithology or stratigraphy or both. It may be due to depositional occurrences, such as a reef formation, alluvial deposits, or submarine turbidite deposits, among others. This type of trap may also occur due to diagenesis, facies change and unconformities. *See also lithology, sequence stratigraphy, diagenesis, facies, and unconformity.*

stringer. In the context of petroleum geology, a very thin zone containing hydrocarbons in a stratigraphic sequence. Special logging tools are required to identify stringers in certain cases.

structural trap. A type of hydrocarbon accumulation trap that occurs after deposition of the rock due to tectonic activity, such as faulting or folding of the rock units. Often multiple tectonic events have occurred related to the hydrocarbon accumulation. Anticlinal traps, caused by an upward folding of the rock, are a type of structural trap responsible for most hydrocarbon accumulations worldwide.

styolite. A geologic feature characterized by a wavy surface, which is often encountered between hydrocarbon-bearing layers in carbonate reservoirs. Styolites have very limited porosity, which may lead to unpredictable results following water and gas injection into a carbonate reservoir. *See also porosity.*

sweep efficiency. Water or other fluids injected to augment oil recovery usually cannot reach every location in a reservoir to completely sweep in-situ oil towards the producers. Sweep efficiency is influenced by rock heterogeneities and the location of injectors and producers, among other factors. Sweep efficiency is classified into areal and vertical sweep. Furthermore, overall recovery during secondary or tertiary flooding is the product of displacement efficiency and volumetric sweep efficiency. *See also displacement efficiency and in-situ fluid.*

synthetic crude oil (SCO). Hydrocarbons derived by a chemical process referred to as upgrading, including the production of bitumen from oil sands and kerogen from oil shales. *See also bitumen, oil sands, and upgrading.*

tar sands. *See oil sands.*

TDT log. *See thermal decay time log.*

teamwork. Relates to reservoir management by a multidisciplinary team of geoscientists and engineers working as a team. Tasks related to data acquisition and analysis, reservoir development and monitoring, and others, are integrated across various disciplines.

tertiary recovery. Thermal, chemical, or any other enhanced oil recovery processes. Implemented in reservoirs once secondary recovery methods are no longer effective and viable. Tertiary recovery involves chemical and miscible flooding, among other processes. *See also enhanced oil recovery, miscible displacement, and recovery.*

thermal decay time (TDT) log. A thermal (neutron) decay time log works on the principle of capturing thermal neutrons, which is influenced by the chloride content of the in-situ fluids. Thermal decay time logs are routinely used in waterflood surveillance to characterize the relative movement of the injected water through the various zones in a reservoir. *See also waterflood surveillance.*

thermal recovery. Frequently implemented in heavy oil reservoirs, thermal recovery includes all of the enhanced oil recovery processes that utilize thermal energy to augment the production of petroleum. Typical examples of thermal recovery processes include steam injection, hot waterflooding, and combination of forward combustion and waterflooding (COFCAW). *See also combination of forward combustion and waterflooding, steam-assisted gravity drainage, and waterflooding.*

thief zone. A relatively thin section in the geologic formation having high conductivity. In certain cases, the injected water during waterflooding is lost as a consequence of a thief zone. The thief zone can be detected by running a downhole temperature survey, among other methods.

tight reservoir. Reservoirs having low to ultra low permeability. A tight reservoir usually requires well stimulation and horizontal or infill drilling, among other measures, to produce commercially. Certain tight gas reservoirs with rock permeability in microdarcies are developed and produced successfully. *See also mobility and permeability.*

tortuosity. Certain configuration of a pore network in the rock that may lead to highly tortuous pathways through which reservoir fluids must flow for production. The degree of tortuosity of the rock pores influences rock permeability. *See also permeability.*

transient flow. A condition in which the fluid pressure changes with time and location during flow through porous media. *See also pseudosteady-state flow and steady state flow.*

transient pressure response. Response obtained from a well as a result of a change in flow or injection rate. Following a well test, the transient pressure response is interpreted to obtain well and reservoir characteristics. *See also derivative plot.*

transition zone. A zone of changing fluid saturation at the interface of two immiscible fluids in the reservoir, where the saturation of the denser fluid increases with depth. For example, depending on rock and fluid properties, an oil/water contact may not be a sharp interface but a transition zone of measurable height. In this zone, both oil and water phases are mobile. In the case of low permeability formations, where the density difference between oil and water is relatively small, the transition zones are usually longer. Some transition zones could be hundreds of feet long. In a fractured formation, however, the transition zone is usually nonexistent. Transition zones are described in chapter 2. *See also capillary pressure, gas/water contact, oil/water contact, and wettability.*

transmissibility. An indicator of productiveness of a geologic formation, obtained as product of rock permeability and formation thickness over fluid viscosity. Transmissibility indicates the ability of a geologic formation to transmit reservoir fluids. The unit of transmissibility is millidarcies-feet per centipoise (mD-ft/cp). Well test interpretation and laboratory core studies lead to estimation of the formation transmissibility. *See also drillstem test, drawdown test, radial flow, radius of investigation, reservoir characterization, reservoir quality, sandstone, and stratified reservoir.*

transmissivity ratio. Characterizes the flow of fluids from a naturally fractured reservoir. The transmissivity ratio is a direct function of the ratio of matrix permeability over fracture permeability. It can be estimated by well test interpretation. *See also permeability and storativity ratio.*

trap. A geologic feature and necessary requisite for hydrocarbon accumulation in porous rocks. Traps prevent the migration of petroleum fluids outside the reservoir, resulting in accumulation. Traps could be either stratigraphic or structural. Examples of geologic traps are shown in chapter 1. *See also hydrodynamic trap, stratigraphic trap, and structural trap.*

true vertical depth (TVD). The vertical distance from a reference point at the surface to the bottom of a well. In a deviated hole, the measured depth (MD) is greater than true vertical depth. Knowledge of the true vertical depth is required in calculating the hydrostatic head and reservoir pressure. *See also measured depth.*

turbidite reservoirs. Hydrocarbon reservoirs that result from the deposition of sediments from turbidity currents, usually located offshore in deepwater marine environments. The horizontal and vertical variations in lithology, size, and shape of these reservoirs add to their complexity from a production point of view.

type curve analysis. Method of well test interpretation where the pressure response obtained from a test is matched against a set of available curves exhibiting pressure responses from formations of known properties. Once a match with the new curve is found, results sought from the well test are obtained. Type curve analysis is illustrated in chapter 5.

ultimate recovery. Indicates the maximum amount of hydrocarbons that can be recovered economically by utilizing primary, secondary, and tertiary methods. *See also recovery.*

unconformity. A geologic feature in which the younger rock is positioned adjacent to much older rock, indicating missing geologic continuity through the ages. This is usually due to erosion and folding. Many hydrocarbon accumulations are trapped by unconformities. Types of unconformities include angular unconformities, nonconformities, and disconformities, among others. *See also nonconformity.*

unconventional resources. According to the Society of Petroleum Engineers, unconventional resources are petroleum accumulations that are pervasive over a large area and are not significantly affected by hydrodynamic forces. They may also be distinguished by trap type, reservoir quality, extraction process, and the amount of processing required. Examples of unconventional resources are some parts of oil sands, coalbed methane, and oil shale deposits.

undersaturated oil reservoir. A reservoir in which the prevailing pressure is above the bubblepoint of the oil. No free gas is present in the reservoir until and unless the reservoir pressure declines to the bubblepoint. Experience has shown that relatively high oil recovery is achieved when the reservoir is produced above the bubblepoint by implementing a pressure maintenance scheme. *See also saturated oil reservoir.*

unitization. Process related to grouping in a reservoir or field owned by multiple organizations or individuals. Reserves and production costs are shared among owners in accordance with their respective entitlement.

upgrading. Prior to refining, bitumen extracted from oil sands requires upgrading to synthetic crude oil having similar viscosity and gravity as conventional crude. Certain extra heavy oils also require upgrading. *See also bitumen, crude oil, oil sands, and synthetic crude oil.*

upscaling. Earth models based on geosciences data are usually developed in fine resolution comprised of millions of cells. These require upscaling to a coarser grid in a dynamic model in order to perform reservoir simulations predicting pressure and rate, among other factors. In a static model, the resolution could be a few feet or less, while in a dynamic model, a single grid may represent an area of thousands of square feet.

viscosity. Measure of resistance of a fluid to flow. Viscous oils are relatively difficult to produce, and recovery efficiency is rather low. The unit of viscosity is centipoise. Viscosity of light oil could be less than 1 cp, while bitumen can have a viscosity greater than 50,000 cp. Correlation of oil and gas viscosities with other properties is provided in chapter 3. *See also displacement efficiency, mobility, pressure-volume-temperature properties, and pseudopressure.*

viscous fingering. *See fingering.*

viscous forces. Forces that arise from the pressure differential between two points in a porous medium. The pressure differential arises as the bottomhole pressure in the producing wells is less than the reservoir pressure. In most reservoirs, viscous forces, distinguished from capillary and gravity forces, are the driving force to produce oil and gas. *See also capillary number and gravity drainage.*

visualization. A task closely associated with reservoir simulation, characterization, and virtual reality. With the advent of digital technology, reservoirs are visualized in 3-D, leading to better conceptualization and formulation of development strategy. Time-lapse studies of fluid flow are also visualized. An application of the technology is drilling of a virtual well in a 3-D visualization model to evaluate its performance. *See also reservoir characterization and reservoir simulation.*

vugular porosity. A secondary porosity that may develop due to certain geologic or geochemical processes. *See also depositional environment, heterogeneous formation, and porosity.*

water-alternating-gas (WAG) injection. An improved oil recovery process whereby alternate slugs of water and gas are injected into the reservoir to maximize recovery. *See also improved oil recovery.*

water cut. Fraction of water produced in the oil stream. Following injected water breakthrough, the water cut increases with time and eventually reaches a very high value to the point that the well may be abandoned. Various corrective actions are available to combat high water cut, including plugging of high permeability streaks and shutting off certain injectors. *See also breakthrough and high permeability streaks.*

water drive. Petroleum production is supported by energy resulting from water influx into the reservoir. Primary recovery from water drive reservoirs is expected to be better than reservoirs producing by certain other mechanisms, as seen in Figure 8–1 (chapter 8). *See also primary recovery.*

waterflooding. Secondary recovery method whereby water is injected into the reservoir to produce additional quantities of oil once natural energies are spent following primary production. A successful waterflooding project requires a detailed reservoir description, determination of remaining oil saturation, and development of injection wells and injection pattern. It also requires optimization of injection rates, reservoir surveillance, and efforts to maximize sweep efficiency. *See also enhanced oil recovery, fractional flow, improved oil recovery, pattern balancing, pressure maintenance, secondary recovery, and wettability.*

waterflood surveillance. Relates to wide-ranging tasks to monitor, evaluate, and enhance waterflood performance at both well and reservoir levels. These include, but are not limited to, measurement and analysis of well rates, pressure and fluid composition, zonal distribution of injected and produced fluids, and evaluation and improvement of sweep efficiency. It also includes profile modification of the injected and produced fluids, diagnosis of potential or existing well problems, pattern balancing, and update of the reservoir simulation model in light of new information obtained from surveillance. *See also reservoir simulation and thermal decay time (TDT) log.*

water influx. Movement of water from an adjacent aquifer located at the bottom or the periphery of a reservoir. The water moves into the reservoir as it depletes due to the production of oil and gas. When a strong water drive is present, the decline in reservoir pressure is comparatively less with time. However, the wells are likely to experience water breakthrough following certain periods of dry production. *See also aquifer, bounded reservoir, and constant pressure boundary.*

water/oil contact. *See oil/water contact.*

water-wet reservoir. A reservoir is said to be water-wet when the water has a greater tendency to adhere to the pore surface than does the oil. The majority of the reservoirs are believed to be water-wet. *See also drainage, oil-wet reservoir, and wettability.*

well allocation factor (WAF). Factor that leads to the calculation of injection efficiency of an injector during waterflooding as the ratio of injected water volume to the volume of oil produced at the offset wells. When water injection efficiencies are known over the entire field, appropriate adjustments can be made in injection rates in order to improve sweep and oil recovery. *See also recovery.*

wellbore. Drilled hole for a well in the field. *See also fracture half-length, gravel pack, limited entry, logging, and sand control.*

wellbore storage. When a well is opened for flow, fluid stored in the wellbore is produced first, not the fluid from rock pores. The resulting pressure response, as monitored during a pressure transient test, is influenced by wellbore storage effects rather than by the formation characteristics. When a well is shut in following production, similar storage effects are observed on the pressure response initially. Immediately after shut in, reservoir fluids continue to flow from the formation until stabilization is reached. *See also after flow.*

well completion. Once a new oil or gas well is drilled and successfully tested, downhole tubulars are installed, and the well is perforated for production. Depending on the characteristics of the formation, the wells are completed in many different ways, including openhole, conventional, multiple, and no-tubing completions. Some other completions are specially designed to control sand production. Due to mechanical or productivity issues, wells can be recompleted at later stages. Horizontal wells are usually openhole or slotted liner completions.

well deliverability plot. Also called an inflow performance relationship curve. *See inflow performance relationship.*

wellhead pressure. *See pressure—wellhead pressure.*

well productivity index. Well productivity index, or simply productivity index (PI), is a measure of how efficiently an oil or gas well can produce under a specific pressure drawdown. It is usually expressed as the rate of oil produced per unit of drawdown pressure (stb/d/psi). The drawdown of a producing well can be determined by the difference in the flowing bottomhole pressure and the pressure at the external reservoir boundary or well drainage area. The productivity index of a well may change with time due to changes in reservoir pressure and also when the formation surrounding the well is either damaged or stimulated. The productivity index of a horizontal well is significantly higher than that of a vertical well in most cases. *See also flow efficiency.*

well spacing. Conventionally expressed in acres, it is an area delineated by wells located at the corners of the pattern area in oil and gas fields. Typical well spacing is found to range between 20 acres and 160 acres. Examples of well spacing are illustrated in chapter 16. A reservoir may initially be developed with a relatively large spacing. During waterflooding at later stages, infill wells are drilled, resulting in reduced well spacing in the field.

well test. Method to obtain and analyze the pressure response in a well when a perturbation in the well flow (shut-in or drawdown) is created. Virtually all wells, injectors, and producers undergo some type of well testing. Common well test types include drawdown, buildup, falloff, interference, step rate, flow-after-flow, and drillstem tests. Well testing is a potent tool available to reservoir engineers to characterize the reservoir, as it is capable of detecting geologic features, boundaries, and fluid fronts located a significant distance away from the test well. Productivity of a well is routinely evaluated from well tests as part of reservoir management. Various applications of well testing in reservoir engineering are described in chapter 5. Well tests are also known as pressure transient tests. *See also buildup test, drillstem test, drawdown test, falloff test, flow-after-flow test, flow regime, interference test, storativity ratio, and step rate test.*

well test interpretation. Analytical and digital procedures that attempt to determine reservoir and well characteristics from transient pressure versus time data obtained from well tests. Classical interpretations, which traditionally depended on manual plotting and computation, have been replaced by digital interpretations in high-speed computers. The capabilities of well test software are described in chapter 5, with illustrative examples.

wet combustion. *See combination of forward combustion and waterflooding.*

wet gas. Natural gas containing relatively heavy hydrocarbons compared to methane and ethane, which can be extracted in surface facilities for various purposes. The heavier hydrocarbons condense out as liquids under surface conditions.

wettability. Property of hydrocarbon-bearing rocks that determines whether water or oil will adhere to the pore surface preferentially. It is dependent on the mineral content of the rock and the composition of the fluids in the pores. It is a function of the interfacial tension that exists between the oil phase and the pore surface, between the water phase and the pore surface, and between the two fluid phases. Rocks can be either water-wet or oil-wet, the former being more likely. Wettability characteristics can change as the pore surface comes into contact with certain hydrocarbon compounds. In an oil-wet reservoir, waterflood performance may be less than satisfactory, as in-situ oil is likely to adhere to the pore walls during water drive. Certain rocks are of mixed or intermediate wettability. Rocks in transition zones can be of mixed wettability. *See also core, displacement efficiency, drainage, interfacial tension, J function, oil-wet reservoir, petrophysical model, transition zone, waterflooding, and water-wet reservoir.*

wildcat. An exploratory well drilled in a location where very little downhole information is available, including formation pressure and hydrocarbon accumulation. The decision to drill more wells in the area hinges on the success of an exploratory well. Many wildcats result in dry holes, while others strike oil or gas. *See also exploration.*

wireline formation tester. *See formation tester.*

wireline logging. *See logging.*

z factor. *See gas deviation factor.*

zone. A stratum or body of rock that has distinctly unique properties in comparison to adjacent zones, or can be recognized by its predominant fluid content. For example, a gas zone may be located above an oil zone in a reservoir. *See also field.*

In preparing this section, the contribution of Schlumberger, SPWLA, and SPE through their Web sites is thankfully acknowledged.

Index

D

E

F

S

T

U

V

X–Z

About the Authors

Dr. Abdus Satter, Ph.D.

Dr. Abdus Satter is president of Satter and Associates, which was formed in Houston upon his retirement from Texaco in 1998. His company is involved in engineering consulting and training services. He has participated in and led reservoir management projects in the U.S., UK, Cameroon, Denmark, Indonesia, Nigeria, Trinidad, and elsewhere.

He is a Distinguished Member and a Distinguished Lecturer of the Society of Petroleum Engineers on the topic of Integrated Reservoir Management (1996–1997), and Asset Management through the Reservoir Life Cycle (2002–2003). He teaches SPE and industry courses.

He has 40 years of worldwide, diversified experience and expertise in reservoir engineering, reservoir simulator development and applications, water flood, enhanced recovery techniques, and reservoir studies and operations. He has been published numerous times in these areas.

His previous published books with co-authors are *Integrated Petroleum Reservoir Management*, *Integrated Water Flood Asset Management*, and *Computer-Assisted Reservoir Management*.

He holds a B.Sc. in Mechanical Engineering from the University of Dhaka, P.E., and M.S. degrees in Petroleum Engineering from the Colorado School of Mines, and a Ph.D. in Engineering Science from the University of Oklahoma.

Dr. Ghulam Iqbal, Ph.D., P.E.

Ghulam Iqbal holds M.S. and Ph.D. in Petroleum Engineering from the University of Oklahoma, where he was a recipient of Oklahoma Mining and Minerals Research Institute scholarship. He also holds a bachelor degree in Chemical Engineering from Bangladesh University of Engineering and Technology. He is a registered professional engineer in the state of California. He has decades of combined experience in reservoir engineering and simulation, model development of fluid flow systems, data integration, information technology, environmental issues, multi-disciplinary teamwork and oilfield asset management.

Spanning over continents, his career assignments included management of a giant oil field in the middle-east producing over half a million barrels daily. He worked with a highly dedicated team which pioneered the application of multi-lateral horizontal well technology in the region in 1990s. The integrated team provided innovative solutions to combat issues related to early water breakthrough, low productivity, reservoir characterization and field monitoring, among others, with an emphasis on state-of-the-art techniques. He received the best performance award in 2000 for outstanding services in managing assets for the company.

Author of many technical publications, including contribution to publication of the Society of Exploration Geophysicists, Dr. Iqbal has conducted several hands-on workshops for oil and gas professionals overseas on reservoir engineering, asset management and petroleum basin modeling under the U.S. Agency for International Development program.

Dr. James Buchwalter, Ph.D., P.E.

Dr. Jim Buchwalter is president of Gemini Solutions Inc. (GSI). He and Dr. Ray Calvert started GSI in 1996 after a 15 year career as a reservoir engineer with Texaco. GSI markets a user friendly PC based integrated mapping, simulation, and nodal analysis system throughout the world, and currently the company has licenses in more than 65 companies. GSI also offers reservoir engineering training and consulting simulation services.

He is a member of SPE and has worldwide experience in reservoir engineering and simulation. He maintained and enhanced Texaco's in-house simulator before forming GSI. He was also active in developing easier to use simulation interfaces, which resulted in a faster simulation work flow at Texaco, and expansion of the simulation community from a dozen or so experts to hundreds of engineers in the 1990's.

He has published numerous times on various topics in the reservoir simulation field. He holds B.Sc. and M.S. degrees in Chemical Engineering from Ohio State University, and a Ph.D. in Chemical Engineering from Rice University.